Lecture Notes of the Institute for Computer Sciences, Social Informatics and Telecommunications Engineering 406

More information about this series at https://link.springer.com/bookseries/8197

Honghao Gao · Xinheng Wang (Eds.)

Collaborative Computing: Networking, Applications and Worksharing

17th EAI International Conference, CollaborateCom 2021
Virtual Event, October 16–18, 2021
Proceedings, Part I

Springer

Editors
Honghao Gao
Shanghai University
Shanghai, China

Xinheng Wang
Xi'an Jiaotong-Liverpool University
Suzhou, China

ISSN 1867-8211 ISSN 1867-822X (electronic)
Lecture Notes of the Institute for Computer Sciences, Social Informatics
and Telecommunications Engineering
ISBN 978-3-030-92634-2 ISBN 978-3-030-92635-9 (eBook)
https://doi.org/10.1007/978-3-030-92635-9

This Springer imprint is published by the registered company Springer Nature Switzerland AG
The registered company address is: Gewerbestrasse 11, 6330 Cham, Switzerland

Preface

We are delighted to introduce the proceedings of the 17th European Alliance for Innovation (EAI) International Conference on Collaborative Computing: Networking, Applications and Worksharing (CollaborateCom 2021). This conference has brought together researchers, developers, and practitioners around the world who are interested in fully realizing the promises of electronic collaboration from the aspects of networking, technology and systems, user interfaces and interaction paradigms, and interoperation with application-specific components and tools.

This year's conference attracted 206 submissions. Each submission was reviewed by an average of three reviewers. After a rigorous review process, 69 papers were accepted, including 62 full papers and seven short papers in oral presentation sessions at the main conference tracks. The conference sessions covered optimization, UAV and traffic systems, recommendation systems, network and security, IOT and social networks, image handling and human recognition, edge computing, collaborative working, and deep learning and its applications. Apart from the high-quality technical paper presentations, the technical program also featured two keynote speeches and two technical workshops. The two keynote speeches were delivered by Jie Wu from Temple University and Dan Feng from Huazhong University of Science and Technology. The two workshops organized were Securing IoT Networks (SITN) and Data-driven Fault Diagnosis with Collaborative Computing. The SITN workshop aims to bring together expertise from academia and industry to build secure IoT infrastructures for smart society. The workshop on Data-driven Fault Diagnosis with Collaborative Computing aims to provide a forum for researchers and industrial practitioners to exchange their latest results on data-driven fault diagnosis techniques with collaborative computing.

Coordination with the steering chair, Imrich Chlamtac, and the steering committee members was essential for the success of the conference. We sincerely appreciate their constant support and guidance. It was also a great pleasure to work with such an excellent organizing committee team for their hard work in organizing and supporting the conference. In particular, we are grateful to the Technical Program Committee who completed the peer-review process for technical papers and helped to put together a high-quality technical program. We are also grateful to the conference manager, Karolina Marcinova, for her support and all the authors who submitted their papers to the CollaborateCom 2021 conference and workshops.

We strongly believe that the CollaborateCom conference provides a good forum for all researchers, developers, and practitioners to discuss all science and technology aspects that are relevant to collaborative computing. We also expect that the future

CollaborateCom conferences will be as successful and stimulating as this year's, as indicated by the contributions presented in this volume.

October 2021

Honghao Gao
Xinheng Wang
Yuanyuan Yang
Kaizhu Huang
Tun Lu

Organization

Steering Committee

Chair

Imrich Chlamtac   University of Trento, Italy

Members

Song Guo   University of Aizu, Japan
Bo Li   Hong Kong University of Science and
                                       Technology, China
Xiaofei Liao   Huazhong University of Science and Technology,
                                       China
Xinheng Wang   Xi'an Jiaotong-Liverpool University, China
Honghao Gao   Shanghai University, China

Organizing Committee

General Chairs

Eng Gee Lim   Xi'an Jiaotong-Liverpool University, China
Yuanyuan Yang   State University of New York, USA
Honghao Gao   Shanghai University, China
Xinheng Wang   Xi'an Jiaotong-Liverpool University, China

Technical Program Committee Co-chairs

Kaizhu Huang   Xi'an Jiaotong-Liverpool University, China
Tun Lu   Fudan University, China

Web Chair

Fei Cheng   Xi'an Jiaotong-Liverpool University, China

Publicity and Social Media Chairs

Li Kuang   Central South University, China
Andrei Tchernykh   CICESE Research Center, Mexico

Workshops Chairs

Yusheng Xu   Xidian University, China
Shahid Mumtaz   Instituto de Telecomunicações, Portugal

Publications Chairs

Youhuizi Li   Hangzhou Dianzi University, China
Azah Kamilah Binti Draman   University Teknikal Malaysia Melaka, Malaysia
Xiaoxian Yang   Shanghai Polytechnic University, China

Panels Chair

Kyeongsoo (Joseph) Kim   Xi'an Jiaotong-Liverpool University, China

Tutorials Chair

Yuyu Yin   Hangzhou Dianzi University, China

Local Chairs

Junyao Duan   Xi'an Jiaotong-Liverpool University, China
Xiaoyi Wu   Xi'an Jiaotong-Liverpool University, China

Technical Program Committee

Anwer Al-Dulaimi   University of Toronto, Canada
Amjad Ali   Korea University, South Korea
Junaid Arshad   University of West London, UK
Zhongqin Bi   Shanghai University of Electric Power, China
Bin Cao   Zhejiang University of Technology, China
Buqing Cao   Hunan University of Science and Technology,
                                       China
Liang Chen   University of West London, UK
Shizhan Chen   Tianjing University, China
Yihai Chen   Shanghai University, China
Ying Chen   Beijing University of Information Technology,
                                       China
Fei Dai   Yunnan University, China
Yucong Duan   Hainan University, China
Xiaoliang Fan   Fuzhou University, China
Shucun Fu   Nanjing University of Information Science and
                                       Technology, China
Fekade Getahun   Addis Ababa University, Ethiopia
Jiwei Huang   China University of Petroleum, Beijing, China

Tao Huang	Silicon Lake University, China
Congfeng Jiang	Hangzhou Dianzi University, China
Malik Ahmad Kamran	COMSATS University Islamabad, Pakistan
Li Kuang	Central South University, China
Rui Li	Xidian University, China
Youhuizi Li	Hangzhou Dianzi University, China
Wenmin Lin	Hangzhou Dianzi University, China
Jianxun Liu	Hunan University of Science and Technology, China
Shijun Liu	Shandong University, China
Xihua Liu	Nanjing University of Information Science and Technology, China
Xuan Liu	Southeast University, China
Yutao Ma	Wuhan University, China
Lin Meng	Ritsumeikan University, Japan
Shunmei Meng	Nanjing University of Science and Technology, China
Elahe Naserianhanzaei	University of Exeter, UK
Yu-Chun Pan	University of West London, UK
Shanchen Pang	China University of Petroleum, China
Lianyong Qi	Qufu Normal University, China
Kuangyu Qin	Wuhan University, China
Stephan Reiff-Marganiec	University of Leicester, UK
Imed Romdhani	Edinburgh Napier University, UK
Changai Sun	University of Science and Technology, China
Xiaobing Sun	Yangzhou University, China
Wenda Tang	Lancaster University, UK
George Ubakanma	London South Bank University, UK
Shaohua Wan	Zhongnan University of Economics and Law, China
Dongjing Wang	Hangzhou Dianzi University, China
Jian Wang	Wuhan University, China
Junhao Wen	Chongqing University, China
Yiping Wen	Hunan University of Science and Technology, China
Yu Weng	Minzu University of China, China
Yirui Wu	Hohai University, China
Yunni Xia	Chongqing University, China
Haolong Xiang	University of Auckland, New Zealand
Jiuyun Xu	China University of Petroleum, China
Xiaolong Xu	Nanjing University of Information Science and Technology, China

Yueshen Xu   Xidian University, China
Xiaoxian Yang   Shanghai Polytechnic University, China
Yuyu Yin   Hangzhou Dianzi University, China
Li Yu   Hangzhou Dianzi University, China
Yuan Yuan   Michigan State University, USA
Jun Zeng   Chongqing University, China
Gaowei Zhang   Nanyang Technology University, China
Jie Zhang   Nanjing University, China
Yanmei Zhang   Central University of Finance and Economics,
                                 China
Yiwen Zhang   Anhui University, China
Zijian Zhang   Beijing Institute of Technology, China
Xuan Zhao   Nanjing University, China
Zhuofeng Zhao   North China University of Technology, China
Yu Zheng   Nanjing University of Information Science and
                                 Technology, China
Ao Zhou   Beijing University of Posts and
                                 Telecommunications, China
Guobing Zou   Shanghai University, China

Contents – Part I

Optimization for Collaborate System (Workshop Papers)

Chinese Named Entity Recognition Based on Dynamically Adjusting
Feature Weights 3
  Qing Lv, Limin Zheng, and Miao Wang

Location Differential Privacy Protection in Task Allocation for Mobile
Crowdsensing Over Road Networks 20
  *Mohan Fang, Juan Yu, Jianmin Han, Xin Yao, Hao Peng, Jianfeng Lu,
  and Ngounou Bernard*

"Failure" Service Pattern Mining for Exploratory Service Composition 38
  *Yunjing Yuan, Jing Wang, Yanbo Han, Qianwen Li, Gaojian Chen,
  and Boyang Jiao*

Optimal Control and Reinforcement Learning for Robot: A Survey 54
  Haodong Feng, Lei Yu, and Yuqing Chen

KTOBS: An Approach of Bayesian Network Learning Based on K-tree
Optimizing Ordering-Based Search 67
  *Qingwang Zhang, Sihang Liu, Ruihong Xu, Zemeng Yang,
  and Jianxiao Liu*

Recommendation Model Based on Social Homogeneity Factor and Social
Influence Factor 83
  Weizhi Ying, Qing Yu, and Zuohua Wang

Attention Based Spatial-Temporal Graph Convolutional Networks
for RSU Communication Load Forecasting 99
  Hang Zheng, Xu Ding, Yang Wang, and Chong Zhao

UVA and Traffic System

Mobile Encrypted Traffic Classification Based on Message Type Inference 117
  Yige Chen, Tianning Zang, Yongzheng Zhang, Yuan Zhou, and Peng Yang

Fine-Grained Spatial-Temporal Representation Learning with Missing
Data Completion for Traffic Flow Prediction 138
  Shiqi Wang, Min Gao, Zongwei Wang, Jia Wang, Fan Wu, and Junhao Wen

Underwater Information Sensing Method Based on Improved
Dual-Coupled Duffing Oscillator Under Lévy Noise Description 156
Hanwen Zhang, Zhen Qin, and Dajiang Chen

Unpaired Learning of Roadway-Level Traffic Paths from Trajectories 171
Weixing Jia, Guiling Wang, Xuankai Yang, and Fengquan Zhang

Multi-UAV Cooperative Exploring for the Unknown Indoor Environment
Based on Dynamic Target Tracking 191
Ning Li, Jiefu Tan, Yunlong Wu, Jiachi Xu, Huan Wang, and Wendi Wu

Recommendation System

MR-FI: Mobile Application Recommendation Based on Feature
Importance and Bilinear Feature Interaction 213
Mi Peng, Buqing Cao, Junjie Chen, Jianxun Liu, and Rong Hu

The Missing POI Completion Based on Bidirectional Masked Trajectory
Model 229
Jun Zeng, Yizhu Zhao, Yang Yu, Min Gao, and Wei Zhou

Dual-Channel Graph Contextual Self-Attention Network for Session-Based
Recommendation 244
Teng Huang, Huiqun Yu, and Guisheng Fan

Context-aware Graph Collaborative Recommendation Without Feature
Entanglement 259
Tianyi Gu, Ping Li, and Kaiwen Huang

Improving Recommender System via Personalized Reconstruction
of Reviews 277
*Zunfu Huang, Bo Wang, Hongtao Liu, Qinxue Jiang, Naixue Xiong,
and Yuexian Hou*

Recommendation System and Network and Security

Dynamic Traffic Network Based Multi-Modal Travel Mode Fusion
Recommendation 299
*Nannan Jia, Mengmeng Chang, Zhiming Ding, Zunhao Liu,
Bowen Yang, Lei Yuan, and Lutong Li*

Improving Personalized Project Recommendation on GitHub Based
on Deep Matrix Factorization 318
Huan Yang, Song Sun, Junhao Wen, Haini Cai, and Muhammad Mateen

An Intelligent SDN DDoS Detection Framework 333
Xiang Zhang, Chaokui Zhang, Zhenyang Zhong, and Peng Ye

Inspector: A Semantics-Driven Approach to Automatic Protocol Reverse
Engineering 348
Yige Chen, Tianning Zang, Yongzheng Zhang, Yuan Zhou, Peng Yang,
and Yipeng Wang

MFF-AMD: Multivariate Feature Fusion for Android Malware Detection 368
Guangquan Xu, Meiqi Feng, Litao Jiao, Jian Liu, Hong-Ning Dai,
Ding Wang, Emmanouil Panaousis, and Xi Zheng

Network and Security

PSG: Local Privacy Preserving Synthetic Social Graph Generation 389
Hongyu Huang, Yao Yang, and Yantao Li

Topology Self-optimization for Anti-tracking Network via Nodes
Distributed Computing 405
Changbo Tian, Yongzheng Zhang, and Tao Yin

An Empirical Study of Model-Agnostic Interpretation Technique
for Just-in-Time Software Defect Prediction 420
Xingguang Yang, Huiqun Yu, Guisheng Fan, Zijie Huang, Kang Yang,
and Ziyi Zhou

Yet Another Traffic Black Hole: Amplifying CDN Fetching Traffic
with RangeFragAmp Attacks 439
Chi Xu, Juanru Li, and Junrong Liu

DCNMF: Dynamic Community Discovery with Improved Convex-NMF
in Temporal Networks 460
Limengzi Yuan, Yuxian Ke, Yujian Xie, Qingzhan Zhao, and Yuchen Zheng

Network and Security and IoT and Social Networks

Loopster++: Termination Analysis for Multi-path Linear Loop 479
Hui Jin, Weimin Ge, Yao Zhang, Xiaohong Li, and Zhidong Deng

A Stepwise Path Selection Scheme Based on Multiple QoS Parameters
Evaluation in SDN 498
Lin Liu, Jian-Tao Zhou, Hai-Feng Xing, and Xiao-Yong Guo

A Novel Approach to Taxi-GPS-Trace-Aware Bus Network Planning 520
Liangyao Tang, Peng Chen, Ruilong Yang, Yunni Xia, Ning Jiang,
Yin Li, and Hong Xie

Community Influence Maximization Based on Flexible Budget in Social
Networks 534
Mengdi Xiao, Peng Li, Weiyi Huang, Junlei Xiao, and Lei Nie

An Online Truthful Auction for IoT Data Trading with Dynamic Data
Owners 554
Zhenni Feng, Junchang Chen, and Tong Liu

**IoT and Social Networks and Images Handling and Human
Recognition**

Exploiting Heterogeneous Information for IoT Device Identification Using
Graph Convolutional Network 575
Jisong Yang, Yafei Sang, Yongzheng Zhang, Peng Chang,
and Chengwei Peng

Data-Driven Influential Nodes Identification in Dynamic Social Networks 592
Ye Qian and Li Pan

Human Motion Recognition Based on Wi-Fi Imaging 608
Liangliang Lin, Kun Zhao, Xiaoyu Ma, Wei Xi, Chen Yang, Hui He,
and Jizhong Zhao

A Pervasive Multi-physiological Signal-Based Emotion Classification
with Shapelet Transformation and Decision Fusion 628
Shichao Zhang, Xiangwei Zheng, Mingzhe Zhang, Gengyuan Guo,
and Cun Ji

A Novel and Efficient Distance Detection Based on Monocular Images
for Grasp and Handover 642
Dianwen Liu, Pengfei Yi, Dongsheng Zhou, Qiang Zhang,
Xiaopeng Wei, Rui Liu, and Jing Dong

Images Handling and Human Recognition and Edge Computing

A Novel Gaze-Point-Driven HRI Framework for Single-Person 661
Wei Li, Pengfei Yi, Dongsheng Zhou, Qiang Zhang, Xiaopeng Wei,
Rui Liu, and Jing Dong

Semi-automatic Segmentation of Tissue Regions in Digital
Histopathological Image 678
    Xin He, Kairun Chen, and Mengning Yang

T-UNet: A Novel TC-Based Point Cloud Super-Resolution Model
for Mechanical LiDAR 697
    Lu Ren, Deyi Li, Zhenchao Ouyang, Jianwei Niu, and Wen He

Computation Offloading for Multi-user Sequential Tasks in Heterogeneous
Mobile Edge Computing 713
    Huanhuan Xu, Jingya Zhou, and Fei Gu

Model-Based Evaluation and Optimization of Dependability for Edge
Computing Systems 728
    Jingyu Liang, Bowen Ma, Sikandar Ali, and Jiwei Huang

Author Index 749

Contents – Part II

Edge Computing

Energy-Efficient Cooperative Offloading for Multi-AP MEC in IoT
Networks 3
  Zhihui Cao, Haifeng Sun, Ning Zhang, and Xiang Lv

Multi-truth Discovery with Correlations of Candidates in Crowdsourcing
Systems 18
  Hongyu Huang, Guijun Fan, Yantao Li, and Nankun Mu

D2D-Based Multi-relay-Assisted Computation Offloading in Edge
Computing Network 33
  Xuan Zhao, Song Zhang, Bowen Liu, Xutong Jiang, and Wanchun Dou

Delay-Sensitive Slicing Resources Scheduling Based on Multi-MEC
Collaboration in IoV 50
  Yan Liang, Xin Chen, Shengcheng Ma, and Libo Jiao

An OO-Based Approach of Computing Offloading and Resource
Allocation for Large-Scale Mobile Edge Computing Systems 65
  Yufu Tan, Sikandar Ali, Haotian Wang, and Jiwei Huang

Edge Computing and Collaborative Working

Joint Location-Value Privacy Protection for Spatiotemporal Data
Collection via Mobile Crowdsensing 87
  Tong Liu, Dan Li, Chenhong Cao, Honghao Gao, Chengfan Li,
  and Zhenni Feng

Hybrid Semantic Conflict Prevention in Real-Time Collaborative
Programming 104
  Wenhua Xu, Yiteng Zhang, Brian Chiu, Dong Chen, Jinfeng Jiang,
  Bowen Du, and Hongfei Fan

Supporting Cross-Platform Real-Time Collaborative Programming:
Architecture, Techniques, and Prototype System 124
  Yifan Ma, Zichao Yang, Brian Chiu, Yiteng Zhang, Jinfeng Jiang,
  Bowen Du, and Hongfei Fan

Collaborative Computing Based on Truthful Online Auction Mechanism
in Internet of Things 144
   Bilian Wu, Xin Chen, and Libo Jiao

A Hashgraph-Based Knowledge Sharing Approach for Mobile Robot
Swarm 158
   Xiao Shu, Bo Ding, Jie Luo, Xiang Fu, Min Xie, and Zhen Li

Collaborative Working and Deep Learning and Application

CASE: Predict User Behaviors via Collaborative Assistant Sequence
Embedding Model 175
   Fei He, Canghong Jin, and Minghui Wu

A Collaborative Optimization-Guided Entity Extraction Scheme 190
   Qiaojuan Peng, Xiong Luo, Hailun Shen, Ziyang Huang,
   and Maojian Chen

A Safe Topological Waypoints Searching-Based Conservative Adaptive
Motion Planner in Unknown Cluttered Environment 206
   Jiachi Xu, Jiefu Tan, Chao Xue, Yaqianwen Su, Xionghui He,
   and Yongjun Zhang

Multi-D3QN: A Multi-strategy Deep Reinforcement Learning for Service
Composition in Cloud Manufacturing 225
   Jun Zeng, Juan Yao, Yang Yu, and Yingbo Wu

Transfer Knowledge Between Cities by Incremental Few-Shot Learning 241
   Jiahao Wang, Wenxiong Li, Xiuxiu Qi, and Yuheng Ren

Deep Learning and Application

Multi-view Representation Learning with Deep Features for Offline
Signature Verification 261
   Xingbiao Zhao, Changzheng Liu, Benzhuang Zhang, Limengzi Yuan,
   and Yuchen Zheng

Backdoor Attack of Graph Neural Networks Based on Subgraph Trigger 276
   Yu Sheng, Rong Chen, Guanyu Cai, and Li Kuang

A UniverApproCNN with Universal Approximation and Explicit Training
Strategy 297
   Yin Yang, Yifeng Wang, and Senqiao Yang

MS-BERT: A Multi-layer Self-distillation Approach for BERT
Compression Based on Earth Mover's Distance   316
    Jiahui Huang, Bin Cao, Jiaxing Wang, and Jing Fan

Smart Contract Vulnerability Detection Based on Dual Attention Graph
Convolutional Network   335
    Yuqi Fan, Siyuan Shang, and Xu Ding

Crowdturfing Detection in Online Review System: A Graph-Based
Modeling   352
    Qilong Feng, Yue Zhang, and Li Kuang

Attention-Aware Actor for Cooperative Multi-agent Reinforcement
Learning   370
    Chenran Zhao, Dianxi Shi, Yaowen Zhang, Yaqianwen Su,
    Yongjun Zhang, and Shaowu Yang

Geographic and Temporal Deep Learning Method for Traffic Flow
Prediction in Highway Network   385
    Tianpu Zhang, Weilong Ding, Mengda Xing, Jun Chen, Yongkang Du,
    and Ying Liang

How are You Affected? A Structural Graph Neural Network Model
Predicting Individual Social Influence Status   401
    Jiajie Du and Li Pan

Multi-order Proximity Graph Structure Embedding   416
    Wang Zhang, Lei Jiang, Huailiang Peng, Qiong Dai, and Xu Bai

Deep Learning and Application and UVA

PATR: A Novel Poisoning Attack Based on Triangle Relations Against
Deep Learning-Based Recommender Systems   435
    Meiling Chao, Min Gao, Junwei Zhang, Zongwei Wang, Quanwu Zhao,
    and Yulin He

Low-Cost LiDAR-Based Vehicle Detection for Self-driving Container
Trucks at Seaport   451
    Changjie Zhang, Zhenchao Ouyang, Lu Ren, and Yu Liu

Author Index   467

Optimization for Collaborate System
(Workshop Papers)

Chinese Named Entity Recognition Based on Dynamically Adjusting Feature Weights

Qing Lv, Limin Zheng$^{(\boxtimes)}$, and Miao Wang

College of Information and Electrical Engineering, China Agriculture University,
Beijing, China
{S20193081363,zhenglimin,S20193081360}@cau.edu.cn

Abstract. Named entity recognition is a basic task in NLP, and it is an important basic tool for many NLP tasks such as information extraction, parsing, question answering system and machine translation. The extraction of sequence features of datasets directly affects the recognition effect of named entities, and only the accumulation of local sequence features cannot capture the long distance dependencies. The extraction of global sequence features improves this problem, but loses some local features. Long entities are nested within short entities and have different entity attributes from short entities, resulting in identification errors. To solve these problems, a Chinese named entity recognition algorithm based on Bert +FL-LGWF+CRF is proposed. In this method, the text is encoded into a word vector matrix by Bert as the input to FL-LGWF (Entity Level-Local And Global Weighted Fusion). FL-LGWF utilizes CNN (Convolutional Neural) to extract the local sequence features of the text vector, and use BISTM (Bidirectional Long Short-Term Memory) to extract contextual global sequence features, and perform dynamic weight fusion on the extracted sequence features. Then the score matrix of the tag is obtained according to the entity attribute level. Finally, the global optimal tag sequence is obtained through the CRF layer. Experimental results show that the proposed Bert +FL-LGWF+CRF model has higher F1 value on both public data sets and self-created data sets.

Keywords: Named entity recognition · Dynamic weight fusion · Entity level local · CNN · BILSTM

Supported by the National Key Research and Development Program of China (2017YFC1601803) and Beijing innovation team project of modern agricultural industrial technology system(BAIC02-2020).

1 Introduction

Named entities are the entities with specific meaning or strong reference in the text, usually including the name of the person, place name, organization name and so on. Named Entity Recognition (NER) refers to the extraction of Named entities from unstructured input text [1]. Named entity recognition has been widely applied in various fields, such as knowledge graph construction [2], knowledge base construction [3,4], network search [5], machine translation [6], automatic question and answer, etc. [7].

Initially, named entity recognition was a rule-based and dictionary-based approach, which adopted specific rule templates or special dictionaries built by linguists manually according to the characteristics of data sets, and used matching methods to process the text to realize named entity recognition. On this basis, Rau et al. [8] proposed for the first time to combine manual writing rules with heuristic ideas and realize automatic extraction of named entities of company name type from text. However, it consumes a lot of manpower and is not easy to expand in other data sets or types, so there are very obvious limitations. In the machine learning approach, named entity recognition is treated as a sequential labeling problem. Yoshua [9] et al. proposed the Hidden Markov Model (HMM) to directly Model transition probability and performance probability, and to calculate co-occurrence probability, so as to carry out sequence labeling. Sutton et al. [10] proposed a named entity recognition method based on Conditional Random Field (CRF), and manually annotated the feature template of the entity, and annotated the sequence. CRF can use internal and contextual characteristic information to label a location. The above methods not only consume time and effort, but also have a certain error in manual annotation because feature selection relies on the prior experience of human, which leads to a low accuracy. In recent years, Deep Neural Network has been widely concerned in the field of natural language processing, and named entity recognition based on Deep Neural Network (DNN) [11] has become a research hotspot. This technology mainly utilizes the strong computing power of neural network to automatically extract the context features at the sentence level and realize entity recognition. Collobert et al. [12] proposed the named entity recognition method based on neural network for the first time. In this method, each word has a fixed size window, but the effective information between long words is ignored. Lample [13] proposed a named entity recognition method based on Long Short-Term Memory (LSTM), which effectively solved the problem of Long time dependence effect. However, these methods do not take into account the sequence of words in sentences. Huang [14] proposed a Bidirectional Long Short-Term Memory (BILSTM) that can better capture Bidirectional semantic dependence by combining forward and backward LSTM. However, because the training is fixed word embedding, there is a problem that the polysemy cannot be expressed. Devlin et al. [15] proposed the Bidirectional Encoder Representation from Transformers (Bert), which represents the context through pre-training and fine-tunes to solve the problem of inadequate context representation in NER. In order to make full use of the advantages of each model, scholars put forward a series of combination models. Dong [16] proposed a combination model of LSTM+CRF to extract sequence features by LSTM, and input the extracted sequence features into the CRF layer to add constraints to the predictive labels. Li [17] proposed Bert +BILSTM+CRF model, in which Bert was used for text encoding, and BILSTM was used for sequence feature extraction. CRF layer added some constraints for the final predicted tags to ensure that the predicted tags were legitimate. Bert +BILSTM+CRF model has excellent performance in accuracy and recall rate, and is one of the mainstream models for named entity recognition at present. However, BILSTM can not take into account local features when extracting global features of sequences. For the existence of entity nesting problem, it will also produce errors in the identification of named entities.

In summary, existing models cannot extract sequence features locally or globally according to sequence features, and the problem of entity nesting in named entity recognition task is not taken into account. For the above problems. In this paper, a hybrid sequence feature extraction and fusion entity attribute priority classification method are proposed to judge entity tags. The contributions of this paper are as follows:

(1) A weighted fusion method of CNN and BILSTM is proposed. CNN is used to extract the local features of the sequences. BILSTM extracts global sequence features related to text context. The contribution of local features to each word in the sequence was extracted. The contribution of global features to each word in the sequence was extracted. Multiply local and global sequence features with corresponding extracted contribution degrees to get the weights of local and global features to the words in the sequence, and add the weights.

(2) An EL is proposed based on the entity nesting problem, and the priority level of the attribute containing the nested entity is set to be higher than that of the nested entity attribute.

(3) The named entity recognition based on Bert+EL-LGWF+CRF is proposed. Bert is used to encode the text as a sequence composed of word vector matrix; CNN and BILSTM are used to extract sequence features. The generated sequence features fuse the attribute level value of the entity; The CRF layer calculates the optimal solution of the whole sequence for global optimization, and obtains the optimal prediction result of the whole sequence. The BERT pre-training model directly calls the model trained by Y [17].

2   Model Approach

2.1   BERT

Bert (Bidirectional Encoder Representation from Transformers) addresses the polysemy problem by dynamically encoding each word in text into a low-dimensional representation of a real-valued vector. This enhances the semantic representation of words. The network architecture of BERT is shown in Fig. 1. Bert is composed of multiple layers of Transformer (TRM in Fig. 1) [18]. The input encoding vector of Bert is the unit sum of three embedded features, and the output is the trained word encoding matrix $T_1, T_2..., T_n$ Fig. 1). The Transform coding unit is the most important part of BERT, and its core is to model a piece of text based on self-attention mechanism, as shown in Fig. 2. Input X1 and X2 of the Transform coding unit, add relative position coding information (Positinal Encoding in Fig. 2), and code through self-attention layer (self-attention in Fig. 2) and feedforward neural network (FNN in Fig. 2). Residual network and layer normalization (Add & Normal in Fig. 2) are added to solve the deep learning degradation problem Fig. 2.

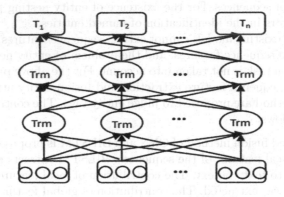

Fig. 1. The network architecture of BERT.

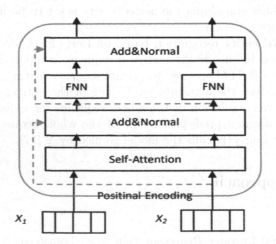

Fig. 2. Transform encoding unit.

2.2 CNN

CNN [19] is a neural network capable of representation learning and parallel computing. Convolutional neural networks can process multi-dimensional data, as shown in Fig. 3. The convolution layer is composed of convolution kernel (Conv) and activation function (ReLU). The convolution layer performs feature extraction on input data. After feature extraction in the convolutional layer, the output features will be transferred to the pooling layer for feature selection and information filtering. Full-connection layer (FC) is located in the last part of Convolutional Neural Network hidden layer. Full-connection layer is the nonlinear combination of extracted features to get the output Fig. 3).

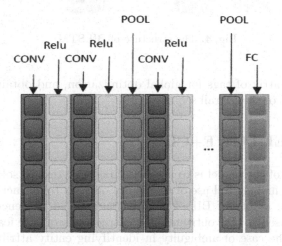

Fig. 3. Convolutional neural network diagram.

2.3 BILSTM

BILSTM is a bidirectional cyclic neural network, which can process long sequences of sentences and extract the global features of sentences [20]. The structure of BILSTM is shown in Fig. 4. BILSTM consists of a forward LSTM and a backward LSTM, and combines a set of sequences $X_1, X_2, ..., X_n$ as input, returns a set of sequences $H_1, H_2, ..., H_n$, these sequences contain information about each step of the input sequence; BILSTM can deal with the remote dependence problem, and takes into account the information before and after the current moment. BILSTM comprehensively considers all the information of the whole sequence Fig. 4).

2.4 CRF

CRF uses the dependency information between adjacent tags for sentence-level labeling, calculates the optimal solution of the overall sequence by adding the

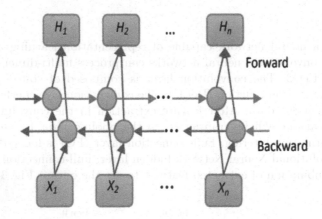

Fig. 4. The structure of BILSTM.

transfer score matrix of tags for global optimization, and obtains the optimal prediction result of the overall sequence [21].

3   BERT+EL-LGWF+CRF

The design idea of the model is to use Bert to encode text to solve the problem of polysemy. By making full use of the ability of CNN to extract local sequence features and the ability of BILSTM to extract global sequence features, the weight value is used as the contribution of extracting sequence features from the same data. In the case of ambiguity in identifying entity attributes, the rank of entity attributes is taken as the condition of priority judgment. CRF was optimized globally to get the final result. A named entity recognition model based on Bert+EL-LGWF+CRF was constructed, and an optimal sequence feature was finally obtained through continuous training and adjustment of the model.

3.1   Weighted Fusion According to CNN and BILSTM

CNN extracts the local features of the sequence, inputs the word vector matrix encoded by BERT into the CNN network, and uses the convolutional neural network to generate the local features of fixed size and independent from the input. This is shown in Fig. 5.

Let $(Pad, Pad, x1, x2, ..., xn, Pad, Pad)$ be the input of the CNN network, where $(x1, x2, ..., Xn)$ represents the input sentence, Pad (element 0) represents the data filling part, xi represents the vector coding of the ith word in the sentence, n represents the number of words in the sentence sequence. After passing through the CNN neural network, n outputs are obtained, $y1, y2..., yn$ is shown in formula (1).

$$y_i = \left([f\theta]_{[w]_t, i}\right) \tag{1}$$

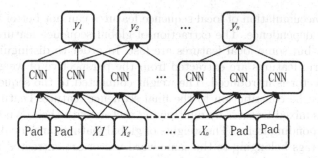

Fig. 5. CNN extracts local features of data.

where, f_θ is the fractional matrix with parameter θ, I is the sequence number of words in the sequence, [w] stands for the tag set, $[[w]_4t$ stands for the tag number of 0 t in the tag set, and t is the last tag number in the tag set.

CNN extract local features affect the accuracy of the named entity recognition, such as "poison eggs" entity attributes as involved food name, CNN first to identify the word "poison", and then identify the word "chicken", finally identify the word "egg", before and after such dependencies will be missing, causing the attribute recognition as independent entities, "poison" Identify the "egg" attribute as the name of the food in question.

BILSTM extracts the global characteristics of the data. The character coding sequence is input into the BILSTM layer, and the global feature extraction is performed to obtain the representation fraction matrix of the sequence. This is shown in Fig. 6. The model consists of input layer W, hidden layer H and output layer Y. W0, W1,... , Wn represents the features of input matrix, the hidden layer H is composed of the combination of forward LSTM and backward LSTM, and the output layer Y obtains the score matrix of each label in the word vector Fig. 6). Although BILSTM can ensure the long distance dependency, the accuracy of BILSTM is lower than that of CNN in the identification of independent entity attributes.

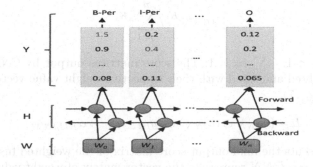

Fig. 6. BILSTM extracts global characteristics of the data.

Only the accumulation of local sequence features can not better capture the long-distance dependence. The extraction of global sequence features improves this problem, but some local features are lost. In order to distinguish different features, hybrid features are extracted from the perspective of weighted fusion and weight vector is introduced, The weight connected to the sequence feature is regarded as the contribution to the final sequence feature. Feature weighted fusion obtains mixed features by multiplying vectors with their own weights and adding corresponding items. The weights of global features and local features on the sequence tags belonging to the same word vector are extracted respectively for weighted fusion, as shown in Fig. 7.

CNN and BILSTM output each tag in the word vector score matrix (y1, y2... , yn), denoted by y, and the representation fraction matrix of the sequence output by the CNN model is denoted by y-CNN; The representation fraction matrix of the sequence output by the BILSTM model is denoted as Y-BILSTM; Y-CNN and Y-BILSTM pass through the network of full connection layer, and the calculation process of full connection layer is shown in the following formula.

$$F_{CNN} = g\left(W_{CNN}y_{CNN} + b\right) \tag{2}$$

$$F_{BILSTM} = g\left(W_{BILSTM}y_{BILSTM} + b\right) \tag{3}$$

where, g(·) is the activation function, b is the offset quantity, W-CNN and W-BILSTM are the weights of the connection in extracting sequence features. The weight changes dynamically with the training.

$$K_{CNN} = Dense_{unit=1}\left(y_{CNN}\right) \tag{4}$$

$$K_{BILSTM} = Dense_{unit=1}\left(y_{BILSTM}\right) \tag{5}$$

Y obtains the weight vector matrix (K1, K2, ... , Kn), the weight vector matrix is referred to as K. In which, $Dense_{(unit = 1)}$ is a fully connected function whose weight matrix is 1 dimension.

Pass K through Softmax layer, the α weight vector matrix $(\alpha 1, \alpha 2, ..., \alpha n)$, and the value of α i is shown in Formula (6).

$$\alpha_i = e^{K_i} / \sum_{j=1}^{n} e^{K_j} \tag{6}$$

where, $0 < \alpha_i < 1$, $\sum_i \alpha_i = 1$; The score matrices output by CNN and BILSTM are weighted and fused with their respective weight value vector matrices, as shown in Eq. (7)

$$L = \left(y_{CNN} \otimes \alpha_{CNN}\right) \oplus \left(y_{BILSTM} \otimes \alpha_{\bar{B}ILSTM}\right) \tag{7}$$

where L represents the final output score matrix of the weighted fusion of CNN and BILSTM, $\alpha - CNN$ represents the vector matrix of weight value extracted from Y-CNN, $\alpha_B ILSTM$ represents the vector matrix of weight value extracted from Y-BILSTM, $\otimes$ represents matrix multiplication, and $\oplus$ represents matrix

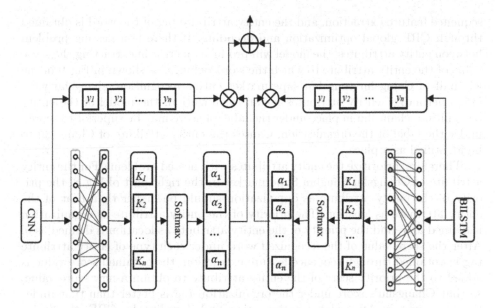

Fig. 7. Schematic diagram of feature extraction for weighted fusion.

addition Fig. 7). The advantage of mixed features is that the local sequence features extracted by CNN and the global sequence features extracted by BIL-STM can be taken as the proportion of features extracted by word vectors in the same sequence according to the weight value. By adopting the weighted fusion method, the information contained in mixed features can be made more comprehensive through the training of the network.

3.2   EL

The process of named entity recognition is that the model scores the category of entity attributes to which each word in the input text belongs. The attribute tag with a higher score is the entity category to which the word belongs. To solve the problem of entity nesting, the priority of entity properties is set.

$$EL = EL_1 + EL_2 + \cdots + EL_n \tag{8}$$

$$S = L \oplus EL \tag{9}$$

As shown in Eqs. (8) and (9). Where, ELi represents the priority ranking matrix of the ith entity attribute, and I is the number of entity attributes defined in the named entity recognition task. L is the final output score matrix of the weighted fusion of CNN and BILSTM; EL is the priority grade matrix composed of all entity attribute grade matrices; $\oplus$ represents matrix addition; $+$ represents matrix stitching operation.

In the case that the priority level of entity attribute is not set, the score value of the recognized word under each type of entity attribute tag is obtained through sequence feature extraction, and the entity attribute tag of the word is obtained through CRF global optimization and decoding. If there is a nesting problem between entity attributes, the model will produce an error in extracting the score value of the entity attribute to which the word belongs, as shown in Fig. 9 in the left half of changshu yong hui supermarket entity recognition, changshu yong hui supermarket entity attributes for an organization, the model of the entity recognition, changshu in place under the label of score may be superior to scored under the label of the organization, Causes the entity attribute of Changshu to be identified as a place.

Therefore, prioritize the entity attributes with nested problems. For the entity attribute of Changshu Yonghui Supermarket in the right part of Fig. 8, the priority of the entity attribute as organization should be higher than that of the entity attribute as location. The priority of the entity attribute as organization is defined as 1, and the priority of the entity attribute as location is defined as 0. After the score value of the recognized word under each type of entity attribute tag is obtained through sequence feature extraction, the obtained score value is added to the priority score of the entity attribute to obtain a new score value, so that Changshu's score under the organization tag is better than that under the location tag. After global optimization and decoding by CRF, the entity attribute tag of the word is obtained. Fig. 8).

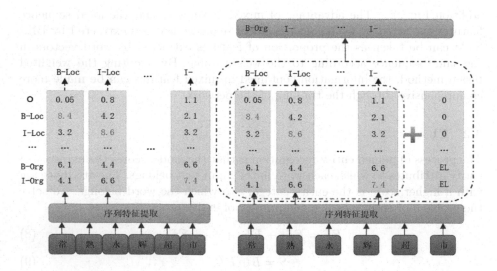

Fig. 8. Fusion El diagram.

Therefore, the proposed Chinese named entity model is shown in Fig. 9.

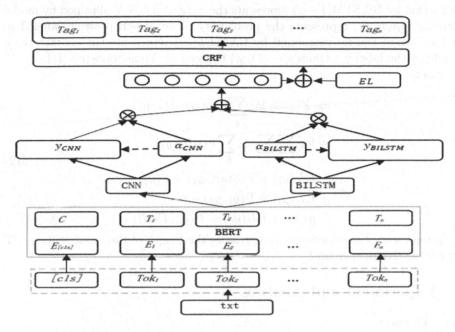

Fig. 9. A Chinese named entity recognition model based on BERT+EL-LGWF+CRF.

As shown in Fig. 9, the text is first partitioned to get TOK1, TOK2, ... ,Tokn; The obtained word segmentation sequence extracts text features through Bert, obtains word granularity vector matrix, and obtains the vector representation of each word in the sentence T1, T2,... Tn; Bert outputs the word vector encoding matrix as CNN and BILSTM inputs. The generated representations of CNN and BILSTM are dynamically weighted and fused, and the score value of the category label obtained are added to the priority score of the entity attribute to obtain the final score value. Finally, the global optimal sentence level tag sequence is obtained by learning the dependency between tags through CRF layer.

3.3   Loss Function

The maximum likelihood method is used for training, and the corresponding loss function is shown as the formula.

$$- \log p(y \mid x) = -score(x, y) + log Z(x) \tag{10}$$

where

$$score(x, y) = \sum_i \alpha p_{B(x_i y_i)} + (1 - \alpha) p_{P(x_i y_i)} + EL(y_i) + Trans(y_{i-1}, y_i) \tag{11}$$

α represents the weight of BILSTM obtained by model training, $p_{B(x_i,y_i)}$ represents the probability score of i-th position labeled y_i in the case of feature extraction by BILSTM$(1-\alpha)$ represents the weight of CNN obtained by model training, $p_P(x_i, y_i)$ represents the probability score of i-th position labeled y_i in the case of feature extraction by CNN.$EL(y_i)$ represents the characteristic weight of the label y_i. Trans$(y_(i-1), y_i)$ is given by TransitonScore [16].

where

$$
\log Z(x) = \log \sum_y \exp(score(x, y))
$$
$$
= \log \sum_{y_1} \sum_{y_2} \cdots \sum_{y_n} \exp\left(p\left(y_1\right)\right.
$$
$$
+ EL\left(y_1\right) + T\left(\text{Start}, y_1\right) + p\left(y_2\right) \quad\quad (12)
$$
$$
+ EL\left(y_2\right) + T\left(y_1, y_2\right) + \cdots
$$
$$
+ p\left(y_n\right) + EL\left(y_n\right) + T\left(y_n, End\right))
$$

p represents EmissionScore, n represents the length of the input sentence. T represents TransitonScore?

4   Experiment and Analysis

4.1   Dataset

People's Daily, MSRA, Golden-Horse data sets are adopted to self-labeled food additive related food safety event news report data set, referred to as food data set, as the experimental data set. The size of the data set is shown in Table 1. The validity of Chinese named entity recognition is verified for the proposed model.

In this paper, 10,680 news reports about food safety incidents related to food additives in the past five years are retrieved from the Food Partner website, and the entity attributes are screened out. These entity attributes are divided into 10 categories, and the corpus is marked with Yedda. The data set of news reports on food safety events related to food additives, referred to as food data set, is obtained. The entity attributes and labels of the food data set as well as the number of details are shown in Table 2.

The entity attribute was marked with BIO-style labeling strategy. For example, the name of the Food concerned is labeled as "FOOD", and a nine-season shrimp (I-food) of the Food concerned appeared in the data set, that is, the shrimp (I-food) is labeled as "B-food" and the additive is seaweed gelatin. The words that do not belong to any one of the 10 categories are marked as O. Examples of BIO labeling are shown in Table 3.

4.2   Evaluation Indices

Set of evaluation index for P (accuracy), R (a) recall rate, F1 value of P to correctly identify the number of entities of identify the percentage of the number

Table 1. Experimental dataset.

Dataset type	Data	Dataset size
Raw dataset	MSRA	42452
Raw dataset	People's Daily	20864
Raw dataset	Golden-horse	11731
Labeled datasets	Food datasets	625518 words

Table 2. Entity attributes and labels of food datasets.

Entity attribute	Label	Number of entities
Time	Time	4260
Location	Loc	5292
Person	Per	1164
Name of the food involved	Food	9042
Production enterprise	Org	5244
Additive	Add	4458
Trace elements, heavy metals	Elements	1044
Hazards, first aid measures	Influences	1326
Involved in the file	Document	648
Reason	Reason	2016

Table 3. An example BIO tag.

九	节	虾	注	有	海	藻	明	胶	。
B-Food	I-Food	O	O	O	B-Add	I-Add	I-Add	I-Add	O

of entities, R for the correct identification number of named entities standards results in the percentage of the number of entities, the weighted harmonic mean formula one is P and R is the influence of balance P and R composite indicator, The definitions are shown in Eqs. (10), (11) and (12).

$$P = \frac{TP + TN}{TP + FN + FP + TN} \times 100\% \tag{13}$$

$$R = \frac{TP}{TP + FN} \times 100\% \tag{14}$$

$$F1 = \frac{2 \times P \times R}{P + R} \times 100\% \tag{15}$$

Where, TP – Predicts positive cases to be true Fn – Predicts the positive example to be false FP –Predicts the counterexample to be true TN – Predicts the counterexample to be false.

4.3 Experimental Results and Analysis

Bert+BILSTM+CRF and Bert+CNN+CRF were selected as comparison models. Where: (1) BILSTM extracts global features of sequences, but some local features are lost due to the consideration of long distance dependencies; (2) CNN extracts the local features of the sequence and splices them together for sequence tag recognition. However, only the accumulation of local sequence features cannot capture the long-distance dependencies well. (3) This model gives consideration to both global and local features, and the theoretical effect is better than that of the comparative model.

All model parameters remain consistent. Experiments are carried out on the open data set and the food data set respectively, and the obtained experimental results are shown in Table 4. By analyzing Table 4, it can be seen that the proposed weighted fusion model obtained optimal results in P, R and F1 of all data sets on the common data sets. Compared with Bert + Bilstm +CRF, Bert +CNN+CRF and other single mode structures, the weighted fusion model significantly improves P, R and F1 of the three data sets. Compared with the weighted fusion model, the weighted fusion model has significantly improved P, R and F1 on the three data sets, indicating that the mixed feature extraction model has better performance in performance. In MSRA data set, the weighted fusion model improves the F1 value by 0.88% compared with Bert+CNN+CRF, which is the minimum value in the table. In the People's Daily data set, the weighted fusion model is 1.78% higher than that of Bert+CNN+CRF in the F1 value, which is the maximum value in the table. In the food dataset, the F1 value of the weighted fusion model reaches 82.69, which is 4.46% higher than that of the single-mode extraction method Bert+CNN+CRF with the lowest F1 value. This is because the entity attribute of the food data set is more nonstandard than that of the public data set, and the entity attribute has different lengths and obvious gaps. By comparison, it is shown that the weighted fusion of local features extracted by CNN and global features extracted by BILSTM constitutes a mixed feature extraction model, which has a great improvement in the performance of the named entity recognition task. It is fully proved that the mixed feature extraction model can improve the accuracy of named entity recognition compared with the single-mode.

Table 4. Experimental results of each method on datasets.

	MSRA			People's Daily			Golden-horse			Dataset about food		
	P	R	F1	P	R	F1	P	R	F1	P	R	F1
Weighted fusion model	96.77	96.92	96.84	95.85	95.74	95.79	92.98	92.23	93.12	83.19	82.19	82.69
BERT+BILSTM+CRF	95.96	95.96	95.96	95.04	93.99	94.51	92.12	92.52	92.32	82.39	80.41	81.39
BERT+CNN+CRF	96.50	96.53	96.52	94.03	93.99	94.01	91.86	92.89	92.37	78.68	77.78	78.23

There is no problem of entity nesting in the public data set, so the food data set labeled by oneself is used to determine the effectiveness of entity attribute level. To reduce named entity identification errors due to entity nesting problems, a priority level is defined for each entity attribute. 10% of the attributes of each entity are randomly selected for observation, in which there is no nested relationship between entity attributes such as a person, time, place, name of related food, additive, trace element, and heavy metal, so the priority level of these entity attributes is classified as 0. In the production enterprises with physical attributes, there are entities with nested locations with physical attributes, such as "Zhangjiagang Oshang Supermarket", "Suzhou Food and Drug Administration", etc., so the priority level of the production enterprises with physical attributes is classified as EL. Entities with nested entity attribute heavy metals and additives, such as "excessive lead", "excessive Fempni", etc., are caused by entity attribute events. Therefore, the priority level of entity attribute events is classified as EL. For the harm caused by physical attributes, the first aid measures have entities nested with trace elements of physical attributes, such as "affecting the absorption of calcium in the human body", etc. Therefore, the priority level of the harm caused by physical attributes is classified as EL. The entity attribute refers to the entity whose documents have nested the entity attribute manufacturer, place, and time, such as "Regulations on the Protection of Consumer Rights and Interests of Liaoning Province", "Notification of Food Safety Sampling Inspection of Shenzhen Market Supervision Administration in 2020", etc. Since the priority level of the entity attribute manufacturer is EL, the priority level of the entity attributes involved in the file is 2EL. The priority levels of each entity's attributes are shown in Table 5.

Table 5. Attribute priority of each entity

Entity attribute	Priority
Time	0
Location	0
Person	0
Name of the food involved	0
Production enterprise	EL
Additive	0
Trace elements, heavy metals	0
Hazards, first aid measures	EL
Involved in the file	2EL
Reason	EL

On the basis of the experiment of weighted fusion model (using CNN) in Table 4, EL is defined as different values respectively, and the experimental data obtained are shown in Table 6. From the analysis of Table 6, it can be seen that the performance of the model is constantly changing with different EL. When EL is between 0 and 0.05, F1 value fluctuates little, the difference is less than

Table 6. Experimental results of different El values.

EL	P	R	F1
0.025	83.48	82.19	82.83
0.05	83.90	81.79	82.83
0.1	85.29	83.11	84.18
0.2	84.34	83.20	83.77
0.4	85.82	84.27	85.05
0.8	86.14	84.54	85.27
1.3	85.95	83.20	84.55
1.6	84.86	82.82	83.83

1%, indicating that the level weight assigned to entity attributes is too small to affect the effect of named entity recognition. When EL is 0.8, F1 value reaches a maximum of 85.27 which is 2.58% higher than that when EL is 0. When EL is 1.3, F1 value is 84.55, which is lower than the peak value of 84.55. When EL is 1.6, F1 value drops to 83.83. The experimental results show that defining EL values for entity attributes can improve the performance of entity recognition in named entity recognition tasks with nested entities.

5 Summary

This paper proposes a sequential labeling method based on Bert+EL-LGWF-CRF model. This method mainly solves two problems in Chinese named entity recognition. (1) The recognition accuracy of irregular entity attributes by a single model is low. (2) Long entity attributes nested short entity attributes of different categories.

The method is evaluated on the Chinese named entity recognition task. The Experimental results show that the hybrid feature extraction model with weighted fusion of CNN and BILSTM is superior to the model with single CNN or BILSTM feature extraction in accuracy, recall value and F1. Especially in the small sample data set, the convergence speed is fast, and it is easier to learn the characteristics of the data set. On the basis of mixed feature extraction, the priority level of entity attributes is increased according to the nested features of the identified entity attributes, which can effectively solve the nested problem between Chinese entities and has excellent performance.

Use a large number of data sets to train the model. The trained model can extract entities from related text without any tags, which not only saves a lot of manpower and time, but also the accuracy and speed of manpower when the amount of text is too large. Incomparable, it is of great significance for the statistics and analysis of information needed in large-scale texts.

References

1. Sun, P., Yang, X., Zhao, X., et al.: An overview of named entity recognition. In: 2018 International Conference on Asian Language Processing (IALP). IEEE (2019)
2. Xie, R., Liu, Z., Jia, J., et al.: Representation learning of knowledge graphs with entity descriptions. In: Thirtieth AAAI Conference on Artificial Intelligence (2016)
3. Riedel, S., Yao, L., McCallum, A., et al.: Relation extraction with matrix factorization and universal schemas. In: Proceedings of the Conference of the North American Chapter of the Association for Computational Linguistics: Human Language Technologies 2013, pp. 74–84 (2013)
4. Shen, W., Wang, J., Luo, P., et al.: Linden: linking named entities with knowledge base via semantic knowledge. In: Proceedings of the 21st International Conference on World Wide Web, pp. 449–458 (2012)
5. Zhu, J., Uren, V., Motta, E.: ESpotter: adaptive named entity recognition for web browsing. In: Althoff, K.-D., Dengel, A., Bergmann, R., Nick, M., Roth-Berghofer, T. (eds.) WM 2005. LNCS (LNAI), vol. 3782, pp. 518–529. Springer, Heidelberg (2005). https://doi.org/10.1007/11590019_59
6. Babych, B., Hartley, A.: Improving machine translation quality with automatic named entity recognition. In: Proceedings of the 7th International EAMT Workshop on MT and Other Language Technology Tools, Improving MT Through Other Language Technology Tools, Resource and Tools for Building MT at EACL 2003 (2003)
7. Bordes, A., Usunier, N., Chopra, S., et al.: Large-scale simple question answering with memory networks. arXiv preprint arXiv:15060.02075 (2015)
8. Rau, L.F.: Extracting company names from text. In: IEEE Conference on Artificial Intelligence Application. IEEE (1991)
9. Bengio, Y., Frasconi, P.: An input output HMM architecture. Adv. Neural Inf. Process. Syst. **7**(4), 427–434 (1995)
10. Sutton, C.: Dynamic conditional random fields: factorized probabilistic models for labeling and segmenting sequence data. In: Proceedings of the 21st International Conference on Machine Learning 2004 (2007)
11. Bishop, C.M.: Neural Networks or Pattern Recognition (2005)
12. Collobert, R.: Natural language processing from scratch (2011)
13. Lample, G., Ballesteros, M., Subramanian, S., et al.: Neural architectures for named entity recognition (2016)
14. Huang, Z., Xu, W., Yu, K.: Bidirectional LSTM-CRF Models for Sequence Tagging (2015)
15. Devlin, J., Chang, M.W., Lee, K., et al.: BERT: pre-training of deep bidirectional transformers for language understanding (2018)
16. Dong, C., Zhang, J., Zong, C., et al.: Character-based LSTM-CRF with radical-level features for Chinese named entity recognition (2016)
17. Cui, Y., Che, W., Liu, T., et al.: Pre-training with whole word masking for Chinese BERT (2019)
18. Vaswani, A., Shazeer, N., Parmar, N., et al.: Attention is all you need. arXiv (2017)
19. Gu, J., Wang, Z., Kuen, J., et al.: Recent advances in convolutional neural networks. Pattern Recogn. (2015)
20. Ye, J., Zou, B., Hong, Y., Shen, L., Zhu, Q., Zhou, Q.: Negation and speculation scope detection in Chinese. J. Comput. Res. Dev. **56**(7), 1506–1516 (2019)
21. Pinto, D., McCallum, A., Wei, X., et al.: Table extraction using conditional random fields. In: 26th Annual International ACM SIGIR Conference on Research and Development in Information Retrieval, pp. 235–242 (2003)

Location Differential Privacy Protection in Task Allocation for Mobile Crowdsensing Over Road Networks

Mohan Fang$^{(\boxtimes)}$, Juan Yu, Jianmin Han, Xin Yao, Hao Peng,
Jianfeng Lu, and Ngounou Bernard

Zhejiang Normal University, Jinhua 321004, China
{fangmohan, yujuan, yaoxin, hpeng}@zjnu.edu.cn,
{hanjm, lujianfeng}@zjnu.cn

Abstract. Mobile Crowdsensing (MCS) platforms often require workers to provide their locations for task allocation, which may cause privacy leakage. To protect workers' location privacy, various methods based on location obfuscation have been proposed. MCS over road networks is a practical scenario. However, existing work on location protection and task allocation few considers road networks and the negative effects of location obfuscation. To solve these problems, we propose a Privacy Protection Task Allocation framework (PPTA) over road networks. Firstly, we introduce Geo-Graph-Indistinguishability (GeoGI) to protect workers' location privacy. And then we model a weighted directed graph according to the road network topology and formulate a linear programming to generate an optimal privacy mechanism, which aims to minimize the utility loss caused by location obfuscation under the constraint of GeoGI. We also improve the time-efficiency of the privacy mechanism generation by using a δ-spanner graph. Finally, we design an optimal task allocation scheme based on obfuscated locations via integer programming, which aims to minimize workers' travel distance to task locations. Experimental results on Roma taxi trajectory dataset show that PPTA can reduce average travel distance of workers by up to 23.4% and increase privacy level by up to 21.5% compared to the existing differential privacy methods.

Keywords: Mobile Crowdsensing · Location privacy · Task allocation · Road network · Linear programming

1 Introduction

Mobile Crowdsensing (MCS) [1] is an emerging paradigm that combines of crowd-sourcing and mobile devices, it engages workers to collect urban-scale sensing data with sensor-equipped smartphones. MCS has the advantages of low sensing cost, wide coverage, flexible deployment and strong computing capability, and enables a large number of applications in real-life, such as air quality monitoring [2], traffic information mapping [3], and feature description of interest points [4]. Task allocation [5] is an important part in MCS that can significantly impact the efficiency of MCS. It can be

H. Gao and X. Wang (Eds.): CollaborateCom 2021, LNICST 406, pp. 20–37, 2021.
https://doi.org/10.1007/978-3-030-92635-9_2

distinguished into two scenarios: *Worker Selected Task* (WST) and *Server Allocated Task* (SAT). In this paper, we focus on the SAT.

In the SAT, workers need to upload their locations to MCS platforms for task allocation. When an adversary observes the location, he/she may infer worker's religion, home/working address, interest preference [6], etc. Therefore, it is necessary to protect worker's location privacy in task allocation.

In recent years, lots of work has been proposed for location privacy protection, and most of them focused on the obfuscation-based methods which allow workers to upload obfuscated locations instead of actual locations to MCS platforms. However, these solutions still have the following limitations.

(1) Most work on location protection and task allocation [7, 8] does not consider road networks. They assume that the location can be obfuscated to any point in a 2-dimensional distribution. As shown in Fig. 1(a), if the location is obfuscated to an unreasonable place, such as on a lake, in a forest, or on a railroad. An adversary can infer that this location is not true and use an attack model (such as Bayesian inference attack) to predict the worker's actual location, which may increase the risk of privacy leakage. Moreover, these methods assume that the distance between locations is measured by the Euclidean distance. When workers' mobility (such as driving) is restricted by road networks, this assumption may cause high utility loss. As shown in Fig. 1(b), the locations u_1 and u_2 are obfuscated to the locations u'_1 and u'_2. The Euclidean distance from u'_1 to t and from u'_2 to t on 2D are 650 m and 620 m, respectively. For optimal task allocation, task t should be assigned to worker u_2. However, the shortest distance from u'_1 to t and from u'_2 to t are 940 m and 1780 m (unavoidable detour) over road networks. The utility loss caused by u'_2 reaches 1160 m, which is much higher than that of u'_1 (290 m). Hence, the road network is a factor that cannot be ignored in privacy protection and task allocation.

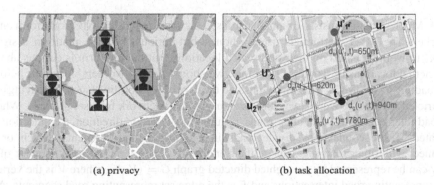

(a) privacy   (b) task allocation

Fig. 1. Influence of road networks on privacy and task allocation

(2) Existing work based on obfuscated methods [7, 9] usually takes the obfuscated locations provided by workers as the actual locations for task allocation. This means that they did not consider the influence of location obfuscation on task allocation efficiency, which is not reasonable and may cause high utility loss.

Recently, Geo-Graph-Indistinguishability (GeoGI) [10] has been presented in Location-Based Services (LBS) and an implementation method-Graph Exponential Mechanism (GEM) is proposed to solve the problem of privacy protection over road networks. However, task allocation is not considered in LBS. Therefore, GEM is not suitable to solve the problems raised in this paper. To tackle the above problems, we design a new privacy mechanism satisfying GeoGI and propose a Privacy Protection Task Allocation framework (PPTA) over road networks. In a nutshell, our contributions can be summarized as follows.

1) To protect workers' location privacy while preserving high utility in task allocation over road networks, we first propose to introduce Geo-Graph-Indistinguishability (GeoGI) to MCS and model a weighted directed graph according to the road network topology. Based on the graph, we formulate a linear programming to generate an optimal privacy mechanism, which can minimize the utility loss caused by location obfuscation under the constraints of GeoGI. We also improve the time-efficiency of the privacy mechanism generation by using δ-spanner graph.

2) To reduce the impact of location confusion on task allocation, we take the privacy mechanism proposed in PPTA as a key parameter to generate the task allocation scheme. With this idea, we formulate the problem of the optimal task allocation scheme generation as an integer programming, which aims to minimize workers' travel distance to task locations.

3) Experimental results on Roma taxi trajectory dataset show that PPTA can reduce the average travel distance of workers by up to 23.4% and increase privacy level by up to 21.5% compared to the existing differential privacy methods.

2 Preliminary

The MCS system consists of three parties, i.e., task requester, MCS platform and worker. The MCS platform is usually honest-but-curious, i.e., the platform assigns tasks to workers based on their obfuscated locations, but the platform is curious about the actual locations of workers. Therefore, we should protect workers' locations before uploading to the MCS platform. Consider the workers' mobility (such as driving) is restricted by road networks, we can represent the road network by a set of roads. When a road intersects, joins with other roads, or turns in a different direction, a connection is created. These connections divide roads into multiple road segments, which only connect with other road segments at their endpoints. Therefore, the road network of a city can be represented by a weighted directed graph $G = (V, E)$, where V is the vertex set representing road intersections and E is the edge set representing road segments. All vertices in V are on the road network, and for any pair of vertices $v_k, v_l \in V$, the weight of edge (v_k, v_l) is the shortest distance $d_g(v_k, v_l)$. Next, we introduce the privacy model and the adversary model.

2.1   Privacy Model

Geo-Graph-Indistinguishability (GeoGI) [10] is a novel location differential privacy model originally proposed for LBSs over road networks. It ensures that an adversary could not infer users' true locations from their released obfuscated locations. The location obfuscation process is to input an actual location u and output an obfuscated location u' through the probability matrix P. The probability matrix P is the key to realize GeoGI, which encodes the probability of obfuscating from a location to any location. GeoGI provides a feasible way to solve the problem of privacy protection for MCS over road networks.

The multiplicative distance between two distributions σ_1, σ_2 on some set $\mathcal{S}$ as $d_p(\sigma_1, \sigma_2) = sup_{S \in \mathcal{S}} |\ln \frac{\sigma_1(S)}{\sigma_2(S)}|$, with the convention that $|\ln \frac{\sigma_1(S)}{\sigma_2(S)}| = 0$ if both σ_1, σ_2 are zero and ∞ if only one of them is zero. Then, given $\varepsilon \in R^+$, Geo-Graph-Indistinguishability is defined as follows:

Definition 1. *(Geo-Graph-Indistinguishability)* [10]. *A probability matrix P on the road network $G = (V, E)$ satisfies Geo-Graph-Indistinguishability iff $\forall u_1, u_2, u'$ in V,*

$$d_p(P(u'|u_1), P(u'|u_2)) \le \varepsilon d_g(u_1, u_2) \tag{1}$$

where $P(u'|u_1)$ is the probability of obfuscating u_1 to u', $d_g(u_1, u_2)$ is the shortest distance from u_1 to u_2 on the road network, and privacy budget ε is a parameter of GeoGI. The smaller ε, the higher privacy.

The definition can be also formulated as $\forall u_1, u_2, u' \in V$, $\frac{P(u'|u_1)}{P(u'|u_2)} \le e^{\varepsilon d_g(u_1, u_2)}$. This formulation implies that GeoGI is an instance of d_x-privacy [11] proposed by Chatzikokolakis et al. The authors showed that an instance of d_x-privacy guaranteed strong privacy. Intuitively, this definition guarantees that if the obfuscation location is u', for any two locations u_1 and u_2 in V, the obfuscating probability of them to u' is approximate. Even though an adversary knows the obfuscated location u' and the probability matrix P, he/she cannot distinguish which one is the actual location. It is worth noting that GeoGI relies on a city's road network topology, and this results in the privacy protection level and utility varying depending on the road network even if the privacy parameter remains the same.

Specifically, in contrast to consider individual people's location [12] in LBS, task allocation efficiency in MCS depends on all the workers' locations. As shown in Fig. 2 (a), we assume a scenario with one worker u_1 and four tasks t_1, t_2, t_3, t_4. Due to task t_1, t_2 are closer to worker u_1, task t_1, t_2 will be allocated to worker u_1. When a new worker u_2 is added, as shown in Fig. 2(b), where $d_g(u_1, t_3) + d_g(u_2, t_2) < d_g(u_1, t_2) + d_g(u_2, t_3)$. For optimal task allocation, MCS platform will select worker u_1 perform task t_1, t_3 and worker u_2 perform task t_2, t_4. Therefore, the workers' location distribution must be considered in the generation of the probability matrix.

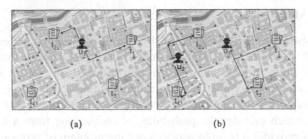

(a)   (b)

Fig. 2. Influence of workers distribution in task allocation

2.2 Adversary Model

We assume that workers' locations in MCS may suffer from Bayesian inference attack [6], i.e., if an adversary knows probability matrix P, worker's obfuscated location u' and location distribution $\pi(u)$, he/she can estimate the posterior distribution $\sigma(\hat{u}|u')$ of the actual location u by resorting to the Bayes' Equation [13]. It is defined as follows:

$$\sigma(\hat{u}|u') = \frac{P(u'|\hat{u}) \cdot \pi(\hat{u})}{\sum_{u* \in V} P(u'|u*) \cdot \pi(u*)} \tag{2}$$

where u' is the obfuscated location, $\hat{u}$ is the location inferred by an adversary when observing the location u', $u*$ is any location in V, and $P(u'|\hat{u})$ is the probability of obfuscating $\hat{u}$ to u'.

We use the inference error (IE) proposed by Shokri et al. [14] to quantify the privacy level of a mechanism. The researchers translated location privacy into IE by measuring how accurately an adversary could infer the worker's actual location. Formally, IE can be formulated as follows:

$$IE(\pi_a, P, \sigma, d_g) = \sum_{u \in V} \sum_{u' \in V} \sum_{\hat{u} \in V} \pi_a(u) P(u'|u) \sigma(\hat{u}|u') d_g(u, u') \tag{3}$$

where $\pi_a(u)$ is the prior knowledge of workers' location distribution by an adversary, $P(u'|u)$ is the probability of obfuscating u to u', $\sigma(\hat{u}|u')$ is the posterior distribution of the actual location u, and $d_g(u, \hat{u})$ is the shortest distance from u to $\hat{u}$ on the road network.

Due to a probability matrix P satisfies GeoGI, it can limit the promotion of an adversary's posterior knowledge $\sigma(u|u')$ about workers' distribution over the prior knowledge π_a, i.e., $\sigma(u|u')/\pi_a \leq e^{\varepsilon D(R)}$ where $D(R)$ is the maximum distance of any two locations in region R. Please refer to [15] for the theoretical proof.

Consider an extreme situation, if an adversary knows the exact location of the worker through some ways in advance, then IE will always be zero. Therefore, we assume that the prior knowledge of the adversary $\pi_a(u)$ is equivalent to the public knowledge of workers' location distribution $\pi(u)$ (e.g., leaked by public check-ins [16]).

3   PPTA Framework

In this section, we introduce the PPTA framework. Figure 3 shows the overview of PPTA which consists of two modules: Location Obfuscation and Task Allocation Based on Obfuscated Locations. In the first module, the MCS platform needs to obtain workers' location distribution based on the historical sensing data and uses it as a parameter to generate a probability matrix via linear programming. After the platform generates the matrix, workers can download it into their smartphones, and then obfuscate their actual locations according to the probabilities encoded in the matrix. The obfuscated locations are uploaded to the platform for task allocation. In the second module, after receiving workers' obfuscated locations, the platform will assign tasks to proper workers, attempting to minimize the total traveling distance to the task locations. Since workers' uploaded locations are obfuscated, directly seeing them as actual locations for task allocation is not reasonable. Therefore, the probability matrix should be considered in the generation of a task allocation scheme for better allocation efficiency.

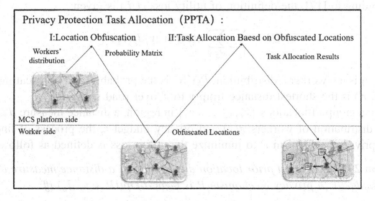

Fig. 3.   PPTA framework

3.1   Location Obfuscation

In PPTA, each worker obfuscates his/her location through the probability matrix and uploads the obfuscated location to the MCS platform. Figure 4 shows an example of probability matrix, where workers' possible location is discrete into five vertices $\{v_1, v_2, v_3, v_4, v_5\}$. In this case, the probability matrix generated by the MCS platform is a 5×5 matrix. If the worker's actual location is v_2, according to the matrix, the probabilities that the worker outputs v_1, v_2, v_3, v_4, v_5 as the obfuscated location are 0.1, 0.3, 0.1, 0.1 and 0.4, respectively. Because the generation of probability matrix satisfies GeoGI, even if an adversary knows the obfuscated location and probability matrix, he/she cannot infer the actual location of the worker.

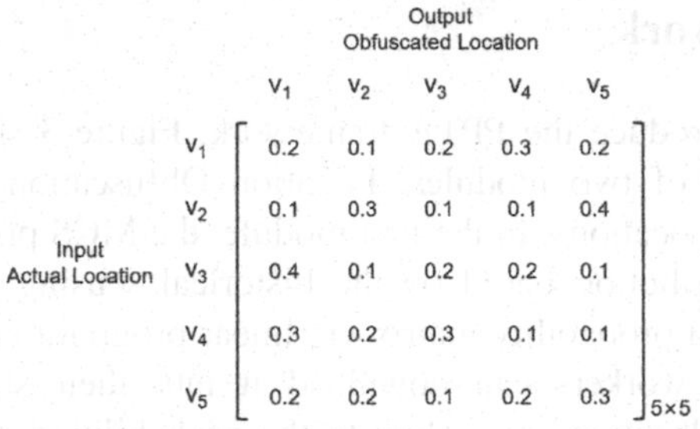

Output
Obfuscated Location

$$
\begin{array}{c}
\text{Input} \\
\text{Actual Location}
\end{array}
\begin{array}{c}
V_1 \\ V_2 \\ V_3 \\ V_4 \\ V_5
\end{array}
\begin{array}{ccccc}
V_1 & V_2 & V_3 & V_4 & V_5 \\
\left[\begin{array}{ccccc}
0.2 & 0.1 & 0.2 & 0.3 & 0.2 \\
0.1 & 0.3 & 0.1 & 0.1 & 0.4 \\
0.4 & 0.1 & 0.2 & 0.2 & 0.1 \\
0.3 & 0.2 & 0.3 & 0.1 & 0.1 \\
0.2 & 0.2 & 0.1 & 0.2 & 0.3
\end{array}\right]_{5\times5}
\end{array}
$$

Fig. 4. Example of probability matrix

Although traditional privacy mechanisms provide a simple way to achieve privacy protection, these methods are independent of the prior knowledge of workers' location distribution which may cause high utility loss. To reduce this loss, we take the workers' distribution as a key parameter to generate the probability matrix P, which can be calculated from the workers' historical sensing data.

According to [17], the definition of utility loss (UL) is given:

$$
UL = \sum_{v \in V} \sum_{v' \in V} \pi(v) P(v'|v) d_g(v, v') \tag{4}
$$

where $\pi(v)$ is the workers' distribution, $P(v'|v)$ is the probability of obfuscating v to v', and $d_g(v, v')$ is the shortest distance from v to v' over road networks.

Given a group of locations $\{v_1, v_2, \ldots, v_n\}$ in region, a distance measure d_g, overall location distribution of workers $\pi(v)$ and privacy budget ε, the process of finding an optimal privacy mechanism P to minimize the utility loss is defined as follows:

Definition 2 [17]. *Given a prior location distribution π, a distance measure d_g and a privacy budget ε, a privacy mechanism P is GeoGI-OptUL (π, d_g) iff:*

1. *P is geo-graph-indistinguishability and*
2. *for all mechanisms P', if P' is geo-graph-indistinguishability then $UL(P, \pi, d_g) \leq UL(P', \pi, d_g)$.*

Note that GeoGI-OptUL (π, d_g) optimizes UL given a privacy constraint of GeoGI. Now, we can formulate the problem of privacy mechanism generation (PMG) as a linear programming, which can minimize the utility loss:

$$
\min_P \sum_{v \in V} \sum_{v' \in V} \pi(v) P(v'|v) d_g(v, v') \tag{5}
$$

$$
subject\ to : P(v'|v_1) \leq e^{\varepsilon d_g(v_1, v_2)} P(v'|v_2), \quad v_1, v_2, v' \in V \tag{6}
$$

$$
\sum_{v' \in V} P(v'|v) = 1, \quad v \in V \tag{7}
$$

$$P(v'|v) \geq 0, \quad v, v' \in V \tag{8}$$

where constraint (6) is the definition of GeoGI, the shortest distance d_g from workers to tasks is calculated by Dijkstra's algorithm on the road network and updated by the MCS platform before each round of task allocation. Therefore, the road network is considered in PMG. Constraints (7–8) are the basic requirements of probability. Obviously, PMG satisfies GeoGI.

It should be noted that PPTA does not need workers to upload locations frequently, only once in PPTA. The location obfuscation runs completely in the worker's smartphone, so no one else knows the worker's actual location.

3.2 Task Allocation Based on Obfuscated Locations

We consider a scenario where M tasks need to be allocated to N workers ($N < M$). After the locations of N workers have been obfuscated and uploaded to the MCS platform, the task allocation scheme designs by the MCS platform can be represented by an indicator matrix $X = \{x_{i,j}\}_{N \times M}$, where the matrix element $x_{i,j}$ indicates whether task j is allocated to worker i, i.e., $x_{i,j} = 1$ if task j is assigned to worker i; otherwise, $x_{i,j} = 0$. The matrix X needs to satisfy the constraints: (1) $\sum_i x_{i,j} \geq p, \forall j (j = 1, \ldots, M)$, i.e., at least p workers are required to perform a task. (2) $\sum_j x_{i,j} \geq q, \forall i (i = 1, \ldots, N)$, i.e., each worker needs to perform at least q tasks. For the MCS platform, considering the phenomenon of malicious uploading data by workers, constraint (1) is to improve the quality of data. For workers, constraint (2) is to accept more tasks in a single task allocation without uploading location multiple times can not only improve the revenue, but also reduce the risk of privacy leakage.

The goal of task allocation is to ensure each task should be assigned to workers and the total travel distance of workers is minimized, which is also known as a common metric to measure the task allocation efficiency. Hence, we formulate the optimal task allocation scheme as follows, and solve it with an integer programming.

$$\min_x \sum_{u' \in V} \sum_{t \in V} d_g(u', t) x(u', t) \tag{8}$$

$$subject\ to : x(u', t) \in \{0, 1\}, \forall u', t \in V \tag{9}$$

$$\sum_{t \in V} x(u', t) \geq q, \quad \forall u' \in V \tag{10}$$

$$\sum_{u' \in V} x(u', t) \geq p, \quad \forall t \in V \tag{11}$$

where $d_g'(u', t)$ represents the travel distance from the obfuscated location u' to the task t over road networks. x is the task allocation scheme generated by the MCS platform according to the obfuscated locations.

Due to the MCS platform receives the obfuscated locations, it is unreasonable to directly regard them as the actual locations for task allocation. Although the privacy mechanism designed in this paper has taken it into account, it still brings high utility loss. According to the definition of probability matrix P, any two locations are obfuscated to the same location is approximate. That means, for an obfuscated location, all locations in region may be its actual location. Therefore, we combine the probability matrix into the generation of task allocation scheme and use the distance from all possible locations to the task instead of the distance from the obfuscated location to the task. The mathematical relationship is as follows:

$$d_g'(u', t) = \frac{\sum_{u \in V} \pi(u)P(u'|u)d_g(u, t)}{\sum_{u \in V} \pi(u)P(u'|u)} \tag{12}$$

From the above formulation, we can find that the higher the obfuscated probability of a location, the more likely it is to be the actual location, and the larger its distance weight, the lower the effect of location obfuscation on task allocation. Finally, by plugging Eq. (12) into Eq. (8), we can get a final objective function:

$$\min_x \sum_{u' \in V} \sum_{t \in V} \frac{\sum_{u \in V} \pi(u)P(u'|u)d_g(u, t)}{\sum_{u \in V} \pi(u)P(u'|u)} x(u', t) \tag{13}$$

The value range of the variable x in the model is limited to integer, thus this is an integer programming. The above two mathematical models can be solved by standard LP approaches, such as the simplex methods, or the advanced program solver (e.g., *CPLEX, Lingo*).

3.3   Speed-Up with δ-Spanner Graph

In the process of PMG, the number of constraints (6) is $O(|V|^3)$, which makes the method proposed in this paper difficult to extend to large-scale regions in real-life. Considering the total number of constraints to generate the task allocation scheme is $O(|N||M|)$, which is far less than $O(|V|^3)$. Thus, we only need to optimize PMG, which is the most time-consuming part in PPTA. Some common approaches, such as the dual form of linear programming, can be used to speed up PMG. However, although there are fewer constraints, the constraints in the dual problem will become more complex, so it is not practical.

We speed up PPTA by using δ-spanner graph. It ensures that for a given obfuscated location, it compares whether the obfuscating probability of adjacent locations in region satisfies GeoGI, rather than any two locations. According to [17], we construct a δ-spanner graph, which contains all the vertices in a weighted directed graph but reduces the number of edges. Stretch factor δ is an important parameter of δ-spanner graph, which represents the maximum ratio of the distance between any two vertices in two graphs. The definition is as follows:

Definition 3. (Dilation) [17]. *Let $G_\delta = V, E_\delta$ be a spanner graph. The dilation of G_δ is calculated as:*

$$\delta = \max_{v \in V, v' \in V, v \neq v'} \frac{d_{g_\delta}(v, v')}{d_g(v, v')} \tag{14}$$

A spanner with dilation δ is called a δ-spanner graph.

Now, to speed up PPTA, we introduce δ-spanner graph to GeoGI and the following theorem holds:

Theorem 1 [18]. *If $G_\delta(V, E_\delta)$ is a δ-spanner graph, and a probability matrix P satisfies:*

$$P(v'|v_1) \leq e^{\frac{\varepsilon}{\delta} d_{g_\delta}(v_1, v_2)} P(v'|v_2), \quad (v_1, v_2) \in E_\delta, v' \in V \tag{15}$$

Then, P satisfies Geo-Graph-Indistinguishability.

Proof. According to Eq. (14), we can obtain

$$d_{g_\delta}(v, v')/\delta \leq d_g(v, v'), \quad \forall v, v' \in V \tag{16}$$

By using Eq. (6) and Eq. (16), we can derive

$$P(v'|v_1) \leq e^{\frac{\varepsilon}{\delta} d_{g_\delta}(v_1, v_2)} P(v'|v_2) \leq e^{\varepsilon d_g(v_1, v_2)} P(v'|v_2), \quad \forall (v_1, v_2) \in E_\delta, v' \in V \tag{17}$$

This concludes the proof.

According to Definition 3 and Theorem 1, the mathematical model of PMG is updated:

$$\min_P \sum_{v \in V} \sum_{v' \in V} \pi(v) P(v'|v) d_{g_\delta}(v, v') \tag{18}$$

$$subject\ to: P(v'|v_1) \leq e^{\frac{\varepsilon}{\delta} d_{g_\delta}(v_1, v_2)} P(v'|v_2), \quad \forall (v_1, v_2) \in E_\delta, v' \in V \tag{19}$$

$$\sum_{v' \in V} P(v'|v) = 1, \ v \in V \tag{20}$$

$$P(v'|v) \geq 0, \ v, v' \in V \tag{21}$$

The number of constraints (19) is $O(|E_\delta||V|)$ by using δ-spanner graph. For a δ-spanner graph [17], $|E_\delta| = \frac{|V|}{\delta - 1}$, thus the number of constraints in PMG can be reduced from $O(|V|^3)$ to $O(|V|^2)$ approximately. Following previous work [17], when δ equal to 1.08, the experimental effect is the best.

4 Evaluation

In this section, we first evaluate the performance of the proposed PPTA framework in terms of privacy and utility with a real-world dataset. Then, we evaluate the time-efficiency of PMG before and after the optimization of δ-spanner graph.

4.1   Experiment Configurations

Evaluation Scenario

We conduct experiments by using a publicly real-world taxi trajectory dataset in Roma [23]. The dataset contains GPS coordinates of approximately 320 taxis collected over 30 days and some of them are selected for the experiments. The longitude range of the selected dataset is (12.418, 12.574) and the latitude range is (41.859, 41.947). Most of the data is in the central of Roma and a small percentage in the suburbs. We select to use a taxi dataset since taxi services can be regarded as a MCS application type (taxi driver can be considered as a worker, passenger can be considered as a task).

To evaluate the PPTA, we set up the experiments with different parameters. The privacy budget ε ranges from ln (2) to ln (8). ε is usually chosen by workers, for simplicity, we set the same ε for each worker in the experiment. In each round of task allocation (a round is set to 1 h), the number of workers (N) ranges from 15 to 40, and the number of tasks (M) ranges from 45 to 120. Note that before each round of task allocation, we learn workers' location distribution $\pi(u)$ according to workers' historical sensing data. We also conduct experiments for different task distributions (Fig. 5), which are compact, scattered and hybrid. Finally, we change the size of the region to evaluate the time-efficiency of PMG before and after the optimization of δ-spanner graph.

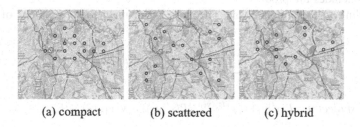

(a) compact   (b) scattered   (c) hybrid

Fig. 5.   Three types of task distribution

In this paper, we use Lingo to solve two linear programming problems, and python is used for experiments. All experiments are conducted on inter (R) core (TM) i7-4710 hq CPU@2.5 GHz, 8 GB RAM, win10 OS.

Evaluation Metrics

ATD (Average Travel Distance). Lots of work [7, 9] considers workers' travel distance to task locations as an important factor in task allocation. Following previous work, we use ATD as the utility metric, which can be calculated as:

$$ATD = \sum_{(u,t)\in X} d_g(u,t)/|x| \tag{22}$$

where $|x|$ is the total number of tasks allocated to workers. The smaller the ATD, the lower the utility loss is.

IE (Inference Error). An adversary can infer worker's actual location by resorting to the Bayes' Equation, if he/she knows the worker's upload location, workers' location distribution and probability matrix. We use IE (Eq. (3)), i.e., the expected distortion from the inferenced location (by adversary) to the actual location, to quantify the privacy level of PPTA. If IE is smaller, the privacy level of the mechanism is lower.

Baselines

Laplace Mechanism (LAP) [7]. Laplace is a traditional differential privacy mechanism and tends to obfuscate a location to its nearby location with high probability. Its probability distribution is derived from a two-dimensional version of the Laplace distribution as follows.

$$P_{lap}(u'|u) \propto e^{-\varepsilon \frac{d_e(u,u')}{D(R)}} \tag{23}$$

where d_e is the Euclidean distance. $D(R)$ is the maximum distance between any two locations in the region R.

Exponential Mechanism (EXP) [19]. Exponential mechanism is also widely used to achieve differential privacy. In the design of Exponential mechanism, a scoring function needs to be modeled to obtain high utility. Given a location, a better obfuscation should be assigned a higher score. In MCS, a higher score is preferred for the location obfuscation that leads to lower utility loss. With this idea, we design the following Exponential mechanism.

$$P_{\exp}(u'|u) \propto e^{\frac{\varepsilon}{2} \cdot (1 - \frac{d_g(u,u')}{\max_{u* \in R} d_g(u,u*)})} \tag{24}$$

Graph Exponential Mechanism (GEM) [10]. Graph exponential mechanism employs the idea of exponential mechanism. This mechanism considers road networks so that high utility can be expected. Moreover, since this mechanism satisfies GeoGI, strong privacy based on differential privacy is guaranteed. The obfuscated probability is as follows.

$$P_{gem}(u'|u) \propto e^{-\frac{\varepsilon}{2}d_g(u,u')} \tag{25}$$

No Privacy Protection. No privacy protection means that the MCS platform knows all workers' actual locations, which can be regarded as the lower bound of ATD for task allocation based on obfuscated locations.

4.2   Experimental Results

Utility

In the experiment, we first evaluate PPTA in term of utility (task allocation efficiency). If ATD is closer to No Privacy, the utility loss is lower. It can be seen from Fig. 6(a) that all the methods (except No Privacy) will lead to the continuous decrease of ATD, as the privacy budget ε increases. According to the definition of differential privacy, a larger ε denotes the lower privacy level, which leads to a location will be obfuscated to its nearby location with a high probability. Thus, ATD shows a downward trend. In addition, privacy budget ε cannot influence on the No Privacy, so the ATD of No Privacy stays the same. In comparison of the four privacy mechanisms, the ATD generated by Laplace mechanism is the largest. This is because Laplace mechanism does not consider the characteristics of road networks and use Euclidean distance, resulting in traditional differential privacy mechanisms are not suitable to provide location protection over road networks. Graph exponential mechanism and Exponential mechanism add road network constraints in the design process, so their utility loss is lower than Laplace mechanism. For the Exponential mechanism, we redesign its scoring function so that a location will be obfuscated with a higher probability to a location with smaller utility loss. As for Graph exponential mechanism, it is a privacy mechanism proposed in LBS, which does not consider task allocation. Therefore, compared with the Exponential mechanism, the application of Graph exponential mechanism to MCS will result in a larger utility loss. From the experimental results, we can find that these methods still have a certain gap with PPTA in term of utility. The main reason for this situation is that PPTA considers the influence of location obfuscation on task allocation and reduces the influence in the optimization model.

Figure 6(b) describes the relationship between the number of tasks and ATD. It can be seen from Fig. 6(b) that with the increase of number of tasks, ATD shows an upward trend. This is because when the number of tasks increases, each worker needs to perform more tasks to complete the task requirements of each round. From the experimental results, we can find that ATD generated by PPTA is always smaller than the other three privacy mechanisms. Figure 6(c) describes the relationship between the number of workers and ATD. As the number of workers increases, ATD shows a downward trend. This is because workers who upload their locations will be assigned tasks (otherwise privacy will be sacrificed in vain) and there are more choices to assign tasks for the platform. Compare with other privacy mechanisms, PPTA generates the smallest ATD regardless of the number of workers or tasks and has a stable performance.

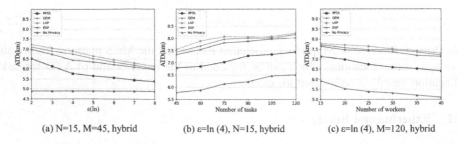

(a) N=15, M=45, hybrid   (b) ε=ln (4), N=15, hybrid   (c) ε=ln (4), M=120, hybrid

Fig. 6. Varying privacy budget, task number, and worker number

We also evaluate the effect of different task distributions on utility: compact distribution, scattered distribution, and hybrid distribution. From previous experiments, we can find that Exponential mechanism performed better than Laplace mechanism and

Graph exponential mechanism in term of utility, so we only compare Exponential mechanism and PPTA. As shown in Fig. 7, we can find that the scattered distribution generates the largest ATD of three task distributions, and the compact distribution has the smallest ATD. This is because the distance between tasks under the scattered distribution is longer than other two task distributions, and workers need to travel a longer distance to accomplish all tasks. From the experimental results, we can also find that no matter compact, scattered or hybrid, PPTA generates much smaller ATD than Exponential mechanism. This means that our proposed method can achieve stable performance across different distributions. Noted that compared with the scattered and hybrid distribution, ATD generated by PPTA and Exponential mechanism under the compact distribution is the closest to no privacy, which indicates that the compact distribution has the lowest influence on utility. On the contrary, the scattered distribution has the greatest influence on utility.

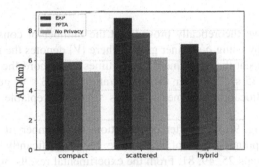

Fig. 7. Varying task distribution (ε = ln (4), N = 15, M = 45)

Privacy

We next evaluate PPTA in term of privacy. If IE is smaller, the privacy level of mechanism is lower. As shown in Fig. 8, as ε increases, all methods lead to the continuous decrease of IE. According to Eq. (3), we can find that a larger ε means the adversary's inferenced location is closer to the actual location. Thus, IE shows a downward trend. With the comparison of other three mechanisms (LAP, GEM, EXP), the IE generated by PPTA is the largest which indicates PPTA provides the best privacy protection under the same privacy budget ε. From the experimental results, when ε is ln (5), the largest difference in IE between PPTA and other privacy mechanisms. It means that when consider both privacy and utility in MCS, the privacy budget ε set as ln (5) has a best effect. As shown in Fig. 6(a) and Fig. 8, when the privacy budget ε is constantly increasing, ATD of PPTA is closer to No Privacy and the task allocation efficiency is improved. However, IE is constantly decreasing and the privacy level will be lower.

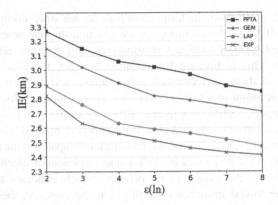

Fig. 8. Varying privacy budget (N = 15, M = 45, hybrid)

Time-Efficiency

In Sect. 3.3, we have theoretically proved that the number of constraints in PMG is reduced to $O(|V|^2)$ by using δ-spanner graph, where $|V|$ denotes the number of vertices in the graph. We repeat the experiments many times and record the mean time, which takes about 1 min 45 s (19 s after the optimization) and 3 s to generate probability matrix and task allocation scheme, which is totally acceptable in real-life MCS applications.

As shown in Fig. 9(a) (b), after optimization, the number of iterations and the corresponding computation time of PMG have been significantly reduced when the location number equals 25, 49, 81. From the experimental results, we can find that the larger the location number, the more significant the optimization effect. When the locations' number reaches 81, the computation time is shortened by five times and the number of iterations is reduced by nearly half after optimization.

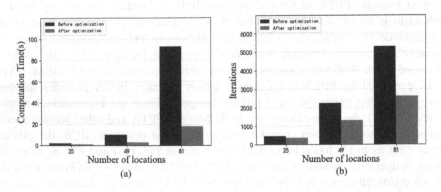

Fig. 9. Varying region size ($\varepsilon = \ln (4)$, N = 15, M = 45, hybrid)

5   Related Work

In recent years, a variety of location privacy protection approaches have been proposed, such as anonymity [20], encryption [21], etc. However, these works have the drawback of dependence on trustful platforms and high cost. To address these problems, we focus on obfuscation-based methods, and differential privacy has been applied to address location privacy issues in MCS.

In the context of sparse MCS, Wang et al. [18, 22] propose a privacy protection framework, which takes into account the level of privacy, the prior knowledge about workers' location distribution, and the data quality loss due to location obfuscation. Particularly, the framework can provide a guaranteed level of differential and distortion privacy with reduced data quality loss in Sparse MCS applications. However, they only evaluate the data quality loss and do not consider the privacy level of the proposed framework. Yang et al. [7] analyzes the shortcomings of existing works and propose a mixed integer nonlinear programming model which aims at minimizing workers' travel distance. It uses differential geo-obfuscation to protect workers' location privacy regardless of adversaries' prior knowledge, without the involvement of any trustful third-party. However, it does not consider the adversary attack and use Euclidean distance as a metric, which still has some defects in privacy and utility. Liu et al. [8] consider two kinds of task allocation scenarios, multi-task with few workers and few tasks with multi workers. They design the multi-objective optimization model and propose the W-ILP and C-ILP algorithms to select workers with the minimum total incentive payments and minimum total travel distance. However, they did not consider the road networks, which may cause insufficiencies in terms of privacy and utility.

6   Conclusion

In this paper, we have proposed a Privacy Protection Task Allocation framework (PPTA) to protect workers' location privacy for MCS over road networks. We introduce GeoGI and model a weighted directed graph according to the road network topology. Then, we formulate a linear programming to generate an optimal privacy mechanism. It considers the level of privacy protection, the prior knowledge about workers' location distribution and the utility loss due to location obfuscation. We also improve the time-efficiency of the privacy mechanism generation by using δ-spanner graph. Finally, we design an optimal task allocation scheme based on obfuscated locations by using an integer programming, which aims to minimize workers' travel distance to task locations. Experimental results on Roma taxi trajectory dataset show that PPTA can reduce average travel distance of workers by up to 23.4% and increase privacy level by up to 21.5% compared to the existing differential privacy methods.

In this paper, we assume that the workers are not related to each other. However, this assumption may not be reasonable. Adversary may learn about workers' correlation from their check-in records, which may cause unexcepted privacy leakage. In the future, we will aim at researching this correlation attack and will propose a corresponding solution.

Acknowledgments. The authors would also like to appreciate the anonymous reviewers for their valuable suggestions, which lead to a substantial improvement of this paper. This research has been funded by the National Natural Science Foundation of China (Grant No. 61672468, 61702148).

References

1. Capponi, A., Fiandrino, C., Kantarci, B., Foschini, L., Kliazovich, D., Bouvry, P.: A survey on mobile crowdsensing systems: challenges, solutions, and opportunities. IEEE Commun. Surv. Tutorials **21**(3), 2419–2465 (2019)
2. Fiandrino, C., et al.: CrowdSenSim: a simulation platform for mobile crowdsensing in realistic urban environments. IEEE Access **5**, 3490–3503 (2017)
3. Tong, Y., Chen, L., Shahabi, C.: Spatial crowdsourcing: challenges, techniques, and applications. Proc. VLDB Endow. **10**(12), 1988–1991 (2017)
4. Chon, Y., Lane, N.D., Li, F., Cha, H., Zhao, F.: Automatically characterizing places with opportunistic crowdsensing using smartphones. In: Proceedings of the 2012 ACM Conference on Ubiquitous Computing, New York, USA, pp. 481–490 (2012)
5. Wang, J., Wang, L., Wang, Y., Zhang, D., Kong, L.: Task allocation in mobile crowd sensing: state-of-the-art and future opportunities. IEEE Internet Things J. **5**(5), 3747–3757 (2018)
6. Chen, J., Ma, H., Zhao, D., Liu, L.: "Correlated differential privacy protection for mobile crowdsensing. IEEE Trans. Big Data 1 (2017)
7. Wang, L., Yang, D., Han, X., Wang, T., Zhang, D., Ma, X.: Location privacy-preserving task allocation for mobile crowdsensing with differential geo-obfuscation. In: Proceedings of the 26th International Conference on World Wide Web, Perth Australia, pp. 627–636 (2017)
8. Liu, Y., Guo, B., Wang, Y., Wu, W., Yu, Z., Zhang, D.: TaskMe: multi-task allocation in mobile crowd sensing. In: Proceedings of the 2016 ACM International Joint Conference on Pervasive and Ubiquitous Computing, New York, USA, pp. 403–414 (2016)
9. Qiu, C., Squicciarini, A., Li, Z., Pang, C., Yan, L.: Time-efficient geo-obfuscation to protect worker location privacy over road networks in spatial Crowdsourcing. In: CIKM 20: Proceedings of the 29th ACM International Conference on Information & Knowledge Management. CIKM (2020)
10. Takagi, S., Cao, Y., Asano, Y., Yoshikawa, M.: Geo-Graph-indistinguishability: protecting location privacy for LBS over road networks. In: Foley, S.N. (ed.) DBSec 2019. LNCS, vol. 11559, pp. 143–163. Springer, Cham (2019). https://doi.org/10.1007/978-3-030-22479-0_8
11. Chatzikokolakis, K., Andrés, M.E., Bordenabe, N.E., Palamidessi, C.: Broadening the scope of differential privacy using metrics. In: Privacy Enhancing Technologies, Berlin, Heidelberg, pp. 82–102 (2013)
12. Qiu, C., Squicciarini, A.C., Pang, C., Wang, N., Wu, B.: Location privacy protection in vehicle-based spatial Crowdsourcing via geo-indistinguishability. IEEE Trans. Mobile Comput. **99**, 1 (2020)
13. Yang, B., Sato, I., Nakagawa, H.: Bayesian differential privacy on correlated data. In: Proceedings of the 2015 ACM SIGMOD International Conference on Management of Data - SIGMOD '15, Melbourne, Victoria, Australia, pp. 747–762 (2015)
14. Shokri, R., Theodorakopoulos, G., Boudec, J.L., Hubaux, J.: Quantifying location privacy. In: 2011 IEEE Symposium on Security and Privacy, pp. 247–262 (2011)

15. Andrés, M.E., Bordenabe, N.E., Chatzikokolakis, K., Palamidessi, C.: Geo-indistinguishability: differential privacy for location-based systems. In: Proceedings of the 2013 ACM SIGSAC Conference on Computer & Communications Security, New York, USA, pp. 901–914 (2013)
16. Dingqi, Y., et al.: Participatory Cultural Mapping Based on Collective Behavior Data in Location-Based Social Networks. https://dl.acm.org/doi/abs/10.1145/2814575. Accessed 21 July 2021
17. Bordenabe, N.E., Chatzikokolakis, K., Palamidessi, C.: Optimal geo-indistinguishable mechanisms for location privacy. In: Proceedings of the 2014 ACM SIGSAC Conference on Computer and Communications Security, New York, USA, pp. 251–262 (2014)
18. Wang, L., Zhang, D., Yang, D., Lim, B.Y., Han, X., Ma, X.: Sparse mobile crowdsensing with differential and distortion location privacy. IEEE Trans. Inform. Forensic Secur. **15**, 2735–2749 (2020)
19. McSherry, F., Talwar, K.: Mechanism design via differential privacy. In: 48th Annual IEEE Symposium on Foundations of Computer Science (FOCS'07), pp. 94–103 (2007)
20. Duckham, M., Kulik, L.: A formal model of obfuscation and negotiation for location privacy. In: Pervasive Computing, Berlin, Heidelberg, pp. 152–170 (2005)
21. Xiong, J., et al.: A personalized privacy protection framework for mobile crowdsensing in IIoT. IEEE Trans. Industr. Inf. **16**, 4231–4241 (2020)
22. Wang, L., Zhang, D., Yang, D., Lim, B.Y., Ma, X.: Differential location privacy for sparse mobile crowdsensing. In: 2016 IEEE 16th International Conference on Data Mining (ICDM), Barcelona, Spain, pp. 1257–1262 (2016)
23. CRAWDAD dataset. https://crawdad.org/roma/taxi/20140717. Accessed July 2014

"Failure" Service Pattern Mining for Exploratory Service Composition

Yunjing Yuan, Jing Wang(✉), Yanbo Han, Qianwen Li,
Gaojian Chen, and Boyang Jiao

Beijing Key Laboratory on Integration and Analysis of Large-scale Stream Data,
School of Information Science and Technology, No.5 Jinyuanzhuang Road,
Shijingshan District, Beijing, China
wang_jing@ncut.edu.cn, yhan@ict.ac.cn

Abstract. To adapt to uncertain and dynamic requirements, exploratory service composition enables business users to construct service composition processes in a trial-and-error manner. A large number of service composition processes are generated, which can be learned to improve the reusability of the service composition processes. By mining these service composition processes and abstracting the mining results to service patterns, the efficiency of service composition can be effectively improved. At present, there has been some research on the methods of service pattern mining. Most of the work focuses on successful service composition processes, but the ones that fail are also valuable. For example, by using the mining results of failure service composition processes, the accuracy of service recommendations can be improved. To solve this problem, this paper proposes a "failure" service pattern mining algorithm (FSPMA) for exploratory service composition, which extends the gSpan algorithm, and can mine "failure" service patterns from service composition processes for further reuse. Meanwhile, the exploratory service composition model and the service pattern model are explained for the FSPMA. The prototype implementation of the exploratory service composition environment is introduced, which integrates the FSPMA. The experimental evaluation is explained to verify the algorithm, and the result shows that the efficiency of the FSPMA has a significant improvement in mining "failure" service patterns compared with the gSpan algorithm and the TKG algorithm. Finally, the application of "failure" service patterns in service recommendations is given.

Keywords: Exploratory service composition · Service pattern mining · gSpan

1 Introduction

In the big data era, people's lives rely more and more on the services delivered via the Internet, and servitization becomes one of the most important trends [1, 2]. User requirements are increasingly complicated, and no single service could completely fulfill a coarse-grained requirement [3]. Service composition has become increasingly popular in both business and scientific domains [4, 5]. In some situations, end users can't make a thorough experiment plan in advance and cannot decide on which steps to take next without reviewing the results of the previous steps [6]. They may want to

H. Gao and X. Wang (Eds.): CollaborateCom 2021, LNICST 406, pp. 38–53, 2021.
https://doi.org/10.1007/978-3-030-92635-9_3

dynamically adjust the composition logic at runtime. Therefore, exploratory service composition [6] is paid great attention in academics these years. There are a large number of service composition process instances generated at runtime, which may contain valuable knowledge. The knowledge refers to previous experiences and is valuable for reuse. If it can be fully utilized, the efficiency of exploratory service composition will be improved.

Through the analysis of service composition process instances generated by exploratory service composition, it is found that users' selection of services has certain commonalities. Some services often appear together in one service composition process track, as a form of a fragment. The fragment shows the characteristic of large granularity and high reusability. Related research refers to this service fragment which frequently appears as a service pattern [7]. There are two kinds of tracks for service composition process instances generated by exploratory service composition which are successful track and failure track. So, the corresponding service patterns can be divided into two kinds, one is the "success" service pattern, and the other is the "failure" service pattern. Applying these two kinds of service patterns to exploratory service composition, on the one hand, can improve the efficiency of constructing service composition processes, on the other hand, can make full use of previous knowledge to improve the reusability of service composition processes. Recently, most of the work is only focused on the mining of "success" service patterns, and there is little research on the mining of "failure" service patterns. The value of the corresponding historical dataset has not been fully utilized. In addition, if the method of mining "success" service patterns is used in mining "failure" service patterns, it will cause unnecessary waste of time and resources. Because if a track is a failure track, it does not mean the entire track is failed, but only part of the track is failed. Mining can be limited to the scope of failure.

To solve this problem, this paper proposes a method for modeling exploratory service composition instance and its corresponding service pattern and designs a "Failure" Service Pattern Mining Algorithm (FSPMA) which extends the gSpan algorithm. Through the proposed models and algorithm, "failure" service patterns in the failure process tracks can be efficiently mined. Compared with the gSpan algorithm and the TKG algorithm, the mining efficiency is significantly improved.

The rest of the paper is organized as follows. Section 2 of this paper introduces the analysis of related research status and the problems that need to be solved. Section 3 gives the formal definitions of exploratory service composition instance and service pattern. Section 4 elaborates on how to mine the "failure" service patterns based on the FSPMA algorithm that focuses on probe points and failure process tracks. Section 5 designs and implements the prototype. Section 6 gives experimental analysis and performance evaluation of the FSPMA algorithm. Section 7 introduces the application of the "failure" service patterns that are used in service recommendations. Finally, summarize the full paper and look forward to future work.

2 Related Work

2.1 Log-Based Service Pattern Mining

Currently, there are mainly four kinds of log-based service pattern mining algorithms. One is the α algorithm [8]. This algorithm scans all instances in the log, abstracts the basic relationship between activities, and directly constructs the processes according to the type of basic relationships. It can handle various control flow structures, but cannot deal with the noise in the log. The second is the heuristic algorithm [9], which mainly considers the frequency of process instances in the log. It can mine the main behaviors of the processes and deal with the log noise, but it ignores the details of the processes and cannot handle log diversity and quality monitoring. The third is the genetic algorithm [10], which is a search technology that simulates the evolution of biological processes. It can handle various types of control flow structures and log noise at the same time, but the algorithm may not get the optimal process model in the end. The last is based on the log classification algorithm. It clusters the execution instances which are saved in the log [11], divides them into multiple sub-logs, and uses existing mining algorithms on the sub-logs. This algorithm can handle the log diversity well, but it relies on the existing specific mining algorithms.

2.2 Process-Based Service Pattern Mining

Process-based service pattern mining can be abstracted as frequent subgraph mining (FSM). A survey done by literature [12] presents several significant FSM algorithms. According to the survey, no "new" algorithms were proposed recently but there had been much work on developing variations of existing algorithms [13]. FSM algorithms mainly adopt two algorithm ideas, the Apriori algorithm and the FP-growth algorithm.

The mining algorithms that apply Apriori's idea include the AGM algorithm [14] and the improved algorithms based on the AGM algorithm. The AGM algorithm is based on recursive statistics. It can mine all frequent subgraphs, but its execution efficiency is lower for large databases. Its improved algorithms include the FSG algorithm [15] and AcGM [16] algorithm, their execution efficiency is higher than that of the AGM algorithm. In the current research, the representative work of service pattern mining algorithms that use Apriori's idea is shown in the literature [17].

Another algorithm idea is the FP-growth algorithm [18, 19], which compresses the data into a frequent pattern tree, stores the association relationship of the items, and finally generates frequent sets for the pattern tree. Since it does not need to generate candidate frequent sets repeatedly, its execution efficiency is higher than that of the Apriori algorithm. FSM algorithms based on the idea of this algorithm such as gSpan algorithm [20] and FFSM algorithm [21]. The rightmost path expansion and frequent pruning strategy of gSpan algorithm greatly reduce the running time of the algorithm. FFSM algorithm only scans the embedding set when calculating the support degree, which outperforms gSpan. But FFSM cannot be used in the context of directed graphs; while gSpan, with some minor changes, can accommodate directed graphs [12]. Besides, TKG [22], which extends gSpan algorithms, utilizes a dynamic search procedure to always explore the most promising patterns first. Experiments show it has almost excellent performance as well as gSpan, but its memory use is high. According to the survey, the representative work of applying the idea of the FP-growth algorithm in service pattern mining is present in literature [23].

The related work of service pattern mining above is aimed at the mining of successful service composition process. They mine the entire service composition process. For failed tracks, they will produce unnecessary waste of time and resources. Therefore, this paper comprehensively considers the characteristics of current service pattern mining and its existing problems, adopts a process-based service pattern mining algorithm, considers the mining efficiency and feasibility of these optional algorithms, and finally chooses to extend the gSpan algorithm so that the algorithm can support the mining of "failure" service patterns.

3   Model Definition

This section mainly introduces the exploratory service composition instance model and the service pattern model.

3.1   Exploratory Service Composition Instance Model

Exploratory service composition can support business users to incrementally construct service composition processes at runtime. During the composition, if the intermediate result is incorrect or unsatisfactory, probe points can be added at the corresponding activity, so that new tracks can be derived from the original track until a successful track is explored. The exploratory service composition instance model is shown in Fig. 1.

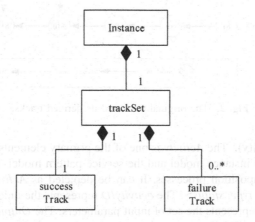

Fig. 1. Exploratory service composition instance model.

Definition 1 (Instance). The *Instance* describes the exploratory execution for certain goals through service composition and is regarded as a package of tracks. It can be depicted as a 3-tuple: *Instance = <instanceID, name, trackSet>*. The *InstanceID* represents the unique identifier of the exploratory service combination instance. The *name* represents the name of the exploratory service composition instance. The *trackSet* represents the track set contained in the instance.

Definition 2 (Track). The *Track* is the ultimate executable process unit, and can be depicted as *Track = <trackProfile, instanceID, status, activities, transitions, direct-Deriv, dataPocket, exploredstate>*. The *trackProfile* represents the basic track information, like *trackID, name, createTime*, etc. The *InstanceID* indicates the instance to which the track belongs, a track must belong to one certain instance, so the *instanceID* cannot be null. The *status* ∈ *{init, running, suspend, complete, terminated}* [4], which represents the execution status of the track. The *activities* and *transitions* represent the activities and the transition relationships of the tracks respectively. The *directDeriv* locates the original track in the derived relationship, which is shown as Fig. 2, defined as *directDeriv = <originalTrack, probePoint>*, where *originalTrack* marks the parent track of the current derived track, *probePoint* marks the probe point in the original track when it is derived. And the operation of adding a probe point is shown as Definition 5. The *dataPocket* is the data information generated by the track. The *exploredstate* ∈ *{success, failure}*, represents the track state after explored, which can be successful or failed.

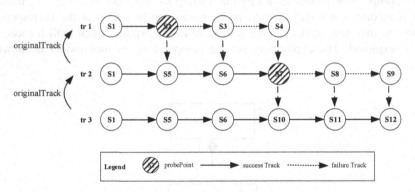

Fig. 2. The original track and its derived tracks.

Definition 3 (Activity). The *Activity* is one of the primary elements of the exploratory service composition instance model and the service pattern model. It is the executable task in service composition processes. It can be depicted as *Activity = <activityID, Input, Output, QoS, type, status>*. The *activityID* represents the unique identifier of the activity. The *Input* represents the set of input parameters. The *Output* represents the set of output parameters. The *QoS* means the quality of the activity. The *type* ∈ *{service, start, end, orSplit, orJoin, andSplit, andJoin}* [4] represents the type of the activity. The *status* ∈ *{init, running, suspend, executed}* is the execution status of the activity.

Definition 4 (Transition). The *Transition* is the other primary element of the models and is a mapping from one activity to another activity. It can be depicted as *Transition = <tranID, fromAct, toAct, dataMapping>*. The *tranID* represents the unique identifier of the transition relationship. The *fromAct* represents the source activity of the transition relationship. The *toAct* represents the target activity of the transition relationship. The *dataMapping* ∈ *{Exact, Plugin, Container, Disjoint}* represents the collection of data mapping relationships between the source activity and the target activity.

Definition 5 (the Operation of Adding a Probe Point). The *Operation of Adding a Probe Point* can be recorded as *addProbePoint (pt, a)*, where *pt* is the track and *a* is the activity. If the *status* of *pt* is not *terminated* and the *status* of *a* is *executed*, then the operation of adding a probe point is allowed. This operation will generate a new track *pt'*. The *dataMapping* of the transition relationship generated by cloning from the original track remains unchanged, so are the predecessor activities of probe point *a*. The probe point *a* and the successor activities are reset and initialized after being cloned.

3.2 Service Pattern Model

After the execution of the exploratory service composition instance, both the successful tracks and the failed tracks can be generated. Two kinds of service patterns can be gained by mining these two kinds of tracks. Considering the characteristics of a service pattern, it can be defined as follows:

Definition 6 (Service Pattern). A service pattern is a frequent subgraph mined from a set of process tracks, which can be depicted as *SP = <spID, activities, transitions, type>*. The *spID* represents the unique identifier of the service pattern. The *activities* are the activities contained in the service pattern. The *transitions* are the transitional relationships that exist between activities. The *type* ∈ *{success, failure}* represents which kind of tracks the service pattern is mined from.

4 FSPMA

Our problem is defined as: mining the tracks of exploratory service composition instances and obtaining service patterns. And we mainly focus on mining the failure tracks to obtain the "failure" service patterns. To solve the problem, this paper proposes a Failure Service Pattern Mining Algorithm (FSPMA) which extends the gSpan algorithm.

Compared with the gSpan algorithm, the extension of the FSPMA is that it limits the mining scope to the vicinity of the probe points. In the following, the algorithm is introduced in detail. It takes a set of failure tracks as input. Firstly, the input called T is abstracted as a graph set *D*. The predecessor activities of the probe points on the tracks are marked as success points (can also be said as vertices). The successor activities of the probe points are marked as failure points (Algorithm 1 line 1). Secondly, vertices and edges are sorted based on their frequencies. The infrequent vertices and edges will be removed (Algorithm 1 line 2–3), and the remaining frequent edges will be sorted in DFS lexicographical order (Algorithm 1 line 5). Thirdly, to discover "failure" service patterns, frequent subgraphs must contain at least one probe point or failure point. So, the algorithm only performs Subgraph_Mining for three kinds of frequent edges which are shown in Fig. 3. This step is the main part where FSPMA extends the gSpan algorithm (Algorithm 1 line 8). These three types of frequent edges can grow based on the gSpan algorithm's search strategy and growth strategy which are the depth-first search and the rightmost path expansion strategy (Subprocedure 1 line 4). For the strategies, details can be seen in [20] or [22], we will not introduce them further. After mining the frequent subgraphs, remove the mined edge from the original graph (Algorithm 1 line 11), continue with the expansion mining of the next frequent edge which must be one of the three kinds of edges in Fig. 3, and repeat these procedures until all frequent subgraphs are found. Finally, after filtering the frequent subgraphs (Algorithm 1 line 14), we can get the "failure" service patterns.

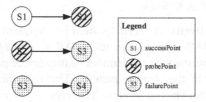

Fig. 3. Three kinds of frequent edges that can grow.

An example is given below to illustrate the operation of the algorithm, as shown in Fig. 4. Given the 8 graphs which are converted from tracks and the minimum support threshold *minSup* = 2, the algorithm sorts vertices and edges according to the frequency, removes infrequent ones, and obtains all frequent edges. Then, only three types of frequent edges which are shown in Fig. 3, can grow based on the depth-first search and the rightmost path expansion. When no new frequent subgraphs are generated, the growth for the next edge starts. Repeat these procedures and get all frequent subgraphs. Finally, remove the repeated frequent subgraphs, and obtain the final result.

Algorithm 1: FSPMA

Input: Track set T, *minSup*

Output: "Failure" service pattern collection *S*

1: Abstract T to graph set D and label success points and failure points;

2: sort the labels in D by their frequency;

3: remove infrequent vertices and edges;

4: $S^1 \leftarrow$ all frequent 1-edge graphs in D;

5: sort S^1 in DFS lexicographic order;

6: $S \leftarrow S^1$;

7: **for each** edge *e* in S^1 **do**

8:     **if** both e.from and e.to are failure points

                or e.from is probePoint and e.to is a failure point

                or e.to is probePoint;

9:         Initialize s with e, set s.D by graphs which contain e;

10:        $S \leftarrow$ Subgraph-Mining (D, S, s);

11:        $D \leftarrow D - e$;

12:        **if** $|D| <$ minSup;

13:            break;

14: remove the repeated frequent subgraphs from S;

15: **return** S;

Subprocedure 1: Subgraph-Mining

Input: *D*, "Failure" service pattern collection *S*, frequent fragment *s*, *minSup*

Output: *S*

1: **if** $s \neq \min(s)$

2:     **return** S;

3: $S \leftarrow S \cup \{s\}$;

4: enumerate s in each graph in D and count its children;

5: **for each** *c*, *c* is s' child **do**

6:     **if** support(c) $\geq$ *minSup*;

7:         $s \leftarrow c$;

8:         $S \leftarrow$ Subgraph-Mining (D_s, S, s);

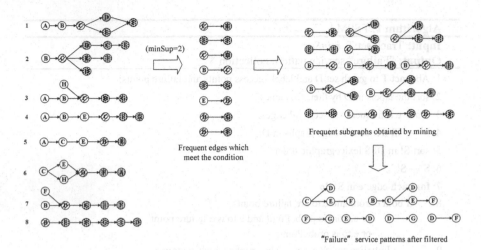

Frequent edges which meet the condition

Frequent subgraphs obtained by mining

"Failure" service patterns after filtered

Fig. 4. An example of the algorithm.

5 Prototype Implementation

Exploratory service composition environment is an online service composition system developed to support the FSPMA algorithm. Figure 5 shows its architecture. The environment has five modules, which are the modules of exploratory service orchestration tool, runtime user interaction, execution engine, service pattern mining, and service library.

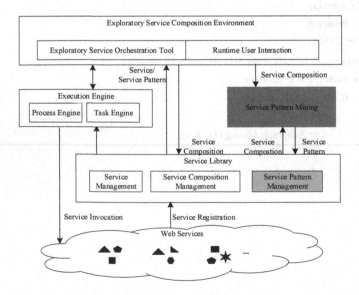

Fig. 5. The architecture of exploratory service composition environment.

The modules of exploratory service orchestration tool, runtime user interaction, execution engine, and service library are the fundamental modules of the environment. The module of exploratory service orchestration tool supports users to build the service composition processes. And users can execute and adjust the processes in the runtime user interaction module. Both the module of exploratory service orchestration tool and the module of runtime user interaction can interact with the execution engine module. There are two engines contained in the execution engine, which are the process engine and the task engine. This module can provide the functions of application parsing, service collaboration, service invocation, and exception handling. The service library module provides services and service patterns to exploratory service orchestration tool and runtime user interaction and saves the service composition processes which are generated by these two modules.

For the mining part, the module of service pattern mining has two data sources, which are the service composition process instances generated by runtime user interaction and the historical service composition process instances saved in the service library. Because the tracks of the process instances can be divided into two kinds, success and failure, this module has two submodules, one of them integrates the FSPMA algorithm to mine "failure" service patterns from failure tracks. The other integrates the gSpan algorithm to mine "success" service patterns from success tracks. These service patterns are written back to the service library in the end. Therefore, the function of service pattern management is developed for the module of the service library, to manage the service patterns.

6 Experiment

6.1 Dataset and Environment

This section uses simulation experiments to evaluate the performance and effectiveness of the FSPMA algorithm. To improve the credibility of the experiment, this paper uses 1405 processes crawled from the myexperiment (www.myexperiment.org) research community as the experimental dataset. The data is in the XML or t2flow format with the highest visits and downloads in the community. This community is a collaborative environment for global biocomputing researchers to publish and share information about biocomputing processes and experimental plans. Scientists can safely publish their processes and experiments, share them with groups, and find other researchers' processes. It can reduce follow-up experiment time, share professional knowledge, and avoid reengineering.

To meet the needs of the algorithm, this paper analyzes the collected biological processes, abstracts services as vertices, sets service names as vertex labels, and abstracts the relationships between services as edges. To obtain more frequent service patterns, we model the most sixteen frequent services as probe points. Since the processes crawled from the website can be executed successfully, failure process tracks are constructed through simulation. A total of 10,490 failure tracks are constructed for the mining algorithm.

The simulation program is developed using Anaconda 3.6 and PyCharm, runs on a PC with a Windows operating system, CPU of AMD Ryzen 7 2.90 GHz, and memory of 16G.

6.2   Experimental Verification

The experiment mainly observes mining efficiency improvement of the FSPMA algorithm compared with the gSpan algorithm and the TKG algorithm. The impacts of *minSup*, the number of tracks, and the number of mining results on the runtime of the algorithms are analyzed while the experiments' mining accuracy is the same. The TKG algorithm can find the top-k frequent subgraphs, where the only parameter is k, the number of patterns to be found. It is also an extension of gSpan. Since TGK can generate the optimal minSup according to the value of k, we only compare the runtime with the TKG algorithm in terms of the number of tracks and the number of mining results.

The algorithm parameters are mainly the *minSup* and the number of tracks. As shown in Fig. 6, we use 10,000 tracks and set the *minSup* from 3 to 15. The result shows that when the *minSup* is adjusted to 10, the runtime of the FSPMA reduces the most compared with the gSpan algorithm. As can be seen from the histogram, the number of mining results that the two algorithms produce is the same. Besides, by comparing the contents of the output files, the frequent subgraphs obtained by the two algorithms are exactly the same. The mining accuracy of other subsequent experiments is also exactly the same.

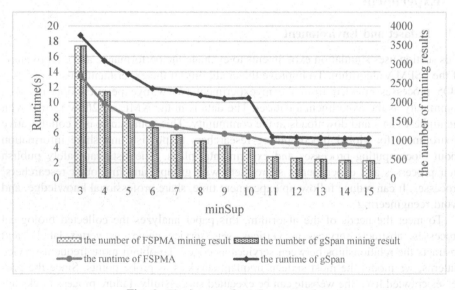

Fig. 6. The impact of *minSup* on runtime.

Then, we select the *minSup* = 10 and analyze the runtime under different numbers of tracks. By changing the number of tracks from 1000 to 10,000, Fig. 7 illustrates the runtimes of these three algorithms and shows the FSPMA achieves better performance compared with the gSpan algorithm and the TKG algorithm. The histogram can show their accuracy is the same.

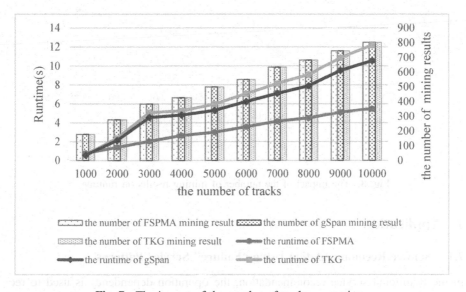

Fig. 7. The impact of the number of tracks on runtime.

In addition, the number of mining results may also affect the runtime of the algorithm. When the *minSup* = *10*, the runtimes of the three algorithms under different numbers of mining results are shown in Fig. 8. As the number of mining results increases, the runtime of the FSPMA algorithm is greatly lower than that of the gSpan algorithm and the TKG algorithm.

In summary, the FSPMA algorithm can effectively focus mining on the generation of "failure" service patterns, avoiding the mining of the entire tracks of service composition process instances which the gSpan algorithm and the TKG algorithm do. The efficiency is improved under different *minSup* and the number of tracks. When the *minSup* is set to 10 under the current dataset, the efficiency of FSPMA is improved the most. And under this *minSup*, as the number of tracks increases, the efficiency improvement of the FSPMA algorithm is stable at about 43% compared with the gSpan algorithm, and about 50% compared with the TKG algorithm.

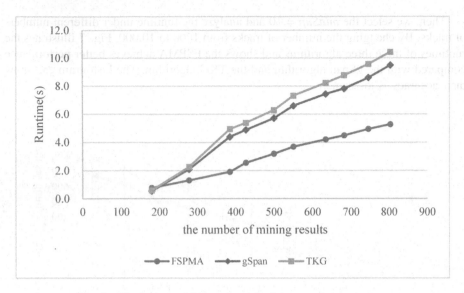

Fig. 8. The impact of the number of mining results on runtime.

7 Application

7.1 Service Recommendation Using "Failure" Service Patterns

In the traditional service recommendation, the operation dependency is used to recommend services that are highly constrained by current services. The operation dependency can be identified by combining the semantic matching of inputs and outputs interfaces between service operations and the analysis of process instances [5]. Through the semantic matching of inputs and outputs interfaces between service operations, the semantic matching degree μ can be calculated. The analysis of process instances can generate the reuse degree λ. These two values are used for recommending the Top N services with high strength of dependencies as candidate services through pattern matching.

However, this recommendation method also recommends the services where their strengths of dependency are high, but there are a large number of error process instances during the analysis of process instance. Therefore, we improve the traditional service recommendation method by using the "failure" service patterns, which can improve the accuracy of service recommendations.

The strengths of dependency (τ) between services considers the frequency degree (γ) of "failure" service patterns which is generated by FSPMA, besides the semantic matching degree (μ) and the reuse degree (λ).

The calculation formula is as follows:

$$\tau = w_1 \times \mu + w_2 \times \lambda - w_3 \times \gamma \tag{1}$$

Among them, w_1 is the weight of the semantic matching degree, w_2 is the weight of reuse degree, w_3 is the weight of the frequency degree of the "failure" service pattern, $w_1 > 0$, $w_2 > 0$, $w_3 > 0$, and $w_1 + w_2 + w_3 = 1$.

When the strength of dependency satisfies $\tau > \chi$, where χ is the minimum dependency strength, it is considered that there may be a relationship of dependency between services. Otherwise, there is no relationship of dependency. Finally, the recommendation can be given by selecting Top N services that satisfy the relationship of dependency.

7.2 An Example

In this section, we use an example of a bioinformatics process to illustrate how the use of the "failure" service patterns can improve the accuracy of service recommendations. As shown in Fig. 9. The traditional method does not consider the "failure" service patterns. The correctness of the process instances cannot be guaranteed during process instance analysis. In this case, *remove_entrez_duplicates, remove_Nulls2, create_report* are recommended. After considering the "failure" service patterns, the fragment *hsapiens_gene_ensembl -> remove_entrez_duplicates* which is used to recommend by the traditional method is a sub-segment of the "failure" service pattern. Although its reuse is high, the wrong process instances are mainly used in the analysis of process instances. This recommendation may cause the execution of the current process to fail. After considering the "failure" service pattern, this problem can be effectively solved, and the recommended results are *remove_Nulls2, create_report, split_for_duplicates*.

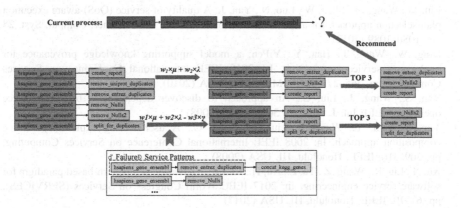

Fig. 9. Comparison of service recommendation using different methods.

8 Conclusion

This paper focuses on the issues related to mining "failure" service patterns from failure tracks of exploratory service composition instances. To make full use of historical service composition process instances, an exploratory service composition instance model and a service pattern model are defined at first, then the FSPMA algorithm is proposed, which extends the gSpan algorithm and focuses the mining on the vicinity of the probe points in the failure tracks to get "failure" service patterns. The prototype is introduced to support the FSPMA algorithm in the paper. Evaluation experiments are given to compare the FSPMA algorithm with the gSpan algorithm and the TKG algorithm, which proves the effectiveness and practicability of the FSPMA algorithm. Finally, we use the "failure" service patterns to improve the accuracy of service recommendations.

In future research, the incremental mining method will be studied to support the dynamic evolution and timely update of the service patterns in the service library, and the evolutionary trend of service patterns will be analyzed. In addition, the experimental verification of the "failure" service pattern applied to service recommendations will be realized.

Acknowledgements. The research work was supported by the International Cooperation and Exchange Program of National Natural Science Foundation of China (No.62061136006).

References

1. Xu, H., Wang, X., Wang, Y., Li, N., Tu, Z., Wang, Z., Xu, X.: Domain priori knowledge based integrated solution design for internet of services. In: 2020 IEEE International Conference on Services Computing (SCC), pp. 446–453. IEEE, Beijing, China (2020)
2. Liu, R., Wang, Z., Xu, X.: Parameter tuning for S-ABCPK: an improved service composition algorithm considering priori knowledge. Int. J. Web Serv. Res. (IJWSR) **16**(2), 88–109 (2019)
3. Liu, M., Wang, M., Shen, W., Luo, N., Yan, J.: A quality of service (QoS)-aware execution plan selection approach for a service composition process. Futur. Gener. Comput. Syst. **28** (7), 1080–1089 (2012)
4. Ding, W., Wang, J., Han, Y.: ViPen: a model supporting knowledge provenance for exploratory service composition. In: 2010 IEEE International Conference on Services Computing, pp. 265–272. IEEE, Miami, FL, USA (2010)
5. Yan, S., Wang, J., Liu, C.: An approach to discover dependencies between service operations. IEEE Int. J. Software **3**(9), 36–43 (2008)
6. Yan, S., Han, Y., Wang, J., Liu, C., Wang, G.: A user-steering exploratory service composition approach. In: 2008 IEEE International Conference on Services Computing, pp. 309–316. IEEE, Honolulu, HI, USA (2008)
7. Xu, X., Liu, R., Wang, Z., Tu, Z., Xu, H.: RE2SEP: a two-phases pattern-based paradigm for software service engineering. In: 2017 IEEE World Congress on Services (SERVICES), pp. 67–70. IEEE, Honolulu, HI, USA (2017)
8. Fang, X., Gao, X., Yin, Z., Zhao, Q.: An efficient process mining method based on discrete particle swarm optimization. Inf. Technol. J. **10**(6), 1240–1245 (2011)
9. Sangaiah, A.K., Hosseinabadi, A.A.R., Shareh, M.B., Bozorgi Rad, S.Y., Zolfagharian, A., Chilamkurti, N.: IoT resource allocation and optimization based on heuristic algorithm. Sensors **20**(2), 539 (2020)
10. Bratosin, C., Sidorova, N., van der Aalst, W.: Discovering process models with genetic algorithms using sampling. In: Setchi, R., Jordanov, I., Howlett, R.J., Jain, L.C. (eds.) KES 2010. LNCS (LNAI), vol. 6276, pp. 41–50. Springer, Heidelberg (2010). https://doi.org/10.1007/978-3-642-15387-7_8

11. Song, M., Günther, C.W., van der Aalst, W.M.P.: Trace clustering in process mining. In: Ardagna, D., Mecella, M., Yang, J. (eds.) BPM 2008. LNBIP, vol. 17, pp. 109–120. Springer, Heidelberg (2009). https://doi.org/10.1007/978-3-642-00328-8_11
12. Jiang, C., Coenen, F., Zito, M.: A survey of frequent subgraph mining algorithms. Knowl. Eng. Rev. **28**(1), 75–105 (2013)
13. Wijesinghe, C.R., Weerasinghe, A.R.: Mining frequent patterns in bioinformatics workflows. Int. J. Biosci. Biochem. Bioinf. **10**(4), 161–169 (2021)
14. Zhou, G., et al.: An improved method of AGM for high precision geolocation of SAR images. ISPRS Int. Arch. Photogrammetry Remote Sens. Spat. Inf. Sci. **42**(3), 2479–2485 (2018)
15. Akoglu, L., Tong, H., Koutra, D.: Graph-based anomaly detection and description: a survey. Data Min. Knowl. Disc. **29**(3), 626–688 (2015)
16. Huynh, B., Nguyen, D., Vo, B.: Parallel frequent subgraph mining on multi-core processor systems. ICIC Express Lett. **10**(9), 2105–2113 (2016)
17. Meng, H., Wu, L., Zhang, T., Chen, G., Li, D.: Mining frequent composite service patterns. In: 2008 Seventh International Conference on Grid and Cooperative Computing, pp. 713–718. IEEE, Shenzhen, Guangdong, China (2008)
18. Shafiq, M., Alhajj, R., Rokne, J.: Reducing search space for web service ranking using semantic logs and semantic FP-tree based association rule mining. In: Proceedings of the 2015 IEEE 9th International Conference on Semantic Computing (IEEE ICSC 2015), pp. 1–8. IEEE, Anaheim, CA, USA (2015)
19. Labbaci, H., Medjahed, B., Aklouf, Y.: Learning interactions from web service logs. In: Benslimane, D., Damiani, E., Grosky, W.I., Hameurlain, A., Sheth, A., Wagner, R.R. (eds.) DEXA 2017. LNCS, vol. 10439, pp. 275–289. Springer, Cham (2017). https://doi.org/10.1007/978-3-319-64471-4_22
20. Yan, X., Han, J.: gSpan: graph-based substructure pattern mining. In: 2002 IEEE International Conference on Data Mining, 2002, Proceedings, pp. 721–724. IEEE, Maebashi City, Japan (2002)
21. Fan, Z., Peng, Y., Choi, B., Xu, J., Bhowmick, S.: Towards efficient authenticated subgraph query service in outsourced graph databases. IEEE Trans. Serv. Comput. **7**(4), 696–713 (2014)
22. Fournier-Viger, P., Cheng, C., Lin, J.-W., Yun, U., Kiran, R.U.: TKG: efficient mining of top-K frequent subgraphs. In: Madria, S., Fournier-Viger, P., Chaudhary, S., Reddy, P.K. (eds.) BDA 2019. LNCS, vol. 11932, pp. 209–226. Springer, Cham (2019). https://doi.org/10.1007/978-3-030-37188-3_13
23. Liu, R., Xu, X., Wang, Z., Sheng, Q., Xu, H.: Probability matrix of request-solution mapping for efficient service selection. In: 2017 IEEE International Conference on Web Services (ICWS), pp. 444–451. IEEE, Honolulu, HI, USA (2017)

Optimal Control and Reinforcement Learning for Robot: A Survey

Haodong Feng[1,2], Lei Yu[1,2], and Yuqing Chen[1(✉)]

[1] School of Advanced Technology, Xi'an Jiaotong-Liverpool University,
Suzhou, China
Yuqing.Chen@xjtlu.edu.cn

[2] Department of Electrical Engineering and Electronics, University of Liverpool,
Liverpool, UK

Abstract. Along with the development of systems and their applications, conventional control approaches are limited by system complexity and functions. The development of reinforcement learning and optimal control has become an impetus of engineering, which has show large potentials on automation. Currently, the optimization applications on robot are facing challenges caused by model bias, high dimensional systems, and computational complexity. To solve these issues, several researches proposed available data-driven optimization approaches. This survey aims to review the achievements on optimal control and reinforcement learning approaches for robots. This is not a complete and exhaustive survey, but provides some latest and remarkable achievements for optimal control of robots. It introduces the background and facing problem statement at the beginning. The developments of the solutions to existed issues for robot control and some notable control methods in these areas are reviewed briefly. In addition, the survey discusses the future development prospects from four aspects as research directions to achieve improving the efficiency of control, the artificial assistant learning, the applications in extreme environment and related subjects. The interdisciplinary researches are essential for engineering fields based on optimal control methods according to the perspective; which would not only promote engineering equipment to be more intelligent, but extend applications of optimal control approaches.

Keywords: Data-driven optimization · Optimal control · Reinforcement learning · Model bias

1 Introduction

Optimal control is to make the performance index achieve optimal state, which can find an optimal channel to achieve predetermined destination [1]. The performance indexes are various based on real systems and missions. Optimal control

This work was supported by the Research Development Fund RDF-20-01-08 provided by Xi'an Jiaotong-Liverpool University.

H. Gao and X. Wang (Eds.): CollaborateCom 2021, LNICST 406, pp. 54–66, 2021.
https://doi.org/10.1007/978-3-030-92635-9_4

has two major principles on solving such issues which are the Pontryagin's minimum principle (PMP) and the dynamic programming (DP) [2]. PMP provides a requisite condition for optimality. Meanwhile, DP provides an ample condition for optimality via solving a differential equation, which is known as the Hamilton-Jacobi-Bellman (HJB) equation. They are the basis of optimal control theory and most algorithms were derived and designed based on PMP and DP approaches. In the following mentioned controllers, the ideas of PMP and DP were involved and applied. As one of the key components in optimal control algorithms, the cost functions or performance index functions are essential for optimization process. In the design of optimal controllers, some performance indexes are described in quadratic problems with destination state, process state, and action sections which are called linear quadratic regulator problems (LQR). However, along with the development of industry and technology, preceding conventional optimal control approaches are offline and they need complete model of system dynamics [3], it is hard to employ these kinds of optimal control solutions with dynamics uncertainties and changes as a result.

Currently, the widely utilized data-driven optimization algorithms of robot control are developed according to reinforcement learning (RL) which is known as a goal-oriented learning method. The agent with RL can learn an action control strategy to modify a long-term reward via interacting with outside environment [4]. At each iteration, an agent with RL evaluate feedback about the performance used state and action, to improve the performance in next step action. The purpose of strategies is to maximum the expected reward. Model-free reinforcement learning (MFRL) method has been employed to solve a wide list of robot tasks, such as playing graphical games and learning locomotion works. MFRL is generally applicable, require relatively fine tuning, and do not try to construct environment model [5]. However, such real-time model-free learning algorithm requires higher complexity of sampling and millions of samples are necessary if great performance must be achieved [6]. On the other hand, model-based reinforcement learning (MBRL) has much more sample efficiency and it uses a learned model via sample to assist learning [7]. To achieve great efficiency, MBRL has often utilized either function approximators or Bayesian models which can fit dynamics using few samples [8].

An advanced form of RL is called approximate DP which is also a solution methodology of optimal control. The purpose of MBRL is to find maximum reward or minimum cost in iterations as same as optimal control. MBRL connects the disparity between classical optimal control theory and adaptive control strategies. The objective is to learn the optimal strategy and cost function for an unknown dynamics. Unlike classical optimal control, MBRL finds the solution to the HJB equation online in real time. And MBRL algorithms are optimal, unlike traditional adaptive controllers that are not usually designed to be optimal in the sense of minimizing cost function. As a result, MBRL which can also be called optimal control has become an efficient and available approach on robot control [2].

There are three issues on robot control. Firstly, the purpose of optimal control, which is to achieve the optimal performance, is the major limitation of it, many researches focus on the precision and efficiency of control [6,8–11]. The second is capacity of computation which limits the algorithm application. The real time MFRL requires a large amount of computation and strong processor, so that it is hard to be employed on real-time robots [12–14]. Finally, the preceding model bias while model learning is also a problem, which is emphasized by many researchers who attempt to reduce the model bias by policy optimization [3,15–18].

The main objective of this paper is to investigate the advantages and disadvantages of optimal control and RL approaches for robots based on our understanding of present and recent advanced automatic techniques in the references. This paper can be applied to guide the design of efficient controllers from data-driven learning aspect for the automatic robots. The rest of this paper is organized as follows: Sect. 2 presents the general statement of optimal control problems' formations. Section 3 reviews the previous published solutions facing for robot optimal control design, regarding the improvement of precision and system complexity, the reduction of model bias, and the decrease of computation. Section 4 discusses the future researches directions to improve the performance of the robot optimal control applications, and the interdisciplinary development among optimal control and other engineering areas, different from the previous published related reviews. In Sect. 5, the main conclusions and contributions are described.

2   Optimal Control Problem Statement

The optimal control problem has fixed formation, this section will discuss cost function defining certain optional control problems. The optimal algorithms aim to solve such constrained optimization problem with cost function [1]

$$\min_{u(\cdot)} I[u(\cdot)] = \int_0^T L(x(t), u(t), t)dt \qquad (1)$$

Subject to : $\dot{x}(t) = f(x(t), u(t), t)$ and $x(0) = x_0$.

where $t \in T = [0, T]$ represents the time interval, $T \in R_+$ denotes the finally terminal time, $x(t) = [x_1(t), \ldots, x_n(t)]^T \in R^n$ denotes the state and $u(t) = [u_1(t), \ldots, u_m(t)]^T \in R^m$ denotes the control input at time t, $u(\cdot)$ denotes the control function, $I[u(\cdot)]$ denotes the total cost, $L(x(t), u(t), t)$ represents the running cost, $f(x(t), u(t), t)$ denotes the dynamics of the system while x_0 denotes the initial state.

From a predefined initial state $x(\cdot)$, optimal policy can search action $u(\cdot)$ with bounded constrains to make cost index minimum. The Lagrange function is often employed as quadratic form in LQR problem [9]

$$L(x(t), u(t), t) = \frac{1}{2}\left(x^T Q x + u^T R u\right) \qquad (2)$$

where $x \in \mathrm{R}^n, u \in \mathrm{R}^m, u \in \mathrm{R}^m, Q \in \mathrm{R}^{n \times n}, R \in \mathrm{R}^{m \times m}$.

If the terminal is predefined, the cost function often has a terminal cost weight ϕ

$$\min_{u(\cdot)} I[u(\cdot)] = \min_{u(\cdot)} \phi(x(T), T) + \int_0^T L(x(t), u(t), t)dt \tag{3}$$

ϕ is terminal cost weight. It would be predefined bigger than Q and R, if the destination is expected to achieve rapidly. The aim of optimal control algorithms is to minimum the cost using optimal control sequences, although they pay attention to different positions.

3   Solutions of Optimal Control for Robot

Several successful stories presented available solutions for problems caused by model bias, complex system, and large computation, which are developments of optimal control.

3.1   Overview the Related Approaches

The efficiency is of core important in optimal problem and several success stories are performed [19,20]. Supervised learning, the precise of which is better with preset labels, sometimes is combined with MBRL to improve efficiency and evaluate outside environment for DP [6]. A crucial choice for any practical utilization is advanced policy search algorithms which can be used to find optimal policy for complex system. Along with the system efficiency, the complexity of dynamic system is always a fundamental task in the field of robotics and engineering, and a variety of methods to solve it have been presented [6,21]. For the simple linear dynamic system, it is easier to find optimal solution, the dynamics formation is $\dot{x}(t) = Ax(t) + Bu(t)$ [22]. However, the nonlinear system is facing large tough issues as the dynamics function is uncertain. [15] uses linear formation to represent the unknown nonlinear dynamic system, but the formation of linear function has bias when it is utilized to express nonlinear dynamics. For many physical applications, the degree of difficulty is increasing along with the degree of freedom in nonlinear system.

Aforementioned applications in nonlinear systems with unknown dynamics, the first step of algorithms is often to learn unknown dynamics model called model learning. There are a large number of stories involved model leaning for optimal control and MBRL using multi-layer neural networks and Gauss mixed model (GMM) [6,9,10,23], however precise model of nonlinear high-dimensional dynamics cannot be learned directly. As learned model of dynamics leads to harmful error which influences policy improvement largely [3], the method of model learning can be optimized so that policy evaluation and policy update can be hold based on precise dynamics model architecture. Another method is to modify policy improvement process. Algorithms can interact with environment each iteration and continuously optimize policy with measured data to reduce model bias. In several success stories, a large amount of measured samples of real system are employed to learn precise model [5, 24, 25] which improves computation of processor and adds burdens on operation. Improving sample efficiency has become an essential task in practice.

Some algorithms needing large computation capacity are available in simulation environment with high performance processors. However, they may meet low efficiency and poor performance in practice on robot system. Conventional model predictive control (MPC) [26] needs to predict predefined steps actions at each iteration [16] and compared with MPC several algorithms reduce computational complexity to reduce optimization time and improve the efficiency. The core problems of optimal control and MBRL on robot has been reviewed from three aspects, which hinder the developments and applications of robot in practice. Some excellent algorithms will be discussed in the following sections about the solutions of aforementioned optimal control issues.

3.2 Improve Precision and System Complexity

The degree of freedom (DoF) is often an essential factor which influences the performance of robot. It is hard to model and control a high-dimensional robot whose system is much more complex. [9] proposed an optimal control algorithm with learned local model, which can be employed to complete dexterous manipulation using a tendon-drive 24 DoF robot hand. The nonlinear system with high DoF faces challenging issue with less sample data and its dexterous manipulation needs higher dimension states and actions compared with single works. At the beginning of algorithm, time-varying linear-Gaussian model is learned and fitted from little size sample applying linear regression approach by a Gaussian mixture model, as the control law is locally linear at each time step although the controllers are time-varying. Time-varying linear-Gaussian dynamics are formed as Eq. (4)

$$p\left(x_{t+1} \mid x_t, u_t\right) = \mathrm{N}\left(f_{xt} x_t + f_{ut} u_t + f_{ct}, \mathrm{F}_t\right) \tag{4}$$

KL-constrained optimization with LQR method and line search method are used to update and optimize controller at each iteration. This solution can complete complex work in high-dimension system and the precision of algorithm is great as same as doing the dexterous manipulation by a human being.

[6] used model-based deep reinforcement learning method to represent MFRL to reduce the value test sample. It utilizes a medium-sized deep neural network which was trained by gradient decent, to learn the system model. Then MPC method is employed to complete the task. To overcome the bias of model and improve the efficiency of MFRL, it combines these two learning strategies. The MBRL is used to initial the model free learner so that MFRL (policy gradient) algorithm just need fewer trials to achieve the destination. The algorithm was verified that it has larger rewards and higher precision than that using model-based approach for four different agents.

Approach in [10] is quiet similar as [6] which used the same model learning method (deep neural network). However, it merged computer vision into the RL. At the begin of the algorithm, two channels are input, one is the state and action channel to learn system model, another is the image which is input to the convolutional neural network (CNN) so that the algorithm can judge various surfaces outside. This approach can make the legged millirobots perform better in different surfaces. The image-conditioned approach can achieve less cost and high precision in different surfaces compared with other MBRL and optimal control methods.

Based on the learning system model, a neural-optimal control structure was proposed by [27] for the unknown discrete nonlinear system with discount factor in the cost function. For nonlinear discrete systems with additive disturbances, a robust model predictive control (MPC) method by self triggering was proposed in [28]. The method has an adaptive predictive time-domain scheme and can stabilize the system while ensuring suboptimal convergence. A novel error bound was designed by work of [29] based on the work in [30], it showed that a model's ability to self-correct is more tightly related to MBRL performance. The model of MBRL can guarantee the robust to model class limitations. A fault-tolerant control method of actuator based on tube-based MPC and set-based fault detection was proposed in [31]. It implemented an active fault isolation method after fault detection with the constraint-handling ability of MPC. The aforementioned researches tried and proposed several available approaches to improve the accurate of the state-of-the-art algorithms from various aspects and directions.

3.3 Overcome Model Bias

As solutions in which are used to improve precision of controllers, several algorithms were proposed to overcome model bias of the nonlinear systems. There are two directions on solving this problem. By using an advanced and efficient model learning approach, more exact model can be trained and learned via deep neural network or linear-Gaussian approximation. Another approach is to employ precision policy iteration algorithms which can optimize strategies according to interaction with environment at each iteration.

In work of [32], a new internal model control (IMC) structure based on fuzzy model was proposed to provide efficient and robust control achievement. A command filtered robust controller was proposed in [33]. In the presence of parameter uncertainty and interference, the given reference signal is tracked by adjusting the aircraft attitude angle. The method uses command filtering inversion to compensate the dynamic error and filtering error of the actuator. The improved stable linear filter is used to deal with the measurement error. [34] presented an approach to find area control error signal based on the frequency biased estimation, so that the performance of load-frequency control system can be improved. Aiming at the speed control problem of constrained nonlinear electric vehicles, a new nonlinear optimal model predictive controller design method was proposed in [35]. In terms of system dynamics, the proposed method can be regarded as a general case of the obtained results.

In order to realize iterative control, an improved embedded reinforcement learning algorithm was proposed by research [36]. The algorithm used off-policy learning to make the dynamics completely unknown, so as to reduce the influence of unknown disturbances and add disturbance compensation controller. For nonlinear continuous systems with state and input constraints, an event model-based predictive control approach was proposed by [37] to solve the uncertain problems such as random time delay caused by communication medium instability.

The effective learning of robot (humanoid) physical model was studied in work [38]. Aiming at the learning problem of body model, an active learning method was proposed. As the learning of serial robot kinematic model, the recursive least square method (RLS) is used to complete the learning process online, which is better than the commonly used gradient method.

Although the model learning methods have been developed by some success stories, the models of complex and high-dimension nonlinear systems are still hard to learn or need tremendous effort to complete. Several researches paid attention on using optimal control approaches to achieve expected objective. [3, 16] were representatives of optimal feedback control algorithms. [16] presented an iterative online optimal feedback approach to solve optimal control problem for the continuous linear or nonlinear systems. A model can be employed to compute the variables for cQP using offline calculation. And the measured data are collected to compute optimal action set online. This method can overcome model bias caused by inexact model and have high efficiency to find optimal control policy. In addition, it can also be employed on discrete system after improvement. If model should be learned from the system discretely using precision model learning method, it would reduce the model bias and be more rapidly find the optimal control solution.

3.4   Reduce Computation

[16] tried to reduce computational complexity during iteration process, which is lower than the nonlinear MPC controller. Computation is an essential factor when implement algorithms in robot equipment caused by limitations of processor. MFRL approaches need much computation as high performance computer, which face limitations on real-time robot.

[12] proposed an adaptive moment estimation (Adam) approach, which is an effective random optimization method requires little memory and first-order gradients. It is computationally efficient and has little memory requirements. This algorithm combines two methods AdaGrad [39], which is very suitable for sparse gradient and RMS Prop, and has good on-line and non-stationary effects.

Conventional MPC method can be optimized to improve efficiency and reduce computational complexity. [13] showed an adaptive MPC algorithm for the linear systems with parameter uncertainty constraints. The proposed approach uses RLS based estimator to learn the unknown systems. And it extended the robust MPC controller in [40] to allow online model adaptation while ensuring closed-loop stability and recursive feasibility. The computational complexity was concerned by research and an adaptive method is given with the reduced computational complexity as well.

[41] presented an arbitrary time control algorithm with limited processor resources, which calculates the components of the control input vector in order to maximize the available processing resources at each time step. To reduce the computational complexity, in the optimal feedback control system, two heuristic algorithms based on greedy search and multiplier alternating direction were proposed by work of [42]. [43] provided a problem solution from initial state to final state for stochastic linear systems controlling linear noise, and minimized the mean square deviation of the target state. The ability of offline computation in this algorithm can reduce computation during process. The aforementioned approaches are the solutions to reduce computational quantity and complexity.

4 Future Prospects and Discussion

Although the approaches in optimal control and MBRL are facing the aforementioned issues, the development is still prosperous as the prospect is bright, and their applications are very essential and wide. Various methods are proposed to overcome these numerous problems and adapt to different application scenarios. Some surveys about reinforcement learning based robotics were reviewed by works of [44–49]. According to these surveys, this paper discusses four directions which can be foreseen in the development of optimal control at present.

To overcome the bias caused by inexact model, aforementioned algorithms were presented in Sect. 3. However, more efficient and optimal policies are looked forward to explore by researches. This is the trend on the development of optimal control and MBRL algorithms. Meanwhile, less sample size and high sample efficiency are pursued as it can reduce the computational complexity and improve the efficiency of controller. If less sample are necessary during model learning, the controller would be easier to be employed in different environments widely. Therefore, more frequent interaction between the agent and environment will be necessary during optimization process to learn and know the parameters better. Optimal control policy can be learned via small sample size by robot as same as that via large sample size. On the other hand, model bias can be reduced using efficient model learning methods and policy update algorithms.

Let robot learn to complete works is hard to apply in real-time practice as numerous limitations in technique. However, if the state-of-the-art algorithms would be used in industrial practice, the efficiency of production would be highly improved and cost would be reduced a lot. Thus, combining actuality with the state-of-the-art control methods has demand prompt solution. It is unnecessary to make intelligent robot complete the whole complex missions in recent years. The help from human being is available when optimal control algorithms are employed in complex nonlinear industrial systems and artificial assistant learning method can be developed in several applications. Just like teachers at school when humans are young, humans act as teachers of robot at the beginning of learning certain works. Apprenticeship learning [50] is one of the example.

Humans can also do partial works and make robot learn another easy and tedious part of works. By this way, human and robot would complete works together and the efficiency of production would be improved.

Although the researches on optimal control theory applied in common environment have been held frequently, the applications of optimal control algorithms in complex, extreme, and special environment are also needed to be explored such as applications in marine technology, deep-sea equipment, and astronautics fields. The environments are more nonlinear complex and special compared with that in the ground. More parameters and factors are necessary to consider when algorithms of controllers are designed. Meanwhile, robot arm and other automatic devices are urgent needs during the scientific investigation in deep-sea and space. For example, Remotely Operated Vehicle (ROV), which is widely used in deep-sea investigation, needs cable to connect through water that constrains the range of ROV activity and increases cost of the equipment. Intelligent Autonomous Underwater Vehicle (AUV) can represent ROV in several suitable works, with high efficient MBRL algorithms it can learn complicated environment in deep-sea [51,52]. However, novel approaches of optimal control are necessary to develop so that extreme external environment can be considered. With optimal control optimization on robot, these types robots would work more efficient, which is helpful for conduct of research in extreme environment.

Optimal control algorithm is not an independent application, whose development has been influenced by computer science. Although novel optimal control methods are often used on robot system to show their availability, they can be utilized by other fields with data-driven optimization as well, like similar control methods. These applications involve smart grid [15], sustainable architecture [53], energy generation [54], and multidisciplinary optimization [55]. Meanwhile, optimal control is widely used in fault prediction and diagnosis [31,56–58] for industrial process control. Researches in interdisciplinary subjects can prompt the development in engineering and extend applications of optimal control algorithms. For instance, optimal control approaches can help power system keep stable voltage and frequency states when faults happening on any nodes via state evaluation and decision. Applications in related fields are one of the directions of optimal control development for existing algorithms and novel algorithms.

The aforementioned four research directions are about bias reduction, approach for applications on industrial robots, applications on complex environmental robots, and applications on other related fields, which are based on different aspects. The direction for bias reduction is a mission on general algorithms themselves, which can promote the generation and improvement of advanced optimal control algorithms. The second direction is an approach to apply the state-of-the-art algorithms on industrial practice by the way of cooperation. The third and forth directions are the applications in different environment and subjects. Although they seem to have little to do with the optimal control algorithm, the application is also an essential part of the development of optimal control. The applications in aforementioned fields are not sample transplant, they need to be tuned according to control objectives and different cases. Trying different applications can also help adjust the performance of optimal controllers, which has a positive impact for the development of optimal control. Thus, this survey regards them as the future prospective directions.

5 Summary and Conclusions

This survey describes the development and applications of the previous optimal control algorithms on robot technology, and the comparison and analysis of different kinds of control approaches using for the automatic robots. In addition, three main problems of today's algorithms are discussed and several successful stories are reviewed to solve these kinds of problems. The survey discusses the future prospective directions of optimal control and MBRL which include overcoming limitations, artificial assistant learning, applications in complex environment, and interdisciplinary development in other data-driven optimization fields. With the development of advance optimization approaches, more problems on optimal control and RL will be addressed, which will be applied more widely in real-time industrial practice as well. Through this survey, the following conclusions can be drawn.

1. Researches will pay attention on the reduction of model bias and the improvement of the optimization efficiency. More precise modeling methods and more efficient control algorithms with little computational complexity will be developed for high-dimensional robot systems.
2. The optimal control approaches will be used in the other areas' control widely, and their applications will make the systems achieve the optimal solution and have better performance.
3. The applications on complex and extreme environmental robots will help to carry out researches in extreme environment regarding the deep-ocean and space for natural science exploration.
4. Artificial assisted learning applications for the industrial robots are available ways to apply the advanced optimal control algorithms in practice.

Acknowledgment. This work was supported by the Research Development Fund RDF-20-01-08 provided by Xi'an Jiaotong-Liverpool University.

References

1. Chen, Y., Roveda, L., Braun, D.J.: Efficiently computable constrained optimal feedback controllers. IEEE Rob. Autom. Lett. 4(1), 121–128 (2019)
2. Kober, J., Bagnell, J.A., Peter, J.: Reinforcement learning in robotics: a survey. Int. J. Rob. Res. 32(11), 1238–1274 (2013)
3. Chen, Y., Braun, D.J.: Hardware-in-the-loop iterative optimal feedback control without model-based future prediction. IEEE Trans. Rob. 35(6), 1419–1434 (2019)
4. Rastogi, D., Koryakovskiy, I., Kober, J.: Sample-efficient reinforcement learning via difference models. In: 3rd Machine Learning in Planning and Control of Robot Motion Workshop at ICRA (2018)

5. Silver, D., et al.: Mastering the game of Go with deep neural networks and tree search. Nature **529**(7587), 484–489 (2016)
6. Nagabandi, A., Kahn, G., Fearing, R.S., Levine, S.: Neural network dynamics for model-based deep reinforcement learning with model-free fine-tuning. In: 2018 IEEE International Conference on Robotics and Automation (ICRA), pp. 7559–7566 (2018)
7. Kupcsik, A., Deisenroth, M.P., Peters, J., Loh, A.P., Vadakkepat, P., Neumann, G.: Model-based contextual policy search for data-efficient generalization of robot skills. Artif. Intell. **247**, 415–439 (2017)
8. Deisenroth, M., Rasmussen, C.E.: PILCO: a model-based and data-efficient approach to policy search. In: 28th International Conference on Machine Learning (ICML), pp. 465–472 (2011)
9. Kumar, V., Todorov, E., Levine, S.: Optimal control with learned local models: application to dexterous manipulation. In: 2016 IEEE International Conference on Robotics and Automation (ICRA), pp. 378–383 (2016)
10. Nagabandi, A., et al.: Learning image-conditioned dynamics models for control of underactuated legged millirobots. In: 2018 IEEE/RSJ International Conference on Intelligent Robots and Systems (IROS), pp. 4606–4613 (2018)
11. Schulman, J., Levine, S., Abbeel, P., Jordan, M., Moritz, P.: Trust region policy optimization. In: 31th International Conference on Machine Learning (ICML), pp. 1889–1897 (2015)
12. Kingma, D.P., Ba, J.: ADAM: a method for stochastic optimization. In: 2015 International Conference for Learning Representations (ICLR) (2015)
13. Zhang, K., Shi, Y.: Adaptive model predictive control for a class of constrained linear systems with parametric uncertainties. Automatica **117**, 108974 (2020)
14. Rottmann, A., Burgard, W.: Adaptive autonomous control using online value iteration with gaussian processes. In: 2009 IEEE International Conference on Robotics and Automation (ICRA), pp. 2106–2111 (2009)
15. Vrabie, D., Pastravanu, O., Abu-Khalaf, M., Lewis, F.L.: Adaptive optimal control for continuous-time linear systems based on policy iteration. Automatica **45**(2), 477–484 (2009)
16. Chen, Y., Braun, D.J.: Iterative online optimal feedback control. IEEE Trans. Autom. Control **66**(2), 566–580 (2021)
17. Losey, D.P., McDonald, C.G., O'Malley, M.K.: A bio-inspired algorithm for identifying unknown kinematics from a discrete set of candidate models by using collision detection. In: 6th IEEE International Conference on Biomedical Robotics and Biomechatronics (BioRob), pp. 418–423 (2016)
18. Saputra, A.A., Wi Tay, N.N., Toda, Y., Botzheim, J., Kubota, N.: Bézier curve model for efficient bio-inspired locomotion of low cost four legged robot. In: 2016 IEEE/RSJ International Conference on Intelligent Robots and Systems (IROS), pp. 4443–4448 (2016)
19. Morton, J., Witherden, F.D., Jameson, A., Kochenderfer, M.J.: Deep dynamical modeling and control of unsteady fluid flows. In: 2018 Conference on Neural Information Processing Systems (NIPS) (2018)
20. Corneil, D., Gerstner, W., Brea, J.: Efficient model-based deep reinforcement learning with variational state tabulation. In: 35th International Conference on Machine Learning (ICML), pp. 1049–1058 (2018)
21. Lioutikov, R., Paraschos, A., Peters, J., Neumann, G.: Sample-based informationl-theoretic stochastic optimal control. In: 2014 IEEE International Conference on Robotics and Automation (ICRA), pp. 3896–3902 (2014)

22. Yaghmaie, F.A., Braun, D.J.: Reinforcement learning for a class of continuous-time input constrained optimal control problems. Automatica **99**, 221–227 (2019)
23. Levine, S., Wagener, N., Abbeel, P.: Learning contact-rich manipulation skills with guided policy search. In: 2015 IEEE International Conference on Robotics and Automation (ICRA), pp. 156–163 (2015)
24. Goedhart, M., Van Kampen, E.J., Armanini, S.F., de Visser, C.C., Chu, Q.P.: Machine learning for flapping wing flight control. In: 2018 AIAA Information Systems-AIAA Infotech @ Aerospace (2018)
25. Jordan, M.I., Rumelhart, D.E.: Forward models: supervised learning with a distal teacher. Cogn. Sci. **16**(3), 307–354 (1992)
26. Åkesson, B.M., Toivonen, H.T.: A neural network model predictive controller. J. Process Control **16**(9), 937–946 (2006)
27. Liu, D., Wang, D., Zhao, D., Wei, Q., Jin, N.: Neural-network-based optimal control for a class of unknown discrete-time nonlinear systems using globalized dual heuristic programming. IEEE Trans. Autom. Sci. Eng. **9**(3), 628–634 (2012)
28. Sun, Z., Dai, L., Liu, K., Dimarogonas, D.V., Xia, Y.: Robust self-triggered MPC with adaptive prediction horizon for perturbed nonlinear systems. IEEE Trans. Autom. Control **64**(11), 4780–4787 (2019)
29. Talvitie, E.: Self-correcting models for model-based reinforcement learning. In: 31 Conference on Artificial Intelligence (AAAI), pp. 1–12 (2017)
30. Talvitie, E.: Model regularization for stable sample rollouts. In: the 30th Conference on Uncertainty in Artificial Intelligence, pp. 780–789 (2014)
31. Xu, F., Ocampomartinez, C., Olaru, S., Niculescu, S.I.: Robust MPC for actuator-fault tolerance using set-based passive fault detection and active fault isolation. Int. J. Appl. Math. Comput. Sci. **27**(1), 43–61 (2017)
32. Kumbasar, T., Eksin, I., Guzelkaya, M., Yesil, E.: Adaptive fuzzy internal model control design with bias term compensator. In: 2011 IEEE International Conference on Mechatronics, pp. 312–317 (2011)
33. Li, X., Cao, L., Hu, X., Zhang, S.: Command filtered model-free robust control for aircrafts with actuator dynamics. IEEE Access. **7**, 139475–139487 (2019)
34. Daneshfar, F., Mansoori, F., Bevrani, H.: Multi-agent reinforcement learning design of load-frequency control with frequency bias estimation. In: The 2nd International Conference on Control, Instrumentation and Automation (ICCIA), pp. 310–314 (2011)
35. Vafamand, N., Arefi, M.M., Khooban, M.H., Dragicevic, T., Blaabjerg, F.: Nonlinear model predictive speed control of electric vehicles represented by linear parameter varying models with bias terms. IEEE J. Emerg. Sel. Topics Power Electron. **7**(3), 2081–2089 (2019)
36. Song, R., Lewis, F.L., Wei, Q., Zhang, H.: Off-policy actor-critic structure for optimal control of unknown systems with disturbances. IEEE Trans. Cybern. **46**(5), 1041–1050 (2016)
37. Varutti, P., Findeisen, R.: Event-based NMPC for networked control systems over UDP-like communication channels. In: 2011 American Control Conference, pp. 3166–3171 (2011)
38. Martinez-Cantin, R., Lopes, M., Montesano, L.: Body schema acquisition through active learning. In: 2010 IEEE International Conference on Robotics and Automation (ICRA), pp. 1860–1866 (2010)
39. Duchi, J., Hazan, E., Singer, Y.: Adaptive subgradient methods for online learning and stochastic optimization. J. Mach. Learn. Res. **12**(7), 2121–2159 (2011)

40. Fleming, J., Kouvaritakis, B., Cannon, M.: Robust tube MPC for linear systems with multiplicative uncertainty. IEEE Trans. Autom. Control **60**(4), 1087–1092 (2015)
41. Gupta, V., Luo, F.: On a control algorithm for time-varying processor availability. IEEE Trans. Autom. Control **58**(3), 743–748 (2013)
42. Demirel, B., Ghadimi, E., Quevedo, D.E., Johansson, M.: Optimal control of linear systems with limited control actions: threshold-based event-triggered control. IEEE Trans. Control Netw. Syst. **5**(3), 1275–1286 (2017)
43. Jenson, E.L., Chen, X., Scheeres, D.J.: Optimal control of sampled linear systems with control-linear noise. IEEE Control Syst. Lett. **4**(3), 650–655 (2020)
44. Nguyen, H., La, H.: Review of deep reinforcement learning for robot manipulation. In: 2019 3nd IEEE International Conference on Robotic Computing (IRC), pp. 590–595 (2019)
45. Khan, S.G., Herrmann, G., Lewis, F.L., Pipe, T., Melhuish, C.: Reinforcement learning and optimal adaptive control: an overview and implementation examples. Ann. Rev. Control. **36**(1), 42–59 (2012)
46. Polydoros, A.S., Nalpantidis, L.: Survey of model-based reinforcement learning: applications on robotics. J. Intell. Rob. Syst. **86**(2), 153–173 (2017)
47. Bhagat, S., Banerjee, H., Ho Tse, Z.T., Ren, H.: Deep reinforcement learning for soft, flexible robots: brief review with impending challenges. Robotics **8**(1), 4 (2019)
48. Khan, M.A.M., et al.: A systematic review on reinforcement learning-based robotics within the last decade. IEEE Access **8**, 176598–176623 (2020)
49. Zhao, W., Queralta, J.P., Westerlund, T.: Sim-to-real transfer in deep reinforcement learning for robotics: a survey. In: 2020 IEEE Symposium Series on Computational Intelligence (SSCI), pp. 737–744 (2020)
50. Abbeel, P., Coates, A., Ng, A.Y.: Autonomous helicopter aerobatics through apprenticeship learning. Int. J. Rob. Res. **29**(13), 1608–1639 (2010)
51. Cui, R., Yang, C., Li, Y., Sharma, S.: Adaptive neural network control of AUVs with control input nonlinearities using reinforcement learning. IEEE Trans. Syst. Man Cybern. Syst. **47**(6), 1019–1029 (2017)
52. Refsnes, J.E., Sorensen, A.J., Pettersen, K.Y.: Model-based output feedback control of slender-body underactuated AUVs: theory and experiments. IEEE Trans. Control Syst. Technol. **16**(5), 930–946 (2008)
53. Eller, L., Siafara, L. C., Sauter, T.: Adaptive control for building energy management using reinforcement learning. In: 2018 IEEE International Conference on Industrial Technology (ICIT), pp. 1562–1567 (2018)
54. Avila, L., De Paula, M., Carlucho, I., Sanchez Reinoso, C.: MPPT for PV systems using deep reinforcement learning algorithms. IEEE Lat. Am. Trans. **17**(12), 2020–2027 (2019)
55. Nguyen, T., Mukhopadhyay, S.: Multidisciplinary optimization in decentralized reinforcement learning. In: 16th IEEE International Conference on Machine Learning and Applications (ICMLA), pp. 779–784 (2017)
56. Dan, H., et al.: Error-voltage-based open-switch fault diagnosis strategy for matrix converters with model predictive control method. IEEE Trans. Ind. Appl. **53**(5), 4603–4612 (2017)
57. Yu, B., Zhang, Y., Qu, Y.: MPC-based FTC with FDD against actuator faults of UAVs. In: 15th International Conference on Control, Automation and Systems (ICCAS), pp. 225–230 (2015)
58. Kim, K., Raimondo, D.M., Braatz, R.D.: Optimum input design for fault detection and diagnosis: model-based prediction and statistical distance measures. Control Conference. In: 2013 European Control Conference (ECC), pp. 1940–1945 (2013)

KTOBS: An Approach of Bayesian Network Learning Based on K-tree Optimizing Ordering-Based Search

Qingwang Zhang, Sihang Liu, Ruihong Xu, Zemeng Yang, and Jianxiao Liu[✉]

College of Informatics, Hubei Key Laboratory of Agricultural Bioinformatics, Huazhong Agricultural University, Wuhan 430070, China

Abstract. How to construct Bayesian Networks (*BN*) efficiently and accurately is a research hotspot in the era of artificial intelligence. By limiting the tree-width of the network, the Bayesian network learning based on *k*-tree can be used to process large-scale of variables. However, this method has the problems of low accuracy, further to optimize the order of adding nodes, *etc*. In order to solve these problems, this work proposes a Bayesian learning method based on k-tree optimizing ordering-based search (*KTOBS*). Firstly, the local learning search strategy is adopted to obtain the candidate parent sets of each variable efficiently and accurately. Then it selects $k + 1$ nodes based on the obtained candidate parent node sets, and constructs the corresponding initial sub-network. Then the heuristic evaluation strategy is used to add subsequent nodes successively, and thus to get the initial network. Finally, it optimizes the network iteratively through switching nodes until the score of network no longer increases. The experimental results show that *KTOBS* can learn a network structure with higher accuracy than other *k*-tree algorithms in a given limited time.

Availability and implementation: codes and experiment dataset are available at: http://122.205.95.139/KTOBS/.

Keywords: Bayesian network · Candidate parent node · *k*-tree · Ordering-based search

1 Introduction

Bayesian network (*BN*), also known as causal networks or probabilistic networks, describes the relationship between variables through probability graph model. Bayesian network has the function of visualizing multiple knowledge diagrams, and has unique advantages in solving uncertainty and reasoning. Bayesian network is currently one of the most effective theoretical models in the field of uncertain knowledge expression and reasoning [1]. At present, Bayesian network has been applied in many fields, including medical diagnosis, language understanding, speech recognition, pattern recognition,

---

Q. Zhang and S. Liu—These authors contributed equally to this article.

H. Gao and X. Wang (Eds.): CollaborateCom 2021, LNICST 406, pp. 67–82, 2021.
https://doi.org/10.1007/978-3-030-92635-9_5

data mining, and artificial intelligence, etc. How to construct Bayesian network efficiently and accurately has attracted many scholars and become a research hotspot in the field of artificial intelligence.

Bayesian network structure learning is the process of finding a network to fit the sample dataset as much as possible. As the number of nodes increase, the complexity of Bayesian network learning increases exponentially. Bayesian network learning methods can be divided into three categories: constraint-based methods, scoring search-based methods and the hybrid methods. Constraint-based methods use statistics or information theory to quantitatively analyze the dependence relationship between variables firstly, and then find a network that is independent of these conditions and consistent with the dependence relationships. The representative constraint-based methods mainly include *grow*-shrink (*GS*), parents & children (*PC*) [2, 3], three-phase dependency analysis (*TPDA*) [4], incremental association Markov blanket (*IAMB*) [5], interleaved incremental association (*Inter-IAMB*), *etc.*

The scoring search-based method is a commonly used Bayesian network structure learning algorithm. The first step in this method is to determine the candidate parent sets for each variable. The second step is to select a parent set from the candidate parent sets of each variable, and then obtain the *DAG* graph of the network. The classical search scoring-based algorithms include *K2* [1], *K3* [6], hill climbing (*HC*) [7], tabu search (*Tabu*), etc. Gao and Ji combined the Markov Blanket and *BN* scoring function to improve the efficiency of existing score-based learning algorithms [8, 9]. In the view of modeling dependencies between groups of variables rather than between individual variables, Parviainen *et al.* used the groupwise faithfulness assumption and conditional independencies to learn the network between variables in different groups [10]. Later, Niinimäki *et al.* used metropolis-coupled Markov chain Monte Carlo and annealed importance method into sampling partial order to further enhance the *BN* learning efficiency [11]. From the aspect of structure constraint, Campos *et al.* used the decomposable properties of score functions to reduce the time and memory costs without losing global optimality guarantees [12, 13]. Regarding searching the Bayesian Network with the largest score as a constrained optimization problem, Bartlett used the integer linear programming (*ILP*) to get the global optimization Bayesian Network [14, 15].

Nie *et al.* proposed a sampling method to generate representative k-trees, and thus to learn a Bayesian Network with bounded tree-width [16, 17]. In 2016–2018, Mauro Scanagatta *et al.* proposed a k-tree optimizing Bayesian learning method of bounded tree width from the perspective of tree decomposition of graph structure [18–20]. However, this algorithm has the problems of low accuracy, needing to further optimizing the order of adding nodes, *etc.* In order to solve the above problems, this paper proposes a Bayesian network structure learning method based on k-tree optimizing ordering-based search (*KTOBS*).

- It uses dynamic programming and local learning methods to obtain the candidate parent node sets of each node efficiently and accurately. Then it selects $k + 1$ nodes and constructs initial corresponding sub-network on the basis of the obtained candidate parent sets.

- The heuristic evaluation strategy is used to add subsequent nodes successively, and thus to obtain the initial network. Then it exchanges the position of the adjacent node pairs in the initial ordering iteratively until the *BN* score of the corresponding network no longer increases.
- The classical datasets of *Alarm1*, *Alarm3*, *Alarm5*, *Insurance1*, *Insurance3*, *Insurance5*, are used for experimental verification. The experimental results show that the network constructed using *KTOBS* has the highest *BN* score. It can obtain a network structure with higher accuracy than other *k*-tree algorithms in given limited time.

2 Bayesian Network and k-tree

2.1 Bayesian Network

The scoring search-based method mainly includes two parts: scoring function and search strategy. The most commonly used scoring functions include *K2* (also known as *CH* score) [21], *BD* score [22], *BIC* (Bayesian Information Criterion) [23], *etc.* The *BIC* scoring function is an approximate calculation method of the marginal likelihood function within a large number of samples. This work mainly uses the *BIC* scoring function. The *BIC* scoring is decomposable, that is, the score of the entire network can be obtained by the sum of all the nodes. The *BIC* scoring function is shown in Eq. (1).

$$\mathrm{BIC}(G) = \sum_{i=1}^{n} \mathrm{BIC}(X_i, \Pi_i) =$$

$$\sum_{i=1}^{n} (\mathrm{LL}(X_i | \Pi_i) + Pen(X_i, \Pi_i)) = \quad (1)$$

$$\sum_{i=1}^{n} \left(\sum_{\pi \in |\Pi_i|, x \in |X_i|} N_{x,\pi} \widehat{\theta}_{x|\pi} - \frac{\log N}{2} (|Xi| - 1)(|\Pi_i|) \right)$$

In Eq. (1), $\theta^{\wedge}_{x|\pi}$ represents the conditional probability $P(X_i = x \mid \prod_i = \pi)$ of the maximum likelihood estimation. $N_{x,\ \pi}$ represent the number of occurrences satisfying $(X = x \wedge \prod_i = \pi)$ in the dataset. $|X_i|$ represents the number of values that X_i can take and $|\prod_i|$ represents the product of the value number of the parent node of X_i.

2.2 Tree Width and k-tree

$H = (V, E)$ represents the undirected graph, V represents the node set and E represents the edge set. The tree decomposition of the undirected graph H is expressed as (C, T), in which $C = \{C_1, C_2,...,C_m\}$ represents the subsets of V.

Tree Width: The tree width is defined as $\max(|C_i|)-1$, where $|C_i|$ represents the number of vertices in C_i. The tree width of H is the smallest width among all possible tree decompositions of G.

k-tree: If the undirected graph $T_k = (V, E)$ represents the largest graph whose tree width is k, then it is a k-tree, and adding any edge to T_k will increase its tree width. The definition of k-tree is shown as follows: for a $(k + 1)$-clique, it is a complete graph with $k + 1$ nodes, and $(k + 1)$-clique is a k-tree. A $(k + 1)$-clique can be decomposed into multiple k-cliques. Z represents a node that has not been included in V. The graph obtained by connecting Z to each node in the k-clique of T_k is also a k-tree.

3 BN Learning Based on K-tree Optimizing Ordering-Based Search

3.1 Obtaining Candidate Parent Set

We adopt the algorithms of local learning and dynamic programming to determine the candidate parent sets for all the nodes. It can help to enhance the calculation efficiency and improve the accuracy of the candidate parent sets. Supposing the node set $X = \{X_1, X_2, ..., X_n\}$ and the number of nodes is n, the process of obtaining candidate parent nodes for node X_i is shown in Algorithm 1.

```
Algorithm 1. Algorithm of getting CandidateParentSets
Input: Xᵢ: variable, n: the number of variables
Output: Π: CandidateParentSet
1:  P(Xᵢ) ←∅
2:  for j←1, …, n
3:      if Xⱼ!=Xᵢ
4:          insert Xⱼ into T(Xᵢ)
5:      endif
6:  endfor
7:  insert Xᵢ into P(Xᵢ)
8:  while (T(Xᵢ)!=∅)
9:      Remove v from T(Xᵢ)
10:     Insert v into P(Xᵢ)
11:     S←FindCandidateParents(Xᵢ, P(Xᵢ))
12:     P(Xᵢ) ←S∩P(Xᵢ)
13: endwhile
14: Π←GetSubsets(S)
15: return Π
```

In Algorithm 1, n represents the number of all nodes. $T(X_i)$ represents the candidate parent variable set to be searched of X_i. S represents the local candidate parent set has been found, and it can be obtained using $FindCandidateParents(X_i, P(X_i))$ [24]. Π represents the candidate parent sets of X_i. Initially, we set $P(X_i)$ to $\varnothing$ and add all the nodes to $T(X_i)$ except for X_i, as shown in Step 1–6. Then it adds X_i to $P(X_i)$, and takes out a node from $T(X_i)$ and adds it to $P(X_i)$ in each iteration, Then it uses *FindCandidateParents*$(X_i, P(X_i))$ to get S, as shown in Step 7–11. Then it assigns $P(X_i)$ to $S \cap P$ (X_i). The above operations are repeated until $T(X_i)$ is $\emptyset$, and it adds all the subsets of S to Π, as shown in Step 12–15.

Through the above steps, we can get the candidate parent sets of nodes A, B, C, D, E, F. The candidate parent node set of each node is arranged in descending order according to the Bayesian network score, as follows, A: $\{D, F\}$, $\{D\}$, $\{F\}$, $\{\}$; B: $\{C, D\}$, $\{C\}$, $\{D\}$, $\{\}$; C: $\{B\}$, $\{E\}$, $\{\}$; D: $\{A, B\}$, $\{A, E\}$, $\{B, E\}$, $\{A\}$, $\{B\}$, $\{\}$; F: $\{A\}$, $\{E\}$, $\{\}$.

Firstly, it determines $T(X_i)$ that means the candidate parent set to be searched of node X_i, takes a node from $T(X_i)$ and adds it to the local operation set $P(X_i)$ each iteration. Then it determines X_i's local candidate parent set S by calling *FindCandidateParents*$(X_i, P(X_i))$ [24]. Then it updates $P(X_i)$ to be the intersection of $P(X_i)$ and S. When $T(X_i)$ is null, it takes all the subsets of S as the candidate parent sets of X_i.

3.2 Initial Network Construction

Firstly, it selects $k + 1$ nodes according to the initial sub-network nodes selection rule. The accurate Bayesian network learning algorithm [25] is used to get the initial sub-network of the $k + 1$ nodes. Then the heuristic evaluation rule is used to add subsequent nodes iteratively, and it uses the k-tree constraint condition to reduce the complexity of processing large number of variables.

Select the Initial k + 1 Nodes: Supposing node set $R = \emptyset$, which expresses the node set that have been selected to construct the initial sub-network. The node set $Z = \emptyset$, it expresses the node set that can be selected to construct the initial sub-network. Firstly, it randomly selects a node to join R, and adds all the candidate parent nodes of R to Z. Then it selects a node in Z, and adds all the candidate parent nodes of the node to R. Repeated the above operations until R includes $k + 1$ nodes. In the process of selecting nodes from Z, if $Z = \emptyset$, it selects a node from all the nodes that have not been added in Z and updates Z. When $k = 2$, $2 + 1 = 3$ nodes are selected to construct the initial sub-network. The specific process of selecting $k + 1$ nodes from the 6 nodes and constructing the initial sub-network is shown in Fig. 1.

In Fig. 1, $R = \emptyset$, $Z = \emptyset$, it first randomly selects node A to join R, and adds all candidate parent nodes $\{D, F\}$ of node A to Z, as shown in Fig. 1(2). Then it selects node D to join R in Z, and adds all its candidate parent nodes $\{B, E\}$ to Z, as shown in Fig. 1(3). Then it selects node B to join R in Z, and adds all its candidate parent nodes $\{C, D\}$ to Z, as shown in Fig. 1(4). Then the number of nodes in R is $k + 1$, and ends the algorithm.

Generate Sub-Network of the Initial k + 1 Nodes: The accurate Bayesian learning algorithm [25] is used to construct the initial sub-network G_{k+1} and k-tree K_{k+1} of $k + 1$ nodes.

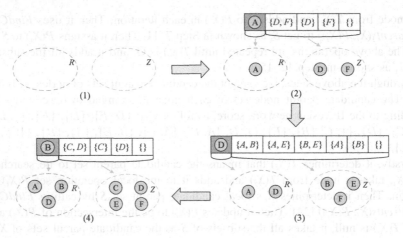

Fig. 1. Select $k + 1$ nodes to construct the initial sub-network

Heuristic Evaluation Method: It uses Eq. (2) to calculate the heuristic score $e(X_i)$ for the remaining other nodes X_i, and it selects nodes with larger BN scores to join the network structure. There are two parts in Eq. (2): the first part is the relative size measurement between the BN score of the feasible parent node set that each node can currently reach and the BN scores of all its candidate parent node sets. The latter part is the absolute size measurement between the BN score of the feasible parent node set that each node can currently reach and the BN scores of all its candidate parent node sets.

$$e(X_i) = \frac{sc^C(X_i) - sc^W(X_i)}{sc^B(X_i) - sc^W(X_i)} + \frac{sc^C(X_i) - sc^Q(X_i)}{sc^P(X_i) - sc^Q(X_i)} \tag{2}$$

$$sc^C(X_i) = \max_{\Pi \in L_i^*} score(\Pi) \tag{2-1}$$

$$sc^B(X_i) = \max_{\Pi \in L_i} score(\Pi) \tag{2-2}$$

$$sc^W(X_i) = \min_{\Pi \in L_i} score(\Pi) \tag{2-3}$$

$$sc^P(X_i) = \max_{i \subset X} \left(\max_{\Pi \in L_i^*} score(\Pi) \right) \tag{2-4}$$

$$sc^Q(X_i) = \min_{i \subset X} \left(\max_{\Pi \in L_i^*} score(\Pi) \right) \tag{2-5}$$

In the above equations, L_i represents all the feasible parent sets of node X_i. L^*_i represents a subset of all feasible parent sets of node X_i. Π represents a set of candidate parent nodes of X_i, and score(Π) represents the *BN* score of parent node set Π.

Add Subsequent Nodes to Build the Network: G_{i-1} and K_{i-1} represent the *DAG* and *k*-tree respectively before node X_i updated. Adding node X_i to G_{i-1}, it can get the updated *DAG* G_i. It connects node X_i to the nodes corresponding to *k*-tree and constrains its parent Π_i to be the *k*-clique of K_{i-1}, and thus to obtain the updated *k*-tree K_i. Compared with K_{i-1} before updated, K_i has an additional $(k + 1)$-clique. Figure 2 shows the change process of the network and *k*-tree structure after adding nodes iteratively.

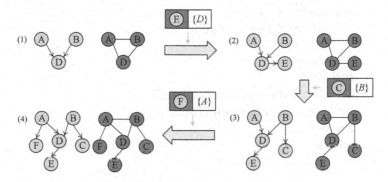

Fig. 2. Add subsequent nodes to build the network

In Fig. 2, when to add the first subsequent node, $e(E) > e(C)$ and $e(E) > e(F)$, node *E* is added to the current network. The candidate parent set with the highest score of *E* is $\{D\}$, and the updated *DAG* and *k*-tree are shown in Fig. 2(2). When to add the second subsequent node, $e(C) > e(F)$, it adds node *C* to the network. The candidate parent set with the highest score for node *C* is $\{B\}$, and the updated *DAG* and *k*-tree are shown in Fig. 2(3). When to add the last node, only node *F* has not yet been joined the network. Then node *F* is directly added to the network, and the parent set with the largest score is $\{A\}$. The updated *DAG* and *k*-tree are shown in Fig. 2(4). The specific process of constructing the initial network is shown in Algorithm 2.

```
Algorithm 2. Algorithm of initial network construction
Input: X: variable set, Π: CandidateParentSet, k:
treewidth
Output: Graph, order: variable addition order, T: Sub-
network node set
1: Graph←∅, order←∅, T←∅
2: Randomly pick k+1 nodes in X and put them in T
3: Graph=Exact algorithm(T)
4: X←X-T
5: while X!=∅
6:      e_max←∞, index=0
7:      for each X_i ∈X do
8:           if e(X_i)>e_max then
9:           e_max=e(X_i)
10:              index=i
11:      end for
12:      Graph←Update(Graph, Π_index)
13:      order←order.append(X_index)
14:      X←X-X_index
15: end while
16: return order, T, Graph
```

In Algorithm 2, X represents the nodes set, Π represents the candidate parent set of a node, k represents the tree width, $order$ represents the order in which subsequent nodes are added into the network, T represents the $k + 1$ nodes to be constructed the initial sub-network, and $Graph$ represents the obtained BN network. Firstly, the initial sub-network node selection rule is used to select the node set T, and it constructs the corresponding network and k-tree structure through the exact BN learning algorithm, as shown in Step 2–3. Then it removes T from X to avoid adding the repeated nodes, as shown in Step 4. Then it selects the node X_{index} with the largest evaluation index $e(*)$ from X. Then it adds node X_{index} to $Graph$ and gets the node adding order. Then it removes X_{index} from X, as shown in Step 6–14. It repeats the above operations iteratively until X is empty.

3.3   Optimizing Network Using Ordering-Based Search

Based on the ordering obtained by adding nodes, we optimize the network structure through switching the nodes order iteratively until the score of the network not increases. The nodes in the ordering satisfy the following property: only the node sets and its subsets in front of X_i can be regarded as the parent node set of X_i. Through the strategy of exchanging adjacent nodes in ordering, we can get a network with higher BN score through multiple iterations. The operator of exchanging adjacent nodes in ordering is shown in Eq. (3).

$$Swap(i) : (X_1, ..., X_i, X_{i+1}, ..., X_m) \mapsto (X_1, ..., X_{i+1}, X_i, ..., X_m) \qquad (3)$$

In Eq. (3), $order[i]$ represents the i-th node X_i in ordering. Firstly, it selects the node pair of $order[i]$ and $order[i + 1]$ that increases the network score the most, and then exchanges them to update ordering. The above operation is executed iteratively until the exchange operations fail to increase the network score. The relative positions of nodes that are not involved in the exchange operation have not changed, and only the supersets of $order[i]$ and $order[i + 1]$ need to be recalculated after the exchange operation. We use the following two rules to improve calculation efficiency.

(1) If the parent node of $order[i]$ contains $order[i + 1]$, the parent node set of $order$ $[i]$ needs to be recalculated.
(2) If the parent node of $order[i + 1]$ contains $order[i]$, the parent node set of $order$ $[i + 1]$ needs to be recalculated.

The specific process of using ordering-based search to optimize the initial network is shown in Algorithm 3.

Algorithm 3. Algorithm of optimizing network using order-
ing-based search
Input: *order*: variable addition order, *T*: variable set,
Graph: initial network, *L*: best parent set, *k*: tree width
Output: *G*: Bayesian network

```
1:  forbidden[n], index←0, maxScore←-∞
2:  for each Tᵢ∈T do
3:       forbidden[Tᵢ]=false
4:  end for
5:  while maxScore not changing
6:       for each order[i] do
7:           oldScore←getScore(Graph, or-
der[i])+getScore(Graph, order[i+1])
8:           newScore←0
9:           forbidden[order[i+1]]←false
10:          for each Π∈L_order[i] do
11:               flag←true
12:               for each x∈Π do
13:                   if forbidden[x]==true then
14:                       flag←false, break
15:               if flag==true then
16:                   newScore←score(Π)
17:          for each Π in L_order[i+1] do
18:               flag←true
19:               for each x in Π do
20:                   if forbidden[x]==true then
21:                       flag←false, break
22:               if flag==true then
23:                   newScore←newScore+score(Π)
24:          forbidden[order[i+1]]←true
25:          if newScore>oldScore and newScore>maxScore
26:               maxScore←newScore
27:               index=i
28:      end for
29:      if maxScore change then
30:          forbidden[order[index+1]]←false
31:          Swap(order[index], order[index+1])
32:          Update(G)
33: end while
34: return G
```

In Algorithm 3, *forbidden* expresses whether the node can be the parent node of the
current node or not. *index* indicates the number of the node that currently needs to be

exchanged, and *maxScore* indicates the largest score that the current exchange operation can achieve. In Step 2–4, it sets the *forbidden* of all nodes in *T* to false, which means that all nodes in *T* can be regarded as parent nodes of the subsequent nodes. In Step 5–33, it optimizes the network by repeatedly traversing the ordering, and ends the traversal when all the exchange operations fail to increase the network score. In Step 7, it calculates the sum of scores of the current parent node set of *order*[i] and *order* [i + 1], and records it as *oldScore*. In Step 8, *newScore* is used to record the sum of the scores of the parent node set after exchanging *order*[i] and *order*[i + 1]. In Step 11, it sets *forbidden* of *order*[i + 1] to false to ensure that *order*[i] can pick *order*[i + 1] as its parent node. In Step 12–16, it recalculates the best score that *order*[i] can reach after the exchanging, and denotes it as *newScore*. In Step 17–23, it recalculates the best score that *order*[i + 1] can achieve after switching nodes, and denotes it as *newScore*. The Step 24 is used to undo the exchange operation of *order*[i] and *order*[i + 1]. The Step 25–37 is used to judge whether the current exchange operation is the best exchange operation, and record the position information (*index*) of the exchange operation. In Step 29–32, it performs the exchange operation of the best node pair, and updates *order* and *forbidden*. After traversing *ordering* several times, it gets the network *G* with the largest *BN* score, as shown in Step 34.

Figure 3 shows an example of using ordering to optimize the initial network.

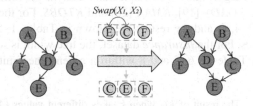

Fig. 3. Switching node order to optimize the network

In Fig. 3, it traverses the node order and gets the best switching node pair of *E*, *C*. Then it exchanges the position of *E*, *C* and updates the network and *order*.

4 Experiment

4.1 Dataset and Evaluation Method

The commonly used datasets of *Alarm1*, *Alarm3*, *Alarm5*, *Insurance1*, *Insurance3*, *Insurance5* are used for experiment comparison and analysis. We use the *Precision(P)*, *Recall rate(R)* and *F1-score* to do the experiment evaluation and analysis, as shown in the Eq. (4)–Eq. (6).

$$P = \frac{TP}{TP + FP} \tag{4}$$

$$R = \frac{TP}{TP + FN} \tag{5}$$

$$F1 = \frac{2 * P * R}{P + R} \tag{6}$$

In the equations, TP refers to the number of correctly identified edges, TN refers to the number of correctly identified non-linked edges, FP refers to the number of incorrectly identified edges, and FN refers to the number of incorrectly identified non-linked edges.

We carry the experiment on a computer with the configuration of Inter(R) Core (TM) i5-8300H CPU @ 2.30 GHz 2.30 GHz and 8G memory. The number of samples in each dataset is 5000. In order to ensure the accuracy of the experiments, we generate each dataset 10 times, and take the average of 10 datasets for comparison and analysis.

4.2 Experiment Results and Analysis

Experiments of Tree Width

In the k-tree related algorithms, the tree width k has a greater impact on the accuracy and efficiency of these algorithms. We compare and analyze the following k-tree related algorithms: KG [18], $KGADV$ [19], $KMAX$ [20] and $KTOBS$. For the $Alarm$ dataset, the tree width is set to 2, 4, 6, and the results are shown in Tables 1, 2, 3 and 4 within a given time limit of 10s. For the $Insurance$ dataset, the tree width is set to 2, 4, 6, and the results are shown in Tables 5, 6, 7 and 8 within a given time limit of 10 s.

Table 1. The result of KG when k takes different values ($Alarm$).

KG	Iterations	Accuracy	Recall	Precision	F1-score
k = 2	391588.3	0.957	0.341	0.259	0.294
k = 4	82304.2	0.953	0.352	0.422	0.383
k = 6	55292.1	0.953	0.352	0.435	0.389

Table 2. The result of $KGADV$ when k takes different values ($Alarm$).

KGADV	Iterations	Accuracy	Recall	Precision	F1
k = 2	525762.4	0.956	0.319	0.243	0.276
k = 4	110756.1	0.9496	0.309	0.383	0.342
k = 6	68878.3	0.9496	0.309	0.376	0.339

Table 3. The result of $KMAX$ when k takes different values ($Alarm$).

KMAX	Iterations	Accuracy	Recall	Precision	F1
k = 2	65149	0.958	0.361	0.283	0.317
k = 4	11299.2	0.947	0.289	0.372	0.325
k = 6	10347.3	0.952	0.338	0.417	0.373

Table 4. The result of *KTOBS* when *k* takes different values (*Alarm*).

KTOBS	Iterations	Accuracy	Recall	Precision	F1
k = 2	126199.4	0.962	0.434	0.337	0.381
k = 4	33068.6	0.992	0.884	0.885	0.884
k = 6	22526.9	0.995	0.917	0.939	0.928

Table 5. The result of *KG* when *k* takes different values (*Insurance*).

KG	Iterations	Accuracy	Recall	Precision	F1
k = 2	552799.3	0.930	0.558	0.263	0.358
k = 4	152056.2	0.944	0.655	0.515	0.577
k = 6	95244.4	0.940	0.617	0.504	0.5575

Table 6. The result of *KGADV* when *k* takes different values (*Insurance*).

KGADV	Iterations	Accuracy	Recall	Precision	F1
k = 2	727985.1	0.930	0.560	0.269	0.364
k = 4	194096.6	0.941	0.630	0.510	0.563
k = 6	108880.7	0.946	0.671	0.540	0.599

Table 7. The result of *KMAX* when *k* takes different values (*Insurance*).

KMAX	Iterations	Accuracy	Recall	Precision	F1
k = 2	133164.3	0.936	0.637	0.308	0.415
k = 4	30496.5	0.936	0.588	0.458	0.515
k = 6	31004.6	0.943	0.650	0.494	0.562

Table 8. The result of *KTOBS* when *k* takes different values (*Insurance*).

KTOBS	Iterations	Accuracy	Recall	Precision	F1
k = 2	206243.3	0.939	0.683	0.331	0.446
k = 4	50120.4	0.969	0.880	0.673	0.763
k = 6	32039.2	0.973	0.910	0.698	0.790

In Table 1-Table 8, the number of iterations of the four kinds of algorithms decreases gradually as the tree width *k* increases. It means that the efficiency gradually decreases as tree width *k* increases. In addition, it can be seen that the accuracy of the four kinds of algorithms gradually increases as the tree width *k* increases. On the whole, the four indicators (*Accuracy*, *Precision*, *Recall rate*, *F1-score*) of *KTOBS* are higher than those of *KG*, *KGADV* and *KMAX*. Considering the efficiency and accuracy together, we set tree width *k* to 4 in the following experiments.

Comparison and Analysis of Accuracy

Given a limited time (10s), this experiment compares and analyzes the accuracy of *KG* [18], *KGADV* [19], *KMAX* [20], *OBS* [26] and *KTOBS* in the datasets of *Alarm(Alarm1, Alarm3, Alarm5)* and *Insurance(Insurance1, Insurance3, Insurance5)*. The experimental results are shown in Fig. 4.

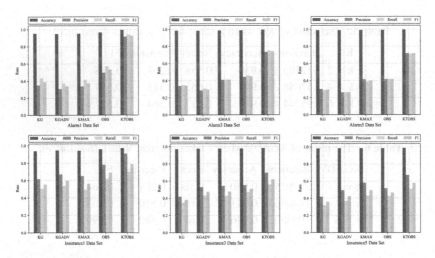

Fig. 4. Accuracy comparison on *Alarm* and *Insurance* dataset

In Fig. 4, the learning accuracy (*Accuracy, Precision, Recall, F1*-score) of *KTOBS* is significantly higher than that of *KG, KGADV, KMAX* and *OBS*. The accuracy of the five kinds algorithms is relatively high, basically above 0.95. The *Precision, Recall, F1-Score* of *KG, KGADV, KMAX* is lower than *OBS* and *KTOBS* apparently. But the *Precision, Recall, F1-Score* of *OBS* is slightly worse than *KTOBS*.

Efficiency Comparison and Analysis

This experiment compares and analyzes the running time of one iteration of *KG* [18], *KGADV* [19], *KMAX* [20], *OBS* [26] and *KTOBS* in the datasets of *Alarm(Alarm1, Alarm3, Alarm5)* and *Insurance(Insurance1, Insurance3, Insurance5)*. The experimental results are shown in Fig. 5.

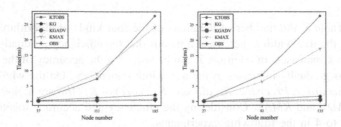

Fig. 5. Efficiency comparison on *Alarm* and *Insurance* dataset

In Fig. 5, the running time of one iteration of all algorithms increases with the increasing of the variable number in the dataset. It is consistent with the law of efficiency decreasing as the complexity of the problem increases. The time spent in one iteration of *KTOBS* is slightly higher than that of other four kinds of algorithms. In the *Alarm* dataset, the learning efficiency of *KTOBS* and *KMAX* is not much different.

In summary, *KTOBS* can obtain a network structure with a higher accuracy within the given limited time. *KTOBS* does not have obvious advantages in the view of the running time of a single iteration. But *KTOBS* has a greater advantage than other methods in the same limited running time in the view of the accuracy.

5   Conclusions

This work proposes a Bayesian network structure learning method (*KTOBS*) based on k-tree optimizing ordering-based search. Firstly, the strategy of dynamic programming and local learning is adopted to get the candidate parent sets of each node efficiently and accurately. Then it constructs the initial sub-network of $k + 1$ nodes, and adds the subsequent nodes successively according to the heuristic evaluation strategy, At the same time, it reduces the complexity of processing large number of nodes according to the k-tree structure. Based on the node order obtained by adding the subsequent nodes, it exchanges the adjacent nodes in ordering iteratively until the score of the network no longer increases. The classical datasets of *Alarm* and *Insurance* are used in experiment verification and analysis. Experiment results show that *KTOBS* can obtain a network structure with higher accuracy than other k-tree related algorithms in a given limited time.

Acknowledgement. This research is supported by the Fundamental Research Funds for the Central Universities under grant No. 2020BC211, S202010504276.

References

1. Cooper, G.F., Herskovits, E.: A Bayesian method for the induction of probabilistic networks from data. Mach. Learn. **9**(4), 309–347 (1992)
2. Spirtes, P., Glymour, C.: An algorithm for fast recovery of sparse causal graphs. Soc. Sci. Comput. Rev. **9**(1), 62–72 (1990)
3. Spirtes, P., Glymour, C., Scheines, R.: Causality from probability. In: Evolving Knowledge in Natural and Artificial Intelligence, pp. 181–199. Pitman, London (1989)
4. Cheng, J., Bell, D., Liu, W.: Learning Bayesian Networks from Data: An Efficient Approach Based on Information Theory (1997)
5. Tsamardinos, I., Aliferis, C.F., Statnikov, A.R.: Algorithms for Large Scale Markov Blanket Discovery. International Flairs Conference, pp. 376–380 (2003)
6. Bouckaert, R.R.: A stratified simulation scheme for inference in Bayesian belief networks. Uncertainty Proc. **5**(1), 110–117 (1994)
7. Tsamardinos, I., Brown, L.E., Aliferis, C.F.: The max-min hill-climbing Bayesian network structure learning algorithm. Mach. Learn. **65**(1), 31–78 (2006)

8. Gao, T., Ji, Q.: Efficient score-based Markov blanket discovery. Int. J. Approximate Reasoning **80**, 277–293 (2017)
9. Gao, T., Fadnis, K., Campbell, M.: Local-to-global Bayesian network structure learning. In: Proceedings of the 34th International Conference on Machine Learning, pp. 1193–1202 (2017)
10. Parviainen, P., Kaski, S.: Learning structures of Bayesian networks for variable groups. Int. J. Approximate Reasoning **88**, 110–127 (2017)
11. Niinimäki, T., Parviainen, P., Koivisto, M.: Structure discovery in Bayesian networks by sampling partial orders. J. Mach. Learn. Res. **17**(1), 2002–2048 (2016)
12. Campos, C.P., Ji, Q.: Efficient structure learning of Bayesian networks using constraints. J. Mach. Learn. Res. **12**(3), 663–689 (2011)
13. Li, A., Beek, P.: Bayesian network structure learning with side constraints. In: International Conference on Probabilistic Graphical Models, pp. 225–236 (2018)
14. Bartlett, M., Cussens, J.: Integer linear programming for the Bayesian network structure learning problem. Artif. Intell. **244**, 258–271 (2017)
15. Bartlett, M., Cussens, J.: Advances in Bayesian network learning using integer programming. arXiv preprint arXiv:1309.6825 (2013)
16. Nie, S., de Campos, C.P., Ji, Q.: Efficient learning of Bayesian networks with bounded tree-width. Int. J. Approximate Reasoning **80**, 412–427 (2017)
17. Nie, S., De Campos, C.P., Ji, Q.: Learning bounded tree-width Bayesian Networks via sampling. In: European Conference on Symbolic and Quantitative Approaches to Reasoning and Uncertainty. Springer, Cham, pp. 387–396 (2015).https://doi.org/10.1007/978-3-319-20807-7_35
18. Scanagatta, M., et al.: Learning treewidth-bounded Bayesian networks with thousands of variables. In: Advances in Neural Information Processing Systems, pp. 1462–1470 (2016)
19. Scanagatta, M., Corani, G., de Campos, C.P., Zaffalon, M.: Approximate structure learning for large Bayesian networks. Mach. Learn. **107**(8–10), 1209–1227 (2018). https://doi.org/10.1007/s10994-018-5701-9
20. Scanagatta, M., et al.: Efficient learning of bounded-treewidth Bayesian networks from complete and incomplete data sets. Int. J. Approximate Reasoning **95**, 152–166 (2018)
21. Cooper, G., Hersovits, E.: A Bayesian method for the induction of probabilistic networks from data. Mach. Learn. **9**, 309–347 (1992)
22. Heckerman, D., Geiger, D., Chickering, D.M.: Learning Bayesian networks: the combination of knowledge and statistical data. Mach. Learn. **20**(3), 197–243 (1995)
23. Schwarz, G.: Estimating the dimension of a model. Ann. Stat. **6**(2), 461–464 (1978)
24. Silander, T., Myllymaki, P.: A simple approach for finding the globally optimal Bayesian network structure. In: Proceedings of the Twenty-Second Annual Conference on Uncertainty in Artificial Intelligence (2006)
25. Cussens, J.: Bayesian network learning with cutting planes. In: Proceedings of the 27th Conference Annual Conference on Uncertainty in Artificial Intelligence, Proceedings of the Twenty-Seventh Conference on Uncertainty in Artificial Intelligence, Barcelona, Spain, 14–17 July, pp. 153–160 (2011)
26. Teyssier, M., Koller, D.: Ordering-based search: a simple and effective algorithm for learning Bayesian networks. In: Proceedings of the 21st Conference on Uncertainty in Artificial Intelligence. Arlington, USA, pp. 584–590 (2005)

Recommendation Model Based on Social Homogeneity Factor and Social Influence Factor

Weizhi Ying[1] , Qing Yu[1], and Zuohua Wang[2(✉)]

[1] Tianjin Key Laboratory of Intelligence Computing and Network Security,
Tianjin University of Technology, Tianjin 300384, China
[2] China Everbright Bank CO. LTD., Beijing 100035, China

Abstract. In recent years, more and more recommendation algorithms incorporate social information. However, most social recommendation algorithms often only consider the social homogeneity factor between users and do not consider the social influence factor. To make the recommendation model more in line with the real-life situation, this paper proposes a novel graph attention network to model the homogeneity effect and the influence effect in the user domain. Besides, we also extended this idea to the item domain, using information from similar items to alleviate the problem of data sparsity. Also, considering that there will be interactions between the user domain and the item domain, which together affect the user's preference for the item, we use a contextual multi-armed bandit to weigh the interaction between the two domains. We have conducted extensive comparative experiments and ablation experiments on two real public datasets. The experimental results show that the performance of our proposed model in the rating prediction task is better than other social recommendation model.

Keywords: Graph attention network · Social recommendation · Multi-armed bandit

1 Introduction

With the rapid development of the Internet, people have entered an information overload era [1]. Since the traditional collaborative filtering recommendation methods usually have problems with the data-sparse and cold start. Therefore, many scholars propose to use social networks to establish a novel recommendation system (i.e., social recommendation) to alleviate the problems mentioned above.

In the social recommendation, the user's decision is often affected by various factors. As shown in Fig. 1, users are affected by the user homogeneity factor and user influence factor when making decisions in the user domain. User homogeneity factor means that users and social friends often have similar preferences [2]. The user influence factor implies that the user may recommend the item to his friends after they purchase (or click) an item [3]. Besides, the item homogeneity factor and item influence factor also exist in the item domain. Item homogeneity factor means that similar items have similar attractiveness. Also, the item influence factor means that the user clicks on

H. Gao and X. Wang (Eds.): CollaborateCom 2021, LNICST 406, pp. 83–98, 2021.
https://doi.org/10.1007/978-3-030-92635-9_6

an item, and its similar items are more likely to be clicked by the other user. Finally, the four factors will affect the user's decisions together.

Previous research on social recommendation has adopted many ways to simulate social effects [4–6]. Converts social information into the regularization term [7, 8]. Uses network embedding methods to map each user into a low-dimensional vector [9, 10, 19]. Uses graph neural networks to obtain the social diffusion process.

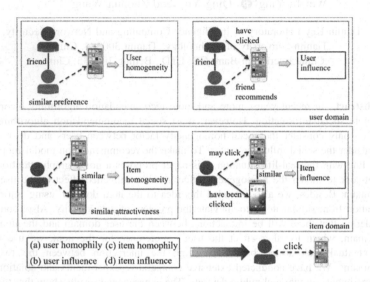

Fig. 1. Social homogeneity factor and social influence factor.

However, the studies mentioned above have the following shortcomings: First, most research on social recommendation only considers the effect of social homogeneity among users. There is not only a social homogeneity but also exists social influence among users. Secondly, many social recommendation studies assume that restricts social friends' influence through constant weights. This way of restraint does not conform to the social situation in real life.

Therefore, in this paper, we focus on the rating prediction task and propose RMHFIF (Recommendation Model Based on Social Homogeneity Factor and Social Influence Factor), a novel social recommendation model based on the Novel Graph Attention Network (NGAT). Specifically, the RMHFIF model consists of four NGAT modules. In the user domain, an NGAT is used to aggregate the embedding vectors of neighboring users that reflect user preferences to obtain user homogeneity factor; an NGAT is used to convolve the context-aware preferences of neighboring users to obtain user influence factor. Similarly, in the item domain, we also use NGAT to model item homogeneity factor and item influence factor. To verify the effectiveness of the model proposed in this paper, we conducted experiments on two public datasets. In summary, the main contributions of this paper are as follows:

- We propose a novel graph attention weight calculation method. And we apply it to model the social homogeneity factor and social influence factor.
- We propose the RMHFIF model, which models the four factors in the user domain and the item domain. It enables the model to weigh the homogeneity factor and influence factor according to the environment.
- We conducted extensive experiments on public datasets. The experimental results showed that our proposed model is better than the other recommendation model.

2   Related Work

In recent years, the integration of social relationships into recommendation algorithms has received a lot of attention [18, 20]. The social recommendation model assumes that the preferences of users and his/her friends are similar. Many scholars have proposed many social recommendation models based on this assumption. Ma H et al. proposed the SoRec method, which is a social recommendation algorithm based on feature sharing [21]. In this method, the user's feature vector is learned by simultaneously decomposing the rating matrix and the social relationship matrix, so that the learned user feature vector can take into account the user's rating habits and social characteristics. Jamali et al. proposed the SocialMF method, which assumes that the user's behavioral preferences should be similar to the average preferences of the user's social neighbors [6]. This method adds the trust propagation mechanism to the model, which can improve the accuracy of the recommendation for cold-start users. Lin et al. proposed the CSR method [22]. This method models the characteristics of social relationships and models the influence characteristics of social relations. Rafailidis D proposed the SDPL method, which is a ranking model that performs social deep pairwise learning with users' trust and distrust relationships [23].

Graph neural networks can learn graph structure data. The model related to our work is GraphRec [18]. GraphRec uses the graph neural network to learn the hidden factors of users and items in the user-user social graph and user-item feedback graph, and finally completes the task of rating prediction. Although the previous work has achieved good results, the previous work only modeled the homogeneity in social relationships and did not model the influence effects in social relationships. In this paper, we propose the RMHFIF model to fill this gap.

3   Preliminary and Problem Definition

3.1   Novel Graph Attention Network

The setting of the attention weight in the traditional graph attention network [11] is shown in Eq. (1). In Eq. (1), $\overrightarrow{h}_i$ represents the characteristics of the input node, W represents the weight matrix, N_i is the neighborhood of node i in the graph, and $\overrightarrow{a}$ is the weight vector.

$$\alpha_{ij} = \frac{\exp(LeakyReLU(\vec{a}^T[W\vec{h_i}\|W\vec{h_j}]))}{\sum_{k\in N_i}\exp(LeakyReLU(\vec{a}^T[W\vec{h_i}\|W\vec{h_k}]))} \qquad (1)$$

Inspired by [24], the graph attention network uses the attention mechanism to aggregate the information of neighbor nodes. Applying it to the social network graph can be understood as each user node in the social network graph aggregates the information of neighboring user nodes. However, the traditional graph attention network weight calculation formula does not add a bias term. In real life, everyone's interests are different, and people with different interests can also become friends, which affects the user's interests. Therefore, in this section, a novel calculation method for the attention weight of the graph attention network is proposed. The bias term is added to the original weight calculation formula. Using this calculation method can provide a better social graph representation, the specific expression is as follows:

$$\alpha_{ij} = \frac{\exp(\tanh(\vec{a}^T[W\vec{h_i}\|W\vec{h_j}]+b_{ij}))}{\sum_{k\in N_i}\exp(\tanh(\vec{a}^T[W\vec{h_i}\|W\vec{h_k}]+b_{ik}))} \qquad (2)$$

3.2 Notations

Table 1 shows the mathematical notations and their definitions used in this paper.

Table 1. Notation.

Symbols	Definitions
R	user-item rating matrix
r_{ui}	user u's rating for item i
M	the number of users
N	the number of items
$C_I(u)$	the set of items rated by user u
$C_U(i)$	the set of users who have rated item i
V_U	the set of users
E_U	the set of edges connecting users
SN_U	the social network between users, $SN_U = (V_U, E_U)$
e_{uv}	the frequency of interaction between user u and user v
$F_U(u)$	the set of the user u's social friends
$F_I(i)$	the set of item i's similar items

3.3 Problem Definition

The definition of the social recommendation problem for this paper is as follows: Given a rating matrix R and the social network SN_U, predict the user's rating for items that have not yet been rated.

4 The Proposed Model

The architecture of the proposed model is shown in Fig. 2. In this section, we will introduce in detail each module of the model and the method of model training.

4.1 Model Details

4.1.1 Original Input and Similar Item Network

First, we require rating matrix R and social network SN_U as input. The similarity coefficient between items Sim_{ij} is defined as the number of users who rate item i and item j at the same time. Specifically, given a threshold τ, if $Sim_{ij} > \tau$ then item i is similar to item j. We define the similar item set as V_I and connect similar items to form an edge set E_I. V_I and E_I together form a similar items network $SN_I = (V_I, E_I)$.

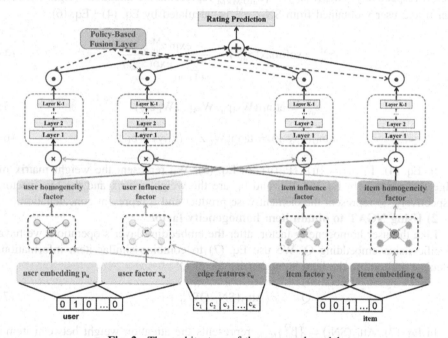

Fig. 2. The architecture of the proposed model.

4.1.2 Embedding Layer

Because each user or item's original input is a high-dimensional and sparse one-hot vector. We first need to map each user vector or item vector to a low-dimensional dense vector representation through the embedding operation. Inspired by [12], each user vector can be mapped into a specific user embedding $P = \{p_u\}_{D \times M}$ and a rating-based user embedding factor $X = \{x_u\}_{D \times N}$. The specific user embedding reflects the user's interest, and the rating-based user embedding factor reflects the implicit influence of the user's past rating on the current decision. Similarly, each item can be mapped to a specific item embedding $Q = \{q_i\}_{D\times N}$ and rating-based item embedding factor $Y = \{y_i\}_{D\times M}$. The specific item embedding reflects the item's attributes, and the rating-based item embedding factor reflects the influence of the item's past rating on the user's decision-making.

4.1.3 NGAT Layer
1) Using NGAT to obtain user homogeneity factor
After the embedding layer's operation, we have a specific user embedding P. We use Eq. (3) to calculate and output the user homogeneity factor M^*:

$$M^* = \sigma(Att_M(SN_U)PW_M^T + b_M) \tag{3}$$

In Eq. (3), σ, W_M and b_M are the activation function, weight matrix, and bias vector, respectively. $Att_M(SN_U) = \{\alpha_{uv}^M\}_{M\times M}$ represents the attention weight between user u and user v obtained from SN_U. α_{uv}^M is calculated by Eq. (4)– Eq. (6).

$$\alpha_{uv}^M = softmax(\alpha_{uv}^{M*}) = \frac{\exp(\alpha_{uv}^{M*})}{\sum\limits_{w\in\Gamma_U(u)} \exp(\alpha_{uw}^{M*})} \tag{4}$$

$$\alpha_{uv}^{M*} = attn_U(W_M p_u, W_M p_v, W_E e_{uv}) \tag{5}$$

$$attn_U(x, y, z) = \tanh(W_U^T z \otimes (x||y) + b_U) \tag{6}$$

In Eq. (4), $\Gamma_U(u) = \{u\} \cup F_U(u)$. In Eq. (5), W_E represent the weight matrix of edge features e_{uv}. In Eq. (6), W_U and b_U are the weight matrix and the bias vector, respectively. $\otimes$ represent the element-wise product and $||$ represent concatenation.

2) Using NGAT to obtain item homogeneity factor
Like the user homogeneity factor, after the embedding layer's operation, we have specific items embedding Q. We use Eq. (7) to combine similar item information, calculate and output the item homogeneity factor Q^*.

$$Q* = \sigma(Att_Q(SN_I)QW_N^T + b_N) \tag{7}$$

In Eq. (7), $Att_Q(SN_I) = \{\beta_{ij}^Q\}_{N\times N}$ represents the attention weight between item i and item j obtained from SN_I. β_{ij}^Q is calculated by Eq. (8)–Eq. (10).

$$\beta_{ij}^Q = softmax(\beta_{ij}^{Q*}) = \frac{\exp(\beta_{ij}^{Q*})}{\sum_{k\in\Gamma_I(i)} \exp(\beta_{ij}^{Q*})} \tag{8}$$

$$\beta_{ij}^{Q*} = attn_I(W_Q q_i, W_Q q_j), \quad j \in \Gamma_I(i) \tag{9}$$

$$\text{attn}_I(x, y) = \tanh(w_I^T(x||y) + b_I) \tag{10}$$

In Eq. (9), $\Gamma_I(i) = \{i\} \cup F_I(i)$. It's worth noting that the attention weight calculated by Eq. (7) remains unchanged after item i and item j are determined, which means Q_* is fixed for the item.

3) Using NGAT to obtain user influence factor

Unlike the user homogeneity factor, the user influence factor is often context-aware. User influence factor is different when facing different items and different friends. Through the embedding layer, we can get the item embedding X_u, which is clicked by user u. Use Eq. (11) to make the item that clicked by the user u interact with the item i^+ that has not yet been clicked.

$$X_u^{i^+} = \{x_j \otimes x_{i^+} | j \in C_I(u)\} \tag{11}$$

Equation (9)'s product operation can be helpful for modeling in a dynamic user influence environment. Define the rating-based user embedding $P_{i^+} = \{p_u^{i^+}\}_{D \times M}$. Next, we use the maximum pooling (MP) operation to select the most important D-dimensional features. The specific process is shown as follows:

$$p_{ud}^{i^+} = \underset{j \in C_1(u)}{\text{MP}} \{x_{jd} \cdot x_{i^+d}\}, \, d = 1, \ldots, D \tag{12}$$

In Eq. (12), MP stands for maximum pooling operation. We use Eq. (13) to calculate the user influence factor $P_{i^+}^*$.

$$P_{i^+}^* = \sigma(\text{Att}_P(SN_U)PW_P^T + b_P) \tag{13}$$

In Eq. (13), $\text{Att}_P(SN_U) = \{\alpha_{uv,i^+}^P\}_{M \times M}$, $\alpha_{uv,i^+}^{P^*}$ is calculated by Eq. (14)−Eq. (15).

$$\alpha_{uv,i^+}^P = \text{softmax}(\alpha_{uv,i^+}^{P^*}) = \frac{\exp(\alpha_{uv,i^+}^{P^*})}{\sum_{w \in \Gamma_U(u)} \exp(\alpha_{uw,i^+}^{P^*})} \tag{14}$$

$$\alpha_{uv,i^+}^{P^*} = \text{attn}_U(W_P p_u^{i^+}, W_P p_v^{i^+}, W_E e_{uv}), v \in \Gamma_U(u) \tag{15}$$

The $\text{attn}_U(\bullet)$ used in Eq. (15) is the same as Eq. (6). It's worth noting that the attention weight α_{uv,j^+}^P depends on the user's rating history and the specific candidate item i^+. It means that $P_{i^+}^*$ will continue to change with the context. Such a design conforms to the real situation of social influence among users.

4) Using NGAT to obtain item influence factor

Like user influence, the calculation of the rating-based item represents $N_{u^+} = \{n_i^{u^+}\}_{D \times N}$ is shown in Eq. (16)−Eq. (17).

$$Y_i^{u^+} = \{y_v \otimes y_{u^+} | v \in C_I(i)\} \tag{16}$$

$$n_{id}^{u^+} = \underset{v \in C_U(i)}{MP} \{y_{vd} \cdot y_{u^+d}\}, \forall d = 1, ..., D \tag{17}$$

In Eq. (17), $n_{id}^{u^+}, y_{u^+d}, y_{vd}$ are the d-th features of $n_i^{u^+}, y_{u^+}, y_v$ respectively. Next, we calculate the item influence by Eq. (18).

$$N_{u^+}^* = \sigma(Att_N(SN_U)N_{u^+}W_N^T + b_N) \tag{18}$$

In Eq. (18), $Att_N(SN_U) = \{\beta_{ij,u^+}^N\}_{N \times N}$, and β_{ij,u^+}^N is calculated by Eq. (19)–Eq. (20).

$$\beta_{ij,u^+}^N = softmax(\beta_{ij,u^+}^{n^*}) = \frac{exp(\beta_{ij,u^+}^{n^*})}{\sum_{k \in \Gamma_I(i)} exp(\beta_{ik,u^+}^{n^*})} \tag{19}$$

$$\beta_{ij,u^+}^{N^*} = attn_I(W_N n_i^{u^+}, W_N n_j^{u^+}), \quad j \in \Gamma_I(i) \tag{20}$$

4.1.4 Pairwise Neural Interaction Layer

After using NGAT to obtain homogeneity factor and influence factor, inspired by [13], we make homogeneity factors and influence factors interact with each other in pairs. Input the four factors into different neural networks.

$$z_0 = [m_u^* \otimes q_i^*, m_u^* \otimes n_i^*, p_u^* \otimes q_i^*, \quad p_u^* \otimes n_i^*] \tag{21}$$

$$g_k^a(z_{k-1}) = tanh(W_k^a z_{k-1}^a + b_k^a), k \in [1, K-1] \tag{22}$$

$$h_a = g_K^a(...g_2^a(g_1^a(z_0[a]))), a \in \{1, 2, 3, 4\} \tag{23}$$

We use the tower structure. The higher layer has fewer neurons [13].

4.1.5 Policy-Based Fusion Layer

Here, the four interactive features h_a are further merged into a synthetic feature. The homogeneity factor and the influence factor can work together, but these factors are different for different users and items. Thus, we modeled the weight distribution of the homogeneity factor and influence factor as a contextual multi-armed bandit problem. The action is represented by $\gamma \in \{1, 2, 3, 4\}$, indicates which factor is selected, and the environment is a user-item pair. The random strategy can be represented by conditional probability $p(\gamma | p_u, q_i)$, representing the probability of choosing different social effects given a specific user-item pair (u, i).

We use Eq. (24)–Eq. (25) to calculate the conditional probability $p(\gamma|p_u, q_i)$:

$$e_\gamma = w_\gamma[p_u||q_i||(p_u \otimes q_i)] + b_\gamma \tag{24}$$

$$p(\gamma|p_u, q_i) = \frac{\exp(e_\gamma)}{\sum\limits_{s=1}^{4} \exp(e_s)} \tag{25}$$

Next, comprehensive representation can be expressed as follows:

$$h = E_{\gamma \sim p(\gamma|p_u, q_i)}(h_\gamma) = \sum_{\gamma=1}^{4} p(\gamma|p_u, q_i) \times h_\gamma \tag{26}$$

4.1.6 Output Layer and Loss Function

The model final outputs the user u's rating for unrated item i, which is specifically calculated by the following equation:

$$\hat{r}_{ui} = w_o h + b_o \tag{27}$$

We follow the research of [18] and also use the mean square error loss function:

$$H_1 = \sum_{(u,i)\in O} (\hat{r}_{ui} - r_{ui})^2 \tag{28}$$

In Eq. (28), O represents the set of user-item ratings in which user u has rated item i, r_{ui} is the actual rating of user u on item i and $\hat{r}_{ui}$ represents the predicted rating.

4.2 Model Training

4.2.1 Mini-Batch Training

In each epoch, since the number of friends of each user is different, we performed a sampling operation: First, set a threshold F, and then perform sampling operations. If the number of friends exceeds F, then F friends are randomly sampled as input data. If the number of friends is less than F, 0 is filled so that the vector with the number of dimensions F is reached.

4.2.2 Alleviate Overfitting

To alleviate the over-fitting problem, we adopted L1 regularization. The following equation represents the regularization loss term:

$$H_2 = \sum_u (||p_u|| + ||y_u||) + \sum_i (||q_i|| + ||x_i||) \tag{29}$$

In summary, the final loss function of our proposed model is as follows:

$$L_{final} = H_1 + \lambda H_2 \tag{30}$$

In the experimental part, we will further discuss the influence of the choice of regularization parameter λ on the model's performance. Besides, we adopt the dropout strategy to alleviate overfitting.

5 Experiments

To comprehensively evaluate the performance of our proposed model, we conducted extensive experiments to answer the following questions:

RQ1 Compared with other recommendation models, how does our social recommendation model perform?

RQ2 The proposed model uses both the homogeneity factor and the influence factor. Only use one of them, how will the performance of the model change?

RQ3 The proposed model modifies the traditional attention weight calculation method in the graph attention network. If only the attention weight calculation method in the traditional graph attention network is used, how will the performance of the model change?

RQ4 How does the change of hyperparameters affect the performance of the model?

5.1   Dataset Introduction

We choose two public data sets for the experiment. The statistics of the datasets are shown in Table 2.

- **Ciao**: This dataset was obtained by Tang Jiliang [14] crawled from the article review website Ciao in May 2011.

Table 2. Dataset statistics.

Statistics	Ciao	Epinions
Users	7,375	49,290
Items	105,114	139,738
Ratings	284,086	664,824
Rating range	[1,5]	[1,5]
Relations	111,781	487,181

- **Epinions**: On the Epinions website, visitors can read various item reviews to determine their purchase behavior and write reviews to get rewards. This dataset is Paolo Massa crawled from the Epinions.com website [15].

5.2   Experimental Setup

5.2.1   Experimental Environment Setting

To prove the effectiveness of our proposed model, we select 80% of the user-item interaction data as the training set and the remaining 20% as the test set. All experimental environments are based on Python 3.7 and use TensorFlow 1.15. The training and testing of the model are based on the NIVIDA TESLA T4 GPU.

5.2.2   Evaluation Metrics

We use Mean Absolute Error (**MAE**) and Root Mean Square Error (**RMSE**) to evaluate our proposed model's rating prediction quality.

The definition of MAE and RMSE are as follows:

$$MAE = \frac{1}{T} \sum_{i,j} |R_{ij} - \hat{R}_{ij}| \tag{31}$$

$$RMSE = \sqrt{\frac{1}{T} \sum_{i,j} (R_{ij} - \hat{R}_{ij})^2} \tag{32}$$

In Eq. (31), R_{ij} represents the rating of user i to item j, R_{ij} represents the rating of user i to item j predicted by the model, and T represents the number of ratings in the test set. We can find that the smaller MAE and RMSE, the better the model's performance.

5.2.3   Compare Models

We choose the following comparison models for comparison to verify the effectiveness of the model proposed in this paper:

- **SVD++** [12]: This is a recommendation model that considers explicit feedback information and implicit feedback information.
- **SoReg** [4]: This is a classic social recommendation model that transforms social information into social regularization terms.
- **TrustSVD** [16]: This model integrates rating implicit feedback and social implicit feedback information.
- **CUNE** [17]: This is a recommendation model that alleviates the sparsity problem of social information by generating top-k potential friend information.
- **GraphRec** [18]: This is a social recommendation model using a graph neural network. This model uses the social graph and user-item interaction graph.

- **SREPS** [8]: This is a collaborative algorithm based on the essential preference space, which combines explicit feedback, implicit feedback, and social relationships.

5.3   Comparative Experiments: RQ1

We report the comparative experiment results in Table 3. The percentages in Table 3 are improvements to our RMHFIF model over other models. It can be seen from Table 3 that our proposed RMHFIF model is always better than other models. The

Table 3. Performance comparison with other models

Model	Ciao		Epinions	
	RMSE	MAE	RMSE	MAE
SVD++	1.034	0.761	1.077	0.832
SoReg	0.995	0.752	1.072	0.824
TrustSVD	1.033	0.759	1.049	0.814
CUNE	1.028	0.766	1.063	0.819
GraphRec	0.979	0.759	1.057	0.817
SREPS	0.955	0.722	1.039	0.801
RMHFIF	**0.941**	**0.706**	**1.027**	**0.775**
Improvement	1.47%	2.22%	1.15%	3.35%

experimental results verify the effectiveness of our proposed model in the social recommendation. We have some findings from the experimental results. First, in most cases, the traditional recommendation model's performance based on matrix factorization is not as good as the recommendation model that considers social information. This result is consistent with previous studies' results. Secondly, the performance of GraphRec is significantly better than SoReg, TrustSVD, and CUNE. This result shows that graph neural network applies to social recommendations can dramatically improve the model's recommendation performance.

5.4   Ablation Experiments: RQ2

To verify whether each component in the model is effective, we conducted an ablation experiment. The experimental results are shown in Figs. 3 and 4. The model we proposed uses both the homogeneity factor and the influence factor. Therefore, we removed the homogeneity factor and the influence factor in the model to generate two variants, and we named them RMHFIF-α and RMHFIF-β, respectively. In addition, since this paper modifies the attention weight calculation method in the traditional graph attention network, to verify whether the modification is effective, we will use the traditional graph attention network to replace it, and name the variant model RMHFIF-γ. The definitions are as follows:

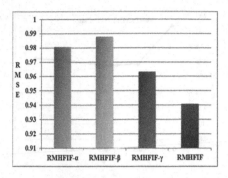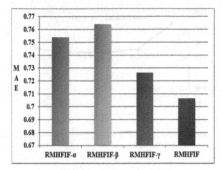

Fig. 3. Performance comparison of different variants in the Ciao dataset.

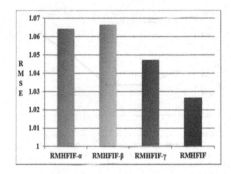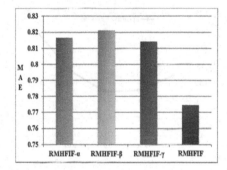

Fig. 4. Performance comparison of different variants in the Epinions dataset.

- **RMHFIF-α**: This variant removes the social homogeneity factor.
- **RMHFIF-β**: This variant removes the social influence factor.
- **RMHFIF-γ**: Modify the original graph attention network attention weight calculation method in the NGAT layer to the traditional graph attention network weight calculation method.

From the experimental results, we can find that: i) the performance of the RMHFIF model is always better than the other variants. It is enough to prove that both the homogeneity factor and the influence factor can play a role in the model, and neither is indispensable. ii) the homogeneity factor and the influence factor have similar effects on the performance of the model. Using the homogeneity factor and the influence factor at the same time can significantly improve the model performance. iii) The performance of RMHFIF is better than RMHFIF-γ, which is enough to prove that the use of a novel attention network attention weight calculation method can significantly improve the performance of the model.

5.5 Parameter Sensitivity Experiments: RQ3

This subsection studies the impact of some hyperparameter changes on performance, including dropout rate ρ, embedding dimension D, and regularization parameters λ.

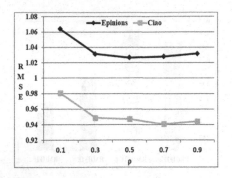
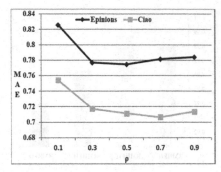

Fig. 5. The effect of dropout rate ρ on model performance.

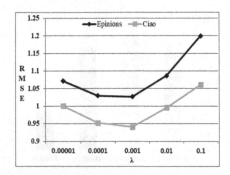
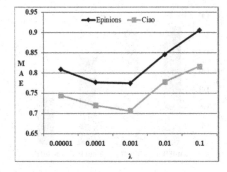

Fig. 6. The effect of regularization parameter λ on model performance.

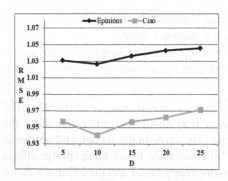
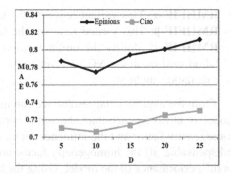

Fig. 7. The effect of embedding dimension D on model performance.

The results are shown in Figs. 5, 6 and 7. From Figs. 5, 6 and 7 we can see that: i) The choice of dropout rate ρ impacts the model's performance, and the best ρ value is different for different datasets. If the value of ρ is too large, too many neurons discarded will hinder the model's training process. If the value of ρ is too small, it will lose the ability to alleviate overfitting. ii) The choice of the regularization parameter λ is very important. If λ is too large or too small, the performance of the model will decrease. iii)

The proper embedding dimension is very important. If the embedding dimension is too small, the expressiveness of the model will decrease. If the embedding dimension is too large, the complexity of the model will increase and the representation vector will become sparse. Therefore, we need to find a suitable embedding dimension.

6 Conclusion

In this paper, we propose the RMHFIF model, which uses novel graph attention networks to obtain the homogeneity factor and the influence factor. We also model the four factor's weight distribution as a contextual multi-armed bandit problem for training. The RMHFIF model solves the problem that most of the current social recommendation models only consider social homogeneity factors without considering social influence factors. We conducted extensive comparative experiments and ablation experiments in two public datasets. The experimental results show that the performance of RMHFIF is better than the other recommendation model.

References

1. Huang, L.W., et al.: Survey on deep learning-based recommender systems. Chinese J. Comput. **41**(7), 1619–1647 (2018)
2. Mcpherson, M., Lovin, L.S., Cook, J.M.: Birds of a feather: homophily in social networks. Ann. Rev. Sociol. **27**(1), 415–444 (2001)
3. Marsden, P.V,, Friedkin, N.E.: Network studies of social influence. Sociol. Methods Res. **22** (1), 127–151 (1993)
4. Ma, H., Zhou, D., Liu, C., Lyu, M.R., King, I.: Recommender systems with social regularization. In: Proceeding of the 4th ACM International Conference on Web Search Data Mining, pp. 287–296 (2011)
5. Wang, X., He, X.N., Nie, L.Q., Chua, T.S.: Item silk road: Recommending items from information domains to social users. In: Proceeding of the 40th SIGIR Conference, pp. 185–194 (2017)
6. Jamali, M., Ester, M.: A matrix factorization technique with trust propagation for recommendation in social networks. In: Proceeding of the 4th ACM conference on Recommender systems(RecSys), pp.135–142 (2010)
7. Wen, Y.F., Guo, L., Chen, Z.M., Ma, J.: network embedding. based recommendation method in social networks. In: Proceeding of the 27th WWW Conference, pp. 11–12 (2018)
8. Liu, C.Y., Zhou, C., Wu, J., Hu, Y., Guo, L.: Social recommendation with an essential preference space. In: Proceeding of the 32rd AAAI Conference, pp. 346–353 (2018)
9. Wu, L., Sun, P., Hong, R., Fu, Y., Wang, X., Wang, M.: SocialGCN: an efficient graph convolutional network based model for social recommendation (2018). arXiv:1811.02815. https://arxiv.org/abs/1811.02815?context=cs.SI
10. Ying, R., He, R., Chen, K., Eksombatchai, P., Hamilton, W.L., Leskovec, J.: Graph convolutional neural networks for web-scale recommender systems. In: Proceeding of the 24th ACM SIGKDD International Conference on Knowledge Discovery & Data Mining (KDD), pp. 974–983 (2018)
11. Velikovi, P., et al.: Graph attention networks (2017). arXiv:1710.10903. https://arxiv.org/abs/1710.10903

12. Koren, Y.: Factorization meets the neighborhood: a multifaceted collaborative filtering model. In: Proceeding of the ACM SIGKDD International Conference on Knowledge Discovery Data Mining, pp. 426–434. ACM, New York, USA (2008)
13. He, X., Liao, L., Zhang, H., Nie, L., Hu, X., Chua, T.: Neural collaborative filtering. In: Proceeding of the 26th WWW Conference, pp. 173–182 (2017)
14. Tang, J., Gao, H., Liu, H.: mTrust: discerning multi-faceted trust in a connected world. In: Proceeding of the 5th ACM International Conference on the Web Search Data Mining, pp. 93–102 (2012)
15. Massa, P., Avesani, P.: Trust-aware recommender systems. In: Proceeding of the 1st ACM Conference on the Recommender System, pp. 17–21. Minneapolis (2007)
16. Guo, G., Zhang, J., Yorke-Smith, N.: TrustSVD: collaborative filtering with both the explicit and implicit influence of user trust and of item ratings. In: Proceeding of the 29th AAAI Conference on Artificial Intelligence, pp. 123–129 (2015)
17. Zhang, C., Yu, L.: Collaborative user network embedding for social recommender systems. In: Proceedings of the 17th SIAM International Conference on Data Mining (SDM), pp. 381–389 (2017)
18. Fan, W., Ma, Y., Li, Q., He, Y., Zhao, E., Tang, J., Yin, D.: Graph neural networks for social recommendation. In: Proceeding of the 28th WWW Conference, pp.417–426 (2019)
19. Wu, Q., et al.: Dual graph attention networks for deep latent representation of multifaceted social effects in recommender systems. In: Proceeding of the 28th WWW Conference, pp. 2091–2102 (2019)
20. Yang, L., et al.: ConsisRec: enhancing gnn for social recommendation via consistent neighbor aggregation. In: SIGIR 21: The 44th International ACM SIGIR Conference on Research and Development in Information Retrieval. ACM (2021)
21. Ma, H., Yang, H., Lyu, M.R., King, I.: SoRec: social recommendation using probabilistic matrix factorization. In: Proceeding of the CIKM, pp. 931–940 (2008)
22. Lin, T.-H., Gao, C., Li, Y.: Recommender systems with characterized social regularization. In: Proceeding of the 27th ACM International Conference on Information and Knowledge Management, pp. 1767–1770 (2018)
23. Rafailidis, D.: Leveraging Trust and Distrust in Recommender Systems via Deep Learning (2019). arXiv preprint, arXiv:1905.13612
24. Bahdanau, D., Cho, K., Bengio, Y.: Neural Machine Translation by Jointly Learning to Align and Translate (2014). arXiv preprint, arXiv:1409.0473

Attention Based Spatial-Temporal Graph Convolutional Networks for RSU Communication Load Forecasting

Hang Zheng[1], Xu Ding[2], Yang Wang[1], and Chong Zhao[3(✉)]

[1] School of Computer Science and Information Engineering,
Hefei University of Technology, Hefei 230009, China
[2] Institute of Industry and Equipment Technology, Hefei University of Technology,
Hefei 230009, China
[3] Engineering Quality Education Center of Undergraduate School,
Hefei University of Technology, Hefei 230009, China
zhaochong@hfut.edu.cn

Abstract. As a special type of base station, Road Side Units (RSU) can be deployed at low cost and effectively alleviate the communication burden of regional Vehicular Ad-hoc Networks (VANETs). However, because of peak hour communication demands in VANETs and limited energy storage, it is necessary for RSU to adjust their participation in communication according to the requirements and allocate energy reasonably to balance the workload. Firstly, tidal traffic flow is generated according to the information of morning and evening peak in the city, so as to simulate the vehicle distribution around RSU on urban roads. Secondly, by inputting the historical information around RSU and the topological relationship between RSU, a network load prediction model is established by using the Attention based Spatial-Temporal Graph Convolutional Networks (ASTGCN) to predict the future communication load around RSU. Finally, according to the forecast of the future communication load, a RSU working mode alteration scheme is proposed with respect to the safety range amongst vehicles in order to control the corresponding area communication load. Compared with other models, our model has better accuracy and performance.

Keywords: VANETs · RSU · Communication load forecasting · ASTGCN · Energy schedule

1 Introduction

Due to the ability of extending the horizon of drivers which will improve road traffic safety, VANETs are attracting an extensive attention from both academia

Supported by the Fundamental Research Funds for the Central Universities (GRANT NO. PA2021GDSK0095).

H. Gao and X. Wang (Eds.): CollaborateCom 2021, LNICST 406, pp. 99–114, 2021.
https://doi.org/10.1007/978-3-030-92635-9_7

and industry [1,2]. Vehicles in VANETs act as communication nodes, exchanging safety-related information such as location, speed, and brake status with surrounding ones, base stations and infrastructures by means of on-board devices. As an important part of VANETs, Road Side Unit (RSU) provide significant support for regional vehicle communication [3]. TRSUs are with fixed locations and could improve communication speed and reduce the delay in VANETs [4,5].

These studies show that RSU can play an important role in improving the communication quality of VANET. However, it is not necessary to keep the RSU at full power. Under the circumstance of ensuring low communication delay in this area, the online algorithm is used to dynamically adjust the working state of the RSU to arrange the energy consumption wisely. Since the communication delay is closely related to the communication load, we predict the communication load of the RSU for a period of time in the future to assess whether the VANET in the corresponding area is in a safe state and decide whether to continue using the RSU.

With RSU usually being deployed at the intersection of urban roads, the communication load of RSU is closely related to the vehicle information of road interchanges that can be obtained from the urban intelligent transportation system [6], such as the number of vehicles in the area, the occupancy rate of lanes and other factors. Vehicle data are sampled every fixed time interval at certain locations along the road successivley. Apparently, the observed data is not independent but dynamically related to surrounding and historical data. And the correlation of vehicle data on urban roads possesses dynamic features in both spatial and temporal dimensions. It becomes a very challenging problem that exploring nonlinear and complex spatiotemporal data to discover inherent spatiotemporal patterns and make accurate communication loads prediction.

With the development of the VANETs, many cameras, sensors and other information collection devices have been deployed on urban roads. Each device is placed at a unique geospatial location, constantly generating time series data about traffic. These devices store rich traffic time series data with geographic information, providing a solid data foundation for communication load forecasting.

On this basis, we use a deep learning model: Attention based Spatial-Temporal Graph Convolution Network (ASTGCN) to predict communication load at every location on the traffic network. This model can process the traffic data directly on the original graph-based traffic network and effectively capture the dynamic spatial-temporal features. Through the effective evaluation of the future network load, a certain safety threshold is set to determine the working status of the RSU to reduce energy consumption. The main contributions of this paper are listed as follows:

- Attention based Spatial-Temporal Graph Convolution Network has been improved so that it can predict the RSU communication load on each node through the vehicle data on the graph-based transportation network.
- An RSU energy decision algorithm based on future communication load evaluation is proposed to enable RSU to reduce energy consumption.

The remainder of this paper is organized as follows. The related works are introduced in Sect. 2. The system model is presented in Sect. 3. The communication load evaluation is introduced in Sect. 4. The communication load prediction model are introduced in Sect. 5. Simulation and analysis are introduced in Section 6 and a conclusion is made in Section 7.

2 Related Works

At present, in many studies, researchers altered the working modes of the RSU to reduce the global energy consumption. In [7], Wen et al. proposed the RSU could determine to work or not depending on the current communication situation. If there was any vehicle in the network communicating with others, the RSU kept service time, otherwise stopped working. However, the non-differentiated service of RSU without considering network load made the performance of RSU unstable. In [8], Patra et al. proposed an algorithm taking into account the coverage of RSU and the overall energy consumption. The goal of the algorithm was to minimize the energy consumption rate.

Reyhanian et al. proposed two novel online approaches for enabling energy trading in multitier cellular networks with non-cooperative energy-harvesting base stations (BSs) to minimize the nonrenewable energy consumption in a multitier cellular network [9]. Zhang et al. proposed a resource-allocation solution for the heterogeneous CR sensor networks to achieve the sustainability of spectrum sensors and conserve the energy of data sensors. Extensive simulation results demonstrate that the energy consumption of the data sensors can be significantly reduced, while maintaining the sustainability of the spectrum sensors [10].

These studies focused on the global energy consumption allocation problem of RSU. However, the main reason of short-term and inconsistent service provided by RSU is the imbalanced network communication demand brought by the tidal fluctuation of traffic [11]. If the communication load can be predicted, the RSU would be able to adjust energy allocation in advance and be adapt to the network's tidal communication and the stability of the RSU service would be guaranteed.

The prediction of communication load is a spatial-temporal data prediction problem that many researchers have used deep learning to solve in recent years. Especially in the field of traffic flow, many deep learning models have been proposed to predict traffic flow. In [12], Liang et al. designed a ST-ResNet model based on the residual convolution unit to predict crowd flows. In [13], Yao et al. proposed a method to predict traffic by integrating CNN and long-short term memory (LSTM) to jointly model both spatial and temporal dependencies. Although the spatial-temporal features of the traffic data can be extracted by these model, their limitation is that the input must be standard 2D or 3D grid data. In [14], Guo et al. designed a novel spatial-temporal convolution module for modeling spatial-temporal dependencies of traffic data. It consists of graph convolutions for capturing spatial features from the original graph-based traffic network structure and convolutions in the temporal dimension for describing dependencies from nearby time slices.

Motivated by the studies mentioned above, considering the graph structure of the traffic network and the dynamic spatial-temporal patterns of the traffic data, we employ Attention Based Spatial-Temporal Graph Convolutional Networks to predict the communication load. Then specify the energy plan based on the network load forecast so that the RSU can obtain a longer effective working time.

3   System Model

In our research scenario, RSU is arranged at each intersections, as shown in Fig. 1. RSU is equipped with multiple antennas, which can receive signals sent by multiple vehicles at the same time.

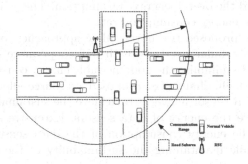

Fig. 1. Typical RSU scenario in VANETs

As the uneven distribution of vehicles have an unbalanced impact on communication requirements, the coverage of each RSU area is divided into several subareas, such as road section, connection points and entrances. The scheduling time of RSU is divided into K time frames, and the traffic flow, lane occupancy rate and number of vehicles in each area are counted every other time frame. D_i^t represents the vehicle information collected by the ith RSU in the tth time range. The vehicle information around RSU includes the number of vehicles in various surrounding areas, traffic flow, and traffic occupancy rate. Limited by length of the road, there is an upper limit for the number of vehicles in each subarea.

The vehicles and RSU can only initiate communication during the time slot assigned to it. Since the length of time frame is very short, during the same time frame, the relative positions among vehicles can be considered unchanged. The RSU is equipped with multiple antennas that can forward messages for multiple vehicles at the same time. But the vehicle can only communicate with only one vehicle in a time slot. Both vehicles and RSU use half-duplex mode which means they cannot transmit or receive messages simultaneously in one time slot.

In this study, RSU is deployed at each node of the entire road network. The collected information of the surrounding vehicles of each RSU is continuous in time, and the information corresponding to the time is used as a node to form a graph, as shown in Fig. 2.

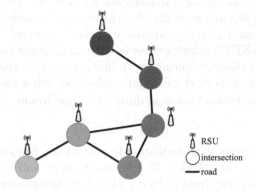

Fig. 2. The spatial structure of traffic data

The straight lines in the figure represent the roads in the city, the circles represent the nodes where the roads intersect, and the RSU is deployed around each intersection. We define a traffic network as an undirected graph $G = (V, E, A)$, where V is a finite set of $| V | = N$ nodes; E is a set of edges, indicating the connectivity between the nodes; $A \in R^{N \times N}$ denotes the adjacency matrix of graph G.

Symbols used in the following discussion are listed in Table 1.

Table 1. Notation table

Symbol	Definition
N	The total number of vehicles on the node
D_i^t	Vehicle information of the tth time segment around the ith RSU
$c_{i,j}^t$	$c_{i,j}^t$ is binary value indicating that vehicle i and vehicle j need to complete an information exchange during t^{th} time frame
$r_{i,j}^t$	$r_{i,j}^t$ represents the distance between vehicle i and vehicle j
R_d, R_v	Communication radius of RSU and vehicles
$D_{i,j}$	Routing path of v_i sends message to v_j
$D_{i,j}(k,l)$	Indicates that v_k sends message to v_l as routing nodes for v_i and v_j information exchange
$C_{i,j}$	The amount of data sent by v_i to v_j during the tth time frame
$R_{i,j}^t$	$R_{i,j}^t$ is the communication speed between v_i and v_j
Y_v^t	Communication load at t^{th} time without RSU
Y_s^t	Communication load at t^{th} time with RSU

4   Communication Load Evaluation

In this section, we will introduce how to evaluate the network load and provide a method for obtaining training data sets based on historical data.

In VANETs, the most important communication content is safety-related information, which has higher requirements for network delay. Therefore, the length of the time slot required for all vehicles to complete this type of message interaction is used as the communication load standard. According to the requirements of VANETs, vehicles must initiate an information interaction with others within safety distance during every time frame. C is the set of communication sessions that element $c_{i,j}^t$ is binary value indicating that v_i and v_j need to complete an information exchange during t^{th} time frame,

$$C = \{c_{i,j}^t\}, (r_{i,j}^t \leq R_v) \tag{1}$$

Vehicle communicates with each other in a multi-hop fashion. Let $D_{i,j}$ denote the routing path and $D_{i,j}(k,l) \in D_{i,j}$ denote that v_k and v_l take part in the data forwarding task sent from v_i to v_j. Therefore, the amount of data sent by v_i is equal to the amount of data received by v_j, i.e.,

$$\sum_{k=1}^{N} D_{i,j}(i,k) = \sum_{k=1}^{N} D_{i,j}(k,j), (r_{i,k}^t \leq R_v, r_{j,k}^t \leq R_v, c_{i,j}^t = 1) \tag{2}$$

Let $F_{i,j}^t$ denote the amount of data sent by v_i to v_j during the tth time frame. $F_{i,j}^t$ could be divided into relayed data and directly transmitted data, and we have:

$$C_{i,j}^t = \sum_{k=1(k\neq i)}^{N} \sum_{l=1(l\neq k)}^{N} D_{k,l}^t(i,j) + \sum_{l=1}^{N} D_{i,j}^t(i,l),$$
$$(r_{k,l}^t \leq R_v, r_{i,l}^t \leq R_v, r_{i,j}^t \leq R_v) \tag{3}$$

If v_i could communicate with v_j directly during t^{th} time frame, the communicate rate can be obtained as following

$$R_{i,j}^t = W \cdot \log_2(1 + \frac{P_i}{N_0}\delta(r_{i,j}^t)^{-\gamma}|h_{i,j}^t|^2), (r_{i,j}^t \leq R_v) \tag{4}$$

$R_{i,j}^t$ is the communication speed between v_i and v_j, W is bandwidth and $h_{i,j}^t$ is the Rayleigh distributed fading magnitude with $E\left[|h_{i,j}^t|^2\right] = 1$. N_0 is the power of additive white Gaussian noise (AWGN). δ is the log-normal shadowing component, with a mean of 0 dB. γ is typically chosen as the path fading exponent for VANETs.

In order to have a unified evaluation standard for the network load carried by the regional VANETs, the required time slots for communication is used as a metric for the communication load represented by Y_v^t, then we have

$$Y_v^t = \min \frac{\sum_{i=1}^{N} \sum_{j=1(i\neq j)}^{N} \frac{C_{i,j}^t}{R_{i,j}^t}}{N} \tag{5}$$

The RSU covers the entire network and the main job it participates in V2V communication is forwarding data for vehicles. Y_s^t is the number of time slots required when RSU participates in communication during t^{th} time frame, which can be calculated as

$$Y_s^t = \kappa(N_a) \cdot Y_v^t \tag{6}$$

κ is the EH-RSH improvement function, related to the number of antennas in RSU denoted by N_a, meaning the RSU can provide service for N_a vehicles simultaneous.

5   The Communication Load Prediction Model

In this section, we will introduce how to make a prediction of the communication load and explain the various modules of the Attention Based Spatial-Temporal Graph Convolutional Networks(ASTGCN) model in detail.

Suppose the fth time series recorded on each node in the traffic network G is the traffic flow sequence, and $f \in (1, \ldots, F)$. We use $x_t^{c,i} \in R$ to denote the value of the cth feature of node i at time t, and $x_t^i \in R^F$ denotes the values of all the faatures of node i at time t. $X_t = (x_t^1, x_t^2, \ldots, x_t^N)^T \in R^{N \times F}$ denotes the values of all the features of all the nodes at time t. $\chi \in (X_1, X_2, \ldots, X_\tau) \in R^{N \times F \times \tau}$ denotes the value of all the features of all the nodes over τ time slices.In addition, we set $y_t^i = x_t^{f,j} \in R$ to represent the communication load of RSU in the node i at time t in the future.

Given χ, all kinds of the historical measurements of all the nodes on the traffic network over past τ time slices, predict future communication load of RSU sequences $Y = (y^1, y^2, \ldots y^N)^T \in R^{N \times T_p}$ of all the nodes on the whole traffic network over the next T_p time slices, where $y^i = (y_{\tau+1}^i, y_{\tau+2}^i, \ldots, y_{\tau+T_p}^i)$ denotes the future communication load of node i from $\tau + 1$. Each node on the traffic network G detects F measurements with the same sampling frequency, that is, each node generates a feature vector of length F at each time slice.

In order to transform this information into our predicted results, we improved the Attention Based Spatial-Temporal Graph Convolutional Networks model. Fig. 3 shows the overall framework of the ASTGCN model proposed in this paper. It is composed of five parts: time attention module, spatial attention module, spatial dimension graph convolution module, time dimension convolution module, and fully connected layer module.

Suppose the sampling frequency is q times per day. Assume that the current time is t_0 and the size of predicting window is T_p. We intercept the time series segment of length T_r along the time axis and use it as the most recent time slice input, where T_r is integer multiples of T_p.

$$\chi_r = (X_{t_0-T_h+1}, X_{t_0-T_h+2}, \ldots, X_{t_0}) \in R^{N \times F \times T_r} \tag{7}$$

A spatial-temporal attention mechanism is used in our model to capture the dynamic spatial and temporal correlations on the traffic network. It contains two kinds of attentions, spatial attention and temporal attention.

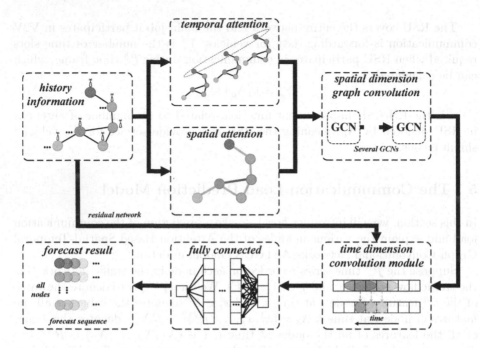

Fig. 3. The framework of ASTGCN

5.1 Temporal Attention

In the temporal dimension, there exist correlations between the traffic conditions in different time slices, and the correlations are also varying under different situations. Thus, we use an attention mechanism [15] to adaptively attach different importance to data:

$$E = V_e \cdot \sigma(((\chi_r^{(r-1)})^T U_1)U_2(U_3\chi_r^{(r-1)}) + b_e) \qquad (8)$$

$$E'_{i,j} = \frac{exp(E_{i,j})}{\sum_{j=1}^{T_{r-1}} exp(E_{i,j})} \qquad (9)$$

where $V_e, b_e \in R^{T_{r-1} \times T_{r-1}}, U_1 \in R^N, U_2 \in R^{C_{r-1} \times N}, U_3 \in R^{C_{r-1}}$ are learnable parameters. The temporal correlation matrix E is determined by the varying inputs. The value of an element $E_{i,j}$ in E semantically indicates the strength of dependencies between time i and j. At last, E is normalized by the softmax function. We directly apply the normalized temporal attention matrix to the input and get $\hat{\chi}_h^{(r-1)} = (\hat{X}_1, \hat{X}_2, \ldots, \hat{X}_{T_{r-1}}) = (X_1, X_2, \ldots, X_{T_{r-1}})E' \in R^{N \times C_{r-1} \times T_{r-1}}$ to dynamically adjust the input by merging relevant information.

In the time dimension, as shown in the Fig. 4, the historical traffic data of different locations has different effects on the communication load of other nodes at different times in the future. We train this weight through the attention mechanism and convert it into a matrix $R^{N \times C_{r-1} \times T_{r-1}}$.

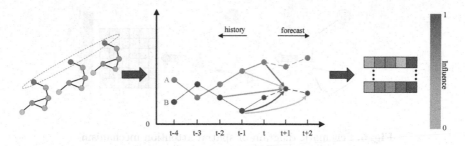

Fig. 4. Schematic diagram of temporal attention mechanism

5.2 Spatial Attention

In the spatial dimension, the traffic conditions of different locations have influence among each other and the mutual influence is highly dynamic. Here, we use an attention mechanism to adaptively capture the dynamic correlations between nodes in the spatial dimension.

$$S = V_s \cdot \sigma((\chi_r^{(r-1)} W_1) W_2 (W_3 \chi_r^{(r-1)})^T + b_s) \tag{10}$$

$$S'_{i,j} = \frac{exp(S_{i,j})}{\sum_{j=1}^{N} exp(S_{i,j})} \tag{11}$$

where $\chi_h^{(r-1)} = (X_1, X_2, \ldots, X_{T_r-1}) \in R^{N \times C_{r-1} \times T_{r-1}}$ denotes the input of the rth spatial module. C_{r-1} is the number of channels of the input data in the rth layer. T_{r-1} is the length of the sequence for forecasting, When $r = 1$, $C_0 = F$, $T_0 = T_r$. $V_s, b_S \in R^{N \times N}, W_1 \in R^{T_r-1}, W_2 \in R^{C_{r-1} \times T_{r-1}}, W_3 \in R^{C_{r-1}}$ are learnable parameters and sigmoid σ is used as the activation function. The attention matric S is dynamically computed according to the current input of this layer.The value of an elements $S_{i,j}$ in S semantically represents the correlation strength between node i and node j. Then a softmax function is used to ensure the attention weights of a node sum to one.

As shown in Fig. 5, when performing the graph convolutions, we will accompany the adjacency matrix A with the spatial attention matrix $S' \in R^{N \times N}$ to dynamic adjust the impacting weights between nodes.

5.3 Graph Convolution in Spatial Dimension

In order to make full use of the topological properties of the traffic network, at each time slice we adopt graph convolutions based on the spectral graph theory to directly process the signals, exploiting signal correlations on the traffic network in the spatial dimension. The spectral method transforms a graph into an algebraic form to analyze the topological attributes of graph, such as

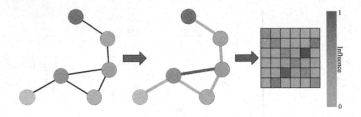

Fig. 5. Schematic diagram of spatial attention mechanism

the connectivity in the graph structure. In spectral graph analysis, a graph is represented by its corresponding Laplacian matrix, and its properties can be obtained by analyzing the Laplacian matrix and its eigenvalues. Therefore, we use Chebyshev convolution polynomial [16] to solve this problem.

$$g_\theta * Gx = g_\theta(L)x = \sum_{k=0}^{K-1} \theta_k T_k(\tilde{L})x \qquad (12)$$

where $*G$ denotes a graph convolution operation, the parameter $\theta \in R^K$ is a vector of polynomial coefficients. $\tilde{L} = \frac{2}{\lambda_{max}}L - I_N$, λ_{max} is the maximum eigenvalue of the Laplacian matric, I_N is a unit matrix. The recursive definition of the Chebyshev polynomial is $T_k(x) = 2xT_{k*-1}(x) - T_{k-2}(x)$, where $T_0(x) = 1, T_1(x) = x$. Using approximate expansion of Chebyshev polynomial to solve this formulation corresponds to extracting information of the surrounding 0 to $(K-1)^{th}-order$ neighbors centered on each node in the graph by the convolution kernel g_θ.

In order to dynamically adjust the correlations between nodes, for each term of Chebyshev polynomial, we accompany $T_k(\tilde{L})$ with the spatial attention matrix $S' \in R^{N \times N}$ [14], then obtain $T_k(\tilde{L}) \odot S'$, where $\odot$ is the Hadamard product. Therefore, the above graph convolution formula changes to:Q

$$g_\theta * Gx = g_\theta(L)x = \sum_{k=0}^{K-1} \theta_k T_k(\tilde{L}) \odot S'x \qquad (13)$$

We can generalize this definition to the graph signal with multiple channels. For example, in the recent component, the input is $\hat{\chi}_h^{r-1} = (\hat{X}_1, \hat{X}_2, \ldots, \hat{X}_{T_{r-1}}) \in R^{N \times C_{r-1} \times T_{r-1}}$, where the feature of each node has C_{r-1} channels. For each time slice t, performing C_r filters on the graph $\hat{X}_t$, we get $g_\theta * G\hat{X}_t$, where $\ominus = \{\ominus_1, \ominus_2, \ldots, \ominus_{C_r}\} \in R^{K \times C_{r-1} \times C_r}$ is the convolution kernel parameter. Therefore, each nodes is updated by the information of the $0 \sim K-1$ neighbors of the node, as shown in Fig. 6.

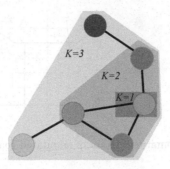

Fig. 6. The receptive field of different K values in Chebyshev convolution

5.4 Convolution in Temporal Dimension

After the graph convolution operations having captured neighboring information for each node on the graph in the spatial dimension, a standard convolution layer in the temporal dimension is further stacked to update the signal of a node by merging the information at the neighboring time slice. Take the operation on the rth layer in the recent component as an example:

$$\chi_h^{(r)} = \Phi * (ReLU(g_\theta * G\hat{X}_h^{(r-1)})) \in R^{C_r \times N \times T_r} \tag{14}$$

where $*$ denotes a standard convolution operation, Φ is the parameter of the time dimension convolution kernel. The activation function uses ReLU to avoid gradient disappearance and gradient explosion problems in back propagation.

5.5 Fully Connected Layer

By maximizing the interaction between the extracted features and the input data, we can preserve the potential correlation between the extracted features and the input data. And in order to establish the relationship between the extracted features and the final output, we need to map them to the same dimension, for which we use a fully connected layer. At the same time, due to the large difference between the dimensions of the input data and the data that needs to be predicted, the introduction of the fully connected layer can accelerate the convergence of our model.

6 Simulation and Analysis

6.1 Dataset

In order to evaluate the performance of our model, we conducted comparative experiments on the data set obtained from the SUMO simulation. The road network simulated by SUMO uses the real road network in a certain area of Hefei, China, as shown in Fig. 7.

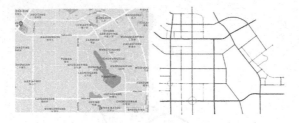

Fig. 7. Schematic diagram of simulated road network

The relevant information is collected every second and aggregated every 5 min. The traffic flow of the road nodes, the vehicle occupancy rate, the number of vehicles in the area and the communication load within the jurisdiction of each RSU. The entire road network contains 105 nodes and 282 edges, and 28 days of vehicle data are simulated (the daily traffic flow data is simulated by morning and evening peak hours). We extract 24-h related information of a certain RSU in the obtained information and draw it, as shown in Fig. 8.

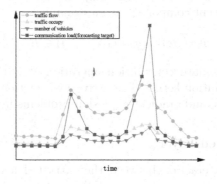

Fig. 8. Three measurements are detected on RSU and the future communication load is the forecasting target

We choose data on the first 23 days as the training set, and the remains as the test set. The communication radius of RSU is selected as 100m, and the safety distance between cars is selected as 50 m. The vehicle needs to transmit information to all vehicles within the safe distance for each calculation.

6.2   Model Parameters

We implemented the ASTGCN model based on the MXNet framework. We tested the number of terms of the Chebyshev polynomial K to changed the size of the receptive field of the graph convolution in the graph. In our model, all the graph convolution layers use 64 convolution kernels. All the temporal convolution layers use 64 convolution kernels and the time span of the data is ad justed by controlling the step size of the temporal convolutions. For the input historical information length, we set it as: the input recent vehicle information $T_h = 18$, the prediction window size $T_p = 6$. The input sequence is 3 times the length of the output sequence, which means that our goal is to predict the next half an hour communication load within. The mean square error is used to calculate the predicted value and the true value as the loss function, and minimized by the back propagation algorithm. In the training process, the batch size is selected as 16, the optimizer selects ADAM optimization, and the learning rate is 0.001.

6.3 Comparison and Result Analysis

Through our model, we predict the communication load of all RSUs in the road network during the next period of time. As shown in the Fig. 9, this is a 24-h communication load forecast formed by splicing the communication load predicted by a certain node.

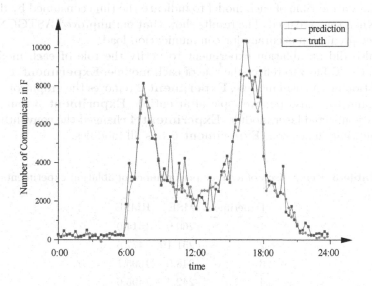

Fig. 9. Communication load estimation simulation

We compare our models with the following four baseline methods on our dataset:

- LSTM [17]: Long Short Term Memory network, a special RNN model.
- GRU [18]: Gated Recurrent Unit network, a special RNN model.
- STGCN [19]:A spatial-temporal graph convolution model based on the spatial method.
- MSTGCN [14]: A multi-layer spatial-temporal graph convolutional neural network

Table 2. Average performance comparison of different approaches on our dataset.

Model	MAE			RMSE			Validation time
	1	3	6	1	3	6	
LSTM	475.9	478.5	485.7	2176	2178	2185	3.128s
GRU	469.4	470.6	473.7	2174	2175	2177	3.070s
STGCN	286.8	295.1	301.8	1755	1763	1779	1.787s
MSTGCN	242.42	255.8	274.7	1499	1560	1627	6.719s
ASTGCN (ours)	**230.7**	**245.3**	**260.9**	**1366**	**1495**	**1589**	**6.632s**

Table 2 shows the average results of communication load prediction performance over the next half one hour. We respectively evaluate the next 1, 3, and 6 prediction sequences. Root mean square error (RMSE) and mean absolute error (MAE) are used as the evaluation metrics. At the same time, we also enumerate the validation time of each model to indicate the time consumed by different models when they are used. The results show that our improved ASTGCN model has better prediction accuracy for communication load.

We also did an ablation experiment to verify the role of each module, as shown in the Table 3 to reflect the role of each module. **Experiment 1** removes the attention mechanism module, **Experiment 2** removes the graph convolution module, and then also removes spatial attention, **Experiment 3** removes the final fully connected layer module. **Experiment 4** changed the convolution step size in the time dimension, **Experiment 5** has all modules,

Table 3. Comparison of average performance of ablation experiments

Experiment	MSE	RMSE
1	260.9	1514.0
2	251.19	1477.5
3	248.9	1505.1
4	242.4	1495.6
5	230.7	1366.0

After removing the attention mechanism module, the performance of the model decreases. In order to more intuitively reflect the role of the attention module, we extract the four points in the road network separately as a sub-image, and we output the trained attention matrix and convert it into a heat map to reflect the difference the correlation between nodes is strong or weak.

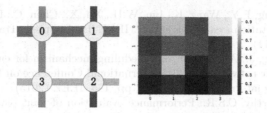

Fig. 10. The attention matrix obtained from the spatial attention mechanism

As shown in the Fig. 10, each row represents the strength of the correlation between the corresponding node and the rest of the nodes. It can be seen that each node has a relatively stronger correlation with itself. And it can be seen that there is a stronger focus on node 0, which is very reasonable, because from the traffic flow point of view, node 0 has more traffic and at the same time bears a greater communication load, which is easily congested. In addition, the relative degree of correlation between node 1 and node 3 is also low, because from the route point of view, there is no direct route connection between the two nodes, but is established through multi-hop Chebyshev graph convolution.

7   Conclusion

In this paper, we construct a traffic flow data set and use the Attention Based Spatial-Temporal Graph Convolutional network communication load estimation model to predict the network load state at the next moment, which is used to guide RSU to make appropriate actions for VANETs to improve the global network communication quality. Experimental simulation shows that model can accurately predict RSU communication load, dynamically adjust working modes, balance tasks between high-load and low-load networks and reduce RSU's energy consumption.

References

1. Hartenstein, H., Laberteaux, K.P.: A tutorial survey on vehicular ad hoc networks. IEEE Commun. Mag. **46**(6), 164–171 (2008)
2. Lee, E., Lee, E.K., Gerla, M., Oh, S.Y.: Vehicular cloud networking: architecture and design principles. IEEE Commun. Mag. **52**(2), 148–155 (2014)
3. Dragicevic, T., Lu, X., Vasquez, J.C., Guerrero, J.M.: Dc microgrids-part ii: a review of power architectures, applications and standardization issues. IEEE Trans. Power Electron. **31**(5), 1–1 (2015)
4. Barrachina, J., Garrido, P., Fogue, M., Martinez, F.: Road side unit deployment: a density-based approach. IEEE Intell. Transp. Syst. Mag. **5**(3), 30–39 (2013)
5. Gao, Z., Chen, D., Cai, S., Wu, H.C.: Optdynlim: an optimal algorithm for the one-dimensional RSU deployment problem with nonuniform profit density. IEEE Trans. Industr. Inf. **15**(2), 1052–1061 (2019)

6. Zhang, J., Wang, F.Y., Wang, K., Lin, W.H., Xu, X., Chen, C.: Data-driven intelligent transportation systems: a survey. IEEE Trans. Intell. Transp. Syst. **12**(4), 1624–1639 (2011)
7. Wen, C., Zheng, J.: An RSU on/off scheduling mechanism for energy efficiency in sparse vehicular networks. In: 2015 International Conference on Wireless Communications & Signal Processing (WCSP), pp. 1–5. IEEE (2015)
8. Patra, M., Murthy, C.S.R.: Performance evaluation of joint placement and sleep scheduling of grid-connected solar powered road side units in vehicular networks. IEEE Trans. Green Commun. Networking **2**(4), 1197–1209 (2018)
9. Reyhanian, N., Maham, B., Shah-Mansouri, V., Tushar, W., Yuen, C.: Game-theoretic approaches for energy cooperation in energy harvesting small cell networks. IEEE Trans. Veh. Technol. **66**(8), 7178–7194 (2017)
10. Zhang, D., et al.: Energy-harvesting-aided spectrum sensing and data transmission in heterogeneous cognitive radio sensor network. IEEE Trans. Veh. Technol. **66**(1), 831–843 (2017)
11. Gupta, L., Jain, R., Vaszkun, G.: Survey of important issues in UAV communication networks. IEEE Commun. Surv. Tutorials **18**(2), 1123–1152 (2016)
12. Liang, Y., Ke, S., Zhang, J., Yi, X., Zheng, Y.: Geoman: multi-level attention networks for geo-sensory time series prediction. In: IJCAI, vol. 2018, pp. 3428–3434 (2018)
13. Yao, H., Tang, X., Wei, H., Zheng, G., Yu, Y., Li, Z.: Modeling spatial-temporal dynamics for traffic prediction. arXiv preprint arXiv:1803.01254, pp. 922–929 (2018)
14. Guo, S., Lin, Y., Feng, N., Song, C., Wan, H.: Attention based spatial-temporal graph convolutional networks for traffic flow forecasting. Proc. AAAI Conf. Artif. Intell. **33**, 922–929 (2019)
15. Feng, X., Guo, J., Qin, B., Liu, T., Liu, Y.: Effective deep memory networks for distant supervised relation extraction. In: IJCAI, vol. 17, pp. 1–8 (2017)
16. Simonovsky, M., Komodakis, N.: Dynamic edge-conditioned filters in convolutional neural networks on graphs. In: 2017 IEEE Conference on Computer Vision and Pattern Recognition (CVPR), pp. 1–10 (2017)
17. Hochreiter, S., Schmidhuber, J.: Long short-term memory. Neural Comput. **9**(8), 1735–1780 (1997)
18. Chung, J., Gulcehre, C., Cho, K., Bengio, Y.: Empirical evaluation of gated recurrent neural networks on sequence modeling. In: NIPS 2014 Workshop on Deep Learning, December 2014, pp. 1–9 (2014)
19. Yan, S., Xiong, Y., Lin, D.: Spatial temporal graph convolutional networks for skeleton-based action recognition. In: Thirty-Second AAAI Conference on Artificial Intelligence, pp. 1–8 (2018)

UVA and Traffic System

Mobile Encrypted Traffic Classification Based on Message Type Inference

Yige Chen[1,2], Tianning Zang[1,2(✉)], Yongzheng Zhang[1,2], Yuan Zhou[3], and Peng Yang[3]

[1] Institute of Information Engineering, Chinese Academy of Sciences, Beijing, China
zangtianning@iie.ac.cn
[2] School of Cyber Security, University of Chinese Academy of Sciences, Beijing, China
[3] National Computer Network Emergency Response Technical Team/Coordination Center of China, Beijing, China

Abstract. With the growing attention to the security and privacy of mobile communications, advanced cryptographic protocols are widely applied to protect information confidentiality and prevent privacy leakage. These cryptographic protocols make it difficult to classify encrypted traffic for network management and intrusion detection. Existing mobile encrypted traffic classification approaches intend to alleviate this problem for TLS 1.2 encrypted traffic through modeling message attributes. However, these approaches are facing tough challenges in classifying TLS 1.3 traffic because most plaintext handshake messages are encrypted in TLS 1.3. To tackle this problem, we propose a mobile encrypted traffic classification approach based on Message Type Inference (MTI). We use a Recurrent Neural Network-Conditional Random Field (RNN-CRF) network to infer the hidden message types of encrypted handshake messages. Moreover, we employ machine learning to integrate three kinds of length features. The experimental results demonstrate that the RNN-CRF network achieves 99.92% message type inference accuracy and 98.96% F1-score on a real-world TLS 1.3 dataset and our proposed approach MTI achieves 96.66% accuracy and 96.64% F1-score on a fourteen application real-world TLS 1.3 dataset. In addition, we compare MTI with existing encrypted traffic classification approaches, which demonstrates MTI performs considerably better than state-of-the-art approaches for TLS 1.3 traffic.

Keywords: Encrypted traffic classification · Message type inference · RNN-CRF

1 Introduction

Mobile encrypted traffic classification is a fundamental research problem for network management and security, which intend to classify mobile encrypted traffic into applications. As the proportion of encrypted traffic on the Internet dramatically increases [12], many network management and cybersecurity technologies

© ICST Institute for Computer Sciences, Social Informatics and Telecommunications Engineering 2021
Published by Springer Nature Switzerland AG 2021. All Rights Reserved
H. Gao and X. Wang (Eds.): CollaborateCom 2021, LNICST 406, pp. 117–137, 2021.
https://doi.org/10.1007/978-3-030-92635-9_8

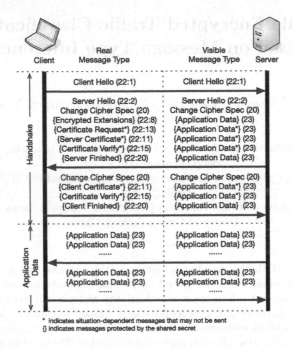

Fig. 1. A schema of TLS 1.3 communication between peers

largely rely on mobile encrypted traffic classification, such as application Quality-of-Service (QoS) [1,8] and network intrusion detection [6]. Existing encrypted traffic classification approaches primarily focus on Transport Layer Security 1.2 (TLS 1.2) [7] since it has been widely used in mobile applications [11]. With the continued development and popularity of Transport Layer Security 1.3 (TLS 1.3) [26], existing works are facing the problem of poor applicability or even inapplicability because of the confidentiality improvement of the cryptographic communication protocol. Figure 1 is a schema of TLS 1.3 communication between peers, including a client on the left and a server on the right. Each row in the figure represents a TLS message in the TLS session, and the numbers in the parentheses are message types. A full TLS 1.3 session consists of handshake messages and application data messages. The handshake messages intend to negotiate the security parameters of a session while the application data messages transfer actual payloads. The left column is a sequence of real message types before being encrypted, while the right column is a sequence of the visible message types after these messages are encrypted. Notice that many real handshake message types are masked by Application Data (23) after encryption, including Encrypted Extensions (22:8), Certificate Request (22:13), Server/Client Certificate (22:11), Certificate Verify (22:15), and Server/Client Finished (22:20). The handshake messages that can still be seen by third-party observers consist of Client Hello (22:1), Server Hello (22:2), and Change Cipher Spec (20). Therefore, the approaches based on message type information will face a significant drop in classification accuracy.

Table 1. A comparison example between named-entity recognition and message type inference

Sentence	Named-entity	Flow	Message type
All	B-event[a]	(22:1, 517)[d]	22:1
Tropical	I-event[b]	(22:2, 127)	22:2
Cyclones	I-event	(20, 6)	20
Are	O[c]	(23, 2519)	22:11
Driven	O	(20, 6)	20
By	O	(23, 58)	22:20
High	B-substance	(23, 92)	23
Heat	I-substance	(23, 292)	23
Content	I-substance	(23, 62)	23
Waters	I-substance	(23, 31)	23

[a] The B- prefix indicates the beginning of a chunk.
[b] The I- prefix indicates that the tag is inside a chunk.
[c] The O tag indicates that the token belongs to no chunk.
[d] The message feature tuple consists of visible message type and message length.

In addition to the difficulties caused by the evaluation of cryptographic protocols, the evolution of the DNS protocol also causes an impact on the attribute-based approaches. The traditional DNS protocol transfers plain DNS records without encryption. Therefore, third-party observers can directly extract the DNS records and exploit the content of these records to facilitate encrypted traffic classification [4]. The DNS over TLS (DoT) [15] and DNS over HTTPS (DoH) [14] can encrypt and protect the DNS traffic between DNS querier and recursively DNS server. Therefore, when DNS traffic is unavailable due to the use of these two secure DNS protocols by the DNS querier, we should consider how to maintain the accuracy of classification.

In this paper, we propose a mobile encrypted traffic classification approach for TLS 1.3 encrypted traffic based on Message Type Inference (MTI), which contains a Recurrent Neural Network-Conditional Random Field (RNN-CRF) network and a feature classifier based on machine learning. Although most handshake messages of TLS 1.3 encrypted traffic are encrypted and masked as application data messages, the message length and some plaintext handshake information are still available, so we can exploit these information to infer the types of encrypted handshake messages as alternative message attributes. This message type inference problem is similar to a Natural Language Processing (NLP) sequence tagging task called Named-Entity Recognition (NER). Table 1 presents a comparison example between the NER and the message type inference problem. The NER example in the table left means to determine the named-entity of the sentence's words while the message type inference in the table right is to predict the real message type of the message feature tuple that contains the visible message type and message length. A fairly common method to solve the NER problem is adopting a Recurrent Neural Network-Conditional Random Field (RNN-CRF) network, where RNN encodes sequential words to predict Named-Entity probability distributions in word-level and CRF integrates the neighbor tag information in sentence-level to modify prediction results [16,22]. Because the protocol specification defines the handshake process, the relatively fixed chronological order of various handshake messages in an independent session can provide the neighbor tag information of TLS 1.3 messages. Take Fig. 1 as an example, for the visible message types in figure right, the ninth message Change Cipher Spec (20) explicitly divides the handshake messages from the client and server, thereby indicating the real type of neighboring messages. Therefore, we can use the RNN-CRF network to learn sequential feature and neighbor tag information to realize message type inference and conduct machine learning to integrate various features to classify mobile encrypted traffic.

We briefly summarize our contributions as follows:

- We propose a mobile encrypted traffic classification approach for TLS 1.3 encrypted traffic based on message type inference (MTI). Our approach overcomes the accuracy decline caused by the handshake message encryption.
- We adopt an RNN-CRF network to realize message type inference for encrypted handshake messages and conduct machine learning to integrate different kinds of features to realize mobile encrypted traffic classification.
- We train the RNN-CRF network upon a TLS 1.3 encrypted traffic dataset with decryption keys to recover real message types. Then, we compare MTI with state-of-the-art approaches on a real-world TLS 1.3 dataset. The experimental results show that MTI can reach 96.66% accuracy and 96.64% F1-score, which outperforms several state-of-the-art approaches.

The rest of this paper is organized as follows. Section 2 summarizes the related work. Section 3 introduces the details of mobile encrypted traffic classification based on message type inference. Section 4 introduces the evaluation and presents the experimental results. Section 5 discusses and concludes this paper.

2   Related Work

In this section, we introduce the related work of the mobile encrypted traffic classification, including traditional unencrypted traffic classification, sequence feature-based encrypted traffic classification, and attribute feature-based encrypted traffic classification.

2.1   Traditional Unencrypted Traffic Classification

Traditional unencrypted traffic-oriented approaches mainly rely on port numbers or payloads to classify unencrypted traffic into applications. The port-based approaches inspect the port numbers of the unencrypted traffic and identify the application to which the traffic belongs by looking up a pre-stated port-to-application list or a public list provided by an authority such as Internet

Assigned Numbers Authority (IANA) [24]. Nevertheless, these approaches will be invalid when the applications use port hopping or random port numbers to hide the real port. The payload-based approaches inspect the content transmitted by both communication paries and infer its application through fingerprint matches. A straightforward approach is to directly extract the feature fragments of the traffic and compare them with the collected feature records to find the most likely application [9]. With the popularity of cryptographic protocols in communications to protect user privacy, these two kinds of approaches are no longer applicable because of the unified port number and confidential payload of the encrypted traffic.

2.2   Sequence Feature-Based Encrypted Traffic Classification

The sequence feature widely used in the classification include message type sequences and packet length sequences. Many existing works adopt developed sequence-oriented models to model sequence features to conduct traffic classification, such as Markov chain models and deep learning-based models. Korczyński et al. (2014) first use first-order Markov chains to model message type sequences for each application and classify encrypted traffic by calculating the similarity between the message type sequences and each application' Markov model [18]. Then, Shen et al. (2017) strengthen the first-order Markov model to the second-order one to promote classification accuracy and propose Second-Order Markov chain fingerprints with application attribute Bigram (SOB) [28]. When the message type sequence-based approaches process encrypted traffic of TLS 1.3, these approaches can only exploit visible message type sequences for Markov chain modeling. Since most handshake messages are masked as application data messages, the diversity of message type sequences will be greatly reduced, which will lead to an inevitable accuracy decline of these approaches.

Liu et al. (2018) try to apply Markov chain models to packet length sequences and propose Multi-attribute Markov Probability Fingerprints (MaMPF) to handle the large number of unique packet lengths [20]. Since the statistical distribution of the packet length frequency obeys the power law, they choose high-frequency packet lengths as length blocks and replace other packet lengths with nearest length blocks in distance priority rules to reduce the number of unique packet lengths. Zhang et al. (2020) focus on the unknown traffic in the encrypted traffic and build an autonomous deep learning-based model on packet length sequences to filter out the encrypted traffic with low classification confidence as unknown encrypted traffic [31]. Then, they label the unknown traffic into new applications through an autonomous clustering model and add this labeled traffic to the dataset to retrain the classifier that supports the new applications. Although the message lengths are independent of the content of the message, TLS 1.3 may combine multiple handshake messages into a new single message and encrypt the single message into an application data message. Thus, some message length information may be omitted and these approaches based on the message length will face a slight accuracy reduction in classifying TLS 1.3 encrypted traffic.

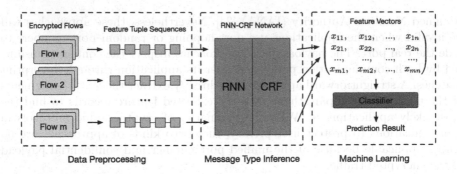

Fig. 2. The system overview of MTI

2.3   Attribute Feature-Based Encrypted Traffic Classification

The attributes commonly used to model TLS 1.2 encrypted traffic include statistic features and string features. Taylor et al. (2017) focus the statistical feature of the bursts in the flow, such as minimum, maximum, mean, median absolute deviation, standard deviation, etc. [30]. Based on the statistical features, they propose an automatic fingerprinting framework called AppScanner, which can identify the mobile application of encrypted traffic in real-time. Chen et al. (2019) propose a string feature-based approach named Multi-Attribute Associated Fingerprints (MAAF) [4]. The MAAF exploits several different attributes, including domain names, certificates, and application data lengths, to train the classifier for mobile encrypted traffic classification. The string attributes adopted by MAAF are extracted from plaintext DNS traffic and certificate messages. Currently, DNS traffic may be encrypted by DoT/DoH and certificate messages will be encrypted in TLS 1.3, which will lead to a severe compatibility challenge of MAAF in dealing with DoT/DoH and TLS 1.3.

3   System Introduction

In this section, we first provide an overview of our proposed mobile encrypted traffic classification system. Then, we describe the detail of the data preprocessing module of the system. Next, We present the design of the RNN-CRF network and the training of the network. Finally, we introduce how to perform machine learning to integrate different features for encrypted traffic classification.

3.1   System Overview

Our mobile encrypted traffic classification system consists of three modules, including Data Preprocessing, Message Type Inference, and Machine Learning, as shown in Fig. 2. The data preprocessing module extracts TLS 1.3 encrypted traffic from online raw traffic as system input and organizes this encrypted traffic into flows. In this paper, we intend to classify the encrypted traffic in units of flows to determine which application these flows belong to. In general, a TLS 1.3 encrypted flow is composed of handshake messages for session key negotiation and application data messages that carrying communication payloads. For each TLS message, we can directly extract two features without modification, namely message length and visible message type, and store these two features as a feature tuple. Therefore, we can represent TLS 1.3 encrypted flows as feature tuple sequences for subsequent processing. We also try to extract related DNS traffic of the flow through IP address matching and adopt it as a classification feature. For all input encrypted flows, the RNN-CRF network tries to recover the real message types of encrypted handshake messages through rigorous analysis of feature tuple sequences and outputs predicted message type sequences. Based on the real message types inferred by the RNN-CRF network, we can format the sequence of feature tuples into a uniform vector of message length, where each dimension stores the length of a specific-type message. Finally, we can train a classifier for the uniform vectors through supervised machine learning and implement classification for the uniform vectors of other encrypted flows.

3.2 Data Preprocessing

Online raw traffic contains raw packets that are captured from the Internet in the order of timeline. For the captured traffic to be classified, we first select out TLS 1.3 encrypted packets and organize these packets into independent flows. The organization of flows generally relies on the IP addresses, port numbers, and TCP sequence and acknowledgment numbers. We consider flows as basic units in the traffic classification since all packets in a flow must belong to the same session initiated by a client and a server and these packets belong to the same application.

The basic transmission units of TLS flows are messages that are reassembled by TCP packets. Compared with TLS 1.2, TLS 1.3 flows encrypt many key handshake messages and mask them with Application Data (23) to protect user privacy in communication, such as Encrypted Extensions (22:8), Certificate Verify (22:15), and Server Certificate (22:11). For each TLS 1.3 flow, we extract the message lengths and the visible message types of all messages to form a feature tuple sequence, where each tuple contains the message length and visible message type of a TLS message.

In general, mobile applications will find the optimal server IP address in the application initialization phase by querying the recursive DNS server for the preset domain name of the application server. Thus, the preset domain name is particularly relevant to the application and the DNS traffic can be used as an important classification feature in the mobile encrypted traffic classification [4]. We can find out the related plain text DNS traffic of a given encrypted flow by comparing the answer record of the DNS traffic and the server IP address of the flow. However, we cannot get the DNS traffic in the following situations and should ignore the related DNS traffic in the traffic classification. 1) The application connects with the server through hard-coded IP addresses. 2) The application obtains server IP addresses through an alive TLS session with the

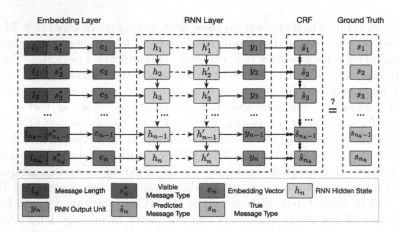

Fig. 3. The architecture of RNN-CRF network

connected server. 3) The client adopts DNS over HTTPS (DoT) or DNS over TLS (DoH) technology to connect a recursive DNS server for domain name querying.

3.3   Message Type Inference

Problem Definition. We first formulate the problem of message type inference. Let the sequential feature tuples of the k_{th} encrypted flow extracted by the data preprocessing module be

$$X^{(k)} = [(l_1^{(k)}, s_1^{*(k)}), (l_2^{(k)}, s_2^{*(k)}), ..., (l_{n_k}^{(k)}, s_{n_k}^{*(k)})] \tag{1}$$

where $l_i^{(k)}$ is the length of the i_{th} message, $s_i^{*(k)} \in S_{type}$ is the visible message type of the i_{th} message, n_k is the number of messages in the k_{th} flow. For the k_{th} encrypted flow, when we obtain the TLS secret of the encrypted flow, we can parse out the ground truth message type sequence

$$Y^{(k)} = [s_1^{(k)}, s_2^{(k)} ..., s_{n_k}^{(k)}] \tag{2}$$

where $s_i^{(k)} \in S_{type}$ is the real message type of the i_{th} message. We aim to build a sequence labeling model $\Phi(X^{(k)})$ to predict the message type sequence $\widehat{Y}^{(k)}$, and try to make the prediction result $\widehat{Y}^{(k)}$ the same as the ground truth message type sequence $Y^{(k)}$.

RNN-CRF Network. To solve this sequence labeling problem, we propose an RNN-CRF network that consists of an embedding layer, a recurrent neural network (RNN), and a conditional random field (CRF), as shown in Fig. 3. The embedding layer embeds the input feature tuple that consists of the message length l_i and the visible message type s_i^* into a integrated embedding vector e_i

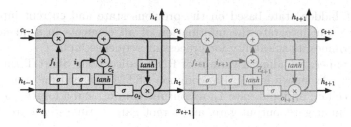

Fig. 4. The architecture of LSTM cells

that contains both the message length and message type information. The RNN layer is designed for processing and encoding sequence inputs and regulate the flow of information along the sequence. As a result, the hidden states h_i is a representation of all previous inputs. The RNN output y_i corresponds to the i_{th} message and represents the feature of the predicted message type. Finally, the conditional random field can reasonably combine the neighbor tag information of the previously predicted message type $\hat{s}_{i-1}$ with the predicted feature y_i provided by the RNN layer to infer the message type $\hat{s}_i$.

Embedding Layer. The input of the RNN-CRF network is feature tuple sequences, so we need to embed feature tuples into embedding vectors for the subsequent numerical calculation. We first create two randomly initialized character embedding matrix $E_l, E_{s^*} \in \mathbb{R}^{K \times d}$ for message length embedding and visible message type embedding, where K is the size of feature element set and d is the dimension of embedding vectors. In essence, the embedding matrixes are embedding vector lookup tables, and we can map the feature tuple $(l_i^{(k)}, s_i^{*(k)}), i \in [1, n_k]$ into a tuple of two independent embedding vectors $(E_l(l_i^{(k)}), E_{s^*}(s_i^{*(k)}))$. Since the subsequent RNN-CRF network only receives numerical vector sequences, and the embedding dimensions of two kinds of features are the same, we can concatenate two kinds of independent embedding vectors into an integrated vector and define the embedding function as $e_i^{(k)} = E(l_i^{(k)}, s_i^{*(k)}) = E_l(l_i^{(k)}) \oplus E_{s^*}(s_i^{*(k)})$.

Compared with scalar message lengths, the embedding vectors have multiple dimensions to form the feature information and can increase the discrimination of different message lengths. We represent the discrete visible message types with embedding vectors so that we can perform numerical calculations on the message type features. Besides, the embedding matrixes are trainable during the model training, and we can learn the task-oriented optimal values of these two embedding matrixes to improve the representation of the embedding vectors.

RNN Layer. The Recurrent Neural Network (RNN) is one kind of neural network where connections between RNN cells form a directed graph along a temporal sequence [27]. The RNN is widely used to process sequence inputs in many research fields, such as text classification in natural language processing [32]. The hidden state of the RNN cell typically stores the information of inputted sequence elements. Along with the input of sequence elements, RNN calculates the current hidden state based on the previous state and current input. As a result, RNN can integrate the information of all elements in sequence order, which determines its suitability for processing sequence input.

In this paper, we adopt a variant of RNN called Long Short-Term Memory (LSTM) [13]. As shown in Fig. 4, a common LSTM unit at time step $i \in [1, n_k]$ contains a cell c_i, an input gate i_i, an output gate o_i, and a forget gate f_i. The outputs of input gate, output gate, and forgot gate at time step i are

$$i_i = \sigma(W_{\mathbf{i}} \cdot [\mathbf{h}_{i-1}, x_i] + b_{\mathbf{i}}) \tag{3}$$

$$o_i = \sigma(W_{\mathbf{o}} \cdot [\mathbf{h}_{i-1}, x_i] + b_{\mathbf{o}}) \tag{4}$$

$$f_i = \sigma(W_{\mathbf{f}} \cdot [\mathbf{h}_{i-1}, x_i] + b_{\mathbf{f}}) \tag{5}$$

where $W_{\mathbf{i}}$, $W_{\mathbf{o}}$, and $W_{\mathbf{f}}$ are trainable weight matrices. $b_{\mathbf{i}}$, $b_{\mathbf{o}}$, and $b_{\mathbf{f}}$ are bias vector parameters need to be learned. $\mathbf{h}_{i-1}$ is the hidden state at time step $i-1$. The sigmoid function $\sigma = 1/(1 + e^{-x})$ is an activation function to introduce non-linearity. The cell input $\tilde{c}_i$, cell state c_i, and hidden state f_i are

$$\tilde{c}_i = tanh(W_{\mathbf{c}} \cdot [\mathbf{h}_{i-1}, x_i] + b_{\mathbf{c}}) \tag{6}$$

$$c_i = f_i \cdot c_{i-1} + i_i \cdot \tilde{c}_i \tag{7}$$

$$\mathbf{h}_i = o_i \cdot tanh(c_i) \tag{8}$$

where $W_{\mathbf{c}}$ and $b_{\mathbf{c}}$ are trainable weight and bias. The forgot gate f_i and input gate i_i jointly determine how to forget the previous cell state c_{i-1} and how to receive the current cell input $\tilde{c}_i$. Finally, the output gate o_i controls how information is processed as the hidden state $\mathbf{h}_i$. The three gates of the LSTM effectively control the flow of information between adjacent LSTM cells and solve the vanishing gradient problem of vanilla RNN [23]. For the hidden state $\mathbf{h}_i$ at time step i, we use a dense layer to map the message type features to the tagging classes. The output score of the RNN layer r_i is

$$r_i = \sigma(W_r \cdot \mathbf{h}_i + b_r) \tag{9}$$

where W_r and b_r are the parameters of the dense layer. The r_i is a classification score vector where each dimension stores the score of the corresponding tag.

Conditional Random Field. The conditional random field (CRF) is a statistical model that considers neighboring information to make structured predictions [19]. Therefore, the CRF is suitable to solve the sequence tagging problem by making use of neighbor tag information. Take the NER task in natural language processing as an example, the linear chain CRFs can efficiently use past tags to predict the current tag [16, 22]. In this paper, we connect a CRF network after the RNN layer to form an RNN-CRF network, which combines the feature of past inputs and neighbor tag information at sentence level to perform message type tagging. The parameter of a CRF layer is a state transition metric.

Suppose the state transition metric $|A|_{s,s'}$ determines the transition score from the message type s to s' in consecutive time steps and the transition metric $|A|_{s,s'}$ is independent of the message type position in the sequence. The parameter of the RNN-CRF network is

$$\tilde{\theta} = \tilde{\theta}_{RNN} + |A|_{s_i, s_j} \tag{10}$$

where $\tilde{\theta}_{RNN}$ if the parameters of the RNN layer. The final score of the flow $X^{(k)}$ with a message type prediction $\widehat{Y}^{(k)}$ is the sum of CRF score and RNN layer score, which is as follows

$$score(X^{(k)}, \widehat{Y}^{(k)}, \tilde{\theta}) = \sum_{i=1}^{n_k} (|A|_{\hat{s}_{i-1}^{(k)}, \hat{s}_i^{(k)}} + r_{i_{\hat{s}_i^{(k)}}}) \tag{11}$$

where $\hat{s}_i^{(k)}$ is the predicted message type of the i_{th} message in k_{th} flow, r_i is score output of RNN layer for i_{th} message. Given a flow input $X^{(k)}$ and RNN-CRF network parameters $\tilde{\theta}$, we can get the optimal message type prediction $\widehat{Y}^{(k)}$ with the highest final score by the Viterbi algorithm [10].

Network Training. In order to train the RNN-CRF network on the collected dataset, we use a negative log-likelihood function $-\log p(Y|X)$ as the loss function, where X, Y indicate the input flow and the corresponding application, respectively. The likelihood function $p(Y|X)$ is defined as the probability of classifying X as Y, so we intend to maximize the $p(Y|X)$ in the training of the RNN-CRF network, which is equivalent to minimize the negative log-likelihood function $-\log p(Y|X)$.

In the previous subsection, we have defined the prediction score $score(X, Y)$ of the flow X with a prediction Y. Therefore, we can define the likelihood function as the quotient of the power of the prediction score $e^{score(X,Y)}$ and the power sum of all potential Y scores $\sum_Y e^{score(X,Y)}$. So, the loss function of the RNN-CRF network can be defined as

$$loss = -\log p(Y|X) = -\log \frac{e^{score(X,Y)}}{\sum_Y e^{score(X,Y)}}$$
$$= -(score(X, Y) - \log \sum_i Z(Y_n = i)) \tag{12}$$

where $\sum_i Z(Y_n = i)$ is the power sum of all n-length potential path score with $Y_n = i$. We define $Z(Y_t = i)$ as

$$Z(Y_t = i) = \sum_{Y_t} e^{score(start \rightarrow Y_t)} \tag{13}$$

We can recursively calculate $\log Z(Y_t = i)$ when we obtain the power sum of all $t - 1$-length potential path score, the state transition metric $|A|$, and the RNN layer score r_t

$$\log Z(Y_t = i) = \log \sum_{Y_t} e^{score(start \to Y_t)}$$

$$= \log \sum_{Y_{t-1}=1}^{m} e^{score(start \to Y_{t-1}) + |A|_{Y_{t-1}, Y_t} + r_t Y_t} \tag{14}$$

where m indicates the number of potential message type tags in the prediction. As a result, we can improve the calculation speed of $\sum_i Z(Y_n = i)$ through dynamic programming algorithm. Based on the loss function, we can use a gradient descent algorithm to train the network upon the collected dataset.

3.4 Machine Learning

Based on the real message type sequence inferred by the RNN-CRF network, we can align the message types of the encrypted handshake messages in different encrypted flows and generate the feature vectors of these encrypted flows. The features we use include the length of the relevant DNS packet, the lengths of the type-tagged encrypted handshake messages, and the lengths of application data messages. For a given encrypted flow, we can obtain these features in chronological order as the communication is established and progressed. Therefore, we can determine the combination of these features according to the performance requirements of real-time prediction and classification accuracy. When we adopt more features in traffic classification, we can generally achieve higher classification accuracy while we need to face a slower classification efficiency and a larger classification latency. In mobile encrypted traffic classification, when we need to predict the application of the encrypted flow before its establishment, we can only exploit the length of DNS traffic to make classifications. When we want to predict the application of the encrypted flow before payload transmission, we can make use of the length of the DNS traffic and the lengths of the type-tagged handshake messages based on the trained RNN-CRF network. When we require a high-accurate encrypted traffic classifier, we should adopt all available features in both the training of the classifier and the classification of the encrypted traffic.

We adopt several supervised machine learning methods to train an effective classifier on these feature vectors for mobile encrypted traffic classification. In this paper, we select three alternative supervised machine learning models, including C4.5 [25], Random Forest [2], and XGBoost [3]. The C4.5 is a landmark algorithm to generate a decision tree for a given set of feature vectors. In detail, the C4.5 adopts the normalized information gain as its splitting criterion to build the decision tree, so the training and prediction of the C4.5 are remarkably efficient. The Random Forest is an ensemble learning method designed for classification or regression tasks. Compared with the single tree generated by C4.5, the Random Forest builds a set of decision trees for random training data and feature sampling results to solve the overfitting problem caused by the single decision tree. The XGBoost is an optimized gradient boosting decision tree implementation designed for both efficiency and effectiveness. The gradient boosting decision tree is an ensemble model of decision trees that are

Table 2. Summary of the experimental dataset

#	Application	Flows	Packets	Domain	Size (MB)
1	Booking.com	3690	695989	42	498.79
2	Breitbart	3515	516367	126	318.38
3	Canva	3762	610755	20	418.77
4	ESPN	3885	408017	140	236.72
5	Facebook	3634	1925147	86	1300.90
6	Fox News	3690	812188	111	500.00
7	Okczone	3736	338559	66	197.27
8	Quizlet	3735	525844	41	324.50
9	Spotify	3712	296209	14	158.48
10	Steam	3771	1004865	24	695.20
11	VK	3851	1329839	137	917.81
12	Wikipedia	3684	588059	6	337.96
13	Yahoo! News	3539	850082	145	562.11
14	Zillow	3886	626698	65	415.16
	Total	52090	10528618	1023	6882.04

recursively created to predict the errors or prior created trees. Therefore, each machine learning method has its specific advantages and applicable scenarios. The C4.5 is suitable for large-scale classification tasks because of its simplicity and efficiency. The feature sampling mechanism of Random forest can effectively avoid the overfitting problem in features learning and provides high classification accuracy. The gradient boosting of XGBoost can effectively deal with complex features and generates a set of cooperating decision trees to achieve high classification accuracy. We can choose the appropriate method according to the performance requirement of the specific scenario.

4   Evaluation

In this section, we first introduce the experimental datasets in the evaluation. Then, we specify the experimental settings and describe the definition of the evaluation metrics. Finally, we present experimental results of the message type inference, our proposed approach, and the comparisons with existing approaches.

4.1   Preliminary

Dataset. To objectively evaluate the effectiveness of our proposed classification approach, we first need to collect a TLS 1.3 mobile encrypted traffic dataset and a TLS 1.3 traffic dataset with ground truths of encrypted handshake message

Table 3. Summary of the dataset with TLS decryption keys and the experimental results of RNN-CRF

Message type	Dataset		RNN-CRF (GRU)		RNN-CRF (LSTM)	
	Count	Ratio	Prec.	Rec.	Prec.	Rec.
20	19988	0.1254	**1.0000**	**1.0000**	**1.0000**	**1.0000**
21	3632	0.0228	**0.9980**	**0.9980**	**0.9980**	**0.9980**
22:1	10225	0.0641	**1.0000**	**1.0000**	**1.0000**	**1.0000**
22:11	3569	0.0224	**0.9889**	0.9944	0.9882	**0.9979**
22:15	3075	0.0193	**0.9992**	0.9992	**0.9992**	**1.0000**
22:2	10233	0.0642	**1.0000**	**1.0000**	**1.0000**	0.9998
22:20	17484	0.1097	**0.9996**	0.9997	0.9991	**0.9999**
22:25:15:20	60	0.0004	**1.0000**	**1.0000**	**1.0000**	**1.0000**
22:4	7685	0.0482	**0.9980**	0.9987	**0.9980**	**0.9997**
22:4:4	2056	0.0129	**0.9988**	**1.0000**	0.9976	0.9988
22:8	7724	0.0485	**0.9977**	0.9994	0.9974	**0.9997**
22:8:20	1639	0.0103	**1.0000**	**1.0000**	**1.0000**	0.9984
22:8:25:15:20	187	0.0012	0.8824	**0.8000**	**0.9524**	**0.8000**
23	71732	0.4500	0.9996	**0.9993**	**0.9999**	0.9992
Total/Acc/F1	159407	1.0000	0.9991	0.9874	**0.9992**	**0.9896**

types. Chen et al. [4] describe two kinds of mobile application traffic collection schemes, namely Active Traceset Collection and Passive Traceset Collection. The active scheme collects mobile application traffic by tracking the mobile phone connected to a controllable workstation and trying to operate mobile applications in the mobile phone to generate online traffic of the specified mobile applications. The passive scheme first collects unlabeled traffic from pre-deployed port mirroring switches and utilizes DNS records to separate the traffic of different applications. To purposefully collect encrypted traffic of TLS-1.3-employed applications, we adopt the Active Traceset Collection scheme and collect a real-world dataset that contains 52,090 encrypted flows of fourteen mobile applications. Table 2 presents the summary of the experimental dataset.

We also collect a TLS 1.3 traffic dataset with TLS decryption keys by assigning a desktop web browser to access TLS-1.3-employed websites and capturing both the encrypted browsing traffic and TLS decryption keys. Based on the TLS decryption keys, we can decrypt TLS handshake messages and recover message type ground truths to support the training of the RNN-CRF network. Table 3 left presents the statistic of the TLS-1.3 traffic dataset with ground truths of encrypted handshake message types.

Experiment Settings. We first consider the parameters of the RNN-CRF network for message type inference. We embed sequential elements to 50 dimension vectors through the embedding layer. We set the number of the RNN layers to 2 and set the dimension of the RNN hidden states to 100. We enhance the robustness of the neural network by setting the dropout [29] to 0.5 to randomly omit network nodes during model training. Besides, we adopt Adam optimizer [17] with a 0.0002 learning rate for the training of the RNN-CRF network.

The machine learning for the feature vectors should consider some encrypted traffic-related parameters. The feature vector of a given encrypted flow includes the length of the relevant DNS packets, the lengths of handshake messages, and the lengths of application data messages. We limit the maximum number of the adopted messages in machine learning to 15. The machine learning model is selected from C4.5, Random Forest, and XGBoost. We conduct a random parameter search and employ 5-fold cross-validation to obtain the optimal parameters of the selected machine learning model on the training data.

We randomly divide the experimental dataset into a training dataset and a test dataset, which respectively account for 70% and 30% of the original dataset. We take the average results of 5 repeated experiments to reduce random errors.

Evaluation Metrics. We employ three common metrics to evaluate the effectiveness of a given approach on the mobile encrypted traffic classification task. The precision and recall of the application i are defined as follows:

$$Precision_i = \frac{TP_i}{TP_i + FP_i}, \quad Recall_i = \frac{TP_i}{TP_i + FN_i} \tag{15}$$

where True Positive TP_i indicates the number of application i's samples that are correctly classified to application i, False Positive FP_i indicates the number of samples that are wrongly classified to application i, and False Negative FN_i indicates the number of application i's samples that are misclassified to other classes. Besides, we use F1-score to represent the overall classification effectiveness, which is defined as:

$$F1_i = \frac{Precision_i * Recall_i}{Precision_i + Recall_i}, \quad F1 = \frac{\sum_{i=1}^{n} F1_i}{n} \tag{16}$$

where the F1-score of application i is defined as the harmonic mean of $Precision_i$ and $Recall_i$, and the overall F1-score is defined as the macro average of all $F1_i$.

4.2 Analysis of the Message Type Inference

We first study the effectiveness of the RNN-CRF network in inferring the real message types of encrypted messages. We evaluate the RNN-CRF network on the TLS 1.3 traffic dataset with ground truths of handshake message types. Since we focus on the accuracy of message type inference, we calculate the evaluation metrics in the granularity of messages instead of flows. Our RNN-CRF network adopts LSTM to build its RNN layer, so we compare it with another RNN variant called Gated Recurrent Unit (GRU) [5].

Table 3 gives the experimental results of using LSTM and GRU RNN-CRF network for message type inference. Both LSTM and GRU perform high accuracy, while LSTM is slightly better than GRU. The F1-scores of LSTM and

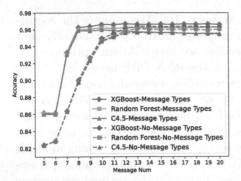

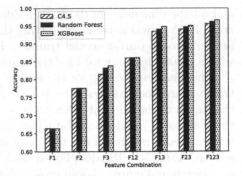

Fig. 5. Comparison results of the inferred message types

Fig. 6. Comparison results of the feature combination

GRU reach 98.96% and 98.74%, respectively. Although the dataset is imbalanced for some message types, the RNN-CRF network can still show high precision and recall for the message types with a small volume, such as 22:25:15:20 and 22:8:25:15:20. Based on the experimental results, we can train an RNN-CRF network upon the whole dataset with TLS decryption keys and use the model to predict the message types of the encrypted messages in other datasets.

In this subsection, we study the effectiveness of the adopted features in machine learning, including the length of relevant DNS packets, the inferred message types of encrypted handshake messages, the lengths of handshake messages, the lengths of application data messages. We first study the impact of using the inferred message types of encrypted handshake messages on classification accuracy. When we do not use inferred message types to align the message types of the encrypted handshake messages in feature vector generation, we directly use the length of relevant DNS packets and message lengths in chronological order to build feature vectors. Besides, we vary the number of adopted message lengths to study the impact of the inferred message types under different numbers of message lengths. We employ three machine learning models to eliminate the random error caused by the machine learning model.

4.3   Analysis of Adopted Features

Figure 5 shows the comparison results of the inferred message types. We can find that when the number of message lengths takes from 5 to 11, the classification accuracy of using massage types is larger than not using it for all three machine learning methods. Besides, the accuracy disparity between using message types and not using message types increases sharply first and then decreases slowly. The classification accuracy of using message types is overall higher than not using message types. This may be because when the number of messages is small, aligning the message lengths of the same message type in constructing feature vectors can reasonably avoid the information deviation caused by the dimension dislocation of the feature vectors. Thus, when the number of messages is limited

Table 4. Experimental results of different machine learning models

Application	C4.5		Random forest		XGBoost	
	Prec.	Rec.	Prec.	Rec.	Prec.	Rec.
Booking.com	0.9646	0.9718	0.9715	0.9788	**0.9765**	**0.9820**
Breitbart	0.8731	0.9145	0.8876	0.9279	**0.8886**	**0.9369**
Canva	0.9445	0.9705	0.9427	0.9697	**0.9478**	**0.9724**
ESPN	0.9190	0.9058	0.9277	0.9127	**0.9351**	**0.9253**
Facebook	0.9867	0.9818	**0.9953**	0.9821	0.9936	**0.9837**
Fox News	0.9319	0.9115	0.9410	0.9299	**0.9417**	**0.9380**
Okezone	0.9488	0.9582	0.9638	0.9732	**0.9702**	**0.9740**
Quizlet	0.9463	0.9427	0.9596	0.9438	**0.9601**	**0.9551**
Spotify	0.9838	0.9658	0.9711	**0.9708**	**0.9862**	0.9674
Steam	0.9829	0.9840	0.9753	0.9893	**0.9849**	**0.9901**
VK	0.9939	0.9910	0.9939	0.9876	**0.9955**	**0.9926**
Wikipedia	0.9979	0.9995	0.9987	**1.0000**	**0.9992**	**1.0000**
Yahoo! News	0.9622	0.9404	0.9717	0.9461	**0.9736**	**0.9504**
Zillow	0.9554	0.9525	0.9713	0.9581	**0.9790**	**0.9627**
Acc/F1	0.9566	0.9564	0.9623	0.9621	**0.9666**	**0.9664**

to a small number, the usage of message types can significantly improve the classification accuracy of encrypted traffic classification.

Then, we study the effect of other features in machine learning, including the length of the relevant DNS packet, the lengths of handshake messages, the lengths of application data messages. We intend to evaluate the classification accuracy of all kinds of feature combinations. We notate the length of the relevant DNS packet, the lengths of handshake messages, the lengths of application data messages as F1, F2, F3, respectively, and use the combination of notations to indicate the combination of features.

Figure 6 shows the experimental results of different feature combinations. As shown in the figure, the classification accuracy increases with the enrichment of the features. The accuracy of the combination of three features is significantly higher than other combinations and the accuracy of the combinations of two features is also higher than single features. In addition, we can find that when we only use one feature for machine learning, the lengths of application data messages can provide more discrimination information than the other two features. This may be because the lengths of application data messages are relevant to the payload transmission pattern of the application. Because of the TLS protocol specification, the number and content of handshake messages are somewhat restricted, so the lengths of handshake messages perform slightly lower classification accuracy than application data messages. The legal lengths of DNS packets are limited by the DNS protocol specification and each flow generally corresponds to one relevant DNS packet, so it performs the worst among the three features. Since the length of application data messages is particularly important for feature learning, when we have high requirements for both real-time performance and classification accuracy, we need to focus on balancing the number of application data messages and the time cost to capture traffic.

We also compare the classification results of using different machine learning models, as presented in Table 4. We find that the accuracy of XGBoost achieves 96.66%, and the F1-score achieves 96.64%, which is a little better than C4.5 and Random Forest. XGBoost also has better precision and Recall for most applications, which demonstrates that when we only focus on the classification accuracy, XGBoost is indeed better than the other two models. Therefore, when the encrypted traffic classification task requires high accuracy and can tolerate other relatively poor performances, such as time complexity, we prefer to choose XGBoost as the machine learning model.

4.4 Comparisons with Existing Approaches

The attribute-based approaches face severe compatibility problems to the encrypted handshake messages, so we only compare our proposed approach with three sequence-based state-of-the-art approaches. The comparison approaches include Second-Order Markov chain fingerprints with application attribute Bigram (SOB) [28], Multi-attribute Markov Probability Fingerprints (MaMPF) [20], and Flow Sequence Network (FS-Net) [21]. SOB employees second-order Markov chain to model message type sequence and use machine learning models to incorporate the lengths of certificate messages and application data messages [28]. Since most handshake messages including the certificate message are encrypted as application data messages, we take the visible message type sequence as model input and leave the length of the certificate empty. MaMPF transfers first message length sequences into length block sequences and then employees Markov chain to model message type sequences and length block sequences [20]. So, we can directly apply MaMPF on the message length sequence and visible message type sequence of TLS 1.3 encrypted traffic. Compared with the former two approaches, FS-Net only uses an auto-encoder neural network to model message length sequences [21], so we can fully apply FS-Net on TLS 1.3 encrypted traffic.

Table 5 presents the experimental results of comparisons with existing approaches. As shown in the table, the accuracy and F1-score of MTI reach 96.66% and 96.64% respectively, which are significantly better than the other approaches. The accuracy and F1-score of SOB are only 61.77% and 61.45%. This is because the message types of most handshake messages are masked by application data messages, which conceals much of the feature information for the encrypted traffic classification. MaMPF performs better than SOB, which is attributed to the usage of message length sequence information. The accuracy and F1-score of FS-Net reach 90.92% and 90.90%. This demonstrates that FS-Net can appropriately deal with the classification of TLS 1.3 encrypted traffic.

Table 5. Experimental results of different approches

Application	SOB		MaMPF		FS-Net		MTI	
	Prec.	Rec.	Prec.	Rec.	Prec.	Rec.	Prec.	Rec.
Booking.com	0.6914	0.5850	0.7372	0.5190	0.9291	0.9321	**0.9765**	**0.9820**
Breitbart	0.5376	0.4520	0.6895	0.5741	0.6963	0.8249	**0.8886**	**0.9369**
Canva	0.8301	0.5877	0.6990	0.7327	0.9231	0.9698	**0.9478**	**0.9724**
ESPN	0.6457	0.5002	0.8221	0.6393	0.8656	0.7916	**0.9351**	**0.9253**
Facebook	0.6158	0.8406	0.6205	0.8059	0.9921	0.9679	**0.9936**	**0.9837**
Fox News	0.4346	0.3820	0.6338	0.5799	0.8685	0.7703	**0.9417**	**0.9380**
Okezone	0.5311	0.6233	0.8478	0.6695	0.8726	0.9214	**0.9702**	**0.9740**
Quizlet	0.6362	0.6589	0.7530	0.8257	0.9030	0.8957	**0.9601**	**0.9551**
Spotify	0.6639	0.8272	0.6206	0.9216	0.9698	0.9588	**0.9862**	**0.9674**
Steam	0.7816	0.6889	0.9029	0.7276	0.9187	0.9455	**0.9849**	**0.9901**
VK	0.7421	0.7673	0.7410	0.8452	0.9686	0.9812	**0.9955**	**0.9926**
Wikipedia	0.7814	0.8563	0.9514	0.5900	0.9936	0.9900	**0.9992**	**1.0000**
Yahoo! News	0.4102	0.5603	0.5048	0.8315	0.9020	0.8765	**0.9736**	**0.9504**
Zillow	0.4088	0.3447	0.7714	0.6595	0.9475	0.8994	**0.9790**	**0.9627**
Acc/F1	0.6177	0.6145	0.7093	0.7084	0.9092	0.9090	**0.9666**	**0.9664**

5 Discussion and Conclusion

In this paper, we design an RNN-CRF network to infer the real message types of TLS 1.3 encrypted handshake messages and classify TLS 1.3 mobile encrypted traffic through the machine learning. The RNN network primarily considers the input visible message type and message length information, while the CRF network focuses on the neighbor tag information of the message type sequence. The evaluation results demonstrate the applicability of MTI to TLS 1.3 traffic, which shows 96.66% classification accuracy and 96.64% F1-score. However, when we encounter encrypted DNS traffic of DoT/DoH, we cannot obtain the relevant DNS information of the encrypted flow but only make use of message type and length information, which would cause slight classification accuracy decreases. In future work, we plan to study how to make use of DoT/DoH traffic as an encrypted traffic classification feature and then maintain the classification accuracy when encountering the DNS traffic of DoT/DoH.

Acknowledgment. This work is supported by the Strategic Priority Research Program of the Chinese Academy of Sciences (No.XDC02030100), the National Key Research and Development Program of China (Grant No.2018YFB0804704), and the National Natural Science Foundation of China (Grant No.U1736218).

References

1. Aceto, G., Ciuonzo, D., Montieri, A., Pescapé, A.: Mobile encrypted traffic classification using deep learning. In: 2018 Network traffic measurement and analysis conference (TMA), pp. 1–8. IEEE (2018)
2. Breiman, L.: Random forests. Mach. Learn. **45**(1), 5–32 (2001)
3. Chen, T., Guestrin, C.: Xgboost: a scalable tree boosting system. In: Proceedings of the 22Nd ACM SIGKDD International Conference on Knowledge Discovery and Data Mining, pp. 785–794. ACM (2016)
4. Chen, Y., Zang, T., Zhang, Y., Zhouz, Y., Wang, Y.: Rethinking encrypted traffic classification: a multi-attribute associated fingerprint approach. In: 2019 IEEE 27th International Conference on Network Protocols (ICNP), pp. 1–11. IEEE (2019)
5. Cho, K., et al.: Learning phrase representations using RNN encoder-decoder for statistical machine translation. In: Proceedings of the 2014 Conference on Empirical Methods in Natural Language Processing (EMNLP), pp. 1724–1734 (2014)
6. Cova, M., Kruegel, C., Vigna, G.: Detection and analysis of drive-by-download attacks and malicious javascript code. In: Proceedings of the 19th international conference on World wide web, pp. 281–290 (2010)
7. Dierks, T., Rescorla, E.: The transport layer security (tls) protocol version 1.2. IETF RFC5246 (2008)
8. Fiedler, M., Hossfeld, T., Tran-Gia, P.: A generic quantitative relationship between quality of experience and quality of service. IEEE Netw. **24**(2), 36–41 (2010)
9. Finsterbusch, M., Richter, C., Rocha, E., Muller, J.A., Hanssgen, K.: A survey of payload-based traffic classification approaches. IEEE Commun. Surv. Tutor. **16**(2), 1135–1156 (2013)
10. Forney, G.D.: The viterbi algorithm. Proc. IEEE **61**(3), 268–278 (1973)
11. Google: Google Online Security Blog: An Update on Android TLS Adoption. https://security.googleblog.com/2019/12/an-update-on-android-tls-adoption. html
12. Google: HTTPS encryption on the web - Google Transparency Report. https://transparencyreport.google.com/https/overview
13. Hochreiter, S., Schmidhuber, J.: Long short-term memory. Neural Comput. **9**(8), 1735–1780 (1997)
14. Hoffman, P., McManus, P.: Dns queries over https (doh). IETF RFC8484 (2018)
15. Hu, Z., Zhu, L., Heidemann, J., Mankin, A., Wessels, D., Hoffman, P.: Specification for dns over transport layer security (tls). IETF RFC7858 (2016)
16. Huang, Z., Xu, W., Yu, K.: Bidirectional lstm-crf models for sequence tagging. arXiv preprint arXiv:1508.01991 (2015)
17. Kingma, D.P., Ba, J.: Adam: a method for stochastic optimization. arXiv preprint arXiv:1412.6980 (2014)
18. Korczyński, M., Duda, A.: Markov chain fingerprinting to classify encrypted traffic. In: 2014 IEEE International Conference on Computer Communications (Infocom), pp. 781–789. IEEE (2014)
19. Lafferty, J., McCallum, A., Pereira, F.C.: Conditional random fields: probabilistic models for segmenting and labeling sequence data. In: Proceedings of the Eighteenth International Conference on Machine Learning, pp. 282–289. Morgan Kaufmann (2001)
20. Liu, C., Cao, Z., Xiong, G., Gou, G., Yiu, S.M., He, L.: Mampf: encrypted traffic classification based on multi-attribute markov probability fingerprints. In: 2018 IEEE/ACM 26th International Symposium on Quality of Service (IWQoS), pp. 1–10. IEEE (2018)

21. Liu, C., He, L., Xiong, G., Cao, Z., Li, Z.: Fs-net: a flow sequence network for encrypted traffic classification. In: 2019 IEEE International Conference on Computer Communications (Infocom), pp. 1–9. IEEE (2019)
22. Ma, X., Hovy, E.: End-to-end sequence labeling via bi-directional LSTM-CNNS-CRF. arXiv preprint arXiv:1603.01354 (2016)
23. McCloskey, M., Cohen, N.J.: Catastrophic interference in connectionist networks: the sequential learning problem. In: Psychology of learning and motivation, vol. 24, pp. 109–165. Elsevier (1989)
24. Qi, Y., Xu, L., Yang, B., Xue, Y., Li, J.: Packet classification algorithms: from theory to practice. In: IEEE INFOCOM 2009, pp. 648–656. IEEE (2009)
25. Quinlan, J.R.: C4.5: programs for machine learning. Elsevier (2014)
26. Rescorla, E.: The transport layer security (tls) protocol version 1.3. IETF RFC8446 (2018)
27. Rumelhart, D.E., Hinton, G.E., Williams, R.J.: Learning representations by back-propagating errors. Nature **323**(6088), 533–536 (1986)
28. Shen, M., Wei, M., Zhu, L., Wang, M.: Classification of encrypted traffic with second-order Markov chains and application attribute bigrams. IEEE Trans. Inf. For. Secur. **12**(8), 1830–1843 (2017)
29. Srivastava, N., Hinton, G., Krizhevsky, A., Sutskever, I., Salakhutdinov, R.: Dropout: a simple way to prevent neural networks from overfitting. J. Mach. Learn. Res. **15**(1), 1929–1958 (2014)
30. Taylor, V.F., Spolaor, R., Conti, M., Martinovic, I.: Robust smartphone app identification via encrypted network traffic analysis. IEEE Trans. Inf. For. Secur. **13**(1), 63–78 (2017)
31. Zhang, J., Li, F., Ye, F., Wu, H.: Autonomous unknown-application filtering and labeling for dl-based traffic classifier update. In: 2020 IEEE International Conference on Computer Communications (Infocom), pp. 1–9. IEEE (2020)
32. Zhou, P., Qi, Z., Zheng, S., Xu, J., Bao, H., Xu, B.: Text classification improved by integrating bidirectional LSTM with two-dimensional max pooling. arXiv preprint arXiv:1611.06639 (2016)

Fine-Grained Spatial-Temporal Representation Learning with Missing Data Completion for Traffic Flow Prediction

Shiqi Wang, Min Gao[(✉)], Zongwei Wang, Jia Wang, Fan Wu, and Junhao Wen

School of Big Data and Software Engineering, Chongqing University,
Chongqing, China
{shiqi,gaomin,zongwei,jiawang,wufan,jhwen}@cqu.edu.cn

Abstract. Spatial-temporal traffic flow prediction is beneficial for controlling traffic and saving traffic time. Researchers have proposed prediction models based on spatial-temporal representation learning. Although these models have achieved better performance than traditional methods, they seldom consider several essential aspects: 1) distances and directions from the spatial aspect, 2) the bi-relation among historical time intervals from the temporal aspect, and 3) missing historical traffic data, which leads to an imprecise spatial-temporal features extraction. To this end, we propose Fine-Grained Features learning based on Transformer-encoder and Graph convolutional networks (FGFTG) to improve the performance of traffic flow prediction in a missing data scenario. FGFTG consists of two components: feature extractors and a data completer. The feature extractors learn fine-grained spatial-temporal representations from spatial and temporal perspectives. They extract smoother representation with the information of distance and direction from a spatial perspective based on graph convolutional networks and node2vec and achieve bidirectional learning for temporal perspective utilizing transformer encoder. The data completer simulates the traffic flow data distribution and generates reliable data to fill in missing data based on generative adversarial networks. Experiments on two public datasets demonstrate the effectiveness of our approach over the state-of-the-art methods.

Keywords: Traffic flow prediction · Generative adversarial network · Graph convolutional neural network · Transformer encoder

1 Introduction

The appearance of cars from the 19th century has brought tremendous changes to people's life. It dramatically helps government agencies to avoid potential catastrophic traffic accidents [48] and brings convenience to daily travel [3]. With the development of technology, the auto industry has been widely used

H. Gao and X. Wang (Eds.): CollaborateCom 2021, LNICST 406, pp. 138–155, 2021.
https://doi.org/10.1007/978-3-030-92635-9_9

in intellectualization since the 1980s. In recent years, traffic flow prediction is consequently becoming a hot issue for researchers because it can predict the state of road traffic and is beneficial in a wide range of applications [12,28,31]. Traffic jams can be significantly reduced due to route planning based on the prediction. The prediction can also provide insights to the regulatory authorities for decision-making, risk assessment, and traffic management.

The traffic flow is easily influenced by multiple factors such as weather, holidays, and traffic accidents, which tremendously aggravates prediction accuracy. Meanwhile, traffic data provided by road sensors are sometimes missing due to sensor damage or network congestion. For example, the missing data in PeMS dataset [44] accounted for 11.3%. Consequently, analyzing and coping with the fast-changing and missing data effectively becomes an urgent problem to solve. Existing studies [13,23,39,43,47] usually consider spatial and temporal information at the same time. Generally, they usually use Graph Convolutional Network to extract spatial features, which only focus on spatial node's flow change (content information) and neighbor information, ignoring the influence of distance and direction information between nodes. Figure 1 shows the real change of traffic flow data on the freeway. Detector 1 and detector 2 detect traffic flow data in the same direction at different locations, while detector 1 and detector 3 are in the same location in different directions. The result shows that the distance between detector 1 and detector 3 is closer while the flow distribution is further, indicating that direction is more critical in this region. Likewise, extracting temporal features also has several shortcomings like error accumulation and one-way learning occlusion, resulting in partial temporal information loss. In alleviating the problem of missing data, early studies [5,14,21] fill in data according to fixed distribution assumptions, such as Gaussian distribution. Nevertheless, these fixed data distributions usually fail to fit the real flow changes due to the insufficient consideration of the actual traffic flow context. In summary, these kinds of methods are too brutal to ensure the robustness of filling data.

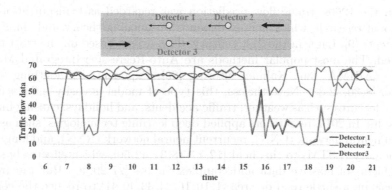

Fig. 1. The line chart of traffic flow (reflect the direction and distance of spatial node).

Facing these problems, we propose a fine-grained spatial-temporal representation learning with missing data completion for accurate traffic flow prediction, which is challenging due to: (1) integrating spatial features like distance and direction information and (2) considering bi-relation among historical time intervals from the temporal aspect and (3) generating reliable traffic flow data for missing values under the high complexity and variability distribution. To tackle the aforementioned challenges, we propose FGFTG (Fine-Grained spatial-temporal Features learning based on Transformer encoder and Graph convolutional networks), which contains two parts. The first is temporal-spatial feature extractors. In particular, spatial feature extractor based on graph convolution neural network (GCN [7]) and node2vec [15] technique is designed for integrating content, neighbour, distances, and directions information. Temporal feature extractor based on transformer encoder [35] is created to consider bi-relations among time intervals. The second part is data completer. We design DCGAN (Data Completion based on Generative Adversarial Networks) to fit the complex distribution of traffic flow data and generate reliable data for missing values. In summary, we present the main contributions as follows:

- We innovatively propose FGFTG, a spatial-temporal traffic flow data prediction framework, to learn fine-grained representations by two feature extractors.
- We design a data completer model DCGAN to simulate the real traffic flow data distribution and generate reliable data to fill in missing data.
- We conduct experiments on two public datasets. The experiments show that our model FGFTG significantly outperforms the state-of-the-art traffic flow prediction models and demonstrate our model's effectiveness in filling in missing data.

2   Related Work

Early in the 1960s, traffic flow prediction was regarded as transportation and operational research, which mainly depends on queuing theory and simulation experiments [9]. Later in the 21st century, data-driven based on the statistics is presented. The most popular methods are Auto-Regressive Integrated Moving Average (ARIMA) [36], Kalman filtering [26], Exponential Smoothing model [45], etc. [8,30,32,38]. Nevertheless, this type of model is easily influenced by dynamic features such as weather, traffic accidents, and holidays, causing inaccurate results. In 2014, researchers applied deep learning technologies, such as convolution neural network (CNN), recurrent neural network (RNN), and long short term memory (LSTM) to this field [12,17,18,37,42,50], which effectively solves the problems of massive data and complex factors [27,29,46]. But researchers only focus on a single road or area [1,10,16,24,33,40,41] to reduce the computation process, while ignoring the spatial dependency between roads or areas. In 2017, Zheng et al. [47] proposed the ST-ResNet model, which is the first time proposing the concept of spatial-temporal traffic flow prediction. In their work, not only the temporal features were calculated, spatial features are also considered. Subsequently, many researchers started this study based on spatial and temporal dependency.

In 2018, Li et al. [23] proposed Graph Convolutional Networks based on Recurrent Neural Network (GCRNN) to deal with the complex spatial dependency on road networks and non-linear temporal dynamics with changing road conditions. Yu et al. [44] proposed Spatio-Temporal Graph Convolutional Networks (STGCN) comprising several spatio-temporal convolutional blocks to model spatial and temporal dependencies. Yao et al. [43] proposed a DeepMulti-View Spatial-Temporal Network (DMVST-Net) framework to model both spatial and temporal relations. Later in 2019, DeepSTN [25] chose the ConvPlus structure to model the long-range spatial dependence among crowd flows in different regions and combine PoI (Point of Interest) distributions and time factors to express the effect of location attributes. Cao et al. [4] analyzed seasonal dependencies based on data analysis and extracted different features based on these dependencies for training the prediction model. Geng et al. [13] proposed a spatial-temporal multi-graph convolution network (ST-MGCN) from three aspects of neighborhood graph, functional similarity graph, and transportation connectivity graph to extract temporal features for traffic prediction. In 2020, Sun et al. [34] divided the urban area into different irregular regions by road network and viewed each region as a node that is associated with time-varying inflow and outflow. Auto-ST [22] designs a novel search space tailored for the spatio-temporal domain, which consists of optional convolution operations and learnable skip connections. However, this work neglects the problem of information loss in the temporal dimension and cannot efficiently integrate the content, neighbors, distance, and direction information of nodes in the spatial dimension.

3   Methodology

Figure 2 illustrates the framework of FGFTG. It consists of two feature extractors and a data completer, where the latter provides data support for the former. Specifically, the traffic spatial-temporal flow graph containing missing data go through the DCGAN model to fill in the incomplete parts. In this way, we gain the traffic flow graph without missing values at T-2, T-1, T+1, and T+2. Then, feature extractors regard these graphs as input, utilizing a spatial feature extractor and a temporal feature extractor to gain fine-grained representations. At last, we can predict the traffic flow graph at time T.

3.1   Feature Extractors

Current approaches for traffic flow prediction are inadequate for extracting spatial features and temporal features. Traditional methods [13,23,43,47] using GCN to extract spatial features fail to account for the distance and direction information between nodes. Likewise, temporal extraction methods based on RNN and GCN also have some problems, such as accumulation of errors,

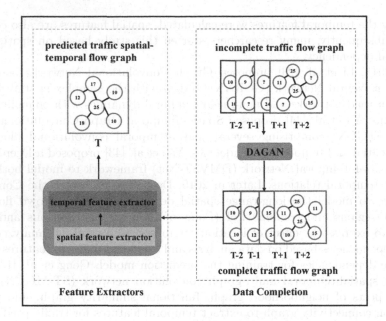

Fig. 2. The framework of FGFTG.

one-way learning occlusion, and limited GCN field of view, which fail to consider bi-relations among historical time intervals. To fill the gaps in existing research, we propose a feature learning method as shown in Fig. 3. From left to right, each red box represents the traffic spatial-temporal flow graph at different time nodes coming from DCGAN. The yellow part is spatial feature extracting layers, which output vectors with spatial information. The blue part is the temporal feature extracting layer composed of n-layer temporal feature extractors and output vectors with temporal information. Finally, feature vectors are mapped to the predicted flow distribution graph through a fully connected layer. Our model can effectively predict traffic flow data based on the spatial feature extractor and the temporal feature extractor.

Spatial Feature Extractor

In this part, we further utilize GCN [7] and node2vec [15] to obtain spatial features, including the content of spatial nodes, neighbour dependency, distance, and direction. GCN obtains content and neighbor features by combining local graph structure and node features. The inputs are composed of a traffic flow data matrix and neighbor information matrix. According to [18], we consider a multi-layer GCN with the following layer-wise propagation rule:

$$H^{(l+1)} = \sigma\left(\hat{D}^{-\frac{1}{2}}\hat{A}\hat{D}^{-\frac{1}{2}}H^l W^l\right) \tag{1}$$

where $\hat{D}$ is the diagonal matrix of traffic flow data, $\hat{A}$ is the adjacency matrix of neighbour information, W^l denotes the trainable weight matrix of layer l, H^l is the matrix of activations in the lth layer, and σ denotes an activation function.

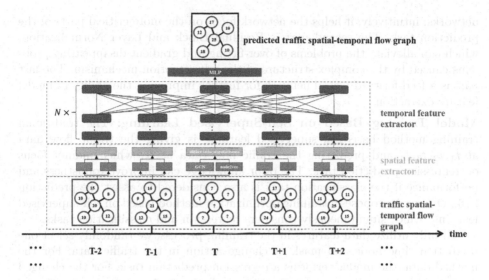

Fig. 3. The framework of feature extractors for traffic flow prediction.

Then, we utilize the node2vec method to obtain distance and direction information by depth-first search and breadth-first search. Finally, we concatenate the two feature vectors, which simultaneously combine the content information, neighbors, distance, and direction. We regard them as the spatial feature, which is the input of the temporal extraction layer.

Temporal Feature Extractor: In this part, we select transformer encoder [35] as temporal feature extractor. Compared with CNN, we expand its view field during the whole training phase. Compared with RNN, our model can avoid dependence on time sequence and considers bidirectional learning. The whole temporal extraction part is composed of multi-layer temporal feature extractors. Figure 4 demonstrates one layer of the extractor. The first part is Multi-head Attention Networks, which consists of multiple self-attention mechanism

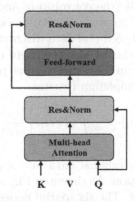

Fig. 4. One layer of temporal feature extractor.

networks; intuitively, it helps the network focus on the more critical parts of the prediction task. The second part is Residual Block and Layer Normalization, which can alleviate the problems of over-fitting and gradient disappearing problems caused by the complex structure of the self-attention mechanism. The last part is a feed-forward neural network for further improving the ability of model feature extraction.

Model Training Based on Self-Supervised Learning: The traditional training method based on supervised learning is giving the traffic flow data at t_1, t_2, t_3, t_4 and predicting the traffic flow data at t_5, which mainly focus on regression prediction task. The prior knowledge can improve robustness and performance if the classification task is also considered to assist the prediction task. Our work proposes a training optimization method based on self-supervised learning, which simultaneously considers regression and classification tasks.

To fully utilize data features in the training process, we randomly select one prediction time node and mask or change action in the traffic data. For the masked data, the model performs a regression prediction task. For the changed data, the model performs a classification prediction task which aims to classify and judge whether the data is changed. Compared with the traditional training method, our method holds several advantages. Firstly, it can observe the complete traffic flow data in the training process, which is beneficial for the prediction task. Secondly, the classification sub-task can enhance model robustness and better judge the reliability of predicted traffic data. Thirdly, we make a regression sub-task for the masked data, ensuring the final traffic flow prediction precision.

3.2 Data Completer

Traditional data completion methods such as random fill or average fill are too brutal to fit the real data distributions. To this end, we propose a data completer DCGAN based on the idea of GAN to fill in the missing data. It consists of a generator and a discriminator, where the generator tries to generate enhanced data similar to real data distribution. The discriminator aims to judge the authenticity of generated data. Specifically, the generator is constructed by the full connected and deconvolution layers, while the convolution and full connected layers construct the discriminator. We introduce the Kullback-Leibler Divergence (KLD) to evaluate the divergence between real data distribution and generate data distribution. Equation 2 defines the KL divergence of distribution $p(x)$ relative to distribution $q(x)$. Our model can be split into three parts: Convert Data Format, Model Training, and Data Completion Process.

$$\mathrm{KL}(\mathrm{p}\|\mathrm{q}) = -\int p(x) \ln \left[\frac{q(x)}{p(x)}\right] dx \tag{2}$$

Convert Data Format

In this part, we transform the original data into a spatial-temporal traffic flow matrix for subsequent calculation. As shown in Fig. 5, we firstly number the spatial node in a single time point. The six spatial nodes in the figure are numbered

1–6. Its corresponding traffic flow data are [15,10,20,10,8,12], which are regarded as a matrix column. Moreover, each spatial node contains a period of temporal nodes. For example, spatial node 1 has seven temporal nodes whose traffic data are [15,13,12,10,9,7,5]. We take these temporal node traffic data as a row of this matrix. After constructing the matrix, the value "0" of the matrix is considered as the missing data.

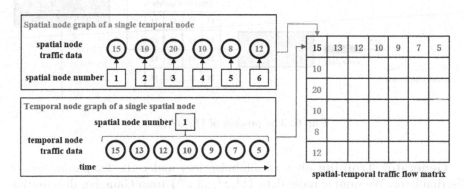

Fig. 5. The transformation of spatial-temporal traffic flow data in matrix form.

Model Training: In the training process as shown in Fig. 6, the generator produces enhanced data based on Gaussian distribution, and the discriminator judges their authenticity. Both of them train and optimize parameters respectively according to the loss function as Eq. 3:

$$\min_{G} \max_{D} V(G, D) = E_{x \sim p_{\text{data}}} [D(x)] + E_{x \sim p_G}[1 - D(G(z))] \tag{3}$$

where x denotes real traffic data, z denotes the noise data generated by Gaussian distribution, $G(\cdot)$ is a mapping function of the generator, and $D(\cdot)$ is the neural network function of the discriminator. The specific steps are as follows.

Discriminator Training aims to maximize loss under the fixed parameters of the generator. Initially, we sample $\{x^1, x^2, ..., x^n\}$ from the real traffic flow data and obtain enhanced traffic flow data $\{\hat{x}^1, \hat{x}^2, ..., \hat{x}^m\}$ created by generator. Then, we take gradient ascent to optimize parameters θ_D to maximize the loss function according to the Eq. 4:

$$\max_{D} V = \frac{1}{m} \sum_{i-1}^{m} D\left(x^i\right) + \frac{1}{m} \sum_{i-1}^{m} \left(1 - D\left(\tilde{x}^i\right)\right),$$

$$\theta_D = \theta_D + \mu \nabla V\left(\theta_D\right). \tag{4}$$

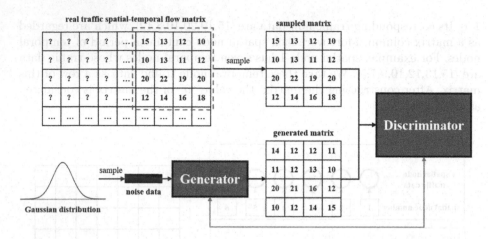

Fig. 6. The process of DCGAN.

Generator Training aims to minimize loss under the fix parameters of discriminator. We sample noise data $\{z^1, z^2, ..., z^m\}$ from Gaussian distribution and take gradient descent to optimize parameters θ_G so as to minimize the loss function according to the Eq. 5:

$$\min_{G} V = \frac{1}{m} \sum_{i-1}^{m} \left(1 - D\left(G\left(z^i\right)\right)\right),$$
$$\theta_G = \theta_G - \mu \nabla V\left(\theta_G\right). \tag{5}$$

Data Completion Process

We use the trained generator for the data completion process. Firstly, we locate missing data and construct a real traffic data matrix by its neighbor data. Then, we randomly sample several noise data based on Gaussian distribution and transform them into multiple traffic spatial-temporal flow metrics by the generator. Finally, we calculate the Euclidean distance between the generated metric and the real matrix (except the missing items) and choose the highest similarity matrix to fill in the missing value.

4   Experiments and Analysis

4.1   Experimental Settings

Datasets. Two datasets showing in Table 1 are used in our experiments.

- **PeMS** was collected from Caltrans Performance Measurement System in real-time, which records 59 days of traffic flow data from January to February in 2018 at important sites of California Highway in the United States. There are 307 sensors in the system, corresponding to 307 spatial nodes. The sensors

Table 1. The information of dataset.

DataSets	Spatial node	Temporal node	Spatial edge	Traffic flow Max/Avg	Record days
PeMS	307	16,992	341	336/186	59
XIAMEN	95	17,856	296	254/112	62

are connected with 341 roads, corresponding to 341 spatial edges. Every five minutes is regarded as a time node, and every day corresponds to 288 time nodes. The whole dataset corresponds to 16,992 time nodes in total.

- **XIAMEN** records 62 days of traffic flow data in Xiamen, China, from July to August in 2016. There are 95 sensors corresponding to 95 spatial nodes. The sensors are connected with 296 roads, corresponding to 296 spatial edges. Every five minutes is regarded as a time node, and every day corresponds to 288 time nodes. The whole dataset corresponds to 17,856 time nodes in total.

In particular, we sort the data by time order, and the first 60% is the training set, 20% are used for testing, and the remaining 20% for validation.

Evaluation Metric: We use Rooted Mean Square Error (RMSE) and Mean Absolute Errors (MAE) to evaluate the proposed model, which are defined as Eq. 6 and Eq. 7:

$$RSME = \sqrt{\frac{\sum_{i=1}^{n}(\hat{y}_i - y_i)^2}{n}} \tag{6}$$

$$MAE = \frac{\sum_{i=1}^{n}|\hat{y}_i - y_i|}{n} \tag{7}$$

where y_i and $\hat{y}_i$ respectively denote the real value and prediction value at the ith time interval. n is the total number of samples in the testing data.

4.2   Performance of Traffic Flow Prediction

Methods for Comparison: We compare our method FGFTG with the following 8 baselines:

- **Linear regression (LR):** We compare our method with different versions of linear regression methods: Ridge Regression (i.e., with L2-norm regularization) and Lasso (i.e., with L1-norm regularization).
- **History Average (HA):** Historical average predicts the future traffic flow data using average values of historical records.
- **Vector Auto Regression (VAR)** [2]: VAR can capture the pairwise relationships among all flows but has massive computational costs due to a large number of parameters.

- **Autoregressive Moving Average Model (ARIMA)** [36]: ARIMA is a combination of Auto Regression (AR) and Moving Average (MA) with a different process.
- **XGBoost** [6]: XGBoost is a mighty boosting tree-based method and is widely used in data mining applications.
- **STGCN** [44]: STGCN is a universal framework for processing structured time series. It is able to tackle traffic network modeling and predicting issues and be applied to more general spatio-temporal sequence learning tasks.
- **MSTGCN** [11]: Each module of MSTGCN uses the GCN model to extract spatial features. It uses a one-dimensional convolution method to extract temporal features to capture the spatial-temporal correlation of traffic data effectively.
- **GMAN** [49]: GMAN uses an encoder-decoder structure to simulate the influence of spatial-temporal factors on traffic conditions. The encoder encodes the input traffic characteristics. The decoder converts the encoded traffic characteristics into the traffic feature vector and utilizes the traffic feature vector to predict the output sequence.

Performance Comparison

Table 2 shows the performances of eight baselines and our model on two datasets with missing values. The results show that our FGFTG method performs the best in terms of all measurements on both datasets. The deep learning models achieve a better performance than the traditional model and tree model. On the PeMS dataset where the missing value accounts for 11.3%, compared with the worst deep learning model STGCN, even the best traditional XGBoost model still holds higher RSME and MAE (an increase of 42.0% and 130.7%, respectively). This result powerfully demonstrates the necessity of studying the deep learning model. Compared with deep learning models, our model FGFTG still achieves better performance.

To verify the effectiveness of the FGFTG structure, we firstly use the traditional training method without Self-Supervised Learning $FGFTG^{withoutSSL}$ to compare with the best deep learning method GMAN. Results show our model achieves a 6.9% and 3.8% lower RMSE and MAE, respectively, indicating that our model performs better prediction performance. Later, we add Self-Supervised Learning in the training process to further improve our performance. Compared with GMAN, FGFTG reduces the RMSE and MAE by 11.3% and 6.7%. Moreover, similar results can be seen on the XIAMEN dataset. Consequently, the results demonstrate that our FGFTG has good generalization performance on flow prediction tasks.

Time Comparison: Table 3 shows the running time (measured by the second) of different baselines on PeMS dataset. To make the comparison fairly, all the experiments are conducted on the same machine with a 10-core 20-thread CPU (Xeon E5-2630 v4, 2.20GHz) 128G RAM. We can easily observe that Ridge and Lasso regression take the shortest running time but present the worst performance on precision. HA, ARIMA and FGFTG differ by running time of 1–2 seconds, but the prediction precision differs by 2–4 times. Due to the complex

Table 2. The comparison of model prediction effects.

Method	PeMS		XIAMEN	
	RMSE	MAE	RMSE	MAE
Ridge	91.52	62.82	66.51	45.47
Lasso	90.49	64.71	64.22	43.16
HA	54.14	36.76	44.03	29.52
ARIMA	68.13	32.11	43.30	24.04
VAR	51.73	33.76	31.21	21.41
XGBoost	34.41	22.75	26.55	18.20
STGCN	24.23	9.86	16.40	7.44
MSTGCN	22.87	9.67	17.01	7.13
GMAN	21.83	8.43	15.84	7.28
$FGFTG^{withoutSSL}$	20.33	8.11	15.23	6.87
FGFTG	**19.36**	**7.86**	**14.71**	**6.44**
Improvement	11.31%	2.35%	7.13%	9.68%

structural design, VAR and XGBoost hold extensive calculations and take more running time. As deep learning methods, STGCN, MSTGCN, and GMAN perform worse than FGFTG, whether in prediction precision or time-consuming. Generally speaking, our proposed FGFTG model achieves the best performance. The same results can be seen on the XIAMEN dataset.

Table 3. The comparison of time consumption.

Method	Cost_time (seconds)	
	PeMS	XIAMEN
Ridge	1.51	0.93
Lasso	1.32	0.75
HA	7.45	6.86
ARIMA	6.45	5.65
VAR	21.45	18.65
XGBoost	34.41	26.26
STGCN	34.26	9.75
MSTGCN	10.88	9.11
GMAN	13.10	10.65
$FGFTG^{withoutSSL}$	8.92	8.62
FGFTG	8.90	8.61

4.3   Effect of Data Completer

To further explore the influence of data missing, we set different scenarios to further show their impact on traffic flow prediction. There are three machine learning algorithms been compared: Ridge, XGBoost, and STGCN. It is evident in Fig. 7 that the more missing data, the worse the performance of the model.

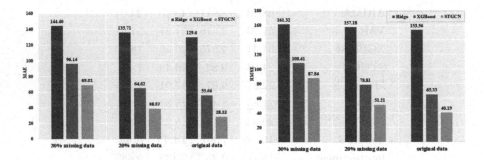

Fig. 7. The effects of missing data in PeMS dataset.

Next, we design two experiments to verify the performance of our proposed data completer: DCGAN.

– **Reliability of Data Generation**: We randomly choose 30% data and set their value to zero, regarding them as missing data. Then, we apply different generation methods to fill the missing data and adopt MAE to evaluate the reliability of different methods.
– **Validity of Promoting Prediction Accuracy**: Different data completion methods are applied to fill the missing data and compare the performance in the spatial-temporal traffic prediction.

Methods for Comparison: We compare our DCGAN with the following five baselines:

– **Randomly fill**: Replace missing data with random values taken from the training set.
– **Average fill**: Replace missing data with average values taken from the training set.
– **Moving average fill**: Replace missing data with average values taken from the neighbor of missing data.
– **Matrix Factorization (MF)** [20]: Decompose the user-item interaction matrix into the product of two lower dimensionality rectangular matrices and replace the missing data with the help of the decomposed matrices.
– **Singular Value Decomposition (SVD)** [19]: Decompose the user-item interaction matrix into the product of three matrices: two lower dimensionality rectangular matrices and one non-negative real diagonal matrix.

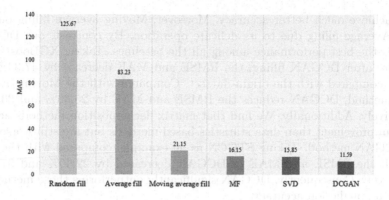

Fig. 8. The results of reliable verification experiment.

Reliability of Data Generation: We use the generated data by five baselines to fill in the missing values and calculate the MAE value. As shown in Fig. 8, in the existing methods, the reliability of generated data by random filling and average filling is larger than moving average filling. Furthermore, the MAE value of DCGAN is 45.20% lower than moving average filling. Compared with methods based on matrix decomposition, DCGAN also outperforms the MF and SVD methods by 28.23% and 26.88%. Results show that our DCGAN outperforms all baseline and is more reliable.

Validity of Promoting Prediction Accuracy: We take five different methods to fill in missing data and regard the results as inputs of three prediction models to promote prediction accuracy. Table 4 shows the performance of the proposed method as compared to all other competing methods. Results show that not every data filling method is helpful for traffic flow prediction. Such as the Random filling method, the RMSE, and MAE values increased after filling operation, indicating the instability of this method. Average filling and Moving average

Table 4. The result of valid promoting prediction accuracy.

Method	RMSE			MAE		
	Ridge	XGBoost	STGCN	Ridge	XGBoost	STGCN
Origin DataSet	153.56	65.33	40.19	129.60	55.66	28.12
Random fill	158.40	67.23	50.18	134.83	59.41	31.62
Average fill	144.29	58.66	39.46	100.98	46.23	26.64
Moving average fill	112.15	46.12	38.42	91.26	34.28	22.32
MF	115.31	40.23	32.62	78.41	32.16	18.62
SVD	101.41	41.48	31.26	76.45	30.51	14.46
DCGAN	**91.52**	**34.41**	**24.23**	**62.82**	**22.75**	**9.86**
Improvement	9.75%	14.47%	22.49%	17.83%	25.43%	31.81%

filling achieve much better accuracy. Moreover, Moving average filling outperforms Average filling due to its delicate operation. By contrast, our DCGAN achieves the best performance among all the baselines. Taking XGBoost as an example, after DCGAN filling, the RMSE and MAE decrease by 47.33% and 59.13% compared with the origin dataset. Compared with the Moving average filling method, DCGAN reduces the RMSE and MAE by 25.39% and 33.63%, respectively. Additionally, We find that matrix decomposition methods achieve better improvement than data statistics-based methods but are still worse than our DCGAN method. Taking STGCN as an example, compared with the SVD method, the RMSE and MAE of DCGAN decreased by 22.49% and 31.81%, respectively. Consequently, DCGAN significantly outperforms those methods in promoting prediction accuracy.

5   Conclusion and Future Work

In this paper, we proposed FGFTG to learn fine-grained spatial-temporal representation for traffic flow prediction. In particular, we first present the spatial-temporal feature extractors to learn better representations, which can fuse content, neighbor, distance, and direction simultaneously and solve the problem of temporal feature loss. Next, to improve the robustness and integrity of data, we propose a novel data completer DCGAN to fill in missing data. Experiments based on two public datasets demonstrate that the proposed FGFTG can lead to better performance than state-of-the-art models. We plan to convert the origin data into graph nodes formula rather than metric to demonstrate the dependency among spatial nodes better for future work. We also plan to explore more temporal dimension features such as weather, holidays, and other factors.

Acknowledgments. This research was supported by the National Key Research and Development Program of China (2020YFB1712903), the Research Program of Chongqing Technology Innovation and Application Development (CSTC2019jscx-zdztzxX0031 and cstc2020kqjscx-phxm1304), and the Overseas Returnees Innovation and Entrepreneurship Support Program of Chongqing (cx2020097).

References

1. Abadi, A., Rajabioun, T., Ioannou, P.A.: Traffic flow prediction for road transportation networks with limited traffic data. IEEE Trans. Intell. Transp. Syst **16**(2), 653–662 (2015)
2. Ang, A., Piazzesi, M.: A no-arbitrage vector autoregression of term structure dynamics with macroeconomic and latent variables. J. Monet. Econ. **50**(4), 745–787 (1999)
3. Beirão, G., Cabral, J.S.: Understanding attitudes towards public transport and private car: a qualitative study. Transp. policy **14**(6), 478–489 (2007)
4. Cao, L., Ma, K., Cao, B., Fan, J.: Forecasting long-term call traffic based on seasonal dependencies. In: Wang, X., Gao, H., Iqbal, M., Min, G. (eds.) CollaborateCom 2019. LNICST, vol. 292, pp. 231–246. Springer, Cham (2019). https://doi.org/10.1007/978-3-030-30146-0_16

5. Chen, C., Wang, Y., Li, L., Hu, J., Zhang, Z.: The retrieval of intra-day trend and its influence on traffic prediction. Transp. Res. Part C Emerg. Technol **22**, 103–118 (2012)
6. Chen, T., Guestrin, C.: Xgboost: a scalable tree boosting system. In: Proceedings of the 22nd ACM SIGKDD international Conference on Knowledge Discovery and Data Mining, pp. 785–794 (2016)
7. Defferrard, M., Bresson, X., Vandergheynst, P.: Convolutional neural networks on graphs with fast localized spectral filtering. arXiv preprint arXiv:1606.09375 (2016)
8. Deng, D., Shahabi, C., Demiryurek, U., Zhu, L., Yan, L.: Latent space model for road networks to predict time-varying traffic. In: ACM Sigkdd International Conference (2016)
9. Drew, D.R.: Traffic flow theory and control. McGraw-Hill Series in Transportation 316 (1968)
10. Fan, Z., Xuan, S., Shibasaki, R., Adachi, R.: Citymomentum: an online approach for crowd behavior prediction at a citywide level. In: the 2015 ACM International Joint Conference (2015)
11. Feng, N., Guo, S., Song, C., Zhu, Q., Wan, H.: Multi-component spatial-temporal graph convolution networks for traffic flow forecasting. J. Softw **30**(3), 759–769 (2019)
12. Fu, R., Zhang, Z., Li, L.: Using LSTM and GRU neural network methods for traffic flow prediction. In: 2016 31st Youth Academic Annual Conference of Chinese Association of Automation (YAC), pp. 324–328. IEEE (2016)
13. Geng, X., Li, Y., Wang, L., Zhang, L., Liu, Y.: Spatiotemporal multi-graph convolution network for ride-hailing demand forecasting. In: Proceedings of the AAAI Conference on Artificial Intelligence, vol. 33, pp. 3656–3663 (2019)
14. Ghahramani, Z., Jordan, M.I.: Supervised learning from incomplete data via an em approach. In: Advances in Neural Information Processing Systems, pp. 120–127 (1994)
15. Grover, A., Leskovec, J.: node2vec: scalable feature learning for networks. In: The 22nd ACM SIGKDD International Conference (2016)
16. Hoang, M.X., Yu, Z., Singh, A.K.: FCCF: forecasting citywide crowd flows based on big data. In: the 24th ACM SIGSPATIAL International Conference (2016)
17. Kang, D., Lv, Y., Chen, Y.Y.: Short-term traffic flow prediction with LSTM recurrent neural network. In: 2017 IEEE 20th International Conference on Intelligent Transportation Systems (ITSC), pp. 1–6. IEEE (2017)
18. Kipf, T.N., Welling, M.: Semi-supervised classification with graph convolutional networks (2017). https://openreview.net/forum?id=SJU4ayYgl
19. Klema, V., Laub, A.: The singular value decomposition: its computation and some applications. IEEE Trans. Autom. Control **25**(2), 164–176 (1980)
20. Koren, Y., Bell, R., Volinsky, C.: Matrix factorization techniques for recommender systems. Computer **42**(8), 30–37 (2009)
21. Li, L., Li, Y., Li, Z.: Efficient missing data imputing for traffic flow by considering temporal and spatial dependence. Transp. Res. Part C Emerg. Technol. **34**, 108–120 (2013)
22. Li, T., Zhang, J., Bao, K., Liang, Y., Li, Y., Zheng, Y.: Autost: efficient neural architecture search for spatio-temporal prediction. In: Proceedings of the 26th ACM SIGKDD International Conference on Knowledge Discovery & Data Mining, pp. 794–802 (2020)
23. Li, Y., Yu, R., Shahabi, C., Liu, Y.: Diffusion convolutional recurrent neural network: data-driven traffic forecasting (2017)

24. Li, Y., Zheng, Y., Zhang, H., Chen, L.: Traffic prediction in a bike-sharing system. In: Proceedings of the 23rd SIGSPATIAL International Conference on Advances in Geographic Information Systems, pp. 1–10 (2015)
25. Lin, Z., Feng, J., Lu, Z., Li, Y., Jin, D.: Deepstn+: context-aware spatial-temporal neural network for crowd flow prediction in metropolis. In: Proceedings of the AAAI Conference on Artificial Intelligence, vol. 33, pp. 1020–1027 (2019)
26. Lippi, M., Bertini, M., Frasconi, P.: Short-term traffic flow forecasting: an experimental comparison of time-series analysis and supervised learning. IEEE Trans. Intell. Transp. Syst. **14**(2), 871–882 (2013)
27. Lv, Y., Duan, Y., Kang, W., Li, Z., Wang, F.Y.: Traffic flow prediction with big data: a deep learning approach. IEEE Trans. Intell. Transp. Syst. **16**(2), 865–873 (2015)
28. Lv, Y., Duan, Y., Kang, W., Li, Z., Wang, F.Y.: Traffic flow prediction with big data: a deep learning approach. IEEE Trans. Intell. Transp. Syst. **16**(2), 865–873 (2014)
29. Ma, X., Zhuang, D., He, Z., Ma, J., Wang, Y.: Learning traffic as images: a deep convolutional neural network for large-scale transportation network speed prediction. Sensors **17**(4), 818 (2017)
30. Moreira-Matias, L., Gama, J., Ferreira, M., Mendes-Moreira, J.: Predicting taxi-passenger demand using streaming data. IEEE Trans. Intell. Transp. Syst. **14**(3), 1393–1402 (2013)
31. Polson, N.G., Sokolov, V.O.: Deep learning for short-term traffic flow prediction. Transp. Res. Part C Emerg. Technol. **79**, 1–17 (2017)
32. Shekhar, S., Williams, B.M.: Adaptive seasonal time series models for forecasting short-term traffic flow. Transp. Res. Rec. **2024**(1), 116–125 (2007)
33. Silva, R., Kang, S.M., Airoldi, E.M.: Predicting traffic volumes and estimating the effects of shocks in massive transportation systems. Proc. Natl. Acad. Sci. United States Am. **112**(18), 5643–8 (2015)
34. Sun, J., Zhang, J., Li, Q., Yi, X., Zheng, Y.: Predicting citywide crowd flows in irregular regions using multi-view graph convolutional networks. IEEE Trans. Knowl. Data Eng. (99), 1 (2020)
35. Vaswani, A., et al.: Attention is all you need. In: Advances in Neural Information Processing Systems, pp. 5998–6008 (2017)
36. Wei, L., Zheng, Y., Chawla, S., Yuan, J., Xing, X.: Discovering spatio-temporal causal interactions in traffic data streams (2011)
37. Wei, W., Wu, H., Ma, H.: An autoencoder and LSTM-based traffic flow prediction method. Sensors **19**(13), 2946 (2019)
38. Li, X., et al.: Prediction of urban human mobility using large-scale taxi traces and its applications. Front. Comput. Sci. **6**(001), 111–121 (2012)
39. Xu, J., Zhang, Y., Jia, Y., Xing, C.: An efficient traffic prediction model using deep spatial-temporal network. In: Gao, H., Wang, X., Yin, Y., Iqbal, M. (eds.) CollaborateCom 2018. LNICST, vol. 268, pp. 386–399. Springer, Cham (2019). https://doi.org/10.1007/978-3-030-12981-1_27
40. Xu, Y., Kong, Q.J., Klette, R., Liu, Y.: Accurate and interpretable bayesian mars for traffic flow prediction. IEEE Trans. Intell. Transp. Syst **15**(6), 2457–2469 (2014)
41. Xuan, S., Zhang, Q., Sekimoto, Y., Shibasaki, R.: Prediction of human emergency behavior and their mobility following large-scale disaster. ACM (2014)
42. Yang, B., Sun, S., Li, J., Lin, X., Tian, Y.: Traffic flow prediction using LSTM with feature enhancement. Neurocomputing **332**, 320–327 (2019)
43. Yao, H., Fei, W., Ke, J., Tang, X., Ye, J.: Deep multi-view spatial-temporal network for taxi demand prediction (2018)

44. Yu, B., Yin, H., Zhu, Z.: Spatio-temporal graph convolutional networks: a deep learning framework for traffic forecasting (2017)
45. Yu, H.F., Rao, N., Dhillon, I.S.: Temporal regularized matrix factorization for high-dimensional time series prediction. In: NIPS, pp. 847–855 (2016)
46. Yu, R., Li, Y., Shahabi, C., Demiryurek, U., Yan, L.: Deep Learning: a generic approach for extreme condition traffic forecasting. In: Proceedings of the 2017 SIAM International Conference on Data Mining (2017)
47. Zhang, J., Zheng, Y., Qi, D.: Deep spatio-temporal residual networks for citywide crowd flows prediction. In: Thirty-first AAAI conference on artificial intelligence (2017)
48. Zhang, N., Wang, F.Y., Zhu, F., Zhao, D., Tang, S.: Dynacas: computational experiments and decision support for ITS. IEEE Intell. Syst **23**(6), 19–23 (2008)
49. Zheng, C., Fan, X., Wang, C., Qi, J.: Gman: a graph multi-attention network for traffic prediction. In: Proceedings of the AAAI Conference on Artificial Intelligence, vol. 34, pp. 1234–1241 (2020)
50. Zheng, Z., Yang, Y., Liu, J., Dai, H.N., Zhang, Y.: Deep and embedded learning approach for traffic flow prediction in urban informatics. IEEE Trans. Intell. Transp. Syst. **20**(10), 3927–3939 (2019)

Underwater Information Sensing Method Based on Improved Dual-Coupled Duffing Oscillator Under Lévy Noise Description

Hanwen Zhang[1(✉)], Zhen Qin[2,3], and Dajiang Chen[2,3]

[1] School of Automation Engineering, University of Electronic Science
and Technology of China, Chengdu, Sichuan, China
201952060908@std.uestc.edu.cn
[2] Network and Data Security Key Laboratory of Sichuan Province, University
of Electronic Science and Technology of China, Chengdu, Sichuan, China
[3] School of Information and Software Engineering, University of Electronic
Science and Technology of China, Chengdu, Sichuan, China

Abstract. Sensing underwater information has become particularly important to obtain information about the marine environment and target characteristics. At present, most interference models for underwater information sensing tasks under substantial interference choose Gaussian noise models. However, it often contains a strong impact and does not conform to the Gaussian distribution. Moreover, in the current research on the sensing of underwater unknown frequency signals, there are problems that the sensing method cannot sufficiently estimate the parameters of the unknown frequency signal, and the signal-to-noise ratio threshold is too high. An underwater environment sensing method is proposed by using the Lévy noise model to describe the underwater natural environment interference and estimate its parameters, which can better describe the impact characteristics of the underwater environment. Then, the intermittent chaos theory and variable step method are leveraged to improve the existing dual-coupled Duffing oscillator method. The simulation results show that the proposed method can sense weak signals in the background of strong Lévy noise and estimate its frequency, with an estimation error as low as 0.1%. Compared with the original one, the minimum signal-to-noise ratio threshold is reduced by 3.098 dB, and the computational overhead is significantly reduced.

Keywords: Underwater information sensing · Lévy noise · Dual-coupled Duffing oscillator

1 Introduction

With the improvement of modern technology, e.g., Internet of Things (IoT) [1–3] and Artificial Intelligence (AI) [4, 5], people have begun to develop and utilize marine resources [6, 7]. Therefore, domestic and foreign personnel have conducted extensive research on underwater target technology to sense the underwater environment and exploit underwater resources [8, 9]. Nowadays, most of the research chooses the Gaussian noise model for discussion. This is because non-Gaussian noise does not have

H. Gao and X. Wang (Eds.): CollaborateCom 2021, LNICST 406, pp. 156–170, 2021.
https://doi.org/10.1007/978-3-030-92635-9_10

Markov characteristics and is extremely difficult to process in weak signal sensing research [10]. Noise interference in the underwater environment often contains intense pulses, which do not conform to the Gaussian distribution [11–13]. The Lévy distribution is a generalized form of the Gaussian distribution. It has broader applicability than the Gaussian distribution, and it is the only distribution that satisfies the generalized central limit theorem among all distributions. The Lévy noise is a typical non-Gaussian noise with long tails, discontinuous jumps, and infinite separability. It can maintain the natural noise process's generation mechanism and propagation conditions and align with the actual situation. The Levy noise model established by the Lévy distribution can describe many symmetrical or asymmetrical noises with different impulse degrees by controlling the selection of different parameters, which can better describe the impact characteristics of underwater environmental noise interference, is of great significance to the research of underwater information sensing. Plus, in the current research on the sensing of underwater unknown frequency signals, there are problems that the sensing method cannot sufficiently estimate the parameters of the unknown frequency signal, and the signal-to-noise ratio threshold is too high [14–18].

This paper proposes a more in line with the weak signal sensing of underwater position in the complex marine environment to solve problems mentioned above. The main contributions of this article are as follows:

1. Propose the Lévy noise model to describe the interference of the underwater natural environment and provide a method to analyze, estimate and select parameters that are closer to the actual underwater natural environment interference.
2. Aiming at sensing weak underwater signals with unknown frequencies, an improved signal sensing method of dual-coupled Duffing oscillators is proposed, which is more effective and intuitive.
3. Designed and established a sensing system for weak signals of unknown frequencies underwater based on the improved dual-coupled Duffing oscillator under the interference of the Lévy noise model and verified its feasibility and supriority.

2 Related Work

2.1 Lévy Noise Model

The Lévy process was proposed by the French mathematician Paul Lévy to study of the generalized central limit theorem. It is a random process with independent and fixed increments, indicating that the movement of a point and its continuous displacement are random. The difference between two disjoint time intervals displacement is independent. The displacement and displacement in different time intervals of the same length have the same probability distribution. It can be regarded as a continuous-time simulation of random walk. Lévy noise, also called alpha noise, obeys the theory of stable alpha distribution. The only distribution satisfies the generalized central limit theorem, and a square law attenuates its tailing. The expression of the characteristic function of Lévy noise [19] is as follows:

$$\log\varphi(t) = \begin{cases} -\sigma^\alpha|t|\{1 - i\beta\text{sign}(t)\tan\left(\frac{\pi\alpha}{2}\right)\} + i\mu t, \alpha \neq 1 \\ -\sigma^\alpha|t|\{1 + i\beta\text{sign}(t)\frac{\pi}{2}\log(|t|)\} + i\mu t, \alpha = 1 \end{cases} \tag{1}$$

$$\begin{cases} X = S_{\alpha,\beta}\dfrac{\sin\left(\alpha\left(V+B_{\alpha,\beta}\right)\right)}{(\cos V)^{\frac{1}{\alpha}}}\left(\dfrac{\cos\left(V-\alpha\left(V+B_{\alpha,\beta}\right)\right)}{W}\right)^{\frac{1-\alpha}{\alpha}}, \alpha \neq 1 \\ X = \frac{2}{\pi}\left[\left(\frac{\pi}{2}+\beta V\right)\tan V - \beta\log\left(\frac{W\cos V}{\frac{\pi}{2}+\beta V}\right)\right], \alpha = 1 \end{cases} \tag{2}$$

In (1), $\alpha \in [0,2]$ is the characteristic index, which determines the decay rate of the distribution tail. When $\alpha = 1$, it is Cauchy distribution. When $\alpha = 2$, It is the Gaussian distribution, and the mean value is μ, the variance is $2\sigma^2$. when $\alpha \neq 2$, the mean value is μ, but the variance does not exist. $\beta \in [-1,1]$ is the skew parameter. When $\beta = 0$, the graph is symmetrical, and when β is a positive number, The graph tilts to the right, on the contrary, the graph tilts to the left, $\sigma > 0$ is the scale parameter, which determines the degree of dispersion of the distribution concerning μ, and $\mu \in R$ is the position parameter. The left and right translation can be achieved by adjusting the value of μ. Rfal-Weron proved the expression of Lévy distribution random variable. In (2), V obeys the uniform distribution in the interval $(-2\pi, 2\pi)$, W follows the exponential distribution with the mean value 1, $S_{\alpha,\beta}$ and $B_{\alpha,\beta}$ The definition expression is as follows:

$$S_{\alpha,\beta} = \left[1 + \beta^2\tan^2\frac{\pi\alpha}{2}\right]^{1/2\alpha} \tag{3}$$

$$B_{\alpha,\beta} = \frac{\arctan\left(\beta\tan\frac{\pi\alpha}{2}\right)}{\alpha} \tag{4}$$

Under the conditions of $\beta = 0, \sigma = 1$ and $\mu = 0$, the Lévy distributions corresponding to different α feature indices are shown in Fig. 1:

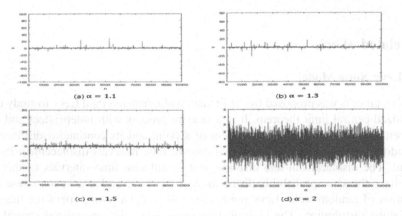

(a) $\alpha = 1.1$ (b) $\alpha = 1.3$

(c) $\alpha = 1.5$ (d) $\alpha = 2$

Fig. 1. Time-domain distribution map of live noise corresponding to different feature index α

It can be seen from Fig. 1 that the smaller the value α, the stronger the impact of noise interference. When $\alpha = 2$, the noise interference has almost no impact. At this time, the Lévy noise degenerates into white noise.

2.2 Chaotic Oscillator Signal Sensing System

Single Duffing Vibrator. The nonlinear dynamic system described by Duffing equation exhibits rich nonlinear dynamic characteristics, including complex oscillation dynamics, bifurcation, and chaos. The specific form of a single Duffing oscillator equation is:

$$\ddot{x}(t) + k\dot{x}(t) - x(t) + x^3(t) = F\cos(\omega t) \tag{5}$$

Where $x(t)$ is the chaotic system variable; k is the damping ratio; t is the time variable; $F\cos(\omega t)$ is the periodic driving force, where F is the amplitude of the periodic driving force, and ω is the angular frequency of the periodic driving force; $-x(t) + x^3(t)$ is the nonlinear restoring force. When k is fixed, the system's state changes regularly with the amplitude of the driving force $F\cos(\omega t)$. Measured by the simulation experiment, when F increases to 0.82673, the system enters the critical state of large-scale periodic motion. At this time, the magnitude of the driving force 0.82673 is the critical threshold for the transition from chaos to a periodic state. The signal sensing method using a single Duffing vibrator can effectively sense the signal, but the immunity to noise interference is low. If the interference is too strong, the phase trajectory will remain in a chaotic state, resulting in an illusion that the signal is not sensed, and misjudgment occurs.

Double Coupling Duffing Vibrator. The double coupling Duffing vibrator is an improved Duffing vibrator, and its specific form is:

$$\begin{cases} \ddot{x}(t) + k\dot{x}(t) - x(t) + x^3(t) + d(y(t) - x(t)) = F\cos(\omega t) \\ \ddot{y}(t) + k\dot{y}(t) - y(t) + y^3(t) + d(x(t) - y(t)) = F\cos(\omega t) + Y(t) \end{cases} \tag{6}$$

Where x is the variable of Duffing oscillator one; y is the variable of Duffing oscillator two; d is the coupling coefficient. When $d = 0$, the coupling effect of the two oscillators disappears. At this time, the dynamic behavior of the double coupled Duffing oscillator is the same as that of the single Duffing oscillator. The dynamic behavior is completely consistent; when $d \neq 0$, the variables x and y will quickly become synchronized under the influence of coupling; $Y(t)$ is the external input signal of the system, including the signal to be sensed $As(t)$ and noise signal $\eta(t)$, A is the amplitude of the signal to be sensed, and the frequency of the signal to be sensed and the driving force frequency are independent of each other. The dynamic characteristics of the dual-coupled Duffing vibrator are similar to that of the single Duffing vibrator. It also changes regularly with the driving force $F\cos(\omega t)$ amplitude when the damping ratio is fixed. However, this signal sensing method does not solve the parameter estimation of unknown frequency signals, and there is still a high signal-to-noise ratio threshold.

3 Approach

3.1 Lévy Noise Model Describes Underwater Natural Environment Interference

The Fokker-Plank equation corresponding to Eq. (1) is:

$$\frac{\partial_p(s,t)}{\partial_t} = \left[\frac{\partial}{\partial_x}A(x) + \frac{\partial^2}{\partial_x}B(x)\right]\rho(s,t) \tag{7}$$

where $A(x) = ax - bx^3 + S(x)$, $B(x) = D$. Since Eq. (7) is a transcendental equation, it cannot be solved directly, but the approximate number of Eq. (1) can be calculated using the finite difference method. Set the system parameters $a = 0.6, b = 0.3, A = 0.3, f = 0.005$, noise parameters $\alpha = 1.5, \beta = 0, \sigma = 1, \mu = 0$, the probability distribution curve of particle density is shown in Fig. 2.

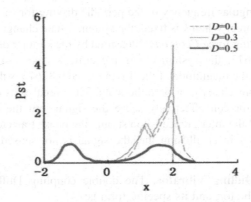

Fig. 2. The particle probability distribution of the system output under the excitation of different noise intensity D

When the Lévy noise intensity is 0, the probability distribution of particle density is only affected by the signal. When the noise intensity D increases from 0.1 to 0.3, the particle density probability shows that the particles are mainly concentrated on the side of the 0 points. When the noise intensity D reaches 0.5, the particles are unevenly distributed on both sides of 0. From the microscopic particle motion law perspective, when the noise intensity D is 0.3, the external noise excitation intensity is not enough. The energy obtained by the particle is not enough to make it cross the potential barrier and can only move left and right in a potential well, so the probability of the particle. The distribution is concentrated on the 0 sides. The resultant force of the signal pull and noise interference that the particle receives at 0 determines whether its distribution is in the left or right potential well. When the noise intensity is 0.5, the particles will be excited by solid noise. They will get enough energy to cross the potential barrier from one potential well into another potential well. At this time, the transfer of particles between potential wells will be affected by noise. Therefore, the distribution of particles on both sides of the 0 point will never be symmetrical. From Fig. 2, we can get a piece of information. When the characteristic index is constant, the sufficient external excitation energy the particles can obtain, the more they can move across the barrier to another potential well. That is to say, the greater the noise intensity, the more the particles are in between the potential wells. The higher the crossover frequency, so when we select the parameters of the Lévy noise model, we need to pay special attention to the selection of the feature index α and the noise intensity D.

In this regard, this paper quotes the method of literature [21] to estimate the characteristic index α and the noise intensity D of the Lévy noise model:

$$E(|X|^{\rho}) = \frac{\rho lg\alpha}{\alpha lgD}C(\rho,\alpha) \tag{8}$$

Where $E(|X|^{\rho})$ is the fractional low-order moment, and ρ is the order, $C(\rho,\alpha) = \frac{2^{\rho+1}\Gamma((\rho+1)/2)\Gamma(-\rho/\alpha)}{\alpha\sqrt{\pi}\Gamma(-\rho/2)}$, $-1 < \rho < \alpha \leq 2$.

Let $Y = lg|X|$ $E(Y) < +\infty$, the moment generating function is:

$$E(|X|^{\rho}) = \lim_{\rho \to 0} \frac{d^q}{d\rho^q} E(|X|^{\rho}), q \in N^* \tag{9}$$

Since Y is only related to α except for the first moment, the first two finite logarithmic moments are listed as:

$$\begin{cases} G_1 = E(|X|^{\rho}) = \Phi_0\left(1 - \frac{1}{\alpha}\right) + lg|\frac{lg\alpha}{lgDcosk}|^{\frac{1}{\alpha}} \\ G_2 = E\left[(lg|X| - E(Y))^2\right] = \Phi_1\left(\frac{1}{2} + \frac{1}{\alpha^2}\right) - \left(\frac{k}{\alpha}\right)^2 \\ \Phi_0 = \frac{dlg\Gamma(t)}{dt} \\ \Phi_1 = \frac{d^2lg\Gamma(t)}{dt^2} \end{cases} \tag{10}$$

Let t = 1, then $\Phi_0 = -0.5772$ and $\Phi_1 = 1.6449$ are obtained. The estimated value can be obtained from the above formula:

$$\begin{cases} \widehat{\alpha} = \left(\frac{G_2}{\Phi_1} - \frac{1}{2}\right)^{-\frac{1}{2}} \\ \widehat{D} = e^{(\Phi_0 - G_1)\widehat{\alpha} - 1} \frac{lg\widehat{\alpha}}{cos\widehat{k}} \\ |\widehat{k}| = [\left(\frac{\widehat{\alpha}^2}{2} - 1\right)\Phi_1 - G_1\widehat{\alpha}^2]^{\frac{1}{2}} \end{cases} \tag{11}$$

To test the effectiveness of the estimation method, the Chambers-Mallows-Stuck (CMS) method is used to generate the Levy noise with $\alpha = 1.5, D = 1$, and the parameters of α, D are estimated. The estimated results are shown in the figure below (Fig. 3):

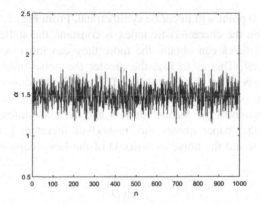

Fig. 3. Estimation of characteristic index α

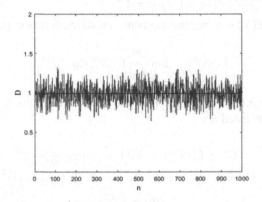

Fig. 4. Estimation of noise intensity D

Where n represents the number of estimates, Fig. 4 and Fig. 5 show that values of α and D are obtained respectively as 1.5026 and 1.1664, which prove that the method can estimate the parameters of interference noise in the actual underwater information sensing environment.

3.2   Improved Signal Sensing Method of Dual Coupling Duffing Oscillator

Although the signal sensing method of the dual-coupled Duffing oscillator improves the speed and accuracy of specific signal-to-noise ratios and threshold solutions, it still has the problem of not being able to estimate the unknown frequency signal parameters, and the signal-to-noise ratio threshold is too high. This paper proposes an improved dual-coupled Duffing oscillator signal sensing method by combining the theoretical knowledge of intermittent chaos and the variable step-size method. The estimated result of the sensed unknown signal frequency is more accurate.

According to the idea of intermittent chaos theory, there is a slight frequency difference between the sensed signal and the built-in driving force of the system, which can make the sensing system produce a chaotic state and periodic state regular changes. Furthermore, the chaotic oscillator has a solid sensitivity to interference and robust noise immunity. So, this paper proposes to sense the signal through the phase change of the chaotic oscillator. The expression of the signal sensing method of the improved dual-coupled Duffing oscillator is as follows:

$$
\begin{cases}
\ddot{x}_1(t) + k\dot{x}_1(t) - x_1(t) + x_1{}^3(t) + d(y_1(t) - x_1(t)) = Fcos(\omega t) \\
\ddot{y}_1(t) + k\dot{y}_1(t) - y_1(t) + y_1{}^3(t) + d(x_1(t) - y_1(t)) = Fcos(\omega t) + Y(t) \\
\ddot{x}_2(t) + k\dot{x}_2(t) - x_2(t) + x_2{}^3(t) + d(y_2(t) - x_2(t)) = \xi \cdot Fcos(\omega t) \\
\ddot{y}_2(t) + k\dot{y}_2(t) - y_2(t) + y_2{}^3(t) + d(x_2(t) - y_2(t)) = \xi \cdot Fcos(\omega t) + Y(t)
\end{cases}
\tag{12}
$$

Where ξ is the influencing parameter of the chaotic oscillator. $Y(t) = Acos(\omega_1 t)$. The other parameters of the two pairs of Duffing oscillators are the same. We generate the differential timing diagrams $x_1(t) - x_2(t)$ of the two pairs of Duffing oscillators to observe whether the signal has been sensed. The timing diagram should be regular. The change of ξ will not affect the waveform but only the magnitude of the phase difference. When $\xi = 1$, the phase difference disappears, and the two pairs of oscillators are the same. The method will degenerate to for ordinary dual-coupled Duffing vibrators. This article has passed many experiments and finally chose $\xi = 1.001$ to obtain a more intuitive effect.

If $f(t)$ is added as the signal to be sensed to the single Duffing vibrator sensing system, the state equation is:

$$
\begin{cases}
\dot{x} = \omega y \\
\dot{y} = \omega[-0.5y + x - x^3 + \gamma cos(\omega t) + f(t)]
\end{cases}
\tag{13}
$$

Where $f(t) = Acos(\dot{\omega}t + \varphi)$, φ is the initial phase of the signal to be sensed, γ is the critical chaos threshold, $\dot{\omega} = \omega + \Delta\omega$, $\Delta\omega$ is the frequency difference between the built-in driving force of the system and the signal to be sensed. The system power is:

$$
L(t) = \gamma cos(\omega t) + Acos(\dot{\omega}t + \varphi) = P(t)cos(\omega t + \phi(t))
\tag{14}
$$

In the formula (14), P(t) represents the polarization force amplitude:

$$
P(t) = \sqrt{\gamma^2 + 2\gamma Acos(\Delta\omega t + \phi) + A^2}, A \ll \gamma
\tag{15}
$$

When P(t) changes periodically, if $P(t) \geq \gamma$, the system is in a large-scale periodic state, and when $P(t) < \gamma$, the system is in a chaotic state.

In this paper, the Runge-Kutta method is used to analyze the Duffing vibrator sensing system numerically. Through calculation, it is obtained that when the intermittent chaotic frequency difference range of the single Duffing vibrator sensing system is $\left|\Delta\omega/_\omega\right| \leq 0.03$, the system is in a state of intermittent chaos. The dual-coupled Duffing oscillator system is $\left|\Delta\omega/_\omega\right| < 0.08$ and the improved dual-coupled Duffing oscillator's intermittent chaotic frequency difference range is $\left|\Delta\omega/_\omega\right| < 0.09$. Compared the intermittent chaotic frequency difference range of the three methods, we can see that the method proposed in this paper detect the step size accurately, which makes it meaningful to use the variable step-size method to improve the sensing system further, and it is easier to realize the unknown frequencies. The intermittent chaotic timing diagram of the dual-coupled Duffing oscillator and the method proposed in this paper is shown in Fig. 5:

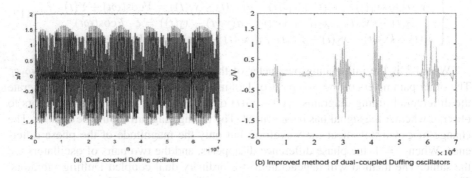

(a) Dual-coupled Duffing oscillator   (b) Improved method of dual-coupled Duffing oscillators

Fig. 5. The intermittent chaotic timing diagram of two methods

Next, we add a variable step-size method based on the above improvements, change the solution step-size, convert the signal into the corresponding discrete sequence of the built-in driving force of the sensing system, and observe the output $x_1(t) - x_2(t)$ Differential timing diagram to determine whether the signal is successfully sensed. The built-in driving force sequence interval of the sensing system is the solution step length of the system, and the sequence interval of the signal to be sensed is T_s ($T_s = \frac{1}{f_s}$, f_s is the sampling rate of the signal). The sensing result is only related to the sampling rate, so the solution step-size of the sensing system can be changed, it can be directly adjusted to the intermittent chaotic state to complete signal sensing. When the variable step intermittent chaos sensing method is applied, the total strategy terms of Eq. (12) are $F\cos(\omega t) + A\cos(\omega_1 t)$ and $\xi \cdot F\cos(\omega t) + A\cos(\omega_1 t)$, and the system is in In the intermittent chaotic state, the solution step-size is $h = \frac{\dot{\omega}}{lwf_s}, \dot{\omega} = \omega + \Delta\omega, l \in (0.94, 1.06)$. The two strategy items are discretized as:

$$
\begin{cases}
L_{n_1} = D \cdot \cos\left(\frac{n\dot{\omega}}{lf_s}\right) + A\cos\left(\frac{n\dot{\omega}}{f_s}\right) \\
L_{n_2} = \xi \cdot D\cos\left(\frac{n\dot{\omega}}{lf_s}\right) + A\cos\left(\frac{n\dot{\omega}}{f_s}\right)
\end{cases}
\tag{16}
$$

If $n \in N^*$, the frequency difference between the built-in driving force and the signal meets the standard of intermittent chaos, the sensing system will appear in the intermittent chaotic state. The specific steps are shown in the figure below:

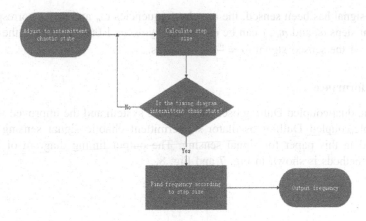

Fig. 6. The specific steps of the sensing system

4   Experiment and Analysis

4.1   Experiment Deployment

To more effectively describe the natural environment interference encountered by underwater information sensing in the real ocean environment, according to the actual underwater data of a reservoir in Hubei, China, measured in the literature [20], through simulation experiments, it is found that when the parameter $\alpha = 1.5, D = 1$, it is closer to the actual situation. The selection of parameters in this paper is determined accordingly.

This paper uses the Chambers-Mallows-Stuck (CMS) method to generate a Lévy noise model with $\alpha = 1.5, D = 1$, and build an intermittent chaotic signal sensing system based on an improved variable-step double-coupling Duffing oscillator under Lévy noise interference. Simulation proves the effectiveness and superiority of the sensing system.

After generating the Lévy noise model with $\alpha = 1.5, D = 1$, take $a_n = 2\pi \frac{1.06^n}{1.06 \omega f_s}$ as the solution step-size of the sensing system, and take $A cos(\omega_1 t) + \eta(t)$ as input signal, $\eta(t)$ is the generated Levy noise. According to Fig. 6, the specific steps of the intermittent chaotic signal sensing system of the improved variable-step double-coupled Duffing oscillator established in this paper are as follows:

Adjust parameters. The intermittent chaotic signal sensing system of the variable-step double-coupled Duffing oscillator sets $F = 0.789, \omega_1 = 1 rad/s, \alpha = 1.001$, $d = 0.2$.

Take $A cos(\omega_1 t) + \eta(t)$ as the input signal into the sensing system, the frequency is set to 1 kHz, and the initial solution step is set to $a_n = 2\pi \frac{1.06^n}{1.06 \omega f_s}$, where $n \in N^*$.

Adjust the solution step size. If the sensing system has an intermittent chaotic state between two adjacent steps a_n and a_{n+1}, then indicates that the signal has been sensed, otherwise go back to the previous step.

If the signal has been sensed, the angular frequencies ω_n and ω_{n+1} corresponding to adjacent steps a_n and a_{n+1} can be calculated by $\omega_n = 1.06^n rad/s$, then the angular frequency of the sensed signal $\ddot{\omega} = \frac{\omega_n + \omega_{n+1}}{2}$ rad/s.

4.2   Performance

We use the dual-coupled Duffing oscillator sensing system and the improved variable-step double-coupled Duffing oscillator's intermittent chaotic signal sensing system established in this paper for signal sensing. The output timing diagram of the two detection methods is shown in Fig. 7 and Fig. 8.

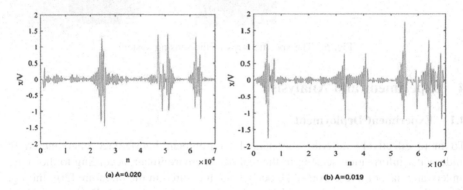

Fig. 7. The output timing diagram of the dual-coupled Duffing oscillator sensing system

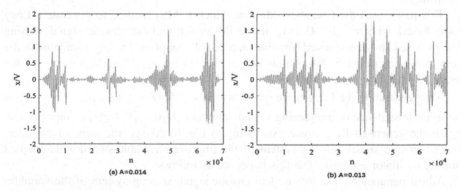

Fig. 8. The output timing diagram of the improved dual-coupled Duffing oscillator sensing system

Figure 7 shows that the dual-coupled Duffing oscillator sensing system appears intermittently chaotic when the adjacent amplitudes A = 0.019 and A = 0.020, indicating that this method can snese periodic signal amplitudes A > 0.020. Figure 8 shows the improved variable step-size double-coupled Duffing oscillator's intermittent chaotic signal sensing system established in this paper when the adjacent amplitudes

A = 0.013 and A = 0.014 appear intermittent chaotic state, indicating that the method can sense periodic signals Amplitude A > 0.014. It shows that the sensing system established in this paper has a broader and more accurate sensing range than the sensing system established in the original method, and comparing Figs. 7 and 8, it can be seen that the output timing diagram of the sensing system established in this paper is clearer and more convenient to observe. The following formula can calculate the minimum signal-to-noise ratio of the two sensing systems:

$$\text{SNR} = 10\frac{A^2}{2\sigma^2} \tag{17}$$

After calculating separately, the SNR of the dual-coupled Duffing vibrator sensing system is −36.98970004 dB. The SNR of the sensing system established in this paper is −40.08773924 dB. It can be seen that the sensing system established in this paper reduces the signal-to-noise ratio threshold of 3.0980392 dB.

The sensing method proposed in this paper can change the standard ratio, form a new solution step size and sensing bandwidth, reduce the number of solution steps, and reduce the amount of calculation. The data of each sensing system is shown in Table 1:

Table 1. Data of each detection system

Method	Common ratio	Number of solving steps	Detection bandwidth
Single Duffing oscillator	1.01	282	$(1.00, 1.01)\omega$
	1.02	197	$(0.99, 1.02)\omega$
	1.03	98	$(0.98, 1.03)\omega$
Double Duffing oscillator	1.05	68	$(0.97, 1.05)\omega$
	1.06	60	$(0.96, 1.06)\omega$
	1.07	57	$(0.95, 1.07)\omega$
Ours	1.06	60	$(0.96, 1.06)\omega$
	1.07	57	$(0.95, 1.07)\omega$
	1.08	35	$(0.94, 1.08)\omega$

It can be seen from Table 1 that the number of solving steps and the amount of calculation of the sensing system established in this paper are obviously the smallest.

Below we verify that the sensing system established in this article can better estimate the frequency of the unknown signal. In the actual underwater signal sensing, multiple signals will be sensed under the same natural environment interference. In this case, when we change the input signal to $0.01cos(10t) + 0.01\cos(20t) + \eta(t)$, the power spectral density is shown in Fig. 9, and the sensing result is shown in Fig. 10.

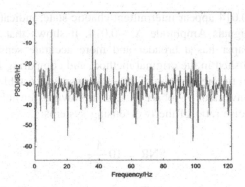

Fig. 9. The power spectral density

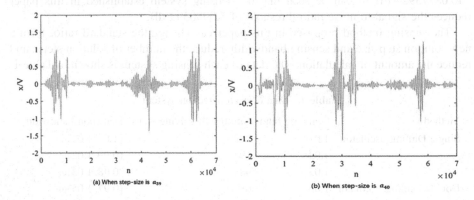

(a) When step-size is a_{39} (b) When step-size is a_{40}

Fig. 10. The sensing results

Figure 10 shows the intermittent chaotic state of the sensing system established in this paper when the adjacent solution steps a = 39 and a = 40. It can be judged that the system senses the signal when the solution steps are 39 and 40, Then can estimate unknown signal frequency. When the signal frequency to be measured is 10 Hz, the adjacent steps are a_{39} and a_{40}, the corresponding sizes of f_{39} and f_{40} are 9.7 Hz and 10.2 Hz. The system's judgment frequency is f_{10} = 9.99 Hz. The calculation error rate is 0.1%; that is, the system can obtain the frequency of the signal to be sensed more accurately under the interference of a healthy natural environment.

5   Conclusion

Aiming at the task of underwater information sensing under solid interference, in this paper, a more general Lévy noise model has been used to better describe the impact characteristics of underwater environmental noise interference in order to describe the natural environment interference of underwater information sensing. Then, an improved dual-coupled Duffing oscillator signal sensing method has been proposed to detect weak underwater signals with unknown frequencies. This method has lower computational overhead, lower signal-to-noise ratio threshold, and can better estimate the unknown signal frequency. Finally, under the Lévy noise interference, an intermittent chaotic signal sensing system based on an improved variable-step double-coupled Duffing oscillator has been established. A large number of simulation experiments have proved the effectiveness and superiority of the system. The results show that, the lowest signal-to-noise ratio threshold can reach −40.0877, and the frequency estimation error rate is 0.1%.

Acknowledgement. This work was supported in part by the National Natural Science Foundation of China (No. 62072074, No. 62076054, No. 62027827, No. 61902054), the Frontier Science and Technology Innovation Projects of National Key R&D Program (No.2019QY1405), the Sichuan Science and Technology Innovation Platform and Talent Plan (No. 2020JDJQ0020), the Sichuan Science and Technology Support Plan (No. 2020YFSY0010), and the Natural Science Foundation of Guangdong Province (No. 2018A030313354).

References

1. Chen, D., et al.: MAGLeak: a learning-based side-channel attack for password recognition with multiple sensors in IIoT environment. In: IEEE Transactions on Industrial Informatics, to Appear (2021)
2. Zhang, N., et al.: Software defined networking enabled wireless network virtualization: challenges and solutions. IEEE Netw. **31**(5), 42–49 (2017)
3. Zhang, N., et al.: Synergy of big data and 5 g wireless networks: opportunities, approaches, and challenges. IEEE Wireless Commun. **25**(1), 12–18 (2018)
4. Alc, L., et al.: Online proactive caching in mobile edge computing using bidirectional deep recurrent neural network. IEEE Internet Things J. **6**(3), 5520–5530 (2019)
5. Ding, Y., et al.: DeepEDN: a deep learning-based image encryption and decryption network for internet of medical things. IEEE Internet Things J. **8**(3), 1504–1518 (2021)
6. Tian, T.: Sonar Technology. Harbin Engineering University Press, Harbin (2010)
7. Liu, X., Qin, Y.: Modem marine power vs state marine strategy. J. Soc. Sci. 73–79 (2004)
8. Cui, F.: Modern sonar technology. Fundam. Defense Technol. 30–33 (2005)
9. Liu, B., Lei, J.: Principles of Hydroacoustic. Harbin Shipbuilding Institute Press, Harbin (1993)
10. Andrew, R.K., Howe, B.M., Mercer, J.A., Dzieciuch, M.A.: Ocean ambient sound: comparing the 1960s with the 1990s for a receiver off the California coast. Acoust. Res. Lett. Online **3**(2), 65–70 (2002)
11. Korakas, A., Hovem, J.M.: Comparison of modeling approaches to low-frequency noise propagation in the ocean. In: IEEE Proceedings of Oceans 2013. Norway (2013)
12. Shi, J., Zhang, X., Hou, T.: Development trend of low-frequency marine environmental noise level caused by ship noise. Torpedo Technol. 112–116 (2010)
13. Siderius, M., Gebbie, J.: Environmental information content of ocean ambient noise. J. Acoust. Soc. Am. **146**(3), 1824–1833 (2019)
14. Li, N., Li, X.K., Liu, C.H.: Detection method of a short-time Duffing oscillator array with variable amplitude coefficients. J. Harbin Eng. **37**, 1645–1652 (2016)
15. Zhou, S., Lin, C.S.: Application of chaos theory for weak signal of ship detecting. J. Wuhan Univ. **33**, 161–164 (2009)

16. Li, S.Q., Wu, X.Z.: Application of ALE based on FTF algorithm in ship-radiated noise detection. Commun Technol. **50**(6), 1175–1180 (2017)
17. Sun, Q.W., Zhang, J.F.: Weak signal detection based on improved chaotic oscillator system with dual coupling. Comput. Mod. **3**, 17–21 (2012)
18. Shi, Z., Yang, S., Zhao, Z.: Research on weak signal detection based on Van der Pol-Duffing oscillator and cross correlation. J. Shijiazhuang Tiedao Univ. **32**, 66–71 (2019)
19. Xu, W., Hao, M., Gu, X.: Stochastic resonance induced by Lévy noise in a tumor growth model with periodic treatment. Mod. Phys. Lett. **28**(11), 1450085 (2014)
20. Ma, S.: Research on Very Low Frequency Seismic Wave Detection Technology Based on Stochastic Resonance under Levi Noise. Northwestern Polytechnical University, Xian (2018)
21. Ma, X., Nikias, C.: Parameter estimation and blind channel identification in impulsive signal environments. IEEE Trans. Signal Process. **43**(12), 2884–2897 (1995)

Unpaired Learning of Roadway-Level Traffic Paths from Trajectories

Weixing Jia[1,2], Guiling Wang[1,2(✉)], Xuankai Yang[1,2], and Fengquan Zhang[1,2]

[1] School of Information Science and Technology, North China University of Technology, Beijing 100144, China
wangguiling@ncut.edu.cn
[2] Beijing Key Laboratory on Integration and Analysis of Large-Scale Stream Data, North China University of Technology, Beijing 100144, China

Abstract. Traffic path data can be used as the basis for traffic monitoring and other technologies, which is essential for developing traffic-related technologies. Traditional methods of traffic path data extraction can no longer meet the needs because they cannot solve the problem of lacking standard benchmark data that may exist in the traffic field. Deep learning-based path extraction methods using large-scale data are a class of promising approaches. However, most of the deep learning-based path extraction methods are supervised and rely on paired training data. This paper proposes an unpaired learning method for fine-grained roadway-level paths from trajectory data based on CycleGAN. The method constructs spatio-temporal features based on HSV color space from trajectories which can enhance the model's ability to recognize the roadway details. It transforms the features using convolutional layers, which can preserve the spatio-temporal information of the features, thus making the extraction results more accurate. We conduct experiments using urban and maritime traffic trajectory data and compare the proposed method with the state-of-the-art methods. The results of our model have more roadway level details, higher precision and F1 score than the other existing unsupervised traffic path learning methods.

Keywords: Deep learning · Unpaired learning · Trajectory mining · Roadway-level path extraction

1 Introduction

Traffic paths are the carriers of transportation and traffic, and traffic paths data can be used as the basis for traffic monitoring and other technologies. Therefore, traffic path extraction is essential for the research in the field of transportation. A traffic path is the route of the traffic facility that a moving object follows

H. Gao and X. Wang (Eds.): CollaborateCom 2021, LNICST 406, pp. 171–190, 2021.
https://doi.org/10.1007/978-3-030-92635-9_11

when traveling or the regular path of many moving objects. The fineness level of the traffic path extraction can be divided into three categories: center line level, roadway level and lane level. The traffic paths of center line level are represented by their center lines; the traffic paths of roadway level are represented by the center lines of their roadways, i.e., the sub-paths of different driving directions; the traffic paths of lane level are represented by their lanes. These three traffic path extraction levels have an increasing level of refinement. The more fine-grained of the traffic path extraction, the more traffic monitoring accuracy we can guarantee. Traditional traffic path extraction methods mainly include manual measurement and extraction methods based on high-resolution remote sensing image processing [1–4]. However, not all traffic paths in traffic-related domains are constructed manually, e.g., maritime traffic paths are the optimal navigation paths inducted from the sailing habits of ships and vessels. Therefore, in order to achieve traffic path extraction in traffic fields, the traditional manual measurement methods and remote sensing image processing methods are no longer applicable.

As presented in Sect. 2, many methods have been proposed to extract traffic paths on land or at sea. Among these methods, the most promising is the deep learning-based approaches because they do not rely on many empirical parameters and make fair use of the prior knowledge in the available map data [5–7]. With the development of computer technology, global positioning technology, GPS, and modern communication technology, a large amount of GPS data collected from moving objects provides an opportunity to extract the traffic paths. In maritime traffic, it is also possible to obtain and collect massive amounts of ship trajectory data with the help of the movement status and spatial position tracking capabilities of maritime transport ships. The Automatic Ship Identification System (AIS) is the successful application of these technologies and makes it possible to extract maritime traffic paths. The large amount of spatio-temporal data generated by the AIS system provides essential raw data for deep learning models to generate maritime traffic paths.

However, most of the deep learning-based methods for traffic path extraction are supervised, which require high-quality paired trajectories and traffic path data for model training. These methods cannot be implemented in many real-world scenarios such as maritime traffic path extraction, which lacks real paths corresponding to the trajectory data for training. In our previous study, we proposed T2I-CycleGAN [5], an unsupervised traffic path learning method based on CycleGAN [8]. To the best of our knowledge, it is the only existing unsupervised traffic path learning method from unpaired data so far to fit in maritime traffic scenarios. But there's a fly in the ointment: it can extract maritime paths of center line level and only skeletons of the traffic paths will be extracted.

In this paper we aim to go further and extract traffic paths of roadway level. The key to extract roadway level paths is constructing features for model training to effectively distinguish different roadways. The trajectory data of a moving object contains spatio-temporal attributes related to its navigation status, such as latitude and longitude coordinates, velocity and direction. If these spatio-temporal attributes can be used to construct features for different roadways, the model's ability to distinguish different roadways can be improved, and the extraction of fine-grained paths can be realized.

The key contributions of this paper can be summarized as follows.

1) In this paper, we propose an unsupervised learning method based on Cycle-GAN for extracting fine-grained paths of roadway level from unpaired data. The cycle consistency of CycleGAN achieves traffic path extraction without paired training data. We construct the point and line features of the trajectory data, use the HSV color space [9] to convert the directions, speed into the color information of the trajectory, and then merge the features in the model. From the results, we can see that our feature construction method and our proposed model can achieve roadway-level path extraction.

2) In the proposed method, we add a convolutional layer to the input part of CycleGAN to get spatio-temporal information of the input trajectory features. Compared with the dense layer added in our previous research [5], the convolutional layer can target the spatially closely connected localities in the trajectory features for extraction, which can be more effective for model training.

The rest of this paper is organized as follows. Section 2 summarizes and analyzes the related work. Section 3 introduces two important preliminary concepts. In Sect. 4, an overview of the approach is presented. Section 5 presents the features to be extracted and the structural matrix for converting the extracted orientation, speed information into RGB colors. In Sect. 6, we describe the selection and modification of the network structure in the CycleGAN-based model. Section 7 gives the parameter details and evaluates the extraction method of this paper and compares it with other methods and Sect. 8 concludes the paper.

2   Related Work

Traffic path extraction methods can be divided into two categories: image-based method and trajectory-based method, according to the different types of data used. The former uses remote sensing images or gray-scale imagery to extract the traffic paths, as in [10]. This type of methods require a lot of computing resources to process image data in the early stages of the extraction process. The latter generates traffic paths based on trajectory data collected by sensors or mobile devices. In recent years, different kinds of trajectory-based methods have been proposed, which can be classified into two main categories: method based on trajectory point clustering and method based on density estimation. There are also some hybrid methods combining more than one strategy.

1) **Method based on trajectory point clustering.** This method takes trajectory points as input, uses clustering methods such as k-means to obtain traffic path segments that are finally connected to form a complete path [11–17]. Ma Wenyao [11] proposed a trajectory similarity measure based on one-way distance. Spectral clustering was used to cluster and analyze the trajectories to obtain the behavior patterns of vessels. Junwei Wu [12] clustered the converging points in the traffic road network and got the intersection positions to solve

the problem of low-quality GPS point extraction. Rade Stanojevic [14] proposed an offline algorithm that clustered GPS points for graph construction and an online algorithm that can create and update the road network. Xingzhe Xie [15] proposed to detect the intersections by finding the common sub-tracks of the GPS traces. Alireza Fathi [16] introduced an intersection detector that uses a localized shape descriptor to represent the distribution of GPS traces around a point. Jing Wang [17] used circular boundaries to separate all GPS traces into road intersections and road segments and then built the road paths. These algorithms improved the tolerance of noise and isolated points in the data. However, because clustering algorithms such as DBSCAN do not consider the overall shape characteristics of the trajectory and divide the trajectory in the clustering process, it is difficult to extract the whole trajectory effectively. And when the density of data distribution varies greatly, the results of point clustering methods tend to be poor.

2) **Method based on geometric morphometrics.** This category of method partitions the region into a set of grids, computes or estimates the density of the trajectory points, converts them into grid densities, classifies the grids into two types (those grids on road areas and those are not) and then constructs the cent lines or road boundaries using geometric morphometrics methods. Wang [18], Ahmed [19,20], Mariescu [21] and Kuntzsch [22] transformed the trajectory data into grid data according to the density. Morphological methods were used to extract grid data, and then vectorization was performed to obtain vector road network data. Tang Luliang [23], Yang Wei [24,25], Tang [26] established a constrained triangulation network to extract the road network through trajectory lines. In the extraction process, because the boundary of the obtained surface data was not smooth, the extracted road had weak folds and excessive redundant data was generated. Li [27] improved the Delaunay triangulation based extraction method by grid merging and sliding local window filtering before triangulation. Among them, the treatment of trajectory density difference has a very big impact on the extraction effect.

The above methods require many empirical parameters and are vulnerable to noisy data. Compared with these methods, the deep learning based road extraction approaches are promising. Jiang Xin et al. [28] extracted roads from high-resolution remote sensing images using the DenseUNet network Ruan et al. [6] proposed DeepMG to extract features from trajectories in both spatial view and transition view and used a convolutional deep neural network to learn road centerlines. Chuanwei Lu et al. [7] used cGAN deep learning network for automatic extraction of road network. This method can extract two-way lanes on the road network. However, it relies on paired labelled training data. To the best of our knowledge, only [5] proposed unsupervised road network extraction methods not relying on paired labelled training data. But it is an extraction method of center line level, there are still some shortcomings that need to be addressed. For example, accuracy and fineness of the results are still not so satisfactory.

Algorithm 1. HSV to RGB

Given: h on domain [0,360], s and v each on domain [0,1].
Desired: The equivalent r, g, and b, each on range [0,255]
1: $h := h/60$;
2: **Let** $i := \text{floor}(h)$; $f := h - i$;
3: **Let** $m := v*(1-s)$;
4:      $n := v*(1-(s*f))$;
5:      $k := v*(1-(s*(1-f)))$;
6: **Switch on** i **into**
7:      **case** 0: $(r, g, b) := (v, k, m)*255$;
8:      **case** 1: $(r, g, b) := (n, v, m)*255$;
9:      **case** 2: $(r, g, b) := (m, v, k)*255$;
10:     **case** 3: $(r, g, b) := (m, n, v)*255$;
11:     **case** 4: $(r, g, b) := (k, m, v)*255$;
12:     **case** 5: $(r, g, b) := (v, m, n)*255$;

Remarks: 1) $\text{Floor}(x)$ is the integer just less than or equal x. 2) Only one case is executed in the switch statement. 3) The expression $(r, g, b) := (x, y, z)*255$ abbreviates $r := x*255$; $g := y*255$; $b := z*255$.

---

3 Preliminary Concepts

The HSV color model is used in our proposed approach to preserve the "direction" information of trajectories. And as introduced in Sect. 1, our proposed learning method is based on CycleGAN [8]. Therefore, in this section, we will introduce two preliminary concepts: the HSV color model and the CycleGAN model.

The HSV color model is a color space in which colors are expressed in terms of hue (h), saturation (s), and value (v). Hue is the dimension with points on it normally called red, yellow, blue-green, etc. Saturation measures the departure of a hue from achromatic, i.e., from white or gray. Value measures the departure of a hue from black, the color of zero energy. In the original literature [9], the range of h is from 0 to 1, but we need to map the orientation of trajectories to h, so the range of h is set from 0° to 360°. In this paper, the HSV representation of trajectory features should be converted to RGB model for later processing. The Algorithm 1 [9] shows how the HSV color space can be converted to an RGB model.

CycleGAN can model the data distributions in two domains separately by using a consistent cycle structure and the cycle consistency loss along with the adversarial loss of GAN [29], as shown in Eqs. 1, 2, 3, so it does not require data to be pairs of samples corresponding to each other at the same location. In Eq. 1, where G tries to generate images $G(x)$ that look similar to images from domain Y, while D_Y aims to distinguish between translated samples $G(x)$ and real samples y. In Eq. 2, for each image x from domain X, the image translation cycle should be able to bring x back to the original image, i.e. $X \rightarrow G(x) \rightarrow F(G(x)) \approx x$. Unlike other GANs, CycleGAN consists of two generators and two discriminators. The generator is used to generate target distribution, and the discriminator is used to determine the difference between the generated target distribution and the real one. Through the adversarial process between the generator and the discriminator, the dynamic balance between the models is realized, which means that the generator can generate data that can make it difficult for the discriminator to distinguish between real and fake.

$$L_{GAN}(G, D_Y, X, Y) = E_{y \sim P_{data}(y)}[\log D_Y(y)] + E_{x \sim P_{data}(x)}[\log(1 - D_Y(G(x)))] \tag{1}$$

$$L_{cyc}(G, F) = E_{x \sim P_{data}(x)}[\| F(G(x)) - x \|_1] + E_{y \sim P_{data}(y)}[\| G(F(y) - y \|_1] \tag{2}$$

$$L_{total} = L_{GAN}(G, D_Y, X, Y) + L_{GAN}(F, D_X, Y, X) + \lambda L_{cyc}(G, F) \tag{3}$$

4   Overview

This section first presents some basic definitions, defines our problem formally, and then analyzes the problems and introduces the overview approach.

4.1   Definition

Definition 1 (Trajectory). Given a moving object v_i, a raw trajectory of v_i is a sequence of positions of it over a period of time. For a moving vessel, v_i can be the Maritime Mobile Service Identify (MMSI). Thus, a raw trajectory can be represented as $T_{v_i} = (v_i, (p_0, p_1, ..., p_n))$, where $p_j = (x_{i,j}, y_{i,j}, t_j)$ indicates the position of the moving object at a certain moment, where t_j is the sampling (or collecting) time of the position, $x_{i,j}$, and $y_{i,j}$ represents the longitude and latitude of v_i at t_j.

Definition 2 (Grid). A grid is represented as $Grid = (Code, Points, Lines, Orientation, Speed)$, where $Code$ is the geographic hash code of the grid, while $Points$ is the point density of the grid, we use one of the three equal quantiles of point densities; $Lines$ represents the number of line segments in the grid ; $Orientation$ is the moving direction of the grid, which is the mean orientation angle of all points in the grid; and $Speed$ is the average moving speed of points in the grid.

Definition 3 (Feature Gridding). Given an area, we divide it horizontally $I - 1$ times and vertically $J - 1$ times. Feature gridding is getting $I \times J$ grids of the target area, where I and J indicates the number of rows and columns of the feature matrix, and $Points$ of all grids in the target area make up to a point feature, same as $Lines$, $Orientation$ and $Speed$ in line feature, orientation feature and speed feature, and these features are denoted as $F_{point} \in \mathbb{R}^{I \times J \times 1}$, $F_{line} \in \mathbb{R}^{I \times J \times 1}$, $F_{orientation} \in \mathbb{R}^{I \times J \times 1}$, $F_{speed} \in \mathbb{R}^{I \times J \times 1}$, respectively.

Definition 4 (Feature Merging). Feature merging concatenates features, i.e. point, line, orientation and speed, into one single feature matrix, with the size of $I \times J \times 5$, to act as the input of the proposed model. Instead of concatenating features directly and simply, feature merging of this paper first transforms $F_{orientation}$ and F_{speed} into a new feature, denoted as $F_{orientation\_speed} \in \mathbb{R}^{i \times j \times 3}$, using HSV color space to preserve the roadway details. The specific experimental procedure is described in Sect. 5.2. Then F_{point}, F_{line} and $F_{orientation\_speed}$ are concatenated into the input of the model as $F_M \in \mathbb{R}^{i \times j \times 5}$.

Definition 5 (Fine-Grained Path). Fine-Grained Paths are the output of the proposed method, and are represented as a grid image with the size of $I \times J$, which is the same size as the trajectory features. Every line in the grid image of the fine-grained paths is a roadway of the traffic rather than the center line of the road. And every pixel in the image can be converted to its corresponding coordinates in the map.

Problem Statement. Given a set of trajectories $T = \{T_{v_1}, T_{v_2}, ..., T_{v_n}\}$, infer its underlying fine-grained paths P.

4.2   Problem Analysis and Approach Overview

In order to learn paths from trajectories without unpaired training data, we design our learning method based on CycleGAN. For urban traffic or maritime traffic, the supervised learning model cannot achieve traffic path extraction when there is no paired training data. In contrast, CycleGAN can handle unpaired training data and thus achieve traffic path extraction. When there is a lack of training data, CycleGAN can expand its training data by certain transformations because it does not need to perform one-to-one matching.

The key to learn fine-grained paths of roadway level is how to extract the roadway level features of trajectories. The directions of the adjacent roadways on the same path are usually opposite or significantly different, so the problem of extracting the roadway level features is how to represent the direction feature of trajectories. It is not appropriate to represent the direction feature directly with the angle value in the range $[0°, 360°)$, for example, the difference between $1°$ and $359°$ is very large but their directions are very close. In the HSV color space, the hue is represented with color wheel, which is consistent with the value range and change pattern of the orientation angle, for example, the HSV colors with the hue of $1°$ and $359°$ at the RGB color wheel are very similar. Inspired by the different representations of color space, if we represent the trajectory direction angle with "hue" dimension in HSV color space, the roadways in the opposite directions have very different colors in RGB color space. Therefore, in our method, we encode the direction angle value using HSV and transform it into RGB values to better represent the trajectory direction feature.

Figure 1 gives the overall process of learning the roadway level traffic paths from the unpaired traffic trajectory data and path data. As shown in the figure, the model consists of two components: the convolutional layers and the Cycle-GAN structure. This structure makes the model able to retain the spatio-temporal information of trajectory points and extract the roadway-level paths without standard paired reference data. In the training process, two kinds of data sets, i.e. Trajectory Data and Path Data (note they can be unpaired), are used as the input of the model, as shown in Fig. 1. To achieve the conversion from trajectory data to roadway-level path data, spatio-temporal information are firstly obtained by feature extraction, and then feature gridding and feature merging are utilized to get the input of the proposed model. And at the same time path data is also gridded to get the sample of the target data distribution for the model.

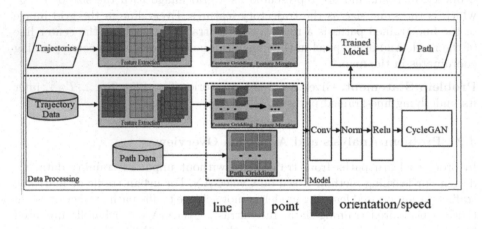

line ⬛ point ⬛ orientation/speed

Fig. 1. The process of extracting fine-grained paths

5   Trajectory Data Transition

5.1   Feature Extraction

Given an area, We partition it into a grid with $I \times J$ cells and for each grid cell, then extract some features from the trajectories. As mentioned in Sect. 4.1, features extracted in this paper include: 1) Point, which is the most straightforward indicator of the underlying roadways; 2) Line, which can help with recovering the roads when points are sparse; 3) Orientation, this feature is helpful for distinguishing two parallel roads in opposite directions. And in Sect. 5.2, how orientation information are transformed into HSV represents is described; 4) Speed, it also helps to model the complete roadway, since, on the same road, the speed usually does not change that much on the same roadway.

5.2   Orientation Converted to Color Information

To get the input of our CycleGAN-based method, feature matrices are organized more effectively rather than concatenated simply and directly. We adopt the HSV color model to transform orientation feature and speed feature of trajectory point into RGB color model.

The orientation of each trajectory point can be calculated by the adjacent trajectory points in a trajectory data, i.e. the azimuth (angle with due north). And then the azimuth value, also denoted as (d), can be converted into the (h) of the HSV color space. As for s and v, if we set them as $s = 1, v = 1$ according to [7], the result of GPS trajectory transformation when $h = d, s = 1, v = 1$ is shown in Fig. 2(c). It is notable that, taking the ground truth of the same area as a reference, there are many noise points when $h = d, s = 1, v = 1$ that may affect the path extraction. Therefore, we encode the normalized value of speed of the trajectory points to the HSV to reduce the saturation of the converted trajectory points outside the normal driving speed range, so that the model ignores these noise points. We set $h = d$, $s = 1$, $v = speed$ and the result is shown in Fig. 2(b), there are obviously less noise points than in (c), and details of the roadways are retained. The comparison of AIS trajectory transformation when $h = d$, $s = 1$, $v = speed$ and $h = d, s = 1, v = 1$ is presented in Fig. 3, it is easy to find that adding the normalized value of speed to HSV can reduce noise points.

We denoted the result of feature transformation as $F_{orientation\_speed} \in \mathbb{R}^{i \times j \times 3}$, and the input of our CycleGAN-based method is achieved by feature merging of $F_{orientation\_speed}$ with F_{point} and F_{line}.

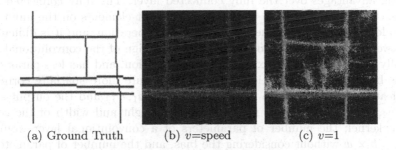

(a) Ground Truth   (b) v=speed   (c) v=1

Fig. 2. Transformation of GPS trajectory features to RGB models

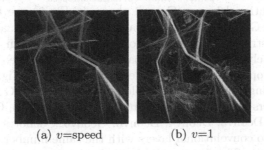

(a) v=speed   (b) v=1

Fig. 3. Transformation of AIS trajectory features to RGB models

6   Training Model

We intend to use the CycleGAN model as a training model for trajectory extraction through the introduction in Sect. 1. As CycleGAN is an image-to-image learning model, we need to convert the input data into three-channel data. In Sect. 5.1 and Sect. 5.2, the F_{point}, F_{line}, $F_{orientation\_speed}$ is a five-channel data, so in order to solve this problem, the model needs to be modified. There are two modification methods:

1) Add a fully connected layer before the input of the generator.

2) Add a convolutional layer before the input of the generator as shown in Fig. 4.

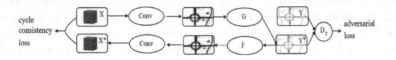

Fig. 4. Training pipeline of the proposed method

When dealing with multi-dimensional data, the convolutional layer has the following advantages over the fully connected layer: The fully connected layer flattens the input data into a vector. The adjacent elements on the input data may no longer be adjacent due to the flattening operation, and it is difficult for the network to capture local information. The design of the convolutional layer naturally has the ability to extract local information, and has less parameters. That is because the convolutional layer can capture more spatially correlated input features; and when the input shape is (c_1, h_1, w_1) and the output shape is (c_2, h_2, w_2), with h and w representing the height and width of the convolutional kernel, the number of parameters of a convolutional layer would be $c_1 \times c_2 \times h \times w$ without considering the bias, and the number of parameters of a fully connected layer would be $c_1 \times c_2 \times h_1 \times w_1 \times h_2 \times w_2$.

After transform the F_{point}, F_{line}, $F_{orientation\_speed}$ into three channels by the added convolutional layers (Gconv1 shown in Fig. 5), we borrow from the design of the CycleGAN [8] network. We adopt ResNet9blocks, a jump-connected encoder-decoder structure. It is able to extract information about data features and accurately reconstruct the road network structure. The overall network structure proposed based on CycleGAN is shown in Fig. 5. The network architectures settings are shown in Tables 1 and 2. In G_x, G_y, D_x, D_y, the convolutional layers are denoted as Gconv1, Gconv2, Gconv3, Gconv4, Gconv8, Dconv1, Dconv2, Dconv3, Dconv4, Dconv5. Gconv5_x denotes a residual block which contains two convolutional layers with the same number of filters in both layers are the same. The fractionally-strided convolutions layers are denoted as Gconv6, Gconv7. For the sake of brevity, the activation function, reflection padding and normalization layers are not shown. Input_x and Input_y represent the source domain data and the target domain data, respectively. And the training process is as follows, where Input_x, Generated_x and Cycli_x represent the transformed feature matrix, and Input_y, Generated_y and Cycli_y represent the roadway-level paths. Input Input_x into G_x to generate Generated_y. Input Input_y into G_y to generate Generated_x. Input Generated_y into G_y to generate Cycli_x. Input Generated_x into G_x to generate Cycli_y. Input_x and Cycli_x are used as training for forward cycle-consistency loss. Input_y and Cycli_y are used as training for backward cycle-consistency loss. Here, k, n, and s denote the kernel size, the number of feature maps, and the stride, respectively.

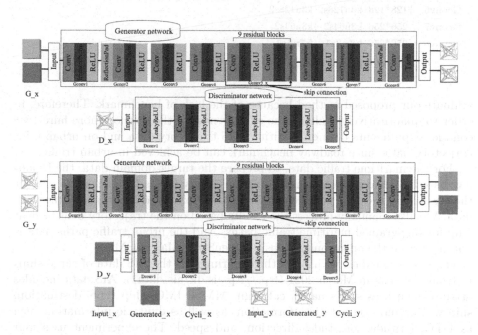

Fig. 5. Network architectures of the generator and the discriminator

7 Experiment and Analysis

7.1 Dataset and Experimental Environment

Our CycleGAN-based model achieve conversion from one data distribution to another, and the two data distributions are trajectory features and roadway-level traffic paths, respectively. We call data sets that have the same data distribution as roadway-level traffic paths target domain data. Note that the target domain data don't need to be paired with the input data.

As mentioned in Sect. 1, it is more challenging to extract traffic paths from maritime AIS trajectory data because there is a lack of standard road network data set in maritime traffic. This also means that it is difficult to quantitatively

Table 1. Model setting of the generator

Layer	Output	G_x	G_y
Gconv1	256*256	k3n3s1	k3n3s1
Gconv2	256*256	k7n64s1	k7n64s1
Gconv3	128*128	k3n128s2	k3n128s2
Gconv4	64*64	k3n256s2	k3n256s2
Gconv5_x	64*64	k3n256s1	k3n256s1
Gconv6	128*128	k3n128s2	k3n128s2
Gconv7	256*256	k3n64s2	k3n64s2
Gconv8	256*256	**k7n3s1**	**k7n5s1**

Table 2. Model setting of the discriminator

Layer	Output	D_x	D_y
Dconv1	128*128	k4n64s2	k4n64s2
Dconv2	64*64	k4n128s2	k4n128s2
Dconv3	32*32	k4n256s2	k4n256s2
Dconv4	31*31	k4n512s1	k4n512s1
Dconv5	30*30	k4n1s1	k4n1s1

evaluate our proposed method because of the lack of benchmark. Therefore, in order to quantitatively evaluate the effectiveness of the method more fairer, we conduct experiments both on maritime AIS trajectory data and on urban GPS trajectory data, since roadway benchmark can be achieved in urban traffic.

For the maritime traffic data, to prepare the target domain data, the Delaunay based method proposed in our previous study [30] is firstly used to extract the paths, and then the path center line data is extracted according to the path data. We also manually process the results based on the trajectory density heat map for fine-grained maritime traffic paths. And the urban traffic paths data is obtained from the open source website OpenStreetMap (OSM).

The maritime data we used in the experiment is the AIS data of cargo ships in Danish waters in March 2021 (ftp://ftp.ais.dk/ais_data/). AIS data includes attributes such as ship's name, call sign, MMSI, IMO, ship type, destination, ship width, and other static information, as well as dynamic information such as UTC, latitude, longitude, direction, and speed. The experiment uses four columns of MMSI, UTC, longitude, and latitude from the data set. The total amount of AIS data for cargo ships in Danish waters is 17.3GB. As shown in Table 3, the data contains travel records for more than 5,930 vessels and more than 7 million trajectory points.

We also used cab GPS(1) trajectory data from February 2, 2008 to February 8, 2008 in Beijing (https://www.microsoft.com/en-us/research/publication/t-drive-trajectory-data-sample/)and GPS(2) trajectory data from Chengdu taxi on August 3, 2014. The dataset contains four attributes, such as cab ID, date, longitude, and latitude. Data details are shown in the Table 3. The land-based trajectory data are introduced here for the later quantitative analysis.

The experiment was run in a Ubuntu 18.04 environment. Model implementation and result analysis based on Python 3.6 and Pytorch 1.4.0 deep learning open source library. The model is trained with a single NVIDIA GeForce RTX 2080Ti GPU and has a memory of 11 GB.

Table 3. Data descriptions

Dataset	Days	Ships/ Vehicles	Points	Lng(min/max)	Lat(min/max)	Amount
AIS	30	5930	7126586	2.5229/17.5107	52.9517/59.1222	17.3 GB
GPS(1)	7	10336	1713331	116.2810/116.4684	39.8451/39.9890	1.40 GB
GPS(2)	1	13605	53045405	103.2696/104.6096	30.2906/31.0324	3.64 GB

7.2 Parameter Setting and Data Division

After obtaining the original data of a rectangular area, it is then divided into 8192×8192 grid areas. The size of each grid area is 256×256, and a total of 1024 grid areas are generated. The number of grids in the test set is 100. The ratio of the training set to the validation set is 9:1, the number of grid regions in the training set is 831, and the number of grid regions in the validation set is 93. Then extract the features in each grid, each grid contains three feature data: F_{point}, F_{line}, and $F_{orientation\_speed}$. Then merge the features in each grid area into an 5 channel feature matrix. Finally, the data of the training set, validation set, and test set are 831, 93, and 100 respectively. AIS data and GPS data are divided in the same way.

During the training phase, we leverage Adam [31] to perform network training with a learning rate $2e-4$ and batch size 1. We keep the same learning rate for the first 100 epochs and linearly decay the rate to zero over the next 100 epochs. The slope of LeakyReLU in the negative part is set to 0.2. More parameters of the model are shown in the Tables 1 and 2.

To show the effectiveness of cycle consistency loss function, we evaluated our model at different λ, as in Eq. 3, settings from 7 to 17. The results are shown in the line graphs in Fig. 6. We find that the network achieves the best performance on both datasets when $\lambda = 16$. A smaller λ or a larger λ reduces the inferred performance of the road network. As shown in Fig. 6.

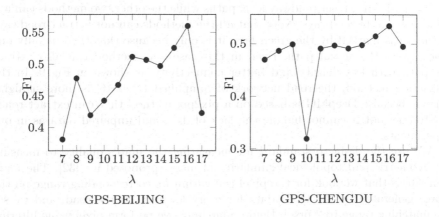

GPS-BEIJING   GPS-CHENGDU

Fig. 6. Effect of λ

7.3   Results and Performance Comparison

Figure 7 shows the extraction result of the fine-grained maritime traffic paths in Danish regional waters. In order to compare and evaluate the extraction results of this experiment, this paper compares and evaluates the extraction methods of T2I-CycleGAN [5], Simplified DeepMG [6].

As introduced in Sect. 1, T2I-CycleGAN [5] is the only comparable unpaired learning method for traffic paths in the existing studies. The DeepMG framework is based on a supervised learning method, it uses OSM data as a reference for model training, its training purpose is more clear than the proposed unsupervised learning. In fact, it is unfair to compare an unpaired learning method with a supervised learning method, because their training sets are different. In general, supervised learning methods should be better than the unpaired learning approach because they the data they use for training must be paired. But here we also compare our approach with DeepMG, the purpose is to show that our method is comparable to the supervised learning method.

Simplified DeepMG: The DeepMG framework uses a CNN-based supervised training method for road network extraction. The trajectory data used in our experiments, when these data are used for T2RNet training, will result in overfitting. In addition, we have not yet implemented the topology construction part of the DeepMG framework, and will implement it in future work. The result is shown in Fig. 7. In the case of using the same test data, the results obtained by the simplified DeepMG are more detailed.

T2I-CycleGAN: By adding a fully connected layer to the CycleGAN generator, the possible sparse features are converted into dense grid images for training. The extracted features include points, lines, speed, direction and spatial features. The final extraction results are shown in the Fig. 7.

In terms of experiments on urban trajectories, we use the road network from OSM map data of Beijing area as the benchmark. The following Fig. 8 and Fig. 9 show the standard road network map and the extracted traffic paths using different approaches.

We can observe that in Fig. 8, for the areas marked by the red dotted box, our method can extract roadway level paths while the other two methods can not achieve accurate roadways. Note that all the methods can not extract roadways in the areas marked by the green box, it is only because that there is only one roadway on the actual path. Even in this case, our method can also extract the path with less density and better connectivity as shown in Fig. 9. In the extraction method, the road network of Simplified DeepMG is more straight. That is because DeepMG is based on a pix2pix method that can extract roads one-to-one, but it cannot handle the lack of data and unpaired data as in our experiments.

We use the de-facto standard for measuring the map's quality to measure the geometric and topological similarity of maps, proposed in [32]. The main idea [20] is that we look for the pixel cell within its corresponding range on the image generated by the OSM data for every location on the road, and we set the matching range to 2 pixel. Before matching, we perform pixel value filtering

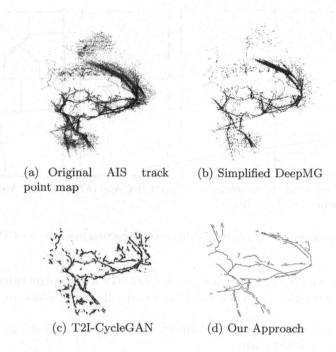

(a) Original AIS track point map

(b) Simplified DeepMG

(c) T2I-CycleGAN

(d) Our Approach

Fig. 7. Comparison of the results of different extraction methods for AIS data

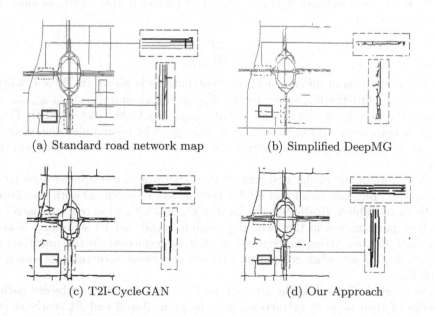

(a) Standard road network map

(b) Simplified DeepMG

(c) T2I-CycleGAN

(d) Our Approach

Fig. 8. Comparison of the results of different extraction methods for GPS data

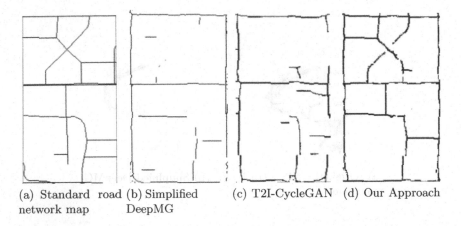

(a) Standard road (b) Simplified   (c) T2I-CycleGAN (d) Our Approach
network map   DeepMG

Fig. 9. Comparison of the results of different extraction methods for GPS data

on all the resulting images to remove dark spots. We set the pixel value filtering range from 150 to 255. We report the F1 score, recall, and precision in the whole result image.

1) Precision, the ratio of the correctly extracted road length to the total length of the extracted paths.

2) Recall, the ratio of the correctly extracted road length to the total length of paths of the benchmark data.

3) F1 score, weighted Harmonic mean of precision and recall, as show in calculation Eq. 4.

$$F_1 = \frac{2P \cdot R}{P + R} \tag{4}$$

The evaluation of different feature combinations is shown in Table 4, where Point, Line, Orientation_Speed means F_{point}, F_{line}, and $F_{orientation\_speed}$ as input, respectively. It can be found that F_{point} has higher metrics than F_{line}, which is consistent with our common knowledge. The results also demonstrate that the $F_{orientation\_speed}$ can improve the Recall, Precision and F1 scores of the model.

We evaluate the model structure of the generator in our approach by comparing the precision, recall and F1 for the test set of seven network structures as shown in Table 5. ResNet9blocks used in CycleGAN and in our approach has the best performance in the form of precision, recall and F1 score and is able to reduce the overfitting problem whether it is performed on the maritime or the urban data set, when comparing with seven network structures as shown in Table 5.

As shown in Table 6, in the unsupervised learning domain, for the evaluation metrics of road network extraction, the Precision, Recall and F1 scores of the extraction method in this paper are higher than those of T2I-CycleGAN. In the supervised and unsupervised learning neighborhoods, the Precision, Recall and

Table 4. Effect of different features

Feature	Precision	Recall	F1 score
Point	0.4498	0.3767	0.4100
Line	0.3739	0.2363	0.2896
Orientation_Speed	0.5597	0.4136	0.4757

Table 5. Generator model comparisons on different datasets

Model	#Params	GPS-BEIJING			GPS-CHENGDU		
		Precision	Recall	F1 score	Precision	Recall	F1 score
ResNet9blocks [33] (adopted in our approach)	11.378M	**0.4779**	0.6767	**0.5602**	0.4101	**0.7643**	**0.5338**
UNet256 [34]	54.410M	0.4458	0.5436	0.4897	0.3756	0.5454	0.4448
FCN [35]	20.104M	0.2784	**0.6809**	0.3952	0.2535	0.6235	0.3604
LinkNet [36]	11.534M	0.2831	0.6782	0.3994	0.2206	0.6124	0.3243
T2RNet [6]	63.082M	0.3174	0.3365	0.3266	0.3011	0.4102	0.3472
D-LinkNet [37]	31.097M	0.3386	0.5131	0.4032	0.2938	0.4803	0.3645
ResNet6blocks [33]	7.838M	0.4501	0.6264	0.5238	0.3875	0.6904	0.4963

Table 6. Comparison of evaluation indicators

Model	GPS-BEIJING			GPS-CHENGDU		
	Precision	Recall	F1 score	Precision	Recall	F1 score
Simplified DeepMG	0.4325	0.6629	0.5235	0.4091	0.5227	0.4587
T2I-CycleGAN	0.3567	0.5613	0.4362	0.3428	0.7289	0.4663
Our approach	**0.4779**	**0.6767**	**0.5602**	**0.4101**	**0.7643**	**0.5338**

F1 Score of the road network extracted by the method in this paper is higher than that of Simplified DeepMG.

8 Conclusion

By proposing an deep learning method based on CycleGAN and using HSV color space to reinforce the spatial features of moving trajectories, this paper achieves roadway-level traffic path extraction without paired standard reference data. In this paper, experiments are conducted with maritime ship AIS data as an example. By referring to the extraction and pre-processing methods of urban traffic networks, we initially processed the collected AIS data and filtered out the noise data. Then based on the CycleGAN model, we proposed an unpaired learning model and realized the extraction of maritime traffic paths by converting the direction feature of the trajectory data into RGB matrix data through HSV color space. We use the OSM data of urban traffic as the reference standard for the results to conduct a quantitative analysis comparing the model and the feature construction method with some baseline methods, and the results show that the proposed method in this paper can extract more accurate traffic paths. For further optimization research of path extraction, on the one hand, we can increase the sample data types sothat the model can learn to generate more path types; on the other hand, the model can be further optimized to integrate more trajectory feature information, so that it can extract richer path information.

Acknowledgements. This work is supported by National Natural Science Foundation of China (Grant No. 61832004) and Projects of International Cooperation and Exchanges NSFC (Grant No. 62061136006).

References

1. Wang, W., Yang, N., Zhang, Y., Wang, F., Cao, T., Eklund, P.: A review of road extraction from remote sensing images. J. Traffic Transp. Eng. (English Edition) **3**(3), 271–282 (2016)
2. Liu, J., Qin, Q., Li, J., Li, Y.: Rural road extraction from high-resolution remote sensing images based on geometric feature inference. ISPRS Int. J. Geo Inf. **6**(10), 314 (2017)
3. Bae, Y., Lee, W.H., Choi, Y., Jeon, Y.W., Ra, J.B.: Automatic road extraction from remote sensing images based on a normalized second derivative map. IEEE Geosci. Remote Sens. Lett. **12**(9), 1858–1862 (2015)
4. Dai, J., Ma, R., Gong, L., Shen, Z., Wu, J.: A model-driven-to-sample-driven method for rural road extraction. Remote Sens. **13**(8), 1417 (2021)
5. Yang, X., Wang, G., Yan, J., Gao, J.: T2I-CycleGAN: a CycleGAN for maritime road network extraction from crowdsourcing spatio-temporal AIS trajectory data. In: Gao, H., Wang, X., Iqbal, M., Yin, Y., Yin, J., Gu, N. (eds.) CollaborateCom 2020. LNICST, vol. 350, pp. 203–218. Springer, Cham (2021). https://doi.org/10.1007/978-3-030-67540-0_12
6. Ruan, S., Long, C., Bao, J., Li, C., Zheng, Y.: Learning to generate maps from trajectories. Proc. AAAI Conf. Artif. Intell. **34**(1), 890–897 (2020)
7. Lu, C., Sun, Q.: Road learning extraction method based on vehicle trajectory data. acta geodaetica et cartographica sinica. J. Surveying Mapp. **49**(06), 26–36 (2020)
8. Zhu, J.Y., Park, T., Isola, P., Efros, A.A.: Unpaired image-to-image translation using cycle-consistent adversarial networks. In: Proceedings of the IEEE International Conference on Computer Vision (ICCV), October 2017
9. Smith, A.R.: Color gamut transform pairs. ACM Siggraph Comput. Graph. **12**(3), 12–19 (1978)
10. Liu, B., Wu, H., Wang, Y., Liu, W.: Main road extraction from zy-3 grayscale imagery based on directional mathematical morphology and vgi prior knowledge in urban areas. PLoS ONE **10**(9), e0138071 (2015)
11. Ma, W., Z., W.: One-way distance-based spectral clustering for ship motion pattern recognition. J. Chongqing Jiaotong Univ. (Natural Sci. Edition) **34**(5), 130–134 (2015)
12. Wu, J., Zhu, Y., Ku, T., Wang, L.: Detecting road intersections from coarse-gained GPS traces based on clustering. J. Comput. **8**(11), 2959–2965 (2013)

13. Aronov, B., Driemel, A., Kreveld, M.V., Löffler, M., Staals, F.: Segmentation of trajectories on nonmonotone criteria. ACM Trans. Algorithms (TALG) **12**(2), 1–28 (2015)
14. Stanojevic, R., Abbar, S., Thirumuruganathan, S., Chawla, S., Filali, F., Aleimat, A.: Robust road map inference through network alignment of trajectories. In: Proceedings of the 2018 SIAM International Conference on Data Mining, pp. 135–143. SIAM (2018)
15. Xie, X., Liao, W., Aghajan, H., Veelaert, P., Philips, W.: Detecting road intersections from GPS traces using longest common subsequence algorithm. ISPRS Int. J. Geo Inf. **6**(1), 1 (2017)
16. Fathi, A., Krumm, J.: Detecting road intersections from GPS traces. In: Fabrikant, S.I., Reichenbacher, T., van Kreveld, M., Schlieder, C. (eds.) GIScience 2010. LNCS, vol. 6292, pp. 56–69. Springer, Heidelberg (2010). https://doi.org/10. 1007/978-3-642-15300-6_5
17. Wang, J., Rui, X., Song, X., Tan, X., Wang, C., Raghavan, V.: A novel approach for generating routable road maps from vehicle GPS traces. Int. J. Geogr. Inf. Sci. **29**(1), 69–91 (2015)
18. Wang, S., Wang, Y., Li, Y.: Efficient map reconstruction and augmentation via topological methods. In: Proceedings of the 23rd SIGSPATIAL International Conference on Advances in Geographic Information Systems, pp. 1–10 (2015)
19. Ahmed, M., Karagiorgou, S., Pfoser, D., Wenk, C.: Map construction algorithms. In: Map Construction Algorithms, pp. 1–14. Springer, Cham (2015). https://doi. org/10.1007/978-3-319-25166-0_1
20. Ahmed, M., Karagiorgou, S., Pfoser, D., Wenk, C.: A comparison and evaluation of map construction algorithms using vehicle tracking data. GeoInformatica **19**(3), 601–632 (2015)
21. Mariescu-Istodor, R., Fränti, P.: Cellnet: inferring road networks from GPS trajectories. ACM Trans. Spatial Algorithms Syst. (TSAS) **4**(3), 1–22 (2018)
22. Kuntzsch, C., Sester, M., Brenner, C.: Generative models for road network reconstruction. Int. J. Geogr. Inf. Sci. **30**(5), 1012–1039 (2016)
23. Furiang, T., Zhang, L., Xue, Y., Zihan, K., Qingquan, L., Kun, D.: A spatio-temporal trajectory fusion and road network generation method in accordance with cognitive laws. J. Surveying Mapp. **44**(11), 1271 (2015)
24. Yang, W., Ai, T.: Road centerline extraction based on multi source trajectory data. Geogr. Geog. Inf. Sci. **32**(003), 1–7 (2016)
25. Yang, W., Ai, T.: Using constrained delaunay triangulation to extract road boundary from multi-source trajectories. J. Surveying Mapp. **046**(002), 237–245 (2017)
26. Tang, L., Chang, R., Zhang, L., Li, Q.: A road map refinement method using delaunay triangulation for big trace data. ISPRS Int. J. Geo Inf. **6**(2), 45 (2017)
27. Li, Z., Wang, G., Meng, J., Xu, Y.: The parallel and precision adaptive method of marine lane extraction based on QuadTree. In: Gao, H., Wang, X., Yin, Y., Iqbal, M. (eds.) CollaborateCom 2018. LNICST, vol. 268, pp. 170–188. Springer, Cham (2019). https://doi.org/10.1007/978-3-030-12981-1_12
28. Xin, J., Zhang, X., Zhang, Z., Fang, W.: Road extraction of high-resolution remote sensing images derived from denseunet. Remote Sens. **11**(21), 2499 (2019)
29. Goodfellow, I., et al.: Generative adversarial networks. Commun. ACM **63**(11), 139–144 (2020)
30. Wang, G., Meng, J., Li, Z., Hesenius, M., Ding, W., Han, Y., Gruhn, V.: Adaptive Extraction and Refinement of Marine Lanes from Crowdsourced Trajectory Data. Mobile Networks and Applications, pp. 1392–1404 (2020)

31. Kingma, D.P., Ba, J.: Adam: a method for stochastic optimization. arXiv preprint arXiv:1412.6980 (2014)
32. Biagioni, J., Eriksson, J.: Inferring road maps from global positioning system traces: survey and comparative evaluation. Transp. Res. Record J. Transp. Res. Board **2291**, 61–71 (2014)
33. He, K., Zhang, X., Ren, S., Sun, J.: Deep residual learning for image recognition. In: Proceedings of the IEEE Conference on Computer Vision and Pattern Recognition (CVPR), June 2016
34. Ronneberger, O., Fischer, P., Brox, T.: U-Net: convolutional networks for biomedical image segmentation. In: Navab, N., Hornegger, J., Wells, W.M., Frangi, A.F. (eds.) MICCAI 2015. LNCS, vol. 9351, pp. 234–241. Springer, Cham (2015). https://doi.org/10.1007/978-3-319-24574-4_28
35. Long, J., Shelhamer, E., Darrell, T.: Fully convolutional networks for semantic segmentation. In: Proceedings of the IEEE Conference on Computer Vision and Pattern Recognition, pp. 3431–3440 (2015)
36. Chaurasia, A., Culurciello, E.: Linknet: exploiting encoder representations for efficient semantic segmentation. In: 2017 IEEE Visual Communications and Image Processing (VCIP), pp. 1–4. IEEE (2017)
37. Zhou, L., Zhang, C., Wu, M.: D-linknet: Linknet with pretrained encoder and dilated convolution for high resolution satellite imagery road extraction. In: Proceedings of the IEEE Conference on Computer Vision and Pattern Recognition Workshops, pp. 182–186 (2018)

Multi-UAV Cooperative Exploring for the Unknown Indoor Environment Based on Dynamic Target Tracking

Ning Li[1], Jiefu Tan[1], Yunlong Wu[2,3(✉)], Jiachi Xu[1], Huan Wang[1], and Wendi Wu[1]

[1] College of Computer, National University of Defense Technology, Changsha 410073, Hunan, China
[2] Artificial Intelligence Research Center (AIRC), National Innovation Institute of Defense Technology (NIIDT), Beijing 100071, China
ylwu1988@nudt.edu.cn
[3] Tianjin Artificial Intelligence Innovation Center (TAIIC), Tianjin 300457, China

Abstract. This paper proposes a method for collaborative exploration adopting multiple UAVs in an unknown GPS-denied indoor environment. The core of this method is to use the Tracking-D*Lite algorithm to track moving targets in unknown terrain, combined with the Wall-Around algorithm based on the Bug algorithm to navigate the UAV in the unknown indoor environment. The method adopts the advantages of the above two algorithms, where the UAV applies the Wall-Around algorithm to fly around the wall and utilizes the Tracking-D*Lite algorithm to achieve collaboration among UAVs. This method is simulated and visualized by using Gazebo, and the results show that it can effectively take the advantages of multiple UAVs to explore the unknown indoor environments. Moreover, the method can also draw the boundary-contour map of the entire environment at last. Once extended to the real world, this method can be applied to dangerous buildings after earthquakes, hazardous gas factories, underground mines, or other search and rescue scenarios.

Keywords: Multi-UAV collaboration · Target tracking · Path planning

1 Introduction

In the past few years, unmanned aerial vehicles (UAVs) have been widely used in military and civilian fields due to their high cost-effectiveness, flexibility, and

This work was supported by the National Key Research and Development Program of China(2017YFB1001901) and the National Natural Science Foundation of China under Grant No. 61906212.

H. Gao and X. Wang (Eds.): CollaborateCom 2021, LNICST 406, pp. 191–209, 2021.
https://doi.org/10.1007/978-3-030-92635-9_12

durability. Due to its flexibility and low risk, UAV has extensively developed in exploring the indoor environments [1]. UAVs have great application scenarios for the exploration of dangerous indoor spaces, such as factories with toxic gas leaks, dangerous buildings after earthquakes, and areas with dangerous nuclear radiation. However, it is very difficult for UAVs to navigate automatically in a GPS-denied unknown indoor environment. Also, A single UAV can use Simultaneous Localization And Mapping(SLAM) method [2] to explore an unknown environment, but it will take up a lot of computing time and storage performance. Therefore, in order to reduce the computational cost of UAVs and the time to search for unknown indoor environments, a collaborative exploration strategy can be adopted through multiple UAVs. This strategy enables them to quickly explore the entire unknown indoor environment at a relatively small cost and provide timely and effective information for subsequent tasks, such as exploration and rescue at accident or disaster sites, building and public facilities inspection, etc.

The collaborative exploration between multi-UAVs needs to provide a flexible and robust exploration strategy since the absence of map information. In a GPS-denied unknown indoor environment, the design of the exploration strategy not only needs to coordinate many UAVs but also needs to be equipped with automatic positioning, flight, and obstacle avoidance navigation algorithms for each UAV. Also, the navigation strategy must ensure that each UAV should search for unknown areas as efficiently as possible and cover the largest range. Moreover, the navigation strategy should avoid repeated exploration between UAVs during searching. Finally, the strategy ought to be robust, the failure of a single UAV will not cause the failure of the whole search process, thereby improving the stability of the entire strategy.

This paper proposes a collaborative exploration strategy for multi-UAVs in unknown indoor environments based on dynamic target tracking. This method uses two path-planning algorithms, one is the Wall-Around algorithm based on Bug algorithms [3–5], which is used to fly around the boundary of environments autonomously; the other is an unknown terrain dynamic target tracking algorithm(Tracking-D*Lite) based on D*Lite [6] and I-ARA* [7], which is used for relay between UAVs.

In summary, the main contributions of this paper are as follows:

- Proposed the Wall-Around algorithm for flying around the boundary of environments. The algorithm is based on the idea of Bug algorithm to cancel the existence of the target point, so that the UAV will navigate around the boundary of environments according to certain rules and finally return to the starting point. The algorithm is mainly used to explore the boundaries of the unknown spaces and provide location information and data for drawing the boundary contour map.
- Proposed a dynamic target tracking algorithm(Tracking-D*Lite for short) for relay between multiple UAVs in unknown terrain. This algorithm is mainly based on the D*Lite algorithm and the I-ARA* algorithm. And the method used for tracking moving target points in the I-ARA * algorithm is extended

to D * Lite algorithm to reduce the search time and the number of expansions in the re-planning process, so as to achieve fast search efficiency and planning quality.
- Built a multi-UAVs exploration system that can navigate autonomously in an unknown indoor environment. The system can coordinate multiple UAVs to explore the unknown indoor environment autonomously. Moreover, the effectiveness of the exploration system is verified by simulation experiments. It shows that the proposed multi-UAVs exploration system can improve exploration efficiency in an unknown indoor environment.

The rest of the paper is organized as follows: Sect. 2 first introduces some methods of exploring the unknown indoor environments by UAVs, then introduces the task scenario of this work. Section 3 will introduce the method used in detail. The first method is the Wall-Around algorithm which used for surrounding the boundary of the unknown environment by UAVs. The second method is the Tracking-D*Lite algorithm that is based on D*Lite and I-ARA* to realize the relay between UAVs. In Sect. 4, we will introduce our experimental setup, including a performance experiment of the Tracking-D*Lite by comparison with repeated-D*Lite, simulation environment and results. Sect. 5 concludes the paper and discusses future work.

2 Related Work and Scenario Description

2.1 Related Work

With the rapid development of UAV technology, more and more researchers have paid attention to the exploration of GPS-denied unknown indoor environments using UAVs. The following three main approaches can be used to solve the problem of autonomous exploration in GPS-denied unknown indoor environments: SLAM, deep reinforcement learning and traditional path planning algorithm.

In exploring unknown indoor environments, SLAM is a comprehensive and accurate method. SLAM enables precise positioning, obstacle avoidance, navigation, and real-time visualization of the flight robot. [8] extended the method suitable for ground robots navigation and mapping to UAVs. It manually built a special drone with a processor and various radars for real-time processing of the acquired spatial information and can build maps and navigate in unknown environments in real-time. [9] equipped with laser radar and a depth camera on the UAV. It uses lidar to sense surrounding obstacles, and the depth camera to avoid obstacles. SLAM-based method is the most classic and accurate method in the field of unknown indoor environments exploration. This method requires the UAV to be equipped with high-performance processing units and storage devices for accurate modeling of complex scenes and autonomous control of the UAV. However, this method cannot be applied to the UAVs with weak computing power and low storage performance, especially some micro UAVs.

The method based on deep reinforcement learning focuses on the learning of positioning, navigation and obstacle avoidance strategies of UAVs in the GPS-denied indoor environment. [10] used deep learning and reinforcement learning methods to recognize video images taken by drones, and designed an application system for search and rescue in an indoor environment. [11] mainly used only on-board sensors for localization within a GPS-denied environment with obstacles through a Partially Observable Markov Decision Process (POMDP) on Robotic Operating System (ROS). [12] proposed a method of using PID + Q-learning algorithm to train a UAV to learn to navigate to a target point in an unknown environment.[13] presented a framework for UAV to explore indoor environments based on deep reinforcement learning methods. The framework is divided into a Markov Decision Process (MDP) and a Partially Observable Markov Decision Processes (POMDP). [14] propose a target discovery framework that combines traditional POMDP-based planning with deep reinforcement learning. Different from [13] is that [14] used multiple UAVs to explore the same indoor environment. The method based on deep reinforcement learning can achieve certain results by using UAVs to explore complex unknown environments, but this method often fails to achieve ideal results in map construction, motion decision-making and planning. In addition, most of the research work of this method is to control a single UAV for exploration, so there is a lack of effective use of multiple UAVs.

There are also some works using traditional path planning algorithms to explore unknown indoor environments. [15] proposed a minimizing navigation method named swarm gradient bug algorithm(SGBA) for tiny flying robots to explore the unknown environment. This work sends a swarm of tiny flying robots to advance in different priority directions, navigates with the Bug algorithm and finally all flying robots return starting position. However, this method has the problem that multiple flying robots may search the same area repeatedly during the exploration process, and due to the necessity to return, the search range of each flying robot cannot be maximized.

In order to explore and navigate UAVs in a GPS-denied unknown indoor environment effectively, this paper proposes an exploration method to coordinate multiple UAVs by combining the Wall-Around algorithm and the Tracking-D*Lite algorithm. This method can be applied to some UAVs(especially micro-UAVs) with weak computing power and small memory, and it is very practical to explore the indoor environment with lower cost.

2.2   Scenario Description

In this paper, we consider using multiple UAVs to explore a GPS-denied unknown indoor environment and draw a boundary map of the entire indoor environment, as shown in the Fig. 1. In this scenario, each UAV starts from the same starting position. The first UAV uses the Wall-Around algorithm to fly around the boundary of environment, the red line indicates the first UAV's flight path. When the first UAV flies low on power or stops working due to an unexpected situation, the second UAV will start to relay. In this process, the Tracking-D*Lite algorithm will be used for tracking relay, as shown by the yellow line. When tracking is completed, the second UAV will continue to fly around the boundary of environment using the Wall-Around algorithm until it returns to the starting position. During the whole exploration process, the path of UAVs around the boundary of environment is drawn in real-time, and the boundary of the entire map is drawn when the exploration is completed.

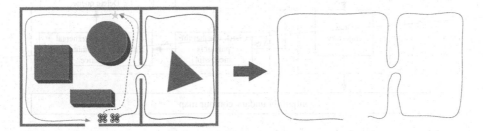

Fig. 1. Schematic figure of two UAVs exploring a GPS-denied unknown indoor environment. The red line represents the trajectory of the first UAV around the wall, the yellow line represents the trajectory tracked by the first UAV during the relay, and the blue dotted line represents the second UAV tracking the first. The two UAVs converge at the pentagram, and the green trajectory indicates the trajectory of the second UAV flying around the wall. (Color figure online)

3   Method

In response to the unknown indoor environment exploration problem described in Fig. 1, we propose a multi-UAV collaborative exploration method for unknown indoor environment based on dynamic target tracking. The method includes three mechanisms: Wall-Around mechanism, relay pursuit mechanism and boundary contour construction mechanism. The specific algorithm scheduling framework of the three mechanisms is shown in Fig. 2, the framework mainly includes four parts: Input, Wall-Around algorithm, Tracking-D*Lite algorithm and Output. The input is the positions of UAVs in the initial state. After inputting it, the Wall Around algorithm is directly called to fly around the wall and draw the flying trajectory of the UAV in real time. The Wall-Around algorithm mainly includes drawing trajectory and flying around the wall. These two work are synchronized. When the UAV stops working (energy exhaustion or unexpected situation), the system will trigger the relay of the next UAV and call Tracking-D*Lite. Tracking-D*Lite mainly includes six parts. When catching up with the last UAV, the Tracking-D*Lite algorithm will return to the Bug algorithm. The output of this framework is the boundary map drawn after multiple UAVs fly around the wall.

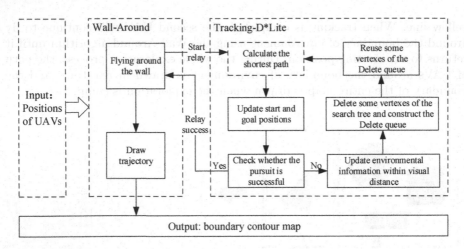

Fig. 2. Diagram of system algorithm framework. It mainly includes two parts of Wall-Around algorithm and Traking-D*Lite algorithm, as well as the conversion process and scheduling process between the two algorithms.

3.1 Wall-Around Algorithm

The Wall-Around algorithm is designed to allow the UAV to fly along the boundary of the indoor environment, namely along walls. For the traditional Bug algorithms [3–5], the UAV constantly advances towards the target point based on understanding the location of the target point. When encountering obstacles, the UAV will keep approaching the target point by surrounding the obstacle to avoid the obstacle. The difference between the Wall-Around algorithm and the Bug algorithm is that it does not have a clear target point, the UAV only needs to surround the boundary of the map and finally return to the starting point. UAVs can navigate in a way similar to the TangenBug [16], sensing obstacles within a certain distance around them through sensors such as vision or lidar, so as to decide to move in a certain direction.

This algorithm divides the whole unknown space into four directions: front, back, left, and right. These four directions are fixed. But in the case of the UAV, it needs to keep choosing the correct direction to fly around the wall. Therefore, the UAV itself has state quantities in four directions, including MD(Main Direction), ND(Next Direction), PD(Previous Direction), and OD(Opposite Direction). MD is the most important, indicating the direction of the current UAV pointing to the wall. The other three directions are calculated by rotating clockwise according to the MD. For example, when MD points to right, ND points to back, PD points to front, and OD points to left. By analogy, four directions can be calculated at any time.

Algorithm 1. Wall Around

1: **procedure** MAIN()
2:      Move towards the closest distance to the wall and set this direction to MD
3:      **while** keep flying **do**
4:          **if** MD is obstacles and ND is free **then**
5:              Go to ND
6:          **else if** MD is free **then**
7:              Go to MD
8:              $MD = $ GETPREDIRECTION(MD)
9:          **else if** MD and ND are obstacles and OD is free **then**
10:             Go to OD
11:             $MD = $ GETNEXTDIRECTION(MD)
12:         **else if** MD, ND and OD are obstacles **then**
13:             Go to PD
14:             $MD = $ GETOPPODIRECTION(MD)
15:         Update ND, OD, PD according to MD

3.2 Tracking-D*Lite

In order to make the relay of UAVs smooth, it is necessary to design an algorithm to track the target in unknown environments. For the grid map used by the traditional path planning algorithms, mainstream path planning algorithms such as A* [17] and JPS [18] are based on the global known map information for path planning between two fixed points. For path planning in unknown spaces, such as D* [19] and D*Lite [6] algorithms can only plan paths between fixed points. As for tracking targets, I-ARA* [7] can quickly re-plan the path of the moving target point, but it is based on the premise that the map information is fully known. Therefore, for two constantly moving UAVs in an unknown environment, under the premise that the position coordinates are known, it is necessary to design a target tracking algorithm to complete the relay task between UAVs. The unknown terrain dynamic target tracking (Tracking-D*Lite for short) algorithm proposed in this paper is mainly modified based on the D*Lite. Since the D*Lite is a path planning algorithm in unknown terrain and cannot be used directly for tracking, the idea of combining the I-ARA* can be used for the purpose of tracking unknown terrain targets.

Before introducing the Tracking-D*Lite, it is necessary to explain the symbols used in the algorithm: S denotes the set of vertexes in the grid map. $s_{start} \in S$ and $s_{goal} \in S$ denotes the start and goal vertex. $Succ(s) \subseteq S$ denotes the set of sub-vertexes of $s \in S$. Similarly, $Pred(s) \subseteq S$ denotes the set of parent vertexes of $s \in S$. $0 < c(s, s') \leq \infty$ denotes the cost of moving vertex s to vertex s'. Since the four-connection expansion method is used in this algorithm, the cost of two neighbouring vertexes is 1. The heuristic function is Manhattan function, denotes by $h(s_{start}, s)$, satisfies $h(s_{start}, s) = 0$ and obeys the triangle inequality $h(s_{start}, s) \leq c(s, s') + h(s_{start}, s')$ for all vertexes $s \in S$ and $s' \in Succ(s)$.

Tracking-D*Lite is similar to D*Lite, it also starts from the goal and extends to the start, and finally gives a path. The calculation of g-value and rhs-value is also involved in the algorithm, $g(s)$ represents the shortest distance from s_{goal} to the current vertex $s \in S$. $rhs(s)$ is calculated based on $g(s)$, the main function of this variable is to find a path vertex with a smaller cost. If current vertex $s = s_{start}$, $rhs(s) = 0$. Instead, if current vertex s is not s_{start}, set $rhs(s)$ to the minimum of the one-step cost plus $g(s)$ among all parent vertexes, shown as Eq.(1).

$$rhs(s) = \begin{cases} 0 & \text{if } s = s_{start} \\ min_{s' \in Pred(s)}(g(s') + c(s, s')) & \text{otherwise} \end{cases} \tag{1}$$

For the vertex $s \in S$, when $g(s) = rhs(s)$, the vertex s is considered to be in a locally consistent state, which means to be able to find the shortest path from s_{goal} to the s. If all vertexes are locally consistent in the graph, the shortest path can be found from s_{goal} to any reachable vertexes. When $g(s) > rhs(s)$, the vertex s is considered to be locally over consistent. It means that a shorter path from s_{goal} to s can be found, which is mainly manifested in a certain area from an obstacle area to a passable area. When $g(s) < rhs(s)$, the vertex s is considered to be locally under consistent. It means that the cost of the shortest path found before from s_{goal} to s becomes larger, and the shortest path needs to be recalculated, which is mainly manifested in a certain area from a passable area to an obstacle area.

Tracking-D*Lite needs to maintain a Frontier priority queue to store vertexes in local inconsistent. These vertexes need to be sorted by a certain rule before selecting some vertexes for expansion and then transform these selected vertexes into locally consistent. The sorting basis of the Frontier priority queue is shown in the key-value $k(s)$ provided by Eq.(2). There are two-part in $k(s)$: $k_1(s) = min(g(s), rhs(s)) + h(s, s_{goal}) + km$ and $k_2(s) = min(g(s), rhs(s))$. And km represents the heuristic compensation value after the start of each move, in order to maintain the strict ascending order of the key values in subsequent searches. The Frontier queue uses the key-value $k(s)$ for sorting. The sorting rules of $k(s)$ are dictionary sorting, which means that the smaller the value of k_1, the higher the priority; if the value of k_1 is the same, the lower the value of k_2, the higher the priority.

$$k(s) = \begin{cases} min(g(s), rhs(s)) + h(s, s_{goal}) + km \\ min(g(s), rhs(s)) \end{cases} \tag{2}$$

Tracking-D*Lite adds the idea of Delete queue in I-ARA* based on D*Lite. Delete queue is used to store the search tree that does not belong to the root vertex of the current search during the current search, and the vertex will be reused for the next search. The execution process of Tracking-D*Lite is shown in Fig. 3.

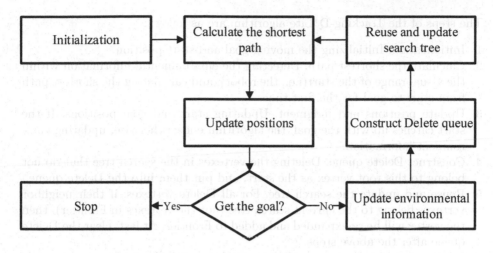

Fig. 3. Flowchart of the Tracking-D*Lite algorithm.

Algorithm 2. Tracking-D*Lite

1: **procedure** MAIN()
2:    $s_{last\_start} = s_{start}$
3:    $s_{last\_goal} = s_{goal}$
4:    INITIALIZE()
5:    COMPUTESHORTESTPATH()
6:    **while** $s_{start} \neq s_{goal}$ **do**
7:        $s_{start} = \text{argmin}_{s' Succ(s_{start})}(c(s_{start}, s') + g(s'))$
8:        Move to s_{start}
9:        $k_m = k_m + h(s_{last\_start}, s_{start})$
10:       $s_{last\_start} = s_{start}$
11:       $s_{goal} = \text{GETNEXTGOAL}()$
12:       $s_{last\_goal} = s_{goal}$
13:       $parent(s_{goal}) = \varnothing$
14:       Scan graph for changed edge cost
15:       **if** any edge costs changed **or** $s_{goal} \neq s_{last\_goal}$ **then**
16:          **if** $s_{goal} \neq s_{last\_goal}$ **then**
17:             **for all** vertex s in the search tree rooted at s_{last_start} but not rooted at s_{start} **do**
18:                $Frontier.\text{remove}(s)$
19:                $rhs(s) = g(s) = \infty$
20:                $parent(s) = \varnothing$
21:                $Delete.\text{insert}(s)$
22:          REUSEDELETEDNODES()
23:       **for all** directed edges (u, v) with changed edge costs **do**
24:          Update the edge cost (u, v)
25:          UPDATENODE(u)
26:       COMPUTESHORTESTPATH()

The steps of the Tracking-D*Lite algorithm are as follows:

1. Initialization. Initializing the moving goal and start position.
2. Calculate the shortest path. Detecting the environmental information within the visual range of the start(i.e., the robot), and calculating the shortest path from start to goal for the first time.
3. Update positions and judgment. Updating start and goal positions. If the start catches up with the goal, the algorithm ends; otherwise, updating environment information.
4. Construct Delete queue. Deleting the vertexes in the search tree that do not belong to this root vertex as the start, and put them into the Delete queue.
5. Reuse and update the search tree. For all Delete vertexes, if their neighbor vertexes belong to the current search tree (not the vertexes in Frontier), then the vertex will be re-expanded and added to Frontier. At last, clear the Delete queue after the above steps.
6. Repeat steps 2 to 5 above until start catch up with the goal.

The pseudocode of Tracking-D*Lite algorithm is mainly shown below. On lines 2–3 of Algorithm 2, the algorithm records the positions of start and goal for subsequent calculations. Then, the MAIN() calls INITIALIZE() to initialize the search problem on line 4. In this progress, the g-value and rhs-value of all vertexes are set according to Eq. (1) and the parent vertex of each vertex is set to empty in the search tree. In the last step in INITIALIZE(), only the vertex of the goal is locally inconsistent, so it is added to the Frontier.

Algorithm 3. Compute Shortest Path

1: **procedure** COMPUTESHORTESTPATH()
2:    **while** $Frontier$.TopKey()$<$CALCKEY(s_{start}) **or** $g(s_{start}) \neq rhs(s_{start})$ **do**
3:        $k_{old} = Frontier$.TopKey()
4:        $n = Frontier$.pop()
5:        **if** $k_{old} <$CALCKEY(n) **then**
6:            $Frontier$.insert(n,CALCKEY(n))
7:        **else if** $g(n) > rhs(n)$ **then**
8:            $g(n) = rhs(n)$
9:            **for all** $s \in Pred(n)$ **do**
10:                UPDATENODE(s)
11:        **else**
12:            $g(n) = \infty$
13:            **for all** $s \in Pred(n) \cup Frontier$ **do**
14:                UPDATENODE(s)

After INITIALIZE() is executed, MAIN() calls the COMPUTESHORTESTPATH() (see Algorithm 3) to search for the shortest path from goal to start. In the process of expanding the search tree's vertexes, for the vertex that is locally over consistent(line 5 of Algorithm 3), set its g-value equal to rhs-value and expand the neighboring vertexes around it. For the vertex that is locally under consistent(line 7 of Algorithm 3), set its g-value to infinity and re-add it to the priority queue Frontier. This step is equivalent to setting it to local over consistent.

Algorithm 4. Reuse Deleted Nodes

1: **procedure** REUSEDELETEDNODES()
2:    **for all** $dn \in Delete$ **do**
3:        **for all** $nn \in Neighbors(nn)$ **do**
4:            **if** $g(nn) \neq \infty$ **and** $rhs(dn) > c(nn, dn) + g(nn)$ **then**
5:                $rhs(dn) = c(nn, dn) + g(nn)$
6:                $parent(dn) = nn$
7:                **if** $dn \in Frontier$ **then**
8:                    $Frontier.$remove(dn)
9:                **if** $g(dn) \neq rhs(dn)$ **then**
10:                   $Frontier.$insert$(dn, $CALCKEY$(dn))$
11:    $Delete = \emptyset$

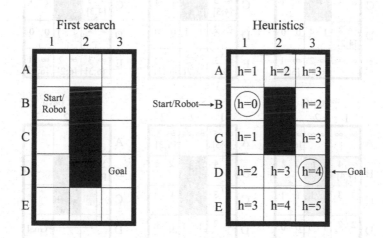

Fig. 4. An example of Tracking-D*Lite (Part 1). As shown in the figure on the left, the robot is located at B1 as the start, and the goal at D3 is the target. In the figure on the right, the robot can only sense the surrounding environment information of one unit distance, so B2 and B3 are obstacles. The h-value in the figure on the right represents the heuristic value, which uses the Manhattan distance from each node on the way to the start.

In the next step, MAIN() updates start and goal's position(line 7–11 of Algorithm 2) and sets goal's parents to null value (line 13 of Algorithm 2). And special attention is needed to calculate the heuristic difference km between different start positions (line 9 of Algorithm 2) to ensure the consistency of Frontier's key values. When the goal's position changes, MAIN()(line 16–22 of Algorithm 2)

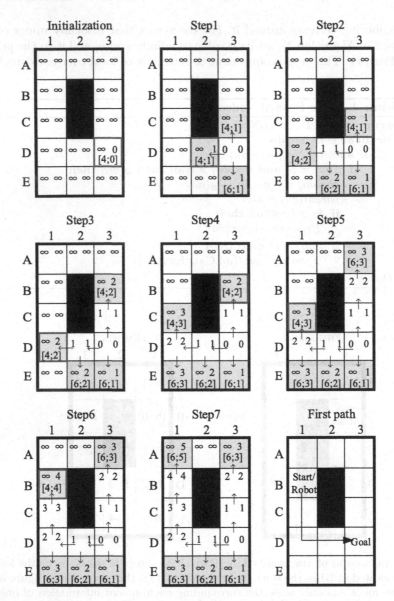

Fig. 5. An example of Tracking-D*Lite (Part 2). Step1 to Step7 represent the process of Tracking-D*Lite in the first expansion process, and finally get the path pointed to by First path. (Color figure online)

deletes the members whose root vertex is not the current start of this search in the search tree and put them into the Delete queue, and then reuse some of the deleted vertexes(see Algorithm 4). On lines 23–25 of Algorithm 2, MAIN()

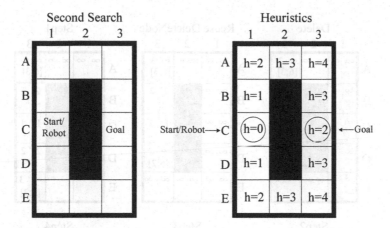

Fig. 6. An example of Tracking-D*Lite (Part 3). When Robot moves one unit distance to C1 according to the path obtained from the first search, Goal also randomly moves to C3. At this time, Robot senses a new obstacle D2, and recalculates the heuristic values of each node in the entire map.

updates the edge cost of the vertexes changed in the grid map and updates Frontier and each vertex's g-value.

The following is an example of Tracking-D*Lite. There is a 5*3 grid map in Fig. 4. The position of the robot (start) at B1 and goal at D3. In Fig. 4, Map is the real environment of the entire map. The robot can sense a map of a unit distance around. Heuristics in Fig. 4 is the heuristic value of the robot at the reachable grid point, and the heuristic function is the Manhattan function.

Figure 5 shows the first time the COMPUTESHORTESTPATH() function is executed by Tracking-D*Lite. The initial search iteration steps of this algorithm are the same as the D*Lite algorithm, which is to search and expand from goal position.

As shown in Fig. 5, the yellow grid represents the vertex in the priority queue, the red box vertex represents the vertex to be expanded next time, the arrow between the vertexes indicates that the parent vertex points to the child vertex. And every vertex has a g-value(in the upper left corner of the grid), rhs-value(in the upper right corner of the grid), and key-value(below in the grid if exist).

The first search is expanded from the goal in the initialization phase. The rhs value of the goal is set to infinity and zero according to Eq. (1), and the priority key value of the vertex is calculated according to Eq. (2). In the example of this algorithm, the four-connection expansion method is used. By selecting the vertex with the highest priority in the priority queue (i.e., the vertex with the smallest Key value), the neighbors of the vertex are expanded and put into Frontier. Then delete the expanded vertexes from the Frontier queue, and finally set the g-value of these vertexes to rhs-value. The path after executing the COMPUTESHORTESTPATH() function for the first time is shown in the First path in Fig. 5.

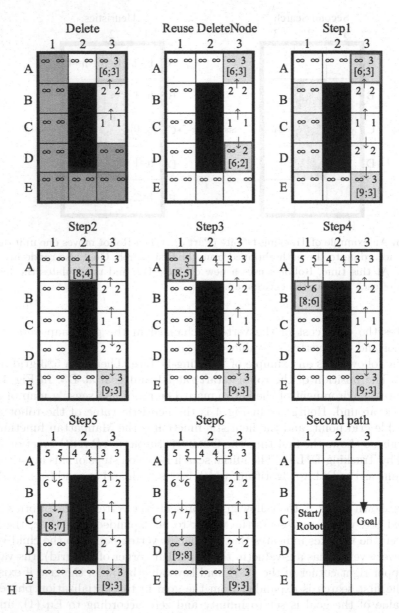

Fig. 7. An example of Tracking-D*Lite (Part 4). This part represents the specific process of reusing part of the Delete node in the re-planning process. Figure of Delete means to delete the search tree that does not take the starting point of the second search (C3) as the root node and store it in the Delete queue (gray grids). Figure of Reuse DeleteNode represents the expansion of the node (D3) in the Delete queue that is close to the new search tree and has not been placed in the priority queue before. Step1 to Step6 represent the process of Tracking-D*Lite in the second expansion process, and finally get the path pointed to by Second path.

After searching for the first path, as shown in Fig. 6, the robot moves one unit distance to reach C1 according to the planned path, and the goal randomly reaches C3 by one unit distance. The robot senses the surrounding environment information and finds that vertex D2 is a new obstacle, so the algorithm updates the map information and set a new heuristic value, and set km to 1 at the same time.

In the second search, it is necessary to delete vertexes that are not related to this search. Therefore, Tracking-D*lite starts from the last goal vertex, uses depth-first search to delete members whose root vertex is not the current goal vertex in the search tree, and puts them into the Delete queue.

In order to use the information effectively after each search. Step 7 in Fig. 5 is the last state after the first search, which contains a search tree with D3 as the root vertex (because D3 is the goal in the first search). And C3 is the goal position in the second search, so we need to cut the search tree. As shown in the Delete graph in Fig. 7, the gray grid represents the vertexes that have been put into the Delete queue. These vertexes are composed of child vertexes with D3 as the root vertex in the first search. We can find that in the Delete graph, the subtree with C3 as the root vertex is retained, so part of the state information of the last search result is retained during the process of deleting vertexes. Then we need to reuse some of the vertexes in the Delete queue.

For all vertexes in Delete queue, if their neighbor vertexes belong to the current search tree (not the vertexes in the priority queue), then the vertexes will be re-expanded and added to Frontier. As shown in Fig. 7 (Reuse DeleteNode), the vertex D3 is added to Frontier.

As shown in steps 1–6 in Fig. 7, the second search process is similar to the first search process. The COMPUTESHORTESTPATH() is called to search for the second path.

The Tracking-D*Lite algorithm proposed in this paper can track moving targets in unknown terrain. The main idea of the algorithm is to re-plan the path of the target after each move. During the re-planning process, part of the vertexes information of the last search results will be reused. The main process is to delete the vertexes that do not belong to the current search tree and keep the subtree with the root vertex in the current search. Then, part of the vertexes in the Delete are reused to reduce the number of vertex expansions in each search process, thereby speeding up the search speed and efficiency and achieving the effect of tracking moving targets.

4   Experiments

4.1   Tracking-D*Lite Algorithm Experiment

The purpose of this experiment is to verify the performance of the Tracking-D*Lite algorithm, which mainly compares the four indicators of the tracking time, the tracking moving distances, the number of expansion and the success rating(when the target is not stopped). This experiment compares the Tracking-D*Lite algorithm with the repeated-D*Lite algorithm (repeated call of D*lite).

The operation platform is Windows 10, and the experiment is developed using the C++ language.

This experiment uses a 60*60 complex grid map to conduct 100 sets of experiments. Each set of experiments is randomly assigned a pair of chaser and target with different positions. The target moving path of each set of experiments is also different. In the comparison test of Tracking-D*Lite and Repeated-D*Lite, the moving speed and path of the target are the same. As shown in Table.1, the experiment has tested the two algorithms when the agent's view ranger is 2, 4, and 6. From the perspective of average tracking time, Tracking-D*Lite can track a target in less time than Repeated-D*Lite, and as the view ranger of the agent increases, the time consumed by both algorithms will increase. However, Tracking-D*Lite takes much less time than Repeated-D*Lite and has a less increase in time. For the average moving distances, the results of different view rangers under the same algorithm are not much different, but Tracking-D*Lite moves shorter than Repeated-D*Lite, which is also the advantage of Tracking-D*Lite. Similar to the result of the average moving distances, the average number of vertexes expansion results of Tracking-D*Lite is much less than that of Repeated-D*Lite. And the number of vertexes expansion under different view rangers is not much different. Finally, for the success rate of the agent tracking the unstopped target. The success rating of the tracking of the two algorithms increases with the increase of the view ranger, but Tracking-D*Lite is much better than Repeated-D*Lite.

Table 1. Comparison of Tracking-D*Lite and Repeated-D*Lite.

Algorithm	View range	Average tracking time per case(ms)	Average moving distances per case	Average number of expansion per case	Success rating
Tracking- D*Lite	2	2863.727	61.250	29596.735	83%
	4	2997.994	64.530	27582.955	85%
	6	3370.850	66.440	26755.190	93%
Repeated- D*Lite	2	6864.102	88.797	78877.591	19%
	4	11421.131	81.950	85398.840	37%
	6	16354.322	84.850	80229.755	40%

4.2   Simulation Experiment

The purpose of simulation experiment is to verify the rationality of method of cooperative exploring for multi-UAV in an unknown indoor environment based on dynamic target tracking. The content of the experiment is to build a simulated indoor environment. Six quadrotor UAV models use the method proposed in this paper to explore the entire environment and draw a boundary contour map.

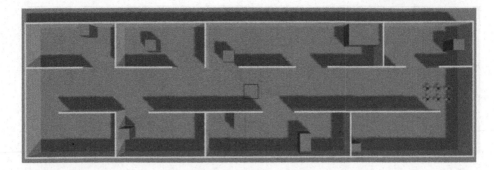

Fig. 8. Simulation experiment scene. 6 UAVs are used to explore an unknown indoor environment in this scene. The unknown indoor environment includes obstacles, small rooms, etc.

The simulation verification experiment is based on the ROS, Gazebo, and Rviz platforms under the Ubuntu16.04 for simulation and development using C++ language, mainly using the hector_quadrotor [20] model toolkit. The specific experimental scene is shown in Fig. 8.

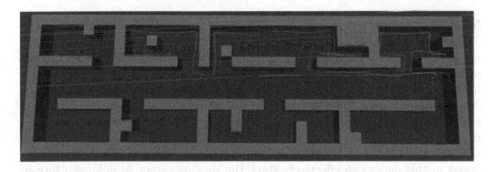

Fig. 9. Track trajectory in real time. The green trajectory is the boundary contour map drawn in real time, and the yellow trajectory is the trajectory of the latter UAV to relay the previous UAV. (Color figure online)

The simulation experiment first integrates the Wall-Around algorithm. To allow UAVs to fly around the boundaries of the map, each UAV can sense environmental information about 1 m around. Then the experiment integrates the Tracking-D*Lite algorithm into the simulation environment. When the power of the last UAV is insufficient, the next UAV will be sent to use Tracking-D*Lite to track the last UAV in the unknown terrain to relay. When the relay is successful, the next UAV continues to use the Wall-Around algorithm to fly around the wall. Finally, in the experiment, the real-time trajectory of each UAV will be drawn and the boundary contour map will be obtained.

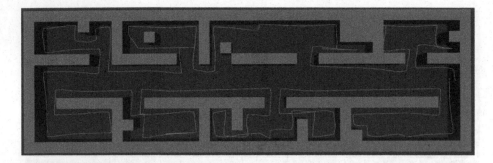

Fig. 10. Boundary contour map. The green trajectory is the final contour map of the entire indoor environment. (Color figure online)

In this experiment, RVIZ is used to draw the trajectory of each UAV in real-time, including the boundary contour map and the path planned during the tracking process. As shown in Fig. 9, the green trajectory is the trajectory of all UAVs flying around the wall, and the yellow trajectory is the trajectory of tracking during the relay. Figure 10 is a boundary contour map drawn after exploring the entire unknown indoor space.

5   Conclusion

In order to improve the efficiency of indoor exploration in a GPS-denied environment, this paper proposes a multi-UAV collaborative exploration method based on dynamic target tracking. This method takes advantages of the Wall-Around algorithm and the Tracking-D*Lite algorithm, using the Wall-Around algorithm to explore the boundary of the unknown indoor environment, and uses the Tracking-D*Lite algorithm to collaborate among UAVs. Finally, completing the task of exploring the entire unknown indoor environment. The method proposed in this paper can effectively track the moving target in unknown terrain and can achieve good results in simulation experiments.

References

1. Floreano, D., Wood, R.J.: Science, technology and the future of small autonomous drones. Nature **521**(7553), 460–466 (2015)
2. Cadena, C., et al.: Past, present, and future of simultaneous localization and mapping: toward the robust-perception age. IEEE Trans. Rob. **32**(6), 1309–1332 (2016)
3. Lumelsky, V., Stepanov, A.: Dynamic path planning for a mobile automaton with limited information on the environment. IEEE Trans. Autom. Control **31**(11), 1058–1063 (1986)
4. Lumelsky, V.J., Stepanov, A.A.: Path-planning strategies for a point mobile automaton moving amidst unknown obstacles of arbitrary shape. Algorithmica **2**(1), 403–430 (1987)

5. McGuire, K.N., de Croon, G., Tuyls, K.: A comparative study of bug algorithms for robot navigation. Rob. Auton. Syst. **121**, 103261 (2019)
6. Koenig, S., Likhachev, M.: Fast replanning for navigation in unknown terrain. IEEE Trans. Rob. **21**(3), 354–363 (2005)
7. Sun, X., Yeoh, W., Uras, T., Koenig, S.: Incremental ara*: an incremental anytime search algorithm for moving-target search. In: Proceedings of the International Conference on Automated Planning and Scheduling, vol. 22 (2012)
8. Grzonka, S., Grisetti, G., Burgard, W.: A fully autonomous indoor quadrotor. IEEE Trans. Rob. **28**(1), 90–100 (2011)
9. Bi, Y., et al.: An autonomous quadrotor for indoor exploration with laser scanner and depth camera. In: 2016 12th IEEE International Conference on Control and Automation (ICCA), pp. 50–55. IEEE (2016)
10. Sampedro, C., Rodriguez-Ramos, A., Bavle, H., Carrio, A., de la Puente, P., Campoy, P.: A fully-autonomous aerial robot for search and rescue applications in indoor environments using learning-based techniques. J. Intell. Rob. Syst **95**(2), 601–627 (2019)
11. Vanegas, F., Campbell, D., Eich, M., Gonzalez, F.: UAV based target finding and tracking in gps-denied and cluttered environments. In: 2016 IEEE/RSJ International Conference on Intelligent Robots and Systems (IROS), pp. 2307–2313. IEEE (2016)
12. Pham, H.X., La, H.M., Feil-Seifer, D., Nguyen, L.V.: Autonomous uav navigation using reinforcement learning. arXiv preprint arXiv:1801.05086 (2018)
13. Walker, O., Vanegas, F., Gonzalez, F., Koenig, S.: A deep reinforcement learning framework for UAV navigation in indoor environments. In: 2019 IEEE Aerospace Conference, pp. 1–14. IEEE (2019)
14. Walker, O., Gonzalez, F., Vanegas Alvarez, F., Koenig, S.: Mutli-UAV target-finding in simulated indoor environments using deep reinforcement learning. In: 2020 IEEE Aerospace Conference. IEEE (2020)
15. McGuire, K., De Wagter, C., Tuyls, K., Kappen, H., de Croon, G.C.: Minimal navigation solution for a swarm of tiny flying robots to explore an unknown environment. Sci. Rob. **4**(35) (2019)
16. Kamon, I., Rivlin, E., Rimon, E.: A new range-sensor based globally convergent navigation algorithm for mobile robots. In: Proceedings of IEEE International Conference on Robotics and Automation, vol. 1, pp. 429–435. IEEE (1996)
17. Hart, P.E., Nilsson, N.J., Raphael, B.: A formal basis for the heuristic determination of minimum cost paths. IEEE Trans. Syst. Sci. Cybern. **4**(2), 100–107 (1968)
18. Harabor, D., Grastien, A.: Online graph pruning for pathfinding on grid maps. In: Proceedings of the AAAI Conference on Artificial Intelligence, vol. 25 (2011)
19. Stentz, A., et al.: The focussed d* algorithm for real-time replanning. In: IJCAI, vol. 95, pp. 1652–1659 (1995)
20. Modeling, control and simulation of quadrotor UAV systems (2014). http://wiki.ros.org/hector_quadrotor

Recommendation System

MR-FI: Mobile Application Recommendation Based on Feature Importance and Bilinear Feature Interaction

Mi Peng, Buqing Cao$^{(\boxtimes)}$, Junjie Chen, Jianxun Liu, and Rong Hu

School of Computer Science and Engineering, Hunan University of Science and Technology, Xiangtan, China

Abstract. With the rapid growth of mobile applications in major mobile app stores, it is challenging for users to choose their desired mobile applications. Therefore, it is necessary to provide a high-quality mobile application recommendation mechanism to meet the user's expectation. Although the existing methods make significant results on mobile application recommendation, the recommendation accuracy can be further improved. More exactly, they mainly focus on how to better interact between mobile applications' features, but ignore the importance or weight of these features themselves. Based on squeeze-excitation network mechanism and bilinear function with combining inner product and Hadamard product, this paper proposes a mobile application recommendation method based on feature importance and bilinear feature interaction to solve this problem. First of all, it exploits a SENET (Squeeze-Excitation Network) mechanism to dynamically learn the importance of mobile applications' features and uses a bilinear function with combining inner product and Hadamard product to effectively learn these features interactions, respectively. Then, the user preferences for different mobile applications are predicted through infusing cross-combined features into a deep model via integrating the classic deep neural network component with the shallow model. The real dataset of Kaggle is used to evaluate the proposed method and the experimental results show that the method can achieve the best results in most cases in terms of AUC and Logloss. It can effectively improve the recommendation accuracy of mobile applications.

Keywords: Mobile application · Recommendation · Feature importance · Bilinear feature interaction

1 Introduction

With the rapid development of 5G, industrial Internet and mobile network, mobile devices have become a very common and indispensable "intermediary" in people's daily life. According to the 46th statistical report on Internet development in China, up to June 2020, China's users in mobile phone have reached 932 million, in which 99.2% Internet users use mobile phones to access the Internet. With the popularity of smart phones and other mobile devices, the number of mobile applications has shown

© ICST Institute for Computer Sciences, Social Informatics and Telecommunications Engineering 2021
Published by Springer Nature Switzerland AG 2021. All Rights Reserved
H. Gao and X. Wang (Eds.): CollaborateCom 2021, LNICST 406, pp. 213–228, 2021.
https://doi.org/10.1007/978-3-030-92635-9_13

explosive growth [1]. Facing such large number of mobile applications with rich information, it is difficult for users to find suitable mobile applications [2]. Therefore, it is necessary to recommend personalized mobile applications to users.

Recommender system is now playing an important role in helping users out of the mobile service overload predicament and automatically recommending proper mobile applications for users [3]. During the process of mobile applications recommendation, the historical interaction information, such as users' rating or comments, can be exploited to estimate the possibility of selecting different mobile applications for users [4]. Among this, traditional collaborative filtering has been widely adopted [5]. It discovers users' preferences for mobile applications by mining user's historical behavior data, and the mobile applications that users may like are predicted for recommendation. That is to say, collaborative filtering achieves the function, such as "guess what you like", "the people who used the mobile application also like". Due to the problem of sparsity and scalability, the recommendation performance of collaborative filtering can be improved. MF (Matrix Factorization) is an enhanced collaborative filtering with collective wisdom and implicit semantics, which models the interaction between users and mobile applications with inner product [6]. It maps user-mobile application scoring matrix with high-dimension into two user-mobile application matrices with low-dimension to solves the problem of data sparsity. Moreover, some extended models on top of matrix factorization are proposed. For example, NCF (Neural Collaborative Filtering) exploits the nonlinear neural network to replace the interaction function of inner product in matrix factorization [7], CDL (Collaborative Deep Learning) extends the embedding function of matrix factorization by combining the deep representations of rich side (feature) information in mobile applications [8]. Most recently, fine-grained models in mobile application recommendation on the basis of matrix factorization via discriminating low-order features, high-order features and their importance are investigated. For instance, NFM (Neural Factorization Machine) [9] and DeepFM (Deep Factorization Machine) [10] simultaneously consider the low-order and high-order feature interactions, MLR (Multiple Linear Regression) [11] only exploits the high-order feature interactions, and AFM (Attentional Factorization Machine) emphasizes the importance (weight) of feature interactions between users and mobile applications [12].

Although the above models and methods make significant results in recommender system, the recommendation accuracy can be further improved. Specifically, they mainly focus on how to better interact between features, but ignore the importance or weight of features themselves. In fact, users have different preferences for different features in mobile applications. For example, compared with the price and size of mobile applications, most users care more about the rating of mobile applications. In other words, different features have different importance to the task of mobile application recommendation. In view of this, inspired by the work of Huang et al. [13], this paper proposes a mobile application recommendation method based on feature importance and bilinear feature interaction, abbreviated as MR-FI. It exploits a SENET (Squeeze-Excitation Network) mechanism to dynamically learn the importance of features, and uses a bilinear function with combining inner product and Hadamard product to effectively learn the features interactions [14], respectively. In summary, the contributions of this paper are as follows:

- To our best knowledge, this is the first work to introduce SENET mechanism into mobile application recommendation, which is conducive to mine the importance of features in mobile applications.
- We propose a novel mobile application recommendation method based on feature importance and bilinear feature interaction, via employing the SENET to learn the importance of features dynamically and the three types of bilinear interaction layer to learn feature interactions in a fine-grained way.
- Through the real dataset of Kaggle, we verify the effectiveness and accuracy of the SENET mechanism used in mobile application recommendation. The experimental results indicate the performance of MR-FI is superior to that of all the comparison models in most cases in terms of AUC and Logloss.

The remainder of this paper is arranged as follows: the Sect. 2 is the related work of mobile application recommendation; the Sect. 3 is the specific proposed method; the Sect. 4 is the experimental evaluation and analysis; and the last section is the summary of our work and the follow-up research work.

2 Related Work

It is challenging for users to find their own interesting mobile applications quickly from a large number of mobile applications with rich content. Therefore, the relevant scholars study mobile application recommendation [17, 25], in order to improve the efficiency of mobile application searching and the accuracy of recommendation results. The main method in mobile application recommendation is based on content information and network structure.

Content-based mobile application recommendation mainly recommends mobile applications to users through user's text description, tags, user information, etc. [17, 21]. For example, Chen et al. [17] propose a novel third-party library recommendation method by integrating topic modeling and knowledge graph technology, which extracts topics from text application description, and uses knowledge graph to merge structured information of third-party library and application, as well as interactive information of recommended application and library. Cao et al. [18] use app information and user information of multiple platforms to improve the accuracy of mobile application recommendation. This alleviates the problem of data sparsity and cold start to a certain extent. Zhong et al. [19] design a mobile application recommendation method based on hyperlinks induced topic search (hits) algorithm and association rules, which considers the importance of mobile applications in association rules and the reliability of users. Pai et al. [20] calculate the positive and negative scores of semantic tendency in each comment according to the pointwise mutual information, and consider the subjective factors such as public opinion, anonymous opinion and star rating, and the objective factors such as the number of downloads and reputation. Xu et al. [21] present an effective function extraction method. One of the main features of this work is to extract mobile application functions from user reviews for recommendation.

Mobile application recommendation based on network structure exploits the similarity between mobile applications to build a network and recommend related mobile applications to users [22, 25]. For example, Liang et al. [22] propose to model functional interaction from different views through attention mechanism. The novelty of this method is that it introduces view segmentation for feature interaction and two levels of attention network construction. Xie et al. [23] use user's historical behavior data and application's auxiliary information to design application recommendation to solve the problem of information overload. Cheng et al. [24] utilize a mechanism to model the three important factors that control the application installation of smartphone users. They may classify the application to be installed as one of these factors, and provide application suggestions for users accordingly. Donghwan et al. [25] express the user's usage pattern as a graph. Based on the graph, the user's recommended value for unknown applications is predicted by measuring the similarity.

In recent years, many applications of CTR model based on deep learning to recommendation system are proposed. FNN [15] supported by factor decomposing machine is a forward neural network which uses FM to pre-train embedded layer. However, FNN can only capture higher-order feature interaction. The wide&deep model (WDL) [16] is introduced in google-play for application recommendation. WDL combines training of wide linear model and deep neural network to recommend the system based on the advantages of memory and generalization. However, the input of WDL in wide domain still needs professional feature engineering, which means that cross domain transformation also needs manual design. In order to reduce the manual work in feature engineering, DeepFM [10] replaces the wide part of WDL with FM, and shares feature embedding between FM and deep components. DeepFM is one of the most advanced models in CTR estimation field. NFM [9] combines the second-order linear feature extracted by FM and the higher-order nonlinear feature extracted by neural network, and NFM has more expressive capability than FM. As mentioned in [12], FM can be hindered by its modeling of all feature interactions with the same weight, because not all feature interactions have the same usefulness and predictability. They propose the attention factor decomposing machine (AFM) [12] model, which can learn the importance of each feature interaction automatically from the data through neural attention network, so as to realize the different contribution of feature interaction to prediction. MLR can be regarded as a natural extension of LR. It uses the idea of dividing and governing, and uses piecewise linear model to fit the nonlinear classification surface of high-dimensional space.

3   Proposed Method

The goal of MR-FI is to dynamically learn features and the importance of feature interaction in a more fine-grained way. The structure of the model is shown in Fig. 1. The MR-FI model consists of embedding layer, SENET layer, bilinear-interaction layer, connectivity layer, neural network and prediction layer. Among them, the embedding layer firstly embeds the sparse representation of the original input features into the dense vector. Secondly, the SENET layer transforms the embedding layer into the SENET-like embedded features to improve the distinguishability of the features. Thirdly, the bilinear-interaction layer models the original embedding and the SENET-like embedding by the second-order feature interaction. Fourthly, the connectivity layer integrates the output connection of the bilinear interactive layer into the dense vector. Finally, the cross-combined features are infused into the neural network, and the prediction score is achieved in the prediction layer.

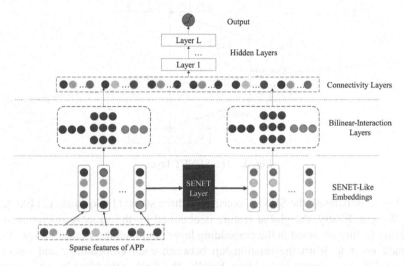

Fig. 1. Structure description of MR-FI model.

3.1 Embedding Layer

The sparse input layer uses sparse representation for the original input features. The embedding layer can embed the sparse features into the low dimensional continuous real valued vector, transform the sparse matrix into the dense matrix by linear transformation, extract the hidden features of the matrix, and improve the generalization ability of the model. The output of the embedded layer is expressed as follows:

$$E = \left[e_1, e_2, .., e_i, \ldots, e_f \right] \tag{1}$$

Where f is the number of domains, $e_i \in \mathbb{R}^k$ is the representation of the i-th domain, and the size of the vector is k. Sparse input layer and embedded layer are widely used in deep learning-based hit rate prediction models, such as NFM and FNN.

3.2 SENET Layer

Different features have different importance to the target task. For example, when predicting a user's preference for a mobile application, the score of the mobile application may be more important than the price of the mobile application. Inspired by the successful application of SENET [14] in the field of computer vision, this paper introduces the SENET mechanism to make the model pay more attention to the importance of features. For a specific mobile application recommendation task, we can dynamically increase the weight of important features and reduce the weight of features with insufficient information through the SENET mechanism.

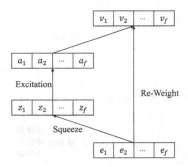

Fig. 2. The SENET layer.

As Fig. 2 illustrated, the SENET consists of three steps: (1) Squeeze, (2) Excitation, (3) Re-Weight. Firstly, the global feature is obtained by the squeeze operation on the embedding feature obtained in the embedding layer, and then the excitation operation is performed on it to learn the relationship between each embedding, and obtain the weight of different domain embedding. Finally, the final embedding result is obtained by multiplying the original embedding. The specific steps are as follows:

Squeeze. The first step is to compress the original embedding E into the statistical vector $Z = [z_1, \ldots, z_i, \ldots, z_f]$ by using the average pooling operation. z_i can be calculated by the following formula:

$$z_i = \frac{1}{k} \sum_{t=1}^{k} e_i^{(t)} \qquad (2)$$

Where, z_i is the global information about the i-th feature representation, and k is the embedded dimension size.

Excitation. The second step is to learn the embedding weights of each domain based on the statistical vector Z, and use two fully connected (FC) layers to learn the weights. The first fully connected layer is the dimension reduction layer of parameter W_1, which uses the σ_1 as a nonlinear function. The second fully connected layer restores the original dimension by increasing the dimension with parameter W_2. Formally, the weight of domain embedding can be calculated as follows:

$$A = \sigma_2(W_2\sigma_1(W_1Z)) \qquad (3)$$

Where $A \in \mathbb{R}^f$ is the vector, σ_1 and σ_2 is the activation function.

Re-Weight. The last step is to re-weight, that is, each field of the embedding layer is multiplied by the corresponding weight, and the final embedding result is $V = \{v_1, \ldots, v_f\}$. The whole operation can be seen as learning the weight co-efficient of each domain embedding, which makes the model more discriminative to each domain embedded features. The embedded V of class SENET can be calculated as follows:

$$V = [a_1 \cdot e_1, \ldots, a_f \cdot e_f] = [v_1, \ldots v_f] \qquad (4)$$

In short, the SENET uses two fully connected layers to dynamically learn the importance of features. For the recommendation task of mobile applications, it increases the weight of important features and reduces the weight of features with insufficient information.

3.3 Bilinear-Interaction Layer

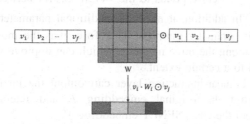

Fig. 3. The method to calculate the feature interactions.

The interaction layer is a layer to calculate the second-order feature interaction. The classical methods of feature interaction are inner product and Hadamard product. The inner product ($\cdot$) is widely used in shallow models such as FM and FFM, while Hadamard product ($\odot$) is commonly used in deep models such as NFM. For example, the inner product is $[a_1, a_2, \ldots, a_n] \bullet [b_1, b_2, \ldots, b_n] = \sum_{i=1}^{n} a_i b_i$. The Hadamard product is $[a_1, a_2, \ldots, a_n] \odot [b_1, b_2, \ldots, b_n] = [a_1 b_1, a_2 b_2, \ldots, a_n b_n]$. Because the inner product and Hadamard product of the interaction layer are too simple to model the feature interaction of sparse data sets effectively. Therefore, this paper uses a more fine-grained method, combining inner product and Hadamard product to learn feature interaction with additional parameters. As Fig. 3 illustrated, the interaction vector p_{ij} can be calculated in the following three ways:

Field-All Type

$$p_{ij} = v_i \cdot W \odot v_j \qquad (5)$$

Where $W \in \mathbb{R}^{k \times k}$, all vectors (v_i, v_j) share the W. So, it is called as "Field-All". The additional parameter is $k \times k$.

Field-Each Type

$$p_{ij} = v_i \cdot W_i \odot v_j \tag{6}$$

Among them, $W_i \in \mathbb{R}^{k \times k}$ corresponds to the parameter matrix of the i-th field, so this layer needs to maintain $f \times k \times k$ parameters, which is called as "Field-Each".

Field-Interaction Type

$$p_{ij} = v_i \cdot W_{ij} \odot v_j \tag{7}$$

Among them, $W_{ij} \in \mathbb{R}^{k \times k}$ corresponds to the weight matrix between the i-th domain and the j-th domain. In addition, it needs an additional parameter of $\frac{f(f-1)}{2} \times k \times k$, which involves more parameter matrices. Therefore, it is the most time-consuming calculation method among the three methods, which can improve the accuracy of the recommended model to a certain extent.

In this part, the bilinear-interaction layer can output the interaction vector $p = [p_1, \ldots, p_i, \ldots, p_n]$ from the original embedding E and interaction vector $q = [q_1, \ldots, q_i, \ldots, q_n]$ from the class SENET embedding V.

3.4   Connectivity Layer

The connectivity layer connects the interaction vectors p and q, and infuses the connected vectors into the next layer of MR-FI model, a standard neural network layer. The specific process can be expressed as follows:

$$C = [p_1, \ldots, p_n, q_1, \ldots, q_n] = [c_1, \ldots, c_{2n}] \tag{8}$$

If each element of vector C is summed up and a sigmoid function is used to output a predictive value, a shallow prediction model is gotten. In order to further improve the performance of the model, this paper considers the combination of shallow component and deep neural network (DNN).

3.5   Deep Network

In order to further improve the performance, we combine the classic deep neural network (DNN) component with the shallow model to form a deep model. The input of this layer is the output vector C of the connectivity layer, and the deep network is composed of multiple fully connected layers, which can implicitly capture higher-order features. Let $a^{(0)} = [c_1, \ldots, c_{2n}]$ denotes the initial input. Next, $a^{(0)}$ is poured into the deep neural network, and the feedforward process is as follows:

$$a^{(l)} = \sigma\left(W^l a^{(l-1)} + b^l\right) \tag{9}$$

Where $a^{(l)}$ is the output of layer l of the deep network, σ is a function of sigmoid, W^l represents the weight matrix of the model, and b^l represents the offset of the model.

After L layer, a dense real valued eigenvector is generated and input into *sigmoid* function for mobile application prediction.

$$y_d = \sigma\left(W^L a^{(L+1)} + b^{L+1}\right) \tag{10}$$

Where L is the number of layers of the deep model.

3.6 Prediction Layer

Finally, the output expression of the model prediction layer is as follows:

$$\hat{y} = \sigma\left(w_0 + \sum_{i=0}^{m} w_i x_i + y_d\right) \tag{11}$$

Where $\hat{y} \in (0, 1)$ is the predictive value of the model, σ is the *sigmoid* function, m is the feature size, and w_i is the i-th weight of the linear part.

The whole training process aims to minimize the following objective function:

$$Logloss = -\frac{1}{N}\sum_{i=1}^{N}(y_i log(\hat{y}_i)) + (1 - y_i) \times log(1 - \hat{y}_i)) \tag{12}$$

Where, y_i is the real label of the i-th mobile application, $\hat{y}_i$ corresponds to the prediction tag of the i-th mobile application, and N is the total number of mobile applications.

4 Experimental Result and Analysis

4.1 Data Set and Experiment Setup

This experiment uses the open data set "App Store" of kaggle as the experimental data set. In this data set, a total of 11 features are selected, including 2 sparsity features and 9 continuity features, that are prime_ genre, cont_ rating, price, rating_ count_ tot, rating_ count_ ver, user_ rating, user_ rating_ ver, sup_ devices.num, ipadSc_ urls. num, lang.num and size_ MB. Since there is no label in the original data set, we manually set the label value of mobile applications with a score of more than 3 and a score of more than 800 as 1, otherwise are 0. The proportion of positive samples is about 0.424, so that the experimental results will not be affected too much due to the uneven distribution of samples. In the experiment, the optimizer is Adam and the loss function is binary cross-entropy.

During model training, the batch size is set to 64, and the proportion of verification set is 0.2. In order to better illustrate the experimental results, different proportions of

Table 1. Category number statistics of top20 mobile applications.

Category name	Number	Category name	Number
Games	3381	Lifestyle	98
Entertainment	456	Shopping	82
Education	408	Weather	69
Photo & video	339	Book	59
Utilities	202	Travel	58
Health & fitness	166	News	55
Productivity	164	Business	54
Music	136	Reference	53
Social networking	113	Finance	47
Sports	103	Food & drink	45

training sets are selected to test the experimental results. The distribution details of the top 20 categories with the largest number are shown in Table 1.

4.2 Evaluation Metrics

AUC and Logloss, which are widely used in click through prediction, are selected as evaluation indexes [10].

AUC is the area under the *ROC* curve. When $0.5 < AUC < 1$, the model is better than the random classifier. In particular, the closer the *AUC* is to 1.0, the higher the authenticity is; When it is equal to 0.5, the authenticity is the lowest. The calculation formula is as follows:

$$\text{AUC}_i = \int_0^1 \text{ROC}_i(fpr)d(fpr) \tag{13}$$

Where *fpr* stands for false positive rate and *fpr* stands for true positive rate. In *ROC* space, the coordinate point (fpr, fpr) describes the trade-off between *FP (false positive case)* and *TP (true positive case)*.

Logloss measures the accuracy of a classifier by punishing the wrong classification. Minimizing the logarithmic loss is basically equivalent to maximizing the accuracy of the classifier. Logloss reflects the average deviation of samples, and is often used as the loss function of the model to optimize. The calculation formula is as follows:

$$Logloss = -\frac{1}{N}\sum_{i=1}^{N}(y_i log(\widehat{y}_i)) + (1 - y_i) \times log(1 - \widehat{y}_i)) \tag{14}$$

Where, y_i is the real label of the *i*-th sample, $\widehat{y}_i$ corresponds to the prediction tag of the *i*-th sample, and N is the total number of mobile application samples.

4.3 Baseline Methods

This paper compares the following methods to better illustrate the experimental results.

MLR [11]: MLR model is a generalization of linear LR model, which uses piecewise linear method to fit data, and can learn higher-order feature combination. The basic strategy is to divide and conquer.

AFM [12]: improves the FM model and introduces the attention mechanism into the feature crossover module. An attention network is used to learn the importance of different combination features (second-order crossover).

FNN [15]: FNN (neural network based on factor decomposition machine) uses FM for supervised learning to obtain the embedding layer, which can effectively reduce the dimension of sparse features and obtain continuous and dense features.

NFM [9]: by combining FM's linear crossover of second-order features with neural network's nonlinear crossover of higher-order features, NFM learns feature combinations that never appear in the training set.

DeepFM [10]: combines the first-order and second-order features of FM and the interaction of higher-order features, and reduces the amount of parameters and shares information by sharing the embedding of FM and DNN.

4.4 Experimental Performance

In this section, we select the training set ratio from 0.2 to 0.9 to compare the model performance. The experimental results are shown in Table 2, Figs. 4 and 5, respectively. In general, MR-FI has the best performance. Especially when the train_size is 0.8, the AUC of MR-FI is 19.2%, 20.99%, 1.22%, 0.27% and 1.08% higher than that of MLR, AFM, FNN, NFM and DeepFM, respectively. Compared with the above methods, the logloss of MR-FI decreased by 32.73%, 34.72%, 2.22%, 3.37% and 3.28%. More specifically, we have the following findings:

MR-FI has the best performance in all models, which shows that considering the weight of original features and using Bilinear interactive learning feature interaction can effectively improve the accuracy of the recommended model.

The performance of depth models such as FNN and DeepFM is better than that of MLR. When the training size is 0.8, the performance of FNN and DeepFM is improved by 17.98% and 18.21%, respectively. It shows that the depth model can better model when features are sparse.

In the shallow model, MLR is always better than AFM. When the training size is 0.2, the accuracy of MLR model is 4.73% higher than that of AFM model, which indicates that learning the weight of high-order feature combination is more important than that of low-order feature combination.

Table 2. Performance comparison of different proportion training sets.

Models	Train_size = 0.2		Train_size = 0.5		Train_size = 0.8		Train_size = 0.9	
	AUC	Logloss	AUC	Logloss	AUC	Logloss	AUC	Logloss
MLR	0.7343	0.5933	0.7708	0.5604	0.7821	0.5430	0.7959	0.5409
AFM	0.6870	0.6288	0.7529	0.5853	0.7642	0.5629	0.7773	0.5607
FNN	0.8684	0.4811	0.9265	0.4079	0.9619	0.2379	0.9732	0.2538
NFM	0.8736	0.4516	0.9348	0.3384	0.9714	0.2494	0.9739	0.2382
DeepFM	0.8703	0.4584	0.9267	0.3723	0.9633	0.2485	0.9761	0.2068
MR-FI	0.8749	0.4491	0.9328	0.3959	0.9741	0.2157	0.9807	0.1825

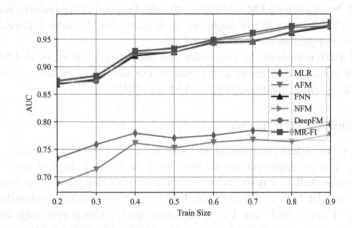

Fig. 4. AUC comparison of different models.

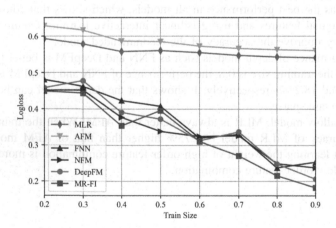

Fig. 5. Logloss comparison of different models.

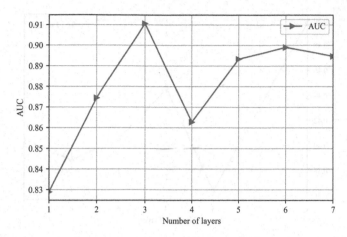

Fig. 6. Influence of neural networks with different layers on AUC.

Starting from FNN, the accuracy of this kind of model is significantly improved by adding neural network. The results show that the neural network can learn more effectively from FM representation, and can get better results than shallow model.

The overall performance of NFM model and DeepFM model is also good, which indicates that the interaction of low-order features and high-order features can improve the accuracy of the model to a certain extent.

4.5   Hyperparameters Analysis

In this part, the hyperparameters analysis mainly includes the size of embedding and the number of layers of neural network. However, in the feature modeling of mobile application recommendation, there are more continuity features and less sparsity features. Therefore, the embedding size has little effect on the accuracy of the model, which is not discussed here. This section mainly analyzes the influence of neural network layers on the model. According to the experimental results of Figs. 6 and 7, it can be observed that: (1) when the number of neural network layers increases from 1 to 3, AUC has been improved to a certain extent, from 0.83 to 0.91; (2) when the number of layers increases to 4, AUC decreases sharply; (3) when the number of neural network layers increases from 4 to 7, AUC has been improved to a certain extent, but it has not reached the highest point of layer 3, and has a downward trend. From the influence of different neural network layers on Logloss, we can find the same trend, that is, when the layer number is 3, Logloss reaches the lowest point. Therefore, we can

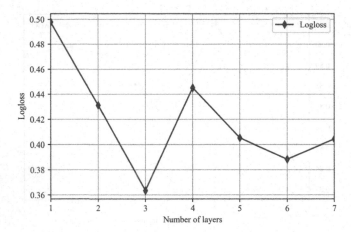

Fig. 7. Influence of neural networks with different layers on logloss.

infer that increasing the number of DNN layers will raise the complexity of the model. Specifically, in a certain number of layers, increasing the number of layers will improve the performance of the model; but if the number of layers increases, subsequently the performance will decline. This is because too complex model can easily lead to over-fitting. In particular, for mobile application recommendation, a good choice is to set the layer size as 3.

5 Conclusion and Future Work

In this paper, a mobile application recommendation method based on feature importance and bilinear feature interaction MR-FI is proposed to dynamically learn the feature importance and fine-grained feature interaction of mobile applications. In this method, on the one hand, a more complex bilinear interaction layer is introduced to learn feature interaction, rather than simply using inner product or Hadamard product to calculate feature interaction. On the other hand, for mobile application recommendation tasks, it uses the SENET module to dynamically learn the importance of features, which will enhance the weight of important features and reduce the weight of unimportant features. By combining SENET mechanism with bilinear feature interaction, the MR-FI model can further improve the accuracy of mobile application recommendation. Through the verification on the open data set of Kaggle, the MR-FI model performs best in most cases. In the future work, we will explore more fine-grained feature interaction and discriminate their importance with different dimensions for further improving the accuracy of mobile application recommendation.

Acknowledgement. Our work is supported by the National Key R&D Program of China (2018YFB1402800), the National Natural Science Foundation of China (No. 61873316, 61872139, 61832014 and 61702181), and the Natural Science Foundation of Hunan Province (No. 2021JJ30274).

References

1. Gao, S., Zang, Z., Gopalakrishnan, S.: A study on distribution methods of mobile applications in China. In: Fong, S., Pichappan, P., Mohammed, S., Hung, P., Asghar, S. (eds.) Seventh International Conference on Digital Information Management, pp. 375–380 (2012)
2. Deng, S., et al.: Toward mobile service computing: opportunities and challenges. IEEE Cloud Comput. **3**, 32–41 (2016)
3. Zhang, D., Lee, W.S.: Question classification using support vector machines. In: Clarke, C. L.A., Cormack, G.V., Callan, J., Hawking, D., Smeaton, A.F. (eds.) SIGIR 2003: Proceedings of the 26th Annual International ACM SIGIR Conference on Research and Development in Information Retrieval, New York, pp. 26–32 (2003)
4. Deerwester, S.C., Dumais, S.T., Landauer, T.K., Furnas, G.W., Harshman, R.A.: Indexing by latent semantic analysis. J. Am. Soc. Inf. Sci **41**(6), 391–407 (1990)
5. Yang, Z., Wu, B., Zheng, K., Wang, X., Lei, L.: A survey of collaborative filtering-based recommender systems for mobile internet applications. IEEE Access. **4**, 3273–3287 (2016)
6. Koren, Y., Bell, R.M., Volinsky, C.: Matrix factorization techniques for recommender systems. Computer **42**, 30–37 (2009)
7. He, X., Liao, L., Zhang, H., Nie, L., Hu, X., Chua, T.S.: Neural collaborative filtering. In: Proceedings of the 26th International Conference on World Wide Web, pp. 173–182 (2017)
8. Wang, H., Wang, N., Yeung, D.Y.: Collaborative deep learning for recommender systems. In: Proceedings of the 21th ACM SIGKDD International Conference on Knowledge Discovery and Data Mining, pp. 1235–1244 (2015)
9. He, X., Chua, T.S.: Neural factorization machines for sparse predictive analytics. In: Kando, N., Sakai, T., Joho, H., Li, H., Vries, A.P. de, and White, R.W. (eds.) Proceedings of the 40th International ACM SIGIR Conference on Research and Development in Information Retrieval, Shinjuku, pp. 355–364 (2017)
10. Guo, H., Tang, R., Ye, Y., Li, Z., He, X.: DeepFM: A Factorization-machine based neural network for CTR prediction. In: Proceedings of the 26th International Joint Conference on Artificial Intelligence, pp. 1725–1731 (2017)
11. Gai, K., Zhu, X., Li, H., Liu, K., Wang, Z.: Learning piece-wise linear models from large scale data for Ad click prediction. CoRR. abs/1704.05194 (2017)
12. Xiao, J., Ye, H., He, X., Zhang, H., Wu, F.: Attentional factorization machines: learning the weight of feature interactions via attention networks. In: Proceedings of the 26th International Joint Conference on Artificial Intelligence, pp. 3119–3125 (2017)
13. Huang, T., Zhang, Z., Zhang, J.: FiBiNET: combining feature importance and bilinear feature interaction for click-through rate prediction. In: Bogers, T., Said, A., Brusilovsky, P., and Tikk, D. (eds.) Proceedings of the 13th ACM Conference on Recommender Systems, pp. 169–177 (2019)
14. Hu, J., Shen, L., Albanie, S., Sun, G., Wu, E.: Squeeze-and-excitation networks. IEEE Trans. Pattern Anal. Mach. Intell. **42**(8), 2011–2023 (2020)
15. Zhang, W., Du, T., Wang, J.: Deep learning over multi-field categorical data. In: Ferro, N., et al. (eds.) ECIR 2016. LNCS, vol. 9626, pp. 45–57. Springer, Cham (2016). https://doi.org/10.1007/978-3-319-30671-1_4
16. Cheng, H.-T., et al.: Wide and deep learning for recommender systems. In: Proceedings of the 1st Workshop on Deep Learning for Recommender Systems, pp. 7–10 (2016)
17. Chen, J., Li, B., Wang, J., Zhao, Y., Yao, L., Xiong, Y.: Knowledge graph enhanced third-party library recommendation for mobile application development. IEEE Access **8**, 42436–42446 (2020)

18. Cao, B., Chen, J., Liu, J., Wen, Y.: A topic attention mechanism and factorization machines based mobile application recommendation method. Mobile Netw. Appl. **25**(4), 1208–1219 (2020). https://doi.org/10.1007/s11036-020-01537-z
19. Pai, H.-T., Lai, H.W., Wang, S., Wu, M.F., Chuang, Y.T.: Recommendations for mobile applications: facilitating commerce in google play. In: Hamdan, H., Boubiche, D.E., and Klett, F. (eds.) Proceedings of the 1st International Conference on Internet of Things and Machine Learning, pp. 10:1–10:6. ACM (2017)
20. Pai, H.T., Lai, H.W., Wang, S., Wu, M.F., Chuang, Y.T.: Recommendations for mobile applications: facilitating commerce in google play. In: Hamdan, H., Boubiche, D.E., and Klett, F. (eds.) Proceedings of the 1st International Conference on Internet of Things and Machine Learning, pp. 10:1–10:6 (2017)
21. Xu, X., Dutta, K., Datta, A.: Functionality-based mobile app recommendation by identifying aspects from user reviews. In: Myers, M.D., Straub, D.W. (eds.) Proceedings of the International Conference on Information Systems - Building a Better World through Information Systems, ICIS, Auckland, New Zealand (2014)
22. Liang, T., Zheng, L., Chen, L., Wan, Y., Yu, P.S., Wu, J.: Multi-view factorization machines for mobile app recommendation based on hierarchical attention. Knowl. Based Syst. **187**, 104821 (2020)
23. Xie, F., Cao, Z., Xu, Y., Chen, L., Zheng, Z.: Graph neural network and multi-view learning based mobile application recommendation in heterogeneous graphs. In: 2020 IEEE International Conference on Services Computing, Beijing, China, pp. 100–107 (2020)
24. Cheng, V.C., Chen, L., Cheung, W.K., Fok, C.-K.: A heterogeneous hidden Markov model for mobile app recommendation. Knowl. Inf. Syst. **57**(1), 207–228 (2017). https://doi.org/10.1007/s10115-017-1124-3
25. Bae, D., Han, K., Park, J., Yi, M.Y.: AppTrends: a graph-based mobile app recommendation system using usage history. In: 2015 International Conference on Big Data and Smart Computing, pp. 210–216. IEEE Computer Society, Jeju, Korea (2015)

The Missing POI Completion Based on Bidirectional Masked Trajectory Model

Jun Zeng[✉], Yizhu Zhao, Yang Yu, Min Gao, and Wei Zhou

School of Big Data and Software Engineering, Chongqing University,
Chongqing, China
{zengjun,zhaoyizhu,yuyang96,gaomin,
zhouwei}@cqu.edu.cn

Abstract. With the development of location-based social networks (LBSNs), users can check in Point-of-Interest (POIs) at any time. However, users do not check in all places they have visited, so the POI trajectory sequence generated through LBSNs is incomplete. An incomplete POI trajectory will have a negative impact on subsequent tasks such as POI recommendation and next POI prediction. Therefore, we complete the missing POI in the user trajectory sequence. Since the POI trajectory sequence is incomplete, it is a challenge to use the pre-order and post-order trajectory sequences with missing POIs. Therefore, we propose a masked POI trajectory model (MPTM) that uses the bidirectionality of BERT to complete the missing POIs in user's behavior sequence. By masking the missing POIs, MPTM fully explores the relationship between the missing POIs and the known POIs to predict the missing POIs. In order to strengthen the relationship between POIs in the user trajectory sequence, we build a graph for each user's incomplete POIs sequence to explore the user's hidden behavior habits. Besides, we design experiments to explore the relationship between the continuity of the number of missing POIs and the predictive ability of the model. The experimental results demonstrate that our MPTM outperforms the state-of-the-art models for completion on missing POIs of user's behavior sequence.

Keywords: The missing POIs · Missing POIs completion · Bidirectional masked trajectory model

1 Introduction

With the rapid development of information technology, human mobility behavior is more easily digitized and shared with friends [1]. Especially with the rapid growth of Location-Based Social Networks (LBSNs), such as Yelp, Gowalla and Foursquare. Point-of-Interest (POI) research has attracted wide attention from both academia and industry [2, 3]. However, users do not check in all places they have visited, there are some missing POIs in the user trajectory data collected by LBSN. In reality, the check-in POIs delivered by users is typically incomplete [4]. An incomplete POI trajectory will have a negative impact on subsequent tasks such as POI recommendation and next POI prediction. Therefore, we complete the missing POI in the user trajectory sequence. Existing studies are mainly focused on next location prediction or POI

© ICST Institute for Computer Sciences, Social Informatics and Telecommunications Engineering 2021
Published by Springer Nature Switzerland AG 2021. All Rights Reserved
H. Gao and X. Wang (Eds.): CollaborateCom 2021, LNICST 406, pp. 229–243, 2021.
https://doi.org/10.1007/978-3-030-92635-9_14

recommendation [5, 6]. POI recommendation is to analyze all the historical check-in data of the user and dig out its internal connections, and predict the user's next check-in location to complete the recommendation. While the missing POI completion is to learn the user's historical check-in data and complete the missing POI. This requires bidirectional learning of the user's POI trajectory sequence. However, discovering and integrating the user's behavior sequence relationship for the completion of the missing POIs in sequence is challenging. The reason is that it is difficult to learn the context of missing positions in the user sequence and establish the connection between POIs due to their incompletion.

The current research is mainly for the completion of GPS trajectory [7, 8], but there is few research on POI trajectory completion. The current POI research focuses on POI recommendation, next POI prediction, etc. Due to the remarkable achievements of deep learning in the field of POI research, deep learning technology such as RNN [9] has gradually replaced simple forms of Collaborative Filtering (CF) [10]. In the missing POI completion problem, we need to learn the pre-order and post-order trajectory sequences with missing POIs. The above methods cannot solve the problem.

To address the limitations mentioned above, we propose a bidirectional model to learn the representations for users' behavior sequences and complete user's missing POIs in sequence. The bidirectional model is more suitable than unidirectional models in modeling user behavior sequences since all POIs in the bidirectional model can leverage the contexts from both left and right side [11]. We use the bidirectionality of BERT [12] to fully mine the before and after information of the missing POI. The pre-training task masked language model (Mask LM) of BERT combines sentence context information through self-attention to predict masked words. Therefore, we apply this idea to the problem of missing POIs completion in trajectory sequence. We regard the user's POI trajectory sequence as a paragraph, and few words or a sentence are predicted through the Masked LM task. Specifically, inspired by the success of BERT4Rec [11] in users' behavior sequences, we propose applying the Masked LM in the problem of missing POIs completion in users' behavior sequences. Transformer [13] encoder is used to learn the sequence relationship and long-distance information dependence around the masked POI.

Besides, in order to strengthen the connection between POIs, we fully explore the implicit features of POIs checked in by user. A graph is created for each user of checked in POIs and used DeepWalk [14] for vector representation learning. The user's POIs network graph contains the precedence and potential contacts of checked in POIs by users. This can be seen as a hidden habit of the user check-ins.

The contributions of our paper are as follows:

- We propose and solve the problem of the missing POIs in the user trajectory sequence and complete them. The incomplete POI trajectory sequence has negative impact on subsequent tasks such as POI recommendation, location prediction, and human mobility. And completing the missing POIs is very important in epidemic prevention and control.
- In order to solve the association between missing POI and the information before and after the missing position in the sequence, we propose a bidirectional mask POI trajectory model (MPTM). It is combined with graph features to mine the

relationship between the missing POI and the known POI in the trajectory sequence and the characteristics of the user's behavior sequence.

- In order to strengthen the relationship between POIs, we build a graph structure for each user to learn the graph features of POI, and dig out the user's hidden behavior habits.

The rest of the paper is organized as follows. Section 2 summarizes the related work, which is highly relevant to our research. Section 3 is the problem statement of POI trajectory sequence completion. Section 4 provides detailed methodology of our proposed model. Section 5 presents experiments and the results, and Sect. 6 concludes this paper and outlines prospects for future study.

2 Related Work

The current research mainly focuses on POI recommendation, next location prediction, etc. In the POI recommendation problem, Han et al. [15] divided the context information of POI into two groups, namely global and local contexts, and developed different regularization terms to merge them for recommendation. In the next location prediction problem, Zhao et al. [16] propose a new spatio-temporal gated network by enhancing long-short term memory network, where spatio-temporal gates are introduced to capture the spatio-temporal relationships between successive check-ins.

These methods have achieved good results on the corresponding problems in POI research. However, there is currently no way to solve the problem of missing POIs in the user's trajectory sequence and to complete the missing POIs. Zhang et al. [4] elaborate that the check-in information delivered by users is typically incomplete in reality. But their work only alleviates the impact of incomplete POI trajectories on recommendation, and do not solve the problem of missing POIs. Incomplete POI trajectory information has negative impact on subsequent tasks such as POI recommendation, location prediction, and human mobility. Therefore, it is necessary to use mature methods to predict the missing POIs in the POI trajectory sequence and complete it.

Liu et al. [17] are the first to learn from natural language processing, treating each POI as a word, and each user's check-in record as a sentence. Then, train the implicit representation vector of each POI and explore the influence of the temporal implicit representation vector on it. With the successful application of RNN in sequential data modeling, RNN can be used to model the user's access sequence in continuous POI recommendations [16]. And subsequent POI research generally use RNN and its variants [18, 19]. In the missing POI completion problem, bidirectional model is required to learn optimal representations for user POI trajectory sequences. Recently, an attention-based sequence-to-sequence method, Transformer [13], achieved state-of-the-art performance and efficiency on machine translation tasks which had previously been dominated by RNN-based approaches [20, 21]. Specifically, due to the success of BERT [12] Mask LM in text understanding, we consider applying the deep bidirectional multi-headed attention to solve the problem of missing POIs completion.

3 Problem Statement

3.1 Scene Description

When an epidemic breaks out, prevention and control are very important. Information technology can help the country control the epidemic. But when the information is incomplete, it will increase the risk of the epidemic in some areas. As shown in Fig. 1, when a confirmed patient is found, it is necessary to track and investigate the patient's historical trajectory within a week to reduce the risk of the epidemic. The patient has been to the supermarket two days before the diagnosis, but has not checked in location information on the LBSNs. This supermarket is frequently visited by the user and has check-in records in LBSNs within one month. Therefore, there is no supermarket in the patient's historical trajectory in the last week on the LBSNs. When tracking and investigating the patient's historical trajectory in the last seven days through LBSNs, the supermarket will be ignored. People who come into contact with the patient in the supermarket are potentially at risk. The risk of the epidemic in the area where the supermarket is located will increase. Therefore, it is very important to predict and complete the user's missing location information in the prevention and control of the epidemic.

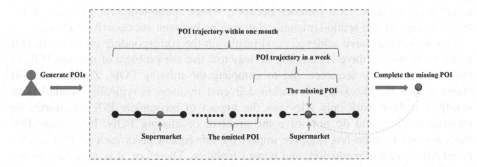

Fig. 1. Scene description

3.2 Problem Definition

In the missing POIs completion problem, the client that uses the LBSNs is called the user, and the generated check-in locations is called the POI. let U and L represent the user and location set respectively. In the POI trajectory sequence generated by the user, the user set is $U = \{u_1, u_2, \ldots, u_{|u|}\}$, the location set is $L = \{l_1, l_2, \ldots, l_{|l|}\}$, and the user POI trajectory sequence $S = \left(l_1^u, l_2^u, \ldots, l_{k-1}^u, l_k^u, l_{k+1}^u \ldots, l_n^u\right)$. User has some missing POIs $l_{missing}^u$. For example, we use $l_{missing}^u$ instead of the missing POI l_k^u. The user's incomplete POI trajectory sequence is $S_m = \left(l_1^u, l_2^u, \ldots, l_{k-1}^u, l_{missing}^u, l_{k+1}^u \ldots, l_n^u\right)$. The model completes the missing supermarkets $l_{missing}^u$ in the last week by learning the user's habit of check-in POI within a month. If the model predicts that the value of $l_{missing}^u$ is the same as the value of l_k^u, the purpose of missing POI completion is achieved.

4   Methodology

4.1   Model Architecture

Different from other POI research problems, the missing POI completion needs to jointly adjust the context in all layers to train deep bidirectional representation. This is because in the problem of missing POI completion, predicting the missing POI needs to combine its surrounding information. Inspired by [12], we use bidirectional masked POI trajectory model (MPTM) to solve the problem of combining the pre-order and post-order trajectory sequences with missing POIs. The Transformer [13] encoder is used to construct a bidirectional MPTM to consider the contextual information around the missing POI. The model structure graph is shown in the Fig. 2.

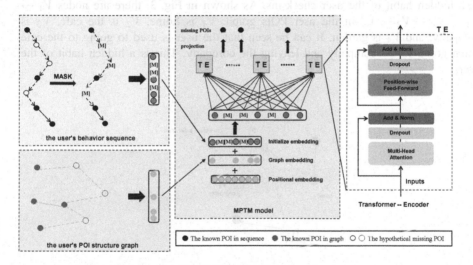

Fig. 2. Model architectures

4.2   Data Processing

For the POI trajectories generated by each user, the POI trajectories are sorted in time to form a sequence. We transform the POI trajectory sequence into a fixed-length n. This fixed-length is determined by the density of the length of the POI checked in by the users. The user's incomplete POI trajectory sequence is $S_m = \left(l_1^u, l_2^u, \ldots, l_k^u, l_{missing}^u, l_{k+1}^u \ldots, l_n^u \right)$, These missing POIs $l_{missing}^u$ are replaced with [M], which are the POI to be predicted by the model. [M] is masked, and refers to mask the missing POIs in the POI trajectory sequence. Therefore, the processed POI trajectory sequence is $S_M = \left(l_1^u, l_2^u, \ldots, l_k^u, [M], l_{k+1}^u \ldots, l_n^u \right)$.

Different from left-to-right language model training, our goal is to let the representation merge the left and right sides of the context to train a deep bidirectional model. Therefore, as same with BERT [12], three methods are used for masking. The three methods are to replace the original POI with [M], original POI and random POI. First, randomly select 15% of the POI in the POI trajectory sequence. Among these selected POIs, 80% is represented by [M], 10% is represented by original POI, and 10% is represented by random POI. This ensures that the POI trajectory sequence of each input is distributional contextual representation. It can not only let the model know which POIs should be predicted, but also weaken the impact of information leakage of the POI replaced by [M].

4.3 POI Graph of Users

To help the model learn more potential relationships between known POIs and missing POIs, we create a POI trajectory structure graph for each user. The graph contains the precedence and potential contacts of the location checked in by users. This can be seen as a hidden habit of the user check-ins. As shown in Fig. 3, there are nodes $V_1 \rightarrow V_2 \rightarrow V_3 \rightarrow V_2 \rightarrow V_4$ in the user POIs graph. V_1 is home, V_2 is the cafe, V_3 is company, and V_4 is garden. It can be seen that the user is used to going to the cafe when going to the company and leaving the company. This is a hidden habit of the user.

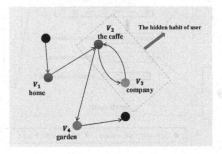

Fig. 3. POI graph of users

We create POI structure graph $G_u = \left(V_{l_1}^u, V_{l_2}^u, \ldots, V_{[M]}^u, \ldots, V_{l_n}^u \right)$ which are directed edge and weightless graphs for each user. DeepWalk [14] is used to learn the representation of POI in the network, including the topological relationship between POI nodes. DeepWalk can learn a network social representation by truncated random walk. Therefore, the relationship between POIs and the feature vector representation of each POI are obtained. DeepWalk generates a sequence of nodes $S_g = \left(\tilde{l}_1, \tilde{l}_2, \ldots, \tilde{l}_{[M]}, \ldots, \tilde{l}_n \right)$ in the graph through random walk. Where the $\tilde{l}_{[M]}$ is the masked POI. Then learn the latent representation $\tilde{l}_g^u \in R^d$ of the node through local information, where $\tilde{l}_g^u \in L_g^u$. $L_g^u \in R^{n \times d}$ contains the hidden relationship between the POIs that the user check-ins, which can be regarded as the user's hidden habit feature. These features can connect missing POIs with known POIs.

4.4   Embedding Layer

In order to learn the relationship between the known POI and the missing POI, it is necessary to initialize the embedding of the POI to establish the feature dimension. We create a location embedding matrix $L^u \in R^{n \times d}$ based on $S_M = \left(l_1^u, l_2^u, \ldots, l_k^u, [M], l_{k+1}^u \ldots, l_n^u\right)$, where n is the sequence length of user and d is the latent dimensionality. In transformer encoder, there is no iterative operation of Recurrent Neural Networks. All POIs of the sequence into the model for parallel processing at the same time. Therefore, it is necessary to provide position information for each POI in order to infer the missing POI through the relationship between other POIs. Like BERT4rec [11], we also use learnable positional embedding $P \in R^{n \times d}$ for position coding. Besides, we introduce the aforementioned graph feature vector L_g^u, which is combined with the P and L_g^u, and sent to the model together. The added vector contains more features, which can help the model to infer the situational information of missing POI. The combined vector is defined as:

$$In = L^u + P + L_g^u \tag{1}$$

4.5   Transformer-Encoder

Multi-Head Attention

In order to learn the expression of multiple meanings in POI trajectory sequence, it is necessary to perform a linear mapping on the input. The multi-headed attention mechanism can be used to extract the meaning of multiple semantics. The attention function is assigned to each POI weight through the three matrices of $Q(Query), K(key), V(value)$, and the correlation between the known POI and the missing POI is calculated according to the weight. The formulas for multi-head attention is as follows:

$$H \quad (In) = Concat(head_1, head_2, \ldots head_i)W \tag{2}$$

$$head_i = Atttention\left(In W_Q^i, In_i W_K^i, In_i W_V^i\right) \tag{3}$$

$$Attention(Q, K, V) = softmax\left(\frac{QK^T}{\sqrt{d/h}}\right)V \tag{4}$$

Where weight matrices W_Q, W_K, W_V are generated based on the input, W is a learnable parameter, $Concat$ is fully connected, and $head_i$ is the $i - th$ of self-attention. W_Q^i, W_K^i, W_V^i is the feature dimension is divided into h.

Position-Wise Feed-Forward Network

Although multi-head attention can use adaptive weights to aggregate the embedding of known POIs and missing POIs, it is still a linear model. In order to improve the nonlinearity of the model, and consider the interaction between the missing POI and the known POI in the dimensions. Like Transformer-encoder [13], we also use Position-wise Feed-Forward Network to improve the nonlinearity of the model. The formula is as follows:

$$FFN(H) = GELU(HW_1 + b_1)W_2 + b_2 \tag{5}$$

Where W_1, W_2, b_1, b_2 are all learnable parameters, and *GELU* is Gaussian Error Linerar Unit.

Stacking Transformer-Encoder Layer
After passing through the multi-head attention module, a network connection is required to learn the potential relationship between the missing POI and other POIs in the input trajectory. We use residual connections to prevent neural network degradation during network training. After each operation of multi-head attention module, the values before and after the operation must be added to obtain the residual connection.

$$Block(H) = LN(A + Drop(FFN(A))) \tag{6}$$

$$A = LN(H + Drop(H)) \tag{7}$$

Where *LN* is the Layer Normalization and *Drop* is the Dropout. Layer Normalization can help stabilize the neural network and speed up its training. Dropout can reduce the complex co-adaptation relationship between neurons, and can avoid the phenomenon of over-fitting.

4.6   The Output Layer

After some Transformer-Encoder blocks that adaptively and hierarchically extract information of previous POIs, we get the final output B for all POIs of the input sequence. We need predict the missing POIs base on B. As same as the BERT4Rec, we apply a two-layer feed-forward network with GELU activation to produce an output distribution over target items:

$$O(l) = softmax(GELU(BW_o + b_o)In^T + b) \tag{8}$$

Where W_o is the learnable projection matrix, B is the output after b transformer blocks, b_o and b are bias terms, In is the embedding matrix for the POI set. And the embedding is a matrix shared between the input and output of the model.

4.7   Network Training

Recall that we convert each user POI trajectory sequence to a fixed length sequence $S_m = \left(l_1^u, l_2^u, \ldots, l_k^u, l_{missing}^u, l_{k+1}^u \ldots, l_n^u\right)$ via truncation or padding locations. And the processed user POI trajectory sequence $S_M = \left(l_1^u, l_2^u, \ldots, l_k^u, [M], l_{k+1}^u \ldots, l_n^u\right)$ is generated as matrix vector L^u, which is used as model input with position embedding P and graph embedding L_g^u. The model outputs the missing POIs representation at the corresponding position, and calculates the loss of the model through the Cross Entropy Loss function. The Cross Entropy Loss is defined as:

$$Loss = -\sum_l (p(l)logq(l)) \tag{9}$$

5 Experiments

5.1 Datasets

We evaluate our proposed method on two real-world LBSN datasets from Foursquare [22], namely NYC and TKY, which have been widely used by previous studies on POI research. This dataset contains check-ins in NYC and Tokyo collected for about 10 months (from 12 April 2012 to 16 February 2013). We remove users with fewer than 10 check-ins and locations which have been visited fewer than 5 times. Table 1 summarizes the statistics of the two datasets.

Table 1. Statistics of datasets.

Datasets	Users	Locations	Check-ins
NYC	1083	38333	227,428
TKY	2293	61858	573,703

Besides, we analyze the distribution and density of user sequence length, as shown in the Figs. 4 and 5. The scatter chart shows that the length of the user's behavior sequence is concentrated within 500 in Fig. 4. And the density map shows the length of user's behavior sequence is concentrated around 140 in Fig. 5. By analyzing these two graphs, we can determine the trend of the centralized distribution of the number of users who checked in the POI. Therefore, the maximum sequence length is set to140. Inspired by SASRec [20], if the sequence length is greater than 140, we consider the most recent 140 actions. If the sequence length is less than 140, we repeatedly add a constant zero vector to the left until the length is 140. We treat the first 50% sequences of each user as training set and validation set, the last 50% as test set.

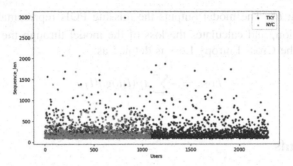

Fig. 4. The scatter chart of users

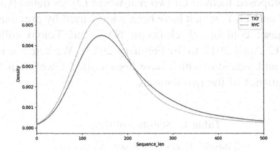

Fig. 5. The density chart of users

5.2 Baselines

To determine the effectiveness of our proposed method, we compare it with the following baselines:

Pop: It is a basic model that only recommends popular POIs.

BPR [23]: It uses the maximum posterior probability obtained by Bayesian analysis to rank and recommend POI.

GCMC [24]: It proposes a graph Auto-Encoder framework to solve the problem of rating prediction in the recommendation system from the perspective of link prediction.

SASRec [20]: It uses a left-to-right Transformer language model to capture users' sequential behaviors, and achieves sequential recommendation.

5.3 Metrics

To evaluate the performance of our proposed method, we employ precision and recall. Precision at a cutoff K, is denoted as $Pre@K$, and Recall at a cutoff k, is denoted as $Recall@K$. Where K is the number of predicted POIs in the result. These are general metrics for POI research used in previous work [25]. The $Pre@K$ is the ratio of recovered POIs to the K predicted POIs, and $Rec@K$ is the ratio of recovered POIs to the groundtruth. Given the user set U. We set the masked POIs as the groundtruth V_u^T, and V_u^P is the set of prediction result. The definitions of *Pre@K* and *Rec@K* are shown as follows:

$$Pre@K = \frac{1}{|U|} \sum_{u \in U} \frac{|V_u^T \cap V_u^P|}{K} \tag{10}$$

$$Rec@K = \frac{1}{|U|} \sum_{u \in U} \frac{|V_u^T \cap V_u^P|}{|V_u^T|} \tag{11}$$

5.4 Settings

In our method, we use four heads of attention modules and two blocks of multi-head attention for the check-in sequence. We train our model using the Adam optimizer with a learning rate of 0.001 and set the dropout ratio to 0.1. The batch size is 16 and the dimension of location embedding is 256. The parameters of the baselines are the default values. The trend of the model's loss with the epoch during the training process is shown in the Fig. 6. The value of loss shows an oscillating decline with the increase of epoch. This is because the set batch size is relatively small, and the error difference of each batch training is large. When the epoch reaches 80, the loss of the model gradually begins to converge. Therefore, the number of training epochs is set to 100 for NYC and TKY.

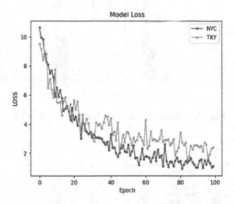

Fig. 6. The model loss with epoch

In the training process, the user behavior sequence are randomly masked to learn the relationship between POIs. In the testing process, in order to explore the influence of the continuity and discontinuity of the missing POIs in the sequence on the predictive ability of the model, we use mask in a fixed position in the user behavior sequence. We conduct multiple tests under both continuous and discontinuous conditions. The conditions are as follows (Table 2):

Table 2. Setting the number of masked POIs

The number of consecutively missing POIs	The total number of missing POIs
0	7
3	6
5	10
7	7

5.5 Comparison with Baselines

We set the parameters of the baselines to the default values. The number of missing POIs is 7 consecutively, and compare our method with baselines on the NYC and TKY datasets. For baselines, take the trajectory from the first POI to the missing POI in the dataset sequence. This is because baselines are all unidirectional models. In this way, we can compare the effect of our model using the information before and after the missing POIs. The comparison results are shown in the Figs. 7 and 8.

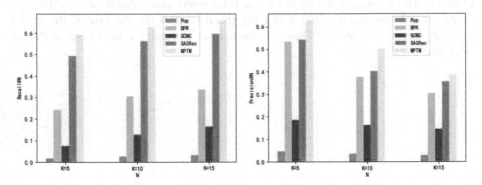

Fig. 7. Comparison with baselines on NYC

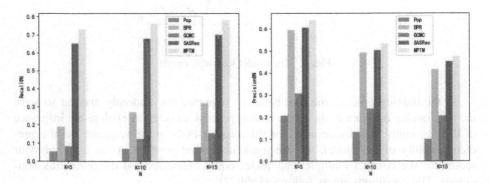

Fig. 8. Comparison with baselines on TKY

Our model MPTM is superior to other methods in recall and precision on both data sets. Pop only predicts based on the popularity of POI without considering the user's behavior habits, so it is much lower than other methods on recall and precision. GCMC constructs a bipartite graph of users and POI through the interaction between users and POI, but does not consider the user's behavior sequence, so the values on recall and precision are low. This also shows that the effect of constructing graph features alone for missing POI completion is not well, and further proves that the combination of graphic features and user behavior sequence is effective. The recall value of BPR on the two datasets is low, but the value of precision is close to SASRec and MPTM. This is because BPR is different from the general ranking model in that it reconstructs a partial order relationship for each user for personalized recommendation to predict. Both SASRec and MPTM use multi-head attention in transformer and consider the potential relationship between user behavior sequence and check-in POIs. But our model MPTM uses a bidirectional structure and takes into account the pre-order and post-order trajectory sequences with missing POIs.

5.6 Impact of POIs in Consecutive Masks

In order to explore the influence of the continuity and discontinuity of the missing POIs in the trajectory sequence on the predictive ability of the model, we use mask in a fixed position in the user behavior sequence. The experimental results are shown in the Tables 3 and 4, where number is the number of consecutive masked POIs. Table 3 is the experiment on the NYC, and Table 4 is the experiment on the TKY. Under all conditions, the recall of the two datasets is high, and the precision is relatively low. This shows that the model recalled many positively correlated POIs in the candidate set, but the number of correct hits is not high. There is not much difference between the recall of NYC and TKY under the four conditions. This shows that the continuity factor of the masked POI does not have much influence when the model recalls the relevant

Table 3. Experiments on NYC

Number	Recall@5	Recall@10	Recall@15	Pre@5	Pre@10	Pre@15
1	0.5974	0.6385	0.6693	0.5224	0.4467	0.3122
3	0.5985	0.6401	0.6657	0.4487	0.3838	0.2681
5	0.6109	0.6460	0.6713	**0.6107**	**0.4305**	**0.3355**
7	0.5916	0.6277	0.6574	0.5174	0.4391	0.3066

Table 4. Experiments on TKY

Number	Recall@5	Recall@10	Recall@15	Pre@5	Pre@10	Pre@15
1	0.7319	0.7643	0.7850	0.6404	0.5350	0.3663
3	0.7287	0.7611	0.7816	0.5465	0.4566	0.3126
5	0.7344	0.7638	0.7838	**0.7344**	**0.5092**	**0.3918**
7	0.7299	0.7599	0.7807	0.6387	0.5319	0.3643

positive samples. However, there is a gap between the precision on datasets, which fluctuates in the range of 0.3 to 0.7. This shows that the continuity factor of the masked POI has an impact on the model's accurate prediction of the missing POI.

In NYC and TKY, when the number of masked POIs is 5 consecutively, the values of recall and precision are higher. This shows that when the number of consecutive missing POIs is too small or too high, it will affect the predictive ability of the model. Especially when k = 5, the value of precision on two datasets is the highest.

6 Conclusions

According to the scene analysis, we propose the missing POI problem in the trajectory sequence. Incomplete user POI trajectory sequence has negative impact on subsequent tasks such as POI recommendation, location prediction, and human mobility. In order to overcome the difficulty of combining the pre-order and post-order trajectory sequences with missing POIs, we propose a bidirectional model MPTM based on the transform encoder. It is combined with graph features to mine the relationship between the missing POI and the known POI in the trajectory sequence and the characteristics of the user behavior sequence. The results demonstrate that our MPTM outperforms the state-of-the-art methods in terms of performance metrics recall and precision. For future work, we consider combining the explicit features of POI such as category and geographic location to solve the problem of user sequence trajectory splicing.

Acknowledgements. This research is sponsored by Natural Science Foundation of Chongqing, China (No. cstc2020jcyj-msxmX0900), the Fundamental Research Funds for the Central Universities (Project No. 2020CDJ-LHZZ-040), and National Natural Science Foundation of China (Grant No. 72074036, 62072060).

References

1. Lian, D., Wu, Y., Ge, Y., Xie, X., Chen, E.: Geography-aware sequential location recommendation. In: Proceedings of the 26th ACM SIGKDD International Conference on Knowledge Discovery and Data Mining, pp. 2009–2019 (2020)
2. Zhao, K., et al.: Discovering subsequence patterns for next POI recommendation. In: IJCAI, pp. 3216–3222 (2020)
3. Zhang, W., Wang, J.: Location and time aware social collaborative retrieval for new successive point-of-interest recommendation. In: Proceedings of the 24th ACM International on Conference on Information and Knowledge Management, pp. 1221–1230 (2015)
4. Zhang, L., Sun, Z., Zhang, J., Lei, Y., Klanner, F.: An interactive multi-task learning framework for next POI recommendation with uncertain check-ins. In: Twenty-Ninth International Joint Conference on Artificial Intelligence and Seventeenth Pacific Rim International Conference on Artificial Intelligence {IJCAI-PRICAI-20} (2020)
5. Zhao, S., Zhao, T., Yang, H., Lyu, M.R., King, I.: STELLAR: spatial-temporal latent ranking for successive point-of-interest recommendation. In: Thirtieth AAAI Conference on Artificial Intelligence (2016)

6. Liao, D., Liu, W., Zhong, Y., Li, J., Wang, G.: Predicting activity and location with multi-task context aware recurrent neural network. In: IJCAI, pp. 3435–3441 (2018)
7. Nawaz, A., Huang, Z., Wang, S., Akbar, A., AlSalman, H., Gumaei, A.J.S.: GPS trajectory completion using end-to-end bidirectional convolutional recurrent encoder-decoder architecture with attention mechanism. Sensors **20**, 5143 (2020)
8. Zheng, K., Zheng, Y., Xie, X., Zhou, X.: Reducing uncertainty of low-sampling-rate trajectories. In: 2012 IEEE 28th International Conference on Data Engineering, pp. 1144–1155. IEEE, (2012)
9. Liu, Q., Wu, S., Wang, L., Tan, T.: Predicting the next location: a recurrent model with spatial and temporal contexts. In: Thirtieth AAAI Conference on Artificial Intelligence (2016)
10. Koren, Y., Bell, R.: Advances in Collaborative Filtering, pp. 77–118. Springer, Boston, MA (2015)
11. Sun, F., Liu, J., Wu, J., Pei, C., Lin, X., Ou, W., Jiang, P.: BERT4Rec: sequential recommendation with bidirectional encoder representations from transformer. In: Proceedings of the 28th ACM International Conference on Information and Knowledge Management, pp. 1441–1450 (2019)
12. Devlin, J., Chang, M.-W., Lee, K., Toutanova, K.: Bert: pre-training of deep bidirectional transformers for language understanding (2018)
13. Vaswani, A., et al.: Attention is all you need. In: Advances in Neural Information Processing Systems, pp. 5998–6008 (2017)
14. Perozzi, B., Al-Rfou, R., Skiena, S.: Deepwalk: online learning of social representations. In: Proceedings of the 20th ACM SIGKDD International Conference on Knowledge Discovery and Data Mining, pp. 701–710 (2014)
15. Han, P., Li, Z., Liu, Y., Zhao, P., Shang, S.: Contextualized point-of-interest recommendation. In: Twenty-Ninth International Joint Conference on Artificial Intelligence and Seventeenth Pacific Rim International Conference on Artificial Intelligence {IJCAI-PRICAI-20} (2020)
16. Zhao, P., Zhu, H., Liu, Y., Xu, J., Zhou, X.: Where to go next: a spatio-temporal gated network for next poi recommendation. Comput. Sci. **33**, 5877–5884 (2019)
17. Liu, X., Liu, Y., Li, X.: Exploring the context of locations for personalized location recommendations. In: IJCAI, pp. 1188–1194 (2016)
18. Lu, Y.S., Shih, W.Y., Gau, H.Y., Chung, K.C., Huang, J.L.: On successive point-of-interest recommendation. Anthology **22**, 1151–1173 (2019)
19. Manotumruksa, J., Macdonald, C., Ounis, I.: A contextual attention recurrent architecture for context-aware venue recommendation. In: The 41st International ACM SIGIR Conference on Research and Development in Information Retrieval, pp. 555–564 (2018)
20. Kang, W.-C., McAuley, J.: Self-attentive sequential recommendation. In: 2018 IEEE International Conference on Data Mining (ICDM), pp. 197–206. IEEE, (2018)
21. Wu, Y., et al.: Google's neural machine translation system: bridging the gap between human and machine translation (2016)
22. Yang, D., Zhang, D., Zheng, V.W., Yu, Z.: Modeling user activity preference by leveraging user spatial temporal characteristics in LBSNs. IEEE Trans. Syst. Man Cybern. Syst. **45**, 129–142 (2014)
23. Rendle, S., Freudenthaler, C., Gantner, Z., Schmidt-Thieme, L.: BPR: Bayesian Personalized Ranking from Implicit Feedback (2012)
24. van den Berg, R., Kipf, T.N., Welling, M.: Graph convolutional matrix completion (2017)
25. Yu, F., Cui, L., Guo, W., Lu, X., Li, Q., Lu, H.: A category-aware deep model for successive poi recommendation on sparse check-in data. In: Proceedings of the Web Conference 2020, pp. 1264–1274 (2020)

Dual-Channel Graph Contextual Self-Attention Network for Session-Based Recommendation

Teng Huang[1], Huiqun Yu[1,2(✉)], and Guisheng Fan[1(✉)]

[1] Department of Computer Science and Engineering,
East China University of Science and Technology, Shanghai, China
{yhq,gsfan}@ecust.edu.cn
[2] Shanghai Key Laboratory of Computer Software Evaluating and Testing,
Shanghai, China

Abstract. The session-based recommendation task is a key task in many online service websites (such as online music, e-commerce, etc.). Its goal is to predict the user's next possible interactive item based on an anonymous user behavior sequence. However, existing methods do not take into account the information implicit in similar neighbor sessions. This paper proposes a new dual-channel graph neural network combined with a self-attention network model. In this model, the graph neural network is used to model the session sequence, and the learning is different through two independent channels between and within the session. The model uses the attention network to learn the weights of different items for the recommendation results. Several experiments are done on two public e-commerce data sets and the results show that the performance of the model proposed in this paper is better than the general methods.

Keywords: Session-based recommendation · Graph neural network · Self-attention network

1 Introduction

With the development of information technology and Internet technology, people have gradually entered the era of information explosion, and the problem of information overload has become one of the urgent problems. In this era, personalized recommendation system came into being [13]. The traditional recommendation system relies on collecting the user's personal information or historical operation records to make recommendations. However, in many network service scenarios (such as online music, online movie websites, e-commerce, etc.), due to privacy protection and other considerations, users may not log in to the website when obtaining network services. To solve such problems, session-based recommendation system has been proposed to predict the next possible interaction (such as clicking, etc.) items [4, 19].

The definition of session in this field is similar to the general concept of web. In general, the session sequence in the session-based recommendation task refers to a sequence of items composed of all items of user interaction within a period of time, such as a few minutes or a few hours, arranged according to time. For example, the

H. Gao and X. Wang (Eds.): CollaborateCom 2021, LNICST 406, pp. 244–258, 2021.
https://doi.org/10.1007/978-3-030-92635-9_15

products that a user browses and clicks on an e-commerce website within one hour constitute a session sequence.

Considering the extremely high value of session-based recommendations, various methods have been proposed. Traditionally, the method based on the Markov chain [9] will predict the user's next behavior based on the previous behavior. In recent years, researchers have introduced neural network models into session-based recommendation tasks and have achieved better results [17]. For instance, GRU4REC [4] uses GRU units to model session sequences. NARM [6] designed a global and local encoder to capture the sequential behavior of the session sequence and the main purpose of user. SR-GNN [15] model the session sequence into a session graph, and use the graph neural network method to model the complex conversions between separate items.

Although the above methods have achieved good recommendation results, these models also have some problems. For example, GRU4REC only considers the sequential behavior information of the session sequence, and does not consider the main purpose of the user. NARM can only model the sequential relationship between consecutive items, ignoring the complex conversion between different items, and these conversions reflect the user's behavior pattern, which is of great significance to the recommendation system. SR-GNN does not consider the similar item conversion implicit in the neighbor sessions, and the hidden layer representation vectors of the item node may not be accurate enough.

In order to further improve the accuracy of the recommendation system, this paper proposes a new model, a dual-channel graph contextual self-attention network for session-based recommendation, which is called DCGC-SAN for short. This model takes into account the differences in item conversion between the current session and neighbor sessions, and uses two channels, intra-session and inter-session channels, to respectively propagate the embedding vectors of the learning graph nodes, then aggregates the features of the two channels through a fusion function, and combines the self-attention network to learn contextual information as the main purpose of the user. Finally, the user's short-term interest and the main purpose are combined to calculate the session representation vector, which is finally used to generate the TOP-N recommendation list.

The main contributions of this paper are summarized as follows:

- To better learn the embedding vectors of graph nodes, a new DCGC-SAN is proposed, which divides the intra-session and inter-session channels learning of accurate graph node embedding vectors.
- The self-attention network is applied to capture the main purpose of the user.
- A large number of experiments and analyses conducted on two public e-commerce data sets show that DCGC-SAN is better than baseline methods in terms of recommendation accuracy.

The rest of this paper is structured as follows. We review related researches and works in Sect. 2. Section 3 presents the proposed method of our model. The experiments and analysis are presented in Sect. 4. Finally, we conclude this paper in Sect. 5.

2 Related Work

2.1 Neural Network Model

Hidasi et al. [4] first proposed that a recurrent neural network with GRU units should be used to extract the features of session sequences. This is the first time that someone introduced a recurrent neural network into session-based recommendation tasks. In order to improve the basic RNN model, Tan et al. [10] used sequence preprocessing, and data enhancement to reduce the occurrence of overfitting and improve the performance of the RNN model. After that, researchers hope to use the implicit collaboration information in similar sessions. For example, KNN-RNN proposed by Jannach et al. [5] uses a heuristic nearest neighbor (KNN) method to sample the nearest neighbors and introduces GRU4REC to capture the sequential relationship. The model based on the graph neural network considers the complex conversion between all items in the item space and changes the pattern dominated by the recurrent neural network models. The most important point of the method based on graph neural networks is the method of graph node embedding vector. Typical methods such as gated graph neural network (GGNN) [7] extend GNN to sequence output. SR-GNN proposed by Wu et al. [15] modeled the session sequence into a session graph, and used the GGNN method to learn intricate item conversions. Recently, inspired by the success of convolution neural networks in the image field, Bruna et al. [2] proposed Spectrogram Convolution Neural Networks (GCN). Wu et al. [14] found that there are many unnecessary calculations in GCN. GCN can be simplified by removing non-linearity and folding the weight matrix to obtain higher efficiency without compromising accuracy.

2.2 Attention Mechanism

The attention mechanism is commonly used in fields such as natural language processing and computer vision. In session-based recommendation tasks, standard soft attention mechanisms have been studied [6], such as NARM and STAMP. Vaswani et al. [11] proposed a self-attention-based model, Transformer, which is a network without any CNN or RNN units. It can model the dependencies between different words and achieve sota performance on machine translation tasks. Anh et al. [1] introduced the Transformer to the session-based recommendation task, which took less time on the premise of achieving better recommendation performance. Xu et al. [16] learned local context information through the attention network and also achieved impressive performance improvements.

3 Proposed Method

3.1 Problem Statement

The goal of the session-based recommendation system is to predict the user's next most likely interactive item based on the current user's interaction sequence. Let $V = \{v_1, v_2, v_3, ..., v_m\}$ denote the collection of items that have appeared in all session sequences. For each session sequence $S = \{v_{s,1}, v_{s,2}, v_{s,3}, ..., v_{s,n}\}$, $v_{s,i} \in V$ represents the user's interactive items in the current session sequence. The goal of the recommendation model is to predict the next possible interactive item $v_{s,i+1}$ for current user. For the session sequence S, the model outputs the probability of all items $\hat{y}$ as the next user interacting item, and the TOP-N recommendation list generated for the user is the TOP-N with the largest score.

3.2 Model Overview

DCGC-SAN consists of three parts. The first part is the graph neural network. The main task of this part is to model the current session sequence and neighbor session sequences, divide different channels to learn accurate graph node embedding vectors. The second part is the self-attention network. The principal task of this part is to capture the main purpose of the user and incorporate rich contextual information into the model. The third part is the prediction layer, which combines the previously generated session sequence representation vector and the graph node embedding vector to calculate the possible interaction probability of all items in the item space. The architecture of DCGC-SAN is shown in Fig. 1.

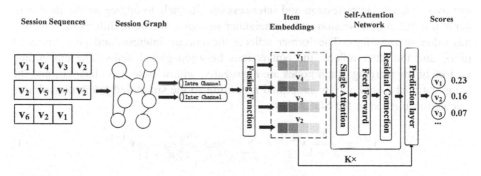

Fig. 1. DCGC-SAN architecture

3.3 Session Graph Construction

First, we need to construct a session graph. For the current session sequence S, extract the n neighbor sessions that are most similar to S. In order to reduce the complexity of calculating the similarity between session sequences, we use the number of repetitive items in different sessions as a measure of similarity. Sort all the session sequences in reverse order, and sample the n session sequences at the top to form a neighbor session set N_s. Model the current session and neighbor session sets as session graph ζ_s. Each node in the session graph represents an item v_i, and each edge (v_i, v_j) represents that the user interacts with the item v_j after interacting with the item v_i in the current session or neighbor sessions.

Figure 2 is the complete process of constructing a session graph, in which the red arrow represents the item conversions in the current session, and the black arrow represents the item conversions in the neighbor sessions.

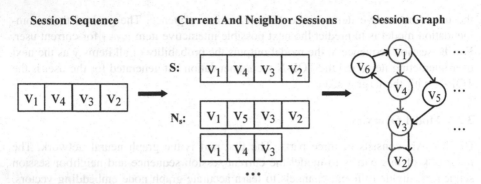

Fig. 2. The construction process of the session graph

3.4 Item Embedding Learning

After the construction of the session graph is completed, the follow-up task is to learn accurate item node embedding vectors from the session graph by graph neural network. The difference between the GGNN used in the DCGC-SAN and SR-GNN is that we use two channels: intra-session and inter-session channels to aggregate the item conversion in the current session and the neighbor sessions. We think different conversion has different meanings. The former reflects the current interests and preferences of users, and the latter reflects the conversions between global items. The embedding vectors of the learning graph nodes are aggregated within these two channels, and the computational complexity is simplified according to the simplified graph convolution calculation method in the SGCN [14].

$$\mathbf{v}_i^{(k)\prime} = \frac{1}{d_i+1}\mathbf{v}_i^{(k-1)} + \sum_{v_j \in V_s} \frac{a_{ij}}{\sqrt{(d_i+1)(d_j+1)}}\mathbf{v}_j^{(k-1)} \tag{1}$$

$$\mathbf{v}_i^{(k)\prime\prime} = \frac{1}{d_i+1}\mathbf{v}_i^{(k-1)} + \sum_{v_j \in V_{N_s}} \frac{a_{ij}}{\sqrt{(d_i+1)(d_j+1)}}\mathbf{v}_j^{(k-1)} \tag{2}$$

Formulas (1) and (2) are two channel learning and node update formulas, where d_i is the degree of the node v_i in the adjacency matrix. a_{ij} indicates whether there is an edge between v_i and v_j in the session graph, if there is an edge, $a_{ij} = 1$, if there is no edge, $a_{ij} = 0$.

Finally, the item graph node embedding vector set learned through the intra-session channel is embedded in the vector set $\mathbf{V}_s = \{\mathbf{v}'_1, \mathbf{v}'_2, \mathbf{v}'_3, ..., \mathbf{v}'_n\}$, where $v'_i = v_i^{(k)\prime}$; the item graph node embedding learned through the inter-session channel represents the vector set $\mathbf{V}_{N_s} = \{\mathbf{v}''_1, \mathbf{v}''_2, \mathbf{v}''_3, ..., \mathbf{v}''_n\}$, where $v''_i = v_i^{(k)\prime\prime}$.

$$\mathbf{v}_i = \text{item\_fusing}(\mathbf{v}'_i, \mathbf{v}''_i) \tag{3}$$

According to formula (3), we aggregate the information in the two channels through the item fusing function, extract effective features, and learn accurate graph node embedding representation vectors. The two vector aggregation methods are used in the following three methods in this article:

- Average pooling (avg): fusion of information from different channels by taking the average of two vectors in each dimension.
- Maximum pooling (max): fusion of information from different channels by taking the maximum value of two vectors in each dimension
- Direct connection (concat): connect two vectors and reduce the dimensionality through a single-layer linear neural network, where the parameter $\mathbf{W} \in \mathbb{R}^{2d \times d}$.

$$\mathbf{v}_i = \sigma(\mathbf{W}[\mathbf{v}'_i, \mathbf{v}''_i]) \tag{4}$$

3.5   Self-attention Network

DCGC-SAN uses a self-attention network to model the main purpose of the current session sequence. The self-attention mechanism is a special attention mechanism. It captures the correspondence between the entire input and output by calculating the conversion weight between each item and all other items in the sequence, which can effectively solve the problem of long-distance items. It is easy to lose the problem of relying on information [11].

For the input session sequence $S = \{v_{s,1}, v_{s,2}, v_{s,3}, ..., v_{s,n}\}$, the d-dimensional embedding vector $\mathbf{S} = [\mathbf{v}_{s,1}, \mathbf{v}_{s,2}, \mathbf{v}_{s,3}, ..., \mathbf{v}_{s,n}]$ is obtained after the embedding layer, and input into the self-attention network. The self-attention network is composed of a single-layer attention layer, a feedforward neural network, and a residual connection layer.

The main function of the single-layer attention layer is to obtain the weight relationship between each item and other items in the session sequence, so that when modelling a single item, the information of other items in the same session sequence can be taken into account, which introduces rich context information to the entire model. The calculation formula of the single attention layer is shown in formula (5).

$$\mathbf{E} = \text{softmax}(\frac{(\mathbf{SW}^Q)(\mathbf{SW}^K)^T}{\sqrt{d}})(\mathbf{SW}^V) \tag{5}$$

The weight matrices $\mathbf{W}^Q, \mathbf{W}^K$ and $\mathbf{W}^V \in \mathbb{R}^{2d \times d}$, $\mathbf{E}$ represents the output of the single-layer attention layer.

The main function of the feedforward neural network is to introduce a nonlinear transformation to the attention network, so that the attention network can obtain information of different spatial dimensions. In order to solve the problems such as the disappearance of the gradient caused by the network level being too deep, a new residual network layer is added after the feedforward neural network [3]. The calculation formulas of the feedforward neural network layer and the residual layer are shown in formula (6).

$$\mathbf{O} = \text{Relu}(\mathbf{EW}_1 + \mathbf{b}_1)\mathbf{W}_2 + \mathbf{b}_2 + \mathbf{E} \tag{6}$$

$\mathbf{W}_1$ and $\mathbf{W}_2 \in \mathbb{R}^{d \times d}$, $\mathbf{b}_1$ and $\mathbf{b}_2$ are the d-dimensional bias vectors, and the integration of the previous three layers is defined as:

$$\mathbf{O} = F(\mathbf{S}) \tag{7}$$

In order to study the most suitable attention network layers for the user's main purpose in modeling session sequences, we repeatedly apply attention layers in the model to form attention networks to capture different types of features. The first layer is defined as $\mathbf{O}^{(1)} = F(\mathbf{S})$, and the kth layer is defined as:

$$\mathbf{O}^{(k)} = F(\mathbf{O}^{(k-1)}) \tag{8}$$

Take the nth row vector of the matrix $\mathbf{O}^{(k)}$ as the output of the attention network and at the same time as the main purpose of the user in the session sequence.

$$\mathbf{C}_t^l = \mathbf{O}_n^{(k)} \tag{9}$$

3.6 Prediction Layer

In order to better predict the user's possible next interactive item, we use the embedding vector of the user's last interactive item in the current session sequence as the user's short-term interest $\mathbf{C}_t^g$:

$$\mathbf{C}_t^g = \mathbf{v}_{s,n} \tag{10}$$

Combining the user's main purpose $\mathbf{C}_t^l$ calculated by the attention network layer, fusion of the two information forms the final session representation vector $\mathbf{C}_t$.

$$\mathbf{C}_t = \sigma(\mathbf{W}_g \mathbf{C}_t^g + \mathbf{W}_l \mathbf{C}_t^g) \tag{11}$$

Then calculate the probability $\hat{y}_i$ of each candidate item $v_i \in V$ as the user's next interactive item. The calculation formula is as follows:

$$\hat{y}_i = \text{softmax}(\mathbf{C}_t^T \mathbf{v}_i) \tag{12}$$

$\mathbf{v}_i$ is the embedding layer representation vector of the item v_i. Finally, the model is trained by minimizing the cross-entropy loss function.

$$Loss(y_i, \hat{y}_i) = -\sum_{i=1}^{n} (y_i \log(\hat{y}_i) + (1 - y_i) \log(1 - \hat{y}_i)) \tag{13}$$

y_i represents the item that the user actually interacts with.

4 Experiments and Analyses

In this section, we first prepare the data needed for the experiments, conduct the experiments, and use the experimental results to answer the following two questions:

Question 1: Can DCGC-SAN proposed in this paper get better recommendation results than the baseline methods?

Question 2: What are the effects of the parameters of DCGC-SAN, such as the number of neighbor sessions sampled and the dimension of the embedding vector, on the recommendation performance?

4.1 Datasets

In order to test the performance, we conducted experiments on two public e-commerce data sets Yoochoose and Diginetica.

The Yoochoose dataset is a public dataset released by RecSys Challenge in 2015. It contains a collection of sessions from a retailer, where each session is encapsulating the click events that the user performed in the session. The Diginetica dataset comes from the competition CIKM 2016. In our experiments, only the data of the transaction part is used.

For fairly comparison, we follow the common data set processing methods in previous studies [4, 6, 8, 10]. All the session sequences are rearranged chronologically, and sessions with length one and items that appear less than five times are deleted. Furthermore, we used the last day of Yoochoose and the last seven days of Diginetica to generate test set. Since it is impossible for the model to generate items that have not appeared before, we delete items from test set that did not appear in the training set. And, moreover, due to the large amount of data in the Yoochoose, the experiments only use the latest 1/64 data, which is called the Yoochoose 1/64.

The statistics of the two data sets after the data preprocessing stage are shown in Table 1.

Table 1. Dataset statistics

Datasets	Yoochoose 1/64	Diginetica
All the items	557,248	982,961
Training sessions	369,859	719,470
Test sessions	55,898	60,858
Items	16,766	43,097
Average length	6.16	5.12

4.2 Evaluation Metric

We use two metrics, P@20 and MRR@20, to evaluate the performance.

P@20: The recommendation accuracy rate of the model, which represents the proportion of correctly recommended items appearing in the Top-20 recommendation list.

MRR@20: The reciprocal average of the correct item's ranking in the Top-20 recommendation list. MRR considers the ranking order of recommended items. The higher the MRR, the higher the ranking of the correctly recommended item in the recommendation list. If the correct item rank is greater than 20, then the value of MRR is 0.

4.3 Experiment Settings

The experiment uses a Gaussian distribution with a mean of 0 and a standard deviation of 0.1 to initialize all parameters in the DCGC-SAN. The optimizer uses the Adam [20] optimizer, initial learning rate is set to 1e-3, and the attenuation is set to 0.1 every 3 training rounds. The batch size is set to 128, and the L2 regularization parameter is set to 1e-5 to reduce overfitting.

4.4 Comparison with Baseline Methods

In order to answer question 1, we conducted experiments on two data sets on the DCGC-SAN and baseline methods. The experimental results are illustrated in Table 2, and the best results are marked in bold.

Table 2. Performance comparison between different methods

Methods	Yoochoose 1/64		Diginetica	
	P@20	MRR@20	P@20	MRR@20
GRU4REC [4]	60.64	22.89	29.45	8.33
NARM [6]	68.32	28.63	49.70	16.17
STAMP [8]	68.74	29.67	45.64	14.32
KNN-RNN [5]	62.36	24.05	31.89	9.65
CSRM [12]	70.79	30.48	50.55	16.38
SR-GNN [15]	70.57	30.94	50.73	17.59
GC-SAN [16]	70.66	30.04	51.34	17.61
TAGNN [18]	71.02	31.12	51.31	**18.03**
DCGC-SAN	**71.37**	**31.73**	**51.60**	17.76

From the experimental results in the table above, we can conclude that:

In the RNN-based recommendation model, GRU4REC uses GRU to capture the sequence information of the session sequences, which is the basic model. Both NARM and STAMP apply the attention mechanism on the final interactive items, in order to capture sequential information. The performance of these two models is better than GRU4REC, which shows the effectiveness of the attention mechanism in session-based recommendation tasks. CSRM introduces the collaboration information in the field into the model. Compared with KNN-RNN, the main difference between the two models is that KNN-RNN uses fixed parameters to combine the two, while CSRM combines more complex nonlinear conversion, so the performance of CSRM is better. Moreover, the performance of KNN-RNN and CSRM is better than that of the basic RNN model, which shows that the collaborative information in the field can improve the performance.

The performance of SR-GNN is better than that based on the RNN model, which shows that the graph node embedding algorithm can effectively learn the representation vector of each graph node. Both GC-SAN and TAGNN are improvements to SR-GNN because of attention mechanism. They all improve the recommendation performance of the graph neural network to a certain extent, and TAGNN has achieved the best performance in the MRR@20 of Diginetica dataset.

DCGC-SAN divides two different channels, intra-session and inter-session channels, disseminates different information within the two channels, and learns more accurate graph node embedding vectors by fusing the information of the two channels. In terms of attention mechanism, our model abandons the simple soft attention mechanism and use a more complex attention network to assign the weight of different items in an adaptive weighting manner. It can be clearly seen from Table 2 that DCGC-SAN achieves the best performance on the P@20 and MRR@20 of the Yoochoose 1/64 dataset and P@20 of the Diginetica dataset.

In order to answer question 2, we modified the parameters of the model and made multiple sets of comparative ablation experiments.

4.5 The Influence of Model Parameters on Experimental Results

The Influence of the Number of Neighbors. DCGC-SAN samples and models similar neighbor session sequences, and the number of samples obviously affects the final recommendation performance. If only considering the item conversion of the current session, the model can dig deeper into the current user's preferences, but it will lose the important item conversions reflected in other similar sessions. If a large number of similar neighbor session sequences are introduced, it is equivalent to expanding the scope of obtaining item conversion, and excessive noise may be introduced. Therefore, on the Yoochoose 1/64 and Diginetica data sets, we increase the number of neighbor session from 0 to 160 to study the impact of different neighbor sample numbers on the recommendation performance. Due to limited space, only P@20 is considered here, and the following experiments only consider P@20 for the same reason.

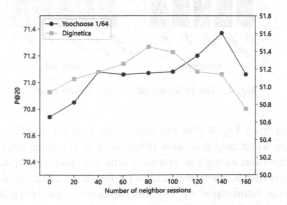

Fig. 3. The influence of the number of sampling neighbors

Figure 3 shows the influence of neighbors with different numbers of neighbors on the recommendation results. As the number increases, the general trend of model performance is to increase first and then decrease. The recommendation performance of DCGC-SAN on the Yoochoose 1/64 at n = 140 and Diginetica at n = 80 reaches its peak. The reason for the decline is that the number of neighbors sampled is too large, resulting in some neighbor sessions that are not very similar to the current session (the larger the number of neighbors, the worse the similarity of the session sequence), which introduces excess noise. Compared with n = 0, on the two datasets, the introduction of item conversion in neighbor sessions (choose the appropriate number of samples) significantly improves the performance of the model, which reflects the rationality of fusing dual-channel item conversion effectiveness.

The Influence of Channel Fusion Method. In order to obtain the most suitable fusion method of the two channel information, we analyzed how different fusion functions affect the experimental results. In the learning stage of the item node embedding vectors in the graph neural network, the following strategies are used. The experimental results are shown in Fig. 4.

- Only use the item node information of the current session (intra-only);
- Only use the item node information in the neighbor session (inter-only);
- The information of the two channels is merged. There are three ways of channel fusion, namely concat, max and avg.

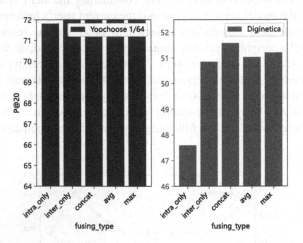

Fig. 4. The influence of channel fusion mode

It can be seen from Fig. 4 that the performance of the model using the current session only is the worst, and it is quite different from the other four methods, which shows that it is far from enough to consider only the item conversion in the current session. In comparison, the recommendation performance of the model that only uses the item conversion information of neighbor sessions is unexpectedly good. This is because although the item conversion in the current session sequence is discarded, the appropriate item conversion can still be learned from a large number of neighbor sessions.

Among the three methods that integrate inter-session and intra-session channels, the concat fusion method works best because it is the most complicated of the three fusion methods and can better integrate the information of the two channels, while max and avg, comparatively speaking, it is too simple.

The Influence of Embedding Layer Dimensions. In Fig. 5, we study the influence of the size d of the embedding vector of the graph node on the experimental results. We increased d from 20 to 160 on the two data sets to explore the influence of the embedding layer dimension on the experimental results.

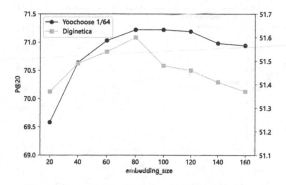

Fig. 5. The influence of embedding layer dimensions

It can be seen from Fig. 5 that ignorantly increasing the dimension of the graph embedding vector does not always improve the performance of the model. For the Yoochoose 1/64, the best embedding size is 80 or 100, and for the Diginetica, the best embedding size is 80. When the dimension of the embedding vector is less than the optimal dimension, the number of dimensions that can be expressed is limited, and the model performance is not too high. When the embedding dimension is greater than the optimal size, the model may be overfitting and the performance will decrease.

The Influence of the Number of Layers of the Attention Network. The attention network can capture the dependency between items that are far away. We first need to determine whether the attention network is effective. We apply the ordinary soft attention mechanism in SR-GNN to our model (DCG-SA), and compare it with the complete DCGC-SAN. Table 3 is the comparison result. From Table 3, it can be seen that the attention network can improve the performance of the recommendation system.

Table 3. Performance comparison of soft attention mechanism and attention network

Methods	Yoochoose 1/64		Diginetica	
	P@20	MRR@20	P@20	MRR@20
DCG-SA	71.02	31.70	50.84	17.32
DCGC-SAN	**71.37**	**31.73**	**51.60**	**17.76**

In order to study the influence of the number of layers of the attention network on the recommendation results, we increased K from 1 to 5 on the two data sets to experiment with the model. The results of the experiment are shown in Fig. 6.

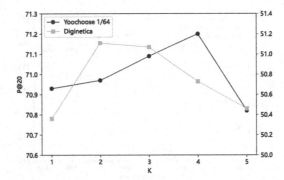

Fig. 6. The influence of K on experimental results

It can be seen from Fig. 6 that increasing K can improve the performance of the DCGC-SAN. After the best K value is exceeded, the larger the K, the worse the performance of the model. This is because too many network layers can obtain more abstract information, but will lose low-dimensional information, while network layers with too few numbers cannot obtain high-dimensional abstract information.

5   Conclusions

This paper proposes the DCGC-SAN model, which divides the intra-session and inter-session channels to learn the information of different item nodes in the learning stage of the graph node embedding vectors, and effectively integrates the two vectors through the fusion function, combined with the attention network capturing the dependency information between items that are far apart. Experimental results on two public e-commerce data sets show that our model is better than baseline methods in terms of recommendation accuracy.

References

1. Anh, P.H., Bach, N.X., Phuong, T.M.: Session-based recommendation with self-attention. In: Proceedings of the Tenth International Symposium on Information and Communication Technology, pp. 1–8 (2019)
2. Bruna, J., Zaremba, W., Szlam, A., LeCun, Y.: Spectral networks and locally connected networks on graphs. ArXiv Prepr. ArXiv13126203 (2013)
3. He, K., Zhang, X., Ren, S., Sun, J.: Deep residual learning for image recognition. In: Proceedings of the IEEE Conference on Computer Vision and Pattern Recognition, pp. 770–778 (2016)
4. Hidasi, B., Karatzoglou, A., Baltrunas, L., Tikk, D.: Session-based Recommendations with Recurrent Neural Networks. ArXiv151106939 Cs (2016)
5. Jannach, D., Ludewig, M.: When recurrent neural networks meet the neighborhood for session-based recommendation. In: Proceedings of the Eleventh ACM Conference on Recommender Systems, pp. 306–310 (2017)
6. Li, J., Ren, P., Chen, Z., Ren, Z., Lian, T., Ma, J.: Neural attentive session-based recommendation. In: Proceedings of the 2017 ACM on Conference on Information and Knowledge Management, pp. 1419–1428. Association for Computing Machinery, New York, USA (2017). https://doi.org/10.1145/3132847.3132926
7. Li, Y., Tarlow, D., Brockschmidt, M., Zemel, R.: Gated graph sequence neural networks. ArXiv Prepr. ArXiv151105493 (2015)
8. Liu, Q., Zeng, Y., Mokhosi, R., Zhang, H.: STAMP: short-term attention/memory priority model for session-based recommendation. In: Proceedings of the 24th ACM SIGKDD International Conference on Knowledge Discovery and Data Mining, pp. 1831–1839 (2018)
9. Rendle, S., Freudenthaler, C., Schmidt-Thieme, L.: Factorizing personalized markov chains for next-basket recommendation. In: Proceedings of the 19th International Conference on World Wide Web, pp. 811–820 (2010)
10. Tan, Y.K., Xu, X., Liu, Y.: Improved recurrent neural networks for session-based recommendations. In: Proceedings of the 1st Workshop on Deep Learning for Recommender Systems, pp. 17–22 (2016)
11. Vaswani, A., et al.: Attention is all you need. ArXiv Prepr. ArXiv170603762 (2017)
12. Wang, M., Ren, P., Mei, L., Chen, Z., Ma, J., de Rijke, M.: A collaborative session-based recommendation approach with parallel memory modules. In: Proceedings of the 42nd International ACM SIGIR Conference on Research and Development in Information Retrieval, pp. 345–354 (2019)
13. Wang, S., Pasi, G., Hu, L., Cao, L.: The era of intelligent recommendation: editorial on intelligent recommendation with advanced AI and learning. IEEE Ann. Hist. Comput. 35, 3–6 (2020)
14. Wu, F., Souza, A., Zhang, T., Fifty, C., Yu, T., Weinberger, K.: Simplifying graph convolutional networks. In: International Conference on Machine Learning, pp. 6861–6871. PMLR (2019)
15. Wu, S., Tang, Y., Zhu, Y., Wang, L., Xie, X., Tan, T.: Session-based recommendation with graph neural networks. In: Proceedings of the AAAI Conference on Artificial Intelligence, pp. 346–353 (2019)
16. Xu, C., et al.: Graph contextualized self-attention network for session-based recommendation. In: IJCAI, pp. 3940–3946 (2019)
17. Yu, F., Liu, Q., Wu, S., Wang, L., Tan, T.: A dynamic recurrent model for next basket recommendation. In: Proceedings of the 39th International ACM SIGIR conference on Research and Development in Information Retrieval, pp. 729–732 (2016)

18. Yu, F., Zhu, Y., Liu, Q., Wu, S., Wang, L., Tan, T.: TAGNN: target attentive graph neural networks for session-based recommendation. In: Proceedings of the 43rd International ACM SIGIR Conference on Research and Development in Information Retrieval, pp. 1921–1924 (2020)
19. Zhu, Y., Li, H., Liao, Y., Wang, B., Guan, Z., Liu, H., Cai, D.: What to do next: modeling user behaviors by time-LSTM. In: IJCAI, pp. 3602–3608 (2017)

Context-aware Graph Collaborative Recommendation Without Feature Entanglement

Tianyi Gu, Ping Li$^{(\boxtimes)}$, and Kaiwen Huang

Southwest Petroleum University, Chengdu, China

Abstract. Inheriting from the basic idea of latent factor models like matrix factorization, current collaborative filtering models focus on learning better latent representations of users and items, by leveraging the expressive power of deep neural networks. However, the settings where rich context information is available pose difficulties for the existing neural network based paradigms, since they usually entangle the features extracted from both IDs and other additional data (*e.g.*, contexts), which inevitably destroy the original semantics of the embeddings. In this work, we propose a context-aware collaborative recommendation framework called CGCR to integrate contextual information into the graph-based embedding process. Our model converts the bipartite graph to a homogeneous one by placing the users and items in the identical feature space. As our method is free of feature crosses, it can preserve the semantic independence on the embedding dimensions and thus improves the interpretability of neural collaborative filtering. We use generalized matrix factorization as the matching function so that the model can be trained in an efficient non-sampling manner. We further give two examples of CGCR: LGC with linear graph convolutional networks and LGC+ with attention mechanism. Extensive experiments on five real-world public datasets indicate that the proposed CGCR models significantly outperform the state-of-the-art methods on the Top-K recommendation task.

Keywords: Collaborative filtering · Top-K recommendation · Matrix factorization · Graph neural network

1 Introduction

As one of the powerful personalized recommendation approaches in modern recommender systems, collaborative filtering makes prediction based on users' past preference embodied in the user-item interactions of similar users as shown in Fig. 1(a). When it comes to extracting users' preference or interest from the observations, matrix factorization [16] plays an important role. In general, matrix factorization finds a common low-dimensional space to describe users and items

© ICST Institute for Computer Sciences, Social Informatics and Telecommunications Engineering 2021
Published by Springer Nature Switzerland AG 2021. All Rights Reserved
H. Gao and X. Wang (Eds.): CollaborateCom 2021, LNICST 406, pp. 259–276, 2021.
https://doi.org/10.1007/978-3-030-92635-9_16

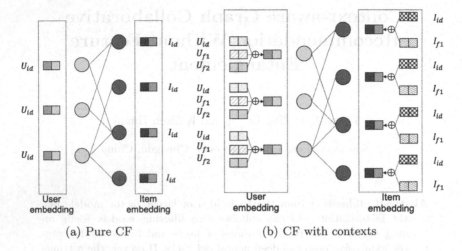

Fig. 1. The traditional collaborative filtering method and a typical treatment of contextual feature fusion. U_{id} and I_{id} denote the embeddings of user ID and item ID, while U_{fi} and I_{fi} represent the embeddings of the i-th feature (contextual information) for user and item, respectively.

wherein latent features (factors) are indicated by the dimensions, and then compare the user/item vectors directly using dot product. To further improve the representations in latent feature space, prior work resorts to deep neural networks [10]. In particular, the very recent work [7,24] shows that the high-order interactions among features can be successfully captured via graph convolution operation on the user-item bipartite graph.

However, in real recommendation processes, it may be inadequate to consider only the user-item interaction behaviors. It is also important to incorporate rich contextual data into the models to adapt the recommender system to specific situation of the user, which challenges current collaborative filtering scheme. An intuitive solution for introducing the auxiliary information (*e.g.*, user's gender, age or item's price) to the existing graph-based collaborative filtering models, is to pool the embedding of the features into one vector (as shown in Fig. 1(b)) for further convolutional embedding learning. The consequence of such feature entangling is the lack of interpretability, that is, the model is not able to learn meaningful and explainable representations for users and items. This problem will be compounded by graph convolution. In fact, the graph defined by the user-item interactions is naturally a heterogeneous graph, implying that different types of nodes may fall in different feature spaces. However, the traditional graph convolution aggregates different types of features from different semantic space. As a result, the original semantics of node embeddings are destroyed [23].

To tackle the problem of semantic damage when introducing contextual information into collaborative filtering, we design a novel **C**ontext-aware **G**raph **C**ollaborative **R**ecommendation framework (**CGCR**), which consists of three parts: *Feature space construction Layer*, *Graph Neural Network Layer*, and *Pre-*

diction Layer. To preserve the embedding semantics of users and items, we first construct a shared feature space that embodies all features from user and item, so that the original user-item bipartite graph can be easily converted to a homogeneous one. To overcome the representation entangling caused by feature crossing, we discriminate information from different feature fields. Then we use graph convolutional network to obtain high-order information contained in the user-item interactions. Finally, we adopt generalized matrix factorization [4] as matching function and train the proposed model in an efficient non-sampling manner. Different from NCF [10], our model endows the representation learning process with transparency and thus can link the feature fields (or factors) to the prediction results. Besides the visualized explanation, we also evaluate the effectiveness of the model on five real-world datasets. The experimental results show the superior performance in context-aware recommendation tasks, compared to prior feature-crossing-based methods.

The main contributions of this work are as follows:

- We propose a new graph-based recommendation framework for context-aware collaborative filtering. By putting the feature representations of users and items into the same subspace, our model is able to incorporate contextual information to the graph convolutional embedding process on the resultant homogenous graph.
- We further propose two examples of CGCR, namely LGC and LGC+, and for the first time use the efficient non-sampling scheme as the optimization solution for the context-aware collaborative filtering.
- Besides visualizing explanations, we also conduct extensive experiments to evaluate the proposed models on five benchmark datasets (two of them are for context-aware recommendation while the rest is for pure collaborative filtering tasks). The results show the significant improvement of our two models over several state-of-the-art models, in terms of model effectiveness and interpretability.

2 Problem Formulation

We formulate the problem of context-aware collaborative filtering as follows: Given the collections of users and items, which are denoted by $\mathbf{U}$ and $\mathbf{V}$ respectively, the user-item interaction pattern is depicted by the matrix $\mathbf{R}$, where $[r_{uv}] \in \{0, 1\}$ indicates whether user u interacts with item v. Then the goal of the collaborative filtering recommendation task is to recommend a list of items that target user u may be interested in, which is formally defined as follows:

Input: Users $\mathbf{U}$, items $\mathbf{V}$ and user-item interaction matrix $\mathbf{R}$.

Output: A ranked item list based on the predicted probability $\hat{r}_{uv}$ that user u would interact with item v (ordered from highest to lowest).

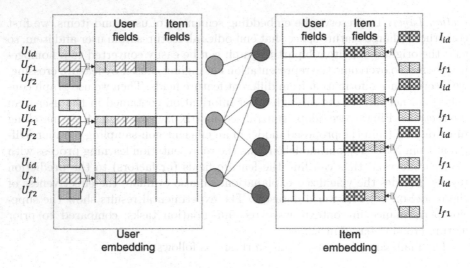

Fig. 2. Feature fusion in the *Feature Space Construction Layer* of the CGCR framework. All colored boxes indicate that the position has been filled with real values, and a blank box indicates that the position is empty and filled with 0.

3   Methodology

In the following, we introduce the proposed context-aware collaborative recommendation framework CGCR, which is based on the aforementioned learning scheme. We first explain the general framework of CGCR. Then, we use generalized matrix factorization with a non-sampling strategy to learn the parameters. Finally, we discussed the model complexity of CGCR.

3.1   General Framework

There are three components in CGCR: (1) a *Feature Space Construction Layer* that initializes the embeddings of users and items and maps them onto the same subspace; (2) a *Graph Neural Network Layer* that aggregates neighbor information onto the target node; and (3) the *Prediction Layer* using generalized matrix factorization scheme.

Feature Space Construction Layer. In the very recent neural collaborative filtering model [10], the embedding layer encodes user IDs and item IDs with multi-layer perceptron to account for the nonlinear interactions. However, in the settings where contextual information is available, how to merge the context data into the embedding should be carefully designed. The reason is that current neural network based collaborative filtering models regard the information about users and items as homogeneous by default (i.e., coming from the same feature space), while different contexts actually have different semantics. To preserve the semantics of various types of features and offer the model explanability, it is necessary to construct a new feature space that contain all factors involved in recommendation. A natural choice is to expand the k dimensional feature space of users (items) with the r dimensional feature space of items (users). Then, the dimension of the derived feature space is $k + r$. Specifically, we construct a new shared feature space for users and items by concatenating all the initial embedding dimensions in the order of *user ID*, *user features*, *item ID*, and *item features* as follows:

$$\mathbf{E} = [\underbrace{\mathbf{e}_{u_1}, \cdots, \mathbf{e}_{u_n}}_{user\ IDs}, \underbrace{\mathbf{e}_{u_{n+1}}, \cdots, \mathbf{e}_{u_N}}_{user\ features}, \underbrace{\mathbf{e}_{v_1}, \cdots, \mathbf{e}_{v_m}}_{item\ IDs}, \underbrace{\mathbf{e}_{v_{m+1}}, \cdots, \mathbf{e}_{v_M}}_{item\ features}]. \tag{1}$$

This operation allows us to convert a heterogeneous bipartite graph to be homogeneous. As is graphically depicted in Fig. 2, there arc only single type of nodes represented by concatenated vectors[1] in the same feature space, which endows the model with the flexibility to leverage the inherent complex relations in the data by graph neural network techniques while keeping each dimension relatively independent.

It is worth noting that the expansion of feature space in the form of vector concatenation also allows each factor (or contextual feature) to have its specific role in the representation learning, which is different from commonly used sum operation in feature fusion. The latter potentially assumes that the summed features have the same weight in forming new representation.

Graph Neural Network Layer. The core to achieve collaborative filtering is to capture users' historical preference as well as the high-order correlations between users and items, which remind us of recent graph neural networks [15,25,28]. The aggregation and message passing involved in graph neural networks can easily resolve those two problems, on top of the converted homogeneous graph. According to graph convolution, the embeddings of users and items can be updated uniformly as:

$$\mathbf{e}_i^{(l+1)} = AGG(\{\mathbf{e}_j^{(l)} : j \in \mathcal{N}_i\}), \tag{2}$$

where the pair (i, j) corresponds to the user-item interaction in a bipartite graph and $\mathcal{N}_i$ is the neighbor of node i. $AGG(\cdot)$ is the aggregation function that can be specified. Here are two aggregation functions, LGC and LGC+ using attention. Their aggregation functions are organized as shown in Table 1.

To make the embeddings more traceable, we restrict each layer of graph convolution with constant dimension. Due to feature space expansion, the initial embeddings will contain many 0s in the extended dimensions (indicated by blanks in Fig. 3). Thus an interesting phenomenon in the learning process is that, the information is updated only on a part of the embedding. Take the user

---

[1] An example of concatenation (*aka.* $\|$) is that $[a_1, a_2, \ldots, a_m]\|[b_1, \ldots, b_n] = [a_1, a_2, \ldots, a_m, b_1, \ldots, b_n]$, where a and b are both scalar.

Table 1. Neighbor aggregation function at a graph convolutional layer for LGC and LGC+.

Model	Neighbor aggregation function
LGC	$e_v^{(k)} = \sum_{u \in \mathcal{N}(v) \cup \{v\}} \frac{1}{\sqrt{|\mathcal{N}(v)| \cdot |\mathcal{N}(u)|}} e_u^{(k-1)}$
LGC+	$e_v^{(k)} = \mathrm{ReLU}(\sum_{u \in \mathcal{N}(v) \cup \{v\}} a_{vu}^{(k)} W^{(k)} e_u^{(k-1)})$

embedding as an example: at the first graph convolution layer, the aggregation for the user is the embeddings of its preferred items, and these item embeddings do not contain any user information. That is, the update occurs only on the item dimensions. Similarly, the update of items' embeddings only involve users' information. From this process one can see that our architecture does not introduce any feature entanglement. The final representations of the users and items are obtained by taking the sum of the embeddings in each layer.

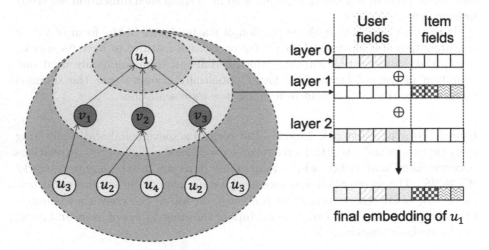

Fig. 3. An illustration of *Graph Neural Network Layer* of the CGCR framework. The result of each layer of graph convolution will leave blanks, and the final embedding representation of u_1 is the element-wise sum of result of each layer.

To implement the model more efficiently, we provide the matrix form of LGC. We first construct the adjacency matrix corresponding to the homogeneous graph using:

$$\mathbf{A} = \begin{bmatrix} \mathbf{0} & \mathbf{R} \\ \mathbf{R}^\top & \mathbf{0} \end{bmatrix}, \tag{3}$$

where $\mathbf{0}$ now is all-zero matrix. The Laplacian matrix $\mathcal{L}$ for the adjacency matrix is formulated as:

$$\mathcal{L} = \mathbf{D}^{-\frac{1}{2}} \mathbf{A} \mathbf{D}^{-\frac{1}{2}}, \tag{4}$$

where $\mathbf{D}$ is the diagonal degree matrix. Let the output of the *feature space construction layer* be $\mathbf{E}^{(0)}$, our goal is $\mathbf{E}^{(l)}$ after l-layer graph convolution. By applying the matrix-form propagation rule, we have:

$$\mathbf{E}^{(l)} = \mathcal{L}\mathbf{E}^{(l-1)} = \mathbf{D}^{-\frac{1}{2}}\mathbf{A}\mathbf{D}^{-\frac{1}{2}}\mathbf{E}^{(l-1)}. \tag{5}$$

As a result, we can simultaneously update the representations for all users and items in a rather efficient way. Similar to LightGCN, we only aggregate the immediate neighbors and do not integrate the target node itself, which means that there is no self-connection in each graph convolution layer.

For the final representations, we combine the results of all graph convolution layers:

$$\mathbf{E} = \alpha_0\mathbf{E}^{(0)} + \alpha_1\mathbf{E}^{(1)} + \cdots + \alpha_l\mathbf{E}^{(l)}. \tag{6}$$

In our experiments, we set α uniformly as $\frac{1}{l+1}$, which is equivalent to the mean pooling of all vectors.

Prediction Layer. Different from the way of employing inner product in the existing graph neural network based collaborative filtering models [4,10], we use the more generalized similarity function to estimate the matching degree as follows:

$$\hat{r}_{uv} = \mathbf{h}^\top(\mathbf{e}_u \odot \mathbf{e}_v), \tag{7}$$

where $\mathbf{e}_u = \mathbf{e}_u^{(1)} \oplus \mathbf{e}_u^{(2)} \oplus \ldots \mathbf{e}_u^{(l)}$ (so is $\mathbf{e}_v$) is the learned embedding representation, and $\mathbf{h}$ is what we call prediction layer.

The above generalized similarity measure provides our model better interpretability. Note that the feature dimensions are never entangled during representation learning and matching, input feature dimensions can be directly associated with the components of $\mathbf{h}$ to manifest their influences on recommending certain items. Since the contexts and IDs are embedded with various dimensions, we evaluate the relative importance of a factor (or feature) as:

$$s_i = \frac{\sum_{k=1}^d \mathbf{h}_{i_k}}{\sum_{p=1}^{|\mathbf{h}|} \mathbf{h}_p}, \tag{8}$$

where i indicates the i-th factor whose embedding dimension is d. This way, our model exhibits not only the transparency in representation learning but the explanability in prediction.

3.2 Optimization

Instead of using the most popular *Bayesian Personalized Ranking* [20] based on negative sampling to learn the model, we train the model on the whole data with the latest proposed *Efficient Non-sampling Matrix Factorization* [4](ENMF). Following this setting, we have the loss of the predictions[2] as:

$$L(\theta) = \sum_{u \in \mathbf{U}} \sum_{v \in \mathbf{V}^+} ((c^+ - c^-)\hat{r}_{uv}^2 - 2c^+ \hat{r}_{uv})$$
$$+ \sum_{i=1}^{d} \sum_{j=1}^{d} ((h_i h_j)(\sum_{u \in \mathbf{U}} e_{u,i} e_{u,j})(\sum_{v \in \mathbf{V}} e_{v,i} e_{v,j})). \tag{9}$$

Note that there are two hyper-parameters, namely, c^+ and c^-, in the objective. They are used to tune the penalty of poor predictions. In practice, we always value the positive samples than negative ones. So the weight of the prediction on the positive samples is often set to 1, and the weight corresponding to the negative samples will be a smaller positive number (lying between 0 and 1). However, different weights on negative sampling could have different impacts on the performance of the model, which will be explored in the experimental part.

3.3   Time Complexity Analysis of CGCR

In addition, compared to negative sampling, we use the efficient non-sampling framework to greatly reduce the training time. It can be found that the time complexity of the whole learning process is primarily determined by the computation of the loss shown in Eq. (9), which is similar to ENMF, namely, $O((|\mathbf{U}| + |\mathbf{V}|)d^2 + |\mathbf{R}|d)$ for d-dimensional embeddings. However, it should be noted that the feature space expansion in the first layer of our model compromises the computation efficiency, compared to ENMF model, as it enlarges the embedding dimensions.

4   Experiments

We now perform experiments to (1) compare CGCR framework to several strong baseline methods in collaborative filtering on five real-world datasets; (2) and analyze the interpretation of our model with visualization. In addition, we explore the effects of negative samples and other hyper-parameters involved in the model.

4.1   Dataset Description

To evaluate the performance of the proposed method, we conduct extensive experiments on the following *implicit feedback* datasets: ***Movielens***[3], ***Zhihu***[4],

---

[2] The proof of Eq. (9) see the previous work [4] for the details.
[3] https://grouplens.org/datasets/Movielens/1m/.
[4] https://github.com/THUIR/CC-CC/tree/master/dataset.

Table 2. Statistics of the datasets.

Dataset	Movielens	Zhihu	Pinterest	Tiktok	KKBOX
User	6,040	9,177	55,187	18,855	24,613
Items	3,706	9,946	9,916	34,756	61,877
User fields	4	4	2	2	2
Item fields	2	3	2	2	2
Instances	1,000,209	810,485	1,500,809	1,493,532	2,170,690
Density	0.0447	0.0088	0.0027	0.0023	0.0014

Pinterest[5]. *Tikiok*[6], *KKBOX*[7]. We summarize the statistics of three datasets in Table 2:

- *Movielens* is widely used as the benchmark for recommendation performance evaluation. In our experiment, we choose the version that contains one million ratings and binarize the data into implicit feedback. User contextual information includes user ID, gender, age, and occupation. The context of items consists of movie ID and movie genre.
- *Zhihu* is obtained from the previous work [21]. Specifically, besides the IDs, for user features, we preserve the number of focusing users, the number of focusing questions, and the city where the user is located; as for item features, the numbers of item likes and item comments are kept.
- *Pinterest, Tiktok, KKBOX* are all used for pure collaborative filtering without contextual data. In this work, we use this dataset to demonstrate that our model can not only be applied to context-aware recommendations but pure collaborative filtering tasks. In order to exploit the characteristics of feature combination of the FM-based models, we added a tag feature after the ID of the user and the item, to indicate that the ID belongs a user or an item.

4.2   Experimental Settings

Baseline Methods. We compare our models with the following state-of-the-art methods:

- **FM** [18]: Though the original FM is written in C++ [19], to make comparisons in the same environment, we use the package built-in TensorFlow framework.
- **AFM** [26]: AFM adds an attention mechanism to the FM framework so that more useful feature combination items will have more positive effects on the results.

---

[5] https://github.com/hexiangnan/adversarial_personalized_ranking/tree/master/Data.

[6] http://ai-lab-challenge.bytedance.com/tce/vc/.

[7] https://www.kaggle.com/c/kkbox-music-recommendation-challenge/data.

- **NFM** [6]: NFM adds MLP to the FM framework to recognize more complex patterns in the data. It can be regarded as the parallel work of AFM.
- **ENMF** [4]: ENMF is a non-sampling framework which learns from the whole training data without sampling by reformulating a commonly used square loss function with rigorous mathematical reasoning.
- **ENSFM** [3]: ENSFM tackles the recommendation problem in context scenarios, which incorporates FM component in the architecture to implement the second-order interactions among the user-self features and item-self features. ENSFM is special in that it uses its own new non-sampling method.
- **LightGCN** [7]: LightGCN is currently the most popular graph-based recommendation model for pure collaborative filtering. In the experiments on Pinterest dataset, LightGCN only uses the IDs of users and items.

Since both LightGCN and ENMF are not designed for context-aware recommendation, to adapt the contextual datasets to these methods, we use sumpooling to combine the features of all users (or items) into a vector as the user's (or item's) ID embedding on Movielens and Zhihu datasets. Although the learning process (graph convolution) of the proposed LGC is quite close to LightGCN, the original LightGCN is learned with BPR loss, which encourages the prediction of an observed entry to be higher than its unobserved counterparts. For a fair comparison, we implement a non-sampling version for LightGCN using the loss depicted in Eq. (9).

Parameter Settings. There is an important parameter in the non-sampling framework we adopt here, namely, the negative sample weight c^-. To see the impact of this parameter on the performance and find the optimal weight for the models, we explore a range of values. In particular, for the baselines (ENMF, ENSFM, LightGCN) and our models (LGC and LGC+), we search the negative sample weight in $\{0.005, 0.01, 0.1, 0.3, 0.5, 1\}$ and learning rate in $\{0.005, 0.01, 0.02, 0.05\}$. Based on the optimal weights, we can choose the proper embedding dimensions for ENMF, ENSFM, LightGCN, LGC and LGC+, while for FM, AFM, and NFM that are based on negative sampling, we fix the embedding dimension to 32. We optimize all models with the Adam [14] optimizer, where the batch size is fixed at 512 for FM, AFM and NFM, and 256 for the rest. Besides, the attention factors for AFM is set as 32, the number of MLP layers for NFM is set as 1 with 32 neurons. For LightGCN, LGC and LGC+, the number of layers of graph convolution is set in $\{1, 2, 3, 4\}$.

Evaluation Metrics. To evaluate the effectiveness of top-K recommendation and preference ranking, we adopt two widely-used evaluation protocols: *HR@K* and *NDCG@K* [13]. To speed up the computation of metrics, the common practice presented in recent work [5, 9, 10] is to use sampled metrics wherein only a smaller set of random items and the relevant items are ranked. Sampled metrics, however, are inconsistent with their exact version [17]. Therefore, in our experiment, we adopt the *leave-one-out* scheme of *all ranking*, that is, we treat all items the user does not interact with as negative samples and rank them. Although this scheme makes the results presented in this work smaller than that of the sampling evaluation scheme, it guarantees the reliability of the evaluation.

4.3 RQ1: Does the Proposed Method Perform Better Than Other Comparison Methods?

We first compare our proposed models with the baseline methods. We investigate the *top-K* performance when K is 20 and number of Graph Convolution Layer is 3. For the purpose of a fair comparison, the embedding size is set as 32 for all approaches. The performance of all methods on five datasets is shown in Table 3, where the percentages of relative improvement on each metric are also presented.

Table 3. Overall performance comparison. NG stand for NDCG. The best result of the state-of-the-art methods is underlined.

	Movielens		Zhihu		Pinterest		Tiktok		KKBOX	
	HR	NG	HR	NG	HR	NG	HR	NG	HR	NG
FM	0.1156	0.0457	0.1064	0.0426	0.0777	0.0291	0.0683	0.0218	0.2101	0.1161
AFM	0.1258	0.0499	0.1519	0.0617	0.0848	0.0322	0.0722	0.0254	0.2643	0.1444
NFM	0.1291	0.0501	0.1696	0.0704	0.0952	0.0373	0.0739	0.0277	0.2681	0.1479
ENMF	0.1459	0.0584	0.2044	0.0876	0.0968	0.0384	0.0753	0.0302	0.2785	0.1516
ENSFM	0.1525	0.0582	0.2074	0.0897	0.1058	0.0412	0.0813	0.0316	0.2822	0.1579
LightGCN	0.1515	0.0593	0.2064	0.0882	0.1011	0.0397	0.0827	0.0319	0.2823	0.1588
LGC	0.1566	0.0611	0.2134	0.0924	**0.1154**	**0.0458**	0.0877	0.0342	0.2920	0.1659
LGC+	**0.1566**	**0.0627**	**0.2178**	**0.0936**	0.1140	0.0457	**0.0881**	**0.0344**	**0.2921**	0.1657

Based on these results, the following observations are drawn:

- Under the non-sampling framework, our model is more effective than Light-GCN, though these two methods use the similar graph network for representation learning. Albeit introducing contextual information into LightGCN, the experimental results prove that our model is superior to LightGCN. Moreover, the comparison between our model and LightGCN sheds light on the positive effect of preserving semantics of the embedded dimensions on recommendation performance, which coincides with the prior conclusion [8,27] that points out the importance of maintaining semantics.
- By using the attention mechanism in LGC+, LGC is further improved in most cases. It is also worth noting that LGC+ surpasses the most competitive baseline method ENSFM, whose performance is better than ENMF on the three datasets (Movielens, Zhihu, Tiktok). The superiority of LGC+ may root in the non-linearity embodied in attention mechanism.

4.4 RQ2: Does the Proposed Method Elucidate the Meanings of Each Dimension of the Embedding?

The prediction layer is an important part of the CGCR model, which not only acts as a matching function to calculate the similarity between user-item pairs, but also offers the model interpretability. Thanks to distinguishing feature domains between users and items and no feature entanglement in every step, the prediction layer can be used to identify the importance of feature domains, which is indicated by the learned parameters h_i. More specifically, we normalize the learned h and use the Eq. (8) to quantify the importance score of each feature domain.

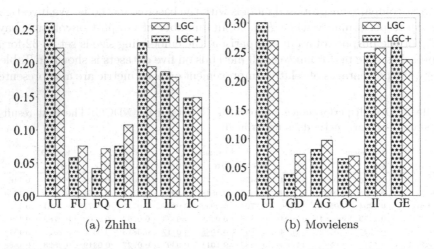

(a) Zhihu   (b) Movielens

Fig. 4. The weight distribution of the prediction layer on Zhihu and Movielens (UI: user id; FU: focusing users; FQ: focusing questions; CT: city; II: item id; IL: item likes; IC: item comments; GD: gender; AG: age; OC: occupation; GE: genre).

We provide the visualization of explainable results in Fig. 4. We can see that

- First and most intuitive, the importance of feature domains is heterogeneously distributed. Among the features, the most important feature is the user IDs and item IDs, followed by the item features like 'Item likes' and 'Item comments' in Zhihu dataset, while several user features are relatively insignificant, fitting well with our intuitions. For example, in Movielens it is shown from Fig. 4(b) that models' recommendation depends mainly on the category of the film, besides the ID features that embed the historical behaviors of the users. It thus give insight into the roles of different feature domains.
- For LGC and LGC+, the importance of different feature domains is marginal, which shows: 1) Introducing attention mechanism into CGCR has little effects on the importance of the feature domain; 2) In both LGC and LGC+, the IDs of users and items are the most important features. That is, historical records play a leading role in the recommendation, which further explain the reason that the proposed models work well on pure collaborative filtering task, even though it is customized for context-aware recommendation.

Note that LightGCN, which also uses the similar prediction layer, fail to find explanations from h. As in LightGCN, the user(item) features are interacted with each other, it is difficult to elucidate the meanings of each dimension of the embedding or separate latent factors.

4.5 RQ3: How Does Number of Graph Convolution Layer Impact?

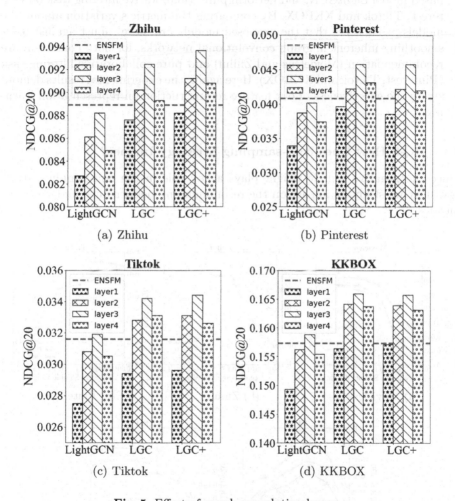

Fig. 5. Effect of graph convolution layers.

To see whether LGC and LGC+ can benefit from multiple embedding propagation layers, we vary the model depth. Figure 5 summarizes the experimental results, from which we have the following observations and analysis:

– It is shown on the four datasets that the increase of the convolutional layer will definitely increase the effectiveness of the model, and when the number of graph convolutional layer reaches 3, the effectiveness of the model will be weakened. The experiments further empirically verify the drawback of graph networks, that is, deeper structure may lead to the oversmoothing that makes all the node embeddings move closer to each other, thereby drastically reducing the recommendation performance.

- We particularly compare our model with the state-of-the-art graph network based model LightGCN, and perform pure collaborative filtering task on Pinterest, Tiktok and KKBOX. By comparing the metric's variation among the models, we observe that the proposed models are more robust against oversmoothing inherent in graph convolutional networks, for both context-aware recommendation (Movielens and Zhihu) and pure collaborative filtering task (Pinterest, Tiktok and KKBOX). Here again, the experimental results demonstrate the effectiveness of our feature reconstruction on representation learning.

4.6 RQ4: How Does Non-sampling Strategy Impact?

Since the negative sample weight plays a key role in the non-sampling strategy, next we study its impact on the recommending performance under various metrics.

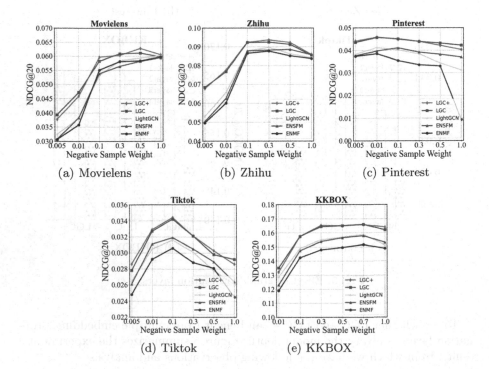

(a) Movielens   (b) Zhihu   (c) Pinterest

(d) Tiktok   (e) KKBOX

Fig. 6. Negative sample weight of each model under three datasets.

We visualize the results in Fig. 6 and reach the following conclusions:

- The shapes of the curves indicate that all models with non-sampling strategy are sensitive to negative sample weights, though the extent of the impacts differs on different datasets. This suggests that in practical applications, the first priority is to search for an optimal weight for certain dataset.
- Note that although the models have different optimal negative sample weights for different datasets, their optimal values on the same dataset are almost the same (corresponding to the peaks of the curves in each plot). This phenomenon reflects the fact that there exists a unique optimal weight in ENMF framework based models on specific dataset, which inspires us a way of fast finding the proper negative weight for the proposed models (e.g., LGC/LGC+) in practice: one could first search for the optimal negative sample weight using ENMF or ENSFM model, and then use it directly in LGC's hyper-parameter setting. This approach can largely improve the learning efficiency.

5 Related Work

5.1 Collaborative Filtering

Recently, graph neural networks (GNNs) [15] have received increasing attention due to their great capacity in various tasks. Inspired by the success in other fields such as node classification and link prediction, researchers also investigate the suitability of GNNs for the task of recommendation. Representative works like NGCF [24] and LightGCN [7] have proven that GNN can significantly enhance collaborative filtering. In addition, side information like the features or attributes of users and items can also play a non-negligible role in graph-based collaborative filtering. However, there is no ideal solution to model graph-based collaborative filtering with additional attributes and features. This is our original intention of proposing CGCR.

5.2 Non-sampling Learning for Top-K Recommendation

It is critically important to design recommendation algorithms that can work with implicit feedback data. To learn from a sparse data, there are generally two optimization strategies: 1) negative sampling strategy and 2) non-sampling (whole-data based) strategy. The first strategy samples negative instances from missing entries, which has been widely adopted for efficient training in previous work [10,20]. However, some studies [2,11,30] have shown that sampling can limit the performance of recommendation, as it is not robust but highly sensitive to the sampling distributions. In contrast, non-sampling strategy sees all the missing data as negative and leverages the whole data with a potentially better coverage, but computation efficiency can be an issue. Recently, some efforts [1,4,11,12] have been devoted to resolving the inefficiency of non-sampling learning. Among them, we point out that ENMF [4] framework is most relevant to our work. ENMF is to learn neural recommendation models from the whole training data without sampling. By reformulating a commonly used square loss function with rigorous mathematical reasoning, each parameter can be updated in a manageable time complexity without sampling, which makes it consistently and significantly outperform the state-of-the-art recommendation models in terms of both recommendation performance and training efficiency.

6   Conclusion and Future Work

In this work, we tackle the problem of semantic separation when integrating contextaul and user-item interaction information in collaborative filtering recommendation. We have proposed a simple but interpretable framework CGCR for context-aware Top-K recommendation. CGCR not only connects graph neural networks and matrix factorization, but also preserves the semantic independence of each feature domain. As a result, CGCR achieves two remarkable advantages: 1) offers a solution to convert a heterogeneous graph to the homogeneous one; 2) seamlessly incorporates contextual features to collaborative filtering with better explainability. Extensive experiments on five real-world datasets have shown that the proposed CGCR consistently and significantly outperforms the state-of-the-art methods including feature-crossing-based models NFM, AFM, and CFM, in terms of both recommendation performance and efficiency.

The proposed CGCR framework is not limited to the recommendation task presented in this paper. We believed that this framework could benefit many other tasks where heterogeneous graph is used. In the future, we will explore our CGCR method on the related tasks like social recommendation [29] and knowledge graph-based recommendation [22].

References

1. Chen, C., Zhang, M., Ma, W., Liu, Y., Ma, S.: Jointly non-sampling learning for knowledge graph enhanced recommendation. In: Proceedings of the 43rd International ACM SIGIR Conference on Research and Development in Information Retrieval (2020)
2. Chen, C., et al.: An efficient adaptive transfer neural network for social-aware recommendation. In: Proceedings of the 42nd International ACM SIGIR Conference on Research and Development in Information Retrieval (2019)
3. Chen, C., Zhang, M., Ma, W., Liu, Y., Ma, S.: Efficient non-sampling factorization machines for optimal context-aware recommendation. In: Proceedings of The Web Conference 2020 (2020)
4. Chen, C., Zhang, M., Zhang, Y., Liu, Y., Ma, S.: Efficient neural matrix factorization without sampling for recommendation. ACM Trans. Inf. Syst. (TOIS) **38**, 1–28 (2020)
5. Deng, Z.H., Huang, L., Wang, C.D., Lai, J., Yu, P.S.: Deepcf: a unified framework of representation learning and matching function learning in recommender system. In: AAAI (2019)

6. He, X., Chua, T.S.: Neural factorization machines for sparse predictive analytics. In: Proceedings of the 40th International ACM SIGIR Conference on Research and Development in Information Retrieval (2017)
7. He, X., Deng, K., Wang, X., Li, Y., Zhang, Y., Wang, M.: Lightgcn: simplifying and powering graph convolution network for recommendation. In: Proceedings of the 43rd International ACM SIGIR Conference on Research and Development in Information Retrieval (2020)
8. He, X., Du, X., Wang, X., Tian, F., Tang, J., Chua, T.S.: Outer product-based neural collaborative filtering. In: IJCAI (2018)
9. He, X., He, Z., Du, X., Chua, T.S.: Adversarial personalized ranking for recommendation. In: The 41st International ACM SIGIR Conference on Research & Development in Information Retrieval (2018)
10. He, X., Liao, L., Zhang, H., Nie, L., Hu, X., Chua, T.S.: Neural collaborative filtering. In: Proceedings of the 26th International Conference on World Wide Web (2017)
11. He, X., Zhang, H., Kan, M.Y., Chua, T.S.: Fast matrix factorization for online recommendation with implicit feedback. In: Proceedings of the 39th International ACM SIGIR conference on Research and Development in Information Retrieval (2016)
12. Hu, Y., Koren, Y., Volinsky, C.: Collaborative filtering for implicit feedback datasets. In: 2008 Eighth IEEE International Conference on Data Mining, pp. 263–272 (2008)
13. Järvelin, K., Kekäläinen, J.: Cumulated gain-based evaluation of IR techniques. ACM Trans. Inf. Syst. **20**, 422–446 (2002)
14. Kingma, D.P., Ba, J.: Adam: a method for stochastic optimization. In: ICLR (2015)
15. Kipf, T.N., Welling, M.: Semi-supervised classification with graph convolutional networks. In: ICLR (2017)
16. Koren, Y., Bell, R.M., Volinsky, C.: Matrix factorization techniques for recommender systems. IEEE Comput. **42**(8), 30–37 (2009)
17. Krichene, W., Rendle, S.: On sampled metrics for item recommendation. In: Proceedings of the 26th ACM SIGKDD International Conference on Knowledge Discovery & Data Mining (2020)
18. Rendle, S.: Factorization machines. In: 2010 IEEE International Conference on Data Mining, pp. 995–1000 (2010)
19. Rendle, S.: Factorization machines with libfm. ACM Trans. Intell. Syst. Technol. **3**, 57:1–57:22 (2012)
20. Rendle, S., Freudenthaler, C., Gantner, Z., Schmidt-Thieme, L.: BPR: bayesian personalized ranking from implicit feedback. In: UAI, pp. 452–461 (2009)
21. Shi, S., et al.: Adaptive feature sampling for recommendation with missing content feature values. In: Proceedings of the 28th ACM International Conference on Information and Knowledge Management (2019)
22. Wang, X., He, X., Cao, Y., Liu, M., Chua, T.S.: Kgat: knowledge graph attention network for recommendation. In: Proceedings of the 25th ACM SIGKDD International Conference on Knowledge Discovery & Data Mining (2019)
23. Wang, X., He, X., Chua, T.S.: Learning and reasoning on graph for recommendation. In: Proceedings of the 28th ACM International Conference on Information and Knowledge Management (2019)
24. Wang, X., He, X., Wang, M., Feng, F., Chua, T.S.: Neural graph collaborative filtering. In: Proceedings of the 42nd International ACM SIGIR Conference on Research and Development in Information Retrieval (2019)

25. Wu, F., Zhang, T., Souza, A., Fifty, C., Yu, T., Weinberger, K.Q.: Simplifying graph convolutional networks. In: ICML (2019)
26. Xiao, J., Ye, H., He, X., Zhang, H., Wu, F., Chua, T.S.: Attentional factorization machines: Learning the weight of feature interactions via attention networks. In: IJCAI (2017)
27. Xin, X., Chen, B., He, X., Wang, D., Ding, Y., Jose, J.: Cfm: convolutional factorization machines for context-aware recommendation. In: IJCAI (2019)
28. Xu, K., Hu, W., Leskovec, J., Jegelka, S.: How powerful are graph neural networks? In: ICLR (2018)
29. Yu, J., Gao, M., Li, J., Yin, H., Liu, H.: Adaptive implicit friends identification over heterogeneous network for social recommendation. In: Proceedings of the 27th ACM International Conference on Information and Knowledge Management (2018)
30. Yuan, F., Xin, X., He, X., Guo, G., Zhang, W., Chua, T.S., Joemon, J.: fBGD: Learning embeddings from positive unlabeled data with BGD. In: UAI (2018)

Improving Recommender System via Personalized Reconstruction of Reviews

Zunfu Huang[1,2], Bo Wang[1,2(✉)], Hongtao Liu[1], Qinxue Jiang[3], Naixue Xiong[4], and Yuexian Hou[1]

[1] College of Intelligence and Computing, Tianjin University, Tianjin, China
{huangwenjie,bo_wang,htliu,yxhou}@tju.edu.cn
[2] State Key Laboratory of Communication Content Cognition, People's Daily Online, Beijing, China
[3] School of Engineering, Newcastle University, Newcastle upon Tyne, UK
b9064217@Newcastle.ac.uk
[4] Department of Mathematics and Computer Science, Northeastern State University, Tahlequah, OK, USA

Abstract. Textual reviews of items are a popular resource of online recommendation. The semantic of reviews helps to achieve improved representation of users and items for recommendation. Current review-based recommender systems understand the semantic of reviews from a static view, i.e., independent of the specific user-item pair. However, the semantic of the reviews are personalized and context-aware, i.e., same reviews can have different semantics when they are written by different users or towards different items. Therefore, we propose an improved recommendation model by reconstructing multiple reviews into a personalized document. Given a user-item pair, we design a cross-attention model to build personalized documents by selecting important words in the reviews of the given user towards the given item and vice versa. A semantic encoder of personalized document is then designed using a cross-transformer mechanism to learn document-level representation of users and items. Extensive experiments on three real-world datasets demonstrate the effectiveness of the proposed model.

Keywords: Recommender system · Personalized reconstruction · Cross attention · Cross transformer

1 Introduction

Recommender systems have played an increasingly important role in online collaboration by helping people establish connections with interested items, e.g., information pieces, products or other users. Towards more effective items recommendation, an essential problem of recommender systems is to understand the users and items on semantic level. Therefore, auxiliary texts, e.g., reviews are

© ICST Institute for Computer Sciences, Social Informatics and Telecommunications Engineering 2021
Published by Springer Nature Switzerland AG 2021. All Rights Reserved
H. Gao and X. Wang (Eds.): CollaborateCom 2021, LNICST 406, pp. 277–295, 2021.
https://doi.org/10.1007/978-3-030-92635-9_17

widely used to improve the representation of users and items. Auxiliary reviews are an excellent source of understanding users' interests and items' characteristics, which can improve the performance of predicting users' rating on items and alleviate the issue of data sparsity and cold start. In this direction, one main challenge is to obtain valuable semantic information from large-scale auxiliary reviews and another consequent challenge is to make the information acquisition personalized, i.e., depending on each special user and item.

To improve the semantic recommendation, in this work, we focus on rating prediction, which is a main task of recommender systems. On predicting users' rating on items, many current works have been proposed. As one of the most widely used recommendation techniques, collaborative filtering has achieved successful results based on the use of users' ratings of items [1–3]. For example, Probabilistic Matrix Factorization (PMF [3]) uses probability matrix factorization technology to learn the potential factors among the user's rating matrices for items. However, this type of method has a serious problem of sparsity, and the predicted score can only reflect the user's overall satisfaction with a certain product, and lacks interpretability. With the development of e-commerce, users are more and more willing to post their reviews on purchased products on e-commerce platforms. Textual reviews contain rich information that can describe the characteristics of users and products, and the use of review information is proved to be able to alleviate data sparseness and cold start problems. Therefore, a lot of work using review information to enhance the recommender system has been proposed in the task of rating score prediction [4–12]. The initial approaches of using reviews are mainly to obtain potential features in reviews through topic modeling [4,11]. For example, the RMR [13] proposes an interpretable LDA [14] model to extract potential features in item review documents. The shortage of existing topic-based methods is dealing with textual reviews as bag-of-words, which loses the information of word order, resulting in the inability to fully capture the semantic information of the reviews.

Recently, in order to extract potential semantic features from reviews more comprehensively, many neural network-based rating score prediction models have been proposed, such as DeepCoNN [15], NARRE [5], DAML [16] and MRCP [17]. Since the convolutional neural network(CNN) can captures the local features and contextual semantic information of the review text, these methods usually use CNN to extract features of the reviews. For example, the DeepCoNN model learns the feature representation of users and items through the parallel CNN on the semantic features of user and item review documents, and performs rating score prediction.

Current methods of using review information can be mainly divided into two types: document-based methods and review-based methods. Although review-based methods have achieved significant improvement in rating score prediction, they still lack effective interpretability for complex scoring behaviors. Document refers to connecting all reviews of users/items into a long text, and document-based methods aim to combine the semantic of single reviews to a more advanced global semantic. Liu et al. [18] has proved that document-based and review-based methods are complementary in the task of rating score prediction. That is, the document-based method and the review-based method can capture the coarse-grained and fine-grained features of reviews, respectively.

For current document-based methods, we indicate two problems: (1) In order to avoid the long tail effect of review documents, all current document-based methods obtain static partial review documents through data preprocessing, e.g., DeepCoNN [15]. As input, partial review documents lose a lot of review information, and may not be able to fully obtain the coarse-grained characteristics of users and items. (2) Current document-based methods only model the word-level relationship between the documents of users and items. However, the document-level matching is believed to be more effective to utilize global semantic, which is the essential advantage of document-based strategy [16,19].

Therefore, to solve these two problems, we propose to reconstruct reviews into compressed personalized documents which can be dealt with as a unit instead of being cut into pieces. The compressed personalized documents are personalized built by identifying the most informative words from reviews of the certain user towards a certain item or vice versa. The identified words are organized into a personalized document in proper semantic order. A cross attention strategy is designed to effectively identify informative words. Then we encode the global semantics of compressed personalized documents of users and items with improved CNN layers. Furthermore, to learn more comprehensive features, we also incorporate document-level and review-level modeling into a unified framework.

In summary, our major contributions are as follows:

(1) We propose a neural recommendation model that reconstructs reviews into personalized documents and extracts coarse-grained features of users and items by modeling the semantic of personalized documents. To the best of our knowledge, we are the first to build personalized review documents for the recommender system.
(2) We propose a cross-attention mechanism to identify important words for personalized document construction. Personalization is achieved by obtaining in-depth word-level interaction features and context-aware interaction features. We design a novel transformer mechanism to learn the interaction feature representation between user-item pairs.
(3) Experimental results on three real-world datasets show that our model is more accurate in predicting users' rating scores than best-performing baselines. At the same time, it also reveals that personalized reconstruction of reviews helps to better capture coarse-grained features.

2   Related Work

In this section, we will review the recent studies which are most related to the works of review-based recommender system.

2.1   Review-based Recommendation with Topic Modeling

Due to the "data sparsity" and "cold start" problems of the collaborative filtering methods based on user-item interaction, researchers have introduced review information to the recommender system. Initial works use topic modeling techniques to learn potential topic features from review texts [4, 6, 11, 13]. For example, HFT [11] uses a LDA-like topic model to extract potential topic features of reviews for scoring prediction. TopicMF [4] uses MF technology to jointly model the user-item rating data and topic features of reviews. RBLT [20] utilizes ratings to promote reviews, and then combines review text and ratings to model user and item features in the shared space theme. These methods outperform models that rely solely on the user-item rating matrix. However, these bag-of-words-based models ignore the word order in the reviews, and cannot learn the local context information.

2.2   Document-Level Recommendation

Recently, document-level methods have been proposed to improve review-based recommendation. These coarse-grained recommender system methods directly combine reviews into a long document for learning the representation of users and items. For example, DeepCoNN [15] uses parallel CNN to learn the characteristics of users or items from review documents. D-Attn [21] uses local attention to learn the document-level feature representation of users and items. CARL [7] learns the interaction between user-item pairs of potential features based on context awareness, and uses it to make rating score predictions. DAML [16] uses the local and mutual attention of CNN to jointly learn the characteristics of reviews. A common problem of current document-level methods is to ignore the diverse and complex interactions between users and items.

2.3   Review-Level Recommendation

With the introduction of attention mechanism, in order to further improve the performance and interpretability of recommender systems, many fine-grained review-level methods have been proposed. For example, NARRE [5] introduces an attention mechanism to obtain the importance of each review on users or items, and provides review-level explanations for rating score prediction. In order to further filter reviews and words, MPCN [19] uses a common attention network based on Gumbel-softmax to dynamically select reviews and words that are important to target users or items. Dong et al. [22]. believes that user content is heterogeneous while item content is homogeneous. Therefore, they proposed the AHN model, which uses co-attention at the sentence-level and review-level to guide the representation learning of the user/item related to the target item/User. In order to increase the interpretability of fine-grained semantic features, MRCP [17] dynamically learns the feature representation of users and items through a three-layer attention framework of word-level, review-level and aspect-level. Although these review-level-based methods can capture more fine-grained features, they cannot effectively understand the coarse-grained features of users and items.

Compared with current methods, our proposed model can not only reconstruct ordered personalized review documents but also uses a transformer mechanism to capture the in-depth interaction characteristics between users and items.

3   Methodology

In this section, we present our proposed method Recommendation with Personalized Reconstruction of Reviews (PPRR) in detail. The overview of PPRR is shown in Fig. 1. The model has three stages: review document reconstruction network(Re-Doc-Net), document-level compiler(Doc-Net), and review-level compiler(Review-Net). The interactive attention characteristics of users and items are captured from word-level and review-level respectively. We will introduce the details of our model (PPRR) in detail below.

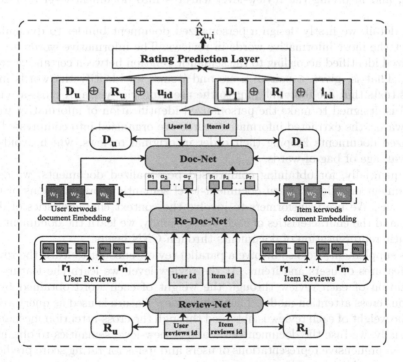

Fig. 1. In the framework of our PPRR method, the three main components are: review document reconstruction network (Re-Doc-Net), document-level encode network (Doc-Net), and review-level encode network (Review-Net).

3.1 Probem Definition

Suppose there is user set U, item set I and the rating matrix $\boldsymbol{R} \in \mathrm{R}^{|U| \times |I|}$, where the entry R_{ui} indicates the rating of user $u \in U$ towards item $i \in I$. For a user u, the reviews written by u can be noted as $r_u = \{r_{u,1}, \cdots, r_{u,n}\}$ where n is the number of reviews. For an item i, all reviews towards i can be represented as $r_i = \{r_{i,1}, \cdots, r_{i,m}\}$ where m is the number of reviews. For a single review r, we represent it as $r = \{w_{r,1}, \cdots, w_{r,t}\}$ where t is the length of r. Additionally, we reconstruct each r_u or r_i into document. We denote the document of r_u as $d_u = \{w_1^u, \cdots, w_T^u\}$, and denote document of r_i as $d_i = \{w_1^i, \cdots, w_T^i\}$ where w and T are the words in the document and the number of words, respectively.

3.2 Overall Framework of PPRR

The proposed PPRR model has three main stages: dynamically selecting informative words by cross attention to build compressed personalized documents of users and items, encoding the compressed personalized documents with CNN layers, and involving the review-level features into document-level representation.

In detail, we firstly design a personalized document builder to dynamically extract the most informative words in reviews. The informative words are personalized identified according to the semantic relation between certain users and items. That is, given a pair of users and items, we identify the words in the item's texts that are more relevant to the user and vice versa. A cross-attention model is designed to make the personalized identification of informative words. Afterwards, the extracted informative words are organized into compressed personalized documents keeping their order in original reviews, which avoids the disadvantage of bag-of-words.

Sequentially, for obtaining compressed personalized documents, we design a document encoder to learn document-level semantic representations of users and items. We use transformers to involve the context characteristics of documents and the characteristics of each word. Finally, we learn the document-level semantic representation of documents through CNN layers.

As supplement, we also involve a parallel review encoder to learn the review-level features of users and items. At the review-level, we learn the feature representation of each review through the weight of each word obtained by the previous cross attention model. Then, the review feature is used as query vector, and the weight of each review is obtained through the cross-attention mechanism.

Finally, we fuse the document-level and review-level semantics to obtain the final comprehensive representations of users and items for rating score prediction in recommendation.

3.3 Review Document Reconstruction Network (Re-Doc-Net)

Word Embedding Layer. Given a review $r = \{w_1, w_2, \cdots, w_t\}$ which has t words, we fill all words into an embedding matrix $\boldsymbol{W}^{d \times |V|}$ where V is the

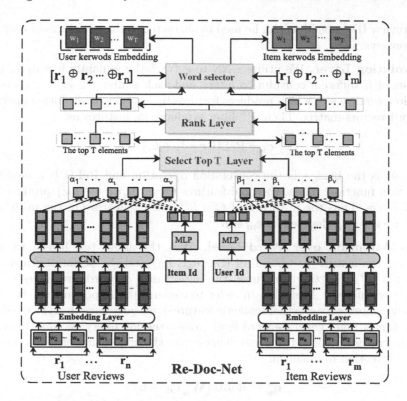

Fig. 2. The document reconstruction network of our PPRR approach, which uses the target ID embedding as the query vector, and obtains the weight of each word through the cross-attention mechanism. In addition, the three operations of **Select Top T Layer**, **Rank layer**, and **Word selector** all sort and select words on the vocabulary without any parameters.

vocabulary of words, d is the dimension of each word vector. Then the embedding vector of review r is $r \in R^{t \times dw}$.

ID Embedding Layer. IDs are usually regarded as identity information of the corresponding user and item in recommender system. Therefore, by encoding the IDs of users and items into low-dimensional vectors $\mathbf{u_{id}}$ and $\mathbf{i_{id}}$, the ID embeddings are denoted as

$$\mathbf{ID_r^u} = \left\{ \mathrm{ID}_{r,1}^u \cdots, \mathrm{ID}_{r,n}^u \right\}, \mathbf{ID_r^i} = \left\{ \mathrm{ID}_{r,1}^i \cdots, \mathrm{ID}_{r,m}^i \right\} \tag{1}$$

where $\mathbf{ID_r^u} \in R^{n \times z}$ and $\mathbf{ID_r^i} \in R^{m \times z}$ are the matrix of user's reviews ID embedding and item's reviews ID embedding, respectively. z is the dimension of ID embedding.

All ID embeddings are initialized randomly, where the ID embeddings of users and items can not only index the identity of each user, but also learn the attention query vector of the words of each user or item. In addition, user and item review ID embedding can be used to characterize the usefulness of user and item reviews.

Convolution Layer. We utilize CNN to extract the semantic feature of r. It consists of K different convolution filters, and each a filter $f \in \mathbb{R}^{l \times dw}$ where l is the filter window size which produces features by applying convolution operator on word vectors matrix. Then, j^{th} filter produces its features as:

$$c_j = \text{ReLU}(r * f_j + b_j) \tag{2}$$

where b_j is the bias, $*$ is the convolution operation and ReLU is a nonlinear activation function. Then, the final features $C = \{c_1, c_2, \cdots, c_K\}$ produced by the K filters. Thus the m-th row of $C \in R^{t \times K}$ is the feature of the m-th word in the review $r_{u,i}$, denoted as $c_m \in R^K$.

Cross Attention over Word Level. After the above two layers, we have obtained the feature vectors of all IDs (u_{id}, i_{id}) and the semantic features of all words (C). For the same user, the importance of words in different item reviews is different. Therefore, in order to capture the importance of words in different contexts (contextual semantic features), we introduce the ID embedding of the target item, generate word-level cross-attention query vectors through a Multilayer Perceptron (MLP), and then capture the importance of words through a cross-attention mechanism:

$$\mathbf{q_w^u} = \text{ReLU}\left(\mathbf{W_w^u} \mathbf{i_{id}} + \mathbf{b}\right) \tag{3}$$

$$g_x = \mathbf{q_w^u} \mathbf{A} \mathbf{c_x}, \quad \alpha_x = \frac{\exp(g_x)}{\sum\limits_{j=1}^{t} \exp(g_j)}, \alpha_x \in (0,1) \tag{4}$$

where $\mathbf{q_w^u}$ is the user word-level query attention vector derived from the corresponding item, $\mathbf{W_w^u}$ is parameter matrix, $\mathbf{A}$ is the harmony matrix in attention, α_x is the attention weight of the m-th word of a review. ReLU is a nonlinear activation function. Then, we aggregate the reviews according to the weight of the words in each review to obtain the feature vector of the review: $\mathbf{h_{u,x}} = \sum\limits_{j=1}^{t} \alpha_{x,j} \mathbf{c_{x,j}}$, then the review feature vector matrices of users and items are $\mathbf{h_u} \in R^{n \times z}$ and $\mathbf{h_i} \in R^{m \times z}$ respectively.

In order to obtain the user's coarse-grained characteristics more comprehensively and to capture the full contextual awareness information (contextual semantic information) to the greatest extent, we dynamically selected the X most important words in the user's reviews under the current target item and reconstructed ordered personalized review documents as shown in Fig. 2. The specific implementation method is as follows:

$$\text{top\_}\alpha_u, \text{top\_idx}_u = \text{TopX}\left(\text{Rank}\left(\alpha\right)\right) \tag{5}$$

$$\text{idx}_u, \text{idx\_idx}_u = \text{Rank}\left(\text{top\_idx}_u\right) \tag{6}$$

$$\mathbf{d_u} = \text{Select}\left(\left[\mathbf{r_1} \oplus \mathbf{r_2} \cdots \oplus \mathbf{r_n}\right], \text{idx}_u\right), \quad \alpha_{dw}^u = \text{Select}\left(\text{top\_}\alpha_u, \text{idx\_idx}_u\right) \quad (7)$$

where Rank$(\cdot)$ is the sorting method, which sorts the internal parameters in descending order, α is the weight of all words in user reviews, TopX$(\cdot)$ is the method of selecting the top X from the internal parameters, $\text{top\_}\alpha_u, \text{top\_idx}_u$ is the weight and position subscript of the first x words with the highest previous importance, respectively, $\text{idx}_u, \text{idx\_idx}_u$ a represents the weight subscripts and the subscripts of the subscripts after reordering the first X weight subscripts. Select$(\cdot)$ is to select important words and weights in user reviews based on subscripts, where $\mathbf{d_u}$ is a newly constructed vector of ordered review documents, α_{dw}^u is the weight of each word vector in $\mathbf{d_u}$. The above process only sorts the data, and does not add any parameters to the model. Similarly, we can obtain new ordered item review-documents $\mathbf{d_i}$ and word weights α_{dw}^i through the above methods.

3.4 Document-Level Encode Network(Doc-Net)

From the review document reconstruction network, we have constructed a new ordered user review document $\mathbf{d_u}$ and item review document $\mathbf{d_i}$. The document-based compilation network is shown in Fig. 3, in order to further obtain user-item complex interaction features and contextual information, We integrate ID embedding information and word position embedding information into each word embedding, and then use the transformer to capture the interaction features at the word-level. The specific calculation method is as follows:

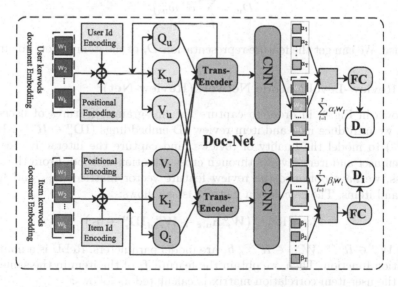

Fig. 3. The Doc-Net of our PPRR approach, which is mainly for feature extraction of personalized review documents, where the cross-transformer mechanism is mainly to dynamically enable word-level information to include target word-level and contextual features.

$$\mathbf{d_w^u} = \text{ReLU} \left(\mathbf{W_w^u u_{id}} + \mathbf{b} \right), \quad \mathbf{d_w^i} = \text{ReLU} \left(\mathbf{W_w^i i_{id}} + \mathbf{b} \right) \tag{8}$$

$$\mathbf{p_u} = \mathbf{d_u} + \mathbf{d_w^u} + \mathbf{pos_u}, \quad \mathbf{p_i} = \mathbf{d_i} + \mathbf{d_w^i} + \mathbf{pos_i} \tag{9}$$

where $\mathbf{W_w^u}$ and $\mathbf{W_w^u}$ are parameter matrices, $\mathbf{d_w^u}$ and $\mathbf{d_w^u}$ are the word-level query attention vector of the review-document, $\mathbf{pos_u}$ and $\mathbf{pos_i}$ are user and item position encodings with the same dimensions as the word embedding, and the encoding method is the same as that of $Transformer$.

$$\mathbf{Q_u} = \mathbf{W_u^Q p_u}, \quad \mathbf{K_u} = \mathbf{W_u^K p_u}, \quad \mathbf{V_u} = \mathbf{W_u^V p_u}$$
$$\mathbf{Q_i} = \mathbf{W_i^Q p_i}, \quad \mathbf{K_i} = \mathbf{W_i^K p_i}, \quad \mathbf{V_i} = \mathbf{W_i^V p_i} \tag{10}$$

$$\mathbf{d_u^T} = Trans(\mathbf{Q_u}, \mathbf{K_i}, \mathbf{V_i}), \quad \mathbf{d_i^T} = Trans(\mathbf{Q_i}, \mathbf{K_u}, \mathbf{V_u}) \tag{11}$$

where $\mathbf{W_u^Q}$, $\mathbf{W_u^K}$, $\mathbf{W_u^V}$ are parameter matrices related to the user. $\mathbf{W_i^Q}$, $\mathbf{W_i^K}$, $\mathbf{W_i^V}$ are parameter matrices related to the item. $Trans(\cdot)$ is the encoder of the transformer mechanism. $\mathbf{d_u^T}$ and $\mathbf{d_i^T}$ are the encoded review-document vectors.

Then, we capture the feature matrix $\boldsymbol{w_u}$ of all words of $\mathbf{d_u^T}$ through the convolutional neural network. Through the above process, we have obtained the weight α_{dw}^u of each word embedding in the user review document. Finally, we obtain the representation vector of the review document $\boldsymbol{D_u}$ of user u based on the item i via weighted summation of all words:

$$\boldsymbol{D_u} = \sum_{j=1}^{T} \alpha_u^j \boldsymbol{w_{u,j}} \tag{12}$$

Likewise, We can get the feature representation $\boldsymbol{D_i}$ of the item review document.

3.5 Review-Level Encode Network(Review-Net)

As shown in Fig. 4, in order to capture the fine-grained features of users and items, we introduce user and item review ID embeddings ($\mathbf{ID_r^u} \in R^{n \times z}$, $\mathbf{ID_r^i} \in R^{m \times z}$) to model the quality of reviews, and capture the interactive features between user and item reviews through cross attention method. From the above process, we have obtained the review feature vector matrices $\mathbf{h_{u,r}}$ and $\mathbf{h_{i,r}}$ of users and items. The calculation method is as follows:

$$f_u = \text{ReLU} \left(W_r^u \mathbf{h_{u,r}} + W_{rid}^u \mathbf{ID_r^u} + b_1 \right) \tag{13}$$

where $W_r^u \in R^{k \times z}$, $W_{rid}^u \in R^{z \times z}$, b_1 are model parameters. ReLU is a nonlinear activation function. Then, we obtain the feature f_i of the item in the same way. Then the user-item correlation matrix is calculated as follows:

$$e_j = W \left(f_u f_i^T \right) + b, \quad \alpha_j = \frac{\exp(e_j)}{\sum\limits_{k=1}^{n} \exp(e_k)}, \alpha_j \in (0, 1), \tag{14}$$

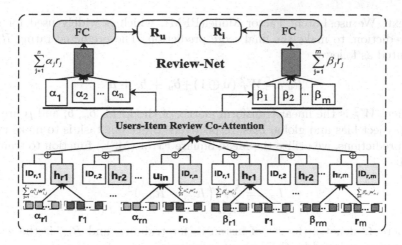

Fig. 4. The Review-Net of our PPRR approach, which mainly extracts the features of each review, and the cross-attention mechanism is mainly to dynamically obtain the weight of each review.

where $W \in R^m$ is the weight matrix of the fully connected layer. $\alpha \in R^n$ is the weight matrix of user reviews, which represents the importance of each review, then the fine-grained feature of users based on reviews is R_u:

$$R_u = W_1 \left(\sum_{j=1}^{n} \alpha_j \mathbf{h}_{\mathbf{u,j}} \right) + b. \tag{15}$$

where $W_1 \in R^{k \times z}$ and b are the weight matrix of the fully connected layer. Then, We use the same method to obtain the final fine-grained feature vector R_i of the item based on reviews.

3.6 Rating Score Prediction Layer

After the above process, we have obtained the characteristics of users and items at the document-level and review-level. Since document-level features and review-level features can describe the coarse-grained features and fine-grained features of users/items, respectively, and the ID embedding of users/items can identify their identity information, we can better perform rating score predictions by combining these three features. Then the final latent features of users and items are denoted as $\mathbf{u}$ and $\mathbf{i}$ respectively. The calculation method is as follows:

$$\begin{aligned} \mathbf{u} &= D_u \oplus R_u \oplus \mathbf{u_{id}} \\ \mathbf{i} &= D_i \oplus R_i \oplus \mathbf{i_{id}} \end{aligned} \tag{16}$$

where $\oplus$ is the concatenation operator.

Next, We use Latent Factor Model(PFM), which is widely used for scoring prediction, to make the final score prediction. The predicted rating $\hat{R}_{u,i}$ is computed as follows:

$$\hat{R}_{u,i} = W_T^2\left(\mathbf{u} \oplus \mathbf{i}\right) + b_u + b_i + \mu \tag{17}$$

where W_T^2 is the linear transform matrix of the LFM, b_u, b_i and μ are user bias, project bias and global bias, respectively. Since our task is to make rating score predictions, we utilize the Mean Square Error (MSE) function to train our model:

$$L_{sqr} = \sum_{u,i \in \Omega} \left(\hat{R}_{u,i} - R_{u,i}\right)^2 \tag{18}$$

where Ω denotes the set of instances for training, and $R_{u,i}$ is the ground truth rating assigned by the user u to the item i.

4   Experiments and Analysis

4.1   DataSets and Experiments Settings

DataSets. In order to prove the effectiveness of our method, we selected real datasets of three fields of *Digital Music*, *Office Products*, and *Tools Improvement* from Amazon[1] core datasets as the data for our experiment. These datasets contain the actual ratings (1–5) and review texts made by users on the items. Following the preprocessing steps used in [5,17], we solve the long-tail effect of reviews by maintaining the length and number of reviews that can cover $p(p = 0.85)$ percentage of users and items. The detailed information of the datasets is shown in Table 1.

Table 1. Comparison of effects between various models.

Datasets	# Users	# Items	# Rating	# Avg. review length	# Words per user	# Words per item	Density(%)
Digital music	5,540	3,568	64,666	69.57	216.21	266.51	0.327
Office products	4,905	2,420	53,228	48.15	197.93	229.52	0.448
Tools improvement	16,638	10,217	134,345	38.75	162.53	212.48	0.079

For evaluation, for each data set, we randomly split 80% of the data as the training set and 10% as the validation set and test set where the validation set for hyper-parameter validation. We ensure that the training set contains at least one user-item interaction, and does not contain any reviews in the validation set and test set.

---

[1] http://jmcauley.ucsd.edu/data/amazon/.

Baselines. We compare the proposed PPRR with the following state-of-the-art rating score prediction methods:

- **PMF** [3]: Probabilistic matrix factorization is a standard matrix factorization that only utilizes rating data to model the characteristics of users and items.
- **RBLT** [20]: This method utilizes a rating-boosted method to combine reviews and ratings to learn the item recommended and user preference distribution in a shared topic space.
- **DeepCoNN** [15]: This method utilizes two parallel CNN networks to learn the feature representations of users and items from user review documents and item review documents, and then performs rating score predictions through FM.
- **D-Attn** [21]: In order to increase the interpretability of recommendations, the model uses local and global dual attention to capture the interpretable features of users and items.
- **NARRE** [5]: This method utilizes a neural attention mechanism to capture the importance and potential features of different reviews, and then combines the reviews and items to make rating predictions.
- **MPCN** [19]: This method introduces the Gumbel-softmax pointer mechanism into the neural network to jointly learn and pay attention to the user-item in-depth interaction.
- **DAML** [16]: The model uses the dual mechanism of local and mutual attention to jointly learn the features of reviews, and combined the rating features to complete the final rating score prediction.
- **CARP** [7]: CARP uses the capsule network to learn the feature representations of users and items from review documents and infer the corresponding emotions to increase the interpretability of rating score predictions.
- **MRCP** [17]: MRCP comprehensively learns the feature representation of users/items from word-level, review-level, and aspect-level by using a three-layer attention framework.
- **NRCA** [18]: This model proposes a neural network recommendation method under the cross-attention framework, which combines document-level and review-level features of users and items, which can more comprehensively capture the representation of users and items.

In addition to the methods mentioned above, there are many well-known methods, such as HFT [11], ConvMF [23], TARMF [10], ANR [24]. These approaches were not involved because they did not perform as well as other baseline methods in our experiments.

Hyper-parameter Settings. In experiments, we used pre-trained word embeddings with 300-dimensions which are trained on more than 100 billion words from Google News to initialize our word embedding. The dimension of the user or item ID embedding (i.e., d) is 32, the number of convolution filters neurons k is 100 (tuning in 50, 100, 150, 200), and the window size of CNN (i.e., l) is 3. The number of words T we choose to construct the review document is set to 200. For the *Transformer* encoding part, we set the overall default number of layers to 1, and the number of multi-head attention layers to 10. We show the performance of the model under different length of review document in Sect. 4.4. In addition, we utilize dropout technology to alleviate the overfitting problem of the model, and the internal dropout ratio of the transformer mechanism of the model is set to 0.5, and the rest of the dropout ratio is set to 0.75. All weight matrices in the model are initialized with a normal distribution with a mean value of 0.0 and a standard deviation of 0.1, and all biases are initialized to 0.1. The non-linear activation function is *ReLU*.

For the optimization of the model, we utilize the Adam optimization strategy, set the learning rate to 0.004, and the weight decay to 0.001. At the same time, using the validation set to tune the overall hyper-parameters of the model.

Evaluation Metric. For the task of rating score prediction in this work, we utilize the well-known Mean Square Error (MSE) as the evaluation metric. Lower MSE indicates the predicted rating $\hat{R}_{u,i}$ is closer to the ground truth rating $R_{u,i}$:

$$MSE = \frac{1}{|\Omega_t|} \sum_{(u,i)\in\Omega_t} \left(\hat{R}_{u,i} - R_{u,i} \right)^2 \tag{19}$$

where Ω_t is the set of the user-item pairs in the testing set.

4.2   Performance Evaluation

The experimental results of all methods over the three datasets are shown in Table 2, in which we have the following observations and analysis:

(1) The review-based method outperforms the rating-based method (i.e., PFM [3]), which demonstrates the advantage of reviews by involving semantic information.

(2) The neural network method outperforms the topic-based method (i.e., RBLT [20]). The main reason is that the neural network can well capture the semantic information in the review and better learn the characteristics of the user/item.

(3) The performance of the method using attention mechanism (i.e., D-Attn, DML, NARRE, MRCP) is generally better than the method without attention (i.e., DeepCoNN), which is mainly because the utility of each word or review may be different, and the attention mechanism can pay attention to these differences.

(4) There is no obvious difference between the document-based method and the review-based method, but our PPRR method achieves the best performance in all data sets by reconstructing the review document at the review-level to capture the fine-grained and coarse-grained features. This experimental result proves the effectiveness of the proposed PPRR model.

Table 2. Comparison of our model **PPRR** and each baseline method on MSE.

Type	Method	Datasets		
		Digital music	Office products	Tools improvement
Rating-based	PFM [3]	1.206	1.092	1.566
Topic-based	RBLT [20]	0.870	0.759	0.983
Document-based	DeepCoNN [15]	1.056	0.860	1.061
	D-Attn [21]	0.911	0.825	1.043
	DAML [16]	0.813	0.705	0.945
	CARP [7]	0.820	0.719	0.960
Review-based	NARRE [5]	0.812	0.732	0.957
	MPCN [19]	0.903	0.769	1.017
	MRCP [17]	0.801	0.702	0.928
Review-and-documnet	NRCA [18]	0.795	0.691	0.929
Proposed method	**PPRR**	**0.779**	**0.678**	**0.923**

4.3 Discussion

In our model, we use word and review level attention to indicate the personalized importance of different words and reviews, and then build a new ordered review document by selecting the most important words, and finally combine the fine-grained features of the review level and coarse-grained features of the document level to learn a comprehensive representation of users and items. In this section, we use ablation experiments to study the effectiveness of the important components of the model, which mainly include three components: review document reconstruction network (Re-Doc-Net), review compiler (Rev-net), and cross-transform (Transformer). Therefore, we designed multiple variants for ablation experiments as follows:

- **PPRR-IT:** the model removes the review-level module, and the input of document-level module is an externally preprocessed static review document(IT).
- **PPRR-ST:** the model removes the review-level module, and the input of the document-level module is reconstructed by the document reconstruction (ST).
- **PPRR-R:** the document-level modules are removed from the model.
- **PPRR-RS:** the model removes the document-level transform mechanism.
- **PPRR-RIS:** the model removes the review-document reconstruction, and uses an external static review document as input in the document-level module.

The experimental results are shown in Table 3. We set all the ablation experimental model parameters to the default values of PPRR. First of all, from the experimental results of the review-level based module (PPRR-R) and document-level module (PPRR-ST), which verified the review-level module and document-level module of the model are effective in capturing the coarse-grained and fine-grained features of users/items. Secondly, the experimental effect of the

Table 3. MSE comparison of different components of model **PPRR**.

Variant	Datasets		
	Digital music	Tools inprovement	Office products
PPRR-IT	0.7806	0.9251	0.6835
PPRR-ST	0.7796	0.9246	0.6808
PPRR-R	0.7823	0.9246	0.6819
PPRR-RS	0.7809	0.9234	0.6796
PPRR-RIT	0.7832	0.9236	0.6813
PPRR	**0.7791**	**0.9230**	**0.6776**

static review-document model (PPRR-IT) is worse than that of the personalized review-document level model (PPRR-ST), which can be seen that the personalized reconstruction of the review-document can more comprehensively capture the user/item Coarse-grained features. In addition, we suppose that each word in the review-document can better learn the user/item features after capturing each word of the target document and the contextual features, so we set up the PPRR-RS model, and the experimental results prove our conclusion.

4.4   Hyper-Parameters Analyses

In this section, we analyze three key hyper parameters in the model: the length of the reconstructed review document, the number of layers of the transformer mechanism, and the number of convolution filters to explore the effectiveness of the hyper parameters for our model. Through experiments, we found that when the length of the reconstructed review document is $T = 200$, the number of transform mechanism layers is 1, and the number of convolution kernels k $= 100$, the model has the best effect. Here we only show the length of the reconstructed review-document.

Effect of Review-document Length. Considering that the length of the review document plays a very important role in the overall performance of our model PPRR, we analyze the influence of the length of the review document on {100,200,300,400,500,600} through experiments on all datasets. The experimental results are shown in Table 4. From the experimental results, we find when the length of the review document $T = 200$, the model works best, which shows that the document length is not as long as possible.

Table 4. The result of different review-document length in PPRR.

Review-document length	Datasets		
	Digital music	Tools inprovement	Office products
T = 100	0.7820	0.9248	0.6811
T = 200	**0.7791**	0.9231	**0.6776**
T = 300	0.7820	0.9304	0.6802
T = 400	0.7838	0.9253	0.6812
T = 500	0.7806	**0.9212**	0.6810
T = 600	0.7795	0.9263	0.6800

5 Conclusion

In this article, we propose a neural recommendation model that can reconstruct reviews into personalized documents and integrate review-level features. In the reconstructed document, we represent the personalized semantic of the reviews of a certain user towards a certain item and vice versa by modeling the personalized importance of each word in the reviews. We propose a cross-transform mechanism to achieve the personalized importance calculating by integrating contextual semantic features of each word. Experimental results on three Amazon public data sets show that the proposed model can effectively improve recommendation performance compared with state-of-the-art baselines. And the reconstruction of personalized documents of reviews is verified to be the essential advantage of the proposed model.

Acknowledgement. This work was supported by a grant from the National Key Research and Development Program of China (2018YFC0809804), State Key Laboratory of Communication Content Cognition (Grant No. A32003), the Artificial Intelligence for Sustainable Development Goals (AI4SDGs) Research Program, National Natural Science Foundation of China (U1736103, 61976154, 61402323, 61876128), the National Key Research and Development Program (2017YFE0111900).

References

1. Koren, Y.: Factorization meets the neighborhood: a multifaceted collaborative filtering model. In: Proceedings of the 14th ACM SIGKDD International Conference on Knowledge Discovery and Data Mining, KDD 2008, pp. 426–434. New York, NY, USA. Association for Computing Machinery (2008)
2. Linden, G., Smith, B., York, J.: Amazon.com recommendations: item-to-item collaborative filtering. IEEE Internet Comput. **7**(1), 76–80 (2003)
3. Salakhutdinov, R., Mnih, A.: Probabilistic matrix factorization. In: Platt, J.C., Koller, D., Singer, Y., Roweis, S.T. (eds.), Advances in Neural Information Processing Systems 20, Proceedings of the Twenty-First Annual Conference on Neural Information Processing Systems, Vancouver, British Columbia, Canada, 3–6 December 2007, pp. 1257–1264. Curran Associates Inc (2007)

4. Bao, Y., Fang, H., Zhang, J.: Topicmf: simultaneously exploiting ratings and reviews for recommendation. In: Brodley, C.E., Stone, P. (eds.), Proceedings of the Twenty-Eighth AAAI Conference on Artificial Intelligence, 27–31 July 2014, Québec City, Québec, Canada, pp. 2–8. AAAI Press (2014)
5. Chen, C., Zhang, M., Liu, Y., Ma, S.: Neural attentional rating regression with review-level explanations. In: Proceedings of the 2018 World Wide Web Conference, WWW 2018, p. 1583C1592, Republic and Canton of Geneva, CHE 2018. International World Wide Web Conferences Steering Committee (2018)
6. Diao, Q., Qiu, M., Wu, C-Y., Smola, A.J., Jiang, J., Wang, C.: Jointly modeling aspects, ratings and sentiments for movie recommendation (JMARS). In: Macskassy, S.A., Perlich, C., Leskovec, J., Wang, W., Ghani, R. (eds.), The 20th ACM SIGKDD International Conference on Knowledge Discovery and Data Mining, KDD 2014, New York, NY, USA - 24–27 August 2014, pp. 193–202. ACM (2014)
7. Li, C., Quan, C., Peng, L., Qi, Y., Deng, Y., Wu, L.: A capsule network for recommendation and explaining what you like and dislike. In: Piwowarski, B., Chevalier, M., Gaussier, É., Maarek, Y., Nie, J.-Y., Scholer, F. (eds.), Proceedings of the 42nd International ACM SIGIR Conference on Research and Development in Information Retrieval, SIGIR 2019, Paris, France, 21–25 July 2019, pp. 275–284. ACM (2019)
8. Liu, H., et al.: Hybrid neural recommendation with joint deep representation learning of ratings and reviews. Neurocomputing **374**, 77–85 (2020)
9. Liu, H., et al.: NRPA: neural recommendation with personalized attention. In: Piwowarski, B., Chevalier, M., Gaussier, É., Maarek, Y., Nie, J.-Y., Scholer, F. (eds.), Proceedings of the 42nd International ACM SIGIR Conference on Research and Development in Information Retrieval, SIGIR 2019, Paris, France, 21–25 July 2019, pp. 1233–1236. ACM (2019)
10. Lu, Y., Dong, R., Smyth, B.: Coevolutionary recommendation model: mutual learning between ratings and reviews. In: Champin, P-A., Gandon, F., Lalmas, M., Ipeirotis, P.G. (eds.), Proceedings of the 2018 World Wide Web Conference on World Wide Web, WWW 2018, Lyon, France, 23–27 April 2018, pp. 773–782. ACM (2018)
11. Julian J. McAuley and Jure Leskovec. Hidden factors and hidden topics: understanding rating dimensions with review text. In Qiang Yang, Irwin King, Qing Li, Pearl Pu, and George Karypis, editors, Seventh ACM Conference on Recommender Systems, RecSys '13, Hong Kong, China, October 12–16, 2013, pages 165–172. ACM, 2013
12. Wang, X., et al.: Neural review rating prediction with hierarchical attentions and latent factors. In: Li, G., Yang, J., Gama, J., Natwichai, J., Tong, Y. (eds.) DASFAA 2019. LNCS, vol. 11448, pp. 363–367. Springer, Cham (2019). https://doi.org/10.1007/978-3-030-18590-9_46
13. Ling, G., Lyu, M.R., King, I.: Ratings meet reviews, a combined approach to recommend. In: Kobsa, A., Zhou, M.X., Ester, M., Koren, Y. (eds.), Eighth ACM Conference on Recommender Systems, RecSys 2014, Foster City, Silicon Valley, CA, USA - 06–10 October 2014, pp. 105–112. ACM (2014)
14. Blei, D.M., Ng, A.Y., Jordan, M.I.: Latent dirichlet allocation. J. Mach. Learn. Res. **3**, 993C1022 (2003)
15. Zheng, L., Noroozi, V., Yu, P.S.: Joint deep modeling of users and items using reviews for recommendation. CoRR, abs/1701.04783 (2017)

16. Liu, D., Li, J., Du, B., Chang, J., Gao, R.: DAML: dual attention mutual learning between ratings and reviews for item recommendation. In: Teredesai, A., Kumar, V., Li, Y., Rosales, R., Terzi, E., Karypis, G. (eds.), Proceedings of the 25th ACM SIGKDD International Conference on Knowledge Discovery & Data Mining, KDD 2019, Anchorage, AK, USA, 4–8 August 2019, pp. 344–352. ACM (2019)
17. Liu, H., Wang, W., Peng, Q., Wu, N., Wu, F., Jiao, P.: Toward comprehensive user and item representations via three-tier attention network. ACM Trans. Inf. Syst. **39**(3), 1–22 (2021)
18. Liu, H., Wang, W., Xu, H., Peng, Q., Jiao, P.: Neural unified review recommendation with cross attention. In: Huang, J., et al. (eds.), Proceedings of the 43rd International ACM SIGIR conference on research and development in Information Retrieval, SIGIR 2020, Virtual Event, China, 25–30 July 2020, pp. 1789–1792. ACM (2020)
19. Tay, Y., Luu, A.T., Hui, S.C.: Multi-pointer co-attention networks for recommendation. In: Guo, Y., Farooq, F., (eds.), Proceedings of the 24th ACM SIGKDD International Conference on Knowledge Discovery & Data Mining, KDD 2018, London, UK, 19–23 August 2018, pp. 2309–2318. ACM (2018)
20. Tan, Y., Zhang, M., Liu, Y., Ma, S.: Rating-boosted latent topics: understanding users and items with ratings and reviews. In: Kambhampati, S. (ed), Proceedings of the Twenty-Fifth International Joint Conference on Artificial Intelligence, IJCAI 2016, New York, NY, USA, 9–15 July 2016, pp. 2640–2646. IJCAI/AAAI Press (2016)
21. Seo, S., Huang, J., Yang, H., Liu, Y.: Interpretable convolutional neural networks with dual local and global attention for review rating prediction. In: Cremonesi, P., Ricci, F., Berkovsky, S., Tuzhilin, A. (eds.), Proceedings of the Eleventh ACM Conference on Recommender Systems, RecSys 2017, Como, Italy, 27–31 August 2017, pp. 297–305. ACM (2017)
22. Dong, X., et al.: Asymmetrical hierarchical networks with attentive interactions for interpretable review-based recommendation. In :The Thirty-Fourth AAAI Conference on Artificial Intelligence, AAAI 2020, The Thirty-Second Innovative Applications of Artificial Intelligence Conference, IAAI 2020, The Tenth AAAI Symposium on Educational Advances in Artificial Intelligence, EAAI 2020, New York, NY, USA, 7–12 February 2020, pp. 7667–7674. AAAI Press (2020)
23. Kim, D.H., Park, C., Oh, J., Lee, S., Yu, H.: Convolutional matrix factorization for document context-aware recommendation. In: Sen, S., Geyer, W., Freyne, J., Castells, P. (eds.), Proceedings of the 10th ACM Conference on Recommender Systems, Boston, MA, USA, 15–19 September 2016, pp. 233–240. ACM (2016)
24. Chin, J.Y., Zhao, K., Joty, S.R., Cong, G.: ANR: aspect-based neural recommender. In: Cuzzocrea, A., et al. (eds.) Proceedings of the 27th ACM International Conference on Information and Knowledge Management, CIKM 2018, Torino, Italy, 22–26 October 2018, pp. 147–156. ACM (2018)

Recommendation System and Network and Security

Dynamic Traffic Network Based Multi-Modal Travel Mode Fusion Recommendation

Nannan Jia[1], Mengmeng Chang[1], Zhiming Ding[2(✉)], Zunhao Liu[1],
Bowen Yang[1], Lei Yuan[1], and Lutong Li[1]

[1] Beijing University of Technology, Beijing 100124, China
{jianannan,changmengmeng,bovin.y,yuanlei,
lilutong}@emails.bjut.edu.cn
[2] Institute of Software, Chinese Academy of Sciences, Beijing 100190, China
zhiming@iscas.ac.cn

Abstract. The travel problem is a challenge to the city's overall development. With the increase of the number of cars and the increase of urban population density, intelligent travel mode provides new solutions to solve these problems. However, the existing research on the choice of travel modes for residents only considers the current traffic conditions, and the preferences of individual users for travel modes are poorly considered, which cannot meet the personalized travel needs of users. From this perspective, a heterogeneous information network based on users' spatial-temporal travel trajectories is proposed in this paper. Considering the dynamic traffic network that is constantly changing during the travel process, and using the graph neural network guided by the meta-path to dynamically model the user and travel mode. Features embedding with rich interactive information, so as to fully learn the users' preferences for travel modes in the time-space travel trajectory, and recommend travel modes that meet personalized needs to users. Finally, the effectiveness of the proposed method is demonstrated by experimental evaluation on real-world datasets.

Keywords: Heterogeneous information network · Graph neural network · Meta-path · Multi-modal travel mode · Fusion recommendation

1 Introduction

With the continuous development of urbanization and the improvement of economic level, people's travel demands become increasingly diverse. When demand exceeds the carrying capacity of the transport system, various traffic and even environmental problems will arise. The complex selection behavior of people in multi-mode transportation network [26] determines the distribution of travel demand on the transportation network. From the perspective of balancing the distribution of traffic demand [3], personalized travel mode recommendation has great research value.

Nowadays, in the era of big data, as an information filtering technology, recommendation system is particularly important for users to recommend information that meets their individual needs. Heterogeneous information networks [4, 5] consider different types of objects, different connection relationships between objects, and

© ICST Institute for Computer Sciences, Social Informatics and Telecommunications Engineering 2021
Published by Springer Nature Switzerland AG 2021. All Rights Reserved
H. Gao and X. Wang (Eds.): CollaborateCom 2021, LNICST 406, pp. 299–317, 2021.
https://doi.org/10.1007/978-3-030-92635-9_18

attribute information, fully reflecting the interaction between different objects in the recommendation system, from which implicit information is mined and deeper regularities are learned, greatly improving recommendation accuracy and opening up a new path for personalized recommendation technology. Besides, it can be seen that traditional user-item recommendations are themselves heterogeneous bipartite graphs, so the use of HIN (Heterogeneous information networks, HIN) in recommendation techniques is inevitable. Common heterogeneous information networks such as social networks, biological neural networks, academic networks, etc., but there is more to it than that - the real world is rich in data, and information networks are everywhere. For example, construct a heterogeneous information network based on the types of movies, directors, and user attributes to express user preferences in a more granular manner [15, 16]. There are many types of research on recommendation by constructing heterogeneous information networks, such as malicious account detection [17] and search intent recommendation [10]. This paper finds that the travel behavior of users on a multi-mode transportation network is also an embodiment of information network, and then proposed to apply the representation learning technology in heterogeneous information network to transportation travel, and combined with dynamic transportation network to recommend multi-mode travel mode.

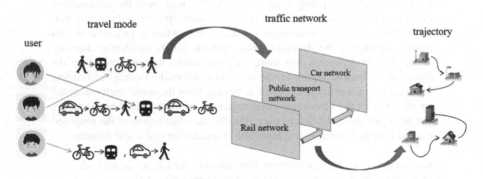

Fig. 1. Interaction of users' Spatial-temporal travel trajectories and transport networks.

Figure 1 illustrates the interaction between the user's Spatial-temporal travel trajectory and the transport network, where the user uses multiple modes of travel to continuously shift between this complex transport network, eventually forming a chain of trajectories. Among them, the complex transportation network is composed of multimodal travel modes. The travel mode is the basic attribute of understanding the user's mobile behavior, and the transportation mode used by the user in the trajectory can reflect the regularity of the user's mobile state. Therefore, most scholars have always been concerned about travel Research fields related to methods have attracted much attention. YuZheng [18–20] et al. used methods such as change-point segmentation based on the identification of travel patterns in GPS trajectories to understand users' travel patterns. Dabiri S [21] et al. used convolutional neural networks to train models for automatic inference of travel patterns. Yao [1] et al. proposed a travel mode choice model that considers the combined effects of rational decision making and inherent choice preferences to analyze users' choice of travel mode [6, 13]. Used multiple sources of data to build a model of residents' travel mode choice and to analyze their travel mode choice behavior. This paper proposes to apply heterogeneous information network representation learning to transport travel to complete personalized recommendations of travel modes to fill the gap to a certain extent. Gunjan Kumar [2] et al. used a sequence-based approach to personalize travel style recommendations to users, but it did not sufficiently take into account the contextual information of the recommended objects and the interaction between them, which is well remedied by our study. Personalized recommendation of travel modes remains an underexplored task, with the following main contributions:

1. Considering the more abundant characteristics of learning users and modes of travel, we innovatively propose building heterogeneous information networks with users, modes of travel, Spatial-temporal attributes (e.g. time, etc.) strongly related to modes of travel, and starting and ending locations.
2. Using the Graphic Neural Network guided by the meta-path to dynamically model users and travel modes, we can get the embedded features with rich interactive information, and fully learn user's preferences for travel modes from HIN.
3. The dynamic traffic network that changes with time during the trip is considered to increase the probability of selecting more time-saving and energy-saving trip modes. Finally, the feasibility and effectiveness of the proposed method are verified by real datasets.

The remainder of this paper is structured as follows. In Sect. 2 some basic concepts used in the article are introduced. Section 3 presents our fusion recommendation model and the main techniques used. Section 4 presents a comprehensive evaluation of our model on a real dataset. Finally, conclusions are drawn in Sect. 5.

2 Concepts Used in the Paper

In this section, we introduce the heterogeneous information networks relevant to the fused recommendation content of this paper and some basic concepts used in the recommendation model.

2.1 Definition of the Fusion Recommendation Problem

Define a set $M = \{U, V, T, L\}$, where U denotes the set of users; V denotes the travel mode used by the user in the travel trajectory, mainly including metro, bus, driving, taxi, since car, walking; T denotes the start time when the user uses a certain travel mode; L denotes the starting and ending location of the user at a certain time when using a certain travel mode; all these belong to the node type in the heterogeneous transportation travel network. By analyzing the user's historical travel trajectory and learning the user's personalized preference for travel mode, combined with the impact of dynamic traffic network changes over time on travel mode, this fusion recommendation finally recommends more time-saving and energy-saving multimodal travel mode $v \in V$ in line with personal preference for users $u \in U$.

2.2 Heterogeneous Transport Travel Networks

In a multi-modal transport network, users take different modes of travel to reach their destination and there is a certain connection between them. In this paper, we construct a heterogeneous information network (see Fig. 2(a)) with the temporal attributes of a user, travel mode, origin and destination, and travel mode to obtain more potential features by learning the semantic relationships between objects such as user and user, user and travel mode, travel mode and its corresponding origin and destination (note: at this point, the temporal attributes are closely connected to the user), and to fully express the user's personalized preference for travel mode. To enable a better understanding of heterogeneous information networks, the construction of a network schema [4] is shown in Fig. 2(b), which specifies the type constraints on the set of objects and the connections between them: there is a connection between users and travel modes, indicating a use-and-used relationship; there is a connection between locations and travel modes, indicating a guide-and-directed relationship, etc.

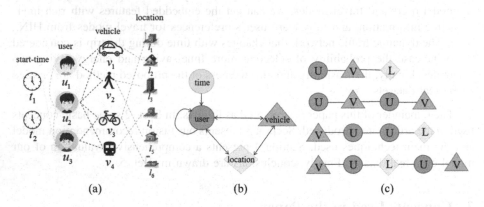

(a)   (b)   (c)

Fig. 2. Interaction of users' Spatial-temporal travel trajectories and transport networks.

2.3 Meta-paths Extraction Based on User Trajectory

A meta-path [7] is a path defined over a network schema linking two types of objects, representing a composite relationship between object types, and is used to extract information about the interaction between objects. Figure 2(c) shows the four meta-paths used in this paper, where user and time are closely linked. The meta-path "user-vehicle-location" indicates that the user has traveled from a certain location to a certain location at a certain time. The meta-path from "user-vehicle-user-vehicle" indicates that a user has used a travel mode at a certain time and that similar users have also used this mode of travel and have also used other travel modes travel. "Vehicle-user-user-location" indicates that a similar user used a travel mode when arriving at a similar starting and ending location; "vehicle-user-location-user-vehicle" indicates that a similar user used a travel mode when arriving at a similar location at a similar time. Meta paths are used in this recommender system to obtain semantic information between objects in a heterogeneous information network.

2.4   Meta-path Guided Neighbors

This paper focuses on integrating rich information through aggregated meta-path-guided neighborhoods. A detailed description of the meta-path-guided neighborhood, as exemplified by the user u_2 in Fig. 2(a): The first-order neighbors of user u_2 under the meta-path $\Phi : U - V - L$ are that $N_{u_2}^1 = \{v_1, v_2, v_3, v_4\}$, The second-order neighbors of u_2 are the first-order neighbors of all nodes in the first-order neighborhood, so its second-order neighbors are $N_{u_2}^2 = \{l_1, l_2, l_3, l_4, l_5, l_6\}$. Therefore all neighbors of the user u_2 are denoted as $N_{u_2} = \left\{ N_{u_2}^0, N_{u_2}^1, N_{u_2}^2 \right\} = \{u_2, v_1, v_2, v_3, v_4, l_1, l_2, l_3, l_4, l_5, l_6\}$.

3   Heterogeneous Transport Travel Network Recommendation Model

In this section, we will introduce in detail the multi-modal travel mode fusion recommendation model (DTN-MMTMRec) based on meta-path under the condition of dynamic transportation network. The idea of this fusion recommendation is to construct the user's Spatial-temporal travel trajectory into a heterogeneous information network, and select different meta-paths, using the different semantic information they express, to build a graph neural network, and to enrich the node embedding of the user and travel mode by aggregating the information of neighboring nodes guided by the meta paths, i.e. to better learn the representation of the user and travel mode, and to recommend a travel mode for the user that meets personalized needs. Figure 3 illustrates the overall framework of this fusion recommendation model. Heterogeneous information network as input, the embedding layer preprocesses the model data with content features to generate an initial node embedding. Then, in the meta-path aggregation layer, the selected meta-path is aggregated with its guided neighbor nodes for dynamic modeling, i.e. the semantic information between the target node and its neighbors is aggregated to obtain rich embeddings of users and travel modes, and finally, the feature embeddings of users and travel modes are fused for multimodal travel mode recommendation. The application of our model to travel mode recommendations is expected to yield better performance, and the use of meta-paths will also improve the interpretability of the recommendation results. The details are described in detail in the following subsections.

3.1   Initial Embedding

Following previous work [7, 8], different types of node features in heterogeneous graphs were mapped to the same vector space, applying a specific type of linear transformation to each type of node:

$$E_v = W_A \cdot x_v^A, E_v^0 = x_v \tag{1}$$

where $x_v \in R^d$ is the original feature vector, $E_v \in R^{d'}$ is the projection latent vector of node v. $W_A \in R^{d' \times d}$ is the parameter weight matrix of type A nodes. After applying this

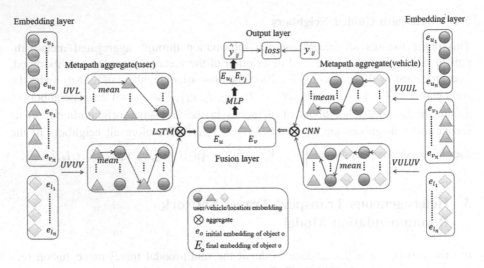

Fig. 3. Overall framework for fusion recommendations.

operation, the projected features of all nodes have the same dimensionality, facilitating the aggregation process for the next model component.

3.2 Practice of Meta-path

An important step in existing recommender systems is to model the interactions between objects to learn more information about the characteristics of the target object. For recommender systems with only two objects, user and item, traditional methods such as matrix decomposition and the more popular [8] method of obtaining higher-order semantics to enrich object embeddings based on a two-part interaction graph of user-item have achieved good results in recent years. For heterogeneous graphs consisting of multiple types of nodes and multiple types of edges, traditional learning embedding methods only involve nodes and do not capture information about neighbouring nodes and the interaction information between neighbouring nodes. With the widespread use of graph neural networks, this deficiency is well filled, where a fixed size number of neighbouring nodes is selected using random wandering for heterogeneous graphs in [11], and then a suitable aggregation function is selected to aggregate the neighbouring vertex feature information to the target node. Similarly, this paper uses different multi-hop meta-paths to determine the neighbors of the target node and aggregates the feature information of the neighboring nodes to derive the final embedding of the target node.

Take Fig. 2(a) as an example to introduce in detail the application of meta-paths in heterogeneous graph neural networks. The feature aggregation process for the target node u_2 is described here. Define two meta-paths $\Phi_{uvl} : U - V - L$, $\Phi_{UVUV} : U - V - U - V$, The initial feature vector of each node is known, and Fig. 4 illustrates the two processes of meta-path aggregation.

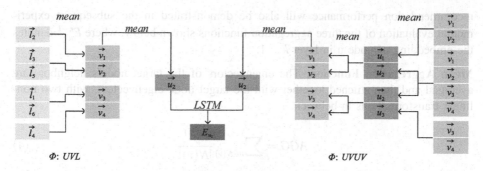

Fig. 4. The meta-path aggregation process.

Internal aggregation of meta paths: choose a suitable aggregation function to aggregate the embeddings of second-order neighbor nodes to obtain the embeddings of first-order neighbor nodes, and then aggregate the embeddings of first-order neighbor nodes to obtain the embeddings of target nodes, as follows:

$$E_{u_2}^{\Phi_{uvl}} = f(e_{v_1}, e_{v_2}, e_{v_3}, e_{v_4}) = f(e_{l_1}, e_{l_2}, e_{l_3}, e_{l_4}, e_{l_5}, e_{l_6}) \tag{2}$$

In formula 2, f is the aggregation function that aggregates the first-order neighbor, second-order neighbor embeddings of u_2 on the specified meta-path. $(e_{l_1}, e_{l_2}, e_{l_3}, e_{l_4}, e_{l_5}, e_{l_6})$ is the second-order neighbor node embedding of user u_2 on the meta-path, $(e_{v_1}, e_{v_2}, e_{v_3}, e_{v_4})$ is the first-order neighbor node embedding, which is updated by aggregating the second-order neighbor node embeddings, and then aggregating the first-order neighbor embeddings to obtain the final embedding of user u_2 on the meta-path Φ_{uvl}.

Aggregation between meta paths: aggregate user u_2 embeddings from different meta paths to get the final embedding of u_2:

$$E_{u_2} = f(E_{u_2}^{\Phi_{uvl}}, E_{u_2}^{\Phi_{uvuv}}, ...) \tag{3}$$

3.3 Meta-path Aggregation Functions

In the meta-path aggregation layer, a basic graph-based neural network is used. The basic idea is to embedding nodes based on their local neighbor information [9]. In this paper, a neural network is used to aggregate the information of target nodes and their neighbors [14, 27].

The aggregation process is divided into two steps, aggregation within a single meta-path and aggregation between different meta-paths. The former mainly aggregates the neighboring nodes of a node, and the average aggregation function is used in this paper. The latter is a feature embedding that aggregates different meta-paths. As different meta-paths express different semantics, the average aggregation function is no longer used, and the LSTM aggregation function or CNN aggregation function is chosen here according to the specific data characteristics. The different effects on recommendation performance will also be demonstrated in the subsequent experimental evaluation of the three aggregation functions shown below, where E_u^{k-1} denotes the embedding of node u in layer $k - 1$:

Mean Aggregation Function. The eigenvectors of the target node's neighbors are averaged and then stitched together with the target node eigenvectors, with two nonlinear transformations in between.

$$AGG = \sum_{u \in N(v)} \frac{E_u^{k-1}}{|N(v)|} \qquad (4)$$

LSTM Aggregation Function. The neighboring nodes of the target node are randomly disordered as the input sequence, and the resulting vector representation and the vector representation of the target node are separately stitched together after a nonlinear transformation to obtain the vector representation of the target node at that layer.

$$AGG = LSTM\left([E_u^{k-1}, \forall u \in N(v)]\right) \qquad (5)$$

CNN Aggregation Function. Unlike the LSTM aggregation function, it does not have to consider time series conditions and can be used directly as an input sequence to generate new features by convolution operations.

$$AGG = CNN\left([E_u^{k-1}, \forall u \in N(v)]\right) \qquad (6)$$

3.4 Semantic Aggregation

The fusion recommendation model in this paper updates the features of nodes in heterogeneous information networks by aggregating neighbors guided by different meta paths to obtain node embedding with rich semantics. In this section, the process of embedding users and modes of travel as target nodes are described in detail.

As represented by the meta-path aggregation layer in Fig. 3. Firstly, a user is regarded as the target node (it can be a multi-hop meta-path or single-hop meta-path), and then the neighbor node features are aggregated one by one according to the selected meta-path. Using the meta-path UVL as an example, the initial embedding of the travel mode and start/stop location is known and the aggregation process is as follows.

- Aggregation between nodes within a single meta-path

The average aggregation function is used to aggregate the embedding of the second-order neighbors, and the aggregation result is spliced with the first-order neighbors to update the embedding of the first-order neighbors:

$$E_{v_j}^{\Phi_{uvl}} = f(E_{l_1}^{\Phi_{uvl}}, E_{l_2}^{\Phi_{uvl}}, E_{l_3}^{\Phi_{uvl}}, ...) \tag{7}$$

The feature embedding of the target node is updated using the same average aggregation function to aggregate the first-order neighbor node embedding:

$$E_{u_i}^{\Phi_{uvl}} = f(E_{v_1}^{\Phi_{uvl}}, E_{v_2}^{\Phi_{uvl}}, ...) \tag{8}$$

- Aggregation between different meta-paths

In this paper's recommendation model, time plays a key role in recommending to users a travel mode that meets their personalized needs, so when modeling users dynamically, users are temporal, and the LSTM aggregation function is used here to better aggregate the embedding of users under different meta-paths. where Φ_i is a different meta-path with the user as the target node:

$$E_{u_i} = f(E_{u_i}^{\Phi_1}, E_{u_i}^{\Phi_2}, E_{u_i}^{\Phi_3}, ..., E_{u_i}^{\Phi_k}) \tag{9}$$

The final embedding of all user nodes is obtained by dynamically modeling all users in the above way:

$$\{E_{u_1}, E_{u_2}, E_{u_3}, ..., E_{u_n}\} \tag{10}$$

The dynamic modeling of travel mode is similar to this. Based on the different meta-paths with travel mode as the target node, the average aggregation function is first used to aggregate the features of the neighboring nodes within the meta-paths one by one, and then the CNN aggregation function is used to complete the feature aggregation between different meta-paths to obtain the final embedding of all travel mode nodes. where Φ_i is the different meta-paths with travel mode as the target node:

$$E_{v_i} = f(E_{v_i}^{\Phi_1}, E_{v_i}^{\Phi_2}, E_{v_i}^{\Phi_3}, ..., E_{v_i}^{\Phi_k}), \{E_{v_1}, E_{v_2}, E_{v_3}, ..., E_{v_n}\} \tag{11}$$

3.5   Evaluation Prediction

In this recommendation model, we predict the probability $\hat{y}_{ij}$ of a user choosing a travel mode at a given time, which is in the range [0, 1]. The final embeddings of users and travel modes are obtained by aggregating meta-path-guided neighbor embeddings, and then we fuse the user, travel mode node embeddings for connectivity and finally input them into the MLP to predict the score $\hat{y}_{ij}$:

$$\hat{y}_{ij} = sigmoid(f(U_i \otimes V_j)) \tag{12}$$

In formula (12), f is the MLP layer with only one output, sigmoid is the activation layer and $\otimes$ is the embedded link operator.

The loss function in this model uses a point-by-point loss function, and the model is adjusted by the loss function to produce optimal results, $\mathcal{Y}$ and $\mathcal{Y}^-$ are the instance sets of positive samples and negative samples respectively:

$$J = \sum_{i,j \in \mathcal{Y} \cup \mathcal{Y}^-} \left(y_{ij} \log \hat{y}_{ij} + (1 - y_{ij}) \log(1 - \hat{y}_{ij}) \right) \tag{13}$$

4 Experiments and Analysis

We perform experimental evaluations on real-world datasets to verify the effectiveness of the model DTN-MMTMRec on multimodal travel mode recommendations.

Table 1. The selected meta-paths used in dataset and the meaning expressed.

Meta-paths	The meaning expressed
U-V-L	User preferences for travel modes
U-V-U-V	Preferences for travel modes between similar users
V-U-U-L	Add location to express travel preferences at a more granular level
V-U-L-U-V	Travel preferences of users arriving at similar locations

4.1 Dataset

Description of the Dataset. To evaluate the effectiveness of the DTN-MMTMRec model in this paper, we conducted experiments on a dataset from the Microsoft Geolife project [19, 20, 22, 23]. The dataset consists of two parts, one is the user's track log, and each track is a sequence of time stamp points, each of which contains its associated longitude, dimension, time, and so on. The other part is the travel mode label file corresponding to the user's track log, which contains the travel mode and starts time used by the user in the track.

The dataset contains 10 different modes of travel, namely bicycle, bus, car, metro, taxi, train, walking, plane, boat, and running. Table 1 shows the percentage of days that each mode of travel occurred at least once for all users, which shows that some modes of travel were used frequently and some were used occasionally. For example, walking is one of the most frequently used by users, while airplanes, sailing boats, and running are hardly ever used. Besides, we observe that users in the dataset use between two and five different modes of travel per day. This was filtered by retaining data for only six modes of travel - walking, cycling, bus, subway, driving, private car, and taxi - and then fusing the data based on the start time of using a particular mode of travel in the labeled data.

Table 2. Percentage of days each mode of travel occurs at least once among all users.

Travle mode	Percent	Rank
Walk	38%	1
Bus	24%	2
Subway	12%	3
Bike	12%	4
Taxi	7%	5
Car	6%	6
Airplane	0.8%	7
Train	0.2%	8
Boat	–	9
Run	–	10

Data Fusion. The main task of data fusion is to incorporate time attributes with key impacts into the data and establish the interaction among users, modes of travel, and starting and ending locations. In the last section, we will make statistics and analysis of the user's space-time travel trajectory data, and ultimately select the six travel modes commonly used by urban residents. After a series of cleaning, filtering, and integrating data, the final results are as follows: [User, mode of travel, start time, start and end location]. Previously, the user data of the recommendation system only contained the user and then established a relationship with the recommendation object, even if the time attribute was added, the relationship between the user, the time, and the recommendation object was also established. In this paper, the data structure is improved as shown in Fig. 5, combining the user and time together, from the original user using a certain travel mode to a certain user using a certain travel mode at a certain time, which can not only capture the user's travel behavior data more accurately but also refine the data on the basis of the original data, which provides a great advantage to the subsequent model training work. The paper then establishes the interaction between the three objects. The user uses a certain mode of travel at a certain time: $\{u_i - t_j : v_k\}$,

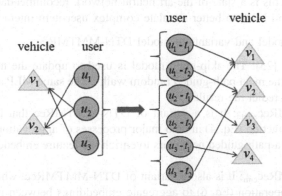

Fig. 5. Improvement of data structure.

Integration of the start and end positions of a user using a certain mode of travel: $\{v_i : l_j - > l_k\}$, The user departs at a certain time and arrives at a certain destination: $\{u_i : l_j - > l_k\}$. Concrete portrayal of the trajectory information of the user using the mode of travel and the construction of a heterogeneous information network.

4.2 Experimental Setup

Baselines. To verify the validity of our proposed methods, three different representatives recommended methods are considered in this paper. The first is based on a collaborative filtering approach: item KNN, MF, SVD. Then there are the more advanced neural network-based models: GMF, MLP, NeuMF. Finally, the HIN-based model: metapath2vec. As well as several variants of model DTN-MMTMRec, the following are specific descriptions of the various methods.

Classical recommendation models based on collaborative filtering:

- itemKNN [28]: This is a classic collaborative filtering approach that evaluates a user's preferences based on their history of interaction with an item, and then makes recommendations for them.
- MF [12, 29]: This is a standard matrix factorization method that projects users and items into the same latent space and uses latent feature vectors to represent them. Thereafter, the user's interaction with the item is modeled as the inner product of its latent vector.
- SVD [30]: This is a feature-based matrix decomposition method, where the object relationships of the heterogeneous information network in this paper are used as feature inputs.

Neural network-based collaborative filtering models:

- GMF: An example of neural collaborative filtering, applying a linear kernel to model latent feature interactions.
- MLP: An example of neural collaborative filtering, using a non-linear kernel to learn interaction functions from data.
- NeuMF [7]: This is a state-of-the-art neural network recommendation method that fuses GMF and MLP to better simulate complex user-item interactions.

HIN-based model and variants of model DTN-MMTMRec:

- metapath2vec [25]: The skip-gram model is used to update the node embeddings generated by the meta-path-guided random walks. The same MLP as in this paper is used here to predict the results.
- DTN–MMTMRec$_{mean}$: It is a variant of DTN-MMTMRec that uses the average aggregation function (Eq. 4) in both major processes of aggregation in the model to aggregate meta-path-guided neighbors to enrich the feature embedding of users and travel modes.
- DTN–MMTMRec$_{cnn}$: It is also a variant of DTN-MMTMRec, where we apply the convolution operation (Eq. 6) to aggregate embeddings between meta-paths.

- DTN–MMTMRec$_{lstm}$: It is a variant of DTN-MMTMRec, and in this model, we use LSTM (Eq. 5) to aggregate embeddings between meta-paths.
- DTN-MMTMRec: This is our complete model.

Evaluation Indicators. In this experiment, to evaluate the recommendation performance, we randomly divide the entire user travel trajectory data into a training set (80%) and a test set (20%), using the K-th precision (Prec@K) and the K-th recall (Recall). @K) and K-th normalized discounted cumulative gain (NDCG@K) are used as evaluation indicators. The larger the NDCG value, the better the performance. For stability, we use different random split training/test sets to run multiple times and average the results.

Implementation Details. We use the python library of Keras to implement the DTN-MMTMRec model. Firstly, the nodes in the heterogeneous information network of this article are initially embedded, and the embedding size is set to 64 dimensions, and then the users, as well as the travel mode sequences, are dynamically modeled and the different effects produced are analyzed by aggregating meta-path-guided neighborhood features using different aggregation functions. In the training phase, the model parameters were randomly initialized with a Gaussian distribution and the model was optimized using a small batch Adam [24], setting the batch size to 256 and setting the learning rate to 0.001. Besides, the number of sampled meta-path instances was four (as shown in Table 2). The other comparison methods followed the appropriate configuration and architecture accordingly and set the same evaluation metrics to facilitate the comparison of effectiveness.

4.3 Result Analysis

Table 3 shows the experimental results of the model as well as the comparison methods on the dataset, with the following main findings.

Table 3. Experimental results for each type of recommendation method on the dataset. The models with an '*' in the data are those with the best recommendation performance in each category, and we use bold to indicate the experimental results of the models. By comparing the metrics, the recommendation performance of the model proposed in this paper is more significant.

Method category	Model	Evaluating indicator		
		Precision@5	Recall@5	NDCG@5
CF	itemKNN	0.2372	0.3513	0.4821
	MF	0.2594	0.3750	0.5012
	SVD	0.2628	0.3757	0.5095*
NCF	GMF	0.2798	0.3951	0.5253
	MLP	0.2810	0.4030	0.5326
	NeuMF	0.2970	0.4127	0.5655*
HIN	Metapath2vec	0.3102	0.4134	0.5701
	DTN-MMTMRec	**0.3189**	**0.4279**	**0.6027**

According to the results of various experimental metrics, the a-model proposed in this paper is very effective for the multimodal travel mode recommendation task, and also demonstrates that the meta-path-guided graph neural network can well enrich the feature embedding of users and travel modes and improve the overall recommendation performance.

Based on the experimental results it can be observed that the meta-path-based model proposed in this paper outperforms the other recommendation methods compared. three recommendation methods based on collaborative filtering, where item KNN has the weakest performance. MF, SVD performance is not very different, which suggests that it is not sufficient to assess user preferences for travel modes based on the interaction between users, travel modes alone. The three examples of neural collaborative filtering frameworks in which a has performed well are based on the idea of modeling complex interactions between users and modes of travel using multi-layer perceptrons by replacing the inner product with a neural structure that can learn arbitrary functions from the data, with experimental results demonstrating the superiority of deep neural networks played in this regard. The DTN-MMTMRec model shows the best recommendation performance, indicating that it is feasible to apply the user's Spatial-temporal travel trajectory to a heterogeneous information network, and the meta-path-guided graph neural network is used to capture heterogeneous information, which produces a good recommendation effect.

The model metapath2vec, also based on a heterogeneous information network and with the same meta-path set up in the experiment, has a lower recommended performance compared to our model. The DTN-MMTMRec model performs aggregation within meta-paths as well as aggregation between meta-paths along existing meta-paths, whereas metapath2vec uses a random wandering strategy to embed node features without learning semantic information between different meta-paths, so its recommendation performance is poor, and it also proves that our model has better recommendation results.

Table 4. Data on evaluation indicators for several variants of the DTN-MMTMRec model using different aggregation functions.

Model	Evaluating indicator		
	Precision@5	Recall@5	NDCG@5
DTN − MMTMRec$_{mean}$	0.3065	0.4051	0.5611
DTN − MMTMRec$_{cnn}$	0.3112	0.4125	0.5893
DTN − MMTMRec$_{lstm}$	0.3162	0.4190	0.5972
DTN − MMTMRec	**0.3189**	**0.4279**	**0.6027**

4.4   Result Analysis

We select four meta paths with different semantics based on the relationships between objects in heterogeneous information networks and then use the aggregate meta paths of graph-neural networks to guide neighbors to improve recommendation performance. Because different meta paths have different semantics, we analyze the impact of different meta paths on the recommended performance by gradually incorporating existing meta paths into the DTN-MMTMRec model. The meta paths in this article are UVL, UVUV, VUUL, and VULUV, which are added to the model sequentially.

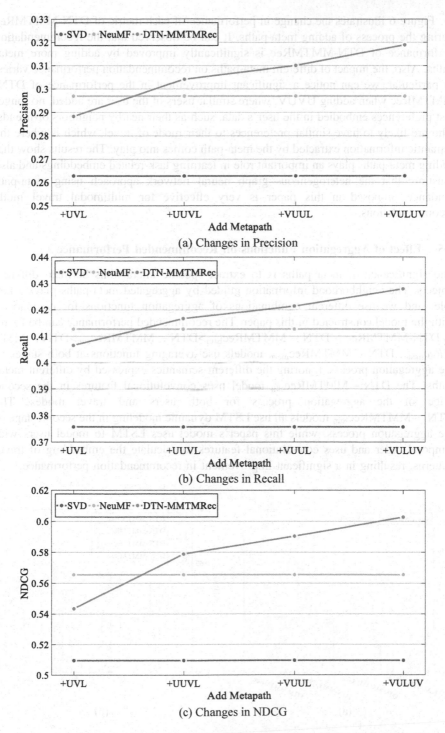

(a) Changes in Precision

(b) Changes in Recall

(c) Changes in NDCG

Fig. 6. Changes in each metric and comparison when adding meta-paths incrementally.

Figure 6 illustrates the change in performance of each metric of DTN-MMTMRec during the process of adding meta-paths. It can be observed that the recommendation performance of DTN-MMTMRec is significantly improved by adding more meta-paths. Also, the impact of different meta-paths on recommendation performance varies. In particular, we can notice a significant improvement in the performance of DTN-MMTMRec when adding UVUV, where similar users of the user are added, no longer just preferences embodied in the user's data, such as their nearby neighbors or friends, who are likely to have similar preferences to their mode of travel, which is where the semantic information extracted by the meta-path comes into play. The results show that adding meta-paths plays an important role in learning task-related embeddings and also illustrate that the heterogeneous graph neural network approach using meta-path guidance proposed in this paper is very effective for multimodal travel mode recommendations.

4.5   Effect of Aggregation Functions on Recommended Performance

The significance of meta paths is to extract specific interactions between different objects. The neighborhood information guided by aggregated meta-paths plays a key role, and we use different combinations of aggregation functions for comparison with the model constructed in this paper. The recommended performance can be found as DTN-MMTMRec > DTN $-$ MMTMRec$_{lstm}$>DTN $-$ MMTMRec$_{cnn}$>DTN $-$ MMT $MRec_{mean}$. DTN $-$ MMTMRec$_{mean}$ models use averaging functions at both stages of the aggregation process, ignoring the different semantics expressed by different meta-paths. The DTN $-$ MMTMRec$_{cnn}$ model uses convolutional features in the second stage of the aggregation process for both users and travel modes. The DTN $-$ MMTMRec$_{lstm}$ models all use LSTM dynamic modeling in the second stage of the aggregation process, while this paper's model uses LSTM to model users with temporal order and uses convolutional features to calculate the embedding of travel patterns, resulting in a significant improvement in recommendation performance.

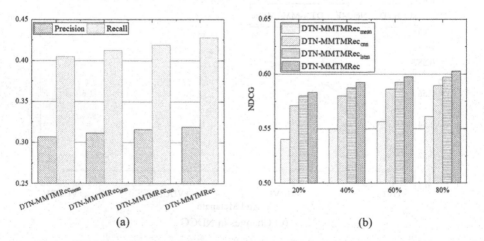

Fig. 7. Impact of different aggregation functions on recommendation performance.

To observe more clearly the effect of aggregation function on the performance of recommendations, this paper will illustrate it from two perspectives. Under the same conditions as the training set and the test set, the precision and recall performance indicators are compared using different aggregation function models (see Fig. 7(a)). We divide the data into five equal parts, one of which is used as a test set, the remaining data is stacked as a training set according to 20%, 40%, 60%, 80% and compared with different aggregators (see Fig. 7(b)). Here we can see that with the increase of training set data, the improvement of recommended performance decreases gradually, and the impact of different aggregators is also compared with Table 4. Echo.The results show that using heterogeneous information networks to obtain richer semantic information is very effective to improve the recommendation performance. The DTN-MMTMRec model proposed in this paper takes full advantage of the meta-path in heterogeneous information networks.

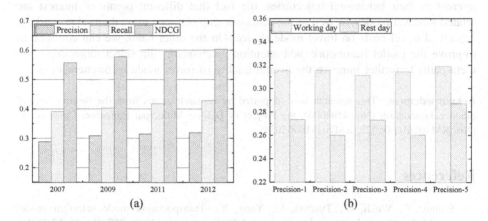

(a)   (b)

Fig. 8. Variation in performance metrics for model DTN-MMTMRec in specific data.

4.6   Performance of the Model on Specific Datasets

Over time, urban population density continues to increase and residents' travel needs are diversified. Our model has played an important role in the complex transportation network. Figure 8 shows the recommended effects of DTN-MMTMRec on different specific datasets. Figure 8 (a) shows that the recommended performance of the model has been improving over the data change from 2007 to 2012, which also proves that our model is more effective as travel demand increases. Figure 8 (b) compares the recommendation effect of DTN-MMTMRec on five different working days and rest days. It is not difficult to see that the recommendation performance of rest days is slower and less stable than that of working days, because the residents' travel on working days is more stable, and the travel demand on rest days varies greatly. The model DTN-MMTMRec is feasible in the recommendation of multimodal travel modes, can meet the personalized needs of daily travel of residents, and is more effective on weekdays.

5 Conclusion

In this paper, our proposed multimodal travel mode recommendation model based on dynamic traffic networks is feasible and effective. The model constructs the user's historical Spatial-temporal travel trajectory into a heterogeneous information network and provides insight into the user's preference for travel modes. Meta-path-guided graph neural networks for dynamic modeling of recommended objects play a key role in the overall recommendation process. At the same time, the real-world dataset is used and the data set is cleaned, filtered, and integrated with new data structure, and experimental evaluation shows that the proposed method in this paper produces good recommendation performance when applied to the recommendation of multimodal travel modes. However there is still relatively little work in this area, the aspects considered are not comprehensive enough and there is much room for improvement. On the one hand, the current work does not take into account users' interest in points of interest in their behavioral trajectories, the fact that different points of interest are related to travel modes, as well as the economic gap between users themselves and the impact of travel costs on travel mode choice. On the other hand, we can continue to improve the model framework, add attention mechanism, and select more important meta-paths to further improve the overall quality of travel mode recommendation.

Acknowledgment. This research was supported by research grants from the Beijing Natural Science Foundation (No. 4192004), the Project of Beijing Municipal Education Commission (No. KM201810005023, KM201810005024).

References

1. Enjian, Y., Weidi, C., Tianwei, L., Yang, Y.: Transportation mode selection model considering traveler's personal preferences. J. Beijing Jiaotong Univ. **209**(01), 46–52 (2020)
2. Kumar, G., Jerbi, H., Mahony, M.: Personalised recommendations for modes of transport: a sequence-based approach. In: International Workshop on Urban Computing at ACM SIGKDD. ACM (2016)
3. Bing-Feng, S., et al.: Urban multimodal traffic assignment model based on travel demand. China J. Highw. Transport **23**(6), 85–91 (2010)
4. Shi, C., et al.: A survey of heterogeneous information network analysis. IEEE Trans. Knowl. Data Eng. **29**(1), 17–37 (2016)
5. Sun, Y., et al.: RankClus: integrating clustering with ranking for heterogeneous information network analysis. In: ACM SIGKDD International Conference on Knowledge Discovery and Data Mining. ACM, Saint Petersburg, Russia (2009)
6. Zhi-Yan, F., et al.: Influence model of social network traffic information on the travel mode choice behavior. J. Transp. Syst. Eng. Inf. Technol. (2019)
7. Xiangnan, H., Lizi, L., Hanwang, Z., Liqiang, N., Xia, H.,Tat-Seng, C.: Neural collaborative filtering. In: 26th International Proceedings on World Wide Web, pp. 173–182 (2017)
8. Xiang, W., et al.: Neural Graph Collaborative Filtering, pp. 165–174 (2019)
9. Shi, C., et al.: Heterogeneous information network embedding for recommendation. In: IEEE Transactions on Knowledge and Data Engineering, pp. 1–1 (2017)

10. Fan, S., et al.: Metapath-guided heterogeneous graph neural network for intent recommendation. In: 25th ACM SIGKDD International Conference. ACM (2019)
11. Zhang, C., et al.: Heterogeneous graph neural network. In: 25th ACM SIGKDD International Conference. ACM (2019)
12. Xiangnan, H., HanWang, Z., Min-Yen, K., Tat-Seng, C.: Fast matrix factorization for online recommendation with implicit feedback. In: 39th International ACM SIGIR conference. ACM (2016)
13. Luan, K., Zhicai, J., Fang, Z.: Research on commuter's choice behavior between travel mode and trip chain. J. Highw. Transp. Res. Dev. (2010)
14. Kipf, T.N., Welling, M.: Semi-Supervised Classification with Graph Convolutional Networks (2016)
15. Binbin, H., et al.: Leveraging meta-path based context for top- N recommendation with a neural co-attention model. In: 24th ACM SIGKDD International Conference. ACM (2018)
16. Shi, C., et al.: Semantic path based personalized recommendation on weighted heterogeneous information networks. In: ACM International on Conference on Information and Knowledge Management. ACM (2015)
17. Liu, Z., et al.: Heterogeneous graph neural networks for malicious account detection. In: ACM International Conference, pp. 2077–2085. ACM (2018)
18. Zheng, Y., et al.: Learning transportation mode from raw GPS data for geographic applications on the web. In: 17th International Conference on World Wide Web, WWW '08, New York, USA, 2008, pp. 247–256 (2008)
19. Zheng, Y., Yukun, C., Quannan, L., Xing, X., Wei-Ying, M.: Understanding transportation modes based on GPS data for web applications. ACM Transactions on the Web (2010)
20. Zheng, Y., Quannan, L., Yukun, C., Xing, X., Wei-Ying, M.: Understanding mobility based on GPS data. In: 10th International Conference on Ubiquitous Computing, UbiComp'08, New York, USA, 2008, pp. 312–321 (2008)
21. Dabiri, S., Heaslip, K.: Inferring transportation modes from GPS trajectories using a convolutional neural network. Transp. Res. Part C Emerg. Technol. **86**, 360–371 (2018)
22. Zheng, Y., Xing, X., Wei-Ying, M.: Geolife: a collaborative social networking service among user, location and trajectory. In: IEEE Database Engineering Bulletin (2010)
23. Yu, Z., Lizhu, Z., Xing, X., Wei-Ying, M.: Mining interesting locations and travel sequences from GPS trajectories. In: 18th International Conference on World Wide Web, WWW '09, New York, USA, 2009, pp. 791–800 (2009)
24. Kingma, D., Ba, J.: Adam: a method for stochastic optimization. Comput. Sci. (2014)
25. Dong, Y., Chawla, N.V., Swami, A.: Metapath2vec: Scalable Representation Learning for Heterogeneous Networks. ACM (2017)
26. Xiaohua, Y., et al.: A study of multimodal traffic assignment based on a multi-level network. J. Transp. Inf. Saf. **36**(001), 103–110 (2018)
27. Hamilton, W.L., Ying, R., Leskovec, J.: Inductive Representation Learning on Large Graphs (2017)
28. Sarwar, B., et al.: Item-based Collaborative Filtering Recommendation Algorithms. ACM (2001)
29. Koren, Y., Bell, R., Volinsky, C.: Matrix factorization techniques for recommender systems. Computer **42**(8), 30–37 (2009)
30. Chen, T., et al.: SVDFeature: a toolkit for feature-based collaborative filtering. J. Mach. Learn. Res. **13**(1), 3619–3622 (2012)

Improving Personalized Project Recommendation on GitHub Based on Deep Matrix Factorization

Huan Yang[1], Song Sun[1], Junhao Wen[1], Haini Cai[1(✉)],
and Muhammad Mateen[2]

[1] School of Big Data and Software Engineering,
Chongqing University, Chongqing, China
{huanyang,sun2007song,jhwen}@cqu.edu.cn
[2] Department of Computer Sciences, Air University, Multan, Pakistan
muhammad.mateen@aumc.edu.pk

Abstract. GitHub is a hosting platform for open-source software projects, where developers can share their open-source projects with others in the form of a repository. However, as the software projects hosted on the platform increase, it becomes difficult for developers to find software projects that meet their need or interest. Considering the practical importance of software project recommendations, we propose a recommendation method based on deep matrix factorization and apply it to GitHub, which is used to recommend personalized software projects in GitHub. With the use of deep neural network, we learn a low dimensional representation of users and projects from user-project matrix in a common space, in which we can capture the user's latent behavior preference of each item, and automatically recommend the top N personalized software projects. The experiments on use-project data extracted from GitHub shows that the proposed recommendation method can recommend more accurate results compared with other three recommendation methods, i.e., UCF (user collaborative filtering), ICF (item collaborative filtering) and PPR (a personalized recommendation method).

Keywords: GitHub · Project recommendation · Personalized recommendation · User behavior · Deep matrix factorization

1 Introduction

GitHub [22] is a large and popular hosting platform for software projects. As of September 2020, GitHub reported that there are more than 56 million developers and 60 million software code repositories active on their websites around the world [1]. These large-scale software projects have increased the difficulty for developers to search for target software projects. In the actual development process, developers will spend a significant amount of time to browse popular

© ICST Institute for Computer Sciences, Social Informatics and Telecommunications Engineering 2021
Published by Springer Nature Switzerland AG 2021. All Rights Reserved
H. Gao and X. Wang (Eds.): CollaborateCom 2021, LNICST 406, pp. 318–332, 2021.
https://doi.org/10.1007/978-3-030-92635-9_19

software libraries or consult different resources, such as project documents, mailing lists and forums, to find software projects they are interested in [13]. What more, the traditional keywords search engines using in consulting are usually relying on text-matching like similarity measures [17–19], but a small number of keywords may not accurately describe all the characteristics of a software project.

In this situation, traditional recommendation methods seem to be hard to work, because that the previous studies mainly focused on collaborative filtering methods and content-based recommendation methods [2]. However, this kind of recommendation methods do not consider the individual need of the developers, while only rely on the similarity of the project description and the project code [17,24]. Besides, they extract the code through the suffix of the project file when calculating the code similarity, which greatly leads to the inaccuracy.

During the past few decades, deep learning has achieved great success in various fields such as computer vision, natural language processing, pattern recognition and so on [6,16]. Since the excellent performance of deep learning in non-linear relationships capturing and data representation for sparse high-dimensional vectors, its high potential in rating prediction has attracted more and more attention in the recommendation system. Besides, to support the collaborative development of software projects, GitHub has implemented various functions, such as create, fork, and star. Hence, it is achievable to capture the projects that meet user's personal preference and interest with the help of deep learning by collecting a mass of user behavior data when they use these functions [23].

In this paper, we propose a personalized recommendation method: we score user behavior based on their operations generate a user-project matrix firstly, then learn a low dimensional representation of users and projects in a common space through neural networks, finally we predict score for each pair of user and project and recommend a list of software projects with high predicted score to developers. It is worthy to note that our method does not focus on any specific software project, but analyzes the behavior data of developers. Therefore, we can avoid many errors caused by similar projects (such as the software projects with the same programming language, and the software projects with the same configuration file that is easy to misjudged as similar projects) and achieve the real personalized recommendation. In summary, our contributions are as follows:

- We propose a recommendation method based on deep matrix factorization to recommend personalized software projects for developers in GitHub. We use user behavior matrix and neural networks to predict users' potential preferences
- The experimental results show that our proposed recommendation method is more effective than other three baseline methods.

The remainder of this paper is organized as follows. In Sect. 2, we introduce some related work of software project recommendation system. In Sect. 3, we present the proposed method of architecture and details. In Section 4, we illustrate our experimental setup. In Sect. 5, we report the experimental results. In

Sect. 6, we list three types of threats to the validity of our method. Section 7 makes a conclusion and give details of the future work.

2   Related Work

2.1   GitHub Project Recommendations

A search of the literature reveal few studies which is about GitHub project recommendations. Some previous researches can be roughly divided into two aspects: CF-based (collaborative filtering) and network-based. The CF-based method is to determine whether the target project has similar software projects by calculating the similarity between projects. The network-based method uses the user's historical operation information [14] to build an information network depending on users and projects. Li et al. [24] proposed a method to detect similar repositories on GitHub, which mainly evaluates the similarity between two repositories by calculating star-based correlation and readme-based correlation. Koskela et al. [11] proposed a hybrid open-source software recommendation method. Specifically, they combined three different similarity measurement methods on three different feature sets to form a recommendation list. Han et al. [8] proposed an approach to predict the popularity of GitHub project, which rely on the number of stars of a project. These studies focused on the features of the project itself. They recommend projects depending on the similarity or popularity of projects, which ignored the interest of users and their real need. In contrast, our study focuses on personalized project recommendation, i.e., we consider developer behavior to recommend relevant projects.

In study [23], the authors used different user behaviors on GitHub to study which types of user behavior data are suitable for recommending related projects. On this basis, Guendouz et al. [7] proposed and discussed a GitHub repository recommendation system based on collaborative filtering, which models user behavior as a user-project matrix for calculating the similarity between users (developers) and items (repositories). Tadej et al. [14] proposed a recommendation system. They constructed a network graph relying on user data from GitHub (such as fork and pull-request), then an unsupervised learning model was used to predict the relationship of items and users. However, only considering the explicit ratings of developer behavior seems not sufficient, so we also take the developer's underlying preference into account to improve the recommendation accuracy.

2.2   Deep Learning in Recommendation Systems

The application of deep learning in recommendation systems has attracted more and more attention. Badiâa Dellal et al. [3] realized a recommendation system based on MLP deep learning adapted to data already defined by their characteristics. He et al. [9] proposed the NCF model, which is short for neural collaborative filtering. They used a multilayer perceptron to learn user-item interaction functions. The study [15] proved that ensembling NCF and matrix factorization can be helpful. Xue et al. [21] proposed a deep matrix factorization model (DMF), which maps users and items into low-dimensional vectors, making full use of explicit ratings and implicit feedback. Lian et al. proposed CCCFNet [12], namely Cross-domain Content-boosted Software Service Recommendation Base on Collaborative Filtering Neural Network, which combines collaborative filtering and content-based filtering in a unified framework. Liu et al. [13] proposed a learning-to rank model named NNLRank by analyzing the different features of projects and expertise of developers. It extracts nine features as inputs from projects and developer's personal information respectively and predicts the project's preference based on a feedforward neural network. With the help of neural network, we take the developer behavior information as input to predict user preference instead of item preference, which is different from these studies above.

3   Proposed Methods

This section describes the design and implementation of the recommendation method we proposed in detail and Fig. 1 shows the overall architecture of the recommendation process.

3.1   Data Collection

Firstly, we download the data from January 2016 to June 2019[1] on GitHub with the use of GHTorrent[2]. Secondly, according to the entities and relationships[3] of these data, we capture user_ids for each organization from table organization_members.csv and table user.csv. With the help of the user_id, we obtain the project_ids from table projects.csv and generate the table uid-pid.csv. Then the data from GitHub are processed into a form of < user_id, project_id, create, fork, star > by using uid_pid.csv, projects.csv and watchers.csv. Finally, we get the datasets of users, projects and corresponding behavior to generate the user-project matrix. We will explain the task of this part in details in the next content.

3.2   Recommender System

In this part, we assign different scores to the user behavior and generate a user-project matrix. Then, we perform deep matrix factorization on this matrix to obtain the predicted score for each pair of user and project, which make full use of the explicit score and implicit feedback. And finally, a sorted list of projects will be recommended for developers.

---

[1] https://ghtorrent.org/downloads.html.
[2] https://ghtorrent.org.
[3] https://ghtorrent.org/relational.html.

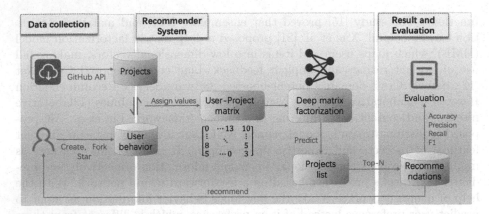

Fig. 1. The architecture of our method.

User-Project Matrix. GitHub provide many functions to support the collaborative development of software, such as create, fork, watch and pull-request. When the developers use these functions, a large amount of user behavior data that record their activities around related software projects will generate. And these data reflect developers' preference and interest [23].

Users' behaviors of repositories on GitHub often contain create, fork and star, different behaviors represent the level of interest in the projects. The create behavior indicates that the user created this project, which means that the project is directly related to the user, so it has the most weight. The fork behavior used by a user implies that the project must meet the need of the user forking it. Finally, the star behavior expresses the interest of a user(s) in a particular project, but the project is not an urgent need of the users.

By scoring each project on account of different user behaviors, we build a user-project matrix. Following [17], we also use the specific construction rules: if the user creates a project, then its score is 10, forks and stars are scored by 5 and 3 points respectively. In addition, if a user creates and stars a project at the same time, the score should be the sum of the two behavior scores, which is 13 points. Similarly, if the project is forked and stared by user at the same time, the score is 8 points. It is worth noting that a project can't have three user behaviors at the same time because create and fork are always opposite and will not appear simultaneously. Therefore, all possible scores for the project are 3, 5, 8, 10, and 13 scores.

Definition: Let $U = \{u_1, u_2 \cdots u_n\}$ denote the set of all developers, $P = \{p_1, p_2 \cdots p_m\}$ denote the set of all software projects, and finally the user-project matrix is expressed as $M_{n \times m}$, where:

$$M_{ij} = \begin{cases} 3, & if \ u_i \ stars \ p_j \\ 5, & if \ u_i \ forks \ p_j \\ 8, & if \ u_i \ stars \ and \ forks \ p_j \\ 10, & if \ u_i \ creates \ p_j \\ 13, & if \ u_i \ stars \ and \ creates \ p_j \end{cases}$$

Recommendations Method. In this paper, we make use of a recommendation method based on deep matrix factorization, which can recommend a list of software projects with high predicted scores to developers. Different from the traditional matrix factorization, in the deep matrix factorization model, two neural networks are used to decompose the rating matrix (User-Project matrix) obtained from the interaction information between the users and the items. By mapping the users and items to a common k-dimensional latent feature space, we can explore the relationship between users and items and learn the potential behavior preference of users in this latent space [21], so we propose the following recommendation model, as shown in the Fig. 2:

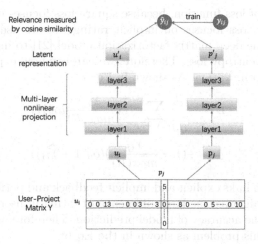

Fig. 2. Deep matrix factorization recommendation model.

From the user-project matrix Y, each user vector u_i represents the i-th user's rating of all items and each project p_j represents the ratings of all users on the j-th project p_j. Let x and y denote the input and output vectors, then the intermediate hidden layer is defined as l_i ($i = 1, 2, \cdots, N - 1$), the weight matrix of layer l_i is represented by W_i, the bias term of layer b_i is represented by bi, finally the output latent representation is represented by h:

$$\begin{aligned} l_1 &= W_1 x \\ l_i &= f(W_{i-1}l_{i-1} + b_i), i = 2, 3, \cdots, N - 1 \\ h &= f(W_N l_{N-1} + b_N) \end{aligned} \quad (1)$$

In the hidden and output layers l_i $(i = 2, \cdots, N - 1)$, we use ReLU as an activation function:

$$f(x) = max(0, x) \tag{2}$$

We use two different multi-layer networks to transform the vectors of users and items, and map them to a common low-dimensional vector space respectively, as shown in Eq. 3.

$$
\begin{aligned}
u'_i &= f_N^U \left(\cdots f_3^U \left(W_{U_2} f_2^U \left(W_{U_1} u_i \right) \right) \cdots \right) \\
p'_j &= f_N^P \left(\cdots f_3^P \left(W_{P_2} f_2^P \left(W_{P_1} p_j \right) \right) \cdots \right)
\end{aligned} \tag{3}
$$

Where U represents the user input matrix and P represents the project input matrix. W_{U_i} and W_{P_j} represent the weight matrix of the user and item in each of the network layer.

Finally, $\widehat{Y}_{ij}$ represents the user's predicted score for the item:

$$\widehat{Y}_{ij} = consine(u'_i, p'_j) = \frac{u'^T_i p'_j}{\|u'_i\| \|p'_j\|} \tag{4}$$

In the choice of loss function, because square loss focuses on explicit ratings and cross-entropy loss focuses on implicit ratings, so we use a new optimization function in the deep matrix factorization model [21] to incorporate explicit ratings into cross-entropy loss. Therefore, we can combine explicit and implicit information for optimization. As shown in Eq. 5.

$$
\begin{aligned}
L = - \sum_{(i,j) \in Y^+ \cup Y^-} &\left(\frac{Y_{ij}}{max(R)} log \widehat{Y}_{ij} \right. \\
&\left. + (1 - \frac{Y_{ij}}{max(R)}) log(1 - \widehat{Y}_{ij}) \right)
\end{aligned} \tag{5}
$$

Although Eq. 5 links explicit and implicit feedback and perform the scale normalization, but it ignores the large difference of rating in user behavior matrix, which will affect the accuracy of model prediction. Therefore, we propose a function to improve this problem as shown in the Eq. 6:

$$S(x) = \frac{1}{1 + e^{-x+3}} \tag{6}$$

By this no-linear function, the score will be mapped compactly into the range from 0.5 to 1, and the gap between the rating is also compressed greatly. On this basis, combined with Eq. 5, a new improved loss function of Eq. 7 is proposed:

$$
\begin{aligned}
L = - \sum_{(i,j) \in Y^+ \cup Y^-} &\left(S(Y_{ij}) log \widehat{Y}_{ij} \right. \\
&\left. + (1 - S(Y_{ij})) log(1 - \widehat{Y}_{ij}) \right)
\end{aligned} \tag{7}
$$

Due to the cross-entropy loss function, the prediction score $\widehat{Y}_{ij}$ may be negative, so we need to correct the predictions with the aid of Eq. 6. When the loss function is less than a predetermined value, it will be equal to a certain minimum value, following [21], we set the minimum value $\mu = 1.0e^{-6}$.

$$Y_{ij}^o = max(\mu, Y_{ij}^o) \tag{8}$$

3.3   Result and Evaluation

To evaluate the performance of our recommendation method, following the literature [5,10], we select all software projects that have not interacted with each user, joint with the test items(interacted with users), to compose the predicted objects, and then a topN recommendation list will be generated for each user depending on the predicted score. We believe that the test items represent the ones users like, so if these items rank higher than others in the test stage, the recommendation shows better performance. For the large data set, since it is too time-consuming to sort each item in the evaluation process, we adopted a common strategy [9], which is to randomly select 30% of the items that users have not interacted with to involve in the predicted objects. The evaluation indicators will be described in detail later.

4   Experimental Setup

4.1   Datasets

Following [17], we extract user behavior and repositories into four groups as our datasets. The first three groups are extracted from three organizations on GitHub: Vim-jp[4], Formidable[5], and Harvestq[6]. In the last group, we extract 2663 active GitHub users and 75417 related repositories.

In GitHub, users have multiple behaviors towards public software projects. We start by extracting all developers of each organizations and get the user_ids, then we extract the project_ids for each user, the result of this operation is a table uid_pid.csv in a form of < user_id, project_id >. Finally, we can get the behavior data(i.e., create, fork, and star) through user_id and project_id. These behavior data are described as < user_id, project_id, create, fork, star >, where the create, fork and star describes the behavior of the user (user_id) for project (project_id). Then, following [17], we randomly selected 60% of the user behavior data as input, and used the remaining 40% of the data for evaluation. The detailed information of each group is shown in Table 1.

4.2   Evaluation and Metrics

We use evaluation Metrics such as Accuracy, Recall, Precision, and F1 [7] to verify the effectiveness of our recommendation method. Accuracy indicates what

---

[4] https://github.com/vim-jp.
[5] https://github.com/FormidableLabs.
[6] https://github.com/harvesthq.

Table 1. Details of four groups of GitHub data

Group name	Users	Projects	Development areas
Vim-jp	47	6262	Vimscript
Formidable	47	2321	Web
Harvesthq	31	944	Android
Large	2663	75417	–

percentage of sample predictions are correct. Recall indicates how many samples among all true samples are predicted to positive. Precision indicates how many predictions are correct in the sample where your prediction is positive. F1 is used to integrate precision and recall as an evaluation indicator. In this paper, the positive sample means the project our method recommends. U represents all users in the test data, and N represents the sum of the number of software projects we recommend for each user. Test represents the number of software projects in the test set that have interacted with users. TopN represents the number of software items recommended for each user. In Table 2, we give the equations of these four metrics.

Table 2. Evaluation metrics

Metric	Formula
Accuracy	$\lvert\{u \mid u \in U, test \cap topN \neq \varnothing\}\rvert / \lvert U\rvert$
Precision	$\lvert test \cap topN\rvert / N$
Recall	$\lvert test \cap topN\rvert / \lvert test\rvert$
F1	$2 * precision * recall / (precision + recall)$

4.3   Statistic Test

When we compare the performance of methods, if the average performance values of multiple methods are different, we still cannot believe that the performance of the higher average method is better than the lower average performance [20]. In this case, we perform Frideman test and Nemenyi post-hoc test, which is widely used in previous studies [20] to analyze the performance difference of the method pairs (i.e., our method and each comparison method) at significant level 0.05. The Frideman test is a non-parametric approach, which compares the overall performance of k algorithms over N data sets. If the p-value [4] calculated by the Frideman test is less than 0.05, the null hypothesis is rejected, that is, all methods perform equally, in other words, there is a significant difference between the methods.

Table 3. The precision and recall of our method, UCF, ICF and PPR

		Precision			Recall		
		Top3	Top5	Top10	Top3	Top5	Top10
Harvesthq	UCF	3.23%	4.52%	3.87%	0.68%	1.53%	2.79%
	ICF	2.15%	3.23%	1.94%	0.45%	1.09%	1.40%
	PPR	5.38%	5.16%	4.52%	3.23%	5.16%	9.66%
	Our	**17.20%**	**26.45%**	**26.45%**	**4.89%**	**12.54%**	**25.08%**
Formidable	UCF	6.82%	5.45%	3.64%	0.98%	1.31%	1.74%
	ICF	2.96%	0.91%	2.67%	0.32%	0.16%	0.97%
	PPR	6.38%	5.53%	6.81%	1.66%	2.44%	5.78%
	Our	**23.40%**	**26.81%**	**27.45%**	**3.60%**	**6.87%**	**14.07%**
Vim-jp	UCF	21.99%	20.43%	15.32%	1.03%	1.60%	2.40%
	ICF	4.96%	3.40%	2.55%	0.17%	0.20%	0.31%
	PPR	5.67%	3.83%	3.40%	0.49%	0.53%	0.96%
	Our	**31.21%**	**34.04%**	**31.49%**	**1.50%**	**2.72%**	**5.04%**
Large	UCF	3.66%	3.07%	2.34%	0.79%	1.10%	1.68%
	ICF	0.55%	0.46%	0.35%	0.12%	0.16%	0.25%
	PPR	4.11%	3.93%	3.91%	0.98%	1.56%	3.10%
	Our	**8.57%**	**9.72%**	**8.26%**	**2.06%**	**3.89%**	**6.61%**

Table 4. The accuracy and F1 of our method, UCF, ICF and PPR

		Accuracy			F1		
		Top3	Top5	Top10	Top3	Top5	Top10
Harvesthq	UCF	10.34%	20.00%	37.93%	1.12%	2.28%	3.24%
	ICF	6.90%	16.67%	20.69%	0.75%	1.63%	1.62%
	PPR	12.90%	25.81%	29.03%	4.03%	5.16%	6.15%
	Our	**29.03%**	**45.16%**	**58.06%**	**7.62%**	**17.01%**	**25.75%**
Formidable	UCF	17.78%	20.00%	24.44%	1.71%	2.11%	2.35%
	ICF	6.82%	4.44%	18.18%	0.58%	0.28%	1.42%
	PPR	17.02%	21.28%	38.30%	2.64%	3.39%	6.25%
	Our	**36.17%**	**38.30%**	**48.94%**	**6.24%**	**10.94%**	**18.60%**
Vim-jp	UCF	44.68%	**55.32%**	**63.83%**	1.97%	2.96%	4.15%
	ICF	12.77%	14.89%	17.02%	0.34%	0.38%	0.55%
	PPR	17.02%	17.02%	31.91%	0.89%	0.93%	1.50%
	Our	**46.81%**	46.81%	53.19%	**2.86%**	**5.05%**	**8.69%**
Large	UCF	10.07%	12.67%	17.47%	1.30%	1.62%	1.96%
	ICF	1.46%	2.15%	3.29%	0.29%	0.24%	0.20%
	PPR	9.04%	11.35%	13.79%	1.58%	2.23%	3.46%
	Our	**16.50%**	**16.63%**	**17.48%**	**3.32%**	**5.56%**	**7.35%**

Table 5. The precision and recall in different dimension

	Precision				Recall			
	k = 50	k = 100	k = 150	k = 200	k = 50	k = 100	k = 150	k = 200
Harvesthq	20.65%	26.45%	23.23%	26.45%	19.57%	25.08%	22.02%	25.08%
Formidable	23.83%	27.45%	23.83%	25.53%	12.21%	14.07%	12.21%	13.09%
Vim-jp	33.83%	31.49%	38.94%	32.55%	5.42%	5.04%	6.23%	5.21%
Large	7.56%	8.26%	7.46%	6.56%	6.05%	6.61%	7.05%	6.35%

Table 6. The accuracy and F1 in different dimension

	Accuracy				F1			
	k = 50	k = 100	k = 150	k = 200	k = 50	k = 100	k = 150	k = 200
Harvesthq	41.94%	58.06%	51.61%	58.06%	20.09%	25.75%	22.61%	25.75%
Formidable	36.17%	48.94%	36.17%	40.43%	16.15%	18.60%	16.15%	17.30%
Vim-jp	57.45%	53.19%	59.57%	55.32%	9.34%	8.69%	10.75%	8.98%
Large	16.16%	17.48%	16.14%	15.16%	6.72%	7.35%	6.72%	6.45%

If the null hypothesis is rejected, it indicates that there are significant differences among the algorithms, then we can continue to conduct follow-up tests. In this paper, Nemenyi post-hoc test is used to find out which algorithms have statistical differences in performance, and the difference of average ranking of each algorithm is correlated with a certain domain value Critical Difference (CD) [4]. If the difference is greater than this domain value, it means that the algorithm with high average ranking is statistically superior to the algorithm with low average ranking; otherwise, there is no statistical difference between the two.

5   Experimental Results

5.1   RQ1:Does the Proposed Method Perform Better Than Other Comparison Methods?

Method: We compare our recommendation results with those of ICF [17,23], UCF [23], and one state-of-the-art recommendation method, a personalized project recommendation method (PPR) [17]:

ICF is project-centric method and it contains two steps: First, it calculates similarity between projects through user behavior. Then, ICF will recommend a list of software projects to users based on the similarity of each items and their historical behavior.

UCF is similar to ICF, UCF is user-centric method and its recommendation process is fallowing: First, it determines the users having similar behavior patterns with the candidate user, which we call similar neighbors, then calculate the candidate user's predicted value for the projects based on the behavior of similar neighbors.

PPR mainly considers the combination of software project characteristics and user behavior. First, it calculates the similarity of readme file and code in each project to obtain the similarity matrices of two projects. Then PPR constructs a user-item matrix, which denotes the historical behavior of users. Finally, a recommendation list is generated for candidate users by examining item similarity matrix through the user behavior.

In addition, we perform statistic significant test to identify whether the differences in performance between our method and other 3 baseline methods are randomness or statistically significant.

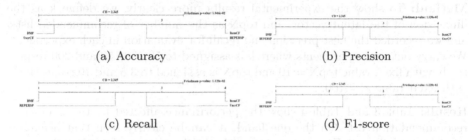

(a) Accuracy   (b) Precision

(c) Recall   (d) F1-score

Fig. 3. Comparison of our method(DMF) against UCF, ICF and PPR with Friedman test and Nemenyi post-hoc test in terms of all 4 indicators.

Result: From the experimental results in Table 3 and Table 4, we can see that our method has better accuracy, precision, recall and F1 than other comparative methods. The main reasons are as follows:

First, the sparsity of the data set is very high, which means that the amount of user behavior is scarce compared with the number of items. So, it is difficult for UCF to accurately find the similar users, and ICF can not make good use of the user behavior data to calculate the similarity of items either. However, on groups of Formidable and Vim-jp, as the number of users and projects increase, we observe a corresponding increase in the accuracy of UCF. The reason behind this correlation may be that the increase of user behavior leads to an increase in user relevance, which makes it easy for UCF to find more similar users.

Second, compared with UCF and ICF, the performance of PPR is improved to a certain extent, but this method may be affected by the same type of suffix when computing the similarity of items through description and source code. Besides, the sparsity of the user-item matrix causes that there are few user behavior data available, good accuracy cannot be achieved in the face of a large amount of data. However, by the aid of neural network, the proposed method can predict unknown user behavior scores based on existing user behavior data, and it alleviate the problem caused by sparsity.

Third, Fig. 3 visualizes the results of Friedman test and Nemenyi post-hoc test for our method and the 3 baseline methods in terms of the 4 indicators. The p-values (all less than 0.05) of Friedman test above the sub-figures show that there exist significant differences among the 4 methods on all indicators. The

Nemenyi test results show that our method significantly performs better than the other baseline methods on all indicators.

To summary, the experimental results show that the proposed method is better than other 3 baseline methods. This means that our method can provide more effective recommendation.

5.2   RQ2:What Is the Effect of the Dimension of the Low-Dimensional Vector and the Number of Recommended Lists on the Performance of the Proposed Method?

Method: To show the experimental results more clearly, we define k as the dimension of the latent vector, and topN as the number of recommended lists, and we recorded the best precision and recall for evaluation in each experiment. We carry out the experiments where k is assigned to 50,100,150 and 200 respectively with fixed value topN = 10 and topN is assigned to 3,5 and 10 respectively with fixed value k = 100.

Result: Table 3 and Table 4 show the performance affected by the number of recommendation list. On the one hand, it can be clearly seen that accuracy, precision, recall and F1 are growing with the increasing of topN. Obviously, when the number of recommended items increases, the target items in the test set are more likely to appear in the recommended list. On the other hand, as the number of items in the data set increases, the recall will decrease due to that we have to recommend topN items. Table 5 and Table 6 show the performance affected by the dimension of latent vectors. When dim increase, there is a slight volatility about accuracy, precision, recall and F1, which may caused by the problem of overfitting.

6   Threats to Validity

6.1   Internal Validity

Internal validity pays attention to the possible faults in the implementation of methods. To minimize this kind of validity, we implement our method by modifying the open source code about DMF model shared by other authors to adapt to our score prediction task. Regarding the reproduction of the baseline methods, we carefully implement them by following the corresponding studies and use the third-party open source code to implement these comparative methods. However, our implementation may not be able to fully reproduce the original methods which may lead to a bias in the comparison between our method and the baseline methods.

6.2   External Validity

External validity focuses on the generalization of the experimental conclusions to other datasets. Our experimental results are derived from user behavior data of four GitHub groups. The first three groups represent three different development areas, which have a relatively small number of users. The fourth group, which include more than 2663 active GitHub users and 75417 related repositories, is less affected by this threat. To further improve the generalization of experimental results, we plan to include more developers and projects.

7 Conclusions

In this paper, we proposed a neural collaborative filtering recommendation method based on user behavior. We extracted users' actions on each related project from different GitHub organizations and generated a user-project matrix by scoring the different actions. Then we utilized a deep neural network to capture user's latent preference, and obtained an abstract data representation from the sparse vector. Finally we computed the predicted scores based on the representation vector, and automatically recommended the topN software projects to developers. The experimental results illustrated that our method obtained better results compared with the other three baseline methods, i.e., UCF, ICF and PPR.

In the future, we plan to extend more user behavior features, such as watch and pull-request, to optimize our method. In addition, we will also consider the impact of time factors.

Acknowledgments. This study was supported by National Natural Science Foundation of China (NSFC): Research on service recommendation of trusted sharing and heterogeneous data fusion in the mobile crowd sensing environment (Grant no.62072060).

References

1. Github: the 2020 state of octoverse report. https://octoverse.github.com (2020)
2. Chen, L., Zheng, A., Feng, Y., Xie, F., Zheng, Z.: Software service recommendation base on collaborative filtering neural network model. In: 16th International Conference, ICSOC 2018, Hangzhou, China, November 12–15, 2018, Proceedings (2018)
3. Dellal-Hedjazi, B., Alimazighi, Z.: Deep learning for recommendation systems. In: 6th IEEE Congress on Information Science and Technology, CiSt 2020, Agadir - Essaouira, Morocco, 5–12 June 2021. pp. 90–97. IEEE (2021)
4. Demšar, J.: Statistical comparisons of classifiers over multiple data sets. J. Mach. Learn. Res. **7**, 1–30 (2006)
5. Elkahky, A.M., Song, Y., He, X.: A multi-view deep learning approach for cross domain user modeling in recommendation systems. In: Proceedings of the 24th International Conference on World Wide Web, pp. 278–288 (2015)
6. Goodfellow, I., Bengio, Y., Courville, A., Bengio, Y.: Deep Learning, vol. 1. MIT Press, Cambridge (2016)
7. Guendouz, M., Amine, A., Hamou, R.M.: Recommending relevant open source projects on github using a collaborative-filtering technique. Int. J. Open Source Softw. Process. **6**(1), 1–16 (2015)

8. Han, J., Deng, S., Xia, X., Wang, D., Yin, J.: Characterization and prediction of popular projects on github. In: 2019 IEEE 43rd Annual Computer Software and Applications Conference (COMPSAC) (2019)
9. He, X., Liao, L., Zhang, H., Nie, L., Hu, X., Chua, T.S.: Neural collaborative filtering. In: Proceedings of the 26th International Conference on World Wide Web, pp. 173–182 (2017)
10. Koren, Y.: Factorization meets the neighborhood: a multifaceted collaborative filtering model. In: Proceedings of the 14th ACM SIGKDD International Conference on Knowledge Discovery and Data Mining, pp. 426–434 (2008)
11. Koskela, M., Simola, I., Stefanidis, K.: open source software recommendations using github. In: Méndez, E., Crestani, F., Ribeiro, C., David, G., Lopes, J.C. (eds.) TPDL 2018. LNCS, vol. 11057, pp. 279–285. Springer, Cham (2018). https://doi.org/10.1007/978-3-030-00066-0_24
12. Lian, J., Zhang, F., Xie, X., Sun, G.: Cccfnet: a content-boosted collaborative filtering neural network for cross domain recommender systems. In: Proceedings of the 26th International Conference on World Wide Web Companion, pp. 817–818 (2017)
13. Liu, C., Yang, D., Zhang, X., Ray, B., Rahman, M.M.: Recommending github projects for developer onboarding. IEEE Access 6, 52082–52094 (2018)
14. Matek, T., Zebec, S.T.: Github open source project recommendation system. arXiv preprint arXiv:1602.02594 (2016)
15. Rendle, S., Krichene, W., Zhang, L., Anderson, J.: Neural collaborative filtering vs. matrix factorization revisited. In: Fourteenth ACM Conference on Recommender Systems, pp. 240–248 (2020)
16. Schmidhuber, J.: Deep learning in neural networks: an overview. Neural Netw. 61, 85–117 (2015)
17. Sun, X., Xu, W., Xia, X., Chen, X., Li, B.: Personalized project recommendation on github. Sci. China Inf. Sci. 61(5), 050106 (2018)
18. Xu, W., Sun, X., Hu, J., Li, B.: Repersp: recommending personalized software projects on github. In: IEEE International Conference on Software Maintenance & Evolution (2017)
19. Xu, W., Sun, X., Xia, X., Chen, X.: Scalable relevant project recommendation on github. In: Proceedings of the 9th Asia-Pacific Symposium on Internetware, pp. 1–10 (2017)
20. Xu, Z., et al.: Tstss: a two-stage training subset selection framework for cross version defect prediction. J. Syst. Softw. 154, 59–78 (2019)
21. Xue, H.J., Dai, X., Zhang, J., Huang, S., Chen, J.: Deep matrix factorization models for recommender systems. In: IJCAI, vol. 17, pp. 3203–3209. Melbourne, Australia (2017)
22. Yu, L., Mishra, A., Mishra, D.: An empirical study of the dynamics of github repository and its impact on distributed software development. In: OnTheMove (OTM 2014) (2014)
23. Zhang, L., Zou, Y., Xie, B., Zhu, Z.: Recommending relevant projects via user behaviour: an exploratory study on github. In: Proceedings of the 1st International Workshop on Crowd-based Software Development Methods and Technologies, pp. 25–30 (2014)
24. Zhang, Y., Lo, D., Kochhar, P.S., Xia, X., Li, Q., Sun, J.: Detecting similar repositories on github. In: IEEE International Conference on Software Analysis (2017)

An Intelligent SDN DDoS Detection Framework

Xiang Zhang[✉], Chaokui Zhang, Zhenyang Zhong, and Peng Ye

University of Electronic Science and Technology of China, Chengdu, China
zhangx@uestc.edu.cn

Abstract. With the development and popularity of computer networks, more and more devices, services and applications are running on the Internet. While it is convenient to the public, more security problems have also brought to the public. Distributed Denial of Attack (DDoS) is just one of the most difficult malicious attacks. It has many different attack forms, causing high damages of services, and is usually hard to detect and defend against. However, the development of Software Defined Networking (SDN) brought new possibilities, due to abilities of global awareness and centralized control. This paper proposes an intelligent SDN DDoS detection framework. In this framework, a security-oriented flow monitoring and sampling algorithm with low-latency is proposed. Meanwhile, we designed a service-oriented recognition model SC-VAE for packet classification. This model combines spectral clustering and variational auto-encoder to detect abnormal traffic by identifying normal streams. It is adaptive to hybrid DDoS attacks, and has a certain predictive effect for unknown DDoS attack not involved in training datasets. Simulation results demonstrate the effectiveness of the proposed framework.

Keywords: SDN · DDoS detection · Network status sampling

1 Introduction

The development of network technology is also accompanied by new threats. With the development of edge computing, Internet of Things (IOT), wireless and other technologies, the demand for network quality and network security are also increasing. The demand for the network is increasing, and many new types of attack are emerging in an endless stream. Many new technologies can also be used in network technology. Time-series content requests and update edge caching accordingly can be predicted by bidirectional deep recurrent neural network (BRNN) model to reduce the network pressure [1]. SDN can also promote the realization of wireless network virtualization to improve the efficiency of resource allocation [12]. However, traditional attacks, DDoS attacks, is rather destructive and difficult to defense. DDoS attack take usage of system software vulnerabilities. It is easy to perform but hard to prevent, quite different

ⓒ ICST Institute for Computer Sciences, Social Informatics and Telecommunications Engineering 2021
Published by Springer Nature Switzerland AG 2021. All Rights Reserved
H. Gao and X. Wang (Eds.): CollaborateCom 2021, LNICST 406, pp. 333–347, 2021.
https://doi.org/10.1007/978-3-030-92635-9_20

from the ordinary infiltration attack. The main goal of this attack is to consume the resources of the network equipment, so that the target equipment cannot provide services normally.

Software Defined Networking (SDN) is a new type of network architecture [8]. SDN separates the control plane and forwarding plane in traditional network, and introduces a network controller to manage the lower-level forwarding equipment generally. SDN uses southbound protocols, as OpenFlow, to dynamically obtain network topology, calculate routing information, send routing instructions to network devices, and eventually obtain network status data such as switch ports and flow tables in real time. Therefore, real-time monitoring and routing of network traffic can be realized at the controller level. Compared with traditional networks, it has a better perception of network traffic changes, what is more, it can control network traffic more flexibly. This can making it easier to detect and protect against DDoS attacks.

The first step to defend against DDoS attacks is to perceive real-time status of the network in SDN. By using OpenFlow, the controller can periodically send requested messages of flow table, and then dynamically obtain flow table information forwarded by the OpenFlow switch. With the analysis of real-time network status, the DDoS attacks can be discovered. The existing methods of detection DDoS attacks are mainly divided into two categories: the first one is attack traffic detection based on entropy or threshold [3], and the second one relies on machine learning and deep learning algorithms to distinguish between attack traffic and normal traffic [4,5,11]. In addition, there is also a multi-level attack detection method that combines entropy and deep learning algorithm [7]. On one hand, by using entropy, the algorithm can achieve a fast recognition speed, which can reduce resource consumption, but with a relatively low accuracy. On the other hand, detection accuracy of using deep learning is very high while it consumes more network resources due to high computational complexity.

These existing approaches that use SDN to defend DDoS attacks are to collect data from all switches in the entire network. However, repeated network flow data will be collected in this way, which will cause a waste of network bandwidth resources. In order to deal with that drawback, the proposed approach in this paper is to collect the flow table data of specific switches by combining the flow sampling algorithm to optimize the resource consumption of the DDoS DDoS detection framework in the early stage of collecting flow information. Also, this approach uses machine learning technology to establish a service-oriented traffic recognition model and a service flow passlist via that model. It becomes possible to recognize more types of DDoS attacks and emerging types of DDoS attacks whit its high generalization ability.

In summary, The main contributions of this paper are as follows:

1. Propose a SDN DDoS detection framework, including:
   (1) An optimization model for traffic monitoring and sampling, in order to reduce the resource consumption when collecting network stream information;

(2) A service flow-oriented classification and identification model, adaptive to hybrid DDoS attacks and potential unknown DDoS attacks.

2. Set up a simulation environment, evaluate and compare the performance of the proposed framework with some latest solutions.

2 Related Works

At present, the main purpose of many researches using SDN for network monitoring is to monitor various link information or message information transmitted in SDN through the ability of centralized control. In SDN, the OpenFlow protocol only provides a message interface to access the flow table of the switch. The paper [9] points out that all approaches of network monitoring are implemented in SDN is based on flow tables, and approaches based on messages haven't been implemented yet. Therefore, an extension package for packet sampling is proposed: FleXam, FleXam extends the OpenFlow protocol and randomly samples a specified number of packet with a certain probability. There are also many network monitoring methods optimized for resource consumption in network monitoring [2,10]. Specific optimization indicators have communication costs. For example, FlowCover [10] established the integer programming problem to minimize the payload of the communication. Using greedy algorithm, it preferentially selected the switch with a small payload consumption load for collection, so as to realize the flow table sampling with low communication payload. The paper [2] proposes a compression algorithm for network monitoring in a hybrid SDN network. Based on the SVD algorithm and linear regression model, this method can replace the link information of all switch links by collecting the link information of a subset of the switches, thereby achieving the purpose of compressing the collected data.

DDoS detection methods in SDN environments use artificial intelligence algorithms such as SVM, deep learning, and random forest to make network attack defense more intelligent [4,5,11]. There is also a simple entropy-based method for identifying abnormal traffic. The former has better results in the accuracy method, while the latter focuses on the minimum resource consumption, For example, Conti M, Lal C, etc. proposed a lightweight method to protect against DDoS attacks in SDN [3], The algorithm proposed by this method can prevent two types of DDoS attacks, routing spoofing and resource consumption. Among them, routing spoofing detection is divided into IP spoofing detection and MAC spoofing detection. Resource consumption detection is divided into two asynchronous detection modules. The first module uses the entropy method to detect random changes in the packet to determine whether there is a DDoS attack. The second module sets up a packet inspection table to determine whether there is a DDoS attack by counting the average number of packets received from a source MAC within a certain time interval. Liu Z, He Y, Wang Wd combined entropy and DNN algorithm to carry out DDoS detection, which was achieved good results [7], the method using entropy detection method for detection firstly. If abnormal flow is found, then the flow will be detected again by particle swarm optimization neural network. Therefore, on the basis of ensuring the rapid detection speed, the accuracy of detection is also improved through the neural network. Kübra, Kalkan, Levent proposed a shared entropy method JESS [6] that can detect and mitigate DDoS attacks. This method can not only detect known DDoS attacks, but also has good accuracy in detecting unknown DDoS attacks. At the same time, this method selects the optimal packet attributes through strategies to reduce the bandwidth resources consumed in the packet collection phase.

3 Security-oriented Flow Monitoring and Sampling

3.1 Performance Analysis of Security-Oriented Flow Table Sampling

In SDN, the most commonly used method for network status awareness is to use the OpenFlow protocol to sample the flow table of forward switches. This can collect current network flow information in the network through the Match field in the flow table. The additional packet payload required to monitor the network in an attack detection framework will have a significant impact on framework resources. Due to this phenomenon, we analyzed the packet payload required for flow table sampling. Suppose the topology of the entire network is graph G, and the set of nodes is V, Each node is mapped to an OpenFlow switch s_i, $s_i \in S$, $i = 1, 2, ..., n$. Define the flow set existing in the network as F, $f_i \in F$, $i = 1, 2, ...m$. Each flow can be uniquely confirmed by the OpenFlow switch sequence P_i that the route passes through, and $P_i = \{s_{i_1}, s_{i_2}, s_{i_3}, ..., s_{i_l}\}$, $i_q \in [1, n]$, $q \in [1, l]$. Let the length of the OpenFlow request flow table packet be L_{req}. The length of each flow table entry is L_{entity}. Because the sampled data is the flow table response, if a switch node s_i has k_i flows through, then at least the packet size $L_{payload_i}$ required to sample the flow table of the switch is:

$$L_{payload_i} = L_{req} + k_i * L_{entity} \tag{1}$$

To get all the flow data in the network, the easiest way is to traverse all the switches to get their flow table data, but this will cause the problem of repeated sampling. Zhiyang Su et al. proposed the FlowCover algorithm [10] to find the smallest set of switches that can collect all flows. This method reduces the number of flow table request messages sent, so that each flow table request brings back as many independent flow table entries as possible. However, from the perspective of monitoring, the polling communication mode of OpenFlow protocol request response to obtain the flow table statistics is indeed a waste of network resources. On the contrary, the subscription notification mode can better realize the periodic flow monitoring tasks. This communication mode only requires the controller to send a subscription message and set the subscribed data type. The OpenFlow switch periodically returns flow table data to the controller, significantly reducing the amount of monitoring packets transmitted in the network. Depending on the subscription notification mode, the flow table sampling and transmission payload of each switch s_i is $L_{new\_payload_i}$:

$$L_{new\_payload_i} = k_i * L_{entity} \tag{2}$$

3.2 Flow Monitoring and Sampling with Low-Latency Based on Optimization Theory

Based on the traffic monitoring service of notification subscription, the traffic is actively automatically reported to the controller, completely avoiding the Open-Flow protocol request message. So it is possible to find out the switches that cover all of the flows in the network during traffic monitoring, and collect the fingerprint on each switch. In this way, the smallest packet sampling payload can be achieved, that is, the sum of the size of the flow table items of all flows.

In the actual network deployment, the controller is deployed in the existing network and relies on the underlying network link. The distance and link bandwidth from each switch to the controller are different, so the transmission delay is also different. Another difficulty in network monitoring is the requirement for real-time network status awareness. We need to be able to perceive changes in the network traffic as quickly as possible, so the latency of monitoring packets transmission in network monitoring must be low enough. Since the sampling is performed on multiple switches, we should set the total latency of packet transmission on multiple switch as the evaluation metric, and establish a model to minimize the total latency required for sampling.

Assuming that the delay vector from the OpenFlow switch to the controller in the network is D, D_i, $i = 1, 2, 3..., n$ represents the network delay from the switch s_i to the controller, and the unit is ms. P_f represents the switches that forward flow f. In order to minimize the total network delay of transmitting traffic packets during sampling, an optimization model is established:

$$\min \quad \sum_{i=1}^{n} D_i * X_i$$

$$s.t. \quad \sum_{i:s_i \in P_f} X_i \geq 1, \forall f \in F \tag{3}$$

$$X_i \in \{0, 1\}, i = 1, 2, 3, ..., n$$

There are two possibilities for each X_i, and there are 2^n in all cases sampled. This problem is an NP problem, and the algorithm complexity is very high to solve directly through polling. Here, heuristic methods can be used to approximate the problem, and the problem can be decomposed into two small goals:

1. The total delay required for one sampling is the smallest
2. Sampled switches can cover all network flows

338   X. Zhang et al.

Then we can adopt a greedy strategy, preferentially select low-latency switches for sampling, and then remove the sampled flows from the flow set F, and repeat this process until all flows are sampled, See the Algorithm 1 for the detailed process.

Data: F, P_f, D
Result: $X = x_1, x_2, ..., x_n$
for x_i in X do
| $x_i \leftarrow 0$
end
while $F \neq \emptyset$ do
| $d \leftarrow$ MAX_INT, $k \leftarrow$ -1 ;
| for d_i in D do
| | if $x_i = 1$ then
| | | continue
| | end
| | if $d \lneq d_i$ then
| | | $d \leftarrow d_i$, $k \leftarrow i$;
| | end
| end
| $x_k \leftarrow 1$;
| for f in F do
| | if $x_k \in P_f$ then
| | | remove f from F ;
| | end
| end
end

Algorithm 1: Solve the set of sampling switches with the smallest total latency in single sampling

---

4   Service Flow-Oriented Attack Recognition Model

4.1   Service Flow Features Required by the Model

The basic features of data mining are directly derived from the flow table collection and packet sampling of the network monitoring module. The flow table collection includes source IP address, destination IP address, source port number (UDP/TCP), destination port number (UDP/TCP), IP message protocol type, flow table matching counter, and flow table matching interval. Packet sampling can obtain the length of the IP message, the partial truncation information of the IP message, and the interval between the arrival of adjacent messages. The source IP address indicates the source address of the message or a forged IP address. In some reflection attacks, it may also be the address of an intermediate server. As a result, the distribution of IP messages may potentially reflect the source and the type of attack of DDoS attacks; The destination IP and port indicate the access address and port provided by the network service. Under normal circumstances, the distribution of packet request a service also has a certain pattern. This feature can reflect the application flow in the network. IP protocol and IP Packet truncation data are used to determine the type of application flow in the network. Basic features are rather limited in representing the pattern of the traffic, but sampling multiple such features continuously can reflect the transaction fragment information of the packets sent by the sender. These transaction fragments allow to infer the sender's intent and determine whether the sender has the intention to attack, so as to identify whether the traffic is normal or abnormal. For example, a sequence of time intervals where traffic matches. When a user accesses a service, multiple packets will be generated on the network. The time interval for the flow matching of these consecutive packets will also reflect the features of the transaction performed by the user to a certain extent. When visiting a portal site, it can be divided into the process of opening a webpage, reading or interacting, and jumping to another webpage to continue reading. There will be a certain dwell time between these actions. These time intervals are also the generation of packets interval.

4.2 DDoS Attack Detection Model Based on Clustering and VAE

Both spectral clustering and variational auto-encoder can be used alone to detect DDoS attacks, but a single model is prone to over-fitting and under-fitting, and cannot achieve good attack recognition results. Therefore, this paper proposes spectral clustering Combine with the variational auto-encoder to build a predictive model. The data used for training is the network flow data X obtained after the network monitoring module data pipeline. First, X is trained through the variational auto-encoder. Both the encoder and decoder of the variational auto-encoder select multi-layer perceptrons (MLPs). After training, we can get the distribution parameters of the potential features of the original data μ and ϵ, Then you can sample a potential feature code X_{code}^i for each sample data X^i, and then concatenate the original data X^i with this hidden feature code, you can get extended data X_{ext}^i, Finally, through spectral clustering, the number of clusters is set to 2, and the new extended data set X_{ext} is trained to obtain a model for detection DDoS attacks. The specific model architecture is shown in Fig. 1.

The main part of model training is the training of the VAE. The VAE only uses normal applications to train, so that the model can learn the potential feature representation of the service flow, and then make the clustering algorithm to find the normal application flow characteristic group. When the model is deployed, the normal application flow data and a small amount of attack application flow data of the automatic application encoder are trained. In order to obtain the cluster numbers of the normal application flow and abnormal attack flow. When new traffic reaches the detection module, the model can be combined with historical traffic data to determine whether the new traffic is DDoS attack traffic, so as to achieve the purpose of attack detection.

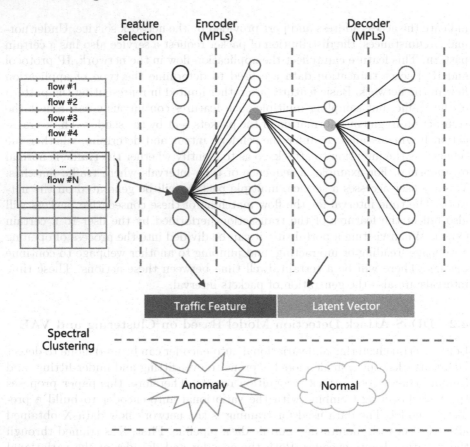

Fig. 1. DDoS attack detection model architecture

4.3 DDoS Attack Defense Based on Recognition Result

When the detection model recognizes a DDoS flow, a series of operations can be used to reduce the damages. First, since we have got the source IP, port and other metadata of the flow, specific flow tables can be sent to block the DDoS traffic on SDN switches in the flow forwarding path. The other way is much gentle, to send specific meter tables to SDN switches, in order to limit traffic flow forwarding rate. However, the IP addresses used by DDoS flow are usually fake, then the real malicious host cannot be located directly through the IP address. But with the global view of SDN, it can be easy to trace the source of malicious traffics. Based on the metadata of traffic flow and the data in switches, malicious traffic can be traced hop by hop to the source. And then, the edge switch can be sent a flow table to block the malicious host, or a meter table to limit the forwarding rate.

5   Simulation and Performance Evaluation

5.1   Simulation Setup

In order to evaluate the approach, we designed a simple SDN-based campus network. The campus network topology is shown in Fig. 2. A Leaf-Spine structure is used to achieve interconnection through the IP network, which is a two-layer network structure. Unlike the traditional three-layer (core layer, convergence layer, access layer), Leaf-Spine is a two-layer network structure. The Leaf layer switches are responsible for accessing all servers and terminal devices. The Spine layer switches act as a backbone switch and is fully connected to the Leaf layer switches. Two edge leaf nodes (Edge Leaf) are configured in the entire network topology, and external networks such as the Internet can be accessed through these two edge nodes.

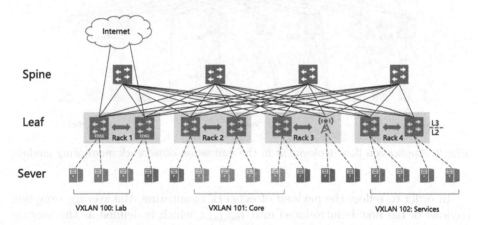

Fig. 2. Campus network topology for simulation

5.2   Network Traffic Sampling Efficiency Evaluation

Three application streams are set up in the test environment, web page information stream, video information stream and background traffic simulated by IPerf tool. Among them, the web page information flow can randomly select three servers on the campus network as the Web server, and access the Web server at a fixed frequency by simulating the client side. Each client uses its own frequency to send request data to server. At the same time, in order to simulate the generation and disappearance of traffic, the client is set to stop for a period of time after every 50 requests to invalidate the flow entry in flow table. The web server has only a fixed-size page, and the client accesses this fixed page every time, so as to facilitate the comparison of the sampling load of different methods. For video stream, we build an Real Time Streaming Protocol (RTSP) video stream server through VLC, and choose two servers in the network as the video stream server, and choose three clients as the consumer end of the video stream. Unlike web streaming, the video streaming is continuous, and the video streaming protocol is set to the RTSP. In addition to the basic application flow, in order to simulate a more realistic environment, we use IPerf to set three background traffic, the bandwidth of the background traffic is set to 1 Mbps, the detailed application flow deployment is shown in Fig. 3.

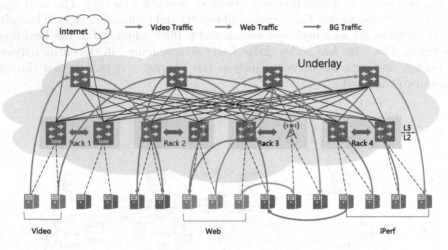

Fig. 3. Application flow deployment in the evaluation of network monitoring module

In order to reflect the payload of network monitoring, the average sampling payload of the flow is introduced into $\overline{f_{payload}}$, which is defined as the average sampling payload of each flow over a period of time. Suppose the payload of each sample in the sampling time is P_i, the flow set of each sample is F_i, and the total number of samples is m, then the average sample payload of the flow is:

$$\overline{f_{payload}} = \frac{\sum P_i{}^m}{\sum |F_i|^m} \tag{4}$$

Regarding the timeliness of network flow sampling, we also defines the metric, total delay D_{simple}, which can better reflect the time cost of acquiring the entire flow status per sample. When the number of flows in the network increases, this metric will increase, so it can reflect how congestion the network behaves. Let the number of packets in a single sampling be N_{simple}, and the end-to-end delay of each sampling is t_i, then D_{simple} is

$$D_{simple} = \sum t_i{}^{N_{simple}} \tag{5}$$

Through the monitoring test in the simulation environment, the network monitoring performance test result is shown in Fig. 4.

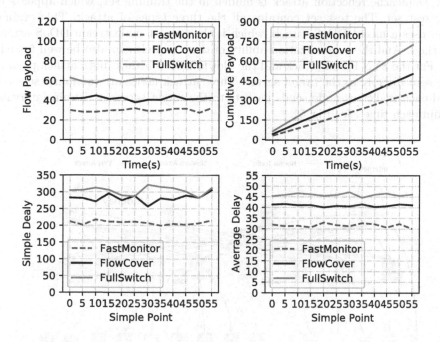

Fig. 4. Network monitoring module performance test results

According to the test results, because of the duplicate sampling problem of FullSwitch, its average message load size is larger than the theoretical load needed per flow. The sampling load of FlowCover algorithm is closer to the theoretical average load for it implements finding the smallest set of switches that can get all flow information, avoiding duplicate sampling. However, the use of polling to obtain information, which causes the request message, will still consume some network resources. The FastMonitor method proposed in this paper adopts a subscription notification mode based on the FlowCover algorithm. In this way, the switch can actively send data to the controller at regular intervals, reducing the bandwidth consumption required for sending request messages and the latency. The average load of FastMonitor method in this test can reduce to only 30 bytes per flow, which decreases the time of each sample by about 60ms compared to FlowCover algorithm. FastMonitor can have a better sense of the changes in the network and thus can react to attacks faster.

5.3 Attack Detection Model Evaluation

To evaluate the model, except the normal flow, we also need to simulate the attack flow. Three common denial-of-service attack flows are selected for simulation, which are the common SYN Flood, the Memcache reflection attack with the most lethality, and the slow attack Slowloris. The data set is 24-h packet capture data that simulates normal application flow data and attack flow. and the Memcache reflection attack is hidden in the training set, which appears in the test set. The test set contains all the three types of attack. This evaluation uses a mixed DDoS attack, which is the current mainstream DDoS attack method. And the generalization ability of these models can also be well tested.

For the SYN Flood attack and the Slowloris slow attack, both attack packets are sent directly to the target server. Three attack nodes are randomly selected and each malicious node sends packets to the target server at 10 min intervals, 3 min each time (See Fig. 5).

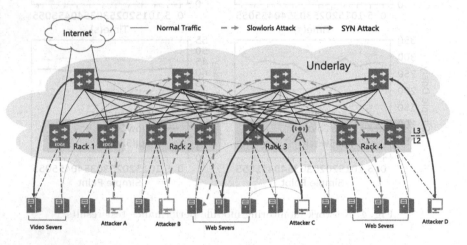

Fig. 5. Deployment of SYN Flood and Slowloris attack simulation

The Memcache reflection attack is different from other types of attacks that send requests directly to the attacked server. This attack forges the source IP of the Memcache request message as the target IP, and then uses the response message of the distributed Memcache servers to achieve an indirect attack. In order to simulate this type of attack, it is necessary to set up two Memcache servers in the network, namely Memcache Sever A and B, and randomly select three attacking nodes to send fake Memcache request messages whose source IP address is the IP address of the attack target. The specific deployment method of Memcache reflection attack is shown in Fig. 6.

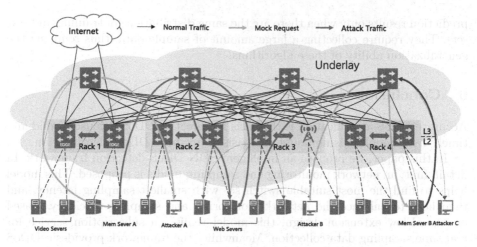

Fig. 6. Deployment of memcache reflection attack simulation

In order to compare the performance of the DDoS attack recognition model proposed in this article, we also test several other common recognition models, such as SVM and deep neural network, and then compare the evaluation scores of different models. Here, multiple evaluation scores are used to reflect the differences of the models. Performance mainly includes commonly used evaluation indicators such as Accuracy, True Postive Rate (TRP), False Postive Rate (FPR) and F1-score (the harmonic mean of precision and recall). For the data used for testing, we collected 24-h data through simulated application flow and attack flow, and use the 24-h data as the test data set to evaluate the model recognition ability. The test results are shown in Table 1.

According to the test results, the SC-VAE attack detection model proposed in this paper has certain advantages over other common models in terms of the accuracy of detection and the ability to correctly determine the anomaly class. Since the SC-VAE distinguishes unknown DDoS attack streams by detecting real application stream features, it has certain ability to detect unknown DDoS attack types. The training set of this test does not contain Memcache reflection attack types, but the test results still maintain a high accuracy. In contrast, common classification-based learning methods such as SVM and DNN can achieve

Table 1. Performance comparison of different models

Model	ACC	FPR	TPR	PPV	F1
SC-VAE	0.9725	0.0373	0.9826	0.9634	0.9728
SVM	0.9668	0.0480	0.9818	0.9533	0.9672
DNN	0.9541	0.0531	0.9615	0.9465	0.9536
VAE	0.9533	0.0712	0.9761	0.9522	0.9583
Spectral clustering	0.9122	0.1035	0.9547	0.8870	0.9311

prediction results only when there are the same distributions of training and test sets. They require collecting a large amount of sample data, thus reducing the generalization ability of these algorithms.

6 Conclusion

Accurate DDoS identification is the basis for effective DDoS defense. Rapid and timely flow sampling is no doubt the basis of accurate DDoS identification.

In this paper, we propose an intelligent SDN DDoS detection framework. In detail, first, an network monitoring and sampling model is proposed. This model helps to find the most suitable switch set with smallest sampling latency, and then help to enhance the network monitoring and sampling efficiency. Based on OpenFlow extension design, this model realizes a subscription service for real-time sampling data collection. Meanwhile, the framework provides a DDoS detection model SC-VAE, consisting of algorithms of spectral clustering and variational auto-encoder. This detection model cluster flows by flow features, then abnormal hybrid DDoS and potential unknown DDoS attack can be detected. Simulation results show that the accuracy of SC-VAE is 97.3%, and the FPR is only 3.7%, which can ensure a good prediction effect, and is better than traditional classification models based on SVM.

Next, we will continue to optimize this intelligent SDN DDoS detection framework, fill in it with defence strategy and implementation.

References

1. Ale, L., Zhang, N., Wu, H., Chen, D., Han, T.: Online proactive caching in mobile edge computing using bidirectional deep recurrent neural network. IEEE Internet Things J. **6**(3), 5520–5530 (2019)
2. Cheng, T.Y., Jia, X.: Compressive traffic monitoring in hybrid SDN. IEEE J. Sel. Areas Commun. **36**(12), 2731–2743 (2018)
3. Conti, M., Lal, C., Mohammadi, R., Rawat, U.: Lightweight solutions to counter DDoS attacks in software defined networking. Wireless Netw. **25**(5), 2751–2768 (2019)
4. Gangadhar, S., Sterbenz, J.P.: Machine learning aided traffic tolerance to improve resilience for software defined networks. In: 2017 9th International Workshop on Resilient Networks Design and Modeling (RNDM), pp. 1–7. IEEE (2017)
5. Arevalo Herrera, J., Camargo, J.E.: A survey on machine learning applications for software defined network security. In: Zhou, J., et al. (eds.) ACNS 2019. LNCS, vol. 11605, pp. 70–93. Springer, Cham (2019). https://doi.org/10.1007/978-3-030-29729-9_4
6. Kalkan, K., Altay, L., Gür, G., Alagöz, F.: Jess: Joint entropy-based DDoS defense scheme in SDN. IEEE J. Sel. Areas Commun. **36**(10), 2358–2372 (2018)
7. Liu, Z., He, Y., Wang, W., Zhang, B.: DDoS attack detection scheme based on entropy and PSO-BP neural network in SDN. China Commun. **16**(7), 144–155 (2019)
8. McKeown, N., et al.: OpenFlow: enabling innovation in campus networks. ACM SIGCOMM Comput. Commun. Rev. **38**(2), 69–74 (2008)

9. Shirali-Shahreza, S., Ganjali, Y.: Flexam: flexible sampling extension for monitoring and security applications in openflow. In: Proceedings of the Second ACM SIGCOMM Workshop on Hot Topics in Software Defined Networking, pp. 167–168 (2013)
10. Su, Z., Wang, T., Xia, Y., Hamdi, M.: Flowcover: low-cost flow monitoring scheme in software defined networks. In: 2014 IEEE Global Communications Conference, pp. 1956–1961. IEEE (2014)
11. Tang, T.A., Mhamdi, L., McLernon, D., Zaidi, S.A.R., Ghogho, M.: Deep learning approach for network intrusion detection in software defined networking. In: 2016 International Conference on Wireless Networks and Mobile Communications (WINCOM), pp. 258–263. IEEE (2016)
12. Zhang, N., et al.: Software defined networking enabled wireless network virtualization: challenges and solutions. IEEE Netw. **31**(5), 42–49 (2017)

Inspector: A Semantics-Driven Approach to Automatic Protocol Reverse Engineering

Yige Chen[1,2], Tianning Zang[1,2], Yongzheng Zhang[1,2], Yuan Zhou[3],
Peng Yang[3(✉)], and Yipeng Wang[4]

[1] Institute of Information Engineering, Chinese Academy of Sciences, Beijing, China
[2] School of Cyber Security, University of Chinese Academy of Sciences,
Beijing, China
[3] National Computer Network Emergency Response Technical Team/Coordination
Center of China, Beijing, China
yp@cert.org.cn
[4] Beijing University of Technology, Beijing, China

Abstract. The automatic protocol reverse engineering for undocumented network protocols is important for many fundamental network security applications, such as intrusion prevention and detection. With the growing prevalence of binary protocols in the network communication to organize data in a terse format and ensure data integrity, the proven reverse approaches for ordinary text protocols face severe challenges in the compatibility. In this paper, we propose Inspector, an automatic protocol reverse engineering approach that exploits semantic fields to infer message formats from binary network traces. Inspector reasonably infers two semantic fields based on the binary content analysis of protocol messages to support clustering messages and message format inference. We evaluate the effectiveness of Inspector on two binary cryptographic protocols (TLS and SSH) and a binary unencrypted protocol MQTT by measuring the accuracy of message clustering and comparing the inferred message formats with the ground truths on a traffic dataset captured from a campus. Our experimental results show that Inspector accurately cluster messages with 100% cluster precision and 100% message recall for TLS, 90% cluster precision and 99.6% message recall for SSH, 100% cluster precision and 92.7% message recall for MQTT. Based on the accurate message clusters, Inspector can correctly infer the format of the messages in the cluster.

Keywords: Protocol reverse engineering · Protocol format inference · Binary protocol · Semantic fields

1 Introduction

This paper concerns protocol reverse engineering for binary protocols based on network traces, which is to automatically infer the application-level specification

Published by Springer Nature Switzerland AG 2021. All Rights Reserved
H. Gao and X. Wang (Eds.): CollaborateCom 2021, LNICST 406, pp. 348–367, 2021.
https://doi.org/10.1007/978-3-030-92635-9_21

of the binary protocol by only accessing the passively collected network traces. The protocol specification stipulates all formats of the legal messages sent by the communicating parties. Consequently, the protocol specification is critical in many network security applications, such as intrusion prevention and detection [21] and unknown application protocol parsing [23]. Taking the botnet as an example [14], when we know the specifications of the malware communication protocol, we can easily recognize the malicious traffic and intercept these traffic to protect vulnerable hosts from being compromised. With the growing attention to network security, the mainstream applications gradually adopt binary cryptographic protocols to ensure data security and ensure data integrity in the network transmission [6]. The Internet of Thing (IoT) devices need a lightweight messaging protocol to minimize bandwidth and power usage in the network communication. Unlike traditional text protocols, such as HTTP, SMTP, and FTP, it is much more difficult for the network administrator to analyze the network traces of binary protocols because the binary byte sequences are designed to be understood by a machine but not a human being.

The classic network-trace-based approaches generally follow the routine of first clustering messages and then inferring the protocol specification based on the clustering results [3,9,23]. Because of the cascade inference routine, the accuracy of the inference results heavily relies on the precision and coverage of the message clustering results. These approaches are incompatible with binary protocol traffic because the binary message structure greatly increases the difficulty of aligning variable position fields and adversely affects the accuracy of message clustering. Thus, a feasible idea to overcome the obstacles of binary protocol reverse engineering is to perceive the binary message structure before conducting message clustering and protocol specification inference.

In this paper, we propose a semantics-driven approach named Inspector to conduct automatic protocol specification inference from network traces. Taking full advantage of the general design methodology of the binary protocol, we reasonably infer two semantic fields, namely the length field and message type field, before conducting the messages clustering and the subsequent protocol specification inference. Based on the inference of the semantic fields, we can accurately cluster messages of the same message format and then directly infer the specification of message formats. Similar to the Discoverer [9], the message format output by the Inspector is a sequence of specification tokens that specify both the semantics and properties of the message fields. To evaluate the effectiveness and accuracy of the Inspector, we collect a dataset of two widely used binary cryptographic protocols, namely Transport Layer Security (TLS) [11] and Secure Shell (SSH) [25], and a lightweight messaging protocol Message Queuing Telemetry Transport (MQTT) [22]. The ground truth specifications of these protocols are public in the Request for Comments (RFC) [12] and OASIS standards [17]. We deploy the Inspector on the collected dataset to infer the specifications without knowing the specification documents in advance. The experimental results show that the Inspector can properly deal with the semantic field inference and message clustering, and then perform accurate protocol specification inference.

We briefly summarize our major contributions as follows:

- We propose a semantic-driven approach named Inspector to implement the specification inference for binary protocols. The Inspector recursively extracts two kinds of semantic fields of the message and then cluster messages based on the semantic fields until each cluster contains messages of the same format.
- We design a hierarchical length field model and a corresponding algorithm to search length field pattern candidates and select the correct one for the messages. We also design a message type field inference algorithm to search for message type fields to support the clustering of messages with the same length field pattern.
- We implement the Inspector and conduct evaluation experiments over three binary protocols. The experimental results show that the Inspector can accurately identify the semantic fields of the messages and partition messages into clusters with 100% cluster precision and 100% message recall for TLS, 90% cluster precision and 99.6% message recall for SSH, 100% cluster precision and 92.7% message recall for MQTT. The Inspector also works well in the format inference and outputs results that are consistent with the ground truths.

The rest of this paper is organized as follows. Section 2 summarizes the related work of the protocol reverse engineering. Section 3 introduces the inference of the length fields and message type fields and then describes the details of the message clustering and format inference. Section 4 introduces the experimental dataset and presents the evaluation details of the Inspector. Section 5 concludes this paper.

2   Related Work

A communication protocol can be regarded as an agreement on how to properly exchange information between communication entities [15]. The specifications of communication protocols are similar to natural languages, including vocabulary and grammar. The vocabulary consists of a set of words or byte values that can compose valid messages. To clearly communicate with each other, the communication entities need to negotiate the semantics of the words or the meaning of the byte values in advance. The grammar specifies the detailed rules of the information exchange procedure. In a communication process, the sender should follow the procedural rules to enclose the raw information into a formatted message. When the receiver observes the formatted message, he can restore the information from the message through inverse operations. The protocol reverse engineering focuses on both the vocabulary and grammar of the protocol specification.

Existing automatic protocol reverse methods can be divided into two families: methods based on execution traces [4,5,8,10] and methods based on network traces [1,3,9,23,27]. The methods based on execution traces include Polyglot [5], Tupni [10], Prospex [8], and Dispatcher [4]. These methods conduct protocol reverse engineering by directly debugging the protocol binary or monitoring how the protocol binary generates and parses messages. Because these methods can dissect the real binary program, the accuracy of the inference results is generally high. However, we must obtain the corresponding protocol binary programs in advance to obtain the execution traces. Besides, some private protocol developers may deliberately use obfuscation techniques to thwart protocol specification inferences such as binary packing, inlining unnecessary instructions, and replacing calls with indirect jumps [16].

In contrast, the methods based on network traces only take the network traces of the target protocol as input and the obtaining of network traces is more straightforward. When the protocol binary programs are available, we can control the binary programs to automatically generate the target traces and then directly capture the traces. When the binary programs are unavailable, we can collect the network traces by passively monitoring the network gateway through port mirroring switches [26] and then sift out the target traces based on the IP address or port number.

Existing methods based on network traces include Discoverer [9], ProDecoder [23], Netzob [3], ProWord [27], and FieldHunter [1]. The basic inference framework of these methods is first clustering messages based on their group characteristics and then performing specification inferences within each cluster. Cui et al. first propose Discoverer [9] to automatically infer the complete protocol format from network traces. Discoverer first tokenizes messages into a sequence of text and binary tokens and then conducts initial clustering messages based on the token pattern. After the initial clustering, Discoverer recursively identifies the format distinguisher (FD) tokens in the messages through a heuristic method and divides existing clusters into subclusters based on the values of FD tokens until each cluster only contains messages of one format. Finally, Discoverer directly infers the protocol specification by making comparisons between the token sequences within each cluster. Wang et al. propose ProDecoder [23] to support the specification inference of asynchronous protocols. ProDecoder first uses n-gram sequences to represents messages and exploits Latent Dirichlet Allocation (LDA) [2] to select the keywords in the messages. Then, ProDecoder uses Information Bottleneck (IB) [20] as the clustering metric to partition messages into clusters based on their keywords. Based on the clustering results, ProDecoder can find the invariant fields among the messages of each cluster and output the message formats in the form of regular expressions. To provide more accurate inference results, Bossert et al. propose Netzob [3] to pre-cluster messages with the same program activity to support the subsequent clustering. Then, Netzob introduces contextual and semantic information as a key parameter to cluster messages and partition the fields in the message. Based on the intermediate inference results, Netzob uses an extended sequence alignment algorithm to align the partitioned fields in different messages to extract the message format. Unlike previous works, the ProWord [27] proposed by Zhang et al. formulates the protocol reverse engineering as an information retrieval problem and focuses on extracting protocol feature words from network traces. They first modify the voting expert algorithm [7] to extract feature word candidates and then select the desired feature words based on the frequency, location, and length of the word candidates. The recent format inference work FieldHunter [1] focus on identifying field boundaries and inferring specific field types for both binary and textual protocols. The FieldHunter successfully infers several semantic fields for binary and textual protocols. However, the FieldHunter is hard to precisely infer the length fields for binary protocols since the message data structure is unknown.

3　System Design

In this section, we first present an overview of the Inspector. Then, we describe the phases of the Inspector in detail, including length field inference, message type field inference, and format inference.

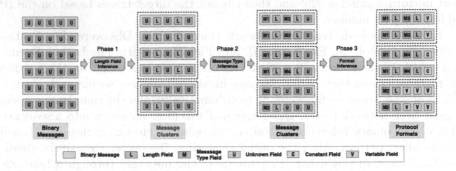

Fig. 1. System architecture of the inspector

3.1　Overview

The system architecture of the Inspector is shown in Fig. 1. The first phase of the Inspector is length field inference, which takes the binary messages as input and searches for the length field pattern candidates based on the proposed hierarchical length field model. After getting all the length field pattern candidates that meet the model requirements, we selected the correct length field pattern based on a reasonable assumption that the correct length field pattern should be more frequently appeared than the wrong ones in the messages and can be well explained. The length field pattern contains both the position information of the length fields and the hierarchical information of the message, so we can cluster messages with the same length field pattern.

The second phase is message type field inference that intends to find the message type fields of the message. Since we have discovered the hierarchy of the messages, we can perform targeted lookups for the message type fields around the diverging positions of the message hierarchy. The main intuition in identifying the message type field is that the number of unique message types should be limited to an appropriate range, and the sizes of the same message type message are relatively close. After inferring the message length field, we can perform finer message clustering so that each cluster only contains messages of one message format.

The third phase format inference focuses on the token properties inference. The output of the message format inference is a sequence of format tokens, including length field, message type field, constant field, and variable field. Since the semantic-driven approach can provide precise and pure clustering results, we can directly infer the token property sequence of the message cluster through a simple and effective sequence alignment method.

3.2 Length Field Inference

The network traces between communication parties can be viewed as consecutive dialogue messages. However, dialogue messages should be divided into independent data packets to meet the transmission protocol of the Internet. Therefore, after collecting network traces from the network link, we first need to recover the original dialogue messages in the same way as the communication parties. For the packets transmitted by Transmission Control Protocol (TCP), we reassemble them into the original message to ensure the integrity [19] and take it as system input. The datagram transmitted by User Datagram Protocol (UDP) is naturally integrated [18], so we directly extract the payloads and regard them as individual messages.

Algorithm 1. Length Field Inference Algorithm

1: **Input:** Message Dataset MD; Bit-Widths BW; Orientations O; Endianness E; Size of the Suffix Field S; Max Field Number NF; Selection Threshold ST; Minimum Number of the Messages in the Cluster NC.

2: **Output:** Message Clusters MCS.

3: **Message** $M(mb, mps, c, p)$: mb = message bytes, mps = message parts to be inferred, c = length field pattern candidates, p = selected length field pattern.

4: **LengthFieldPattern LFP**(ofs, w, ori, e, sub) : ofs = field offset, w = bit-width, ori = orientation, e = endian, sub = sub length fields.

5: **function** CANDIDATESEARCH$(mp, n=1)$

6:     $C = \varnothing$

7:     **if** $n > NF$ **then**

8:         **return** C

9:     **end if**

10:    **for** ofs from 0 to LEN(mp) **do**

11:        **for all** (w, ori, e) in $(BW \times O \times E)$ **do**

12:            $entity \leftarrow$ extract the field bytes by ofs, w

13:            $value \leftarrow$ parse the value of $entity$ by ori, e

14:            $leftlen \leftarrow$ LEN$(mp) - value - w - ofs$

15:            $lengthfield \leftarrow$ **LFP**$(ofs, w, ori, e, \varnothing)$

16:            **if** $n = 1$ **and** $leftlen$ in S **then**

17:                add $[lengthfield, ('suffix', left)]$ to C

18:            **end if**

19:            **if** $left = 0$ **then**

20:                add $[lengthfield]$ to C

21:           **else if** $left > 0$ **then**
22:               $extpart \leftarrow$ Cut out the extra parts
23:               $lf \leftarrow$ CANDIDATESEARCH($extpart, n + 1$)
24:               merge $lengthfield + lf$ to C
25:           **end if**
26:       **end for**
27:   **end for**
28:   **return** C
29: **end function**
30: **function** CANDIDATESELECT($mc, \boldsymbol{ST}, \boldsymbol{NC}$)
31:   **while** mc contains message m with nonempty $m.c$ **do**
32:       $C \leftarrow$ extract all $m.c$ for all m in mc
33:       $lfps \leftarrow$ select the element that exists in most messages from C
34:       $num \leftarrow$ count the messages containing $lfps$
35:       $prop \leftarrow num/\text{LEN}(mc)$
36:       **if** $prop \geq \boldsymbol{ST}$ and $num \geq \boldsymbol{NC}$ **then**
37:           **for** all messages m where $m.c$ contains $lfps$ **do**
38:               $m.lf \leftarrow lfps$ **or** add $lfps$ into related sub length field
39:               $m.c \leftarrow \varnothing$
40:               update new $m.mps$ based on the new $m.lf$
41:           **end for**
42:       **end if**
43:       remove $lfps$ from all $m.c$
44:   **end while**
45: **end function**
46: **procedure** LENGTHFIELDINFER($\boldsymbol{MD}$)
47:   $initmc = \varnothing$
48:   **for** all $message$ in $\boldsymbol{MD}$ **do**
49:       add M($message, [message], \varnothing, \varnothing$) to $initmc$
50:   **end for**
51:   $MCS = [initmc]$
52:   **while** MCS contain m with nonempty $m.mps$ **do**
53:       **for** all mc in MCS **do**
54:           **for** all m containing nonempty $m.mps$ in mc **do**
55:               $m.c \leftarrow$ CANDIDATESEARCH($m.mp$)
56:           **end for**
57:           CANDIDATESELECT(mc)
58:           Split mc into subclusters based on the length field pattern.
59:       **end for**
60:   **end while**
61:   **return** MCS
62: **end procedure**

The primary purpose of the length fields in the message is to specify the boundary of the variable-length payload, so the communication parties can accurately extract the relevant payload from sequential binary messages without explicit delimiters. Thus, when we find the length fields in the message, we can extract the payload just like the communication parties and then perform further format analysis for each message part. There are two main difficulties in seeking the length fields. a) The first difficulty is to deal with the hierarchy of the length fields. From a vertical perspective, only the length fields in the top-level can be directly related to the size of the message, while other length fields are only related to the size of the relevant payload. From a horizontal perspective, there could be multiple parallel length fields in the message, which weakens the correlation between the value of the length field and the size of the message. When we perform length field inference for a message, the only reliable length information is the total size of the message. Without a prior definition of the length field structure, we cannot infer most of the length fields. b) The other difficulty is to deal with the various interpretations of the length field. The length fields in different protocols may have their preferred specifications, such as in terms of position, endianness, and bit-width. To be specific, the length field could be located before or within the relevant payload, the endianness could be little-endian or big-endian, and the bit-width could be 8, 16, 32, or 64 bits. Even in one binary message, the protocol designers may use different bit-width length fields to compress the message size. An intuitive method to deal with the undetermined interpretation of the length field is to conduct a grid search that requires costly computing resources. Therefore, we need a search strategy that can hit the correct interpretation of the length field but only consumes moderate computing resources.

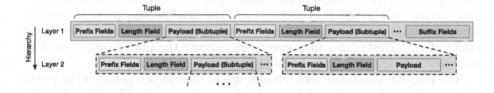

Fig. 2. Hierarchical length field model

In order to solve these two problems, we propose a hierarchical length field model to perform the length field search in a certain routine and develop a corresponding algorithm to implement the length field inference. Figure 2 shows the structure of the hierarchical length field model. In this model, the basic structural units of the message are tuples that arranged in sequence or nested recursively. An ordinary tuple contains prefix fields, a length field, and a payload. The prefix fields typically provide meta information of the message, such as message type, protocol version, and session ID. The value of the length field determines the size of the subsequent payload. The contents of the payload could be sequential bytes or a subtuple that should be parsed in the lower layer. The suffix fields are optional with respect to the protocol format, which may store the message authentication code of the preceding byte sequences. Based on the hierarchical length field model, we develop an algorithm for the search and selection of length field pattern candidates. During the length field inference process, the algorithm is executed sequentially from the top layer to the bottom layer because the inference operation in the lower layer depends on the determination of the payload in the upper layer.

Algorithm 1 shows the details of the length field inference algorithm, which contains a procedure LengthFieldInfer and two functions CandidateSearch and CandidateSelect. We first define the Message and LengthFieldPattern of the inference algorithm. A Message object contains message bytes mb, the message parts to be inferred mps, the length field pattern candidates c, and the selected length field pattern p. A LengthFieldPattern object contains the field offset ofs, bit-width w, orientation ori, endian e, and the length field pattern in the lower layers sub (lines 3–4). The procedure LengthFieldInfer first initializes the message dataset MD to the initial message clusters MCS for the management of the intermediate clustering results. Since the initial Messages have no length field pattern, we consider these messages sharing the same empty length field pattern (lines 47–51). In the process of the hierarchical length field inference, each Message cluster should only contain all messages of the current length field pattern. Then, for all message parts in the message cluster mc, we call the function CandidateSearch to search the length field pattern candidates fitting the hierarchical length field model in the current layer (lines 52–56). The function CandidateSelect intends to pick out the correct length field from numerous candidates based on the relevance between the length field pattern candidates and the messages in a cluster, so we only call it after performing the function CandidateSearch for all Messages in the cluster (line 57). Since the CandidateSelect may produce new length field patterns and the new message parts in the next layer, we should split the current cluster into subclusters to maintain the format purity of each cluster and then iteratively execute the process above until no more new message parts are generated (line 58).

The function CandidateSearch makes use of the characteristics of the length field to perform the search task in an effective strategy to significantly reduce the search space of the length field. Specifically, we clarify the algorithm parameters in advance, including the bit-width of the length field BW, the length field orientation O, the endianness E, the size of the suffix field S, and the maximum field number NF. We first limit the maximum number of length fields in each layer to NF (lines 7–9). Then, we slide the offset ofs on the message part to determine the length field position and traverse the values of the w, ori, e in the Cartesian product $BW \times O \times E$ (line 10–11). Based on the above parameters, we can extract the potential length field entity and parse the value of the selected field entity (lines 12–13). If the value of the selected field exactly matches the size of the left part (with or without a suffix field), we consider this field a candidate (lines 16–20). When the value of the field is smaller than the size of the left message part, we iteratively call the function CandidateSearch to analyze the extra message part and splice the current message field with the recursive results (lines 21–25). The output of the CandidateSearch function is a set of length field pattern candidates that can be applied to the input message part, so most of them are potentially illegal to other message parts and must be filtered out.

The function CandidateSelect in the algorithm processes the numerous length field pattern candidates and selects the correct one from them. The basic principle for selecting the length field pattern is that the correct one is more robust to the messages of the cluster. For the input Message cluster mc, we first extract all length field candidates and select the candidate $lfps$ that exists in most messages (line 32–33). Then, we count the messages containing $lfps$ in the cluster mc and calculate the proportion of the covered messages to the cluster mc (line 34–35). The number of messages in the cluster num should be greater than or equal to the preset minimum number of the message in the cluster NC and the proportion of the covered messages $prop$ should be greater than the preset selection threshold ST (line 36). We may find several different length field pattern candidates meet the selection but cover the same messages because their parsed values of the length fields are all the same, so we simply choose the one with the largest bit-width to maximize robustness. After selecting the length field pattern, we apply the pattern to all covered messages and exclude these messages from subsequent selections (lines 37–39). We also extract new message parts in the next layer since we have obtained the length field pattern in the current layer (line 40). Whether a length field pattern candidate passes the selection or not, we should remove the candidate from all messages before the next selection. Otherwise, these selected candidates would be repeatedly counted and influence the subsequent selections (line 43). Through the collaboration of the procedure LengthFieldInfer and the functions CandidateSearch and CandidateSelect, we can infer the length fields from collected traffic trace without knowing the protocol specification in advance.

3.3   Message Type Field Inference

Similar to the length field, the message type field is also a semantic field that plays a key role in the protocol format inference. The message type field usually indicates the specific format type of the message when exists, so that the communicating parties can reach a consensus on the structure of the relevant payload. Therefore, when we find the message type field in the message, we can further cluster the messages with the same format type. Figure 3 shows an example of the message type fields in the TLS messages. According to the RFC document [11], the byte field at offset 05 is a top-level message type field, which is marked with a light blue background in the figure. In this example, the message type field uses continuous values from 14 to 22 (Hex) while other fields take fixed or random values. This phenomenon only appears in the cluster where the observed messages belong to the same superior format and the message type field is used to distinguish different inferior formats. The message type field typically locates around the layer break position of the message hierarchy since the communicating parties need to know the specification before parsing the transmitted payload. Fortunately, we can exploit the hierarchy of the length field pattern to perform preliminary clustering and make the messages in each cluster shares the

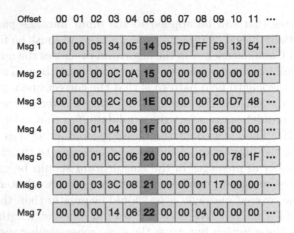

Fig. 3. An example of the message type field in the TLS messages

same length field pattern. We limit the search range of message type field to the unknown fields in each layer to avoid the meaningless searches.

Based on two reasonable intuitions that the message type field typically takes a few distinct values to distinguish different formats and the length of messages in one format tends to be similar, we use the following four indicators to infer the message type field.

– The number of used values num, which measures the unique number of the value appeared in the collected dataset. We estimate this indicator of each unknown field by simply counting the number of unique values.
– The mean of message size variances var, which measures the average message size uniformity of the assumed message type field.

We propose a message type field inference algorithm using these two indicators, as shown in Algorithm 2. The main idea of the algorithm is to limit the values of the first indicator to a predefined value range to make preliminary filtering, and then use the second indicator to select the message type field. Specifically, the procedure MessageTypeInfer first traverses the byte offset ofs from the prefix field start position $Start$ to the prefix field end position End to locate the field of the messages in the given message cluster MC (line 7). For the field at the byte offset ofs, we calculate the number of used values num and the mean of message size variances var for all the messages in the cluster (lines 8–18). We sift out the fields that meet the requirement of the indicator num and consider the one with smallest var as the message type field (lines 19–22). Once the message type field in the message cluster is determined, we can perform a more elaborate message clustering so that each cluster contains all messages of one format.

Algorithm 2. Message Type Field Inference Algorithm

1: **Input:** Message Cluster MC; Prefix Field Start Position $Start$; Prefix Field End Position End; Valid Number of Message Types NT.
2: **Output:** Message Type Field $Msgtype$.
3: **MessageType MT**(ofs, num, var) : ofs = offset, num = the number of unique values, var = the average of message length variances.
4: **procedure** MESSAGETYPEINFER(MC)
5:     $Msgtype$ = None
6:     $minvar = +\infty$
7:     **for** ofs from $Start$ to End **do**
8:         $fieldbytes \leftarrow [m[ofs]$ for m in $MC]$
9:         $byteset \leftarrow$ SET$(fieldbytes)$
10:        $num \leftarrow$ NUM$(byteset)$
11:        $varsum = 0$
12:        **for** $byte$ in $byteset$ **do**
13:            $lens \leftarrow$ the sizes of messages m, where $m[ofs] = byte$
14:            $varsum \mathrel{+}=$ VAR$(lens)$
15:        **end for**
16:        $var \leftarrow varsum/num$
17:        **if** $num \in NT$ **and** $var < minvar$ **then**
18:            $Msgtype \leftarrow$ MT(ofs, num, var)
19:            $minvar \leftarrow var$
20:        **end if**
21:    **end for**
22:    **return** $Msgtype$
23: **end procedure**

3.4 Protocol Format Inference

After obtaining the length fields and the message type fields, we can exploit these fields to cluster messages with the same format and then perform protocol format inference for other fields. Similar to Discoverer, we specify the inferred format as a sequence of token specifications and use token properties to describe field specifications [9]. The token properties describe two field characteristics: binary/text and constant/variable. The first token property focus on the encoding of the field. Although we focus on the format inference of the binary messages, we may still encounter text segments in the messages. A binary field can be parsed through hexadecimal conversion, while a text field is parsed through character decoding. The second token property concerns whether a field is fixed to a certain value in different messages. The inference of this property depends largely on the adequacy of the datasets. When the data source is deficient, we can only observe a few values of a variable token and may misjudge it as a constant field. To avoid this kind of mistake, we need to collect network traces from different sources to enrich the value diversity of variable tokens and improve the accuracy of the token property.

The detailed process of the token attribute inference is as follows. For each unknown token in the message, we collect all the values that have appeared in the cluster's messages. When the token takes the same value across in all messages, we consider it as a binary constant token. When multiple consecutive tokens take variable values within the ASCII coding range in all messages, we consider them as text variable tokens. All tokens outside the above cases will be considered as binary variable tokens.

4   Evaluation

In this section, we first introduce the dataset used in the experiments. Then, we define the evaluation metrics in detail and set the tunable parameters of two inference algorithms. Finally, we present the experimental results and the corresponding analyses.

4.1   Datasets

We evaluate the Inspector on three widely used binary protocols, namely TLS, SSH, and MQTT. The Transport Layer Security (TLS) [11] is a binary application protocol that primarily aims to provide privacy and data integrity between network communicating parties. The Secure Shell (SSH) [25] is a binary application protocol that provides a secure communicating channel for the client and server. Therefore, except for the channel establishment traffic and control fields in the messages, the real payload transmitted by these two protocols is strictly encrypted. The Message Queuing Telemetry Transport (MQTT) is a lightweight unencrypted message protocol that designed for the IoT communications. Since we aim to reverse the format specification of these protocols, we can focus on the message format structure and ignore the content inside the encrypted payload. The ground truth specification details of these three protocols are available in the online RFC documents [12] and OASIS standards [17].

We collect a bunch of protocol traffic traces from a campus network gateway. The summary of the dataset is shown in Table 1. During the trace collection, we try to collect the traffic from different client and server pairs to enrich the diversity of the protocol formats. Some particular messages of the binary protocols may be text messages. For example, the version announcement message of the SSH is a pure text message. Since many previous works have studied format reverse engineering for the text-oriented protocols, we filter out the text messages for individual processes. A simple and effective method to separate the text messages from binary messages is to check the existence of consecutive ASCII printable characters. When a message begins with these kind of characters, we treat this message as a text message.

4.2 Evaluation Metrics

The evaluation metrics are designed to measure the effectiveness of the message clustering and the consistency between the inferred formats and the ground truths. The previous works have proposed many evaluation metrics to evaluate clustering accuracy. Some metrics have been frequently used, such as correctness [3,9,27] and conciseness [3,9], while others are specific to their experiments [23,24]. In our evaluation experiments, we combine the correctness and conciseness to a comprehensive metric named format precision and adopt the message coverage in Discoverer [9] as the message recall.

Table 1. Summary of datasets used in the evaluation

Protocol	Source	Size(B)	# Packets	# Messages	# Formats
TLS	Campus	27,601K	324250	44316	12
SSH	Campus	5,177K	23900	12988	10
MQTT	Campus	3,791K	33164	33879	21

Table 2. Summary of tunable parameters

#	Parameter	Var	Value
1–1	Bit-Widths (Bits)	BW	8/16/32
1–2	Orientation	O	Before/Within
1–3	Endianness	E	Little/big
1–4	Size of the suffix field (Bytes)	S	0/16/32/64
1–5	Maximum field num	NF	5
1–6	Selection threshold	ST	30%, 10%
1–7	Min. Num. of messages in the cluster	NC	10
2–1	Valid Number of message types	NT	$[2, 30]$

- Format Precision: The correctness and conciseness respectively focus on the number of true formats contained in a cluster and the number of clusters required to describe a true format. Since the Inspector can cluster messages in high accuracy, we only consider completely pure and precise clusterings to be the correct inference cases. So, we measure the proportion of inferred message clusters that only contains all messages of one true format.
- Message Recall: We measure the proportion of messages covered by the inferred formats in the dataset as the message recall. Since the format precision can reflect the ability to cover the true formats, we use this metric to evaluate the ability of the inferred protocol formats to recall messages from the dataset.

In order to automatically obtain the ground truth format of the message, we leverage Wireshark [13], a free and open-source network traffic analyzer, to get rid of the tedious manual format annotation works. For the correct message clusters, we further verify the correctness of the semantic inference results in the format by comparing it with the field sequence outputted from Wireshark [13], including the length fields, message type fields, constant fields, and dynamic fields.

4.3 Tunable Parameters

As shown in Table 2, the Inspector has a few tunable parameters. (1–1) The bit-width BW determines the size of the length field. We choose three most commonly used bit-widths for the Inspector. (1–2) We consider length fields that are contiguous before or at the beginning of the relevant payload as valid length field positions. Therefore, the orientation O can significantly restrict the search space of the length field in the message. (1–3) For the endianness E, we consider both little-endian and big-endian in the first level length field search. When searching for length fields at other levels, we directly use the previous orientation and endianness to reduce the search space. (1–4) We determine the size of the suffix field S according to the general conventions of the protocol format. (1–5) The maximum field num NF limits the number of the length fields in the same layer to avoid meaningless searches. We set NF to 5 based on the experimental experience. (1–6) The selection threshold ST controls the minimum proportion of applicative messages for the selected length field pattern. We start the selection from the most general length field pattern, so we assign the first selection threshold to 30% and the subsequent selection thresholds to 5% to cope with the selection order. (1–7) The minimum number of messages in the cluster NC guarantees the minimum number of the target messages of each format. We set NC to 10 to exclude the cluster containing too few messages for the subsequent inference. (2–1) The number of message type NT should be within a proper interval. We set the interval to $[2, 50]$ based on the experimental experience.

4.4 Experimental Results

Table 3. Summary of message clustering results

Protocol	# Inferred formats	Precision	# Covered messages	Coverage
TLS	12/12	100.0%	44316/44316	100.0%
SSH	9/10	90.0%	12972/12988	99.9%
MQTT	21/21	100.0%	31413/33879	92.7%

The presentation of experimental results includes the message clustering and format inference. Except for one format of the protocol SSH, all message clustering results correspond exactly to the ground truths. Table 3 shows the summary of message clustering results. For the protocol TLS, 12 message clusters correspond to all 12 true formats in 100% accuracy and cover 100% messages of the dataset. For the protocol SSH, the Inspector finds 9 message clusters that correspond to 9 true formats in 90% accuracy and cover 99.6% messages of the dataset. For the protocol MQTT, the Inspector finds 21 message clusters that correspond to 21 true formats in 100% accuracy and cover 92.7% messages of the dataset. We manually check the excluded SSH messages and find that these messages are fully encrypted and have no segment for field inferences. These experimental results show that the Inspector can perform very well on the message clustering and can automatically filter out unformatted messages from the clustering.

Table 4 shows the detail of message clustering results. The length field pattern (ofs, w, ori, e, sub) has been introduced in Sect. 3.2. For message type (ofs, v, sub) in the table, ofs means the offset from the current tuple beginning, n means the value of the message type, sub means the message type in the lower layer. All the messages in the cluster have the same message type and their message type is consistent with the official documents. We also find that most of the inferred message types are closely related to the establishment of encrypted communication channels.

The Inspector handles the hierarchical formats as expected and successfully finds the message type in the lower level, such as the 5th–12th TLS formats in Table 4. Besides, the Inspector can still infer the correct length field pattern when the messages contain a suffix field. Notice that the Inspector puts 753 text messages into the 1st SSH format, which is the version announce message format of SSH, such as, 'SSH-2.0-OpenSSH_7.7'. For the text messages in the collected dataset, we can directly identify and filter them before conducting our length field inference. We also noticed that for all the inferred protocols, the orientation is "after" and the endianness is "big". We guess that this orientation is suitable for the design of a hierarchical protocol format, and this endian is convenient for the direct reading and format uniform of the network traffic.

The quality of format inference results greatly depend on the accuracy of the message clustering. Based on our message clustering results, the format inference results are very close to the ground truths. To better understand the results of message format inferences, we present a format inference example of TLS, SSH, and MQTT, as shown in Table 5, Table 7, and Table 6, respectively. For L(x, y) in the table, L means length field byte, x means the minimum legal value, y means the maximum legal value. For M(x), M means message type field byte, x means the message type value. For C(x), C means constant byte, x means the constant value. For V(x, y) and nV(x, y), V means variable byte, nV means an indefinite number of consecutive variable bytes, x and y have the same meaning as in L(x, y). For nT(x, y), nV means an indefinite number of consecutive text bytes, x and y have the same meaning as in L(x, y). The example TLS message is "Client

Table 4. Message clustering results in detail

Protocol	#	Inferred length field pattern	Inferred message type	True message type	# Message
TLS	1	(3, 2, 'after', 'big')	(0, 20)	Change Cipher Spec	5719
TLS	2	(3, 2, 'after', 'big')	(0, 21)	Alert	2487
TLS	3	(3, 2, 'after', 'big')	(0, 22)	Handshake	5714
TLS	4	(3, 2, 'after', 'big')	(0, 23)	Application Data	16459
TLS	5	(3, 2, 'after', 'big', [(2, 2, 'after', 'big')])	(0, 22, [(0, 1)])	Client Hello	2965
TLS	6	(3, 2, 'after', 'big', [(2, 2, 'after', 'big')])	(0, 22, [(0, 2)])	Server Hello	2877
TLS	7	(3, 2, 'after', 'big', [(2, 2, 'after', 'big')])	(0, 22, [(0, 4)])	New Session Ticket	1094
TLS	8	(3, 2, 'after', 'big', [(2, 2, 'after', 'big')])	(0, 22, [(0, 11)])	Certificate	1639
TLS	9	(3, 2, 'after', 'big', [(2, 2, 'after', 'big')])	(0, 22, [(0, 12)])	Server Key Exchange	1689
TLS	10	(3, 2, 'after', 'big', [(2, 2, 'after', 'big')])	(0, 22, [(0, 14)])	Server Hello Done	1691
TLS	11	(3, 2, 'after', 'big', [(2, 2, 'after', 'big')])	(0, 22, [(0, 16)])	Client Key Exchange	1722
TLS	12	(3, 2, 'after', 'big', [(2, 2, 'after', 'big')])	(0, 22, [(0, 22)])	Certificate Status	264
SSH	1	None (Text Message)	None	Version Announce	3168
SSH	2	(0, 4, 'after', 'big')	(5, 20)	Key Exchange Init	3154
SSH	3	(0, 4, 'after', 'big')	(5, 21)	New Keys	2295
SSH	4	(0, 4, 'after', 'big')	(5, 30)	DH[a] Key Exchange Init	1575
SSH	5	(0, 4, 'after', 'big')	(5, 31)	DH Key Exchange Reply	1518
SSH	6	(0, 4, 'after', 'big')	(5, 32)	DH GEX[b] Init	50
SSH	7	(0, 4, 'after', 'big')	(5, 33)	DH GEX Reply	29
SSH	8	(0, 4, 'after', 'big')	(5, 34)	DH GEX Request	49
SSH	9	(0, 4, 'after', 'big'), ('suffix', 32)	None	Encrypted Payload	1134
MQTT	1	(1, 1, 'after', 'big', 0)	(0, 16)	Connect	244
MQTT	2	(1, 1, 'after', 'big', 0)	(0, 32)	ConnAck	310
MQTT	3-9	(1, 1, 'after', 'big', 0)	(0, 48 / 49 / 50 / 51 / 52 / 53 / 56)	Publish (DUP[c] / QoS[d])	10402
MQTT	10	(1, 1, 'after', 'big', 0)	(0, 64)	PubAck	2748
MQTT	11	(1, 1, 'after', 'big', 0)	(0, 80)	PubRec	364
MQTT	12	(1, 1, 'after', 'big', 0)	(0, 98)	PubRel	334
MQTT	13	(1, 1, 'after', 'big', 0)	(0, 112)	PubComp	371
MQTT	14	(1, 1, 'after', 'big', 0)	(0, 130)	Subscribe	305
MQTT	15	(1, 1, 'after', 'big', 0)	(0, 144)	SubAck	355
MQTT	16	(1, 1, 'after', 'big', 0)	(0, 162)	Unsubscribe	6
MQTT	17	(1, 1, 'after', 'big', 0)	(0, 176)	UnsubAck	2
MQTT	18	(1, 1, 'after', 'big', 0)	(0, 192)	PingReq	8178
MQTT	19	(1, 1, 'after', 'big', 0)	(0, 208)	PingResp	7715
MQTT	20	(1, 1, 'after', 'big', 0)	(0, 224)	Disconnect	65
MQTT	21	(1, 1, 'after', 'big', 0)	(0, 240)	Auth	14

[a] Diffie-Hellman   [b] Group Exchange SHA-1   [c] Duplicate Delivery   [d] Quality of Service

Table 5. The format of TLS client key exchange

Token	Ground truth	Token	Ground truth	Token	Ground truth	Token	Ground truth
M (22)	Message type	L (0, 255)	Message length	C (0)	Pubkey length	nV (0, 255)	Pubkey
C (3)	Version	L (0, 255)	Message length	L (0, 255)	Pubkey length		
C (3)	Version	M (16)	Handshake type	L (0, 255)	Pubkey length		

Table 6. The format of MQTT subscribe request

Token	Ground truth	Token	Ground truth	Token	Ground truth	Token	Ground truth
C (130)	Header flags	V (0, 255)	Message identifier	C (0)	Property length	nT (0, 127)	Property
L (0, 255)	Message length	V (0, 255)	Message identifier	V (0, 255)	Property length		

Table 7. The format of SSH DH GEX request

Token	Ground truth	Token	Ground truth	Token	Ground truth	Token	Ground truth
L (0, 255)	Message length	C (0)	DH GEX Min	V (0, 255)	DH GEX num of bits	V (0, 255)	Padding
L (0, 255)	Message length	C (0)	DH GEX Min	C (0)	DH GEX num of bits	V (0, 255)	Padding
L (0, 255)	Message length	C (4)	DH GEX Min	C (0)	DH GEX max	V (0, 255)	Padding
L (0, 255)	Message length	C (0)	DH GEX Min	C (0)	DH GEX max	V (0, 255)	Padding
C (6)	Padding length	C (0)	DH GEX Num of Bits	C (32)	DH GEX max	V (0, 255)	Padding
M (34)	Message code	C (0)	DH GEX Num of Bits	C (0)	DH GEX max	V (0, 255)	Padding

Key Exchange", which sends a public key to the server for key negotiation. The example SSH message is "Diffie-Hellman (DH) Group Exchange SHA-1 (GEX) Request", which specifies the parameters for the subsequent Diffie-Hellman key exchange. The example MQTT message is "Subscribe Request", which is sent from the Client to the Server to register one or more topic subscriptions. We can see that the inferred formats can properly reflect the token properties and semantic fields of the true formats. Notice that the boundary of some fields is imprecise, such as the tokens of the "Pubkey Length" in TLS are "C(0) L(0, 255) L(0, 255)" and the tokens of the "DH GEX Num of Bits" in SSH are "C(0) C(0) V(0, 255) C(0)". This does not mean that our inference results are not applicable, but that we restrict the value range of the field based on the appeared values of the dynamic field in the network. Since the integration of the length field pattern and message type values are specific to the message format, the inferred formats can be clearly distinguished from each other, which guarantees the reliability of the inferred formats when applied to network security applications.

5   Conclusions

In this paper, we propose a protocol reverse engineering approach Inspector, which makes use of the semantic fields for the message clustering and format inference. In order to find the semantic fields in the messages, we propose a hierarchical length field model to search length field pattern candidates and develop two algorithms to infer the length fields and message type fields. The experimental results demonstrate the accurate message clustering and excellent format inference results. However, the Inspector has the limitation that the format inference results are not detailed enough. In future work, we plan to study a more elaborate approach to enrich the properties and semantic fields of the format inference results.

Acknowledgment. This work is supported by the Strategic Priority Research Program of the Chinese Academy of Sciences (No.XDC02030100), the National Key Research and Development Program of China (Grant No.2018YFB0804704), and the National Natural Science Foundation of China (Grant No.U1736218).

References

1. Bermudez, I., Tongaonkar, A., Iliofotou, M., Mellia, M., Munafo, M.M.: Automatic protocol field inference for deeper protocol understanding. In: 2015 IFIP Networking Conference (IFIP Networking), pp. 1–9. IEEE (2015)
2. Blei, D.M., Ng, A.Y., Jordan, M.I.: Latent dirichlet allocation. J. Mach. Learn. Res. **3**, 993–1022 (2003)
3. Bossert, G., Guihéry, F., Hiet, G.: Towards automated protocol reverse engineering using semantic information. In: Proceedings of the 9th ACM Symposium on Information, Computer and Communications Security, pp. 51–62 (2014)
4. Caballero, J., Poosankam, P., Kreibich, C., Song, D.: Dispatcher: enabling active botnet infiltration using automatic protocol reverse-engineering. In: Proceedings of the 16th ACM Conference on Computer and Communications Security, pp. 621–634. ACM (2009)
5. Caballero, J., Yin, H., Liang, Z., Song, D.: Polyglot: automatic extraction of protocol message format using dynamic binary analysis. In: Proceedings of the 14th ACM Conference on Computer and Communications Security, pp. 317–329. ACM (2007)
6. Chen, Y., Zang, T., Zhang, Y., Zhouz, Y., Wang, Y.: Rethinking encrypted traffic classification: a multi-attribute associated fingerprint approach. In: 2019 IEEE 27th International Conference on Network Protocols (ICNP), pp. 1–11. IEEE (2019)
7. Cohen, P., Adams, N.: An algorithm for segmenting categorical time series into meaningful episodes. In: Hoffmann, F., Hand, D.J., Adams, N., Fisher, D., Guimaraes, G. (eds.) IDA 2001. LNCS, vol. 2189, pp. 198–207. Springer, Heidelberg (2001). https://doi.org/10.1007/3-540-44816-0_20
8. Comparetti, P.M., Wondracek, G., Kruegel, C., Kirda, E.: Prospex: protocol specification extraction. In: 2009 30th IEEE Symposium on Security and Privacy, pp. 110–125. IEEE (2009)
9. Cui, W., Kannan, J., Wang, H.J.: Discoverer: automatic protocol reverse engineering from network traces. In: USENIX Security Symposium, pp. 1–14 (2007)
10. Cui, W., Peinado, M., Chen, K., Wang, H.J., Irun-Briz, L.: Tupni: automatic reverse engineering of input formats. In: Proceedings of the 15th ACM Conference on Computer and Communications Security, pp. 391–402 (2008)
11. Dierks, T., Rescorla, E.: The transport layer security (TLS) protocol version 1.2 (2008)
12. Force, T.I.E.T.: Rfc index. https://tools.ietf.org/rfc/index
13. Foundation, W.: Wireshark - go deep. (2020). https://www.wireshark.org/
14. Gu, G., Perdisci, R., Zhang, J., Lee, W.: Botminer: clustering analysis of network traffic for protocol-and structure-independent botnet detection. In: Proceedings of the 17th USENIX Security Symposium, pp. 139–154. USENIX (2008)
15. Holzmann, G.J., Lieberman, W.S.: Design and Validation of Computer Protocols, vol. 512. Prentice hall Englewood Cliffs, Englewood Cliffs (1991)
16. Juan, C., Noah, M.J., Stephen, M., Dawn, S.: Binary code extraction and interface identification for security applications. In: Proceedings of the Network and Distributed System Security Symposium, NDSS 2010. The Internet Society (2010)
17. OASIS: Standards archive - oasis open. https://www.oasis-open.org/standards/
18. Postel, J., et al.: User datagram protocol (1980)
19. Postel, J., et al.: Transmission control protocol (1981)
20. Slonim, N., Tishby, N.: Agglomerative information bottleneck. In: Advances in Neural Information Processing Systems, pp. 617–623 (2000)

21. Sommer, R., Paxson, V.: Outside the closed world: On using machine learning for network intrusion detection. In: 2010 IEEE Symposium on Security and Privacy, pp. 305–316. IEEE (2010)
22. Standard, O.: Mqtt version 5.0. Retrieved 22 Jun 2020 (2019)
23. Wang, Y., et al.: A semantics aware approach to automated reverse engineering unknown protocols. In: 2012 IEEE 20th International Conference on Network Protocols (ICNP), pp. 1–10. IEEE (2012)
24. Wang, Y., Zhang, Z., Yao, D.D., Qu, B., Guo, L.: Inferring protocol state machine from network traces: a probabilistic approach. In: Lopez, J., Tsudik, G. (eds.) ACNS 2011. LNCS, vol. 6715, pp. 1–18. Springer, Heidelberg (2011). https://doi.org/10.1007/978-3-642-21554-4_1
25. Ylonen, T., Lonvick, C., et al.: The secure shell (SSH) protocol architecture (2006)
26. Zhang, J., Moore, A.: Traffic trace artifacts due to monitoring via port mirroring. In: 2007 Workshop on End-to-End Monitoring Techniques and Services, pp. 1–8. IEEE (2007)
27. Zhang, Z., Zhang, Z., Lee, P.P., Liu, Y., Xie, G.: Proword: an unsupervised approach to protocol feature word extraction. In: 2014 IEEE International Conference on Computer Communications (Infocom), pp. 1393–1401. IEEE (2014)

MFF-AMD: Multivariate Feature Fusion for Android Malware Detection

Guangquan Xu[1,2], Meiqi Feng[1], Litao Jiao[2], Jian Liu[1(✉)], Hong-Ning Dai[3],
Ding Wang[4], Emmanouil Panaousis[5], and Xi Zheng[6]

[1] College of Intelligence and Computing, Tianjin University, Tianjin, China
jianliu@tju.edu.cn
[2] Big Data School, Qingdao Huanghai University, Qingdao, China
[3] Faculty of Information Technology, Macau University of Science and Technology,
Taipa, Macau SAR, China
[4] School of EECS, Peking University, Beijing, China
[5] University of Greenwich, London, UK
[6] Department of Computing, Macquarie University, Sydney, Australia

Abstract. Researchers have turned their focus on leveraging either dynamic or static features extracted from applications to train AI algorithms to identify malware precisely. However, the adversarial techniques have been continuously evolving and meanwhile, the code structure and application function have been designed in complex format. This makes Android malware detection more challenging than before. Most of the existing detection methods may not work well on recent malware samples. In this paper, we aim at enhancing the detection accuracy of Android malware through machine learning techniques via the design and development of our system called MFF-AMD. In our system, we first extract various features through static and dynamic analysis and obtain a multiscale comprehensive feature set. Then, to achieve high classification performance, we introduce the Relief algorithm to fuse the features, and design four weight distribution algorithms to fuse base classifiers. Finally, we set the threshold to guide MFF-AMD to perform static or hybrid analysis on the malware samples. Our experiments performed on more than 25,000 applications from the recent five-year dataset demonstrate that MFF-AMD can effectively detect malware with high accuracy.

Keywords: Malware detection · Hybrid analysis · Weight distribution · Multivariate feature fusion

1 Introduction

The proliferation of Android applications has been benefited from the rapid development of portable electronic devices, such as smartphones, tablets, and wearable smartwatches. As a byproduct, the number of malware has been constantly increasing over these years targeting at Android-based smartphones.

H. Gao and X. Wang (Eds.): CollaborateCom 2021, LNICST 406, pp. 368–385, 2021.
https://doi.org/10.1007/978-3-030-92635-9_22

For Android users, there are several indications of this trend of ever-increasing threats of malware [1]. According to the Symantec research report [2], 23,795 Android malware on average were detected daily in 2017, an increment of 17.2% compared to that of 2016. While 360 Beaconlab [3] reported that approximately 12,000 malware samples per day on average were intercepted in 2018. Compared with that of 2017 or even three years ago, the malware detection rate has been reduced substantially. As AV-TEST [4] pointed out, this reduction is due to that malware is being designed and injected in a more complicated way, implying that attackers have focused on "better" quality of malware rather than the quantity. The corresponding malware detection methods should also be improved to adapt to the upgrade and development of malware.

In this work, we propose a model of MFF-AMD to address the above issues. Our model is built via a combination of dynamic and static techniques. Through the analysis of 25,000 samples obtained in the past five years, we extract features that can identify malicious applications from benign ones. More multiscale features can be used to describe an application better so as to detect malware more accurately. Meanwhile, to achieve a balance between efficiency and accuracy, our model automatically determines the analysis method. Specifically, we extract some basic static features including permissions, sensitive API calls, and other related features inferred from the basic features.

We also check the intent to see whether it is used to deliver sensitive messages. To extract more comprehensive dynamic features, we investigate the work related to detecting malware based on dynamic behaviors. We further implement a UI component testing scheme based on Android activity to trigger more malicious behaviors. In short, we extract the malware from more various dimensions, which can improve the malware detection consequently.

When training our model, we propose four weight distribution algorithms for base classifier fusion. We experimentally conclude that the overall performance of MFF-AMD is much better than those of a single-base classifier and other weight distribution algorithms. We also test the Android app samples from 2015 to 2019, to prove the robustness of our model. Experiments show that our model achieves good detection performance for Android malware samples from different years.

In summary, we highlight the major contributions of this paper as follows:

1. In order to solve the problem that existing methods cannot fully extract Android application features, we propose a model named MFF-AMD, which can achieve automated detection of Android malware, and our experimental results show that MFF-AMD can provide 96% accuracy on average.

2. MFF-AMD can automatically control the detection process during sample testing by training models that can adapt to static features or both static and dynamic features. Therefore, our model achieves higher accuracy with lower overhead.

3. In the process of training the model, in order to maximize the overall accuracy of MFF-AMD, we design a lightweight weight distribution algorithm to fuse the base classifiers. Experimental results show that our method can improve the overall accuracy by 3.53% on average.

The rest of the paper is organized as follows: In Sect. 2, we review the related work of Android malware detection based on machine learning. We introduce our feature extraction process in Sect. 3. We then describe our model implementation in Sect. 4. The experiment results and analysis are presented in Sect. 5. Finally, we conclude our paper and outlook future directions in Sect. 6.

2   Related Work

There have been many approaches to detect malware based on machine learning combined with Android static analysis. Wu et al. [5] used the Kmeans and K nearest neighbor (K-NN) algorithms, which combined with static features including permissions, intents, and API calls. Their main contribution is to provide a static analysis paradigm for detecting Android malware and to develop a system called DroidMat. Arp et al. [6] proposed a lightweight detection method based on Support Vector Machine (SVM) running on the mobile terminal.

Unlike static detection, some methods use dynamic features combined with machine learning, for instance, AntiMalDroid [7] is a framework of detecting malware based on the analysis of dynamic behavior via the SVM algorithm, in which the features are extracted from the log behavior sequence. Saracino et al. [8] proposed MADAM, which can combine several classes of features, from distinct Android levels, and applied both anomaly-based and signature-based mechanisms. Afonso et al. [9] proposed a system to dynamically identify whether an Android application is malicious or not, based on the features extracted from Android API calls and system call traces.

The hybrid analysis includes both dynamic analysis and static analysis. The AndroPytool framework [10] is a new hybrid analysis-based work. It proposes a malware detection method based on static and dynamic feature fusion through the combination of ensemble classifiers. They mainly develop their tools based on Flowdroid [11]. Their experiment result shows that AndroPytool can achieve up to 89.7% detection accuracy. TrustDroid [12] is a hybrid approach that can both operate on the phone and on a server. It takes the Android byte code and converts it into a textual description using Jasmin syntax. MARVIN [13] uses machine learning based on hybrid static and dynamic features (SVM and L2 regularized linear classifiers). MARVIN evaluates the risks associated with unknown Android applications in a malicious scoring form from 0 to 10.

Some other works [14–19] combined static analysis and dynamic analysis. We find that many studies calculate dynamic behavior based on logs and some system parameters, which will introduce computational overhead. Different from those works, we utilize a hook framework to directly monitor the triggered dynamic behavior, which has lower overhead. Meanwhile, we parse Android UI components to trigger more behaviors to improve detection as much as possible.

Moreover, we combine more comprehensive dynamic features with static ones used in malware detection to obtain higher accuracy.

3 Multivariate Feature Extraction

The main purpose of extracting features is to distinguish malware from benign functions. According to the related studies, the behavior of the Android application mainly relates to the static code and dynamic behavior; thus, we intend to extract contributive features between malware and benign applications based on these two aspects. In order to improve our model's robustness across diverse malicious samples, we use Android malicious families or Android applications across four years. At the same time, we consider the upgrade of the Android SDK version and the popularity of the 4G network communication with the Android application.

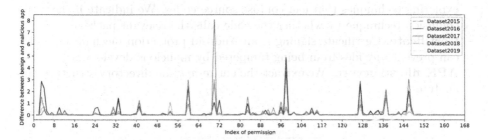

Fig. 1. The different distribution of the same permission on each data set.

3.1 Static Feature Extraction

Static features can be extracted without the need of running an application, which relies on concrete static analysis. We can perform automated extraction of some static features by using the python API provided by *Androguard* to analyze *apk* files. *Androguard* is an open source and supports extensions. It can implement automated reverse *apk* and easily extract static features of applications. We note that many related studies in the feature extraction part only give attention to whether exists a certain feature in the AndroidManifest.xml file or just describe the feature vector in binary code. The possible drawback is that the features of different codes of 1 are the same contribution for malware detection. Therefore, one improvement we made is to increase the frequency of utilization of features based on the original method. Features with large frequency may represent more important features. We also portray as many Android app details as possible.

- **Permissions.** The Android permission mechanism specifies operations that can be performed at different risk levels. With a more than 11,000 separate sample set in the benchmark, our main idea is to extract the permission feature expressed in benchmarks and verify all the samples in our experimental dataset to see if the permission feature is effective in detecting malware. Figure 1 shows the different distribution of the samples we extracted from 2015 to 2019. From Fig. 1 we can see that almost 168 permission features we extracted are identical in distribution on each data set, though there are still some differences. Our assumption is that the greater the difference in permissions, the stronger the ability to distinguish malware.

- **Intents.** Intent can launch activities, create services, and communicate with other Android apps. Figure 2 shows the code that carries sensitive data through an intent object.
- **API calls.** Android officially marks the API's risk level, and those sensitive APIs are often leveraged as a powerful feature in Android malware detection.
- **Components.** We calculate the number of Android components as a continuous feature, including *activity, service, broadcast receiver*, etc.
- **Code features.** Enhancing the robustness of malware detection requires exploring techniques that can confuse source codes. We indicate if there is a confusion technique by whether the code calls the relevant package.
- **Certificate.** Certificate signing is an Android protection mechanism, which can prevent *apk* files from being tampered by malicious developers.
- **APK file structure.** We extract the entire *apk* file directory structure and analyze it.

```
1 button1.setOnclickListenner(new setOnclickListenner){
2     @override
3     public void onClick(View v){
4         String data = 'sensitive message';
5         Intent intent = new Intent(MainActivity.this,TargetActivity.class);
6         intent.putExtra('extra_data',data);
7         startActivity(intent);
8     }
9 }
```

Fig. 2. Easy intent: sensitive data are transformed to TargetActivity by binding explicit intent object.

3.2 Dynamic Feature Extraction

The acquisition of dynamic features relies on the installation and running of the Android application to detect malicious behavior that cannot be detected by static analysis. Through the *Inspeckage* based on the *Xposed* framework, we hook the application to be tested and enable dynamic monitoring of the application's various behaviors. We analyze the log files generated during the dynamic analysis process to achieve batch extraction of dynamic features.

- **Networks**. We capture the network data and analyze the data packet to determine whether the HTTP request contains sensitive data related to the user.
- **File and SQLite operations**. We focus on the file or database operations of the application recorded in the log. Our main research objects are the path and data of the file, which may include sensitive information.
- **Command execution**. Malware can call system programs in sensitive directories to execute commands aiming to achieve camouflage and malicious operations. We count the number of executions of the system command, coded as a continuous feature.

3.3 Application Coverage

Android applications are based on event-driven, and the execution process can be simply summarized as driven by various input events, and the application completes a variety of different logical functions. The key to malware detection based on dynamic analysis is the application's execution coverage. Our model does not intent to insert probes into the application to calculate the code coverage during testing, because modifying code, repackaging, and other operations are not conducive to automated detection. So, in order to improve the effect of dynamic analysis, we design an event testing scheme based on the Android UI view. Our main idea is not to modify the *apk* but to input as many as UI events as possible to the activity component information of the Android application. In this way, potential malicious behavior can be triggered as much as possible. Figure 3 shows the dynamic detection scheme based on the UI view. Android will render the UI view and layout information currently on the screen. We construct the socket script through the activity information obtained by static analysis and try to communicate with the View Server to request the analysis result of the current screen information in real-time. Then, we parse the results on the client and obtain a simplified view tree that can uniquely locate all layout information through the parsing algorithm. Finally, through analyzing the components to formulate corresponding test events, we aim to trigger as many input events as possible. Therefore, we obtain corresponding dynamic features by monitoring the running behavior of the application.

3.4 Feature Selection

To reduce the overhead of training models and improve the detection accuracy, we first select and reduce the dimensionality of both static features and dynamic features. The feature selection is used to filter out the features that are more suitable for detecting malware, so as to have a positive effect on studying malicious samples. We use the Relief algorithm for feature selection, and select features with larger weights to participate in training. The advantage of choosing Relief is to separate the feature selection from the training process, and the algorithm performs well in terms of time complexity.

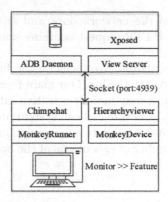

Fig. 3. Testing scheme based on UI view.

4   Implementation

4.1   Architecture

By analyzing the existing studies on Android malware detection, the method of using machine learning is generally divided into several stages: data filtering, feature extraction, model training, and model testing. We analyzed the steps of the malware detection process in detail and optimized it based on the traditional method. As shown in Fig. 4, during the training, our detection model extracts multi-dimensional static features and dynamic features from Android APK samples collected from different sources.

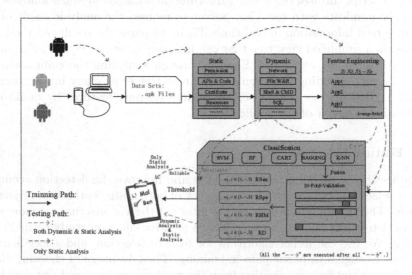

Fig. 4. MFF-AMD's architecture.

In the features pre-processing stage, we use the Relief algorithm to process the features, including the feature selection and feature dimensionality reduction. Based on the processed features, we use five base binary classification algorithms for supervised training, including SVM, RF, CART, BAGGING, and K-NN. Finally, we design four weight distribution algorithms including RSen, RSpe, RHM, and RD to allocate dual weights to the five base classifiers and obtain the combination of a pair of weights distribution strategies with the highest overall accuracy as the final algorithm to merge classifier.

For testing, given an arbitrary Android application, our model first performs static analysis on the application and then compares it with the threshold based on the score given by the classifier. If the score is higher than the threshold, the detection is ended, and the detection result is output. If the score is lower than the threshold, we need to perform hybrid analysis, and finally, compare the two classification results to determine whether the application is malicious. The advantage of this scheme is the ability to ensure accuracy and reduce the detection overhead.

4.2 Weight Distribution Algorithm

In order to improve the detection ability of our model, we adopt a strategy of emphasizing the base classifiers with weight pairs. We focus on how to assign a weight to different base classifiers. We know that different base classifiers have different classification performance. We denote True Positive Rate (TPR) as the detection accuracy of malicious samples by the fused classifier, which we named as sensitivity. We denote True Negative Rate (TNR) as the detection accuracy of benign samples by the fused classifier, which we named as specificity. Then, the overall accuracy of the weighted classifier can refer to the following equations, $P_{overall}$ can be computed as:

$$P_{overall} = \frac{N_M \cdot TPR + N_B \cdot TNR}{N_X} \tag{1}$$

where N_M and N_B represent the number of malicious and benign applications in the sample, respectively, and N_X represents the number of samples. In order to improve the overall generalization ability of the classifier to the sample, for a specific sample set, we can only improve TNR and TPR according to Eq. (1). By observing the performance of the classifier in a great number of malware detection processes, the overall classification accuracy of the classifier is directly proportional to its sensitivity and specificity, while both the sensitivity and the specificity of the most typical classifier are rarely equal in value. If one classifier has a higher specificity and sensitivity as well as a smaller difference between each other, it may have higher overall accuracy and robustness. Based on this, we propose four weight distribution algorithms, which can rank the classifier according to the specificity and sensitivity while assigning weights. These four algorithms are Ranking Algorithm Based on Sensitivity (RSen), Ranking Algorithm Based on Specificity (RSpe), Ranking Algorithm Based on Sensitivity and

Specificity Harmonic Mean (RHM), Ranking Algorithm Based on Sensitivity and Specificity difference (RD).

RSen Algorithm. In order to maximize the accuracy of the model's detection of malware, the simplest idea is to give the greatest weight to the classifier with the highest TPR. Thus, we design the first type of weight distribution algorithm. We use R_1 to represent RSen in this work.

Define e_k as the TPR of classifier k, $k \in \{1, \cdots, 5\}$. Make $E \leftarrow e_k$, we first rank each element of the set E in descending order and obtain $\bar{E}$. We assign weights of e_k according to the following equation:

$$\omega_i = 6 - i, i \in \{1, \cdots, 5\} \tag{2}$$

For instance, the weight of the classifier with the highest TPR is set to 5. The rules of subsequent algorithms are similar.

RSpe Algorithm. Similar to RSen algorithm, we set the maximum weight for the classifier with the highest sensitivity TNR, and the method is the same as the above, so we do not describe the detail. We use R_2 to represent RSpe in this work.

RHM Algorithm. We have known that the harmonic mean of two numbers tends to approach the one with a smaller value. Therefore, when the harmonic mean is large, and the two numbers are large. Classifier ranks are set to directly proportional to the sensitivity and specificity in this algorithm. We use R_3 to represent RHM in this work.

Define m_k as the harmonic mean of the sensitivity and specificity of classifier k, we have the following equation.

$$m_k = \frac{2 \cdot TPR_k \cdot TNR_k}{TPR_k + TNR_k}, \ k \in \{1, \cdots, 5\} \tag{3}$$

Make $M \leftarrow m_k$, and we let a set $M = \{m_k | k = \{1, \cdots, 5\}\}$ be the harmonic mean of classification accuracy of five classifiers. Descending the set M, and the symbol $\bar{M}$ represents the ranked set. We assign the weight for each classifier based on the element sequence in the set $\bar{M}$ according to Eq. (2) and we obtain the ω_i. Finally, the ω_i is going to be combined with classifier k and used to reclassify the category label of instance x.

RD Algorithm. In this algorithm, the weight of the classifier is designed to be inversely proportional to the absolute value of the difference between sensitivity and specificity. As this is a binary classification problem, the smaller the difference between the accuracy of the same classifier for different categories, the more stable the performance and the stronger the robustness. Such a classifier can perform better even on a dataset where the two samples are not balanced. We probably assign the weight to the classifier with stronger robustness by using this algorithm. This time, we use R_4 to represent this algorithm in this work.

Define d_k as the absolute value of the sensitivity and specificity of classifier k, we have:

$$d_k = |TPRe_k - TNR_k|, k \in \{1, \cdots, 5\} \tag{4}$$

Make $D \leftarrow d_k$, and we denote a set $D = \{d_k | k = \{1, \cdots, 5\}\}$ as the difference of classification accuracy of five classifiers. Unlike before, we ascend the set D and the symbol $\bar{D}$ represents the ranked set. Finally, we assign weights for each classifier based on the previous strategy and reclassify instances.

After we assign weights to each classifier using these four weight distribution algorithms, we use the combination of any two algorithms to weight the classifier again to get the optimal weight combination scheme. For example, if R_1 assigns each classifier a weight set as $\{1, 2, 4, 5, 3\}$ and R_2 assigns each classifier a weight set as $\{2, 3, 4, 1, 5\}$. Then the combined scheme $R_1 R_2$ assigns each classifier a final weight set as $\{3, 5, 8, 6, 8\}$. This weight set is used to reclassify the sample. We will evaluate each scheme in the next section to find the optimal solution.

5   Evaluation

The purpose of our experiments is multifaceted. In this section, we analyze the selection of parameters and evaluate the performance and overhead of our model.

5.1   Dataset and Setup

Environment. Our experiments are all done under Windows 10 Enterprise Edition, and the PC is equipped with Intel (R) Core I5-4460 CPU@ 3.20 GHz. We leverage *androguard* (v3.4.0) and python 3.6 to extract static features. For dynamic feature extraction, we used *Google Nexus 5* (Android 5.0, SDK 21) based on the *Xposed* framework using *Genymotion* (v3.0.2). The *MonkeyRunner* and the *HierarchyViewer* that come with Android are used for dynamic event testing. In the feature engineering stage, we use python3 to implement the Relief algorithm and use it for feature extraction and fusion.

Dataset. To perform an effective analysis, we select more than 25,000 samples from multiple sources. It includes benign and malicious Android apps from 2015 to 2019. Table 1 lists all datasets used in our experiments. We describe data set in alphanumeric format for simplicity. For example, samples from 2017 are represented to Dataset-17 respectively. All benign applications were downloaded from *Google Play* (GP) and confirmed by the *Virus Total* platform. We collected applications in Google play using the third-party website *apkCombo*. We did not pay attention to the category of the applications when collecting benign samples, because we aim for a generic solution. Our malicious applications were collected from *VirusShare*. After downloading malicious sample files from the platform, we filter out those non-Android related malicious samples. In order to avoid the biases of the unbalance of samples, the number of benign samples and malicious samples in each dataset is roughly equal.

Metrics. We use common metrics in machine learning to evaluate the classification performance. In our research, precision is expressed as the correct rate of malware detection. The recall rate (TPR) reflects the sensitivity of our model to malware. TNR is also this case for benign samples. F1-score (F1) represents

Table 1. Datasets.

DATASET	Benign apps		Malicious apps		Total
	Source	#App	Source	#App	
Dataset-15	GP	2968	VS	2793	5761
Dataset-16	GP	2992	VS	2837	5829
Dataset-17	GP	2991	VS	2780	5771
Dataset-18	GP	2974	VS	2861	5835
Dataset-19	GP	2784	VS	2963	5747

a comprehensive evaluation indicator of the classifier performance on malware detection. In addition, accuracy (ACC) is defined as the correct classification rate of our model for both benign and malicious samples.

5.2 Results and Analysis

Optimal Combination of Weights Distribution Algorithm. We select five base classifiers in this work, including SVM, Random Forest (RF), Classification and Regression Tree (CART), BAGGING, and K-Nearest Neighbor (K-NN). These classifiers have shown good performance in malware detection, so to demonstrate the effectiveness of our weight distribution algorithms, we still use these base classifiers. We choose to use Dataset-17 to evaluate the five base classifiers. For Dataset-17, we first test the performance of each base classifier in static analysis and hybrid analysis. The results are shown in Table 2. Then, we use four ranking algorithms to assign weights to the five base classifiers. The results are shown in Table 3. Each column represents the weights assigned to the classifier by each algorithm.

Table 2. Performance of base classifier on dataset-17 (%).

Classifier	Static analysis			Hybrid analysis		
	TPR	TNR	ACC	TPR	TNR	ACC
RF	92.81	93.75	93.28	95.35	98.14	96.77
CART	89.87	95.39	92.62	91.47	97.03	94.31
BAG	91.50	92.43	91.97	93.41	96.65	95.07
K-NN	81.70	92.11	86.89	91.86	92.94	92.41
SVM	85.29	92.76	89.02	84.88	94.05	89.56

From Table 2 and Table 3, we can see that in the static analysis, RF is more sensitive to malware, while CART has better performance in identifying benign applications. In the hybrid analysis, the overall performance of RF is better, with

Table 3. Each algorithm combination's performance on dataset-17 (%).

Classifier	Static analysis				Hybrid analysis			
	RSen	RSpe	RHM	RD	Rsen	RSpe	RHM	RD
RF	5	4	5	4	5	5	5	3
CART	3	5	4	3	2	4	3	2
BAG	4	2	3	5	4	3	4	4
K-NN	1	1	1	1	3	1	2	5
SVM	2	3	2	2	1	2	1	1

an accuracy rate of 96.77%. The difference between KNN's TRP and TNR is the smallest, only 1.08%, indicating that the classifier is relatively stable. Overall, the classification performance of SVM and KNN is not good, hence a low weight is assigned when the classifiers are fused. There are performance differences between each classifier, but at the same time, each has its own advantages. Finally, we use a dual weight distribution strategy to combine different algorithms and fuse the classifiers. The results are shown in Table 4 and Table 5, respectively.

Table 4. Each algorithm combination's performance on dataset-17 in the static analysis (%).

Combination	TPR	TNR	F1-score	ACC
R_1R_2	90.85	95.39	95.17	93.11
R_1R_3	92.48	94.74	94.62	93.61
R_1R_4	91.83	95.07	94.90	93.44
R_2R_3	**93.02**	**95.73**	**95.61**	**94.35**
R_2R_4	90.85	95.39	95.17	93.11
R_3R_4	93.27	94.97	94.88	94.10

From the data in Table 4 and Table 5, specifically, in the static analysis, the ACC of the optimal algorithm combination R_2R_3 reaches 94.35%, which is 1.07% higher than the best base classifier RF. After the detection of classifier fusion and hybrid analysis, it can achieve 97.15% accuracy and 98.83% F1-score. Overall, no matter it is static analysis or hybrid analysis, most of the dual weight distribution combinations have played a positive role, indicating the effectiveness of our weight distribution algorithm. For static analysis and hybrid analysis, we selected the optimal algorithm combination that is suitable for each detection method and displayed it in bold in the table.

The Threshold for Static Analysis. In order to achieve a better detection effect with less time overhead, we set a threshold for static analysis. Firstly,

Table 5. Each algorithm combination's performance on dataset-17 in the hybrid analysis (%).

Combination	TPR	TNR	F1-score	ACC
$R_1 R_2$	**95.34**	**98.88**	**98.83**	**97.15**
$R_1 R_3$	93.80	98.88	98.82	96.39
$R_1 R_4$	93.80	98.88	98.82	96.39
$R_2 R_3$	94.95	96.29	96.24	95.63
$R_2 R_4$	93.41	98.88	98.82	96.20
$R_3 R_4$	94.17	96.29	96.21	95.25

the static analysis of the application is performed. If the probability that the model detects the sample as malicious is less than the preset threshold, then the application will be subjected to mixed analysis; otherwise, the detection will be ended. Such a detection method can reduce the overall internal overhead of the model, and at the same time can obtain a higher detection accuracy.

The key issue is how to determine this threshold according to our needs. We noticed using the ROC curve to help us solve the problem. In machine learning, the AUC calculated by the ROC curve is an indicator for evaluating different classifiers. The ROC curve of each base classifier and the fused model during static analysis is shown in Fig. 5. The label in the figure represents the AUC value corresponding to each classifier. It can be seen from the figure that the fused classifier performs better than the base classifier. At the same time, we can calculate the point closest to (0, 1) in the coordinate system, which has a value of 94.83%, and we mark it with an arrow. This point means that when this probability threshold is used to judge as malware, it can make a lower FPR and a higher TPR. At this time, it has the best performance for the entire model, which can reduce the model overhead.

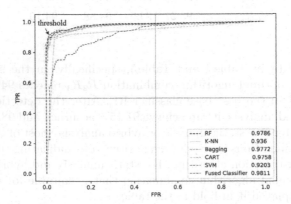

Fig. 5. ROC curve of each classifier.

Performance Evaluation. We evaluate the model performance in terms of execution time. We set up two sets of experiments to count the average detection time of samples that only require static analysis and the average detection time of samples that require static and dynamic analysis. The results are shown in Table 6. We select *apk* files with a size of 5 to 25 MB from the original sample set for evaluation. Due to the large differences between each *apk*, we count the average detection time. As can be seen from the table, for a small sample size (5 to 10 MB), our model only needs to perform static analysis to detect malware, the average time cost is 103.08 s. For small samples that require further dynamic analysis, the average detection increases to 223.33 s. The main time consumed in the process of dynamic analysis consists of two parts, including dynamic test time and feature extraction time. No matter for large samples or small samples, compared to dynamic analysis, the static analysis only needs to perform feature extraction and sample prediction, which can save the detection time.

Table 6. Application's average detection time.

Method	Sample size	App size (MB)	Average detection time (s)			Total (s)
			Testing	Extraction	Prediction	
Static	200	5–10	–	103.08	0.9×10^{-3}	103.08
Static	200	10–25	–	199.86	0.37×10^{-3}	199.86
Hybrid	200	5–10	116	107.33	0.44×10^{-3}	223.33
Hybrid	200	10–25	252	202.27	0.35×10^{-3}	454.27

Model Robustness. In order to verify the robustness of MFF-AMD, we tested different datasets and the results are shown in Table 7. The experimental results show that our model performs well for other datasets. This means that our model can handle almost all Android applications without considering whether the application is up to date. It is worth noting that the accuracy of our model for dataset-15 and dataset-19 is higher than 96%, indicating that our features have a better prevalence. Meanwhile, our average precision and FRP are 97.12%

Table 7. Performance in each dataset (%).

Dataset	TPR	TNR	Precision	ACC	F1
Dataset-15	95.44	97.22	97.16	96.33	96.30
Dataset-16	94.17	96.47	96.23	95.35	95.19
Dataset-17	94.56	96.47	96.24	95.54	95.37
Dataset-18	94.96	98.14	98.00	96.58	96.46
Dataset-19	94.18	98.14	97.98	96.20	96.04

and 2.72%, respectively, indicating that the detection results of our model have a high degree of confidence.

Comparison with Related Work. We selected 5 related studies that have similar features or similar analytical methods. The results are shown in Table 8. The work in the table may use a static method [5], the dynamic method [9,20], or hybrid method [21,22]. We found that static and hybrid methods select features related to permissions and sensitive APIs as their static features, which indicates the prevalence of such features for static Android malware detection. While the difference is that we have added some features such as component features, certificates, etc. At the same time, for some samples, our method can perform dynamic analysis based on static analysis. Most of the dynamic methods are based on dynamic behavior monitoring, by building feature sets from different aspects. Some of these scores in the table look slightly better than us. That is because they were tested in different datasets (our samples are more abundant), which is explained in the following to prove the superiority of our method.

Table 8. Comparison with other approaches (%).

Method	Type		Feature							ACC/F1
	St	Dy	P	A	M	C	N	O	E	
Droidmat	✓	–	✓	✓	✓	–	–	–	–	97.87 91.83
Andromaly	–	✓	–	–	–	–	✓	✓	✓	91.13 –
Afonso	–	✓	–	✓	–	–	✓	✓	–	96.82 96.79
M. Su	✓	✓	✓	✓	–	✓	✓	✓	–	97.4 –
stormDroid	✓	✓	✓	✓	–	–	✓	✓	–	93.80 93.80
MFF-AMD	✓	✓	✓	✓	✓	✓	✓	✓	✓	96.00 95.87

St: static method, Dy: dynamic method, P: permission-based feature, A: API-based feature, M: meta-information-based feature, including component information, intent, package information, file md5, file size, certificate, etc., C: code-feature-based feature, including java reflection, dynamic loading, etc., N: network- information-based feature, O: sensitive-file-based or database-operation-based feature, E: shell-based or command-based feature.

Each accuracy in Table 8 is obtained from its own dataset, because we didn't get all their source code, and some of their datasets are not such comprehensive. Compared with the dataset size of the three works of Andromaly (10800), Afonso (7520), and StormDroid (7970), our dataset samples are more abundant.

Therefore, experiment results show that the advantages exist in comparison with other schemes. The Droidmat's datasets (1738) and M. Su's datasets (1200) have a smaller size, and the obtained accuracy rates are 97.87% and 97.4%, respectively. We also test our scheme on their dataset (Contagio Mobile dataset) separately. In the case of the same sample, our method has an advantage in accuracy, and the detection accuracy reaches 98.37% and 98.34%, which is superior to these two schemes.

6   Conclusion and Future Work

We proposed a high accuracy-oriented detection model of Android malware - MFF-AMD, based on multiscale feature extraction and classifier fusion. It extracts dynamic and static features and is proved to be effective in distinguishing between benign applications and malicious applications. We balanced the contradiction between overhead and accuracy in hybrid analysis via selective dynamic analysis. Our research shows that by using our designed weight distribution algorithm to fuse base classifiers, we can make up the unreliable performance of base classifiers, and effectively improve the overall accuracy of the model. Finally, MFF-AMD performs a better detection rate and robustness based on our data from the past five years. Our work can provide a solid solution to complement current malware detection.

There are some interesting works following this line of research. The process of extracting dynamic features is still costly. Despite we use UI views for dynamic analysis, it may be not able to perform code-level analysis, which may result in the failure to trigger a logical relationship and deeper malicious behavior. One possible solution is to use semantic analysis to enhance our model, which we will study in future work.

Acknowledgment. This work is partially sponsored by National Key R&D Program of China (No. 2019YFB2101700), National Science Foundation of China (62172297, 61902276), the Key Research and Development Project of Sichuan Province (No. 21SYSX0082), Tianjin Intelligent Manufacturing Special Fund Project (20201159).

References

1. SophosLabs 2018 Malware Forecast. https://www.sophos.com/de-de/medialibrary/PDFs/technical-papers/malware-forecast-2018.ashx. Accessed 20 Mar 2019
2. Symantec 2018 Internet Security Threat Report. https://symantec.com/security-center/threat-report. Accessed 2 Mar 2019
3. Core Security. http://blogs.360.cn/post/review_android_malware_of_2018.html. Accessed 20 Mar 2019
4. AV-TEST Antivirus for Android. https://www.av-test.org/en/news/test-20-protection-apps-for-android/. Accessed 26 Sept 2019

5. Wu, D.-J., Mao, C.-H., Wei, T.-E., Lee, H.-M., Wu, K.-P.: DroidMat: Android malware detection through manifest and API calls tracing. In: 2012 Seventh Asia Joint Conference on Information Security, pp 62–69, Tokyo. https://doi.org/10. 1109/asiajcis.2012.18
6. Arp, D., Spreitzenbarth, M., Hübner, M., Gascon, H., Rieck, K.: DREBIN: effective and explainable detection of Android malware in your pocket. In: Proceedings 2014 Network and Distributed System Security Symposium, vol. 14, pp. 23–26 (2014). https://doi.org/10.14722/ndss.2014.23247
7. Zhao, M., Ge, F., Zhang, T., Yuan, Z.: AntiMalDroid: an efficient SVM-based malware detection framework for Android. In: Liu, C., Chang, J., Yang, A. (eds.) ICICA 2011, Part I. CCIS, vol. 243, pp. 158–166. Springer, Heidelberg (2011). https://doi.org/10.1007/978-3-642-27503-6_22
8. Saracino, A., Sgandurra, D., Dini, G., Martinelli, F.: MADAM: effective and efficient behavior-based Android malware detection and prevention. IEEE Trans. Dependable Secure Comput. 15(1), 83–97 (2018). https://doi.org/10.1109/tdsc. 2016.2536605
9. Afonso, V.M., de Amorim, M.F., Grégio, A.R.A., Junquera, G.B., de Geus, P.L.: Identifying Android malware using dynamically obtained features. J. Comput. Virol. Hacking Tech. 11(1), 9–17 (2014). https://doi.org/10.1007/s11416-014-0226-7
10. Martín, A., Lara-Cabrera, R., Camacho, D.: Android malware detection through hybrid features fusion and ensemble classifiers: the AndroPyTool framework and the OmniDroid dataset. Inf. Fusion 52, 128–142 (2019). https://doi.org/10.1016/j.inffus.2018.12.006
11. Arzt, S., Rasthofer, S., Fritz, C., Bodden, E., Bartel, A., Klein, J., et al.: FlowDroid. ACM SIGPLAN Not. 49(6), 259–269 (2014). https://doi.org/10.1145/2666356. 2594299
12. Zhao, Z., Colon Osono, F.C.: TrustDroid: preventing the use of SmartPhones for information leaking in corporate networks through the used of static analysis taint tracking. In: 2012 7th International Conference on Malicious and Unwanted Software. https://doi.org/10.1109/malware.2012.6461017
13. Lindorfer, M., Neugschwandtner, M., Platzer, C.: MARVIN: efficient and comprehensive mobile app classification through static and dynamic analysis. In: 2015 IEEE 39th Annual Computer Software and Applications Conference. https://doi.org/10.1109/compsac.2015.103
14. Bichsel, B., Raychev, V., Tsankov, P., Vechev, M.: Statistical deobfuscation of Android applications. In: Proceedings of the 2016 ACM SIGSAC Conference on Computer and Communications Security (2016). https://doi.org/10.1145/2976749. 2978422
15. Wang, W., Li, Y., Wang, X., Liu, J., Zhang, X.: Detecting Android malicious apps and categorizing benign apps with ensemble of classifiers. Future Gener. Comput. Syst. 78, 987–994 (2018). https://doi.org/10.1016/j.future.2017.01.019
16. Wang, W., et al.: Constructing features for detecting Android malicious applications: issues, taxonomy and directions. IEEE Access 7, 67602–67631 (2019). https://doi.org/10.1109/access.2019.2918139
17. Chen, K., Wang, P., Lee, Y., Wang, X., Zhang, N., Huang, H., et al.: Finding unknown Malice in 10 seconds: mass vetting for new threats at the Google-play scale. In: USENIX Security, vol. 15 (2015). https://doi.org/10.5555/2831143. 2831185

18. Gascon, H., Yamaguchi, F., Arp, D., Rieck, K.: Structural detection of Android malware using embedded call graphs. In: Proceedings of the 2013 ACM Workshop on Artificial Intelligence and Security. https://doi.org/10.1145/2517312.2517315
19. Perdisci, R., Dagon, D., Wenke Lee, Fogla, P., Sharif, M.: Misleading worm signature generators using deliberate noise injection. In: 2006 IEEE Symposium on Security and Privacy (S&P). https://doi.org/10.1109/sp.2006.26
20. Shabtai, A., Kanonov, U., Elovici, Y., Glezer, C., Weiss, Y.: "Andromaly": a behavioral malware detection framework for Android devices. J. Intell. Inf. Syst. 38(1), 161–190 (2011). https://doi.org/10.1007/s10844-010-0148-x
21. Chen, S., Xue, M., Tang, Z., Xu, L., Zhu, H.: StormDroid: a streaminglized machine learning-based system for detecting Android Malware. In: Proceedings of the 11th ACM on Asia Conference on Computer and Communications Security, pp. 377–388, Xi'an (2016). https://doi.org/10.1145/2897845.2897860
22. Su, M.-Y., Chang, J.-Y., Fung, K.-T.: Machine learning on merging static and dynamic features to identify malicious mobile apps. In: 2017 Ninth International Conference on Ubiquitous and Future Networks (ICUFN). https://doi.org/10.1109/icufn.2017.7993923

Network and Security

PSG: Local Privacy Preserving Synthetic Social Graph Generation

Hongyu Huang, Yao Yang, and Yantao Li[✉]

College of Computer Science, Chongqing University, Chongqing 400044, China
yantaoli@cqu.edu.cn

Abstract. Social graph, as a representation of the network topology, contains users' social relationship. In order to obtain a social graph, a server requires users to submit their relationships. As we know, using or publishing social graph will cause privacy leakage to users. For this sake, it is necessary to generate synthetic social graph for various usages. In this paper, we propose PSG, a local Privacy Preserving Synthetic Social Graph Generation method. In order to protect users' privacy, we utilize the local differential privacy model and a truncated Laplace mechanism to allow users to perturb their own data before submission. We then model the graph generation as a combinatorial optimization problem and design a greedy algorithm to maximize the utility of the generated graph. Through theoretical analysis and extensive experiments, we show that our method satisfies local differential privacy as well as maintains attributes of the original social graph.

Keywords: Social networks · Local differential privacy · Synthetic graph generation

1 Introduction

Social graph is often used to represent correlations among all users. By analyzing the social graph, we can obtain much valuable knowledge about the network which can help to improve the performance or quality of services. When generating a social graph, the server requires every user in the network to submit the social connections. Obviously, the information is sensitive which brings serious risk to users' privacy if it is misused by the server or exposed to the third party. Therefore, how to protect the individual privacy of users in social networks by a privacy-preserving way is an emerging research topic. A reasonable solution is to allow each user to add noise in the submission and then the server generates a synthetic social graph. Local differential privacy, as a privacy protection method, has been widely used to protect users' privacy. According to the local differential privacy, the probability that the server infers the real data from the noisy submission can be controlled by the user, which permits users to control their privacy leakage.

© ICST Institute for Computer Sciences, Social Informatics and Telecommunications Engineering 2021
Published by Springer Nature Switzerland AG 2021. All Rights Reserved
H. Gao and X. Wang (Eds.): CollaborateCom 2021, LNICST 406, pp. 389–404, 2021.
https://doi.org/10.1007/978-3-030-92635-9_23

Since the server only has noisy data, it is challenging to guarantee that the synthetic graph has similar attributes to the original social network. However, most of the existing works generate the synthetic graph in a randomized way. That is, the server first requires users to submit social network attributes, such as degree sequences, subgraph counts and degree distributions. Then, the server divides users into different groups based on the information, and calculates the probability of generating an edge between intra-cluster and inter-cluster. Finally, edges are randomly generated according to the probability. The above random model is similar to the BTER model [1]. Nevertheless, the attributes of the social network synthetic graph generated by the random model is indeterminate, and thus we cannot obtain values that are similar to the attributes of the original social network graph.

In this paper, we propose ℙ𝕊𝔾, a local Privacy Preserving Synthetic Social Graph Generation mechanism that generates synthetic graphs under local differential privacy. In ℙ𝕊𝔾, the server first divides users into different groups and requires users to submit the number of friends in each group. Then, users form their degree vectors based on groups, add truncated Laplace based on privacy budget and normalization factor, generate a noise degree vector, and send it to the server. Next, the server uses the noise degree vector to calculate the utility of an edge generation, establishes a utility matrix, and constructs a social network synthetic graph, which aims to maximize the utility. To validate the effectiveness of ℙ𝕊𝔾, we conduct extensive experiments using real social network datasets to verify that the synthetic graph generated by ℙ𝕊𝔾 can maintain attributes of the original social network under the desired privacy level.

The rest of the paper is organized as follows. Section 2 reviews related work. Section 3 defines the problem of synthetic social graph generation under local differential privacy. Section 4 presents the proposed solution, ℙ𝕊𝔾, and proves that it satisfies local differential privacy. Section 5 evaluates the performance of ℙ𝕊𝔾 by extensive experiments. Last, Sect. 6 concludes this work.

2   Related Work

There are a large number of methods proposed to protect social networks and generate social network graphs, which have been investigated for over decades with flourished results.

2.1   Social Network Privacy Protection

Early works on privacy-preserving social network mainly focus on data mining of social network. Chen et al. proposed a privacy policy recommendation model, which aimed to recommend for text-based posts privacy policies to users [2]. Sweeney et al. proposed a k-anonymity model, which made each record have at least the same attribute value as the other $k-1$ records in the data, thereby reducing privacy leakage caused by link attacks [3]. In order to overcome the shortcomings of the k-anonymity model, Machanavajjhala et al. proposed a l-diversity privacy model, where attackers inferred the probability of each k anonymous dataset of private information in social networks was less than $\frac{1}{l}$ [4]. Unfortunately, the aforementioned models are vulnerable to attackers with stronger background knowledge, which has stimulated the use of differential privacy to obtain more rigorous privacy guarantees.

The latest research on applying differential privacy in social networks mainly focuses on: 1) the release of social network statistics, such as degree histograms and subgraph counts [5–7], and 2) the publication of a synthetic graph of a social network, such as BTER [1].

2.2 Synthetic Graph Generation

There are some contributions on formal models for synthetic graph generation. The earliest model is the Erdos-Renyi (ER) model proposed by Paul et $al.$, which assumes that an edge is randomly generated in the network based on the same probability [8]. Aiello et $al.$ proposed the CL model, which was similar to the ER model, but each edge has different probability that satisfies the node degree distributions [9].

The aforementioned graph models ignore to take the protection privacy of social network users in the generation of synthetic graph into account. However, differential privacy is widely used to protect users privacy in social synthetic graph. Qin et $al.$ proposed LDPGen model which was a multi-phase approach to generate synthetic decentralized social graph under local differential privacy [10]. Zhu et $al.$ designed a degree-differential privacy mechanism, and proposed a graph generation approach with field theory [11].

According to previous literature, we find that when the existing social network graph models generate edge relationships between nodes, most of them link nodes by utilizing probability-based methods. However, our method is able to calculate the probability of social network edge relationships for social network graph generation, which is rarely studied in previous works.

3 Preliminaries

3.1 System Overview

We consider a social network with N users. We model the network as an undirected graph $G = (V, E)$, where V and E indicate sets of users and their connections, respectively. We assume that $|V| = N$ and $|E| = M$. We denote $v_i \in V$ in the graph as the i–th user in the network. We say v_i and v_j are neighbors in the graph, i.e., an edge $e_{i,j} \in E$ connects v_i and v_j, iff the two users are friends in the real world. An untrusted data curator, which we call a server in the rest of this paper for simplicity, collects neighbor lists from users to generate a synthetic network. With all neighbor lists, the server uses a three-stage process to generate the network, inspired by [10].

In the first stage, the server randomly divides all users into k disjoint groups, denoted as $R = \{R_1, R_2, ..., R_k\}$. Then, the server distributes these groups to all users. Each user computes friends in each group and then obtains a degree vector $\gamma^i = (\gamma_1^i, \gamma_2^i, ..., \gamma_k^i)$, where γ_k^i denotes the number of friends that user v_i has in the $k-$th group. In order to protect the local privacy, each user perturbs the vector γ^i before submitting it to the server.

In the second stage, the server re-constructs the group R and distributes it to all users again. At this stage, the number of groups and members in each group are different from the first stage. When users obtain R, they do the same as the first stage that compute the degree vector based on the groups, then perturb the vector, and submit it to the server.

In the last stage, the server computes the degree of each user and assigns some other users as his friends. This assignment runs iteratively until all users have enough friends. To this end, the server builds the synthetic social graph.

3.2   Problem Statement

In order to generate a synthetic network in a privacy-preserving way, we need to solve two problems in the stages in previously introduced procedure. The first one is how to find a method to add the noise to the degree vector for the user privacy protection. Specifically, we use the differential privacy [12] as our privacy model. For randomized mechanism $\mathcal{M}$ on a graph, it satisfies $\epsilon-$differential privacy iff for any two neighboring graphs G and G' which only differ in one edge, and any possible $s \in range(\mathcal{M})$, we have $\frac{Pr[\mathcal{M}(G)=s]}{Pr[\mathcal{M}(G')=s]} \leq e^\epsilon$. Generally, we need to add Laplace noise to each entry of the degree vector to satisfy the differential privacy. As we know, however, the degree of each node on the graph must be non-negative and less than the number of nodes. Thus, simply using Laplace distribution may generate out-of-bounds noise. Hence, we design a randomized mechanism that satisfies differential privacy and outputs degrees in the range $[0, N-1]$ to address the first problem.

Our second problem is to generate a synthetic social network graph based on the degree vector submitted by users. Most of the existing works adopt randomized methods to connect nodes with edges. For fixed input parameters, the output network graph of a randomized method may be different. Hence, the performance of synthetic networks cannot be guaranteed. In order to overcome this defect, we try to find a deterministic method to improve the randomness. Suppose we have a metric to measure the utility of connecting v_i and v_j by edge, and then we can generate the synthetic social network graph by solving the following combinatorial optimization problem:

$$max \sum_{i=1}^{N} \sum_{j=1}^{N} u_{i,j} \cdot x_{i,j} \tag{1}$$

$$s.t. \sum_{j=1}^{M} x_{i,j} = d_i, \, i = 1, 2, 3, ..., N \tag{2}$$

$$x_{i,j} = x_{j,i} \tag{3}$$

$$x_{i,j} \in \{0,1\}, i = 1, 2, ..., N, j = 1, 2, ..., M \qquad (4)$$

where $u_{i,j}$ is the utility of $e_{i,j}$, $x_{i,j} = 1$ indicates that $e_{i,j}$ exists and $x_{i,j} = 0$ for otherwise. The d_i is the degree of v_i which can be easily obtained by adding all entries of γ^i. Last, since the graph is undirected, we have $x_{i,j} = x_{j,i}$. Therefore, we propose SNGCO, a $\underline{S}$ynthetic $\underline{N}$etwork $\underline{G}$eneration by $\underline{C}$ombinatorial $\underline{O}$ptimization to address the second problem.

4 Design Details

4.1 Privacy Protection Mechanism Design

As a de facto standard, local differential privacy has been widely used in social networks for user sensitive data protection. It mainly focuses on publishing social network synthetic graph and various types of graph statistics, including social network degree distributions [13], degree sequences [14], subgraph counts [7], and so on. In this context, differential privacy is proposed to avoid the leakage of individual private information in social networks.

Specifically, differential privacy mechanism allows users to submit noisy data. In social networks, users with their own privacy considerations individually utilize traditional Laplace algorithm that satisfies differential privacy to perturb their personal information, and then submit the noisy data to a server. The server use these data to infer the overall statistic information of social networks. However, since the perturbation range of the traditional Laplace algorithm is not restricted, the noise range can be $[-\infty, +\infty]$, which cannot be applied to all scenarios, and may generate erroneous data beyond this range. For example, after the user receives the group result from the server, the user submits the number of friends in each group according to the division result, i.e., the degree vector. This value is non-negative and must be less than the number of nodes in the group. Obviously, the traditional Laplace cannot meet such requirements, so we adopt the truncated Laplace algorithm to control the noisy value within the valid range. By briefly reviewing the algorithm, we notice that the range of the Laplace noise function on the left and right sides of the true value is equal, so we use the distance from the true value to the upper and lower limits to determine the function curves on the left and right sides to obtain the function distribution on the left and right sides. To better illustrate the local differential privacy, it can be defined as:

Definition 1 (*Local differential privacy* [15]). *A randomized mechanism M satisfies ε-local differential privacy, iff for any two input data t and* t*, *we have*

$$\frac{\Pr[M(t) = s]}{\Pr[M(t^*) = s]} \leq e^{\varepsilon}. \qquad (5)$$

Where $s \subseteq$ range (M), and then we say that M satisfies ε-local differential privacy.

Definition 2 (*Local sensitivity* [12]). *Given a graph $G = (V, E)$ containing user nodes $V = \{v_i | 1 \le i \le n\}$, and any function f, the local sensitivity of f can be defined as:*

$$LS_{f(D)} = \max \| f(G) - f(G') \|_1, \tag{6}$$

where G and G' are neighboring graph of users.

Definition 3 (*Laplace algorithm* [12]). *Given a dataset D and function f, if privacy protection mechanism Y satisfies ε-local differential privacy, we have:*

$$Y = f(D) + laplace(\frac{\Delta f}{\varepsilon}), \tag{7}$$

where Δf is the sensitivity.

As mentioned earlier, our goal is to generate a synthetic graph with attributes similar to the original social network graph under user privacy protection. Now we formally describe our solution (referred to $\mathbb{PSG}$), as illustrated in Algorithm 1 (based on the truncated Laplace algorithm). Specifically, the user calculates the number of friends in each group according to the received group results to form a degree vector, and then uses the truncated Laplace algorithm to add noise to the degree vector. Finally, the noise degree vector is returned to a server for data processing.

As described in Algorithm 1, the user mainly uses the truncation mechanism twice in the following two steps. The first step is to add noise to user's node degree (lines 1–4). Specifically, the user initially calculates the distance between himself and the left and right ends of the valid range (line 1). In order to ensure that the range of the probability density function of the truncated Laplace is 1, the user calculates the respective areas of the two ends of the function (line 2). Next, the user calculates the normalization factor based on the previous calculation results (line 3). Then, the truncated Laplace function is obtain though multiplying the normalization factor by the original Laplace function, and the user adds the noise by this function (line 4). In the second step, the user also performs privacy protection for the node degree vector. The user repeats truncated Laplace algorithm on the degree vector, where the difference from the first step is that the valid range varies with the size of each group. To ensure the consistency of user data, the cumulative sum of the user degree vector needs to be equal to the user degree. Therefore, the second step is to add noise to the user degree vector ensuring that the cumulative sum of the user degree vector is equal to user's node degree (lines 6–13). In order to improve the data accuracy, the user initializes the cumulative sum of the degree vector s, and performs in an ascending order processing on each element of the degree vector (lines 5–6). Similarly, as the same step as the first user degree noise addition, the user calculates the valid range in each group, obtains the normalization factor, and calculates the noise addition value in each group (lines 7–10). According to the constraint of the degree, when the cumulative sum of the degree vector is greater than the degree, the i-th element is equal to the absolute value of the difference between the cumulative sum of the previous $(i-1)$-th elements and the degree (lines 11–13). Finally, according to the above operations, the user constructs and returns a noise degree vector (lines 14–15).

Algorithm 1. Truncated Laplace Mechanism for Nodes Degree Vector

---

Input: *group*
Output: Noisy degree vector q
1: $\Delta L = \mu$, $\Delta R = (N-1) - \mu$
2: $L = \frac{e^{\frac{-\Delta R}{\sigma}}}{2}$, $R = \frac{e^{\frac{-\Delta L}{\sigma}}}{2}$;
3: $n = \frac{1}{1-(R+L)}$;
4: $\tilde{d} = d + nLap(\frac{1}{\varepsilon})$;
5: $s = 0$;
6: **for** each d_i in q_i with ascending order :
7:    $\Delta L_i = \mu_i$, $\Delta R_i = |q_i| - 1 - \mu_i$;
8:    $L_i = \frac{e^{\frac{-\Delta R_i}{\sigma}}}{2}$, $R_i = \frac{e^{\frac{-\Delta L_i}{\sigma}}}{2}$;
9:    $n_i = \frac{1}{1-(L_i+R_i)}$;
10:    $\tilde{d}_i = d_i + n_iLap(\frac{1}{\varepsilon})$;
11:    $s = s + \tilde{d}_i$;
12:    **if** $s > \tilde{d}$:
13:       $\tilde{d}_i = |s_{i-1} - \tilde{d}|$;
14: $q = < \tilde{d}_1, \ldots, \tilde{d}_{|q_i|} >$;
15: **return** q

4.2 Privacy Analysis

In this section, we demonstrate that PSG satisfies local differential privacy. Let $\{\tilde{\gamma}_1, \tilde{\gamma}_2, \ldots, \tilde{\gamma}_k\}$ be user's noise degree vector based on k groups, where $\tilde{\gamma}^{v_i} = \gamma_j^{v_i} + n_j^i Lap(\frac{\Delta F}{\varepsilon})$. Denote this mechanism as L.

Theorem 1. *The PSG satisfies ε-local differential privacy.*

Proof. Without loss of generality, any two users v_i and v_j send degree vectors of $\gamma^{v_i} = \{\gamma_1^{v_i}, \gamma_2^{v_i}, \ldots, \gamma_k^{v_i}\}$ and $\gamma^{v_j} = \{\gamma_1^{v_j}, \gamma_2^{v_j}, \ldots, \gamma_k^{v_j}\}$ to a server. If γ^{v_i} and γ^{v_j} differ in one element, we assume $\gamma_k^{v_i} \neq \gamma_k^{v_j}$. Then, we can obtain $|\gamma_k^{v_i} - \gamma_k^{v_j}| = 1$, and $\Delta F = 1$. Given an arbitrary vector $s = (s_1, \ldots, s_k)$, based on traditional Laplace algorithm, the privacy guarantee of differential privacy can be shown in Eq. (8):

$$\frac{\Pr[M(\gamma^u) \in s]}{\Pr[M(\gamma^v) \in s]} = \frac{\Pr[M(\gamma_1^u) = s_1] \ldots \Pr[M(\gamma_k^u) = s_k]}{\Pr[M(\gamma_1^v) = s_1] \ldots \Pr[M(\gamma_k^v) = s_k]}$$
$$= \frac{\Pr[M(\gamma_k^u) = s_k]}{\Pr[M(\gamma_k^v) = s_k]} \tag{8}$$
$$\leq e^{\varepsilon}.$$

Similarly, when using truncated Laplace mechanism, we have $\Pr[L(\gamma_k^v) \in s_k] = n_k^v \frac{\varepsilon}{2\Lambda F} e^{-\frac{\varepsilon|s-\gamma_k^v|}{\Lambda F}}$, $\Pr[L(\gamma_k^u) \in s_k] = n_k^u \frac{\varepsilon}{2\Lambda F} e^{-\frac{\varepsilon|s-\gamma_k^u|}{\Lambda F}}$. It can be shown in Eq. (9):

$$
\begin{aligned}
\frac{\Pr[L(\gamma^u) \in s]}{\Pr[L(\gamma^v) \in s]} &= \frac{\Pr[L(\gamma_k^u) = s_k]}{\Pr[L(\gamma_k^v) = s_k]} \\
&= \frac{n_k^u \frac{\varepsilon}{2\Lambda F} e^{-\frac{\varepsilon|s-\gamma_u|}{\Lambda F}}}{n_k^v \frac{\varepsilon}{2\Lambda F} e^{-\frac{\varepsilon|s-\gamma_v|}{\Lambda F}}} \\
&= \frac{n_k^u}{n_k^v} e^{\frac{\varepsilon|s-\gamma_u|-\varepsilon|s-\gamma_v|}{\Lambda F}} \\
&\leq \frac{n_k^u}{n_k^v} e^{\frac{\varepsilon|\gamma_u-\gamma_v|}{\Lambda F}} \\
&= \frac{n_k^u}{n_k^v} e^{\varepsilon}.
\end{aligned}
\tag{9}
$$

Since the scaling parameters are all the same, and both of L_k^u (i.e., L_k^v) and R_k^u (i.e., R_k^v) have a range of $[0, 0.5)$, we substitute the L_k^v and R_k^v into $L_k^v = L_k^u e^{\varepsilon}$, $R_k^v = R_k^u e^{-\varepsilon}$ to obtain the form shown in Eq. (10):

$$
\frac{n_k^u}{n_k^v} e^{\varepsilon} = \frac{1 - (L_k^u e^{\varepsilon} + R_k^u e^{-\varepsilon})}{1 - (L_k^u + R_k^u)} e^{\varepsilon}
\tag{10}
$$

$$
\leq e^{\varepsilon}.
$$

According to the above proof, we obtain

$$
\frac{\Pr[L(\gamma^u) \in s]}{\Pr[L(\gamma^v) \in s]} \leq e^{\varepsilon}.
\tag{11}
$$

In $\mathbb{PSG}$, users add noise to their degree vector applying the truncated Laplace $n_i Lap(\frac{\Lambda F}{\varepsilon_1})$ and $n_i Lap(\frac{\Lambda F}{\varepsilon_2})$ twice. According to the composability property of differential privacy $\varepsilon = \varepsilon_1 + \varepsilon_2$, $\mathbb{PSG}$ satisfies differential privacy.

4.3   Synthetic Network Generation

As we mentioned in Sect. 3.1, there are two stages of interactions between the server and users. The operations in each stage are the same: the server distributes the partition to all users and they return perturbed vectors to indicate how many friends they have in each group of the partition. Let k_1 and k_2 be the numbers of groups in the first and second stages, respectively.

Before we show how to generate the synthetic network graph, we informally prove that the SNGCO problem is NP-hard. In brief, we transform the SNGCO into another equivalent problem. We know that the degree of v_i must be d_i (see the constraint (2)). We replace d_i with v_i, which we call dummy nodes, denoted as $v_{i,j}, j = 1, 2, ..., d_i$. Now we have two kinds of nodes: the original nodes and the dummy nodes. Please note that the degree of each dummy node is just 1. Obviously, we have $\sum v_i$ dummy nodes in total. Then we can equivalently transform the SNGCO problem into connecting edges between v_i and $v_{k,j}$ under the constraint that $i \neq k$ and at most one of $v_{k,j}, j = 1, 2, ..., d_k$ can connect to v_i. Formally, we add a new constraint, i.e., $\sum_{i=1, i\neq j}^{N} x_{i,j} = 1$, to the SNGCO problem. Next, we re-consider the assignment problem, which is a well known NP-hard problem, to the SNGCO. In the assignment problem, we have M tasks and N workers. Each worker can obtain different gain when he takes different task. And one task can only be taken by one worker. We can simply regard N nodes as workers, and $\sum v_i$ dummy nodes as M tasks. Obviously, this reduction only needs polynomial time. To this end, since the assignment problem is NP-hard, the SNGCO problem is also NP-hard.

Algorithm 2. Utility-based Synthetic Graph Generation

Input: N, $Q = \{q_1, \ldots, q_{|N|}\}$
Output: Edge set E_T
1: Initialize m=a, index=b
2: for any v_i and v_j do :

3: $\quad u[v_i][v_j] = \dfrac{\sum\limits_{t=1}^{k} (q_t^i \times q_t^j)}{\sqrt{\sum\limits_{t=1}^{k} (q_t^i)^2} \times \sqrt{\sum\limits_{t=1}^{k} (q_t^j)^2}}$;

4: Sort(v_i) by $\tilde{d}_i$ in descending order;
5: for each v_i do;
6: $\quad$ s=0;
7: $\quad$ while($s \leq \tilde{d}_i$) do;
8: $\quad\quad$ for j=0 to N-1 do;
9: $\quad\quad\quad$ if $u[v_i][j] \geq m$ and $i \neq j$;
10: $\quad\quad\quad\quad$ index=j, $m = u[v_i][j]$;
11: $\quad\quad\quad$ if e_{ij} not in E_T then;
12: $\quad\quad\quad\quad$ $E_T = E_T \cup e_{ij}$, $s = s + 1$,
13: $\quad\quad\quad\quad$ $u[v_i][j] = min(u[v_i])$;
14: **return** E_T

Comparing our problems with the assignment problem, we observe that our problem is more complicated than the traditional assignment problem. We aim to turn our problem into an assignment-like combinatorial optimization problem to solve. For the assignment problem and its variants, a large number of algorithms are proposed to obtain feasible solution close to the optimal algorithm, and the time complexity is low. Among them, the greedy algorithm is the simplest and most intuitive one.

In this section, we formally describes how to use greedy solution based on utility matrix to generate social graph, i.e., PSG. According to Algorithm 2, the server has two stages: 1) the server calculates a utility matrix based on cosine similarity between nodes, and 2) the server uses a greedy algorithm to generate a social network synthetic graph. Specifically, in the user division phase, the server randomly divides users twice. First, it randomizes the users into k_0 groups. Next, the server collects the user noise degree vector, and then uses k-means to classify users. The optimal k_2 value is calculated by $k_2 = \left\lceil \sum_{\eta=1}^{\eta_{max}} p_\eta(\frac{d}{2} + \frac{(\frac{d}{2})^2 - 2(1+\sqrt{5})(\frac{d}{2})+1}{\varepsilon_2}) \right\rceil$ to obtain the latest user noise degree vector. In real life, we realize that if two people have more friends in common, then they are more likely to become friends. In the same way, if the degree vectors of two nodes are more similar, then they are more likely to be connected. In Algorithm 2, after server collects noise degree vector of users, using correlation with cosine similarity between nodes as the metric, an utility matrix is constructed to express the possibility of connecting between users (lines 2–3). In order to improve the efficiency of data processing, we sort node degree of users as a pre-processing (line 4), and then generate edges in synthetic graph though selecting the maximum value of utility matrix until the constraint conditions are met. Specifically, first, the sum of user degrees is initialized to 0. When the user degree does not satisfy the constraint, the maximum value is continuously selected from the user row corresponding to the utility matrix (lines 5–10). Next, if any two users corresponding to the maximum value do not have an edge, then add an edge to the synthetic graph, and the value of s is increased by 1. Otherwise, the value of s remains. The above operations are continued until the degree of constraint is satisfied (lines 11–13). Finally, we return to the edge set of the synthetic graph of the social network (line 15).

5   Performance Evaluation

In this section, we evaluate the performance of our proposed generating social network graph method. We first describe a large number of experiments to verify that $\mathbb{PSG}$ can maintain the attributes of the original social network graph under the same privacy level. Then, we compare our method with several graph generation models based on real social network datasets to verify that $\mathbb{PSG}$ is more efficient.

5.1   Datasets and Models

We apply our algorithm on real datasets from Stanford Large Network Dataset Collection [16]. In order to facilitate the comparison and versatility of our experiments, we convert the datasets into an undirected graph. The experiments involve two real social graphs and the details are illustrated as follows:

- Facebook: It is an undirected social graph consisting of 4,039 nodes and 88,234 edges.
- Lastfm: It contains 7,624 nodes and 27,806 edges in a social network.

To demonstrate the usability of our proposed algorithm compared with the traditional graph generation models, we focus on three aspects of utility including the mean absolute error (MAE) of clustering coefficient [17], modularity [18], and Adjusted Rand Index (ARI) [19]. The importance of nodes is also an evaluation metric, since different users have different influences in networks. Nodes with high influence are more likely to affect other nodes. For example, a teacher is more influential than students in a teacher-student network. Therefore, the importance of different nodes in the network is not the same. So, we utilize the mean absolute error of eigenvector centrality (EVC) [18] to evaluate influence of users.

In this paper, We compare above metrics in our model with three graph generation models (FCL [20], BTER [1] and LDPGen [10]) based on differential privacy.

5.2 Evaluation Metrics

Clustering Coefficient. Local clustering coefficient, which represents a measure of the tendency of nodes in the graph to cluster together. The local clustering coefficient is defined as follows:

$$C_{\mathrm{i}} = \frac{2\Delta_i}{d_i(d_i - 1)} = \frac{a_{ii}^{(3)}}{a_{ii}^{(2)}(a_{ii}^{(2)} - 1)} \tag{12}$$

Where Δ_i is the number of triangles connected to user v_i in the social network, d_i is v_i degree, $a_{ii}^{(2)}$ represents the second power diagonal element of the neighboring matrix, and $a_{ii}^{(3)}$ is the third power diagonal element. The smaller the MAE value, the higher the data usefulness.

Modularity. It is used to measure the effect of community division results. Generally, the community satisfies that the nodes within the community have high similarity, while the similarity of external nodes is low. Therefore, modularity is proposed to measure community structure.

$$Q = \sum_{c=1}^{N} [\frac{M_c}{n} - (\frac{d_c}{2n})^2] \tag{13}$$

Where n represents the total number of edges in social network graph, M denotes the number of communities, M_c is the sum of edges within the community c, and d_c represents the sum of degrees in c.

Adjusted Rand Index. ARI is also a metric to measure the clustering effect of social networks. The larger the value of the ARI, the more similar the clustering result is with the real original social networks. Given a set of n elements $S = \{O_1, O_2, ..., O_n\}$, if $U = \{u_1, u_2, ..., u_Z\}$ and $Y = \{y_1, y_2, ..., y_R\}$ represent two different divisions of S satisfying $\cup_{i=1}^{Z} u_i = S = \cup_{j=1}^{R} y_j$, then ARI can be denoted:

$$ARI = \frac{RI - E(RI)}{\max(RI) - E(RI)} \tag{14}$$

$$E(RI) = E(\sum_{i,j} \binom{n_{ij}}{2})) = \frac{[\sum_i \binom{n_{i.}}{2} \sum_j \binom{n_{.j}}{2}]}{\binom{n}{2}} \qquad (15)$$

$$\max(RI) = \frac{1}{2}[\sum_i \binom{n_{i.}}{2} + \sum_j \binom{n_{.j}}{2}] \qquad (16)$$

$$RI = \frac{a+d}{a+b+c+d} \qquad (17)$$

Where a (opposite to c) is the logarithm of nodes of the same group in U and the same group in Y. In the same way, b (opposite to d) is the logarithm of nodes that belong to the same group in U but belong to different group in Y. n_{ij} represents the number of nodes in the same group u_i and group y_j, and $n_{i.}$ (resp. $n_{.j}$) is the number of nodes of group u_i (resp. y_j).

Eigenvector Centrality. It measures the influence of nodes in the network, which means that the importance of a node depends not only on the number of its neighboring, but also on the importance of its neighboring nodes. It can be denoted as:

$$x_i = \frac{1}{\lambda} \sum_{j=1}^{N} a_{ij} x_j \qquad (18)$$

Where λ is a constant, and x_i is the importance metric value of node v_i.

5.3   Experimental Results

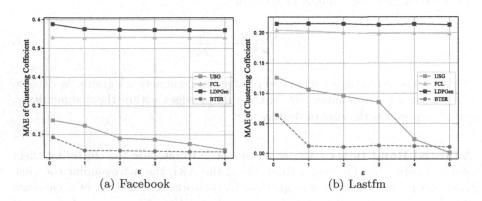

(a) Facebook   (b) Lastfm

Fig. 1. Effect of ε on clustering coefficient

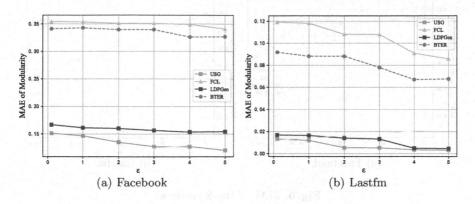

(a) Facebook   (b) Lastfm

Fig. 2. Effect of ε on modularity

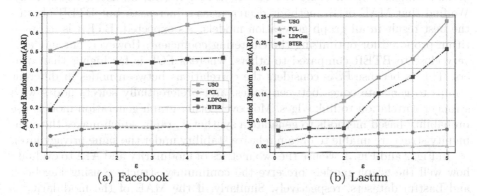

(a) Facebook   (b) Lastfm

Fig. 3. Effect of ε on ARI

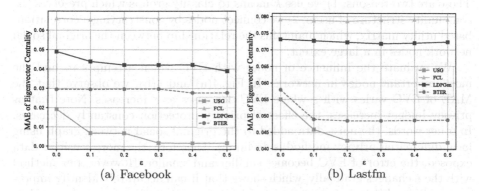

(a) Facebook   (b) Lastfm

Fig. 4. Effect of ε on EVC

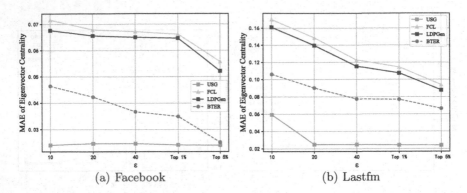

(a) Facebook   (b) Lastfm

Fig. 5. MAE of top-k vertices

Figure 1(a) and 1(b) show that the MAE of clustering coefficient varies with privacy budget ε when using the Facebook and Lastfm datasets, respectively. We find that MAE of all methods decreases as ε increases, but BTER obtains the best result in all graph generation models. As expected, BTER is an algorithm proposed for optimizing the clustering coefficient. However, our model is very close to BTER compared to others, whereas FCL is far higher than others. This is because PSG considers the correlations between nodes in different groups to generate edges between nodes, and preferentially selects nodes with stronger correlation to link edges. Moreover, other graph generation models use probability-based methods to generate a synthetic graph, which makes the community structure unable to obtain relative stability under the same privacy level.

In Fig. 2 and Fig. 3, we use the two metrics of modularity and ARI to evaluate how well the graph models preserve the community structure using Facebook and Lastfm datasets, respectively. Similarly, if the MAE of the modularity is smaller and the value of ARI is more higher, the synthetic social graph is better to preserve community structure of the original social network graph. In Fig. 2 and Fig. 3, we shows PSG significantly outperforms other generation models. These are two reasons: 1) we use k-means to classify groups which preserves the community structure of users; 2) we connect nodes by constructing a correlation-based utility matrix, which maintains the relationship between the original social network nodes in a large extent.

We evaluate the influence of nodes with EVC which attempts to find the most important nodes in networks. In Fig. 4(a) and 4(b), we observe that the MAE of EVC varies with ε where EVC decreases as ε increases. Note that as privacy budget increases, the effect of privacy protection constantly decreases. In other words, the noise data added to the original social network graph using local differential privacy method is declining. Therefore, the more sensitive data exposed, the error of EVC becomes smaller and smaller. However, our method with the ε changes gradually, which shows that it can accurately identify important nodes and better protect the sensitive information of important nodes than other methods. From Fig. 5, we calculate the EVC of common nodes among the most influential top-k nodes in social networks. We select top 10, top 20, top 40, 1% and 5% influential nodes to test. As the number of influential nodes increase, we can find a smaller MAE of EVC. Then, Figs. 4(a) and 4(b) illustrate the effect of PSG is better than other methods. This is because synthetic graph generated by the proposed optimization method is closer to the original graph, and the privacy of the original graph node is better protected based on local differential privacy. That means our methods can better protect privacy while providing accurate node analysis with the deepening of privacy protection.

6 Conclusion

In this paper, we propose PSG, a novel graph generation model that generates a synthetic social network graph with the attributes similar to original graph under local differential privacy. The key idea of this paper is that we turn the problem of graph generation in social networks into a traditional optimal combination problem. PSG proposes an utility matrix based on correlation between nodes and generated edges, which is not used in probability models as the same way as the previous ones. In order to preserve community structure, we utilize k-means to classify groups and according to different groups, users locally perform privacy with truncated Laplace mechanism to protect sensitive information and send noise data to server, so as to better protect the privacy based on the user own privacy consideration. By theoretical analysis and experiments, we verify that PSG can not only protect user privacy, but also improve efficiency of data.

Acknowledgements. This work was partially supported by the National Natural Science Foundation of China under Grants 62072061 and U20A20176, and by the Fundamental Research Funds for the Central Universities under Grant 2021CDJQY-026.

References

1. Seshadhri, C., Kolda, T.G., Pinar, A.: Community structure and scale-free collections of erdős-rényi graphs. Phys. Rev. E **85**(5), 056109 (2012)
2. Chen, L., et al.: A privacy settings prediction model for textual posts on social networks. In: Romdhani, I., Shu, L., Takahiro, H., Zhou, Z., Gordon, T., Zeng, D. (eds.) CollaborateCom 2017. LNICST, vol. 252, pp. 578–588. Springer, Cham (2018). https://doi.org/10.1007/978-3-030-00916-8_53
3. Sweeney, L.: k-anonymity: a model for protecting privacy. Int. J. Uncertain. Fuzziness Knowl. Based Syst. **10**(05), 557–570 (2002)
4. Machanavajjhala, A., Kifer, D., Gehrke, J., Venkitasubramaniam, M.: l-diversity: privacy beyond k-anonymity. ACM Trans. Knowl. Discov. Data (TKDD) **1**(1), 3-es (2007)
5. Zhang, X., Chen, R., Xu, J., Meng, X., Xie, Y.: Towards accurate histogram publication under differential privacy. In: Proceedings of the 2014 SIAM International Conference on Data Mining, pp. 587–595. SIAM (2014)
6. Ding, X., Zhang, X., Bao, Z., Jin, H.: Privacy-preserving triangle counting in large graphs. In: Proceedings of the 27th ACM International Conference on Information and Knowledge Management, pp. 1283–1292 (2018)

7. Sun, H., et al.: Analyzing subgraph statistics from extended local views with decentralized differential privacy. In: Proceedings of the 2019 ACM SIGSAC Conference on Computer and Communications Security, pp. 703–717 (2019)
8. Erdös, P., Rényi, A.: On the evolution of random graphs. In: The Structure and Dynamics of Networks, pp. 38–82. Princeton University Press (2011)
9. Aiello, W., Chung, F., Lu, L.: A random graph model for massive graphs. In: Proceedings of the Thirty-Second Annual ACM Symposium on Theory of Computing, pp. 171–180 (2000)
10. Qin, Z., Yu, T., Yang, Y., Khalil, I., Xiao, X., Ren, K.: Generating synthetic decentralized social graphs with local differential privacy. In: Proceedings of the 2017 ACM SIGSAC Conference on Computer and Communications Security, pp. 425–438 (2017)
11. Zhu, H., Zuo, X., Xie, M.: DP-FT: a differential privacy graph generation with field theory for social network data release. IEEE Access 7, 164304–164319 (2019)
12. Zhu, T., Li, G., Zhou, W., Yu, P.S.: Preliminary of differential privacy. In: Differential Privacy and Applications. Advances in Information Security, vol. 69, pp. 7–16. Springer, Cham (2017). https://doi.org/10.1007/978-3-319-62004-6
13. Day, W.Y., Li, N., Lyu, M.: Publishing graph degree distribution with node differential privacy. In: Proceedings of the 2016 International Conference on Management of Data, pp. 123–138 (2016)
14. Karwa, V., Slavković, A.B.: Differentially private graphical degree sequences and synthetic graphs. In: Domingo-Ferrer, J., Tinnirello, I. (eds.) PSD 2012. LNCS, vol. 7556, pp. 273–285. Springer, Heidelberg (2012). https://doi.org/10.1007/978-3-642-33627-0_21
15. Wang, T., Blocki, J., Li, N., Jha, S.: Locally differentially private protocols for frequency estimation. In: 26th USENIX Security Symposium (USENIX Security 2017), pp. 729–745 (2017)
16. Leskovec, J., Krevl, A.: Snap datasets: Stanford large network dataset collection (2014)
17. Luce, R.D., Perry, A.D.: A method of matrix analysis of group structure. Psychometrika 14(2), 95–116 (1949). https://doi.org/10.1007/BF02289146
18. Rousseau, R., Egghe, L., Guns, R.: Becoming Metric-Wise. A Bibliometric Guide for Researchers. Chandos-Elsevier, Kidlington (2018)
19. Rand, W.M.: Objective criteria for the evaluation of clustering methods. J. Am. Stat. Assoc. 66(336), 846–850 (1971)
20. Pinar, A., Seshadhri, C., Kolda, T.G.: The similarity between stochastic Kronecker and Chung-Lu graph models. In: Proceedings of the 2012 SIAM International Conference on Data Mining, pp. 1071–1082. SIAM (2012)

Topology Self-optimization for Anti-tracking Network via Nodes Distributed Computing

Changbo Tian[1,2], Yongzheng Zhang[1,2], and Tao Yin[1,2(✉)]

[1] Institute of Information Engineering, Chinese Academy of Sciences,
Beijing 100093, China
{tianchangbo,zhangyongzheng,yintao}@iie.ac.cn
[2] School of Cyber Security, University of Chinese Academy of Sciences,
Beijing 100049, China

Abstract. Anti-tracking network aims to protect the privacy of network users' identities and communication relationship. The research of P2P-based anti-tracking network has attracted more and more attentions because of its decentralization, scalability, and widespread distribution. But, P2P-based anti-tracking network still faces the attacks on network structure which can destroy the usability of anti-tracking network effectively. So, a secure and resilient network structure is an important prerequisite to maintain the stability and security of anti-tracking network. In this paper, we propose a topology self-optimization method for anti-tracking network via nodes distributed computing. Based on convex-polytope topology (CPT), our proposal achieves topology self-optimization by each node optimizing its local topology in optimum structure. Through the collaboration of all nodes in network, the whole network topology will evolve into the optimum structure. Our experimental results show that the topology self-optimization method improves the network robustness and resilience of anti-tracking network when confronting to the dynamic network environment.

Keywords: Topology self-optimization · Distributed computing · Node collaboration · Network optimization · Anti-tracking network

1 Introduction

Anti-tracking network [1–4] provides secure, anonymous communication for network users to protect the privacy of their network identities and communication relationships. As a large, scalable and stable network system, the design and implementation of anti-tracking network has faced huge challenges. On the one

Supported by the National Key Research and Development Program of China under Grant No. 2019YFB1005203, the National Natural Science Foundation of China under Grant No. U1736218.

© ICST Institute for Computer Sciences, Social Informatics and Telecommunications Engineering 2021
Published by Springer Nature Switzerland AG 2021. All Rights Reserved
H. Gao and X. Wang (Eds.): CollaborateCom 2021, LNICST 406, pp. 405–419, 2021.
https://doi.org/10.1007/978-3-030-92635-9_24

hand, anti-tracking network is an open network and allows each node join or exit from network freely. Then, malicious nodes can infiltrate in the network and measure network topology and network scale [5-8]. On the other hand, the management and optimization of network topology is essential for the dynamic changed topology. In general, the network optimization is implemented by the network controller which has the global view of the network. But the frequent communication between the controller and network can be easily monitored and traced by the adversary [9-12]. So, a stable, resilient and self-optimizing network structure is the foundation for anti-tracking network to provide secure and reliable communication [13].

Recent researchers [14-17] have developed many approaches to improve the robustness and security of anti-tracking network effectively. From the present researches [18-20] it seems, the topology optimization methods have attracted more and more attentions because an optimum network structure can bring about a vast improvement in the performance of network communication and anti-destroy ability. However, anti-tracking network is an P2P-based network which allows each node joins or leaves the network freely. In general, network optimization is achieved by network controller which is vulnerable to network tracing and monitoring. Some approaches [21-25] have been proposed to achieve network self-optimization, but most of them is focused on the resource allocation, path selection optimization and so on. In the dynamic network environment, network structure oriented self-optimization method is important to improve the security and resilience of anti-tracking network.

To address this problem, we propose a topology self-optimization method for anti-tracking network. Our proposal is based on convex-polytope topology (CPT) [26] and network self-optimization algorithm to improve the security and resilience of anti-tracking network. We make three key contributions in this paper as follows:

- We apply convex-polytope topology in the construction of anti-tracking network topology. Anti-tracking network based on CPT has better robustness.
- We define an optimal topology model based on CPT to achieve the topology optimization of anti-trakcing network.
- We propose a topology self-optimization method based on CPT, which improves the security and resilience of anti-tracking network.

2 Related Works

Network optimization is to improve the performance of anti-tracking network, and the stable, reliable network structure is the prerequisite of efficient communication. Aimed to self-optimization of network topology, Auvinen [27] proposes a topology management algorithm based on neural network which does not predetermine favorable values of the characteristics of the peers. The decision whether to connect to a certain peer is done by a neural network, which is trained with an evolutionary algorithm. Tian [22] proposes smart topology construction method (STon) to provide the self-management and self-optimization of topology for anti-tracking network. By deploying the neural network on each node of the anti-tracking network, each node can collect its local network state and calculate the network state parameters by the neural network to decide the link state with other nodes. With the collaboration of all nodes in the network, the network can achieve the self-management and self-optimization of its own topology. Liu [21] proposes an adaptive overlay topology optimization (AOTO) technique. AOTO is scalable and completely distributed in the sense that it does not require global knowledge of the whole overlay network when each node is optimizing the organization of its logical neighbors. Sun [23] presents THash, a simple scheme that implements a distributed and effective network optimization for DHT systems. THash uses standard DHT put/get semantics and utilizes a triple hash method to guide the DHT clients to choose their sharing peers in proper domains. Liang [24] presents the optimization formulations, and proposes a set of heuristic algorithms for the construction and dynamic management of the multiple sub-stream trees for practical implementation which can significantly improve the delay performance of existing P2P streaming systems. Jelasity [25] proposes a generic protocol for constructing and maintaining a large class of topologies. In the proposed framework, a topology is defined with the help of a ranking function. The nodes participating in the protocol can use this ranking function to order any set of other nodes according to preference for choosing them as a neighboring node. Liao [28] presents a trust-based topology management protocol, which aims to promote the fairness and service quality of P2P system by integrating a trust model into its topology management.

3   Introduction to Convex-Polytope Topology

3.1   Basic Properties

Convex-polytope Topology (CPT) [26] is a structured topology in which all nodes are constructed into a logical structure of convex-polytope as illustrated in Fig. 1. One of the advantages of CPT is the elimination of cutvertex because any two nodes in CPT have at least two non-overlapping paths. In this case, some nodes are removed from CPT, CPT can still keep the convex-polytope structure except the nodes in ring connection.

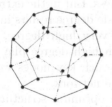

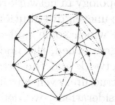

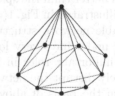

(a) CPT in sparse structure.   (b) CPT in dense structure.   (c) CPT with central point.

Fig. 1. CPT with different structures.

As illustrated in Fig. 2(a), some nodes are removed from CPT, CPT still keep the convex-polytope structure. But in Fig. 2(b), the nodes in ring connection are removed, CPT is split into two parts. But in the practical application, it is a small probability event that the removed nodes happened to be in the ring connection. So, when some nodes are removed, the timely recovery of CPT will keep the robustness and invulnerability of network.

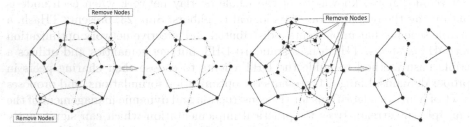

(a) CPT keeps the convex-polytope structure when some nodes are removed.   (b) CPT breaks into two parts when the nodes in ring connection are removed.

Fig. 2. The influence of node removal on the structure of CPT.

However, the connection structure and connection density have a big influence on the performance of CPT. As illustrated in Fig. 1, CPT in sparse topology is susceptible to the network churn, and the network structure is more vulnerable to the disconnection of nodes. CPT in dense topology has better robustness, but key nodes with high degree may appear to become the potential threats. As illustrated in Fig. 1(c), CPT has a central node which has connections with all other nodes. CPT with central node is unstable because the disconnection of central node would cause big damage in the structure of CPT.

3.2   The Optimum Structure of CPT

We consider the optimum structure of CPT based on which network has better robustness and invulnerability. Obviously, network density has a big influence on network connectivity and communication efficiency. The dense topology performs better than the sparse topology in network robustness. But in the extreme case illustrated in Fig. 1(c), the uneven distribution of node degree results in the unstable network structure in which some nodes have very high degree, but other nodes have very low degree. Then, the nodes with high degree play the very important roles in the network and the attack to such key nodes would severely disrupt the network structure, even partition the network.

On the basis of above considerations, we define the optimum structure of CPT as that CPT has the maximum network connectivity while conforming to the convex-polytope structure, and the degree of each node is close to the average of all nodes' degrees. CPT with maximum connectivity(CPT_M) has a special property that each surface of CPT is triangle. If CPT_M has n nodes, the number of edges is $l = 3 \times (n-2)$. So, for CPT_M with n nodes, the number of edges is fixed. Then, we can calculate the average degree $\bar{d}$ of CPT_M as shown in Eq. 1. When N approaches infinity, $\bar{d}$ approximately equals to 6. So, the average degree $\bar{d}$ can be set as the baseline for each node to measure and adjust its local topology.

$$\bar{d} = \lim_{N \to \infty} \frac{2 \times L}{N} = 6 - \lim_{N \to \infty} \frac{12}{N} = 6 \qquad (1)$$

Formally, for CPT with n nodes, its optimum structure can be defined as shown in Eq. 2, in which N_v denotes the number of nodes, N_e denotes the number of edges, $Degree(v_i)$ denotes the degree of node v_i.

$$CPT_{optimum} = \{N_v = n, N_e = 3 \times (n-2), Degree(v_i) \to \bar{d}\}(1 \le i \le n) \qquad (2)$$

4   Topology Self-optimization

The goal of topology self-optimization is to maintain the network topology in the optimum structure of CPT. Topology self-optimization is achieved by nodes' distributed computing. At first, each node calculates the optimum local topology ($T_{optimum}$) according to the situation of its current local topology and $T_{optimum}$ is an optimization objective for each node to adjust its local topology. But the final optimization plan is decided by the collaboration between the current node and its neighboring nodes and make sure the optimized local topology is beneficial to both sides.

Before that, we put some notations used in the following discussion in Table 1 to help readers refer to them conveniently.

Table 1. Notations

Notation	Description
CPT	Convex-polytope topology
$CPT_{optimum}$	The optimum structure of CPT
$T_{optimum}$	The optimum local topology of each node
$T_{original}$	The original local topology of each node
O	The criterion of local topology
$\bar{d}$	The average node degree of $CPT_{optimum}$
d_i	The node degree of node v_i
N_l	The node number of the local topology
r_d	The disconnection request between two nodes
r_c	The connection request between two nodes

4.1 Calculation of Optimum Local Topology

As we have mentioned above, $T_{optimum}$ is the optimization objective for each node to optimize local topology, and $T_{optimum}$ also needs to conform to the property of $CPT_{optimum}$. So, the evaluation standards of $T_{optimum}$ can be concluded as: (1) the distribution of nodes' degree, and (2) the difference between the degree of each node with the average degree $\bar{d}$ in $CPT_{optimum}$.

Here, we define the local topology of node v_i as the topology constructed by node v_i and its neighboring nodes. Then, the criterion of $T_{optimum}$ can be calculated as shown in Eq. 3, in which N_l denotes the node number of the local topology, d_i denotes the node degree of node v_i in the local topology and $\bar{d}$ denotes the degree baseline which has been discussed in Sect. 3.2.

$$O = \frac{\sum_{i=1}^{n}(d_i - \bar{d}))^2}{N_l} \tag{3}$$

Equation 3 computes the variance of all nodes' degree in local topology with the average node degree $\bar{d}$. The deviation of nodes' degree from $\bar{d}$ is lower, the node degree is more close to the average node degree of network. So, $T_{optimum}$ has the minimum value of O. Each node changes its local connection status and calculates O to assess the changed local topology until find the $T_{optimum}$.

In order to keep the network connectivity of $CPT_{optimum}$ unchanged, if one node breaks the link with its neighboring node, it has to instruct the two relevant neighboring nodes to build new connection. As illustrated in Fig. 3, for example, node v_0 breaks the link with node v_4, because two surfaces share the same link (v_0, v_4), the two neighboring nodes v_1 and v_6 need to build new connection with each other. Likewise, node v_0 breaks the link with v_3, then its two relevant neighboring nodes v_2 and v_5 build new connection with each other. In this way, some nodes reduce their degree and the others increase their degree to achieve the equilibrium of network connectivity.

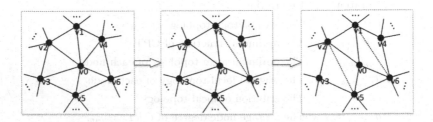

Fig. 3. The adjustment of local topology of node v_0.

The calculation of $T_{optimum}$ is an optimum result search algorithm from all the possibilities that each node adjusts its local topology. In the adjustment of local topology, the parameters in Eq. 3, such as $d_i(1 \leq i \leq m)$ and N_l, also need to be adjusted according to the changed local topology. Algorithm 1 shows the pseudocode of the calculation of $T_{optimum}$ which is implemented by recursive algorithm, the detailed workflow of Algorithm 1 is concluded as follows:

(1) Node v_0 first calculates the criterion O_0 of its original local topology $T_{original}$, and stores it in an array L.
(2) Node v_0 successively disconnects with one of its neighboring nodes v_i to generate a new local topology T_i. Node v_0 calculates the criterion O_i of each new local topology T_i and stores them in the array L.
(3) Based on each T_i, node v_0 recursively executes the step (2) to calculate the criteria of all the new changed topology until node v_0 has no neighboring nodes. All the calculated criteria are stored in the array L.
(4) Find the minimum value O_{min} of criterion in array L, and the local topology related with O_{min} is the optimum local topology for node v_0.

Algorithm 1. Calculation Algorithm of Optimum Local Topology

Input: M: the original connection matrix, C_n: the neighboring nodes set, v_0: current node
Output: T: the connection matrix of $T_{optimum}$
1:
2: **function** $Calculation(M, C_n, L)$ ▷ Recursive algorithm to search all the possibilities that node v_0 changes its local topology
3:    **if** $C_n.size() > 0$ **then**
4:       **for** v in C_n **do**
5:          v_0 disconnects with v
6:          $v_i, v_j \leftarrow Surface(v_0, v)$ ▷ Get the two neighboring nodes in the same surface of node v_0 and v
7:          v_i connects with v_j
8:          $M' \leftarrow Update(M)$ ▷ Get the changed topology
9:          $o_t' = O(M')$ ▷ Calculate O of the changed topology
10:          $L.append(o_t', M')$
11:          $C_n' = C_n.remove(v)$
12:          $Calculation(M', C_n', L)$
13:       **end for**
14:    **else**
15:       **return**
16:    **end if**
17: **end function**
18:
19: **function** $Main(M, C_n, L)$
20:    $o = O(M)$ ▷ Calculate O of the original topology of node v_0
21:    $L.append(o, M)$
22:    $Calculation(M, C_n, L)$
23:    $(o_t, M_t) = L.min()$ ▷ Get the min O and its related local topology
24:    **return** M_t
25: **end function**

According to Algorithm 1, each node searches its $T_{optimum}$ through breaking links. So, the degree of each node in $T_{optimum}$ will not exceed its degree in $T_{original}$. $T_{optimum}$ is just an optimization suggestion for each node to optimize its local topology. The final adjustment of local topology may not completely conform to the structure of $T_{optimum}$ because any adjustment of topology should be confirmed by the relevant nodes.

4.2 Topology Self-optimization via Nodes' Collaboration

The optimum local topology provides the useful information for each node to optimize its local topology. But the optimization of each node's local topology can not totally depend on the optimum local topology. The effect of topology adjustment on other nodes also need to be taken into consideration. So, the self-optimization of topology is achieved by nodes' collaboration.

Each adjustment of topology involves four nodes, two nodes executing disconnection operation and the other two nodes executing connection operation. So, each adjustment of topology should be confirmed by the relevant four nodes.

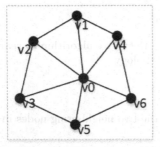

(a) The original topology of node v_0.

(b) The optimum topology of node v_0.

Fig. 4. The original topology of node v_0 and its calculated optimum topology.

Assume the original topology of node v_0 is as shown in Fig. 4(a), and it calculates the optimum topology as shown in Fig. 4(b). To adjust its local topology from $T_{original}$ to $T_{optimum}$, node v_0 needs the following operations: (1) the disconnection of v_0 and v_4, the connection of v_1 and v_6; (2) the disconnection of v_0 and v_1, the connection of v_2 and v_6; (3) the disconnection of v_0 and v_3, the connection of v_2 and v_5. For convenience of discussion, we use r_d and r_c to denote the disconnection request and connection request respectively.

Take the example of the collaboration of node v_0, v_3, v_2 and v_5, the detailed process of these nodes' collaboration in the adjustment of local toppology can be described as follows:

(1) Node v_0 sends r_d to v_3 for disconnection, and sends r_c to v_2 and v_5 to instruct them to build new connection. The criteria of $T_{original}$ and $T_{optimum}$ of node v_0 are denoted by O_0 and O_0' respectively.

(2) After node v_3, v_2, v_5 receive r_c or r_d, they respectively calculate their criteria O_3', O_2', O_5' of their local topologies changed according to the request. The criteria of the original topology of node v_3, v_2, v_5 are denoted by O_3, O_2, O_5 respectively. If $O_3' \leq O_3$, $O_2' \leq O_2$, $O_5' \leq O_5$, node v_3, v_2, v_5 agree with the topology adjustment requested by node v_0. Otherwise, go to Step (3).

(3) For node $v_k (k \in \{2, 3, 5\})$ and $O_k' > O_k$, v_k calculates the difference D_k between O_k and O_k'. Node v_0 calculates the difference D_0 between O_0 and O_0'. If $D_k < D_0$, node v_k agrees with the topology adjustment. Otherwise, go to step (4).

(4) If at least three nodes satisfy the rules shown in step (2) and (3), the topology adjustment has to be implemented. Otherwise, the local topology involes these four nodes stays the same.

In each adjustment of local topology, the relevant nodes can arrive at consensus or not that is decided by effect of changed topology on each node. If the topology adjustment evolves the local topology of relevant nodes to better structure, of course they should agree with the topology adjustment. If some nodes get worse topology structure, they compare the effect of topology adjustment and the node with bigger effect has the decision to adjust topology or not. At last, the local topology has to be adjusted when at least three nodes arrive at consensus.

According to the above adjustment process, each node adjusts its local topology according to its $T_{optimum}$ only when the relevant nodes arrive at consensus. So, the final optimized local topology of each node may not completely accord with its $T_{optimum}$ because some neighboring nodes may not arrive at consensus in topology adjustment.

All nodes adjust their local topology according to their $T_{optimum}$, then the network topology gradually evolves to optimum structure. In case that each node frequently requests for the topology adjustment to affect the performance of network communication, we set the topology stability parameter S as shown in Eq. 4, in which n denotes the node number in the local topology, and O_i denotes the criterion of node v_i' local topology.

$$S = \frac{\sum_1^n (O_r)}{n} (1 \leq r \leq n) \tag{4}$$

The topology stability parameter S is the average value of the topology criterion O of all nodes in the local topology. Each node v_i calculates the parameter S_i in its local topology. Then, the topology stability condition of node v_i can be set as $S_i \leq s_{max}$, in which s_{max} is the upper limit parameter for each node to adjust the sensitivity of the local topology optimization. The parameter s_{max} is smaller, the node implements the local topology optimization more frequently.

5 Performance Evaluation

In this section, the performance of anti-tracking network based on our proposal is evaluated through computer simulations. The simulation computer has a 12-Core

4 GHz CPU and 64 GB RAM. We first evaluate the effectiveness of our proposal (Topology Self-optimization, TS). Then, we compare TS with other two network topology optimization methods: neural network based network optimization method (NN) [22] and distributed hash table based network optimization method (DHT) [23] in network resilience.

5.1   Evaluation of Network Optimization

To evaluate the effectiveness of our proposal, we use ring topology and centralized topology seperately illustrated in Fig. 5(a) and Fig. 5(b) to construct a network with 1000 nodes, and deploy the self-optimization algorithm on this network. We use d_{min}, d_{max} and d_{avg} to denote the minimum node degree, maximum node degree and average node degree of all nodes respectively. We define one round of network self-optimization as that all nodes finish the optimization of its local topology. Then we calculate the above criteria in each round of network self-optimization to analyze the effectiveness of our proposal.

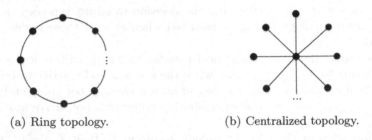

(a) Ring topology.   (b) Centralized topology.

Fig. 5. Two topologies for performance evaluation of network self-optimization.

As shown in Fig. 6(a), the network is constructed in ring topology originally. The node degree of each node is 2. Before each node begins to optimize its local topology, it has to maintain its local topology in maximum connectivity. So, d_{max} increases sharply because the network needs to reach the maximum connectivity at first. Inevitably, the degree of some nodes will get bigger. After all nodes reach the maximum connectivity of their local topology, they begin to optimize their local topology. Then, d_{max} decreases until the network reaches the optimal structure. At last, d_{max} keeps nearly 10, d_{min} keeps nearly 4, and d_{avg} keeps nearly 6 which accords with the property of CPT shown in Eq. 1.

As shown in Fig. 6(b), the network is constructed in centralized topology in which one node has connections with all other nodes. So, the center node has a very high degree. At first, each node needs to maintain its local topology in maximum connectivity, d_{max} keeps unchanged for a few rounds. After the network reaches the maximum connectivity, d_{max} decreases sharply and keeps nearly 9 at last. The change of d_{max} shows the effectiveness of network self-optimization.

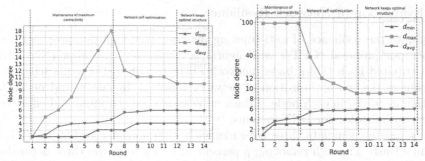

(a) The change of node degree in the self-optimization process of network based on ring topology.

(b) The change of node degree in the self-optimization process of network based on centralized topology.

Fig. 6. The change of β values of CPTs, NN, THash in random-p removal and top-p removal.

In order to present the effectiveness of network self-optimization intuitively, we calculate the node degree distribution after network self-optimization in ring topology and centralized topology respectively. D_r denotes the node degree distribution generated in ring topology, D_c denotes the node degree distribution generated in centralized topology. As illustrated in Fig. 7, the output topology of network self-optimization in both ring topology and centralized topology is almost the same. More than 80% of the nodes have the degree in the interval $[5, 8]$ which proves that the distribution of node degree is approximately close to uniform distribution. So, our proposal is effective to optimize the network into optimal structure.

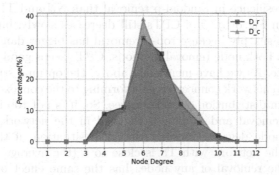

Fig. 7. The node degree distribution after network self-optimization in different topologies.

5.2   Evaluation of Network Resilience

To evaluate the resilience of anti-tracking network based on our proposal, we compare our proposal (CPTs) with neural network based network (NN) [22] and distributed hash tables based network (THash) [23] in the same scenario. NN achieves the self-optimization of network topology by the neural network algorithm depolyed in each node. THash implements a distributed and effective network optimization for DHT systems.

We seperately simulate three networks with 2000 nodes according to CPTs, NN and THash. Through removing p percent of nodes from the three networks each time, we use the node number of maximum connected graph to measure the network resilience when confronted to dynamic network scenario. We use Eq. 5 to quantify the performance of network resilience. $G(p)$ denotes the subgraph after p percent of nodes is removed from the original network, $MCS(G(p))$ denotes the maximum connected subgraph of $G(p)$, $Num(g)$ denotes the node number of a graph g, N_G denotes the node number of the graph G. The metric β measures the maximum connectivity of the network after some nodes are removed from the network. The β is higher, the network resilience is better.

$$\beta = \frac{Num(MCS(G(p)))}{N_G} \tag{5}$$

In the experiments, we use two different ways to remove nodes from network:

- **Random-p Removal:** In each round of nodes removal, we remove p percent of nodes from the network randomly.
- **Top-p Removal:** In each round of nodes removal, we remove p percent of nodes with the highest degree.

As shown in Fig. 8(a), β of CPTs decreases slowly which means CPTs has better network resilience in random-p removal than NN and THash. For top-p removal shown in Fig. 8(b), β of CPTs still decreases slowly, but β of NN and THash decreases sharply because top-p removal has bigger damage to network structure. But in both node removal methods, CPTs keeps good performance in network resilience. As we have mentioned above, the optimal structure of CPT has the maximum network connectivity conforming to the convex-polytope structure, and uniform distribution of node degree. So, it is not too much difference between top-p removal and random-p removal in the network with balanced distribution of node degree. Consider the ideal situation, if the node degree of all nodes in the network with TS is close to 6 (the average node degree of $CPT_{optimum}$), the removal of any nodes has the same effect on the topology structure.

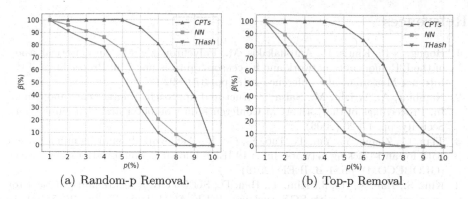

(a) Random-p Removal.   (b) Top-p Removal.

Fig. 8. The change of β values of CPTs, NN, THash in random-p removal and top-p removal.

6 Conclusion

In this paper, we propose a topology self-optimization method for anti-tracking network via nodes distributed computing. Our proposal applies convex-polytope topology (CPT) in the construction of anit-tracking network. Based on CPT, we achieve the topology self-optimization for anti-tracking network. We also define an optimum structure of CPT in which network has maximum network connectivity and balanced distribution of node degree. Each node optimizes its local topology, then the whole network evolves into optimum structure of CPT through the collaboration of all nodes. Each node first calculates its optimum local topology according to its local topology situation. Then, each node negotiates with its neighboring nodes to adjust its local topology according to the calculated optimum local topology. When the relevant nodes arrive at consensus, the local topology will be adjusted according to the optimum local topology. Or, they will keep the local topology unchanged to make sure each adjustment of local topology is beneficial to all the relevant nodes.

In the experiments, we evaluate the network optimization and network resilience of our proposal. The experimental results show that our proposal has a good performance in network optimization. The network based on our proposal can achieve topology self-optimization effectively. Compared with the current network optimization methods, our proposal has better network resilience when confronting to dynamic network environment.

Acknowledgements. The authors would like to thank the anonymous reviewers for their insightful comments and suggestions on this paper. This work was supported in part by the National Key Research and Development Program of China under Grant No. 2019YFB1005203, the National Natural Science Foundation of China under Grant No. U1736218.

References

1. Hoang, N.P., Kintis, P., Antonakakis, M., Polychronakis, M.: An empirical study of the I2P anonymity network and its censorship resistance. In: Proceedings of the Internet Measurement Conference 2018, pp. 379–392 (2018)
2. Nambiar, A., Wright, M.: Salsa: a structured approach to large-scale anonymity. In: Proceedings of the 13th ACM Conference on Computer and Communications Security, pp. 17–26 (2006)
3. Tian, C., Zhang, Y., Yin, T., Tuo, Y., Ge, R.: Achieving dynamic communication path for anti-tracking network. In: 2019 IEEE Global Communications Conference (GLOBECOM), pp. 1–6. IEEE (2019)
4. Kim, S., Han, J., Ha, J., Kim, T., Han, D.: SGX-TOR: a secure and practical tor anonymity network with SGX enclaves. IEEE/ACM Trans. Netw. **26**(5), 2174–2187 (2018)
5. Bauer, K., McCoy, D., Grunwald, D., Kohno, T., Sicker, D.: Low-resource routing attacks against tor. In: Proceedings of the 2007 ACM workshop on Privacy in electronic society, pp. 11–20 (2007)
6. Wright, M.K., Adler, M., Levine, B.N., Shields, C.: The predecessor attack: an analysis of a threat to anonymous communications systems. ACM Trans. Inf. Syst. Secur. (TISSEC) **7**(4), 489–522 (2004)
7. Kang, B.B., et al.: Towards complete node enumeration in a peer-to-peer botnet. In: Proceedings of the 4th International Symposium on Information, Computer, and Communications Security, pp. 23–34 (2009)
8. Loesing, K., Murdoch, S.J., Dingledine, R.: A case study on measuring statistical data in the Tor anonymity network. In: Sion, R., et al. (eds.) FC 2010. LNCS, vol. 6054, pp. 203–215. Springer, Heidelberg (2010). https://doi.org/10.1007/978-3-642-14992-4_19
9. Murdoch, S.J., Danezis, G.: Low-cost traffic analysis of tor. In: 2005 IEEE Symposium on Security and Privacy (S&P 2005), pp. 183–195. IEEE (2005)
10. Chen, T., Cui, W., Chan-Tin, E.: Measuring Tor relay popularity. In: Chen, S., Choo, K.-K.R., Fu, X., Lou, W., Mohaisen, A. (eds.) SecureComm 2019, Part I. LNICST, vol. 304, pp. 386–405. Springer, Cham (2019). https://doi.org/10.1007/978-3-030-37228-6_19
11. Das, D., Meiser, S., Mohammadi, E., Kate, A.: Anonymity trilemma: strong anonymity, low bandwidth overhead, low latency-choose two. In: 2018 IEEE Symposium on Security and Privacy (SP), pp. 108–126. IEEE (2018)
12. Cociglio, M., Fioccola, G., Marchetto, G., Sapio, A., Sisto, R.: Multipoint passive monitoring in packet networks. IEEE/ACM Trans. Netw. **27**(6), 2377–2390 (2019)
13. Shirazi, F., Diaz, C., Wright, J.: Towards measuring resilience in anonymous communication networks. In: Proceedings of the 14th ACM Workshop on Privacy in the Electronic Society, pp. 95–99 (2015)
14. Chen, M., Liew, S.C., Shao, Z., Kai, C.: Markov approximation for combinatorial network optimization. IEEE Trans. Inf. Theory **59**(10), 6301–6327 (2013)
15. Alon, N., Awerbuch, B., Azar, Y., Buchbinder, N., Naor, J.: A general approach to online network optimization problems. ACM Trans. Algorithms (TALG) **2**(4), 640–660 (2006)
16. Chen, J., Touati, C., Zhu, Q.: A dynamic game approach to strategic design of secure and resilient infrastructure network. IEEE Trans. Inf. Forensics Secur. **15**, 462–474 (2019)

17. Shafigh, A.S., et al.: A framework for dynamic network architecture and topology optimization. IEEE/ACM Trans. Netw. **24**(2), 717–730 (2015)
18. Kang, S.: Research on anonymous network topology analysis. In: 2015 International Conference on Automation, Mechanical Control and Computational Engineering, pp. 1416–1421, Atlantis Press (2015)
19. Bouchoucha, T., Chuah, C.-N., Ding, Z.: Topology inference of unknown networks based on robust virtual coordinate systems. IEEE/ACM Trans. Netw. **27**(1), 405–418 (2019)
20. Jansen, R., Johnson, A.: Safely measuring Tor. In: Proceedings of the 2016 ACM SIGSAC Conference on Computer and Communications Security, pp. 1553–1567 (2016)
21. Liu, Y., Zhuang, Z., Xiao, L., Ni, L.M.: AOTO: adaptive overlay topology optimization in unstructured P2P systems. In: GLOBECOM 2003. IEEE Global Telecommunications Conference (IEEE Cat. No. 03CH37489), vol. 7, pp. 4186–4190. IEEE (2003)
22. Tian, C., Zhang, Y.Z., Yin, T., Tuo, Y., Ge, R.: A smart topology construction method for anti-tracking network based on the neural network. In: Wang, X., Gao, H., Iqbal, M., Min, G. (eds.) CollaborateCom 2019. LNICST, vol. 292, pp. 439–454. Springer, Cham (2019). https://doi.org/10.1007/978-3-030-30146-0_31
23. Sun, Y., Richard Yang, Y., Zhang, X., Guo, Y., Li, J., Salamatian, K.: THash: a practical network optimization scheme for DHT-based P2P applications. IEEE J. Sel. Areas Commun. **31**(9), 379–390 (2013)
24. Liang, C., Liu, Y., Ross, K.W.: Topology optimization in multi-tree based P2P streaming system. In: 2009 21st IEEE International Conference on Tools with Artificial Intelligence, pp. 806–813. IEEE (2009)
25. Jelasity, M., Babaoglu, O.: T-man: Gossip-based overlay topology management. In: Brueckner, S.A., Di Marzo Serugendo, G., Hales, D., Zambonelli, F. (eds.) ESOA 2005. LNCS (LNAI), vol. 3910, pp. 1–15. Springer, Heidelberg (2006). https://doi.org/10.1007/11734697_1
26. Tian, C., Zhang, Y., Yin, T.: Modeling of anti-tracking network based on convex-polytope topology. In: Krzhizhanovskaya, V.V., et al. (eds.) ICCS 2020, Part II. LNCS, vol. 12138, pp. 425–438. Springer, Cham (2020). https://doi.org/10.1007/978-3-030-50417-5_32
27. Auvinen, A., Keltanen, Y., Vapa, M.: Topology management in unstructured P2P networks using neural networks. In: 2007 IEEE Congress on Evolutionary Computation, pp. 2358–2365. IEEE (2007)
28. Liao, Z., Liu, S., Yang, G., Zhou, J.: TTMP: a trust-based topology management protocol for unstructured P2P systems. In: He, X., Hua, E., Lin, Y., Liu, X. (eds.) Computer, Informatics, Cybernetics and Applications. LNEE, vol. 107, pp. 271–281. Springer, Dordrecht (2012). https://doi.org/10.1007/978-94-007-1839-5_29

An Empirical Study of Model-Agnostic Interpretation Technique for Just-in-Time Software Defect Prediction

Xingguang Yang[1,2], Huiqun Yu[1,3(✉)], Guisheng Fan[1(✉)], Zijie Huang[1], Kang Yang[1], and Ziyi Zhou[1]

[1] Department of Computer Science and Engineering, East China University of Science and Technology, Shanghai 200237, China
{yhq,gsfan}@ecust.edu.cn
[2] Shanghai Key Laboratory of Computer Software Evaluating and Testing, Shanghai 201112, China
[3] Shanghai Engineering Research Center of Smart Energy, Shanghai, China

Abstract. Just-in-time software defect prediction (JIT-SDP) is an effective method of software quality assurance, whose objective is to use machine learning methods to identify defective code changes. However, the existing research only focuses on the predictive power of the JIT-SDP model and ignores the interpretability of the model. The need for the interpretability of the JIT-SDP model mainly comes from two reasons: (1) developers expect to understand the decision-making process of the JIT-SDP model and obtain guidance and insights; (2) the prediction results of the JIT-SDP model will have an impact on the interests of developers. According to privacy protection laws, prediction models need to provide explanations. To this end, we introduced three classifier-agnostic (CA) technologies, LIME, BreakDown, and SHAP for JIT-SDP models, and conducted a large-scale empirical study on six open source projects. The empirical results show that: (1) Different instances have different explanations. On average, the feature ranking difference of two random instances is 3; (2) For a given system, the feature lists and top-1 feature generated by different CA technologies have strong agreement; However, CA technologies have small agreement on the top-3 features in the feature ranking lists. In the actual software development process, we suggest using CA technologies to help developers understand the prediction results of the model.

Keywords: Software defect prediction · Just-in-time · Classifier-agnostic interpretation · Model interpretation

1 Introduction

Software defects will cause errors and unexpected results of software system [28]. Although software quality assurance activities (such as code review, software

H. Gao and X. Wang (Eds.): CollaborateCom 2021, LNICST 406, pp. 420–438, 2021.
https://doi.org/10.1007/978-3-030-92635-9_25

testing, etc.) can effectively find and repair defects in software system. However, under the constraints of software testing resources and software release time, it is unrealistic for software developers to implement comprehensive quality assurance activities for software systems [43].

Software defect prediction is an effective method to improve the efficiency of SQA activities [21]. By accurately predicting the defect probability of the program modules in the software system, developers can design a reasonable test resource allocation plan, thereby reducing the cost of software development and improving the quality of the software.

Just-in-time software defect prediction (JIT-SDP) is a fine-grained defect prediction method, and its prediction entity is fine-grained code changes. JIT-SDP technology has the following three advantages [19]: (1) Fine granularity. Compared with coarse-grained file, function, or package, code changes contain fewer lines of code. Therefore, once a change is predicted to be defect-prone, developers can efficiently check the code of the change; (2) Immediately. Once the developer submits a code change, the defect prediction model can immediately predict the defect probability of the change. Therefore, developers still clearly remember the process of software development when checking the code for changes. (3) Traceability. There is usually only one developer for a code change. Therefore, project managers can easily find suitable developers to check and fix defect changes. Due to the importance of JIT-SDP for improving software quality, many companies have adopted JIT-SDP technology, such as Avaya [27], Blackberry [35], and Cisco [40].

Currently, JIT-SDP has received extensive research and attention. Most of the research focuses on the prediction performance of the JIT-SDP model [48], data annotation [8], feature extraction [23] and so on. However, few studies have focused on the interpretability of the JIT-SDP model. The interpretability of models is very critical in the field of software engineering [7]. In the research of JIT-SDP, the need for interpretability of prediction models mainly comes from two aspects: (1) Lack of trust. Developers may ask "Why the code change I submitted is predicted to be buggy". In a software analysis system, if the prediction model cannot provide an explanation for the prediction results, the developer may not be able to believe the prediction results of the model [7], which will hinder the application of the JIT-SDP model in actual software development. (2) Data privacy protection. In the application of the JIT-SDP model, if code changes submitted by a developer are often predicted as buggy by the model, the developer's job opportunities, salary, bonus, etc. may be affected. According to the Article 22 of the European Union's General Data Protection Regulation (GDPR) [31], if the data used in the decision-making model affects individuals or organizations, the decision-making model needs to provide an explanation.

Classifier-specific (CS) techniques can provide explanations by analyzing the internal structure of the model. However CS methods are not universal to all prediction models. Many complex classification models such as SVM, deep neural network and other black box models do not have specific model interpretation techniques [5,30]. In the JIT-SDP model, many black-box models have high prediction performance, such as random forest [17], deep neural work [10], etc. Therefore, we propose to use three classifier-agnostic (CA) interpretation techniques to interpret the JIT-SDP model, namely Local *Local Interpretable Model-agnostic Explanations* (LIME) [32], *BreakDown* [9,37] and *SHapley Additive ExPlanations* (SHAP) [24]. These three CA technologies can provide explanations for each instance. The Sect. 3 of this paper describes the basic principles of these three CA technologies. The main contributions of this article are as follows:

1. We introduced three CA technologies LIME, BreakDown and SHAP to explain the prediction results of the JIT-SDP model.
2. We conducted an a large empirical research on six open source projects, and evaluated the distribution of the ranking differences of each feature in different instances;
3. For a given project, we evaluate the agreement of the feature ranking lists generated by the three CA technologies.

The organization structure of the rest of this paper is as follows: Sect. 2 introduces the work related to JIT-SDP and the interpretability of software defect prediction; Sect. 3 briefly introduces the basic principles of the three CA methods; Sect. 4 explains the experimental settings; Sect. 5 introduces the results and analysis of the experiment; Sect. 6 summarizes the research of this paper and explains the future work.

2   Related Work

2.1   Just-in-time Software Defect Prediction

Mockus and Weiss [27] first proposed the JIT-SDP technology. They use the attributes of the code change, such as the number of lines added or deleted, and the number of code modified subsystems, to predict the defect probability of the change, and they apply the prediction model to a 5ESS switch system software. Recently, Kamei et al. [19] conducted large-scale empirical research on six open source projects and five commercial projects. They use 14 change metrics to measure code changes. The results show that their prediction model can achieve 68% accuracy and 64% recall, and can identify 35% of buggy changes when investing 20% of the code inspection effort. Subsequently, Kamei et al. [17] studied the prediction performance of the JIT-SDP model in the cross-project scenario. They used random forest to build models and conducted research on 10 projects. The results show that the method of data merging and ensemble learning can improve the prediction performance of cross-project models. Yang et al. [48] compared the prediction performance of simple unsupervised models and supervised models. They found that in the effort-aware JIT-SDP, the simple unsupervised models are better than the state-of-the-art supervised models. Inspired by the work of Yang et al. [48], Huang et al. [11,12] further compared the prediction performance of supervised and unsupervised models in JIT-SDP.

They found that the unsupervised models have many context switches and high false alarms. To this end, they proposed a supervised model called CBS, which combines the EALR model proposed by Kamei et al. [19], and the LT model proposed by Yang et al. [48]. The results show that the proposed method is significantly better than other benchmark models in precision and F1 indicators. Besides, to improve the performance of the JIT-SDP model, researchers have also proposed many methods, such as multi-objective optimization algorithms [6,44], differential evolution algorithm [45,46], ensemble learning [47], deep learning [10], online learning [3], etc.

2.2 Explainability in Software Defect Prediction

The interpretability of software defect prediction models is a relatively new research area. Jiarpakdee et al. [15] first studied the participants' perceptions of the goal of the software defect prediction model. Research results show that 82% to 84% respondents believe that software defect prediction is useful in the three goals of optimizing test resource allocation, understanding defect-associated characteristics, and understanding the prediction results of prediction models. In addition, the results showed that LIME was the most popular model interpretation technique. Subsequently, Jiarpakdee et al. [14] empirically evaluated three classifier-agnostic interpretation techniques on 32 defect data sets, namely LIME-HPO, LIME, and BreakDown. The results show that it is necessary to use model-agnostic technology to explain the software defect prediction model, and the instance explanations have a high degree of overlap with the global explanations. Wattanakriengkrai et al. [42] proposed a framework called LINE-DP to identify defective lines of code. LINE-DP first uses the code token feature to build a file-level defect prediction model. Then for a file that is predicted to be defective, LINE-DP uses LIME to identify risky tokens. Finally, the code lines with risky tokens are predicted as defective code lines. Rajbahadur et al. [30] analyzed the impact of different model interpretation techniques on defect classifiers. They studied six CS technologies and two CA technologies, and conducted research on 18 software projects. The results show that the feature importance rank list generated by CA and CS methods do not always have a strong agreement. In addition, different CA methods have strong agreement in indicators top-1 overlap and top-3 overlap. To improve the agreement of feature ranks among CS methods, they recommend using the CFS method to remove correlated features. There are also some studies that analyze the impact of imbalance techniques [39] and the correlation of features [16] on model interpretation. Currently, there are relatively few studies on the interpretability of the JIT-SDP model. Pornprasit et al. [29] proposed a JIT-SDP method called JITLine. The method first extracts the tokens features from code changes and builds a commit-level prediction model; then, for each test instance, JITLine uses the model interpretation technology LIME to calculate the feature importance score of each token. Finally, the code lines in the commit are sorted according to the feature importance score.

3   Classifier-Agnostic Interpretation Technique

The interpretability or explainability of a prediction model refers to the degree to which human observers can understand the reasons why the model makes predictions [7,26]. There are two main ways to obtain the interpretability of a machine learning model [22]:

1. **Make the decision-making process of the machine learning model transparent and easy to understand.** This kind of model interpretation method requires the internal structure of the model to be transparent and easy to understand, so it is also called a classifier-specific (CS) interpretation technique. The white box model has a simple internal structure, so it can make the prediction process of the model transparent. For example, the decision-making process of a decision tree can be displayed in a visual tree structure. The path from the root node to the leaf node of the decision tree provides an explanation of the prediction result. Although the white-box model has high interpretability, it has poorer prediction performance than the black-box model.
2. **Provide an explanation for each prediction result of models.** This kind of model interpretation technology can interpret any machine learning algorithm, including complex black-box models (such as random forest, SVM, deep learning, etc.), so it is also called classifier-agnostic (CA) interpretation technology.

Just-in-time software defect prediction is a binary classification task. Most of the current researches use complex black-box algorithms (such as random forest [17], deep learning [10], etc.) to build classification models and obtain high prediction performance. To this end, this paper introduces three state-of-the-art CA technologies LIME [32], BreakDown [9,37], and SHAP [24] to explain the prediction results of the JIT-SDP model. This section details the specific processes of the three CA technologies.

3.1   LIME

LIME is a local surrogate model proposed by Ribeiro et al. [32]. The model uses an interpretable machine learning model (i.e., linear regression, decision tree etc.) to locally mimics the predictions of the black-box model. Therefore, a local surrogate model can be used to explain an individual instance. Assuming that $\hat{f}$ is a black-box model and x is an instance that needs to be explained, the loss function of the LIME method is shown in Eq. 1.

In the Eq. 1, the minimized loss function L reflects the closeness of the prediction results between the black-box model $\hat{f}$ and the proxy model g. G represents the set of interpretable machine learning algorithms; π defines the domain range when sampling around instance x; $Omega(g)$ represents the model complexity of the interpretable model g.

$$explanation(x) = \arg\min_{g \in G} L(\hat{f}, g, \pi_x) + \Omega(g) \tag{1}$$

The key steps of training the local surrogate model are as follows: (1) Perturb the target instance to generate new points; (2) Use the black-box model to predict the new points; (3) According to the closeness of new points to the target instance, set a weight for each point; (4) Use an interpretable machine learning algorithm to train a local surrogate model on the data set with weights. The local surrogate model trained in our experiment is ridge regression. Therefore, the coefficients of the regression model are used to measure the feature importance scores of a change.

3.2 BreakDown

BreakDown is a CA technology. For a target instance, BreakDown uses a greedy strategy to sequentially measure the importance scores of features. Assuming that x is a target instance with m features, $\hat{f}$ is a black box model, and the feature importance score generated by BreakDown satisfies Eq. 2, where $v(j, x)$ represents the feature importance score calculated by BreakDown for the instance on the j-th feature, $\hat{f}(x)$ represents the prediction result of the black-box model on the instance x, and v_0 represents the average value of the prediction results of the black-box model in the training set.

$$\hat{f}(x) = v_0 + \sum_{j=1}^{m} v(j, x) \tag{2}$$

The calculation process of the BreakDown method mainly includes the following four steps: (1) Use the black-box model to predict all the training data, and calculate the average of the prediction results; (2) For each feature, sequentially replace the value of a feature in the training data with the value of the target instance to construct new training data; (3) Use the black box model to predict the new training set. By analyzing the differences in the prediction results on the new data set, the most important features are identified, and the differences in the prediction results are used to measure the importance scores of the features; (4) Repeat the third step until the importance scores of all the features are calculated.

3.3 SHAP

SHAP calculates the contribution of each feature to the predicted result according to the coalition game theory. The model interpretation of SHAP is shown in Eq. 3, where g represents an interpretable model, m represents the size of the coalition, $z' \subseteq \{0, 1\}^m$ and Φ_j represents the Shapley value of the j-th feature, which is used to evaluate the importance score of the feature. In the study of JIT-SDP, since all the features of the target instance x exist, the coalition vector is a vector with all values of 1, so Eq. 3 can be simplified to Eq. 4.

$$g(z') = \Phi_0 + \sum_{j=1}^{m} \Phi_j z'_j \tag{3}$$

$$g(x') = \varPhi_0 + \sum_{j=1}^{m} \varPhi_j \tag{4}$$

4 Experimental Setup

This section introduces the experimental setup from three aspects: data sets, building classification models, and evaluation metrics. We aim to answer the following two research questions through experiments:

- RQ: What are the interpretation results of classifier-agnostic techniques on the JIT-SDP model?
- RQ2: For a given project, what is the difference in the feature importance rank lists generated by different CA methods?

The experiment is run on a workstation, and its configuration is *Intel(R) Xeon(R) Gold 5118 CPU @ 2.30 GHz; RAM 251 GB.*

4.1 Data Sets

The experiment uses public data sets widely used in JIT-SDP research [3,38]. The data set contains the defect data of ten open source projects, which have a development cycle of more than five years, rich historical data and a good defect-inducing changes rate. To extract the metrics values of changes in the projects and identify defective code changes, Cabral et al. [3] used *Commit Guru* [33] to collect the change-level defect data set of projects. Commit Guru is an automated defect data collection tool that can extract code changes from the *Git Hub* and use the SZZ algorithm [36] to identify buggy changes. Table 1 shows the basic information of ten projects, including the name of the project, the number of changes, the number of defective changes, the proportion of buggy changes, the development period of the project, and the programming language. Each instance in the data sets contains 14 metrics, which can be effectively used to build the JIT-SDP model [48]. Table 2 describes the names and definitions of 14 metrics. These metrics can be divided into five dimensions including diffusion, size, purpose, history, and experience. Due to the limitation of the length of the paper, the detailed introduction can refer to Kamei et al. [19]'s study.

4.2 Building Classification Models

Before building the classification model, we use a cross-validation method to divide the data sets of each project into a training set and a test set. K-fold cross-validation [25] randomly divides the data sets into k folds with the same size, each of which has roughly the same proportion of defect changes. Then each fold of data is uses as the test set, and the remaining k-1 fold data is used as the training set. Finally, the average value of k prediction results is used as the final prediction performance. The challenge of k-fold cross-validation is to choose an

Table 1. The Basic Information of Data Sets

Project	#Changes	#Defective changes	%Defective rate	Period	Language
Fabric8	15591	5177	33.2	12/2011–12/2017	Java
JGroups	21468	6305	29.4	09/2003–12/2017	Java
Camel	36753	12488	34.0	03/2007–12/2017	Java
Tomcat	24043	10372	43.1	03/2006–12/2017	Java
Brackets	21347	8082	37.9	12/2011–12/2017	JavaScript
Neutron	24057	9214	38.3	12/2010–12/2017	Python
Spring-Integration	10999	4640	42.2	11/2007–01/2018	Java
Broadleaf	17429	5046	29.0	11/2008–12/2017	Java
Nova	61364	24857	40.5	08/2010-01–2018	Python
NPM	9258	2773	30.0	09/2009–11/2017	JavaScript

Table 2. The description of metrics

Dimension	Metric	Description
Diffusion	NS	Number of modified subsystems
	ND	Number of modified directories
	NF	Number of modified files
	Entropy	Distribution of modified code across each file
Size	LA	Lines of code added
	LD	Lines of code deleted
	LT	Lines of code in a file before the change
Purpose	FIX	Whether or not the change is a defect fix
History	NDEV	Number of developers that changed the files
	AGE	Average time interval between the last and the current change
	NUC	Number of unique last changes to the files
Experience	EXP	Developer experience
	REXP	Recent developer experience
	SEXP	Developer experience on a subsystem

appropriate value of k. We use the widely used 10-fold cross-validation technique to split training and test set [41].

After dividing the training set and the test set, we use the random forest algorithm [2] to build a classification model. Random forest is an ensemble learning algorithm based on bagging. It uses decision trees as the base learners and introduces a random attribute selection scheme. We chose random forest to build the classifier mainly considering the following two reasons: First, compared with traditional machine learning models, random forest has higher robustness, prediction accuracy, and stability [13]; Secondly, the research of Kamei et al. [18] shows that random forest has better prediction performance than other modeling techniques in software defect prediction research. Then they introduced random forest to the research of JIT-SDP [17].

4.3 Evaluation Metrics

This paper adopts three CA methods to explain the JIT-SDP model. To evaluate the differences in the feature importance rank lists generated by different model interpretation methods, similar to the research of Rajbahadur et al. [30], we introduced four evaluation indicators: Kendall's Tao coefficient, Kendall's W coefficient, top-3 overlap, and top-1 overlap. The specific introduction is as follows.

Kendall's Tau coefficient (τ) [20] is a non-parametric statistical coefficient used to evaluate the correlation between two sequences. If the two sequences are completely the same, then $\tau = 1$; if the two sequences are completely opposite, then $\tau = -1$. This paper uses τ to evaluate the agreement of the feature importance rank lists generated by the two CA methods. According to the Akoglu's suggestion [1], the relationship between the results of the τ test and the interpretation of agreement is shown in Eq. 5.

$$Kendall's\ \tau\ Agreement = \begin{cases} weak, & if\ |\tau| \leq 0.3 \\ moderate, & if\ 0.3 < |\tau| \leq 0.6 \\ strong, & if\ 0.6 < |\tau| \leq 1 \end{cases} \tag{5}$$

Kendall's W coefficient is a statistic used to evaluate the agreement of multiple sequences, and its range is between 0 and 1. The larger the value of Kendall's W, the more consistent the feature importance rank lists generated by the three CA methods. The agreement explanation of Kendal's W coefficient is shown in Eq. 6 [4].

$$Kendall's\ W\ Agreement = \begin{cases} weak, & if\ W \leq 0.3 \\ moderate, & if\ 0.3 < W \leq 0.5 \\ good, & if\ 0.5 < W \leq 0.7 \\ strong, & if\ 0.7 < W \leq 1 \end{cases} \tag{6}$$

Top-k overlap [30] is used to evaluate the agreement of multiple sequences on top k values. This paper uses the top-k indicator to evaluate the overlap of the feature importance rank lists generated by LIME, BreakDown, and SHAP on the top-k features. The calculation process of top-k is shown in Eq. 7, where p represents the number of ranking lists. In our experiment, p is set to 3, and k is set to 1 and 3.

$$Top - k\ overlap = \frac{|\bigcap_{i=1}^{p} features\ at\ top\ k\ ranks\ in\ list_i|}{|\bigcup_{i=1}^{p} features\ at\ top\ k\ ranks\ in\ list_i|} \tag{7}$$

Eqs. 8 and 9 give interpretations for the results of top-3 overlap and top-1 overlap, respectively.

$$Top - 3 \; Agreement = \begin{cases} negligible, & \text{if } 0.00 < top - 3 \; overlap \leq 0.25 \\ small, & \text{if } 0.25 < top - 3 \; overlap \leq 0.50 \\ medium, & \text{if } 0.50 < top - 3 \; overlap \leq 0.75 \\ large, & \text{if } 0.75 < top - 3 \; overlap \leq 1.00 \end{cases} \quad (8)$$

$$Top - 1 \; Agreement = \begin{cases} low, & \text{if } 0.00 < top - 1 \; overlap \leq 0.50 \\ high, & \text{if } 0.50 < top - 1 \; overlap \leq 1.00 \end{cases} \quad (9)$$

For example, if the top-3 feature sets generated by the three CA methods are as follows: LIME={LA, NF, Entropy }, BreakDown={NUC, LA, EXP }, SHAP={LA, SEXP, ND}, then top-3 overlap=1/7. Therefore, the top-3 agreement among three CA methods is negligible. If the feature of top-1 generated by three CA methods are: LIME = {LA}, BreakDown − {LA}, and SHAP = {LA}, then top-1 overlap = 1. Therefore, the top-1 agreement among three CA methods is high.

5   Experimental Results and Analysis

In this section, according to the experimental results, we analyze the two research questions proposed in Sect. 4.

5.1   Analysis for RQ1

Motivation. Currently, most of the JIT-SDP researches pay attention to the prediction performance of classifiers while ignoring the interpretability of model when building classification models, which hinders the application of JIT-SDP technology in actual software development. The traditional CS interpretation methods interpret prediction models according to the internal structure of models. However, the CS method cannot explain an individual instance, and cannot explain the black box model. To this end, we introduced three CA interpretation technologies, LIME, BreakDown, and SHAP, to interpret JIT-SDP models.

Approach. To answer RQ1, we counted the differences in the feature importance ranks of the instance explanations generated by the three CA technologies. Specifically, for each project, we first use a 10-fold cross-validation method to divide the data set into a training set and a test set, and then use random forest to build a classification model on the training set. To calculate the difference in the instance explanations in the test set, we first use the CA technology to calculate the feature importance scores for each instance. Then, according to the feature importance scores, we sort the features of each instance. Finally, we counted the differences in feature importance ranks among different instance interpretations, and used box plots to show the distribution of the differences in feature importance ranks. For example, if the feature importance score of one instance c_1 is [LA = 0.108, EXP = 0.049, NUC = 0.259], the feature importance score of another instance $c2$ is [LA = 0.305, EXP = 0.014, NUC = 0.205], the feature importance rank of c_1 is [$1st = NUC, 2nd = LA, 3rd = EXP$], and the feature importance rank of c_2 is [$1st = LA, 2nd = NUC, 3rd = EXP$]. Therefore, the ranking difference between c_1 and c_2 on feature LA is $|2 - 1| = 1$. We apply the calculation process to all instances that are accurately predicted as buggy changes in the test set.

Results

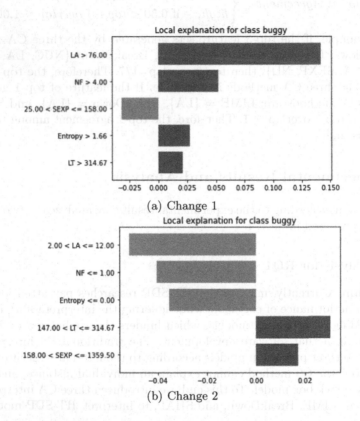

(a) Change 1

(b) Change 2

Fig. 1. An example of instance explanation of LIME

CA technologies can explain the prediction results of the JIT-SDP model and provide visual explains. We selected two instances Change 1 and Change 2 from the project Fabric8. These two instances are accurately predicted as buggy changes by the prediction model, and have a defect probability of 92% and 71%, respectively. Figure 1, Fig. 2, and Fig. 3 show the explanations about Change 1 and Change 2 provided by LIME, BreakDown, and SHAP, respectively.

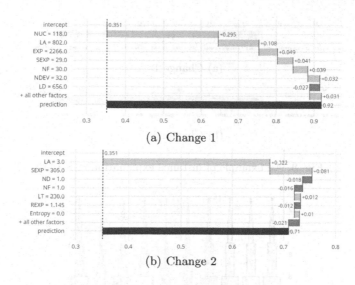

(a) Change 1

(b) Change 2

Fig. 2. An example of instance explanation of BreakDown

Figure 1 shows the visual interpretation generated by LIME for Change 1 and Change 2. The green bar indicates the score that supports the change to become buggy, and the red bar indicates the score that supports the change to be clean. It can be seen from Fig. 1 that the three most important factors for Change 1 to be predicted as buggy are {LA > 76}, {NF > 4} and {Entropy > 1.66}; In the remaining 8% probability, the condition {25 < SEXP ≤ 158} is the main factor for Change 1 to have a clean probability; For Change 2, condition {147 < LT ≤ 314.67} and condition {158 < SEXP ≤ 1359.5} are the main factors that change 2 is predicted to be buggy; In the remaining 29% probability, the condition {2 < LA ≤ 12} is the most important factor for Change 2 to have a clean probability.

Figure 2 shows the explanations of the BreakDown technology on the prediction results of Change 1 and Change 2, in which the green bar and the red bar respectively indicate the feature importance scores that support and oppose the change to be buggy. It can be seen from Fig. 2 that for Change 1, the feature importance scores of the conditions {NUC = 118} and {LA = 802} are 0.295 and 0.108, which are the factors that accounts for the largest proportion of the defect probability; The importance score of the condition {LD = 656} is −0.027, which is the largest factor in the probability of clean in Change 1. For change 2, the feature importance score of condition {LA = 3} is 0.322, which is the largest factor of defect probability; The scores of the conditions {ND = 1} and {NF = 1} are −0.018 and −0.016, which are the largest factors for the existence of clean probability in Change 2.

Figure 3 shows instance explanations generated by the SHAP technology, in which the red and blue areas represent the feature importance scores of supporting and opposing changes that are predicted to be buggy, respectively. The

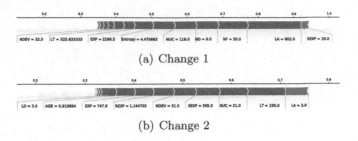

(a) Change 1

(b) Change 2

Fig. 3. An example of instance explanation of SHAP

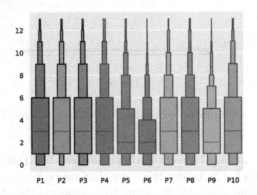

Fig. 4. The distribution of rank differences of each metric generated by LIME

longer the red or blue area is, the larger the absolute value of the corresponding feature importance score is. As can be seen from Fig. 3, for Change 1, LA, NF, and ND are the three most important features that cause Change 1 to be predicted as buggy; SEXP is the most important feature that Change 1 is predicted to be clean. For Change 2, LT, NUC, and SEXP are the three most important features that cause Change 2 to be predicted as buggy; and LA is the main feature of Change 2 with clean probability.

It can be seen from Fig. 1, Fig. 2, Fig. 3 that Change 1 and Change 2 have different instance explanations. To this end, we counted the differences in the feature importance ranks of different instances for each project. Figure 4, Fig. 5, and Fig. 6 respectively show the experimental results of LIME, BreakDown and SHAP technologies on ten projects, where P1, P2,... P10 represents the abbreviations of the names of the ten projects in Sect. 4, and the vertical axis represents the difference value of the feature ranks. As can be seen from Fig. 4, Fig. 5, and Fig. 6, the median of difference of the feature ranks generated by LIME, Break-Down and SHAP is 3 in 7, 10, and 10 projects. Therefore, the most important feature of one instance may rank fourth in another instance. Therefore, there are great differences among instance explanations, which further verifies the importance of introducing CA technology into JIT-SDP model.

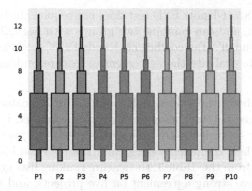

Fig. 5. The distribution of rank differences of each metric generated by BreakDown

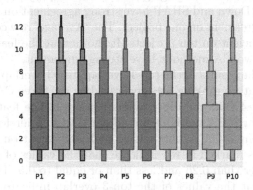

Fig. 6. The distribution of rank differences of each metric generated by SHAP

5.2 Analysis for RQ2

Motivation. The feature importance rank lists generated by CA technologies can provide valuable insights and guide developers to design code changes with low probability of defects. However, the principles and calculation processes of different CA methods are different. Therefore, we further studied the degree of difference in the feature importance rank lists generated by different CA methods.

Approach. For each project, we first use the 10-fold cross-validation method to divide the data set into a training set and a test set. Subsequently, we used LIME, BreakDown, and SHAP to calculate the feature importance scores of each instance in the test set. Finally, the features of the instances are sorted according to feature importance scores. According to the feature importance ranks of all instances in the data set, we use the Scott-Knott effect size difference (ESD) test [41] to calculate feature ranking lists of an project. Scott-Knott ESD test is a variant of Scott-Knott test [34], which uses hierarchical cluster analysis to divide the features of a project into several statistically different groups. Compared with

Scottp-Knott test, Scott-Knott ESD test does not require the input data to obey normal distribution, and can combine two groups with negligible effect size into a group. We use Kendall's Tao coefficient, Kendall's W coefficient, top-3 overlap, and top-1 overlap to evaluate the agreement of feature ranking lists of different CA methods.

Results. Table 3 shows the values of Kendall's Tau indicator for any two CA methods of the three CA technologies. The values marked in red in the table indicate that the τ coefficients are greater than 0.6, and the values marked in blue indicate that the τ coefficients are greater than 0.3 and less than or equal to 0.6. As can be seen from the Table 3, the feature ranking lists generated by LIME and BreakDown have strong agreement on five projects, and there is moderate agreement on the other five projects; The feature ranking lists generated by LIME and BreakDown have strong agreement on ten projects; The ranking lists generated by BreakDown and SHAP have strong agreement on nine projects and have a moderate agreement on one project. The 'Average' row counts the average value of the τ coefficient on ten projects. It shows that the feature ranking lists generated by any two CA methods have strong agreement.

Table 4 shows the results in the top-1 overlap, top-3 overlap and Kendall's W indicators of the feature ranking lists generated by the three CA methods. It can be seen from Table 4 that the top-1 overlap values of the feature ranking lists generated by the three CA methods are all equal to 1. Therefore, in the feature ranking lists, the three CA methods have high agreement on top-1 feature; In addition, the values marked in red indicate that the values of the top-3 overlap indicator are greater than 0.25 and less than or equal to 0.5. The values marked in blue indicate that the values of the top-3 overlap indicator are less than or equal to 0.25. It can be seen that among the top-3 features of the feature ranking lists, the three CA methods have small agreement on 7 projects and negligible agreement on three projects. The Kendall's W coefficient shows that the feature ranking lists generated by the three CA methods have values greater than 0.7 on ten projects, so they have a strong agreement. The 'Average' row counts

Table 3. Kendall's Tau values between two CA methods

Project	τ(LIME, BreakDown)	τ(LIME, SHAP)	τ(BreakDown, SHAP)
Fabric8	0.745	0.711	0.887
JGroups	0.679	0.774	0.748
Camel	0.559	0.701	0.856
Tomcat	0.594	0.671	0.823
Brackets	0.649	0.678	0.894
Neutron	0.561	0.761	0.667
Spring-Integration	0.691	0.810	0.770
Broadleaf	0.607	0.887	0.565
Nova	0.557	0.750	0.758
NPM	0.386	0.724	0.742
Average	0.603	0.747	0.771

Table 4. Top-1 overlap, top-3 overlap, and Kendall's W among three CA methods

Project	Top-1 overlap	Top-3 overlap	Kendall's W
Fabric8	1.000	0.400	0.846
JGroups	1.000	0.400	0.856
Camel	1.000	0.500	0.821
Tomcat	1.000	0.400	0.799
Brackets	1.000	0.500	0.819
Neutron	1.000	0.200	0.851
Spring-Integration	1.000	0.200	0.833
Broadleaf	1.000	0.500	0.828
Nova	1.000	0.500	0.853
NPM	1.000	0.200	0.734
Average	1.000	0.380	0.824

the average of the three evaluation indicators on ten projects. As can be seen from the 'Average' row, the feature ranking lists of the three CA methods have strong agreement and high agreement on top-1 feature; However, they have small agreement on top-3 features.

6   Conclusion and Future Work

We introduced three CA interpretation technologies LIME, BreakDown, and SHAP to interpret the prediction results of the JIT-SDP model. The experiment conducted large-scale empirical research on six open source projects. The empirical results show that: (1) The CA technology has different explanations for different instances. On average, the difference of feature ranks for two instances is 3. (2) The feature ranking lists and the top-1 feature generated by different CA technologies have a strong agreement. However, the top-3 features in the feature lists of different CA technologies have a small agreement. Therefore, in the actual software development process, we recommend using CA technology to provide practical explanations for the prediction results of the JIT-SDP model.

Our research provides knowledge about the JIT-SDP model interpretation. In the future, we will further carry out the following research: (1) This paper uses three CA technologies to explain the prediction results of the JIT-SDP model. In the future, we will study the agreement of the explanations between CA technologies and CS technologies. (2) Different CA technologies have different principles and calculation processes. Therefore, we will explore the agreement of explanations generated by different CA technologies for a given instance. (3) The running time of the interpretation technology is one of the factors that determine whether the participants will adopt the interpretation technology. Therefore, we will further analyze the running time of different CA technologies.

In order to ensure the reproducibility of experimental results, we provide all experimental data and experimental codes, which could be download at https://github.com/yangxingguang/jit_explain

Acknowledgment. This work was supported by the National Natural Science Foundation of China (No. 61772200), the Project Supported by Shanghai Natural Science Foundation (No. 21ZR1416300).

References

1. Akoglu, H.: User's guide to correlation coefficients. Turk. J. Emerg. Med. **18**(3), 91–93 (2018)
2. Breiman, L.: Random forests. Mach. Learn. **45**(1), 5–32 (2001)
3. Cabral, G.G., Minku, L.L., Shihab, E., Mujahid, S.: Class imbalance evolution and verification latency in just-in-time software defect prediction. In: Proceedings of the 41st International Conference on Software Engineering, pp. 666–676 (2019)
4. Cafiso, S., Di Graziano, A., Pappalardo, G.: Using the Delphi method to evaluate opinions of public transport managers on bus safety. Saf. Sci. **57**, 254–263 (2013)
5. Chakraborty, S., et al.: Interpretability of deep learning models: a survey of results. In: IEEE SmartWorld, Ubiquitous Intelligence & Computing, Advanced & Trusted Computed, Scalable Computing & Communications, Cloud & Big Data Computing, Internet of People and Smart City Innovation, pp. 1–6 (2017)
6. Chen, X., Zhao, Y., Wang, Q., Yuan, Z.: MULTI: multi-objective effort-aware just-in-time software defect prediction. Inf. Softw. Technol. **93**, 1–13 (2018)
7. Dam, H.K., Tran, T., Ghose, A.: Explainable software analytics. In: Proceedings of the 40th International Conference on Software Engineering: New Ideas and Emerging Results, pp. 53–56 (2018)
8. Fan, Y., Xia, X., Da Costa, D.A., Lo, D., Hassan, A.E., Li, S.: The impact of changes mislabeled by SZZ on just-in-time defect prediction. IEEE Trans. Softw. Eng. **47**(8), 1559–1586 (2019)
9. Gosiewska, A., Biecek, P.: iBreakDown: Uncertainty of model explanations for nonadditive predictive models. arXiv preprint arXiv:1903.11420 (2019)
10. Hoang, T., Dam, H.K., Kamei, Y., Lo, D., Ubayashi, N.: Deepjit: An end-to-end deep learning framework for just-in-time defect prediction. In: Proceedings of the 16th International Conference on Mining Software Repositories, pp. 34–45 (2019)
11. Huang, Q., Xia, X., Lo, D.: Supervised vs unsupervised models: a holistic look at effort-aware just-in-time defect prediction. In: 2017 IEEE International Conference on Software Maintenance and Evolution, ICSME, pp. 159–170. IEEE Computer Society (2017)
12. Huang, Q., Xia, X., Lo, D.: Revisiting supervised and unsupervised models for effort-aware just-in-time defect prediction. Empirical Softw. Eng. **24**(5), 2823–2862 (2019)
13. Jiang, Y., Cukic, B., Menzies, T.: Can data transformation help in the detection of fault-prone modules?. In: Proceedings of the Workshop on Defects in Large Software Systems, held in conjunction with the ACM SIGSOFT International Symposium on Software Testing and Analysis, pp. 16–20 (2008)
14. Jiarpakdee, J., Tantithamthavorn, C., Dam, H.K., Grundy, J.: An empirical study of model-agnostic techniques for defect prediction models. IEEE Trans. Softw. Eng. 1 (2020)

15. Jiarpakdee, J., Tantithamthavorn, C., Grundy, J.C.: Practitioners' perceptions of the goals and visual explanations of defect prediction models. In: 18th IEEE/ACM International Conference on Mining Software Repositories, pp. 432–443 (2021)

16. Jiarpakdee, J., Tantithamthavorn, C., Hassan, A.E.: The impact of correlated metrics on the interpretation of defect models. IEEE Trans. Softw. Eng. **47**(2), 320–331 (2021)

17. Kamei, Y., Fukushima, T., McIntosh, S., Yamashita, K., Ubayashi, N., Hassan, A.E.: Studying just-in-time defect prediction using cross-project models. Empirical Soft. Eng. **21**(5), 2072–2106 (2016)

18. Kamei, Y., Matsumoto, S., Monden, A., Matsumoto, K., Adams, B., Hassan, A.E.: Revisiting common bug prediction findings using effort-aware models. In: 26th IEEE International Conference on Software Maintenance, pp. 1–10 (2010)

19. Kamei, Y., Shihab, E., Adams, B., Hassan, A.E., Mockus, A., Sinha, A., Ubayashi, N.: A large-scale empirical study of just-in-time quality assurance. IEEE Trans. Softw. Eng. **39**(6), 757–773 (2013)

20. Forthofer, R.N., Lehnen, R.G.: Rank correlation methods. In: Public Program Analysis. Springer, Boston, MA (1981). https://doi.org/10.1007/978-1-4684-6683-6_9

21. Li, Z., Jing, X., Zhu, X.: Progress on approaches to software defect prediction. IET Softw. **12**(3), 161–175 (2018)

22. Lipton, Z.C.: The mythos of model interpretability. Commun. ACM **61**(10), 36–43 (2018)

23. Liu, J., Zhou, Y., Yang, Y., Lu, H., Xu, B.: Code churn: a neglected metric in effort-aware just-in-time defect prediction. In: ACM/IEEE International Symposium on Empirical Software Engineering and Measurement, pp. 11–19 (2017)

24. Lundberg, S.M., Lee, S.: A unified approach to interpreting model predictions. In: Advances in Neural Information Processing Systems 30: Annual Conference on Neural Information Processing Systems, pp. 4765–4774 (2017)

25. Mervyn, S.: Cross-validatory choice and assessment of statistical predictions. J. Roy. Stat. Soc.: Ser. B (Methodol.) **36**(2), 111–133 (1974)

26. Miller, T.: Explanation in artificial intelligence: insights from the social sciences. Artif. Intell. **267**, 1–38 (2019)

27. Mockus, A., Weiss, D.M.: Predicting risk of software changes. Bell Labs Tech. J. **5**(2), 169–180 (2000)

28. Naik, K., Tripathy, P.: Software testing and quality assurance: theory and practice. John Wiley & Sons (2011)

29. Pornprasit, C., Tantithamthavorn, C.: Jitline: A simpler, better, faster, finer-grained just-in-time defect prediction. In: 18th International Conference on Mining Software Repositories, pp. 1–11 (2021)

30. Rajbahadur, G.K., Wang, S., Ansaldi, G., Kamei, Y., Hassan, A.E.: The impact of feature importance methods on the interpretation of defect classifiers. IEEE Trans. Softw. Eng. 1 (2021)

31. Regulation, G.D.P.: Regulation eu 2016/679 of the european parliament and of the council of 27 April 2016. Official Journal of the European Union (2016)

32. Ribeiro, M.T., Singh, S., Guestrin, C.: Why should I trust you?": explaining the predictions of any classifier. In: Proceedings of the 22nd ACM SIGKDD International Conference on Knowledge Discovery and Data Mining, pp. 1135–1144 (2016)

33. Rosen, C., Grawi, B., Shihab, E.: Commit guru: Analytics and risk prediction of software commits. In: Proceedings of the 10th Joint Meeting on Foundations of Software Engineering, pp. 966–969 (2015)

34. Scott, A.J., Knott, M.: A cluster analysis method for grouping means in the analysis of variance. Biometrics, pp. 507–512 (1974)
35. Shihab, E., Hassan, A.E., Adams, B., Jiang, Z.M.: An industrial study on the risk of software changes. In: 20th ACM SIGSOFT Symposium on the Foundations of Software Engineering, p. 62 (2012)
36. Sliwerski, J., Zimmermann, T., Zeller, A.: When do changes induce fixes? In: Proceedings of the 2005 International Workshop on Mining Software Repositories (2005)
37. Staniak, M., Biecek, P.: Explanations of model predictions with lime and break-down packages 10(2), 395 (2018). arXiv preprint arXiv:1804.01955
38. Tabassum, S., Minku, L.L., Feng, D., Cabral, G.G., Song, L.: An investigation of cross-project learning in online just-in-time software defect prediction. In: IEEE/ACM 42nd International Conference on Software Engineering, pp. 554–565 (2020)
39. Tantithamthavorn, C., Hassan, A.E., Matsumoto, K.: The impact of class rebalancing techniques on the performance and interpretation of defect prediction models. IEEE Trans. Softw. Eng. **46**(11), 1200–1219 (2020)
40. Tantithamthavorn, C., McIntosh, S., Hassan, A.E., Ihara, A., Matsumoto, K.: The impact of mislabelling on the performance and interpretation of defect prediction models. In: 37th IEEE/ACM International Conference on Software Engineering, pp. 812–823 (2015)
41. Tantithamthavorn, C., McIntosh, S., Hassan, A.E., Matsumoto, K.: An empirical comparison of model validation techniques for defect prediction models. IEEE Trans. Softw. Eng. **43**(1), 1–18 (2017)
42. Wattanakriengkrai, S., Thongtanunam, P., Tantithamthavorn, C., Hata, H., Matsumoto, K.: Predicting defective lines using a model-agnostic technique. IEEE Trans. Softw. Eng. (2021). https://doi.org/10.1109/TSE.2020.3023177
43. Yang, X., Yu, H., Fan, G., Shi, K., Chen, L.: Local versus global models for just-in-time software defect prediction. Sci. Program. 2384706:1–2384706:13 (2019)
44. Yang, X., Yu, H., Fan, G., Yang, K.: An empirical studies on optimal solutions selection strategies for effort-aware just-in-time software defect prediction. In: The 31st International Conference on Software Engineering and Knowledge Engineering, pp. 319–424 (2019)
45. Yang, X., Yu, H., Fan, G., Yang, K.: A differential evolution-based approach for effort-aware just-in-time software defect prediction. In: Proceedings of the 1st ACM SIGSOFT International Workshop on Representation Learning for Software Engineering and Program Languages, pp. 13–16 (2020)
46. Yang, X., Yu, H., Fan, G., Yang, K.: DEJIT: a differential evolution algorithm for effort-aware just-in-time software defect prediction. Int. J. Softw. Eng. Knowl. Eng. **31**(3), 289–310 (2021)
47. Yang, X., Lo, D., Xia, X., Sun, J.: TLEL: a two-layer ensemble learning approach for just-in-time defect prediction. Information & Software Technology **87**, 206–220 (2017)
48. Yang, Y., et al.: Effort-aware just-in-time defect prediction: simple unsupervised models could be better than supervised models. In: Proceedings of the 24th ACM SIGSOFT International Symposium on Foundations of Software Engineering, pp. 157–168 (2016)

Yet Another Traffic Black Hole: Amplifying CDN Fetching Traffic with RangeFragAmp Attacks

Chi Xu[1,2], Juanru Li[1(✉)], and Junrong Liu[2]

[1] Shanghai Jiao Tong University, Shanghai 200240, China
{sjtu_xuchi,jarod}@sjtu.edu.cn
[2] ZhiXun Crypto Testing and Evaluation Technology Co. Ltd.,
Shanghai 201601, China

Abstract. Content Delivery Network (CDN) has been widely used nowadays as an important network infrastructure to provide fast and robust distribution of content over the Internet. However, an inherent weakness of CDN involved network service is its content fetching amplification issue, that is, the network traffic among the origin server and CDN surrogate nodes is maliciously amplified due to some crafted requests. Such requests can be multiplied by the forwarding of the CDN, posing a serious performance threat to the origin server. Particularly, when the HTTP range request mechanism, which allows the server to respond only a portion of the HTTP message to the request of client, is used, the risk of content fetching amplification is significantly increased. Therefore, defenses against such kinds of traffic amplification have been deployed to protect CDN users from being over charged.

In this paper, we revisited HTTP range request cased content fetching amplification issue and evaluated the deployed defenses of mainstream CDN providers. Specifically, we proposed Range Fragment Amplification (RangeFragAmp) attacks, a new variation of CDN content fetching attack related to HTTP range request mechanism. The proposed RangeFragAmp attacks have concealment and bandwidth consumption capability. Our pentests against five CDN providers with more than 2.5 million users demonstrated that all of their CDNs were vulnerable to RangeFragAmp attacks. Particularly, S-RFA attack, one of the two types of RangeFragAmp attacks, can achieve an amplification factor of 11345 on *Baidu AI Cloud*. We have reported the issues to the involved CDN providers, and expected our study could help CDN designers and developers build more robust systems.

Keywords: CDN security · HTTP range request · Amplification attack · DDoS

This work was partially sponsored by the Shanghai Science and Technology Innovation Fund (Grant No. 19511103900), the National Key Research and Development Program of China (Grant No. 2020AAA0107800), and the National Natural Science Foundation of China (Grant No. 62002222).

H. Gao and X. Wang (Eds.): CollaborateCom 2021, LNICST 406, pp. 439–459, 2021.
https://doi.org/10.1007/978-3-030-92635-9_26

1   Introduction

As an important infrastructure of the Internet, CDN security has received widespread attention. Some resource consumption attacks on the CDN have been found to threaten the security of the CDN itself and servers hosted on it. For example, *CDN loop* attacks constructing loops in CDN aim to reduce the availability of CDN [1,2]. *No-abort* attack, which depletes the bandwidth of the origin server by rapidly dropping the CDN-client connections [3,4]. The amplification attack is an attack in which an attacker can use an amplification factor to multiply his power [5]. The most typical amplification attacks on CDN include: DNS Amplification (*DNS-A*) attack [6], UDP Reflection Amplification (*UR-A*) attack [7], and Range-based Amplification (*RangeAmp*) attacks.

The range request in the HTTP protocol is a mechanism that should be used to improve transmission performance. However, the flaws in the protocol design and the negligence of the CDN in the implementation process allow this mechanism to be used for amplification attacks. For instance, *RangeAmp* attacks [4] disclose the vulnerabilities of HTTP range requests on CDNs.

Although mainstream CDN providers claimed that they have deployed defenses against *RangeAmp* attacks, we found HTTP range requests are complex and it is difficult to implement a comprehensive defense against all variations of malicious range request. Therefore, we revisited those defenses deployed on popular CDNs and tested their effectiveness by proposing a new class of amplification attack against CDN surrogate nodes and origin servers. Our proposed Range Fragment Amplification (RangeFragAmp) attacks leverage the weakness of current fragment based transmission to implement a CDN content fetching amplification, which not only affects the performance of the origin server but also the CDN system. In particular, RangeFragAmp attacks include two kinds of attacks: Small Range Fragment Amplification (S-RFA) attack and Overlapping Range Fragment Amplification (O-RFA) attack. In both attacks, an attacker constructs an HTTP request within a minimum transmission fragment range to the CDN, and drive it to fetch a large range of data from the origin server. Since the size of fragment would not change according to the request range, the attacker only needs to first determine the range size of the fragment, and then requests a specific range that crosses two fragments to produce a series of considerable amplification attacks.

Unlike existing CDN traffic amplification attacks such as *RangeAmp* attacks, our proposed RangeFragAmp attacks rely on multiple range requests to implement traffic amplification. Since the crafted requests comply with RFC 7233 [8], deployed defenses could not effectively detect and block such malicious requests. To the best of our knowledge, we are the first to discuss the security risks posed by malicious range requests against fragment based transmission in CDN application scenarios.

To measure whether mainstream CDN providers adopted a well-protected range forwarding policy against RangeFragAmp attacks, we built various experimental environments and simulated the impact of different amplification attacks on CDNs. We tested five mainstream CDN providers including *Alibaba Cloud*,

Tencent Cloud, *Huawei Cloud*, *Baidu AI Cloud*, and *CloudFront*. As shown in Table 1, we found most CDN products, even though they are immune to existing attacks such as *DNS-A* and *UR-A* attacks, our proposed RangeFragAmp attacks (especially the (S-RFA) attack) could circumvent the protection. This demonstrates CDN providers did not well understand the root cause of range request based traffic amplification attacks, and hence we proposed a better mitigation scheme to comprehensively protect CDNs against such kinds of threats.

Table 1. Summary of CDNs against existing traffic amplification attacks and our proposed RangeFragAmp attacks

CDN provider	DNS-A [6]	UR-A [7]	SBR [4]	OBR [4]	S-RFA	O-RFA
Alibaba Cloud [9]	✗	✗	✗	✗	✓	✗
Tencent Cloud [10]	✗	✗	✗	✗	✓	✓
Baidu AI Cloud [11]	✗	✗	✗	✗	✓	✗
Huawei Cloud [12]	✗	✗	✗	✗	✓	✗
CloudFront [13]	✗	✗	✗	✓	✓	✓

Ethical Consideration. We specially registered experimental accounts for all of our evaluation. In our experiments, we simulated the attacks against our own cloud servers so as not to influence real world network services. Furthermore, when evaluating the feasibility and amplification factor of RangeFragAmp attacks, we used cloud servers with small bandwidth (i.e., 1 Mbps). Therefore, the workloads of tested CDN servers would not increased too much, and our tests would not cause any usability impact on the CDN servers or other cloud servers hosted on them.

Furthermore, we have contacted all influenced CDN providers and reported the vulnerabilities we found to them, and we have received feedback from four of them. For instance, some of the CDN providers planned to adopt a more flexible dynamic fragment mechanism to avoid such amplification.

2   Background

In this section, we first briefly introduce essential features of CDN, then particularly discuss the HTTP range request mechanism and some relevant amplification attacks in CDN applications.

2.1   CDN Overview

CDN is an important Internet infrastructure composed of edge node server clusters distributed in different geographical areas [14]. It not only improves the performance for the websites of its customers but also provides security features such as DDoS (Distributed Denial-of-Service) protection mechanisms [15].

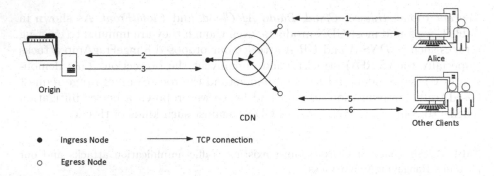

Fig. 1. Multiple segments of connectivity in a CDN environment.

As shown in Fig. 1, the CDN network can be divided into two parts: the central nodes and the edge nodes [16]. The central nodes are usually responsible for global load balancing and content management of the entire system, while the edge nodes play a crucial role in content distribution and caching usually. Edge nodes can be divided into ingress nodes and egress nodes according to their locations and functions. In general, the egress nodes are closer to the client, so they are responsible for the access of client and content distribution. In Fact, a normal client can only establish connections with the egress nodes most of the time. Similarly, the ingress nodes are closer to the origin server, so CDN usually forwards requests through the ingress nodes to obtain the latest contents from the origin. In general, this process is also called *Range Origin Fetch*.

Actually, there are multiple connection paths in the CDN environment between the client and the origin, such as the connections *client-cdn* between the client and the egress nodes, the connections *cdn-origin* between the ingress nodes and the origin, and connections *cdn-cdn* among egress nodes and ingress nodes in CDNs [17]. Besides, a kind of cascade relationship can be established within a CDN or between CDNs [18].

For convenience, we indicated an external CDN near the client as ExCDN, and an internal CDN close to origin as InCDN. Therefore, there are at least 3 TCP connections among client, ExCDN, InCDN and origin as shown in Fig. 1.

Usually, CDN will try to find response content from the caches of edge nodes preferentially when receiving requests from clients [19]. If the cache misses or expires, the CDN will try to forward requests for the latest content data from the origin server through the ingress nodes, and restore it in the caches of the edge nodes. When the client or other users in the neighboring area request the same content again, the request can be fulfilled immediately, through the response from the caches of edge nodes.

For potential consumers who are willing to try using CDN, performance improvement and cost become the things they care about most. Since the network usually expresses fluctuates due to its own features, the extra cost brought by CDN has become an important reference for consumers when choosing a CDN [20].

2.2   HTTP Range Request Mechanism on CDN

With the advent of the mobile Internet era, more and more large-scale media files have been moved to the Internet. Luckily, the range request in the HTTP protocol allows the server to respond only a portion of the HTTP message to the request of client [8]. In fact, HTTP range requests are particularly helpful when sending large media files or used for resume broken transfer and downloads [21].

As an accelerator for content delivery, a reasonable CDN should have the ability of segmented caching for large media files or documents. For example, the client can specify the transmission range of an image to obtain the contents of a specific fragmentation, instead of receiving the whole image. As for the edge nodes, they should be able to cache and respond to the requests of other clients autonomously in a while.

These requirements have been solved when range request from HTTP was introduced into the CDN. In fact, clients often fail to receive a complete file because of a canceled or interrupted HTTP connection [22]. In terms of efficiency, the client expects that it can continue to retrieve the rest of the data in subsequent requests after acquiring a part of the data, rather than retrieving all the data at once. Meanwhile, capturing part of the data on one request is also good for devices that are running out of storage space.

2.3   Differences in CDNs Handling Range Requests

Different CDN providers have different policies for handling range requests. However, there are still no clear definitions or considerations in related protocols or RFCs to help developers hand HTTP range requests in CDN. Different CDN providers choose to implement range forward policies based on different perspectives, including business, operational and technical views.

At present, there are four basic range requests forward policies deployed on CDN, including:

* *Laziness* - Forward the *Range* header without change.
* *Deletion* - Remove the *Range* header directly.
* *Fragment* - Send the *Range* header in fragments.
* *Continuance* - Request all remaining content from specified *Range* header.

Previously, someone also proposed the *Expansion* policy [4], but after research and evaluation, we believe that it has been replaced with *Fragment* and *Continuance* to meet the new security requirements. In practice, the CDN providers use one or a combination of several basic policies. When *Range Origin Fetch* is off by default, most CDNs prefer to adopt the *Deletion* policy [4].

When the *Range Origin Fetch* is enabled, most CDN providers use *Fragment* policy partially or completely. Moreover, The CDNs will divide the range into several fragments according to a preset size when they receive some requests containing a wide range, and then forward the request to the origin server in turn. This heuristic method of fragmentation provides a new opportunity for malicious attackers, who just need to make sure the fragment size of each CDN provider in advance - which is not difficult to do - to implement *SBR* attacks within the scope of each fragment. Although some possible mitigation on the CDN side has been proposed in the paper by Li et al., they suggested adopting *Laziness* policy to complete defend against a *SBR* attack or applying *Expansion* policy but not extend the byte range too much e.g. 8 KB [4]. Unfortunately, we are disappointed to find that most CDNs just ignore them.

2.4 Amplification Attacks

As one of the most popular and effective DDoS attacks, an amplification attack is any attack where an attacker is capable of using an amplification factor to multiply its power [5]. The amplification factor can be protocol vulnerabilities (e.g., *UR-A* attack), security negligence of the specifications (e.g., *DNS-A* attack), or both (e.g., *RangeAmp attacks*). Amplification attacks are "asymmetric", which meanings a relatively small number or low level of resources is required by an attacker to cause a significantly greater number or higher level of target resources to malfunction or fail [5,23]. Examples of amplification attacks on CDN include *DNS-A* attack [6], *UR-A* attack [7], and *RangeAmp*[1] attacks [4].

DNS-A attack is a typical DDoS attack based on reflection. Attacker leverages the functionality of open DNS resolvers in order to overwhelm a target server or network with an amplified amount of traffic, rendering the server and its surrounding infrastructure inaccessible [6]. A *DNS-A* attack based on DNSSEC can leads to an amplification factor of 44 [24].

An *UR-A* attack will send a series of special requests based on the UDP service to the victims, and the identity of the attacker will be hidden by forging IP address. This action will generate much larger response traffic to the victim than the requested data. An *UR-A* attack can reach an amplification factor of 556 [7].

RangeAmp attacks are kind of high efficiency amplification attacks, which allow attackers to exploit the range implement vulnerabilities and damage DDoS protection mechanism of CDNs [4]. *RangeAmp* attacks include two types: Small Byte Range (*SBR*) attack and Overlapping Byte Ranges (*OBR*) attack. A *SBR* attack can lead to an amplification factor of 43093, and an *OBR* attack have ability to reach an amplification factor of 7432, which both pose severe threats to the serviceability of CDNs and availability of websites.

3   RangeFragAmp Attack

In this section, we propose a novel RangeFragAmp attacks. This kind of attacks exploit the vulnerabilities to the upgraded range forwarding mechanism of the

---

[1] *RangeAmp* attack include two types: Small Byes Range (*SBR*) attack and Overlapping Byte Ranges (*OBR*) attack.

CDN, which can break through the existing CDN defense mechanism and bring significant disruptive impacts on the origin server hosted on CDNs and surrogate nodes of CDN itself.

RangeFragAmp attacks include two types: Small Range Fragment Amplification (S-RFA) attack and Overlapping Range Fragment Amplification (O-RFA) attack. The main difference between S-RFA attack and O-RFA attack is on the crafted range of attack requests. A S-RFA attack amplifies on a single Fragment interval, which means it only needs tiny traffic to launch attacks and implement a stable amplification factor. Therefore, a S-RFA attack is more suitable for the attacker with small network bandwidth. However, an O-RFA attack amplifies the traffic across multiple fragment intervals, they need considerable traffic to launch attacks. Meanwhile, the amplification factor of an O-RFA attack depends on the size of the fragment and target resource both. Although O-RFA attack has a better performance when facing a large target resource, it still needs greater bandwidth for attackers.

3.1 Threat Model

In fact, the *Deletion* and *Expansion* policies are beneficial for CDNs to improve service performance [4]. However, improper *Range* request forwarding mechanism will take huge potential security risks to CDNs and origin servers they host, when attackers launch *RangeAmp* attacks. Therefore, CDN providers have upgraded range request forwarding mechanisms – introducing the *Fragment* policy and the *Continuance* policy – to help protect CDN and origin from *RangeAmp* attacks. But we notice that, these fresh policies still require CDN to retrieve many more bytes from origin server than ones requested by client – in a fragmented range.

On the other hand, when the cache of edge nodes is missed, CDN will directly forward fragment range request to origin server. If CDN has no mechanisms to check whether the *Range* header contains several similar parts already stored in the cache, the response sent by CDN can be larger than the one by the origin server. These cases will cause significant traffic differences between different fragment sizes and connections in the path from client to origin server.

The striking traffic differences cause by *Fragment* policy will bring a kind of variant range-based amplification attacks, denoted RangeFragAmp attacks. We also discern two scenarios of a RangeFragAmp attacks and demonstrate them in Sect. 3.2 and Sect. 3.3.

In RangeFragAmp attacks, an attacker can send many maliciously craft but legal range requests to the CDN, as shown in Fig. 2. Different from the previous *RangeAmp* attacks, now the attacker cannot send small range requests naively. Instead, he or she must be prepared before "work". Because the fragment sizes that are usually not the same in different CDN providers, have tremendous affection for amplification effects. "Fortunately", the size of fragment can be easily found in the official documentation (such as *Alibaba Cloud*), or captured on the server of attacker which has been accessed to the same CDN with victim in advance.

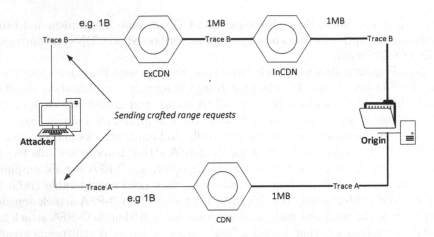

Fig. 2. General construction of RangeFragAmp attacks.

Unlike *RangeAmp* attacks, origin servers are always victims when the attacker launches RangeFragAmp attacks no matter in which scenario of *Trace A* or *Trace B* as shown in Fig. 2. Moreover, the ExCDN and the InCDN in *Trace B* can be other victims when they have been cascaded together by owner or attacker.

3.2   S-RFA Attack

If a CDN adopts the *Fragment* policy to handle range requests, an attacker can craft *Range* header with a few bytes in each fragment range to launch RangeFragAmp attacks. After upgrading the range request forwarding mechanisms, we confirm that all of the CDNs have dropped purely *Deletion* or *Expansion* policy in order to protect from *SBR* attacks. When users enable the *Range Origin Fetch*, they tend to send more fixed-size fragments when they receive range requests. Therefore, an attacker can craft a series of requests that *Range* header with a small byte range in every interval of fragment size to launch RangeFragAmp attacks. We call it S-RFA attack. In a S-RFA attack, the *cdn-origin* connection will transport a much larger traffic than *client-cdn* connection influenced by fragment size. Therefore, an attacker has the ability to consume a large number of bandwidth resources to the origin server through the CDN without determining its real IP address.

As shown in Fig. 3, once the attacker has determined the size of the target file and the provider of CDN that victims used, he or she can launch RangeFragAmp attacks by writing a series of range requests. For example, if a victim has a video of 5 MB on the server and the fragment size of CDN he used is 512 KB (524288 bytes), an attacker is able to craft 10 different requests such as "Range: bytes = 1–1", "Range: bytes = 524289–524289", "Range: bytes = 1048577–1048577" to implement amplification attack. As a result, origin server will return an entire

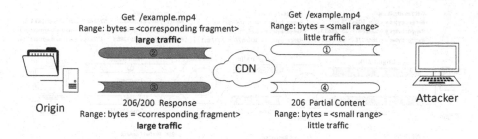

Fig. 3. Flow and example construction of a S-RFA attack.

fragment of target resource like "Range: bytes = 0–524287", "Range: bytes = 524288–1048575", but edge nodes will return only a partial content specified by *Range* header, which can be as small as 1 byte if not considering transmission overhead.

In a S-RFA attack, the size of response traffic in the client-cdn connection is just hundreds of bytes (which is small). When the CDN adopts *Fragment* policy, response traffic between the CDN and origin server completely depends on the size of the fragment adopted by the CDN provider and the size of target resources selected by the attacker. In fact, the bigger the fragment, the larger the amplification factor. On the other hand, the bigger the target resource, the more attack fragments can be crafted, which means the more bandwidth consumption could be done by the attacker.

3.3   O-RFA Attack

If the ExCDN adopts the *Fragment* policy and the InCDN returns corresponding full fragment response without checking whether range contains several similar parts already stored in the cache, an attacker can craft a series of *Range* header that contains a large number of similar but not identical overlapping fragments to launch fresh RangeFragAmp attacks. We name this O-RFA attack. Different from an *OBR* attack, we focus on bandwidth consumption in fragment conditions adopted by CDNs now. In an O-RFA attack, the *ExCDN-InCDN* and *InCDN-origin* connections both transport a much larger traffic than *client-ExCDN* connection, which makes the attacker have significant ability to consume the bandwidth available from ExCDN to origin.

As illustrated in Fig. 4, an attacker launches a series of elaborate range requests that spanning n fragment intervals. For example, when n equals 1, *Range* header could be like "Range: bytes = 524287–524288", "Range: bytes = 10485765–1048576", then sends them to ExCDN. Since CDNs have adopted *Fragment* policy to protect from *OBR* attack now, the fragment range actually forwarded by ExCDN and InCDN covers n complete fragments scope that overrides the minimum range requested by the attacker. In this case shown in Fig. 4, ExCDN and InCDN transmit *Range* header like "Range: bytes = 0–10485765" and "Range: bytes = 524288–1572863" respond to the requests of attackers. As

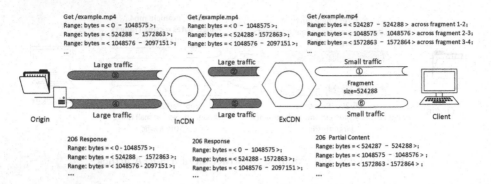

Fig. 4. Flow and example construction of O-RFA attack.

we can see, for an O-RFA attack, the attacker only needs 2-byte requests to consume 1 MB traffic of both *ExCDN-InCDN* and *InCDN-origin* connections in a time.

In an O-RFA attack, when the target resource is fixed, the traffic consumed by the *ExCDN-InCDN* and *InCDN-origin* connections is directly proportional to the fragment size of the deployed CDN and the number of overlapping fragments of the range request crafted by the attacker. Intuitively, the larger the fragment scope covered by the range request, the more traffic will be consumed for *ExCDN-InCDN* and *InCDN-origin* connections.

However, there is a trade-off relation between the amplification factor and the overhead of attacker in an O-RFA attack. The amplification factor of an O-RFA attack will decrease as the number of overlapping fragments in the range request increases until the traffic consumption of attacker is nearly equal to transmission by the origin server.

4   Real-World Evaluation

To evaluate the feasibility and severity of RangeFragAmp vulnerabilities in the wild, we designed and conducted some experiments. Firstly, we have verified whether *RangeAmp* vulnerabilities have been patched up. Then, we examine the five most representatives CDNs in order to find out which are vulnerable to a RangeFragAmp attack. Lastly, we calculate the actual amplification factors and analyze the practical impacts. In all experiments, our origin server is the same Linux server with 1vCPU 2 GiB, CentOS 7.3 64-bit, and 1 Mbps of bandwidth. And our origin website is powered by Nginx/1.16.1 with default configuration deployed.

4.1   Consideration in Selecting CDN Providers

We tested five of the most mainstream CDN providers, including *Ali Cloud, Tencent Cloud, Huawei Cloud, Baidu AI Cloud* and *CloudFront*. According to the previous research of Kashaf et al., CDN services are highly centralized and oligarchic [25]. The five providers we selected have supply service for nearly 75% of websites which have accessed CDN in China [26].

At the same time, all the CDN providers we selected open their services to individual users and provide free trial or large-capacity traffic packages. In fact, we only spent less than 200 RMB to complete all the experiments.

For experimental security and accuracy of results, we deploy our server individually behind these CDNs and apply their default configuration in all subsequent experiments.

4.2 S-RFA Attack Evaluation

Table 2. Range forwarding behaviors vulnerable to S-RFA attack (Considering the readability of the table, we list only three range requests for each CDN.)

CDN	Fragment size (Bytes)	Vulnerable range requests	Forwarded range requests
Alibaba Cloud	524,288	1–1	0–524,287
		524,289–524,289	524,288–1,048,575
		1,048,577–1,048,577	1,048,576–1,572,864
Baidu AI Cloud	1,048,576	5,242,879–5,242,879	0–7,340,031
		1,048,579–1,048,579	7,340,032–11,534,335
		15,728,639–15,728,639	11,534,336–15,728,639
Tencent Cloud	4,096	1–1	0–4,095
		4,097–4,097	4,096–8,191
		8,193–8,193	8,192–16,384
Huawei Cloud	524,288	1–1	0–524,287
		524,289–524,289	524,288–1,048,575
		1,048,577–1,048,577	1,048,576–1,572,864
CloudFront	1,048,576	1–1	0–1,048,575
		1,048,577–1,048,577	1,048,576–2,097,151
		2,097,153–2,097,153	2,097,152–3,145,728

Feasibility of S-RFA Attack. In our experiment, we upload three media files of different sizes (3 MB to 50 MB) to our origin server in advance. In order to ensure the accuracy of experimental results, before each experiment, we will use the *Prerefresh* provided by the CDNs console, which can delete the specified contents in the cache of all edge nodes. Besides, in Table 2 and Table 4 we summarize the existing Range forwarding mechanisms of chosen CDNs through the response header received in the origin server. we try to evaluate its practical advantages and disadvantages. In addition, the monitoring windows watched for traffic on

CDN console is another crucial reference to help us figure out how much economic consumption can be caused by RangeFragAmp attacks.

Our experimental results show that all tested CDNs have upgraded their range forwarding policies. In fact, they have no longer naively adopted *Deletion* or *Expansion* policies when the origin owner chooses to enable *Range Origin Fetch* on CDN, which means a *SBR* attack produces little effect on origin server now. Meanwhile, the *Range Overlap Checking* mechanism makes it difficult for *OBR* attacks to have a performance impact on the origin server too.

As illustrated in Table 2, all of five major CDNs are vulnerable to a S-RFA attack. The second column lists the fragment size when range request have been forwarded by CDNs, and the third column lists attack range we crafted in experiment. Lastly, the forth column presents the policies CDNs handing the corresponding *Range* headers. The details are shown below:

1) **Alibaba Cloud** adopts the *Fragment* policy to hand the *Range* header. Specifically, *Alibaba Cloud* has two ways to forward requests after enabling the *Range Origin Fetch*: *Normal Mode* and *Mandatory Mode* [27]. The *Laziness* and *Fragment* policies are adopted in the *Normal Mode*. In other word, the first range request sent by the client will be forwarded to the origin server directly. Furthermore, the subsequent requests will be sent back to the origin with the size of 512 KB. If the *Mandatory Mode* is enabled, all the requests of clients that should be back to the origin server have to remain a fixed size of 512 KB.

2) **Tencent Cloud** implements *Fragment* and *Laziness* policies after the *Range Origin Fetch* is enabled. However, We find two different fragment sizes during experiments. When the range data requested by the client is bigger than 4 MB, *Tencent Cloud* CDN will forward each range in the size of 1 MB, otherwise, it will forward the request in size of 4 KB.

3) **Baidu AI Cloud** adopts *Fragment* and *Continuance* policies when range request is enabled. In our experiment, we find *Fragment* policy decides range upper boundary of forwarded request, which means the closest fragment covering the request of client will be selected as range start point. As for the lower boundary, CDN adopts the *Continuance* policy that requests all of the remaining content of the specified resource. The connection between CDN and origin will be interrupted as the *client-CDN* disconnects automatically when clients have received all the content they want.

4) **Huawei Cloud** implements the *Fragment* policy when *Range Origin Fetch*. That is to say, all the range requests forwarded to the origin server will be transmitted in the size of 512 KB.

5) **CloudFront** enables the *Range Origin Fetch* by default. Moreover, it adopts a *Fragment* policy. In fact, the requests of clients will be forwarded in a fixed fragment size of 1 MB back to the origin server.

The Amplification Factor of S-RFA Attack. As shown in Table 3, all the CDNs we selected are vulnerable to a S-RFA attack. Our wild experimental results show that the amplification effect of S-RFA attack is positively correlated with the fragment size. The bigger the fragment size on CDN is adopted, the greater the amplification factor under the attack of S-RFA will be. At the same time, we also found that although the fragment size is the same, the amplification factor has different values in different CDNs. For example, the amplification factor of *Baidu AI Cloud* is thousands of times larger than *CloudFront*, but they have the same fragment size 512 KB.

In order to explain this magical phenomenon, we carried out in-depth research. We found differences in the handling of forwarding range requests. *CloudFront* strictly forwards client requests according to the fixed size of the fragment. However, *Baidu AI Cloud* will automatically load the head 3 MB resource content for the first time. In addition, *Baidu AI Cloud* will implement a looser fragment strategy, which makes CDN may not only return to the corresponding fragment. For example, in our experiment we found that it would return at least four fragments at a time. This mechanism that should have improved the *Range Origin Fetch* hit rate causes a great amplification factor.

Table 3. The amplification varies with fragment size and target file size in a S-RFA attack

CDN	Fragment size (Bytes)	Amplification factor		
		3 MB	23 MB	50 MB
Alibaba Cloud[a]	524,288	519	510	507
		531	529	525
Baidu AI Cloud[b]	1,048,576	11,345	6,783	5,952
Tencent Cloud	4,096	15	14	14
Huawei Cloud	524,288	702	700	698
CloudFront	1,048,576	1,347	1,342	1,330

[a]*Alibaba Cloud* has two range request forwarding modes: *Mandatory* and *Normal*. The difference between them is mainly about how to handle the first client request, which leads to a slight difference in the amplification factor.

[b]*Baidu AI Cloud* uses the *Fragment* and *Continuance* policies that make S-RFA attack amplification factor have huge differences depending on the size of target recourse.

On the other hand, we also find that the amplification factors of CDNs (except *Baidu AI cloud*) do not fluctuate significantly with the size change of target resources under a S-RFA attack. We believe a S-RFA attack amplification factor is strongly related to the fragment size. The larger the size of transmitted fragments is, the greater the scope that an attacker can influence.

4.3   O-RFA Attack Evaluation

Feasibility of O-RFA Attack. Table 4 shows that *Tencent Cloud* and *Cloud-Front* are vulnerable to an O-RFA attack when they are deployed as ExCDN

Table 4. Range forwarding behaviors vulnerable to O-RFA attack (Considering the readability of the table, we only list the three range requests when the number of overlapping fragments for each CDN is equal to 1.)

CDN	Fragment size (Bytes)	Vulnerable range requests	Forwarded range requests
Tencent Cloud	4,096	4,095–4,096	0–8,191
		8,191–8,192	4,096–12,287
		12,287–12,288	8,192–16,383
	1,048,576	1,048,575–5,242,880	0–6,291,455
		2,097,151–6,291,455	1,048,576–7,340,031
		3,145,727–7,340,031	2,097,152–8,388,607
CloudFront	1,048,576	1,048,575–1,048,576	0–2,087,151
		2,097,151–2,097,152	1,048,576–3,145,728
		3,145,727–3,145,728	2,087,152–4,194,304

or InCDN. In our experiment we cascade vulnerable CDNs. The second column lists the fragment size which vulnerable CDNs adopted. Then, the third and fourth columns show the range request we crafted and corresponding responses CDNs forwarded.

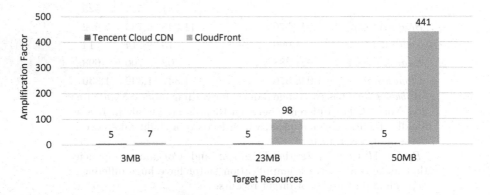

Fig. 5. The amplification varies with fragment size and target file size in an O-RFA attack (The fragment size of *Tencent Cloud* has two different values according to the specific data ranges requested by clients. Here we only present the one that can be successful in an O-RFA attack.)

The Amplification Factor of O-RFA Attack. As shown in Table 5, *Tencent Cloud* and *CloudFront* are vulnerable to O-RFA attack when used as ExCDN or deployed separately like Fig. 2. There are several situations that attackers can perform amplification attacks. 1) When vulnerable CDNs are used as ExCDN, the *ExCDN-InCDN* connection and the *InCDN-Origin* connection will transport much traffic generated by O-RFA attack if the InCDN does not enable the *Range*

Origin Fetch. If InCDN enables the *Range Origin Fetch*, the *InCDN-origin* connection has less impact from O-RFA attack, but *ExCDN-InCDN* connection still suffers a lot of malicious traffic. 2) When a vulnerable CDN is used in a cascading way, an O-RFA attack can produce the same amplification factor between the *ExCDN-InCDN* connection and the *InCDN-origin* connection. That is to say, an O-RFA attack can affect both the CDNs themselves and the original server. 3) In addition, an O-RFA attack has the same impact when the vulnerable CDNs are used alone just shown in Fig. 2 (*Trace B*).

For *Tencent Cloud*, an O-RFA attack works best when the number of overlapping fragments is less than 2, and the increment of amplification factor will begin to decline significantly after the *Threshold*[2] is exceeded. Therefore, for *Tencent Cloud*, the maximum amplification factor of an O-RFA attack is 5, and it does not change with the size of the target resource.

In contrast, when *CloudFront* is attacked by O-RFA, the amplification factor will increase significantly as the target resource becomes larger.

4.4 Severity Assessment

Serious Damage to the Economic Availability of the Website. In the experiment, we find that all the CDNs tested are charged for the actual traffic deliver to client by default, but the victims still have to pay a huge traffic bill after being maliciously attacked. Malicious attackers are able to consume the bandwidth of the origin server and CDN at little cost by means of RangeFragAmp attacks. This condition will become more intractable when mass cloud storage services such as Object Storage Service (OSS) have been wildly introduced into CDNs for acceleration and economy.

A Concealed and Efficient DDoS Attack. Traditional CDN protects the origin server from direct DDoS attacks by masking its IP address [28]. RangeFragAmp attacks does not have to locate the real IP address of the origin server but only needs to send crafted range requests to the CDN to launch amplification attacks that are able to influence the availability of websites. Unfortunately, the vulnerable CDN has no alert under its default configuration when we launch RangeFragAmp attacks.

4.5 CDN Providers Feedback

We reported our findings to CDN providers, and most of them gave us positive feedback. Most CDN providers stated that they had been aware of the potential threat posed by *Small Range Request* since *RangeAmp* attacks were disclosed. In fact, they had already made some corresponding improvements. But it seems that a hasty patch which lacks time for security analysis and experimental verification brings some new troubles. In the future, they will keep evaluating the security

---

[2] Here the *Threshold* is equal to 2.

risk (including our RangeFragAmp attacks) of *Range Origin Fetch* in CDNs. At the same time, they also promise to upgrade and deploy a new range request forward mechanism once they come up with an appropriate method.

As for the size of the fragment, that indicator has the greatest influence on the amplification factor of RangeFragAmp attacks. All providers indicate that, the current value of the fragment size comes from the experience-based setting during the product design stage. In the future, they will refer to our mitigation solutions and keep improving the stability and security of *Range Origin Fetch* in subsequent upgrade.

5   Mitigation

In this section, we will further discuss the root cause of RangeFragAmp vulnerabilities, limitations of our attack, and mitigation solutions.

5.1   Root Cause Analysis

The naive upgrade of *Fragment* policy on CDN without a comprehensive security analysis is the root cause of the RangeFragAmp vulnerability. The paper proposed by Li et al. has fully discussed the flaws in the HTTP range request on RFC 7233 [8]. Moreover, they found there were no additional illustrations on range request in the newest HTTP/2 protocol which made *RangeAmp* can threat both in HTTP/1.1 and HTTP/2 [8,29]. They hold the opinion that, unclear definition and security negligence of the specification constitute the root cause of *RangeAmp* vulnerabilities and the implementation flaw of greatly worsen it.

Through the analysis of the experimental results, we believe that the root cause of RangeFragAmp attacks and *RangeAmp* attacks are similar. There are few considerations about proxy forwarding scenarios like CDN when designing protocol of HTTP range request. Differently, CDN providers have more influence on RangeFragAmp attacks. For a S-RFA attack, the fragment size will affect the amplification factor directly. We notice that, there is a large variation in the size of the fragment among diverse CDNs. To find out the reasons, we have contacted the CDN provider. In the experiment, we find that *Tencent Cloud* performs best in a S-RFA attack test due to its barely 4096 (4K) fragment size. Their engineers respond that, the fragment size they set is based on the need for 4K alignment of the cache system on the edge nodes, which can speed up the cache reading and integration. At the same time, they also set different transmission fragment sizes for different requested files. In contrast, other providers claim that their fragment sizes are based on the experience of developers or just from designs of peers, without sufficient security reasons. Therefore, we believe that the lack of security analysis in planning and designing phase leads to the differences in the performance of defending S-RFA attack among CDN providers.

As for an O-RFA attack, it is caused by the CDN lack of boundary checking and loose forwarding mechanisms for range requests. In our experiments, we find that two vulnerable CDNs are at risk of being exploited by O-RFA attack, though they have different performance features. When the *Threshold* is exceeded, the probability that the request is responded to by the cache in the edge node will be greatly increased. That is why the amplification factor attacked by O-RFA in *Tencent Cloud* is not sensitive to the size of target resource.

As for *CloudFront*, it adopts an extremely loose range request mechanism, and it even automatically corrects most of the illegal requests from clients, which causes the amplification factor of an O-RFA attack to increase significantly according to the size of the target file.

5.2 Limitation

Actually, RangeFragAmp attacks have some limitation and it not suitable for all scenarios. On the one hand, a S-RFA attack has a larger amplification factor and wider victim target. On the other hand, the effect of traffic consumption is not as good as an O-RFA attack. Once all the fragments of the target resource have been transmitted or stored in the cache of edge nodes, a S-RFA attack will become futile. Because the edge node will directly respond to subsequent range request of the attacker from the cache. In fact, a S-RFA attack is more often used to launch attacks from small bandwidth condition such as cellular networks. Whereas, an attacker can launch S-RFA attack free of geographic restriction.

Besides, an O-RFA attack has a better traffic consumption effect. Because its amplification factor is related to the size of the target resource. Therefore, an O-RFA attack can not only attack the origin server but also affect the performance of CDN itself. However, an O-RFA attack has stricter requirements on the bandwidth and range fragment combinations of the attacker compared with a S-RFA attack. Therefore, it is more suitable for the attacker who has a large bandwidth, such as the terminal in high-speed Ethernet.

5.3 Solutions

Actually, previous studies have proposed many mitigation schemes. For example, Li et al. suggested fixing this problem from the server side, CDN side and protocol side [4]. After evaluation, we believe that it too difficult to resist this kind of amplification attack without affecting the normal host on server side. Because it is hard to distinguish the message of the attacker from the normal request, especially for distributed CDN nodes. On the other hand, revising a well-defined and security-aware RFC is indeed a solution able to fix this problem thoroughly. However, the new protocol still requires many researchers to analyze and design, and the protocol upgrade process often takes several years.

Therefore, we believe that improving the range request forwarding policy on the CDN side is the most effective and low-cost approach at present. For example, the edge nodes of CDN should switch to different fragment sizes according to the range of client requests and resource capacity, just like *Tencent Cloud* does. In addition, adopting an incremental caching mechanism on the edge node can reduce the pressure on the origin server effectively when CDN forward request to the origin server. Besides, a better CDN should adopt some more stringent range forwarding policies, including boundary checking and input semantic analysis. Moreover, it is supposed to avoid unnecessary expansion interpretation and reject unreasonable access requests. Last but not the least, we agree with Li et al. on their opinion that CDNs should perform a full security evaluation before supporting new protocol features [4].

6   Related Work

HTTP Range Security. Our work is based on *RangeAmp* attacks [4]. And as far as we know, there is no more academic literature discussing the security risks posed by range requests in CDN the environment before. After *RangeAmp* attacks were published, we revisited the rationale for the attack and tested its latest performance on the most representative CDNs. Our research shows that all the CDNs selected have upgraded the range forwarding policy and fixed the vulnerabilities exploited by *RangeAmp*, but the improved forwarding mechanism has brought new problems. Therefore, we propose a novel range fragment amplification attack based on the existing range request forwarding mechanism to implement the traffic consumption and amplification attack on the CDNs and origin servers.

CDN Security. As an important network infrastructure, CDN is favored by users for its accessibility and security protection [30]. According to reports [31], there is nearly one-fifth of the current Internet traffic transmitted through CDNs. Therefore, the security of CDNs has always been concerned by researchers. For example, Triukose et al. [3] proposed an attack which have abilities to exhaust the bandwidth of origin server by dropping the *front-end* connections rapidly. However, this attack has been proved ineffective in most CDNs [32]. Furthermore, DDoS protection of vulnerable can be directly nullified and abused to attack the origin server through *RangeAmp* attacks proposed by Li et al. [4]. They also mention that, an *OBR* attacker can set a small TCP receive window to make himself only receive little data. In Fact, we evaluate these attacks in our experiment, and find most CDNs tested can mitigate them now. In contrast, our proposed RangeFragAmp attacks can adapt to the new range request forward mechanism on the CDNs, and bypassing the DDoS protection provided by CDNs for the origin server. In addition, we also take economic factors into consideration. Actually, Our RangeFragAmp attacks can be used as a potential scheme of EDoS attack, which can effectively destroy the economic sustainability of websites.

Amplification Attacks. Our research is also a kind of amplification attack which has long been well studied. For example, Booth et al. [7] revealed that, an UDP amplification attack can reach an amplification factor of 556 by recruiting UDP servers on the Internet as reflectors. Mairos et al. [33] introduced a new breed of DNS amplification which has an amplification factor of 44. In fact, the factor of amplification attack is enlarged with the introduction of CDN [34]. For example, *RangeAmp* attacks can even have an amplification factor of 43,300 [4]. Therefore, the harm of amplification attacks in CDN is more serious to a certain extent. RangeFragAmp attacks we proposed have a larger amplification factor than those in the traditional network. Even compared with *RangeAmp* attacks that same to use the range request mechanism of CDN, RangeFragAmp attacks have better concealment and adaptability to the capabilities of attackers. In fact, our experiment shows that the former has been mitigated by most CDNs.

7 Conclusion

In this paper, we revisited HTTP range request cased content fetching amplification issue and evaluated the deployed defenses of mainstream CDN providers. We propose a novel amplification attacks, RangeFragAmp attacks, which take advantage of the defect of the HTTP range request mechanism on CDNs. At the same time, we evaluated the feasibility and severity of the RangeFragAmp attacks in the wild. Lack of security considerations and arbitrary setting of fragment size in the planning stage are the root causes of the vulnerability. In addition, we disclosed our findings to CDN providers and offered our mitigation solutions to them. In the future, we hope to cooperate with CDN providers and academic researchers to discuss CDN security and potential improvement measures.

Acknowledgment. We would like to thank the research and development engineers of the CDNs. The fruitful discussion with them helps us better understand the working mechanism of CDN and the design principle behind it, which provides ideas for our subsequent research At the same time, we are grateful to the reviewers for their comments and suggestions, their feedback helped us build a better job.

References

1. OpenCDN: The idea of traffic amplification attacks (2013). http://drops.wooyun.org/papers/679
2. Chen, J., et al.: Forwarding-loop attacks in content delivery networks. In: NDSS (2016)
3. Triukose, S., Al-Qudah, Z., Rabinovich, M.: Content delivery networks: protection or threat? In: Backes, M., Ning, P. (eds.) ESORICS 2009. LNCS, vol. 5789, pp. 371–389. Springer, Heidelberg (2009). https://doi.org/10.1007/978-3-642-04444-1_23
4. Li, W., et al.: CDN backfired: amplification attacks based on HTTP range requests. In: 2020 50th Annual IEEE/IFIP International Conference on Dependable Systems and Networks (DSN), pp. 14–25. IEEE (2020)
5. Radware: Amplification Attack. https://www.radware.com/security/ddos-knowledge-center/ddospedia/amplification-attack
6. CloudFront: What is a DNS amplification attack? (2020). https://www.cloudflare.com/learning/ddos/dns-amplification-ddos-attack/

7. Booth, T.G., Andersson, K.: Elimination of DoS UDP reflection amplification bandwidth attacks, protecting TCP services. In: Doss, R., Piramuthu, S., Zhou, W. (eds.) FNSS 2015. CCIS, vol. 523, pp. 1–15. Springer, Cham (2015). https://doi.org/10.1007/978-3-319-19210-9_1

8. Fielding, R., Lafon, Y., Reschke, J.: Hypertext transfer protocol (HTTP/1.1): range requests. IETF RFC7233, June 2014

9. Alibaba Cloud: Alibaba Cloud Content Delivery Network. https://www.alibabacloud.com/product/cdn

10. Tencent Cloud: Tencent Cloud Content Delivery Network. https://intl.cloud.tencent.com/product/cdn

11. Baidu AI Cloud: Baidu AI Cloud Content Delivery Network. https://intl.cloud.baidu.com/product/cdn.html

12. Huawei Cloud: Huawei Cloud Content Delivery Network. https://www.huaweicloud.com/en-us/product/cdn.html

13. Amazon Web Services: Amazon CloudFront Content Delivery Network. https://aws.amazon.com/cloudfront/. Accessed 10 Feb 2021

14. Liu, X., Yang, P., Dong, Y., Ahmed, S.H.: A comparative analysis of content delivery capability for collaborative dual-architecture network. In: Romdhani, I., Shu, L., Takahiro, H., Zhou, Z., Gordon, T., Zeng, D. (eds.) CollaborateCom 2017. LNICST, vol. 252, pp. 63–72. Springer, Cham (2018). https://doi.org/10.1007/978-3-030-00916-8_7

15. Tencent: Tencent cloud security white paper. Technical report, Tencent Cloud Security Team & Tencent Research Institute Security Research Center, June 2019

16. Wen, Y., Chen, Y., Shao, M.-L., Guo, J.-L., Liu, J.: An efficient content distribution network architecture using heterogeneous channels. IEEE Access 8, 210988–211006 (2020)

17. Falchuk, B., Żernicki, T., Koziuk, M.: Towards streamed services for co-located collaborative groups. In: 8th International Conference on Collaborative Computing: Networking, Applications and Worksharing (CollaborateCom), pp. 306–315. IEEE (2012)

18. Biazzini, M., Serrano-Alvarado, P., Carvajal-Gomez, R.: Towards improving user satisfaction in decentralized P2P networks. In: 9th IEEE International Conference on Collaborative Computing: Networking, Applications and Worksharing, pp. 315–324. IEEE (2013)

19. Tencent Cloud: Enterprise Content Delivery Network User Guide Product Documentation. Tencent, April 2020

20. GlobalDots: How to Evaluate and Implement a Multi-CDN Strategy. https://www.globaldots.com/content-delivery-network-explained

21. Park, K., Pai, V.S.: Deploying large file transfer on an HTTP content distribution network. In: WORLDS (2004)

22. Hong, K., Kim, Y., Choi, H., Park, J.: SDN-assisted slow HTTP DDoS attack defense method. IEEE Commun. Lett. 22(4), 688–691 (2018)

23. Deshpande, T., Katsaros, P., Basagiannis, S., Smolka, S.A.: Formal analysis of the DNS bandwidth amplification attack and its countermeasures using probabilistic model checking. In: 2011 IEEE 13th International Symposium on High-Assurance Systems Engineering, pp. 360–367. IEEE (2011)

24. Anagnostopoulos, M., Kambourakis, G., Kopanos, P., Louloudakis, G., Gritzalis, S.: DNS amplification attack revisited. Comput. Secur. 39, 475–485 (2013)

25. Kashaf, A., Sekar, V., Agarwal, Y.: Analyzing third party service dependencies in modern web services: have we learned from the Mirai-Dyn incident? In: Proceedings of the ACM Internet Measurement Conference, pp. 634–647 (2020)

26. chinaz.com: CDN cloud real-time observation (2021). https://cdn.chinaz.com/. Accessed 15 Apr 2021
27. Alibaba Cloud: Configure range origin fetch (2021). https://help.aliyun.com/ document_detail/27129.html. Accessed 30 Jan 2021
28. Parno, B., Wendlandt, D., Shi, E., Perrig, A., Maggs, B., Yih-Chun, H.: Portcullis: protecting connection setup from denial-of-capability attacks. ACM SIGCOMM Comput. Commun. Rev. **37**(4), 289–300 (2007)
29. Belshe, M., Peon, R., Thomson, M.: Hypertext transfer protocol version 2 (HTTP/2) (2015)
30. Luglio, M., Romano, S.P., Roseti, C., Zampognaro, F.: Service delivery models for converged satellite-terrestrial 5G network deployment: a satellite-assisted CDN use-case. IEEE Netw. **33**(1), 142–150 (2019)
31. Data Economy: Data economy frontline. Edge computing as the most misunderstood weapon of the IoT world (2019). https://www.clearblade.com/press/data-economy-frontline-edge-computing-as-the-most-misunderstood-weapon-of-the-iot-world/
32. Ife, C.C., Shen, Y., Stringhini, G., Murdoch, S.J.: Marked for disruption: tracing the evolution of malware delivery operations targeted for takedown. arXiv preprint arXiv:2104.01631 (2021)
33. Sieklik, B., Macfarlane, R., Buchanan, W.J.: Evaluation of TFTP DDoS amplification attack. Comput. Secur. **57**, 67–92 (2016)
34. Guo, R., et al.: Abusing CDNs for fun and profit: security issues in CDNs' origin validation. In: 2018 IEEE 37th Symposium on Reliable Distributed Systems (SRDS), pp. 1–10. IEEE (2018)

DCNMF: Dynamic Community Discovery with Improved Convex-NMF in Temporal Networks

Limengzi Yuan[1,2], Yuxian Ke[1], Yujian Xie[1], Qingzhan Zhao[1,2],
and Yuchen Zheng[1,2](✉)🆔

[1] College of Information Science and Technology, Shihezi University, Shihezi, China
[2] Geospatial Information Engineering Research Center, Xinjiang Production
and Construction Corps, Shihezi, China
{ylmz,zqz_inf}@shzu.edu.cn

Abstract. For its crucial importance in the study of temporal networks, techniques for detecting community structures and tracking evolutionary behaviors have been developed. Among these techniques, evolutionary clustering is an efficient method which unveils substructures in complex networks and models the evolution of a system. Most research works in this domain mainly employ Semi-NMF to discover evolving communities. However, in some cases, it can not jointly maintain the quality of community detection and track the temporal evolution infallibly. In this paper, we present a novel community discovery model based on an evolutionary clustering framework using convex non-negative matrix factorization (Convex-NMF), called DCNMF. It is an improvement of Semi-NMF when applied in temporal networks to detect and track evolutionary communities. The proposed model, with temporal smoothness constraint considering the Convex-NMF results, is more accurate and robust both than the evolutionary clustering method based on Semi-NMF and some other existing methods. Specifically, we adopt the gradient descent algorithm to optimize the objective function and prove the correctness and convergence of the algorithm. Experimental results on several synthetic benchmarks and real-world networks show the effectiveness of the proposed method in discovering communities and tracking evolution in dynamic networks.

Keywords: Dynamic community discovery · Temporal networks · Convex non-negative matrix factorization

1 Introduction

Many social, physical, technological, and biological systems can be modeled as networks composed of numerous interacting parts [19]. As an increasing amount of time-resolved data have become available, it has become increasingly

H. Gao and X. Wang (Eds.): CollaborateCom 2021, LNICST 406, pp. 460–475, 2021.
https://doi.org/10.1007/978-3-030-92635-9_27

important to develop methods to quantify and characterize the dynamic properties of temporal networks [8]. Real-world examples of temporal networks include person-to-person communication (e.g., via mobile phones [20]) and one-to-many information dissemination (such as Twitter networks [6]). Analyzing such temporal networks can uncover the important phenomenon and characterize the properties of networks of dynamical units. Dynamic community discovery (DCD) is considered as a sort of effective tool for discovering the structure of complex networks, and ultimately extracting useful information from them. Recently, it is also applied in some edge computing scenarios [14]. A major problem for DCD is to identify stable network decompositions, which comes from the very nature of communities. It lies in the fact that, generally, it is difficult to know if the differences observed on complex networks between consecutive time slices are due to the algorithm's instability or the communities' evolution. A great variety of methods have been proposed to address this problem.

In this paper, we propose a novel model which detects and tracks dynamic communities with convex non-negative matrix factorization (Convex-NMF) based on an evolutionary clustering framework. In most of existing DCD works using non-negative matrix factorization (NMF), it is more common to apply Semi-NMF to address concrete problems. Considering the Semi-NMF form $X = FG^T$, Convex-NMF is obtained when the basis vectors of F are constrained to be convex combinations of the data points [2]. As regards the community membership matrices derived from NMF methods, Convex-NMF solutions are more sparse and orthogonal than Semi-NMF solutions, which is reasonably consistent with the network rules that the node propensities of belonging to communities would not be highly ambiguous. Consequently, we argue that Convex-NMF would be a better option to interpret substructures in evolving real-world networks. Considering the convincing performance of Convex-NMF in clustering problems as well as static community detection, we employ it in reconstructing the network topology task and constraining temporal smoothness for DCD in the proposed model. More concretely, the main contributions of our work are as follows:

- We propose a unified model DCNMF which detects communities and analyzes their evolution. Using probabilistic community membership and temporal smoothness constraints, the proposed model unveils latent community structure and discovers network changes.
- By employing Convex-NMF to obtain community membership matrices, the proposed model achieves community discovery results that are of a better quality than Semi-NMF solutions.
- The proposed model is easily extended to the networks whose number of nodes and communities may change over time, which is a fairly common phenomenon in temporal networks.
- An optimal algorithm is proposed to solve the obtained objective function. We prove that the algorithm is guaranteed to converge and verify the correctness of the algorithm. Analysis of its computational efficiency is also provided.

The rest of the paper is organized as follows. We review related work of dynamic community detection in Sect. 2. In Sect. 3, the proposed gradient descent algorithm and its theoretical analysis are provided. Experimental results performed on synthetic and real-world data are presented in Sect. 4. The conclusion and discussion follow in Sect. 5.

2   Related Work

Recently, considerable research works have been devoted to discover communities in temporal networks. The available methods can be classified into heuristic algorithms and low rank approximation approaches. Since the objective of temporal community detection is typically NP-hard to optimize, some methods employed heuristics to find sets of nodes which can be understood or interpreted as real communities, including two-step methods [21, 24–26] and multi-objective optimization algorithms [3, 4, 10, 16–18]. However, two-step methods extract community structure from each time step independently, which often results in community structure with high temporal variation. In addition, the major drawback of the multi-objective approaches is that the random generation of initial population will greatly increase the search space and hence cause high spatial and temporary complexity.

To overcome the aforementioned problems, low rank approximation based methods [11, 12, 15, 22] (e.g. evolutionary clustering [1]) simultaneously optimized the community discovery accuracy and drift based on the temporal smoothness framework. These methods transformed the time slices of network to the counterparts of each node, which can be used to create an alternative to graph drawings for visualization of node dynamics. Among them, matrix factorizations are widely applied for time-varying community exploration and detection in time-evolving graph sequences. Specifically, NMF is well satisfied with networks, for the reason that most of their edges which commonly correspond to flows, capacity, or binary relationships, are non-negative. Capable of extracting inherent patterns and structures in high dimensional data, the NMF-based methods have become one of the hottest research topics in community discovery.

The NMF-based methods is the most relevant type of approaches to our work, with innate interpretability and good performance in practice. Wu et al. [22] introduced hypergraph in NMF model and utilized the higher-order relationship among the points to promote the clustering performance. Li et al. [11] proposed a method which is based on semi-supervised matrix factorization and random walk to execute community partition. However, these two methods both have the same problem that the number of communities needs to be known as a prior information. Hong Lu et al. [15] used an improved density peak clustering to obtain the number of cores as the pre-defined parameter of NMF and adopted non-negative double singular value decomposition initialization, which can rapidly reduce the approximation error of NMF.

The proposed algorithm DCNMF falls into the category of evolutionary clustering based methods, which discovers communities at time t on the basis of both the topology of the network at t and community structures found previously. It introduces a parameter $\alpha \in [0,1]$ to determine the trade-off between a solution to optimal community detection at t and a solution for maximizing the similarity with the result at $t - 1$. DCNMF hence is capable of coping with the instability problem, while not diverges much from the common community discovery. Moreover, it takes advantage of the partition searched for at time step $t - 1$ to speed up the community discovery at time step t.

3 Algorithm

In this section, we introduce the proposed model, derive the optimization rules, and analyze the complexity of the algorithm.

3.1 Notation

In this paper, bold uppercase letters will donate matrices, e.g. $\mathbf{X}$, bold lowercase letters will donate column vectors, e.g. $\mathbf{x}$, while operators $(\cdot)^\top$ will stand for matrix transposition, e.g. $\mathbf{X}^\top$. Both x_{ij} and $(X)_{ij}$ represent the Entry (i, j) of the matrix $\mathbf{X}$, and the Frobenius norms will be represented by $\|\cdot\|_F$.

Consider a dynamic N-node network whose time-varying structures are captured by the time-series adjacency matrices $\{\mathbf{X}^t \in \mathbb{R}^{N \times N}\}_{t=1}^T$. $x_{ij}^t = 1$ if there is an edge from node i to node j at time t, and $x_{ij}^t = 0$ otherwise. We assume that the dynamic network is undirected, i.e. $x_{ij}^t = x_{ji}^t$, and there are no self-edges, i.e. $x_{ii}^t = 0$.

3.2 The Unified DCNMF Model Formulation

Considering the observed network at time t, denoted by $\mathbf{X}^t$, the non-negative data matrix $\mathbf{X}^t$ can be factorized into $\mathbf{X}^t \mathbf{W}^t \mathbf{G}^{t^\top}$, i.e. $\mathbf{X}^t \approx \mathbf{X}^t \mathbf{W}^t \mathbf{G}^{t^\top}$, with the constraints that $\mathbf{W}^t$ and $\mathbf{G}^t$ are non-negative. In the factorization, $\mathbf{W}^t$ can be considered as the node weight matrix of all nodes and $\mathbf{G}^t$ can be considered to be a community membership matrix with $\mathbf{G}_{ij}^t$ denoting the probability that the node i belongs to the community j. Specifically, $\mathbf{F}^t \approx \mathbf{X}^t \mathbf{W}^t$ can be considered to be a centroid matrix in which each column represents a community central node. As a result, we derive the following function in matrix formulation at time t:

$$\min_{\mathbf{X}^t \geq 0} \|\mathbf{X}^t - \mathbf{X}^t \mathbf{W}^t \mathbf{G}^{t^\top}\|_F^2. \tag{1}$$

Here, we restrict central node column vectors to convex combinations of the columns of $\mathbf{X}^t$ to achieve good interpretability of obtained matrices. For one thing, the basic matrix columns would capture the notion of central nodes whose movements often influence the drifts of nodes who have close relationship with them. For another, restricted constraints lead to the desired NMF solution that community membership matrix $\mathbf{G}^t$ is more sparse and orthogonal, which gives sharper indicators of the community.

We impose the temporal smoothness constraints on community membership matrices to regularize the community structure, so that it is less likely to change dramatically in terms of the community memberships from time $t-1$ to t. The temporal cost is defined as the difference between the community membership matrices at time $t-1$ and that at time t. Regarding both the snapshot cost of modeling network topologies and the temporal cost of smoothness constraint, the cost function can be defined as the sum of community detection quality and historical cost. To achieve evolutionary clustering for DCD, we solve this by maximizing the community detection quality of current time-stamp and minimizing the historical cost, then the obtained cost function is as follows:

$$\min_{\mathbf{W}^t \geq 0, \mathbf{G}^t \geq 0} \|\mathbf{X}^t - \mathbf{X}^t \mathbf{W}^t \mathbf{G}^{t\top}\|_F^2 + \alpha \|\mathbf{X}^{t-1} - \mathbf{X}^{t-1} \mathbf{W}^{t-1} \mathbf{G}^{t\top}\|_F^2, \qquad (2)$$

where, α is a temporal smoothness trade-off parameter.

In experiments, for each time step, we obtain a random initial value of $\mathbf{W}_0^t$ and $\mathbf{G}_0^t$ by setting $\alpha = 0$, then we restart the optimization with $\mathbf{W}^t = \mathbf{W}_0^t$ and $\mathbf{G}^t = \mathbf{G}_0^t$.

Extensions. In this subsection, we introduce two extensions to the proposed unified DCNMF model in order to handle the insertion and removal of nodes and communities.

Assume that at time t, n_1 nodes are removed from and n_2 nodes are inserted into the dynamic network. We first handle the n_1 removed nodes by removing the corresponding n_1 rows from $\mathbf{W}^{t-1}$ and $\mathbf{G}^{t-1}$ in Eq. (2) and from $\mathbf{W}^t$ and $\mathbf{G}^t$ in the last item of Eq. (2). After that, to add n_2 nodes, we pad n_2 rows of zeros to $\mathbf{W}^{t-1}$ and $\mathbf{G}^{t-1}$ in the second item of Eq. (2). Finally, we scale the vector $\mathbf{1}_n$ in Eq. (2) to get $\mathbf{1}_{n'}$. The basic idea behind this heuristic is that we assume that these n_2 nodes already exist as isolated nodes at time $t-1$. Moreover, we just preserve the membership of unchangeable nodes between successive time steps.

Assume that at time t, c_1 communities disappear and c_2 communities emerge in a dynamic network. To handle the c_1 disappearing communities, we remove the corresponding c_1 columns from $\mathbf{W}^t$ and $\mathbf{G}^t$ in Eq. (2). The basic idea of the removal is that considering a disappearing community c_1, we assume that there are no nodes belonging to other communities at time t will join it, and also all those nodes belonging to it at $t-1$ will leave and change to other communities. Even more intuitively, it is equivalent to assuming that the disappearing communities at time t split or combine with other communities. In order to add c_2 new communities, we add c_2 columns to $\mathbf{W}^t$ and $\mathbf{G}^t$ in Eq. (2). The purpose of the addition is that the nodes joining to emerging communities at time t come from previous communities.

3.3   Optimization

To solve the objective function in the Eq. (2), we propose an iterative algorithm using following updating rules which are obtained by using auxiliary functions.

The algorithm iteratively updates $\mathbf{W}^t$ with $\mathbf{G}^t$ fixed and then $\mathbf{G}^t$ with $\mathbf{W}^t$ fixed using following updating rules. And in such a way, the objective function defined in Eq. (2) is monotonically decreased and therefore converges to an optimal solution. The updating rules are as follows,

$$\mathbf{w}_{ij}^t \leftarrow \mathbf{w}_{ij}^t \left(\frac{(\mathbf{X}^{t\top} \mathbf{X}^t \mathbf{G}^t)_{ij}}{(\mathbf{X}^{t\top} \mathbf{X}^t \mathbf{W}^t \mathbf{G}^{t\top} \mathbf{G}^t)_{ij}} \right), \tag{3}$$

then normalize such that $\sum_k \mathbf{w}_{ik} = 1, \forall i,$

$$\mathbf{g}_{ij}^t \leftarrow \mathbf{g}_{ij}^t \left(\frac{(\mathbf{X}^{t\top} \mathbf{X}^t \mathbf{W}^t + \alpha \mathbf{X}^{t-1\top} \mathbf{X}^{t-1} \mathbf{W}^{t-1})_{ij}}{(\mathbf{G}^t \mathbf{W}^{t\top} \mathbf{X}^{t\top} \mathbf{X}^t \mathbf{W}^t + \alpha \mathbf{G}^t \mathbf{W}^{t-1\top} \mathbf{X}^{t-1\top} \mathbf{X}^{t-1} \mathbf{W}^{t-1})_{ij}} \right). \tag{4}$$

The overall procedure of DCNMF can be described as Algorithm 1. Algorithm 1 is able to guarantee that the objective function Eq. (2) converges to a local minimum, and the proof will be presented in the next section. After obtaining $\{\mathbf{W}^t\}_{t=1}^T$ and $\{\mathbf{G}^t\}_{t=1}^T$, we can use $\{\mathbf{G}^t\}_{t=1}^T$ to get the final disjoint communities at each time step and analyze the dynamic behaviors of communities between time intervals.

Algorithm 1. DCNMF

Input: $\mathbf{X}^t, \mathbf{X}^{t-1}, K, \alpha$
    $\mathbf{X}^t$: The adjacency matrix of the network at time step t;
    $\mathbf{X}^{t-1}$: The adjacency matrix of the network at time step $t-1$;
    K: The number of communities;
    α: The temporal smoothness trade-off parameter.
Output: $\mathbf{W}^t, \mathbf{G}^t$
    $\mathbf{W}^t$: The weight matrix of nodes at time step t;
    $\mathbf{G}^t$: The community membership matrix at time step t.
1: Initialize $\mathbf{W}^t, \mathbf{G}^t$ with random values.
2: **repeat**
3:    Update $\mathbf{W}^t$ using Eq. (3) with $\mathbf{G}^t$ fixed.
4:    Update $\mathbf{G}^t$ using Eq. (4) with $\mathbf{W}^t$ fixed.
5:    Normalize such that $\sum_k \mathbf{W}_{ik} = 1, \forall i$.
6: **until** Convergence

Besides, the optimal solution to problem Eq. (2) is constructed from optimal solutions to two subproblems of Eq. (2). We prove the correctness of updating rules Eq. (3) and Eq. (4), and provide the time complexity analysis of Algorithm 1.

$\mathbf{W}^t$**-subproblem.** When update $\mathbf{W}^t$ with $\mathbf{G}^t$ fixed, Eq. (2) is reformulated as:

$$\min_{\mathbf{W}^t \geq 0, \mathbf{G}^t \geq 0} \|\mathbf{X}^t - \mathbf{X}^t \mathbf{W}^t \mathbf{G}^{t\top}\|_F^2. \tag{5}$$

To solve the optimization problem, we introduce the Lagrangian multiplier matrix λ with non-negative values, which constrains the non-negativity of $\mathbf{X}^t$, and then we obtain the following equivalent Lagrangian function:

$$L(\mathbf{W}^t) = tr(-\mathbf{X}^{t\top}\mathbf{X}^t\mathbf{W}^t\mathbf{G}^{t\top} - \mathbf{G}^t\mathbf{W}^{t\top}\mathbf{X}^{t\top}\mathbf{X}^t + \mathbf{G}^t\mathbf{W}^{t\top}\mathbf{X}^{t\top}\mathbf{X}^t\mathbf{W}^t\mathbf{G}^{t\top}).$$
(6)

This function satisfies KKT complementary conditions. By setting the gradient $\frac{\partial L(\mathbf{W}^t)}{\partial \mathbf{W}^t} = 0$, we have the following equation from the complementary conditions:

$$-2\mathbf{X}^{t\top}\mathbf{X}^t\mathbf{G}^t + 2\mathbf{X}^{t\top}\mathbf{X}^t\mathbf{W}^t\mathbf{G}^{t\top}\mathbf{G}^t = \lambda_{ij}\mathbf{W}_{ij}^t = 0.$$
(7)

This is the fixed point equation, and the solution must eventually converge to a stationary point. From Eq. (7), we derive another equivalent equation:

$$(-2\mathbf{X}^{t\top}\mathbf{X}^t\mathbf{G}^t + 2\mathbf{X}^{t\top}\mathbf{X}^t\mathbf{W}^t\mathbf{G}^{t\top}\mathbf{G}^t)(\mathbf{W}_{ij}^t)^2 = 0.$$
(8)

The constrained solution with updating rule in Eq. (3) satisfies Eq. (8), so it satisfies the KKT fixed point condition. The proof of convergence for Eq. (3) is presented in [2].

$\mathbf{G}^t$-**subproblem.** When update $\mathbf{G}^t$ with $\mathbf{W}^t$ fixed, the goal is to solve the optimization problem Eq. (2). Similar to $\mathbf{W}^t$-subproblem, we introduce the Lagrangian multiplier matrix λ with non-negative values which constrains the non-negativity of $\mathbf{G}^t$, and then we obtain the following equivalent Lagrangian function:

$$L(\mathbf{G}^t) = tr(-\mathbf{X}^{t\top}\mathbf{X}^t\mathbf{W}^t\mathbf{G}^{t\top} - \mathbf{G}^t\mathbf{W}^{t\top}\mathbf{X}^{t\top}\mathbf{X}^t + \mathbf{G}^t\mathbf{W}^{t\top}\mathbf{X}^{t\top}\mathbf{X}^t\mathbf{W}^t\mathbf{G}^{t\top}$$
$$- \alpha\mathbf{X}^{t-1\top}\mathbf{X}^{t-1}\mathbf{W}^{t-1}\mathbf{G}^{t\top} - \alpha\mathbf{G}^t\mathbf{W}^{t-1\top}\mathbf{X}^{t-1\top}\mathbf{X}^{t-1}$$
$$+ \alpha\mathbf{G}^t\mathbf{W}^{t-1\top}\mathbf{X}^{t-1\top}\mathbf{X}^{t-1}\mathbf{W}^{t-1}\mathbf{G}^{t\top}).$$
(9)

This function satisfies KKT complementary conditions. By setting the gradient $\frac{\partial L(\mathbf{G}^t)}{\partial \mathbf{G}^t} = 0$, we have the following equation from the complementary conditions:

$$-2\mathbf{X}^{t\top}\mathbf{X}^t\mathbf{W}^t + 2\mathbf{G}^t\mathbf{W}^{t\top}\mathbf{X}^{t\top}\mathbf{X}^t\mathbf{W}^t - 2\alpha\mathbf{X}^{t-1\top}\mathbf{X}^{t-1}\mathbf{W}^{t-1}$$
$$+2\alpha\mathbf{G}^t\mathbf{W}^{t-1\top}\mathbf{X}^{t-1\top}\mathbf{X}^{t-1}\mathbf{W}^{t-1} = \lambda_{ij}\mathbf{G}_{ij}^t = 0.$$
(10)

This is the fixed point equation, and the solution must eventually converge to a stationary point. From Eq. (10), we derive another equivalent equation:

$$(-2\mathbf{X}^{t\top}\mathbf{X}^t\mathbf{W}^t + 2\mathbf{G}^t\mathbf{W}^{t\top}\mathbf{X}^{t\top}\mathbf{X}^t\mathbf{W}^t - 2\alpha\mathbf{X}^{t-1\top}\mathbf{X}^{t-1}\mathbf{W}^{t-1}$$
$$+2\alpha\mathbf{G}^t\mathbf{W}^{t-1\top}\mathbf{X}^{t-1\top}\mathbf{X}^{t-1}\mathbf{W}^{t-1})(\mathbf{G}_{ij}^t)^2 = 0.$$
(11)

The constrained solution with updating rule in Eq. (4) satisfies Eq. (11), so it satisfies the KKT fixed point condition. The proof of convergence for Eq. (4) is presented in [2].

Time Complexity. In Eq. (3), the time to calculate $\mathbf{X}^{t\top}(\mathbf{X}^t\mathbf{W}^t)(\mathbf{G}^{t\top}\mathbf{G}^t)$ and $\mathbf{X}^{t\top}(\mathbf{X}^t\mathbf{G}^t)$ is $\mathcal{O}(4m_1k + 2nk^2)$ and $\mathcal{O}(4m_1k)$, and in Eq. (4), the time to calculate $\mathbf{G}^t((\mathbf{W}^{t-1\top}\mathbf{X}^{t-1\top})(\mathbf{X}^{t-1}\mathbf{W}^{t-1}))$, $\mathbf{X}^{t\top}(\mathbf{X}^t\mathbf{W}^t)$, $\mathbf{X}^{t-1\top}(\mathbf{X}^{t-1}\mathbf{W}^{t-1})$, and $\mathbf{G}^t((\mathbf{W}^{t\top}\mathbf{X}^{t\top})(\mathbf{X}^t\mathbf{W}^t))$ is $\mathcal{O}(4m_2k + 2nk^2)$, $\mathcal{O}(4m_1k)$, $\mathcal{O}(4m_2k)$, and $\mathcal{O}(4m_1k + 2nk^2)$, respectively, where n is the number of nodes, m_1 and m_2 is the number of edges at time t and $t-1$, and k is the number of communities. Therefore, the time to evaluate Eq. (3) and Eq. (4) once is $\mathcal{O}(m_1k + nk^2)$ and $\mathcal{O}(m_1k + m_2k + nk^2)$, respectively, and then the time complexity of DCNMF is $\mathcal{O}(L(m_1k + m_2k + nk^2))$, where L is the iteration number for convergence. The proposed method is applicable to large-scale network, since the number of communities is much smaller than the network size.

4   Experiments and Results

In experiments, we evaluate the performance of the proposed algorithm DCNMF, and compare the results on different types of synthetic and KIT-email real-life datasets with four popular methods: FacetNet [12], DYNMOGA [3], $AFFECT_{kmeans}$ [23], and $AFEECT_{spectral}$ [23]. Since Convex-NMF derives from Semi-NMF, we compare the results of evolutionary clustering method using Semi-NMF, named Semi-NMF, with other methods and DCNMF additionally. DYNMOGA applies local smoothing based multi-objective optimization method to discover dynamic communities. As Facetnet and AFFECT are two classic algorithms among evolutionary clustering methods, we choose them as baselines. We specifically select three synthetic and real-world temporal networks to assess the performance of the proposed method, where the synthetic networks are to prove the accuracy of the algorithm and the real-world one is to validate the feasibility of the algorithm in practice.

In experiments, we set the parameter α of DCNMF and Semi-NMF both as 0.3, and adopt random initialization to NMF solutions at each step for all datasets. Since the compared algorithms converge to local minima, we run 20 times each pair of compared algorithms and finally report the average results into comparison, which ensures the robustness and accuracy of the proposed algorithm as for the community detecting results. Specifically, for Facetnet, we set the parameter of α as 0.8 for three datasets. DYNMOGA can adjust the parameters by itself to search for the global or local optimal solutions. And AFFECT methods also search for the best parameters both in kmeans and spectral based methods on their own.

In this section, we first introduce evaluation measures we employ and then report experiment results of different datasets. The proposed algorithm is not sensitive to the parameter α, and the experimental results of parameter analysis will be shown in Subsect. 4.2.

4.1   Evaluation Measures

In order to measure the DCD performance, we select three evaluation indexes which are widely employed, i.e. Normal Mutual Information (NMI) [9], error rate (CA) [13], and Fscore [9]. Let C_q be the set of the cluster of dataset (the annotated class) and C_p be the set of the cluster detected by the community discovery algorithm. Let $n_p, n_q, n_{p,q}$ be the numbers of the amount of nodes in community C_p, community C_q, and both the communities C_p and community C_q, respectively. The computational processes of three metrics are briefly illustrated as follows.

Normalised mutual information (NMI) is one of the popular evaluation indexes of clustering quality [9], which can be formulated as,

$$NMI = \frac{\sum_{p=1}^{K} \sum_{q=1}^{K} n_{p,q} log(\frac{n \times n_{p,q}}{n_p \times n_q})}{\sqrt{(\sum_{p=1}^{K} n_p log\frac{n_p}{n})(\sum_{q=1}^{K} n_q log\frac{n_q}{n})}}. \tag{12}$$

NMI is a value between 0 and 1, which equals 1 when two partitions are equivalent.

The error rate (CA) [13] can be formulated as

$$CA = \|\mathbf{ZZ}^{\top} - \mathbf{GG}^{\top}\|_F^2, \tag{13}$$

where, $\mathbf{Z}$ is the indicator matrix of clustering result which is computed by a given algorithm, where the i-th row of $\mathbf{Z}$ indicates the community membership of the i-th node (i.e., if the i-th node belongs to the k-th community, then $z_{ik} = 1$ and $z_{ik'} = 0$ for $k' \neq k$). A similar indicator matrix $\mathbf{G}$ is constructed for the ground truth. Then the error rate computed by Eq. (13) measures the distance between the community structure represented by $\mathbf{Z}$ and that represented by $\mathbf{G}$.

Fscore integrates the metrics of precision and recall, which is extensively applied in evaluating the community detection performance [9]. The precision and recall are calculated as

$$Precision(C_q, C_p) = \frac{n_{p,q}}{C_p}, \tag{14}$$

$$Recall(C_q, C_p) = \frac{n_{p,q}}{C_q}. \tag{15}$$

Then the Fscore of the detected community C_p and the real community C_q can be computed as

$$F(C_q, C_p) = \frac{2 \times P(C_q, C_p) \times R(C_q, C_p)}{P(C_q, C_p) + R(C_q, C_p)}. \tag{16}$$

4.2 Synthetic Dataset 1: Dynamic-GN Dataset

GN-network benchmark was previously proposed by Girvan and Newman [5] for static community detection, where each network contains 128 nodes and 4 communities. Each community contains fixed 32 nodes, and the average degree of nodes is 16 fixed. To incorporate evolution into the GN-network, Lin et al. [12] developed dynamic GN-network, in which the membership of three vertices

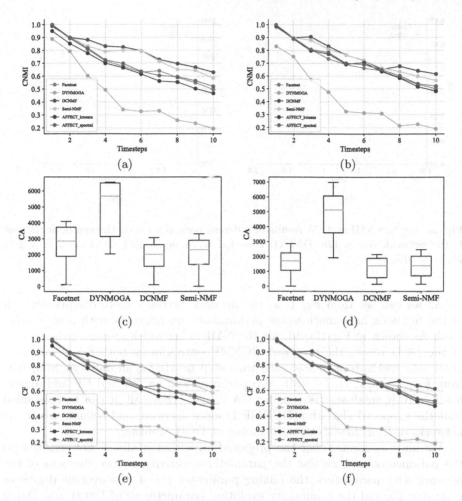

Fig. 1. Average NMI, CA and CF results on the synthetic dataset 1: the network size is 128; (a, c, e): $z = 3, d = 20, nc = 10\%$; (b, d, f): $z = 4, d = 25, nc = 30\%$.

in each community is changed by random assignment to other communities. In experiments, we set the number of time steps $T = 10$, the number of nodes at each timestep $N = 128$, and the number of communities $K = 4$, respectively. The mixing parameter z which controls the noise level of communities, is set to 3 and 4. Moreover, we set the average degree of nodes $d = 20$ and $d = 25$, and the community evolution parameter $nc = 10\%$ and 30%, which is used to control the degree of transition that nodes transfer from their own communities to others randomly between consecutive time slices. For each dataset of this type of temporal network, we run 20 times and take the average NMI, CA, and CF values as the final reported results.

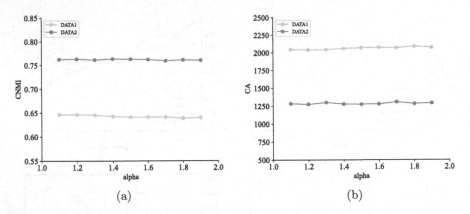

Fig. 2. Average NMI and CA results of different parameter α on the synthetic dataset 1: the network size is 128; DATA1: $z = 4, d = 25, nc = 10\%$; DATA2: $z = 4, d = 25, nc = 30\%$.

As we can see from Fig. 1, as the network becomes more complicated, all of the methods have much worse performance on networks with a high noise level. As shown in Fig. 1(a, b), for the NMI index which measures the quality of the DCD result, the proposed DCNMF outperforms both Semi-NMF and other four methods at almost each time step nearly for all datesets. Similarly, from Fig. 1(c, d, e, f), DCNMF also performs well on CA and CF and always has relatively small variances as for CA values. Above all, it can be concluded that the proposed algorithm DCNMF is more accurate and robust on various networks of Dynamic-GN dataset aiming at DCD problem.

We have mentioned that the proposed algorithm is almost not sensitive to the parameters, and we test the parameter-sensitivity on two datasets of the network with parameters: the mixing parameter $z = 4$, the average degree of node $d = 25$, and the community evolution parameter nc of Data1 and Data2 is 10% and 30%, respectively. Parameter α is an important factor to tune the smoothness of the evolutionary trends of the dynamic networks. Figure 2 shows the results of NMI and CA over the changes of α from $\alpha = 1.1$ to $\alpha = 1.9$. From the results, we can see that the DCD results of DCNMF are stable as for NMI and CA indexes.

4.3   Synthetic Dataset 2: Dynamic-LFR Dataset

To obtain a synthetic network that is more consistent with a real-world network, we employ the extended LFR [7] model to generate dynamic networks, which is based on the embedding of events into synthetic graphs. We test the proposed algorithm on the synthetic datasets which mainly include the node switch event, i.e., nodes switch community membership between consecutive time slices. To evaluate different methods, we construct different synthetic networks by setting different parameters which cover 1000 nodes over 10 snapshots. In each of the synthetic networks, μ is the mixing parameter which controls the level of edges between communities, and p is the probability of the nodes switching among communities. In experiments, we set the number of the time steps $T = 10$, the number of communities $k = 36$, the average degree of nodes $d = 20$, and the mixing parameter $p = 0.5$, $\mu \in [0.1, 0.8]$.

The results of two examples of generated temporal networks are shown in Fig. 3, which involves 1000 nodes, 36 embedded dynamic communities, $p = 0.5$, and $\mu = 0.1$, 0.3, and 0.6, respectively. According to Fig. 3(a, c, e), we observe that the DCNMF performs better when the network has a lower level of noise than the other compared methods. When the probability of switching is high, i.e. the network is more complex, the proposed model DCNMF still has higher community detection quality than other methods as shown in Fig. 3(e). It is also noticed in Fig. 3(b, d, f) that DCNMF has lower errors in most cases, which succeeds to discover dynamic communities accurately. Consequently, these results demonstrate that DCNMF is more accurate and robust than the other methods.

4.4   KIT-Email Data

To validate the feasibility of the proposed algorithm, we compare DCNMF with other methods on KIT-email data which is a large number of snapshots of the e-mail communication network in the Department of Informatics at KIT[1]. In the network, the vertices represent email contacts of the department of computer science at KIT, which evolves during 48 consecutive months from September 2006 to August 2010. If an email is sent to a recipient, there is an edge from the sender vertex to the recipient vertex. We construct the adjacency matrices among 231 active members. In the E-mail network, the department of computer science at KIT is considered as a community. Since the number of communities increases when taking more months as a snapshot, the number of communities is 14, 23, 25, 25, and 27, for the snapshots of 1, 2, 3, 4, and 6 months, respectively.

Table 1 shows the results of compared algorithms on the real world dataset with different resolutions. Correspondingly, we set 2 months, 3 months, 4 months, and 6 months of the KIT email dataset as the length of snapshots to form four temporal networks, on which the results in terms of NMI and CA are presented in Table 1. From the table, it demonstrates that the proposed DCNMF substantially improves the performance of Semi-NMF based method. Besides, although DCNMF does not always perform the best on datsets, it is quite competitive with the best one (i.e. Facetnet) on CNMI and CA. Overall, the proposed method DCNMF shows superior performance on NMI and CA compared with other methods.

---

[1] http://i11www.iti.uni-karlsruhe.de/en/projects/spp1307/emaildata.

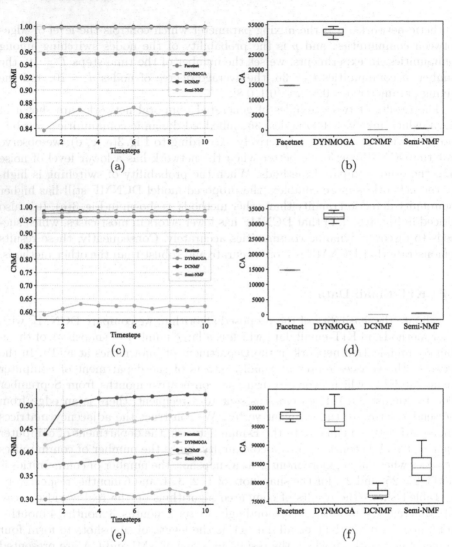

Fig. 3. Average NMI and CA results on the synthetic dataset 2: the network size is 1000, $p = 0.5$; (a, b): $\mu = 0.1$; (c, d): $\mu = 0.3$; (e, f): $\mu = 0.6$.

Table 1. Average NMI and CA values of KIT-email real network data: Data1: taking 2 months as a snapshot; Data2: taking 3 months as a snapshot; Data3: taking 4 months as a snapshot; Data4: taking 6 months as a snapshot.

Method	CNMI			
	$Data_1$	$Data_2$	$Data_3$	$Data_4$
Facetnet	**0.84 ± 0.019**	**0.83 ± 0.018**	0.80 ± 0.018	**0.81 ± 0.018**
DYNMOGA	0.72 ± 0.036	0.68 ± 0.035	0.66 ± 0.029	0.64 ± 0.029
$AFFECT_{kmeans}$	**0.84 ± 0.027**	0.81 ± 0.022	0.80 ± 0.023	0.79 ± 0.021
$AFFECT_{spectral}$	**0.84 ± 0.021**	0.82 ± 0.020	**0.81 ± 0.021**	**0.81 ± 0.016**
SemiNMF	0.83 ± 0.03	0.79 ± 0.036	0.76 ± 0.050	0.76 ± 0.043
DCNMF	**0.84 ± 0.023**	**0.83 ± 0.020**	**0.81 ± 0.021**	**0.81 ± 0.020**

Method	CA			
	$Data_1$	$Data_2$	$Data_3$	$Data_4$
Facetnet	**814.0 ± 77.7**	**1233.4 ± 123.3**	**1621.6 ± 173.2**	**2007.3 ± 175.8**
DYNMOGA	1350.2 ± 251.0	2134.9 ± 369.3	2847.6 ± 404.4	4034.0 ± 585.5
$AFFECT_{kmeans}$	889.0 ± 128.3	1456.7 ± 134.7	1896.6 ± 208.7	2321.9 ± 284.0
$AFFECT_{spectral}$	1017.0 ± 124.8	1431.5 ± 147.3	1708.7 ± 191.0	2239.8 ± 222.6
SemiNMF	911.3 ± 218.0	1473.7 ± 309.5	2131.1 ± 584.9	2778.2 ± 170.1
DCNMF	**847.7 ± 93.3**	**1289.9 ± 103.5**	**1633.3 ± 135.8**	**2117.3 ± 149.8**

5 Discussion and Conclusion

In this paper, we present a unified model named DCNMF which is able to detect communities and track their evolution. Because of the employment of Convex-NMF, a constrained version of Semi-NMF, the proposed algorithm DCNMF has achieved the great success on DCD issues. Compared with several typical methods, it is proved that the proposed method utilizes the previous information more effectively in the task of analyzing dynamics. The experimental results on the synthetic and real data show that DCNMF outperforms other typical DCD methods for various evolutionary networks.

In terms of DCD methods based on temporal trade-off smoothness, one of potential issue is that the long-term coherence of dynamic communities. For the reason that at each iteration the detected community depends on previous ones, temporal trade-off approaches are subject to the risk of an avalanche effect: communities can experience rapid changes. Under such circumstance, it is more suitable to directly employ static community discovery algorithms for solutions at a given time, and it will be our future work to capture such substantial drifts in temporal networks. Furthermore, we also plan to extend the proposed model to networks whose number of nodes and communities may change over time.

Acknowledgements. This work is supported by the Innovation and Cultivation Project for Youth Talents of Shihezi University (Program No. CXPY201905), the Scientific Research Project for High-level Talents of Shihezi University (Program No. RCZK202028), and the Financial and Science Technology Plan Project of Xinjiang Production and Construction Corps (Program No. 2017DB005).

References

1. Chi, Y., Song, X., Zhou, D., Hino, K., Tseng, B.L.: On evolutionary spectral clustering. ACM Trans. Knowl. Discov. Data **3**(4), 1–30 (2009)
2. Ding, C., Li, T., Jordan, M.I.: Convex and semi-nonnegative matrix factorizations. IEEE Trans. Pattern Anal. Mach. Intell. **32**(1), 45–55 (2010)
3. Folino, F., Pizzuti, C.: An evolutionary multiobjective approach for community discovery in dynamic networks. IEEE Trans. Knowl. Data Eng. **26**(8), 1838–1852 (2013)
4. Gao, C., Chen, Z., Li, X., Tian, Z., Li, S., Wang, Z.: Multiobjective discrete particle swarm optimization for community detection in dynamic networks. Europhys. Lett. **122**(2), 28001 (2018)
5. Girvan, M., Newman, M.E.J.: Community structure in social and biological networks. Proc. Natl. Acad. Sci. **99**(12), 7821–7826 (2002)
6. González-Bailón, S., Borge-Holthoefer, J., Rivero, A., Moreno, Y.: The dynamics of protest recruitment through an online network. Sci. Rep. **1**, 197 (2011)
7. Greene, D., Doyle, D., Cunningham, P.: Tracking the evolution of communities in dynamic social networks. In: ASONAM, pp. 176–183 (2010)
8. Holme, P., Saramäki, J.: Temporal networks. Phys. Rep. **519**, 97–125 (2012)
9. Jing, L., Ng, M.K., Huang, J.Z.: An entropy weighting k-means algorithm for subspace clustering of high-dimensional sparse data. IEEE Trans. Knowl. Data Eng. **19**(8), 1026–1041 (2007)
10. Li, Q., Cao, Z., Ding, W., Li, Q.: A multi-objective adaptive evolutionary algorithm to extract communities in networks. Swarm Evol. Comput. **52**, 100629 (2020)
11. Li, W., Xie, J., Xin, M., Mo, J.: An overlapping network community partition algorithm based on semi-supervised matrix factorization and random walk. Expert Syst. Appl. **91**, 277–285 (2018)
12. Lin, Y.R., Chi, Y., Zhu, S., Sundaram, H., Tseng, B.L.: FacetNet: a framework for analyzing communities and their evolutions in dynamic networks. In: WWW, pp. 685–694 (2008)
13. Lin, Y.R., Chi, Y., Zhu, S., Sundaram, H., Tseng, B.L.: Analyzing communities and their evolutions in dynamic social networks. ACM Trans. Knowl. Discov. Data **3**(2), 8:1–8:31 (2009)
14. Liu, F., Lv, B., Huang, J., Ali, S.: Towards mobility-aware dynamic service migration in mobile edge computing. In: CollaborateCom (2020)
15. Lu, H., Zhao, Q., Sang, X., Lu, J.: Community detection in complex networks using nonnegative matrix factorization and density-based clustering algorithm. Neural Process. Lett. (12) (2020)
16. Marler, R.T., Arora, J.S.: Survey of multi-objective optimization methods for engineering. Struct. Multidiscip. Optim. **26**(6), 369–395 (2004)
17. Messaoudi, I., Kamel, N.: A multi-objective bat algorithm for community detection on dynamic social networks. Appl. Intell. **49**(6), 2119–2136 (2019)

18. Mu, C., Zhang, J., Liu, Y., Qu, R., Huang, T.: Multi-objective ant colony opti-
    mization algorithm based on decomposition for community detection in complex
    networks. Soft. Comput. **23**(23), 12683–12709 (2019). https://doi.org/10.1007/
    s00500-019-03820-y
19. Newman, M.: Networks: An Introduction. Oxford University Press, Oxford (2010)
20. Onnela, J.P., et al.: Structure and tie strengths in mobile communication networks.
    Proc. Natl. Acad. Sci. **104**(18), 7332–7336 (2007)
21. Sarswat, A., Jami, V., Guddeti, R.M.R.: A novel two-step approach for overlapping
    community detection in social networks. Soc. Netw. Anal. Min. **7**(1), 1–11 (2017).
    https://doi.org/10.1007/s13278-017-0469-7
22. Wu, W., Kwong, S., Zhou, Y., Jia, Y., Gao, W.: Nonnegative matrix factorization
    with mixed hypergraph regularization for community detection. Inf. Sci. **435**, 263–
    281 (2018)
23. Xu, K.S., Kliger, M., Hero III, A.O.: Adaptive evolutionary clustering. Data Min.
    Knowl. Disc. **28**(2), 304–336 (2013). https://doi.org/10.1007/s10618-012-0302-x
24. Zhang, D., Huang, Y., Wang, Y., Zhu, Y., Zhao, C.: A novel two-step community
    detection approach based on community tree and the n-players cooperative game
    in large-scale social networks. J. Comput. Methods Sci. Eng. **18**(4), 1007–1020
    (2018)
25. Zhang, Y., Yin, D., Wu, B., Long, F., Cui, Y., Bian, X.: Plinkshrink: a parallel
    overlapping community detection algorithm with link-graph for large networks.
    Soc. Netw. Anal. Min. **9**(1), 66 (2019)
26. Zhao, Z., Li, C., Zhang, X., Chiclana, F., Viedma, E.H.: An incremental method
    to detect communities in dynamic evolving social networks. Knowl.-Based Syst.
    163, 404–415 (2019)

Network and Security and IoT
and Social Networks

Loopster++: Termination Analysis for Multi-path Linear Loop

Hui Jin[1], Weimin Ge[1], Yao Zhang[1], Xiaohong Li[1](✉), and Zhidong Deng[2]

[1] College of Intelligence and Computing, Tianjin University, Tianjin, China
{jinh,gewm,xiaohongli}@tju.edu.cn
[2] State Grid Customer Service Center, Tianjin, China

Abstract. Loop structure is widely adopted in many applications, *e.g.* collaborative applications, social network applications, and edge computing. And the termination of the loop is of great significance to the correctness of the program. Most of the previous relative studies focused on determining the termination of a loop program by synthesizing the ranking functions, but not every ranking function can be synthesized. Although a class of linear loop program termination has been proven to be decidable, it is always difficult to analyze the termination of a multi-path linear loop. Xie et al. [20] presented Loopster to quickly check the termination of the multi-path loop program by analyzing the termination of each path and the dependency between paths. But it relies on the monotonicity of variables which is very complicated to check when the variables increase.

To this end, we extend Loopster, named Loopster++, to analyze the termination of multi-path linear loops. In Loopster++, 1) we convert the iterable path into a single path linear loop to analyze its termination. 2) We also propose a novel method to analyze the dependency between linear loop paths. 3) For the cycle constituted by alternate execution between paths, we classify all cycles and give the termination method of the corresponding category cycle. We finally evaluate Loopster++ by analyzing the termination of the benchmarks from the competition on software verification and compare it with the state-of-the-art tools. The empirical results demonstrate the superiority of Loopster++ by achieving high accuracy of 83% in the shortest time.

Keywords: Termination analysis · Multi-path linear loop · Path dependency automaton

1 Introduction

The termination analysis of the program is one of the most important parts of the program verification, and it is of great significance to ensure the correctness of the program. Furthermore, non-termination will cause a variety of program bugs, even incurring denial-of-service attacks [2], and be hardly notified [12]. Therefore, it is imperative to determine the termination of the programs.

© ICST Institute for Computer Sciences, Social Informatics and Telecommunications Engineering 2021
Published by Springer Nature Switzerland AG 2021. All Rights Reserved
H. Gao and X. Wang (Eds.): CollaborateCom 2021, LNICST 406, pp. 479–497, 2021.
https://doi.org/10.1007/978-3-030-92635-9_28

The research on program termination analysis has received a lot of advances. The general approaches are to synthesize the ranking functions [3,7,9,14,17]. Colon and Sipma [7] proposed a method to synthesize the ranking functions based on polyhedral cones and systems of linear constraints. Podelski and Rybalchenko [17] proposed the synthesis of linear ranking functions for linear loop programs except for nested loops. Ben-Amram and Genaim [3] synthesize the ranking functions by integrating polyhedra theory to prove the termination of a loop program which has linear constraints. Leike and Heizmann [14] proposed a ranking template, which covers all methods based on constraint-based synthetic ranking functions. However, having a ranking function is a sufficient condition for termination, which is no longer effective if cannot be founded. Therefore, there are other techniques to analyze the decidable class of loops [5,16,18]. Tiwari [18] showed that the termination of a linear loop of $while(Bx > b)\{x = Ax + c\}$ form is decidable range over $\mathbb{R}$. Braverman [5] generalized the work of Tiwari and proved that the homogeneous form is decidable over integers. If the update matrix A of a simple linear loop program can be diagonalized, Ouaknine et al. [16] proposed how to decide its termination. Note that, these techniques only consider the single path, however, a loop program is normally multi-path in practice.

To analyse loop programs, Xie et al. [19] presented a loop summary framework, namely Path Dependency Automaton (PDA), which summarizes path-sensitive loop on interesting variables. They extracts the properties of each path, and then summarizes the properties of the overall loop based on the dependencies between paths. However, In their approach, all the related variables need to be inductive when building PDA. To efficiently determine the loop termination, Xie et al. further proposed Loopster [20] based on PDA by analyzing the monotonicity of variables, which is not only limited to the inductive variables. But for linear loop programs, the monotonicity of variables is hardly detected.

In this paper, we extend the Loopster to Loopster++ based on the theory of termination of linear loop proposed by Tiwari [18]. Specifically, Loopster++ follows the theory in [18] to analyze the termination of each iterable path so that it is capable of extending the PDA to support linear expression in the loop program. At the same time, we propose a novel method to analyze the dependency between the paths in the multi-path linear loop. Finally, we also present a method to analyze the termination of the cycle generated in this linear loop. To demonstrate the effectiveness of Loopster++, we apply it to analyze the benchmarks from the competition on software verification [1] and we compare it with the top-three tools, i.e., Ultimate Automizer, CPAchecker, and 2LS. The result shows that Loopster++ can correctly handle 100 loop programs in a total of 120 loop programs. Meanwhile, Loopster++ outperforms other tools on efficiency, which is at least 5x faster than the other three tools.

In summary, this paper makes the following contributions:

- We follow the theory of linear loop termination to extend Loopster and propose Loopster++ so that it is capable of supporting multi-path linear loops.

- We extend the algorithm proposed by Tiwari [18] so that the termination of the loop with precondition can be well determined.
- In order to deal with the cycle formed by strong connected components between paths, we make a classification of the cycle derived from Loopster++ and propose the corresponding method to determine the termination of the cycle.

2   Preliminaries

In this section, we define the scope of our work, some professional terms in PDA proposed by Xie et al. [19], the structure of Loopster [20] and the determination algorithm of the termination of linear loop programs.

2.1   Scope of Our Work

We focus on the termination analysis of multi-path linear loops in this paper. Let $X = \{x_1, x_2, ..., x_n\}$ be a finite set of variables ranging over $\mathbb{R}$, and $f(x_1, x_2, ..., x_n)$ be a multivariate linear polynomial. All atomic operations on a loop are in the form of $f(x_1, x_2, ..., x_n) \sim b$, $b \in \mathbb{R}$ and $\sim \in \{<, \leq, >, \geq, =\}$ (= represents the assignment operation). Limited by the linear loop termination analysis algorithm, other operations are not considered in this paper.

2.2   Path Dependency Automaton (PDA)

A control flow graph (CFG) of a loop is a tuple $\mathcal{G} = \{V, E, v_s, V_h, V_e, \iota\}$. Where V is a set of basic blocks, E is a set of directed edges connecting two basic blocks, v_s is the start node of the CFG, executed before entering the loop, V_h and V_e are the set of header blocks and exit blocks of the loop, $\iota(e)$ is the branch condition of the edge $e \in E$.

Example 1. Figure 1(b) shows a CFG of the unnested loop in Fig. 1(a), where $V = \{a, b, c, d, e, f\}$, $E = \{(a, b), (b, c), (b, d), (c, e), (c, f), (e, b), (e, f)\}$, $v_s = a$, $V_h = \{b\}$, $V_e = \{d\}$ and $\iota((b, c)) = 2x + y > 0$. Since there is only one loop, there is only one header block. In Fig. 2(b), since the nested loop has two loops, the head blocks of this loop are b and c.

Given a control flow graph $\mathcal{G} = \{V, E, v_s, V_h, V_e, \iota\}$, the loop path σ is a finite sequence of basic blocks $(v_0, v_1, ..., v_k)$, where $v_0 \in V_h$, $v_k \in V_h \cup V_e$ are the head and tail of σ and are denote as head(σ) and tail(σ), respectively. If $head(\sigma) == tail(\sigma)$, we say σ is an iterable path. The path condition of σ is the conjunction of the branch condition of each edge in the path and denote as θ_σ. We use $\mathcal{V}_\sigma$ to denote the value changes of the variables in the path σ and $\mathcal{V}_\sigma^n$ denote the variables after σ executes n times. $\theta(\sigma_i, \mathcal{V}_{\sigma_j}^n) \mapsto \{true, false\}$ represents if the path condition θ_{σ_i} is satisfiable or not after the path σ_j executes n times.

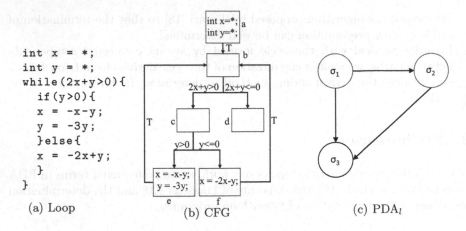

```
int x = *;
int y = *;
while(2x+y>0){
    if(y>0){
        x = -x-y;
        y = -3y;
    }else{
        x = -2x+y;
    }
}
```

(a) Loop   (b) CFG   (c) PDA$_l$

Fig. 1. Unnested loop program

If θ_σ is satisfiable, σ is feasible, otherwise, it is infeasible. Obviously, the infeasible path will not be executed, so in this paper, we only consider feasible paths. In the following sections, we assume that the paths are all feasible.

The precondition of loop denoted as $pre(\mathcal{G})$ constrains the possible valuations for the variables in start node of CFG. The precondition of each path σ which constrains the possible valuations for the variables before executing the path σ.

Example 2. The loop in Fig. 1 has three paths in the CFG: $\sigma_1 = (b, c, e, b)$, $\sigma_2 = (b, c, f, b)$, $\sigma_3 = (b, d)$. σ_1 and σ_2 are the iterable paths. The path condition of σ_1 is $2x + y > 0 \wedge y > 0$. We use Θ_σ to denote the set of conditions of σ, e.g., $\Theta_{\sigma_1} = \{2x + y > 0, y > 0\}$. The loop in Fig. 2 has four paths: $\sigma_1 = (b, c)$, $\sigma_2 = (c, e, c)$, $\sigma_3 = (c, f, b)$, $\sigma_4 = (b, d)$. For each of the iterable path, we can convert it to a linear loop program. For example, in Fig. 1, the path σ_1, we can convert it to:

$$while(2*x + y > 0 \wedge y > 0)\{$$
$$x = -x - y;$$
$$y = -3 * y;$$
$$\}$$

Given a loop with CFG $\mathcal{G} = \{V, E, v_s, V_h, V_e, \iota\}$, the path dependency automaton (PDA$_l$) of this loop is $\mathcal{A} = \{S, T, init, accept\}$, where

- S is a set of states. Each state $\sigma \in S$ corresponds to a path in the loop.
- $T \subseteq S \times S$ is a set of transitions. $(\sigma_i, \sigma_j) \in T$, which represents $\exists n > 0$, s.t. $\theta(\sigma_j, V^n_{\sigma_i}) == true \wedge tail(\sigma_i) == head(\sigma_j)$
- $init$ is a set of initial states in S and $accept$ is a set of accept states in S. An initial state is the firstly executed state and an accept state has no successors.

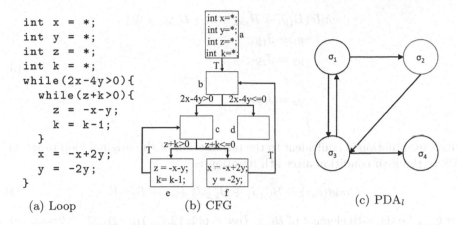

```
int x = *;
int y = *;
int z = *;
int k = *;
while(2x-4y>0){
  while(z+k>0){
    z = -x-y;
    k = k-1;
  }
  x = -x+2y;
  y = -2y;
}
```

(a) Loop   (b) CFG   (c) PDA$_l$

Fig. 2. Nested loop program

Example 3. Figure 1(c) shows the PDA$_l$ of the loop in Fig. 1(a). Where $S = \{\sigma_1, \sigma_2, \sigma_3\}$, $T = \{(\sigma_1, \sigma_2), (\sigma_1, \sigma_3), (\sigma_2, \sigma_3)\}$, $init = \{\sigma_1, \sigma_2\}$, $accept = \{\sigma_3\}$.

2.3 The Structure of Loopster

Loopster [20] uses a divide-and-conquer approach to analyze the termination of a loop program. It is required three steps to determine whether the loop is terminated. 1) Extracting relevant statements from the loop program through the program slicing technique and constructing a control flow graph (CFG). 2) Using CFG to construct PDA and adopting monotonicity to analyze the termination for each path. 3) Analyzing the reachability of the non-terminating path in the loop and the termination of the cycle in PDA to determine the overall termination of the loop.

2.4 Termination of Linear Loop Program

Here we briefly introduce the algorithm proposed by Tiwari [18] to determine the termination of a linear loop program.

Given a homogeneous case of linear loop program

$$while(Bx > 0)\{x = Ax\} \tag{1}$$

We first transform it into

$$while(B'y > 0)\{y = Jy\} \tag{2}$$

where J is a Jordan canonical form of A, P is the transition matrix from A to J, $A = PJP^{-1}$, $x = Py$, and $B' = BP$. Then we only consider the Jordan blocks corresponding to the positive eigenvectors and get the program

$$while(B_1'y_1 + B_2'y_2 + ... + B_r'y_r > 0)\{$$
$$y_1 = J_1y_1;$$
$$y_2 = J_2y_2;$$
$$...;$$
$$y_r = J_ry_r;$$
$$\}$$

$$(3)$$

which termination is equivalent to the termination of the original loop program. Where the k-th condition after i-th iteration is

$$Cond(k, i) = B_{k,1}'J_1^i + B_{k,2}'J_2^i + ... + B_{k,r}'J_r^i \qquad (4)$$

Let $b_{m,n}$ be the n-th element of $B_{k,m}$, $Ind = \{11, 12, ..., 1n_1, 21, 22, ..., 2n_2, ..., r1, r2, ..., rn_r\}$. For all index $ind \in Ind$, if

$$\begin{cases} b_{ind}y = 0 \\ b_{indr}y > 0 \\ Cond(k,i)y > 0, i \in [0, \Pi_2(ind)] \end{cases} \qquad (5)$$

are not satisfiable, the loop is terminating. Where $\Pi_1(ind)$, $\Pi_2(ind)$ are represent the left and right components of ind respectively (for example, $ind = 12$, $\Pi_1(ind) = 1$ and $\Pi_2(ind) = 2$), $b_{ind} = b_{\Pi_1(ind),\Pi_2(ind)}$ and $indr$ represent the right part of ind in Ind (for example if $ind = 12$, $indr \in \{13, 14..., 1n_1, 21, 22, ..., 2n_2, ..., r1, r2, ..., rn_r\}$), and y represent the initial value of each variables in the loop.

For the non-homogeneous case

$$while(Bx > b)\{x = Ax + c\} \qquad (6)$$

we can convert it as the following form and then analyze the terminator of this homogeneous form.

$$while((B\ b)\ x > 0)\left\{x = \begin{pmatrix} A & c \\ 0 & 1 \end{pmatrix} x\right\} \qquad (7)$$

3   Methodology

In this section, we introduce Loopster++ in detail. Loopster++ extends Loopster mainly in the following two parts (marked with red boxes in Fig. 3. 1) When constructing PDA_l, we use linear loop termination analysis to determine the termination of each path. We use the weakest post-condition to determine the dependencies between paths. 2) In the phase of loop overall termination analysis, we extend the linear loop termination algorithm to support the termination of the loop which has preconditions. For the cycle in PDA_l, we propose a new method to analyze the cycle termination.

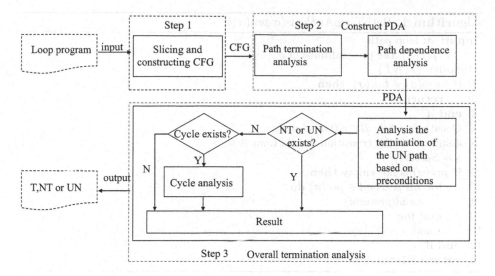

Fig. 3. Overview of Loopster++ (Color figure online)

3.1 Path Termination Analysis

The path can terminate if it can be executed within a limited number of times for any input, otherwise, the path is nonterminating. Obviously, a nonterminating path must be iterable, so we only consider the termination of iterable paths. For each iterable path, we first determine whether the path has a precondition. If there is no precondition, we only need to extract its path condition and value changes, and convert it into a simple linear loop program, and then use the algorithm introduced in Sect. 2.4 to determine the termination of this path. However, for the path with preconditions, we need slightly extend the algorithm introduced in Sect. 2.4. The specific methods are as follows:

From the formula (5) in the algorithm of Sect. 2.4, we can find that the algorithm to determine the termination is searching for whether there is an initial value y that satisfies the condition in formula (5). If it exists, the loop does not terminate. So we can set the range of the initial value of the corresponding variable here so that we can prevent the search for points other than the preconditions so as to determine the termination of the path containing the preconditions. For the precondition $pre(\sigma)$ of the path σ, we assume that all preconditions are in the form of $Cx \sim 0$, where C is the coefficient matrix of the variable x, $\sim \in \{<, >, \leq, \geq, ==\}$ is a relation symbol. We first limit the range of x by $Cx \sim 0$, and then determine the relationship between x and y according to the transformation from formula (1) to formula (2) in Sect. 2.4, so as to limit the search range in formula (5). If the path can be terminated under this restriction, we can determine that the path σ can be terminated under the condition $pre(\sigma)$.

Algorithm 1. PathTermAnalysis(σ,$pre(\sigma)$)

Input: σ: loop path
        $pre(\sigma)$: the preconditions of σ
Output: $\{T, NT\}$
1: **if** $head(\sigma) \neq tail(\sigma)$ **then**
2:     **return** T
3: **end if**
4: Construct $while\ Bx > 0\ do\ x = Ax$
5: assume P is the transition matrix from A to J
6: $s = Solver()$
7: **if** $pre(\sigma)$ is not empty **then**
8:    **for all** $precond \in pre(\sigma)$ **do**
9:        $s.add(preocnd)$
10:   **end for**
11:   $s.add(y = P^{-1}x)$
12: **end if**
13: **return** isTerm(A, B, s)

Algorithm 1 shows how to analyze the termination property of a path. Obviously, it is always terminating if a path σ is not an iterable path ($head(\sigma) \neq tail(\sigma)$) (Lines 1–3). When a path σ is an iterable path, we first extract the path conditions and the variable changes and convert to the $while(Bx > 0) \{x = Ax\}$ form (Line 4). We initialize an empty SMT solver at line 6, and add the precondition in lines 7–11. At line 13, the function "isTerm" is an implementation of the algorithm in Sect. 2.4, this function will return whether the path is terminated.

Example 4. As shown in Fig. 1, for the path $\sigma_1 = (b, c, e, b)$, since $head(\sigma_1) == tail(\sigma_1)$, we construct it into a linear loop to check whether it can be terminated. We first extract its path condition $\theta_{\sigma_1} = 2x + y > 0 \wedge y > 0$ and extract its variable changes $\nu_{\sigma_1} = \{x = -x - y,\ y = -3y\}$. Then convert to the $while(Bx > 0) \{x = Ax\}$ form, where

$$A = \begin{pmatrix} -1 & -1 \\ 0 & -3 \end{pmatrix} \qquad B = \begin{pmatrix} 2 & 1 \\ 0 & 1 \end{pmatrix}$$

For this linear loop program, we can verify that it is terminated, so we return 'T' for the termination of σ_1.

3.2 Inter-Path Analysis

After analyzing the termination of each path, we need further to analyze whether a path can transit to another path. Obviously, for the path σ_i and σ_j if $tail(\sigma_i) \neq head(\sigma_j)$, then σ_i cannot be transferred to σ_j. Therefore, we only analyze the situation of $tail(\sigma_i) == head(\sigma_j)$. If the path σ_i only have one successor node, then σ_i can be transferred to σ_j. For other situations, we divide into two cases according to whether σ_i is iterable.

Algorithm 2. ComputeTran($\mathcal{G}$)

Input: $\mathcal{G} : CFG$
Output: T
1: $T = \{\}$
2: **for all** $(\sigma_i, \sigma_j) \in \{(\sigma_m, \sigma_n) | \sigma_m \in S \wedge \sigma_n \in S \wedge tail(\sigma_m)$ **do**
$$= head(\sigma_n) \wedge m \neq n\}$$
3:     **if** σ_i is termination $\wedge$ σ_i.outdegree $== 1$ **then**
4:         $T = T \cup ((\sigma_i, \sigma_j))$
5:     **else**
6:         $\theta_j{}' = $ substitue$(\theta_j, \sigma_i.updates())$
7:         $\tau_{ij} = \theta_i \wedge \theta_j{}'$
8:         **if** $head(\sigma_i) \neq tail(\sigma_i)$ **then**
9:             **if** τ_{ij} is satisfiable **then**
10:                 $T = T \cup ((\sigma_i, \sigma_j))$
11:             **end if**
12:         **else**
13:             $\theta_i{}' = $ substitue$(\theta_i, \sigma_i.updates())$
14:             $\tau_{ij} = \tau_{ij} \wedge \neg\theta_i{}'$
15:             **if** τ_{ij} is satifiable $\vee$ $\theta(\sigma_j, \mathcal{V}_{\sigma_i}^n)$ is satifiable **then**
16:                 $T = T \cup ((\sigma_i, \sigma_j))$
17:             **end if**
18:         **end if**
19:     **end if**
20: **end for**
21: **return** T

- If the path σ_i is a one-time path, we only need to analyze whether there exists a set of variables that satisfy the path condition θ_i. After execute in σ_i, they satisfy the path condition θ_j. If it exists, we say that σ_i can be transferred to σ_j, otherwise, there is no dependency between the two paths.
- If the path σ_i is iterable, first of all, we analyze whether there exists a set of variables that satisfy the path condition θ_i. After execute in σ_i, they satisfy the path condition θ_j and they not satisfy the path condition θ_i. If it exists, we say that σ_i can be transferred to σ_j, otherwise, we use reachability analysis tools to determine whether these two paths can be transferred.

As shown in Algorithm 2, we present the method to analyze the dependency between every two paths. The necessary condition for σ_i to be transferred to σ_j is that the tail of σ_i is equal to the head of σ_j. Intuitively, if σ_i is termination and whose outdegree is 1, it can definitely be transferred to σ_j (Lines 3–4). At line 6, $\theta_j{}'$ represents the entry conditions after executing σ_i. At lines 8–11, we determine whether σ_i can be transferred to σ_j according to whether $\theta_i \wedge \theta_j$ can be satisfied. That is, after the program is executed, σ_i can satisfy the path condition of σ_j. If σ_i is an iterable path, we analyze whether σ_i can transfer to σ_j at lines 13–16. τ_{ij} is satisfiable means that there is a suitable input so that σ_i can be terminated and transferred to σ_j. $\theta(\sigma_j, \mathcal{V}_{\sigma_i}^n)$ means σ_j can reach after σ_i execute n times.

Algorithm 3. merge($\mathcal{C}$)

Input: $\mathcal{C} = \{\sigma_1, \sigma_2, ..., \sigma_n\}$: the cycle constituted by $\sigma_1, \sigma_2, ..., \sigma_n$
Output: σ_{dummy}
1: valChange = {}
2: pathConditions = []
3: **for all** value $\in \mathcal{C}$.values() **do**
4:     valChange[value] = value
5: **end for**
6: **for all** $\sigma \in \mathcal{C}$ **do**
7:     pathConditions + = substitue(σ.condition, valChange)
8:     **for all** update $\in \sigma$.updates() **do**
9:         valChange[update.left] = substitue(update.right, valChange)
10:     **end for**
11: **end for**
12: σ_{dummy}.setValChange(valChange)
13: σ_{dummy}.setConditions(pathConditions)
14: **return** σ_{dummy}

After analyzing the dependencies between paths, we need to analyze whether there are reachable non-terminating paths. If σ_n is non-terminating, we first collect all the precursors of σ_n, and then for each precursor, we use its weakest post-condition as the precondition of σ_n to determine the termination of σ_n. If σ_n terminates under each precursor, we say that σ_n can be terminated, otherwise, it cannot be terminated.

3.3 Cycle Analysis

If the PDA$_l$ is acyclic and all path is termination, the loop is terminated. But when PDA$_l$ has a cycle (e.g., the nested loop's PDA$_l$), the state in the cycle may be repeatedly executed alternately and result in non-termination. So we need further analysis of the termination of the cycle in PDA$_l$.

Definition 1. *Let* $\mathcal{C} = \{\sigma_1, \sigma_2, ..., \sigma_n\}$, *and it meets the following two conditions* 1) $\mathcal{C} \subset S$, 2) $\sigma_1, \sigma_2, ..., \sigma_n$ *constitute a strongly connected component (SCC) in* PDA$_l$. *We call* $\mathcal{C}$ *is a cycle in* PDA$_l$.

To analysis the cycle $\mathcal{C}$ in PDA$_l$, we divide the cycle into two categories as follows. Our analysis cycle termination method will be divided into type I cycle analysis and type II cycle analysis.

Definition 2. *If all paths in* $\mathcal{C}$ *are one-time paths, the cycle is type I cycle. Otherwise, the cycle is type II cycle.*

Theorem 1. *The termination of type I cycle is equivalent to the termination of the new path composed by each path of this cycle.*

In type I cycle C since all paths are one-time paths and all paths are executed in sequence, we can merge all paths in C one by one into a new path. And then, the termination of the cycle is equivalent to the termination of this new path.

Algorithm 3 merge the cycle C to a dummy path σ_{dummy}. In Theorem 1, we know that the termination of a type I cycle is determined by the path composed of each path in the cycle. Algorithm 3 introduces the specific method of merging each path in the cycle. First, we define an empty path σ_{dummy} and initialize the assignment of each variable to themselves (Lines 1–5). Then, we obtain the path condition of the visited path, and then replace all variables in the condition with the current variable assignment of σ_{dummy}, and add the replaced condition to the path condition of σ_{dummy} (Lines 6–7). Similarly, in lines 8–9, we also substitute the right part of the update statement on the visited path. Finally, we return to the merged path σ_{dummy} (Line 14). Example 5 introduces how to merge the type I cycle.

Example 5. The loop in Fig. 2 has the paths σ_1 and σ_3 which constitute a type I cycle. $\theta_{\sigma_1} = 2x - 4y > 0$ and $\theta_{\sigma_3} = z + k <= 0$, we conjunction of this two path conditions as a new loop condition. The value update in σ_1 is none and in σ_3 is $x = -x + 2y$, $y = -2y$. We construct a new loop as:

$$while(2 * x - 4 * y > 0 \wedge z + k <= 0)\{$$
$$x = -x + 2 * y;$$
$$y = -2 * y;$$
$$\}$$

And then analyze the termination of this new loop, that is the termination of this cycle. If there is an update instruction on block c (example $z = z + 1$), the second condition of this new loop is $z + k + 1 <= 0$.

Definition 3. *All variables in C that can affect the exit path conditions are called key variables in C.*

Theorem 2. *If a type II cycle can be reduced to a type I cycle, and the type I cycle is also a cycle in PDA_l, the termination of the type II type is equivalent to the termination of this type I cycle.*

For the type II cycle, our main idea is to determine whether the cycle can be reduced to a type I cycle, that is, whether all the iterable paths in this cycle can be removed. If it can be reduced, the termination of the type II cycle is the same as the termination of the type I cycle after the reduction. Here we find that there are two types of iterable paths that can be removed:

- For the iterable path σ, if all the variable updates in σ do not involve key variables, we say that σ can be removed.
- For the iterable path σ, if the variable that can affect the conditions of σ is only updated in σ and it is not updated outside σ, the path σ can be removed.

Algorithm 4. CycleAnalysis($\mathcal{C}$)

Input: $\mathcal{C} = \{\sigma_1, \sigma_2, ..., \sigma_n\}$ $\sigma_1, \sigma_2, ..., \sigma_n$ constitute a cycle
    $pre(\sigma_1)$: precondition of σ_1
Output: $\{T, NT, UN\}$
1: **if** $\forall \sigma_i \in \mathcal{C} \wedge \sigma_i.iterable = false$ **then**
2:     $\sigma = \text{merge}(\mathcal{C})$
3:     **return** PathTermAnalysis($\sigma, pre(\sigma_1)$)
4: **end if**
    assume $\mathcal{C}_{II}$ is a set that all iterable path in $\mathcal{C}$
5: **for all** $\sigma \in \mathcal{C}$ **do**
6:     **if** $head(\sigma) == tail(\sigma)$ **then**
7:         $\mathcal{C}_{II}.\text{append}(\sigma)$
8:     **end if**
9: **end for**
10: keyVars $= \mathcal{C}.\text{getKeyVar}()$;
        We use Θ_σ to denote the conditional of σ
11: **for all** $\sigma \in \mathcal{C}_{II}$ **do**
12:     **if** $\exists \theta_i \in \Theta_\sigma$ only change in σ **then**
13:         $\mathcal{C}.\text{del}(\sigma)$
14:         $\mathcal{C}_{II}.\text{del}(\sigma)$
15:     **else if** $\forall$ var $\in \sigma_i.\text{vars}() \wedge$ var $\notin$ keyVars **then**
16:         $\mathcal{C}.\text{del}(\sigma)$
17:         $\mathcal{C}_{II}.\text{del}(\sigma)$
18:     **end if**
19: **end for**

20: **if** $\mathcal{C}_{II}.\text{empty}() \wedge \mathcal{C}.\text{isTypeI}() \wedge \mathcal{C} \in \text{PDA}_l.\text{cycles}()$ **then**
21:     **return** CycleAnalysis($\mathcal{C}$)
22: **end if**
23: **return** UN

In type II cycle $\mathcal{C}$, we first collect all iterable paths in $\mathcal{C}$, and determine whether all iterable paths can be removed according to the above strategy. If the cycle $\mathcal{C}$ is reduced to a type I cycle after removing the iterable path, and we only need to analyze the termination of this cycle as a type I cycle. If there is still an iterable path in $\mathcal{C}$, we say that the termination of the cycle is unknown.

In Algorithm 4, we first check whether the input cycle $\mathcal{C}$ is a type I cycle. If it is a type I cycle, we merge its paths into a new path and then analyze the termination of this new path (Lines 1–3). After line 4 is the algorithm for analyzing the type II cycle. First, we collect all the iterable paths (Lines 5–7) in the cycle $\mathcal{C}$ to $\mathcal{C}_{II}$ and then analyze whether these iterable paths can be removed and remove those removable paths (Lines 10–19). After removing the removable paths, we need to check whether the set of iterable paths $\mathcal{C}_{II}$ is empty; that is, the cycle can be reduced to a type I cycle. If it is empty and $\mathcal{C}$ be reduced to a type I cycle, we return the termination of the new cycle (Lines 20–21); otherwise, we can not analyze the termination of the cycle $\mathcal{C}$ and return "UN" (Line 23).

Table 1. Experimental result

	Loopster++				Ultimate automizer				CPAchecker				2LS			
	CTT	CFF	CUN	COT	CTT	CFF	CUN	COT	CTT	CFF	CUN	COT	CTT	CFF	CUN	COT
Nested	19	2	9	0	26	3	0	1	23	3	3	1	20	3	7	0
Unnested	64	15	11	0	70	15	5	0	57	12	17	4	60	8	22	0
Total	83	17	20	0	96	18	5	1	80	15	20	5	80	11	29	0
Time(s)	344.03				1838.40				9229.49				1864.05			

Example 6. The loop in Fig. 2 has the type II cycle $\mathcal{C}_2 = \{\sigma_1, \sigma_2, \sigma_3\}$. σ_2 is the iterable path where the path condition of σ_2 is $z + k > 0$ and z, k only change in σ_2, so we remove σ_2 in $\mathcal{C}_2$. In fact, in this loop, the key variables are x, y. In this iterable path, the value of the key variable will not be changed, so the path can also be removed by the method in lines 15–17 of the Algorithm 4. After remove, $\mathcal{C}_2 = \{\sigma_1, \sigma_2\}$ is a type I cycle. Finally, we use the method of analysis the type I cycle to analysis the termination of $\mathcal{C}_2$.

4 Implementation and Evaluation

We implement the Loopster++ based on LLVM 8.0 [13], Seahorn [10] and SMT solver Z3 [15]. We also implement Tiwari's linear loops program termination decision algorithm to determine the termination of the path in our tool. To evaluate the effectiveness and performance of loopster++, we compare our tool with the state-of-the-art tools.

4.1 Effectiveness of Loopster++

To evaluate the effectiveness and performance of loopster++, we compare our tool against the other three termination analysis tools, namely Ultimate Automizer [11], CPAchecker [4], and 2LS [6], which are the top three ranking in the termination analysis section of 9th Competition on Software Verification (SV-COMP 2020) [1]. We selected all linear loops with terminating properties in termination-crafted, termination-crafted-lit, and termination–restricted-15 benchmarks from the category of SV-COMP 2020. There are 120 of these loop programs in total, 30 of them are nested loops and 90 are non-nested loops. The termination of these 120 loop programs is known that 99 of them are termination and 21 are non-termination. All experiments run on a Ubuntu 18.04.4 LTS system in a virtual machine with Intel Core i7-4790 CPU @ 3.60 GHz (1 core) and 4.7 GB memory.

Table 1 shows the experimental results, and columns CTT, CFF and CUN represent the number of termination loops, non-termination loops and unknown results separately. COT summarizes the number of programs that analyzed time out and Time represents the total time spent analyzing 120 programs. In this table, we measure the time in second and we set the timeout as 600 s.

In Table 1, the result shows that our tool can correctly handle 100 programs with the shortest time (344.03 s). Ultimate Automizer can correctly analyze 114 programs in 1838.40 s and have 1 program out of time. CPAchecker has correctly analyzed 95 programs, and 5 programs have time out, therefore CPAchecker takes the longest time (9229.49 s). 2LS takes 1864.05 s to analyze all programs and has correctly analyzed 91 programs. However, when analyzing the termination of linear loops, 20 of them cannot get correct results, which have exceeded CPAchecker and 2LS. In summary, Loopster++ is second only to UA in the number of programs handled correctly. But Loopster++ is better than the other three tools in terms of solution time. At least 5 times faster than the most efficient Ultimate Automizer among them.

In these 120 loop programs, we recorded the methods Loopstr++ used when analyzing their termination. Among them, 63 programs can directly determine the overall termination of the loop by analyzing the path termination and the dependence between the paths. There are 26 programs that need to use preconditions to determine the termination of the path in the process of analyzing the termination. There are 37 programs that will generate cycles during the analysis process, of which 20 programs generate loops and we give unknown results.

Table 2. The unknown results given by Loopster++

Loop programs	Is nested
a.01.c	Yes
AliasDarteFeautrierGonnord-SAS2010-loops.c	Yes
AliasDarteFeautrierGonnord-SAS2010-wcet2.c	Yes
java_Nested.c	Yes
McCarthy91_Iteration.c	Yes
NO_03.c	Yes
PastaA1.c	Yes
Urban-WST2013-Fig2-modified1000.c	Yes
Urban-WST2013-Fig2.c	Yes
AliasDarteFeautrierGonnord-SAS2010-cousot9.c	No
AliasDarteFeautrierGonnord-SAS2010-speedpldi2.c	No
AliasDarteFeautrierGonnord-SAS2010-speedpldi3.c	No
c.03.c	No
Flip2.c	No
Gothenburg-1.c	No
Gothenburg_v2-1.c	No
McCarthyIterative.c	No
NO_13.c	No
PastaC3.c	No
UpAndDownIneq.c	No

As shown in Table 2, for the unknown result given by our tool, which has 9 nested loops and 11 unnested loops. The reason why these loops cannot be given

```
int main() {
    int c = 0;
    int x = *;
    int y = *;
    while (x > 0) {
        y = 0;
        while (y < x) {
            y = y + 1;
        }
        x = x - 1;
    }
    return 0;
}
```

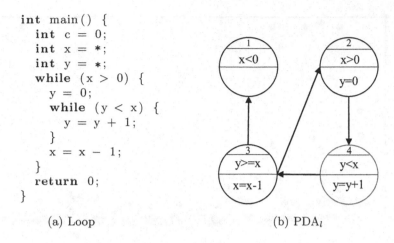

(a) Loop   (b) PDA$_l$

Fig. 4. Loop program a.01.c

the correct results is that there is a cycle in PDA$_l$, and we cannot analyze the termination of the cycle by Algorithm 4 in Sect. 3. We summarize the reasons as follows.

- For unnested loop programs, the cycle in PDA$_l$ is composed of *if...else...*; for example $while(i > 0)\{if(j > 0)\{j = j - 1;\}else\{j = N; i = i - 1;\}\}$, in this loop, the paths formed by the branches of *if* and *else* are all iterable paths, and the interdependence of these two paths forms a type II cycle; and we cannot reduce it to a type I cycle.
- For nested loops, there are still complex dependencies between inner loops and outer loops that constitute the cycle, and we cannot be sure of termination. For example, In Fig. 4, the loop program a.01.c is terminating, but our tool cannot obtain the correct result. Figure 4(b) is the PDA$_l$ of this loop program. From this PDA$_l$, we can see that σ_4 is an iterable path. When we remove σ_4, we cannot get the type I cycle in the PDA$_l$. So we cannot determine whether the cycle is terminated, and the result of the unknown is given. Therefore, if we can find better ways to analyze loops (for example, find more types of cycles that can be analyzed or using the ranking function technique to handle cycle), our tools can be more complete.

4.2 Performance of Loopster++

To evaluate the performance of Loopster++, we compare the performance of our tool with the other three tools in terms of the time it takes to verify the termination of each program. At the same time, we analyzed why our tool has a good performance.

Figure 5 shows the detailed analysis time of each tool. The blue dot is the time distribution of our tool Loopster++. We can see that Loopster++ can verify the termination of the benchmark within 5 s. Note that, 2ls cannot work efficiently

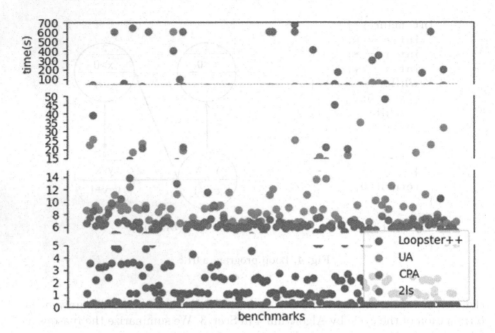

Fig. 5. Detail run time

on all benchamrks, and some benchmarks takes a very long time (which is the main reason for the high total time), even if it can verify most benchmarks in a short time. As shown in Fig. 5, we can also find that our tool is faster than CPAchecker and Ultimate Automizer for the verification of each program. In the following, we analyze and discuss the possible reasons why our tool can steadily prove the termination.

- We separate the analysis of each path and the analysis of dependencies between paths, which simplifies the problem to a certain extent. This determines that our tool analysis of the termination of the program will not take much time.
- When our tool analyzes program termination, the main part of the time is spent on the path termination analysis. As for these benchmarks, the number of path branches is about the same, so it takes us about the same time to analyze the termination of each program.

From Fig. 5, we can find that it takes a long time for Loopster++ to analyze its termination for some loop programs. Our experimental results shows that these loop programs with a sudden increase in time often have the characteristics of more conditions or more variables. For loops with many conditions, the *Ind* Sect. 2.4 will produce more groups of possible non-terminating points when analyzing the termination of the path. For loops with many variables, a higher-dimensional variable update matrix can be generated by Loopster++, which will increase the probability of a greater number of positive eigenvalues and expand the Ind set.

5   Relate Work

In this section, we discuss the related work on the termination proving on the linear loop programs.

Dams et al. first proposed a method to automatically synthesize the ranking functions for linear loops [9]. Michael and Henny [7,8] presented an algorithm to generate the linear ranking functions by manipulating polyhedral cones. Podelski and Rybalchenko [17] presented a complete method for the linear ranking functions through the relationship between linear inequalities. Ben-Amram and Genaim [3] studied the complexity of generating the linear ranking functions for a linear loop and they proved that the complexity is coNP-complete when the variable range over the integers. Leike and Heizmann [14] introduced the notion of linear ranking templates, and based on this to study the constraint-based synthesis of termination arguments for linear loop programs.

However, for the termination of loop programs, the ranking function does not necessarily exist (this is just a sufficiently unnecessary condition). There are other people who pay more attention to study a decidable class of loops. Tiwari [18] used algebraic theory to propose a method for finding nonterminating points (if it does not exist, the program will terminating) and he also proved that it is decidable over the reals $\mathbb{R}$. Braverman generalized the technique presented in [18] and they presented a decision procedure for simple homogeneous linear loop programs over the integers [5]. If the update matrix A of a simple linear loop program can be diagonalized, Ouaknine et al. [16] proposed how to decide its termination.

Compared with the above techniques [3,7–9,14,17], our approach does not need to synthesize the ranking functions. We pay attention to a class of decidable loops to ensure that the termination of each path can be analyzed. Unlike the techniques in [5,16,18], our approach focuses on the termination of multi-path linear loops.

As an extension of Loopster [20], our work is also related to Loopster. Loopster used the monotonicity theory and the dependence between paths to determine the termination of a loop program. The difference is that our approach uses the theory of linear loop programs to determine the termination of each path. At the same time, the way of handling dependencies between paths and the method of analyzing the cycle is also different.

6   Conclusion

In this paper, we extend Loopster as Loopster++ to analyze the termination of multi-path linear loops. We determine the termination of the loop by analyzing the termination of the path, the reachability of each two paths, and the termination of the cycle in PDA_l. Finally, we implemented our approach and designed experiments to compare our tools with the state-of-the-art tools. The empirical results show that our approach is effective to achieve high accuracy and is more efficient with a shorter analysis time. In the future, we plan to extend our approach more complete and make it to support more types of loops, such as nonlinear loop programs.

Acknowledgment. This work has partially been sponsored by the National Science Foundation of China (No. 61872262).

References

1. 9th competition on software verification (2020). https://sv-comp.sosy-lab.org/2020/
2. CVE-2009-1890 (2020). https://cve.mitre.org/cgi-bin/cvename.cgi?name=CVE-2009-1890
3. Ben-Amram, A.M., Genaim, S.: On the linear ranking problem for integer linear-constraint loops. SIGPLAN Not. **48**(1), 51–62 (2013). https://doi.org/10.1145/2480359.2429078
4. Beyer, D., Henzinger, T.A., Théoduloz, G.: Configurable software verification: concretizing the convergence of model checking and program analysis. In: Damm, W., Hermanns, H. (eds.) CAV 2007. LNCS, vol. 4590, pp. 504–518. Springer, Heidelberg (2007). https://doi.org/10.1007/978-3-540-73368-3_51
5. Braverman, M.: Termination of integer linear programs. In: Ball, T., Jones, R.B. (eds.) CAV 2006. LNCS, vol. 4144, pp. 372–385. Springer, Heidelberg (2006). https://doi.org/10.1007/11817963_34
6. Chen, H., David, C., Kroening, D., Schrammel, P., Wachter, B.: Synthesising interprocedural bit-precise termination proofs (t). In: 2015 30th IEEE/ACM International Conference on Automated Software Engineering (ASE), pp. 53–64 (2015)
7. Colón, M.A., Sipma, H.B.: Practical methods for proving program termination. In: Brinksma, E., Larsen, K.G. (eds.) CAV 2002. LNCS, vol. 2404, pp. 442–454. Springer, Heidelberg (2002). https://doi.org/10.1007/3-540-45657-0_36
8. Colóon, M.A., Sipma, H.B.: Synthesis of linear ranking functions. In: Margaria, T., Yi, W. (eds.) TACAS 2001. LNCS, vol. 2031, pp. 67–81. Springer, Heidelberg (2001). https://doi.org/10.1007/3-540-45319-9_6
9. Dams, D., Gerth, R., Grumberg, O.: A heuristic for the automatic generation of ranking functions. In: Workshop on Advances in Verification, pp. 1–8 (2000)
10. Gurfinkel, A., Kahsai, T., Navas, J.A.: SeaHorn: a framework for verifying C programs (competition contribution). In: Baier, C., Tinelli, C. (eds.) TACAS 2015. LNCS, vol. 9035, pp. 447–450. Springer, Heidelberg (2015). https://doi.org/10.1007/978-3-662-46681-0_41
11. Heizmann, M., Hoenicke, J., Podelski, A.: Termination analysis by learning terminating programs. In: Biere, A., Bloem, R. (eds.) CAV 2014. LNCS, vol. 8559, pp. 797–813. Springer, Cham (2014). https://doi.org/10.1007/978-3-319-08867-9_53
12. Larraz, D., Oliveras, A., Rodriguez-Carbonell, E., Rubio, A.: Proving termination of imperative programs using Max-SMT. In: 2013 Formal Methods in Computer-Aided Design, FMCAD 2013, pp. 218–225 (2013)
13. Lattner, C., Adve, V.: LLVM: a compilation framework for lifelong program analysis transformation. In: International Symposium on Code Generation and Optimization, CGO 2004, pp. 75–86 (2004)

14. Leike, J., Heizmann, M.: Ranking templates for linear loops. In: Ábrahám, E., Havelund, K. (eds.) TACAS 2014. LNCS, vol. 8413, pp. 172–186. Springer, Heidelberg (2014). https://doi.org/10.1007/978-3-642-54862-8_12
15. de Moura, L., Bjørner, N.: Z3: an efficient SMT solver. In: Ramakrishnan, C.R., Rehof, J. (eds.) TACAS 2008. LNCS, vol. 4963, pp. 337–340. Springer, Heidelberg (2008). https://doi.org/10.1007/978-3-540-78800-3_24
16. Ouaknine, J., Pinto, J.A.S., Worrell, J.: On termination of integer linear loops. In: Proceedings of the Twenty-Sixth Annual ACM-SIAM Symposium on Discrete Algorithms, SODA 2015, pp. 957–969. Society for Industrial and Applied Mathematics, USA (2015)
17. Podelski, A., Rybalchenko, A.: A complete method for the synthesis of linear ranking functions. In: Steffen, B., Levi, G. (eds.) VMCAI 2004. LNCS, vol. 2937, pp. 239–251. Springer, Heidelberg (2004). https://doi.org/10.1007/978-3-540-24622-0_20
18. Tiwari, A.: Termination of linear programs. In: Alur, R., Peled, D.A. (eds.) CAV 2004. LNCS, vol. 3114, pp. 70–82. Springer, Heidelberg (2004). https://doi.org/10.1007/978-3-540-27813-9_6
19. Xie, X., Chen, B., Liu, Y., Le, W., Li, X.: Proteus: computing disjunctive loop summary via path dependency analysis. In: Proceedings of the 2016 24th ACM SIGSOFT International Symposium on Foundations of Software Engineering, FSE 2016, pp. 61–72. Association for Computing Machinery, New York (2016). https://doi.org/10.1145/2950290.2950340
20. Xie, X., Chen, B., Zou, L., Lin, S.W., Liu, Y., Li, X.: Loopster: static loop termination analysis. In: Proceedings of the 2017 11th Joint Meeting on Foundations of Software Engineering, ESEC/FSE 2017, pp. 84–94. Association for Computing Machinery, New York (2017). https://doi.org/10.1145/3106237.3106260

A Stepwise Path Selection Scheme Based on Multiple QoS Parameters Evaluation in SDN

Lin Liu[1,2], Jian-Tao Zhou[1(✉)], Hai-Feng Xing[3], and Xiao-Yong Guo[1]

[1] College of Computer Science, Ecological Big Data Engineering Research Center of the Ministry of Education, Cloud Computing and Service Software Engineering Laboratory of Inner Mongolia Autonomous Region, National and Local Joint Engineering Research Center of Intelligent Information Processing Technology for Mongolian, Social Computing and Data Processing Key Laboratory of Inner Mongolia Autonomous Region, Big Data Analysis Technology Engineering Research Center of Inner Mongolia Autonomous Region, Inner Mongolia University, Hohhot, Inner Mongolia, China
[2] College of Computer Science and Technology, Inner Mongolia Normal University, Hohhot, Inner Mongolia, China
[3] College of Computer Information and Management, Inner Mongolia University of Finance and Economics, Hohhot, Inner Mongolia, China

Abstract. Nowadays, the best-effort service can not guarantee the quality of service (QoS) for all kinds of services. QoS routing is an important method to guarantee QoS requirements. It involves path selection for flows based on the current network status and the performance criteria of the service requirements. However, it is difficult for proposed solutions to obtain all the available paths owing to not fully considering all the QoS parameters of paths. In this paper, we propose SWQoS, a novel, universal, and stepwise QoS guarantee scheme based on multiple QoS parameter evaluation for selecting the available paths including preferred paths, satisfied paths and reluctant paths in SDN. The experiments show that SWQoS can select all the available paths that meet the performance criteria of the service requirements and have better QoS parameter performance compared with other path selection methods.

Keywords: SDN · QoS routing · Path selection · AHP · Fuzzy synthetic evaluation

1 Introduction

Providing high-quality traffic delivery for various services is a hot topic of current QoS routing research. Currently, some researches including Integrated Services (IntServ)/Resource Reservation Protocol (RSVP) [1], Differentiated Services (Diffserv) [2] and Multi-Protocol Label Switching (MPLS) [3] exist management complicated and are challenging to deploy [4,5]. Path selection

© ICST Institute for Computer Sciences, Social Informatics and Telecommunications Engineering 2021
Published by Springer Nature Switzerland AG 2021. All Rights Reserved
H. Gao and X. Wang (Eds.): CollaborateCom 2021, LNICST 406, pp. 498–519, 2021.
https://doi.org/10.1007/978-3-030-92635-9_29

is another common way used in QoS routing. Path selection is to find the paths that meet the service requirements based on the current network status and the performance criteria of the service requirements. However, that the current researches are based on the conventional network architecture will become extremely difficult owing to the lack of centralized functions and difficultly obtaining the global network status.

SDN [6–9], as a new type of network architecture, effectively makes up for the deficiencies of path selection research in the conventional network architecture. The centralized control and quickly obtaining the global network view function of SDN network provide path selection with the possibility to implement novel, more powerful strategies [5, 10]. However, the existing SDN path selection schemes do not fully consider multiple QoS parameters to determine the path selection, making it difficult to obtain optimal path and all the available paths. Furthermore, some method need dedicated computational resources and training data, yet. Aim to the above deficiencies, we propose SWQoS - a novel, universal and stepwise QoS routing evaluation scheme in SDN.

SWQoS scheme is pre-computing all the candidate paths between a source and a destination in advance. On this basis, the pros and cons of the candidate paths are evaluated by considering the performance criteria of the service requirements and the current network status. More specifically, SWQoS firstly looks for the paths from all the candidate paths where the network status fully meet the performance criteria of the service requirements, called preferred path set. Secondly, if no preferred path, then look for the paths where the network status basically meet the performance criteria of the service requirements, called satisfied path set. Finally, if none of the above paths exist, that is, there are no paths whose status fully meet the performance criteria of the service requirements, AHP (Analytic Hierarchy Process) [11] combined with the fuzzy synthetic evaluation method is used to look for the paths, namely reluctant path set. The contributions of our research are highlighted as follows:

- All the QoS parameters are fully considered and the network service preference is emphasized to ensure the QoS requirements of network services.
- As a universal path selection scheme, SWQoS scheme can be applied to the transmission of various types of service traffic by adjusting parameters and obtains all the available paths according to stepwise path selection.
- SWQoS scheme can obtain the optimal path.

The rest of the paper is organized as follows. In Sect. 2, we discuss the related work on QoS routing. In Sect. 3, we propose SWQoS scheme, including the scheme architecture, finding the candidate paths, determining the QoS weight values of the service requirements, and selecting the available paths. In Sect. 4, three group simulating experiments are implemented with VoIP as an example, and the results are analyzed. Finally, we conclude this work in Sect. 5.

2   Related Work

In the section we review existing researches including QoS routing in the conventional network architecture, QoS routing in SDN, and QoS routing adopting artificial intelligence in SDN.

The solutions of QoS routing, such as IntServ, Differv, and MPLS, mainly depend on more bandwidth capacity in the conventional network architecture. More bandwidth capacity would generate more cost. IntServ utilizes the resource reservation protocol (RSVP) [12] to guarantee the service requirements. The resources are reserved along the packet forwarding path in the IntServ model. Thereby, it has limited scalability. Differv classifies incoming the flows based on the different service types. It only supports per-hop QoS, which reduces the complexity of implementation, but weakens the guarantees. MPLS is another widely used technology that is explicit routed paths.

However, the above routed paths are static, lacking on-demand reconfigurability, difficultly deployed and management complicated.

SDN provides a new perspective for the solution of QoS routing. OpenQoS [13], a novel OpenFlow controller design for multimedia delivery with QoS guarantee, groups the multimedia and data flows. The multimedia flows are dynamically placed on QoS routing path and data flows remain on the conventional shortest-path. FlowQoS [14,15] performs application identifications and flow rate shaping according to the service requirements. HiQoS [16], an SDN-based multipath solution to guarantee QoS, uses the queuing mechanisms of OpenFlow protocol to implement the bandwidth guarantees for different service traffic, and the multipath is generated using a modified Dijkstra algorithm with QoS constrained. ARVS [17] proposes an adaptive routing approach for video traffic with guaranteeing QoS requirements in SDN. It finds out a certain path that is not a good path for the service. Slightly different from the above method, Li et al. [18] propose the fuzzy synthetic evaluation mechanism (FSEM) for path load balancing based on SDN. The paths can be dynamically adjusted according to the global view of the network. This method first uses the Top-K paths selection algorithm to obtain k candidate paths based on the least hop count, and then selects the available paths from the k candidate paths using the fuzzy synthetic evaluation method.

The above schemes in SDN have not fully consider all the QoS parameters and the network service preference, and most of them only obtain an available path and the available path may not be the optimal path.

In recent years, artificial intelligence being trying to introduced into routing mechanisms based on SDN is used as a new potential research direction [19], while supervised learning and reinforcement learning is the most widely used. Bomin Mao et al. [20] propose a supervised deep learning system to directly calculate the paths using traffic patterns. Joao Reis et al. [21] propose a routing framework instead of heuristic algorithm to obtain the best routing decision by training deep neural networks. The characteristics of flows and the network status are taken as the inputs of the deep neural network, a value is an input for each link, and it is converted into a path at last. DROM [22], a deep reinforcement learning mechanism for SDN, achieves a universal and customizable routing optimization. DROM simplifies the network operation and maintenance, thus improving the network performance, such as delay and throughput.

However, artificial intelligence be applied in routing mechanisms need the dedicated GPU devices, a lot of computational resources to train the model, and training is expensive.

3 Proposed Scheme: SWQoS

In this section, we design a path selection scheme adopting a stepwise strategy based on the multiple QoS parameter evaluation in SDN architecture. Next, we will introduce the architecture that we designed, and then elaborate all modules of the SWQoS scheme.

3.1 SWQoS Scheme Architecture

The reference architecture is shown in Fig. 1. SWQoS scheme architecture is built based on SDN architecture. It consists of data, control and application plane. SWQoS lies in the application plane and mainly includes three differentiated modules: path finding, QoS requirements of services, and path selection. The relationship between three modules is shown in Fig. 1, too.

- **Path Finding:** responsible for finding out all the candidate paths between a source and a destination according to the network topology and the hop count.

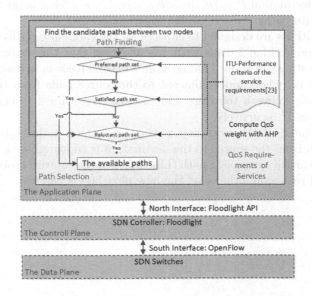

Fig. 1. SWQoS scheme architecture

- **QoS Requirements of Services:** responsible for quantifying the QoS requirement of services by AHP, namely computing the QoS weight values of the service requirements.
- **Path Selection:** responsible for selecting the available path set for the forwarded traffic. The available path set is divided into three classes: preferred path set, satisfied path set and reluctant path set. The preferred path set and satisfied path set are selected according to ITU-performance criteria of the service requirements standards [23]. When the preferred path set and the satisfactory path set do not exist, the reluctant path set is looked up by using the fuzzy synthetic evaluation method combined with the QoS weight values calculated by the QoS requirements of the service.

Path selection as an application is located on SDN application plane. The design does not impose additional functions in the control and data plane of SDN architecture, but makes full use of link discovery, network status statistics, and centralized control function obtained by SDN architecture.

3.2 Path Finding

In order to describe our algorithm clearly, several notions are introduced follows as. Consider an arbitrary network topology $\mathcal{G}$ is composed of the set of nodes $\mathcal{V}$, the set of links $\mathcal{E}$. The nodes indicate all the switches in the network topology, $\mathcal{V} = \{s_1, s_2, ..., s_n\}$, and n is the total number of nodes. $\mathcal{E}$ indicates all the links in the network, $\mathcal{E} = \{e_1, e_2, ..., e_m\}$, and m is the total number of links. In the network topology $\mathcal{G}$, node number starts from 1 to n, and each of the links is represented by (s_i, s_j). We define a set that all the candidate paths are between a source and a destination $\mathcal{P} = \{p_1, p_2, ...p_i, ..., p_k | 1 \leq i \leq k\}$. k is the total number of the candidate paths.

Two algorithms are designed to capture all the candidate paths. Algorithm 1 is the main frame. Line 2 deals with the situation that the source node is the same as the destination node. That is, node will send packets to itself. Line 4 gets the list of the next hop nodes adjacent to the source node from the adjacency table saving the network topology. Line 5 calls Algorithm 2 to capture all the candidate paths $\mathcal{P}$ recursively.

In Algorithm 2, if the next hop is the destination node (Line 3), then a complete path from the source to the destination is captured; if the next hop is not contained in the incomplete path (Line 5), then the capture process continues by the way of recursion; otherwise, the new next hop is checked.

Algorithm 1: Path Finding

Input: sn, dn, *adj_list* //sn is source, dn is destination, and adj_list is the
         network topology.

Output: $\mathcal{P}$ //$\mathcal{P}$ is the set of all the candidate paths.

1   *create_path(path, sn)*;
2   **if** $sn = dn$ **then**
3 | $P = P \cup path$;
4   **else**
5 | *next_hop_list = adj_list[sn]*;
6 | *get_all_paths(path, next_hop_list, $\mathcal{P}$, dn)*
7   **end**

Algorithm 2: get_all_paths

Input: path, *next_hop_list*, $\mathcal{P}$, dn
Output: $\mathcal{P}$

1   **for** *each next_hop in next_hop_list* **do**
2 | *new_path = copy_path(path)*;
3 | **if** *next_path = dn* **then**
4 | | *add_tail(new_path, next_hop)*; $P = P \cup path$;
5 | **else if** *next_hop not in path* **then**
6 | | *add_tail(new_path, next_hop)*;
7 | | *new_next_hop_list = adj_list[next_hop]*;
8 | | *get_all_paths(new_path, new_next_hop_list, $\mathcal{P}$, dn)*;
9 | **else**
10 | | *free_path(new_path)*;
11 | **end**
12   **end**

Path finding module finds out all the paths between two nodes in the network topology, and takes them as the candidate paths, which is the basis for the next path selection.

3.3   QoS Requirements of Services

That different services on the Internet generate flows have the different QoS requirements, which are also called sensitive preferences for services. However, the QoS requirement of services given in Table 1 are qualitative, such as VoIP, video conferencing and online games, etc. [24]. They can not express the QoS requirement of services with accurate QoS weight value. In this subsection, we adopt AHP method to quantify the QoS requirement of services, and obtain accurate QoS weight value of services.

This paper takes VoIP as an example. Using AHP method to determine the QoS weight values of the VoIP service are commonly divided into the following three major steps:

Table 1. The QoS requirement of services [24]

QoS parameters	VoIP	Video conferencing	Online gaming
Bandwidth	Low	Mid	Low
Delay	High	High	High
Loss	Low	Low	Mid
Jitter	High	High	Low

Step 1: Construct a pairwise comparison matrix. In this step, using AHP will be asked to rate the relative importance of QoS weight values instead of the QoS qualitative values be directly used, so as to obtain accurate QoS weight values. We need to construct a pairwise comparison matrix of QoS parameters. We should provide a scale of numbers for the value judgments to construct this comparison at the beginning. That the scales are to judge the relative importance between the QoS parameters of services are shown in the Table 2. Therefore, a pairwise comparison matrix $\mathcal{PM}$ can be obtained as follows:

$$\mathcal{PM} = \begin{matrix} Bandwidth \\ Delay \\ Loss \\ Jitter \end{matrix} \begin{pmatrix} Bandwidth & Delay & Loss & Jitter \\ 1 & 1/5 & 1 & 1/5 \\ 5 & 1 & 5 & 1 \\ 1 & 1/5 & 1 & 1/5 \\ 5 & 1 & 5 & 1 \end{pmatrix}$$

Each element of the matrix $\mathcal{PM}$ represents the importance intensity of between the QoS parameters of VoIP service. The element value is supposed to be an approximation of the relative importance. For example, entering 5 in the (Delay, Bandwidth) position meaning that delay is strongly more important than bandwidth for VoIP.

Table 2. Lineal scale of preferences in the pair-wise comparison process [25]

Intensity of importance	Definition
1	Equal importance
3	Moderate importance
5	Strong importance
7	Very strong or demonstrated importance
9	Extreme importance
2, 4, 6, 8	The importance between 1, 3, 5, 7 and 9 scales

Step 2: Calculate the relative weights. We use the arithmetic mean method to calculate the QoS weight values of the service requirements. Firstly, the pairwise comparison matrix of QoS parameters are normalized by the column. Secondly, to sum the values by row to get the vector. Finally, to

normalize the vector obtained by the second step to get the QoS weight values of VoIP service. The symbol $\mathcal{W}$ denotes the QoS weight value vector of the service requirements. After calculation, the QoS weight values vector of VoIP service is $\mathcal{W} = (0.08333333, 0.41666667, 0.08333333, 0.41666667)$.

Step 3: Check the consistency. This step is very important for the rationality of the judgment. A method is to calculate the Consistency Ratio (CR), namely $CR = CI/RI$. $\mathcal{RI}$ is the average random index, which is computed and tabulated as shown in Table 3. The Consistency Index (CI) is calculated as $CI = (\lambda_{max} - r)/(r-1)$. r is the order of the pairwise comparison matrix $\mathcal{PM}$. λ_{max} represents the maximum eigenvalue of the matrix. If a value of CR is less than 0.1, the numerical judgments will be considered to be acceptable [25]. After the matrix $\mathcal{PM}$ is calculated, $CR = -0.000000$, it meets the consistency requirements.

Table 3. Average random index values according to matrix size [25]

r	1	2	3	4	5	6	7	8	9	10	11
RI	0.00	0.00	0.58	0.90	1.12	1.24	1.32	1.41	1.45	1.49	1.52

QoS Requirements of Services can accurately calculate the QoS weight values of the service, which is the foundation for the application of fuzzy synthetic evaluation method in the reluctant paths.

3.4 Path Selection

In this subsection, we present a stepwise scheme to select the available path set for services among the candidate paths. To determine whether the path is available for services, we need to construct a matrix $\mathcal{A}$ that presents the QoS parameter values of the candidate paths.

QoS parameters of links include delay, jitter, bandwidth, and packet loss rate, but they have different metric character. Bandwidth is a concave metric character, delay and jitter are additive metric characters, and the packet loss rate is a multiplicative metric character. Therefore, we represent bandwidth, delay, jitter, and packet loss rate of the path p_i with symbols bw_i, de_i, jt_i and lo_i, respectively. $(bw_{e_1}, bw_{e_2}, ...)$, $(de_{e_1}, de_{e_2}, ...)$, $(jt_{e_1}, jt_{e_2}, ...)$ and $(lo_{e_1}, lo_{e_2}, ...)$ represent the bandwidth, delay, jitter and packet loss rate on each link along the path, respectively. bw_i is minimum bandwidth constraints over the link of the ith path shown as Eq. (1). de_i and jt_i are the sum of each link delay and jitter over ith path shown as Eq. (2) and (3). lo_i is the multiplication of the loss of each link over ith path shown as Eq. (4).

$$bw_i = \min(bw_{e_1}, bw_{e_2}, ...) \quad e_1, e_2, ... \in p_i \tag{1}$$

$$de_i = \sum (de_{e_1}, de_{e_2}, ...) \quad e_1, e_2, ... \in p_i \tag{2}$$

$$jt_i = \sum (jt_{e_1}, jt_{e_2}, ...) \quad e_1, e_2, ... \in p_i \tag{3}$$

$$lo_i = \prod (lo_{e_1}, lo_{e_2}, ...) \quad e_1, e_2, ... \in p_i \tag{4}$$

The matrix $\mathcal{A}$ is composed of four rows that represent four QoS parameters, and the k columns represent k candidate paths. Each element value in matrix $\mathcal{A}$ is derived from Eqs. (1), (2), (3) and (4). The matrix $\mathcal{A}$ is shown as follows.

$$\mathcal{A} = \begin{array}{c} \\ Bandwidth \\ Delay \\ Jitter \\ Loss \end{array} \begin{pmatrix} p_1 & p_2 & \cdots & p_i & \cdots & p_k \\ bw_1 & bw_2 & \cdots & bw_i & \cdots & bw_k \\ de_1 & de_2 & \cdots & de_i & \cdots & de_k \\ jt_1 & jt_2 & \cdots & jt_i & \cdots & jt_k \\ lo_1 & lo_2 & \cdots & lo_i & \cdots & lo_k \end{pmatrix}$$

We divided into three steps to find out preferred path set, satisfied path set and reluctant path set in order. The International Telecommunication Union-Telecommunication Standardization Sector (ITU-T) G.1010 [23] provides an indication of suitable performance targets, as shown in Table 4.

Table 4. Performance criteria for VoIP services [23]

VoIP	Bandwidth (kbit/s)	Delay (ms)	Jitter (ms)	Loss (%)
Preferred	>64	<150	<1	<3
Limit	>4 & ≤64	≥150 & <400	<1	<3
Not proper	≤4	≥400	≥1	≥3

It is shown that the performance indicators required for VoIP traffic in Table 4. The performance is divided into Preferred, Limit, and Not proper. They correspond to the above mentioned preferred path set, satisfied path set, and reluctant path set, respectively. Next, we describe the details of selecting three path sets.

1) To select preferred path set: We find out preferred path set from the candidate paths, as shown in Eq. (5).

$$\textbf{preferred path set} = \{\, p_i(bw_i, de_i, jt_i, lo_i) \mid max(bw_i) > 64$$
$$\wedge\, min(de_i) < 150 \wedge min(jt_i) < 1 \wedge min(lo_i) < 3 \tag{5}$$
$$\wedge\, (max(bw_i), min(de_i), min(jt_i), min(lo_i)) \in p_i \}$$

If each of the QoS parameter values of the path meets the first line in Table 4, and all QoS parameter values are the preferred values of the current network status, then they belong to the preferred path set.

2) To select satisfied path set: If each of QoS parameter values of the paths meets the second line in Table 4. Namely, all QoS parameter values are the limit values of the current network status, then they belong to satisfied path set, as shown in Eq. (6).

$$
\begin{aligned}
\textbf{satisfied path set} = \{\, p_i(bw_i, de_i, jt_i, lo_i) \mid & max(bw_i) > 4 \\
\wedge\, min(de_i) < 400 \wedge (max(bw_i) < 64 &\vee min(de_i) \geq 150) \\
\wedge\, min(jt_i) < 1 &\wedge min(lo_i) < 3 \\
\wedge\, (max(bw_i), min(de_i), min(jt_i), min(lo_i)) &\in p_i \,\}
\end{aligned}
\tag{6}
$$

3) To select reluctant path set: If no path meets the above two cases, namely the paths that fully meet the service requirements do not exist, the fuzzy synthetic evaluation method is used to select the reluctant paths. The reluctant path refers to that some QoS parameters of the path meet the QoS requirements of the service, or the path is the best QoS parameter value in the current path set, but still can not fully meet the QoS requirements of the service. For example, the delay and jitter of a path meet Preferred, but the bandwidth and packet loss rate belong to Not proper. If the service is sensitive to delay and jitter, the path can meet the QoS requirements.

The fuzzy synthetic evaluation method is based on fuzzy set theory developed by Zadeh [26] for capturing the uncertainties inherent in a system. The whole process is further divided into three steps.

Step 1: Determining the membership functions
The membership functions are established to calculate the membership degrees of each QoS parameter. A membership function (MF) is a curve that defines how each point in the input space is mapped to a membership value between 0 and 1. The symbol $\mathcal{F}$ represents the membership function of the QoS parameters of services. The symbols $\mathcal{B}_{bw_i}$, $\mathcal{B}_{de_i}$, $\mathcal{B}_{jt_i}$ and $\mathcal{B}_{lo_i}$ denote the membership degrees of the service about bandwidth, delay, jitter and packet loss rate, and they are affiliated with Preferred, Limit, or Not proper, respectively. Equations (7), (8), (9), and (10) are shown as:

$$
\mathcal{B}_{bw_i} = \mathcal{F}(bw_1, bw_2, \cdots, bw_k) \tag{7}
$$

$$
\mathcal{B}_{de_i} = \mathcal{F}(de_1, de_2, \cdots, de_k) \tag{8}
$$

$$
\mathcal{B}_{jt_i} = \mathcal{F}(jt_1, jt_2, \cdots, jt_k) \tag{9}
$$

$$
\mathcal{B}_{lo_i} = \mathcal{F}(lo_1, lo_2, \cdots, lo_k) \tag{10}
$$

Step 2: Calculating the fuzzy synthetic evaluation matrix $\mathcal{B}$
The matrix $\mathcal{B}$ is obtained from $\mathcal{A}$ through Eq. (7), (8), (9), and (10) transformation. Each of the elements in matrix $\mathcal{B}$ represents the membership degrees of QoS parameters of the paths.

$$\mathcal{B} = \begin{matrix} Bandwidth \\ Delay \\ Jitter \\ Loss \end{matrix} \begin{pmatrix} \overset{p_1}{\mathcal{B}_{bw_1}} & \overset{p_2}{\mathcal{B}_{bw_2}} & \cdots & \overset{p_i}{\mathcal{B}_{bw_i}} & \cdots & \overset{p_k}{\mathcal{B}_{bw_k}} \\ \mathcal{B}_{de_1} & \mathcal{B}_{de_2} & \cdots & \mathcal{B}_{de_i} & \cdots & \mathcal{B}_{de_k} \\ \mathcal{B}_{jt_1} & \mathcal{B}_{jt_2} & \cdots & \mathcal{B}_{jt_i} & \cdots & \mathcal{B}_{jt_k} \\ \mathcal{B}_{lo_1} & \mathcal{B}_{lo_2} & \cdots & \mathcal{B}_{lo_i} & \cdots & \mathcal{B}_{lo_k} \end{pmatrix}$$

Step 3: Obtaining the available paths
The evaluation results can be obtained by taking into account the QoS weight values of the service requirements, such that a vector $\mathcal{R}$ can be calculated by:

$$\mathcal{R} = \mathcal{W} \bullet \mathcal{B} \tag{11}$$

Where $\bullet$ is a multiplication operator, the evaluation results can be obtained by multiplying the QoS weight values of the service requirements $\mathcal{W}$ and the fuzzy synthetic evaluation matrix $\mathcal{B}$. The conclusion of the fuzzy synthetic evaluation can be obtained by the maximum value principle. The key algorithm for computing $\mathcal{R}$ is shown in Algorithm 3. Lines 1 to 5 are to calculate $\mathcal{B}$, and Line 6 gets $\mathcal{R}$.

Algorithm 3: Computing R

Input: $\mathcal{A}[r][c]$, $\mathcal{W}$// $\mathcal{A}$ is QoS parameter values of $\mathcal{P}$, $\mathcal{W}$ is the weight vector of QoS parameters
Output: $\mathcal{R}$ // the decision value vector of $\mathcal{P}$
1 **for** $i \in [1, r]$ **do**
2   **for** $j \in [1, c]$ **do**
3 | $\mathcal{B}[i][j] = \mathcal{F}(\mathcal{A}[i][j])$;
4   **end**
5 **end**
6 $\mathcal{R} = \mathcal{W} \cdot \mathcal{B}$;

Path Selection module orderly selects preferred path set, satisfied path set and reluctant path set. Even if the candidate paths do not fully meet the QoS requirements of services, a path approximating the QoS requirements of services will be recommended.

4   Experiments and Performance Evaluation

In this section, we conduct three experiments to evaluate the performance of SWQoS in all aspects of QoS parameters and compared with Widest-shortest path (WSP), Shortest-widest path (SWP), shortest-distance path (SDP), and FSEM [18] based on different link status.

4.1   The Experimental Environment and Topology

We have implemented the experiments with Mininet [27,28] on the virtual machine of personal computer. The Floodlight [29] is used as the SDN controller and iperf [30] is used to simulate VoIP flows. The testing results are obtained by iperf and ping tools.

All simulation experiments are performed on the asymmetric network topology including twelve OpenFlow switches and six hosts as shown in Fig. 2. There are fourteen paths from $s1$ to $s12$, which are obtained by Algorithms 1 and 2. Fourteen paths are shown in Table 5. The 3rd, 7th, and 14th paths are the shortest paths based on the hop count, which can be obtained by the Dijkstra algorithm usually.

We ran three groups of experiments to verify the effectiveness of SWQoS in the same topology mentioned above. Link status values assumed by three group

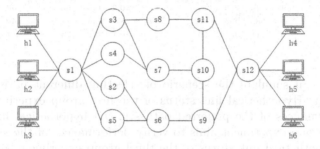

Fig. 2. The network topography

Table 5. All candidate paths from s1 to s12

Path numbers	Paths
1	s1, s2, s6, s9, s10, s7, s3, s8, s11, s12
2	s1, s2, s6, s9, s10, s11, s12
3	s1, s2, s6, s9, s12
4	s1, s3, s7, s10, s9, s12
5	s1, s3, s7, s10, s11, s12
6	s1, s3, s8, s11, s10, s9, s12
7	s1, s3, s8, s11, s12
8	s1, s4, s7, s3, s8, s11, s10, s9, s12
9	s1, s4, s7, s3, s8, s11, s12
10	s1, s4, s7, s10, s9, s12
11	s1, s4, s7, s10, s11, s12
12	s1, s5, s6, s9, s10, s7, s3, s8, s11, s12
13	s1, s5, s6, s9, s10, s11, s12
14	s1, s5, s6, s9, s12

Table 6. Hypothetical link status of three group experiments

LINKS	The first group experiment				The second group experiment				The third group experiment			
	Bandwidth (kb/s)	Delay (ms)	Loss (%)	Jitter (ms)	Bandwidth (kb/s)	Delay (ms)	Loss (%)	Jitter (ms)	Bandwidth (kb/s)	Delay (ms)	Loss (%)	Jitter (ms)
s1, s3	80	3	0.1	0.09	60	37	7	9	65	37	3	0.9
s1, s4	68	9	1	0.3	60	9	0.1	0.3	68	9	1	0.63
s1, s2	70	1	0.6	0.1	63	25	5	1	50	0	1	0.001
s1, s5	80	30	2	1	63	12	0.2	0.38	30	30	2	0.01
s3, s8	52	31	3	0.13	52	31	3	0.13	32	31	3	0.99
s3, s7	80	1	3	0.06	63	1	0.4	0.06	50	32	3	0.632
s4, s7	70	9	1	0.03	63	9	0.1	0.03	48	9	1	0.476
s2, s6	80	5	1	0.4	63	5	1	0.4	80	5	1	0.001
s5, s6	90	3	2	0.3	63	3	0.2	0.3	90	3	2	0.003
s8, s11	60	35	3	0.93	60	35	3	0.93	60	35	3	0.935
s7, s10	90	1	1	0.06	60	1	0.1	0.06	49	10	1	0.594
s6, s9	80	0	1	0.01	60	5	0.1	0.01	80	0	0.1	0.687
s10, s11	80	1	1	1	60	1	1	1	41	11	1	0.1
s10, s9	95	1	2	0.3	60	1	0.2	0.3	50	50	2	0.001
s11, s12	70	40	8	0.3	60	40	8	0.3	47	40	0.8	0.1
s9, s12	70	0	1	0.1	60	0	0.1	0.1	70	1	1	3

experiments can simulate the scenario of selecting different path sets shown as Table 6. (*i*) Hypothetical link status of the first group experiment aims to verify performances of the preferred path set; (*ii*) hypothetical link status of the second group experiment aims to verify performances of the satisfied path set; (*iii*) hypothetical link status of the third group experiment aims to verify performances of the reluctant path set. In the experiments, VoIP simulation traffic of G.711 encoding is first sent along with every path selected respectively from host 1 to host 4 shown as Fig. 2. Then, we capture the throughout, delay, jitter and packet loss rate of the paths selected under the same VoIP traffic load. Last, we implement and analyze three group experiments respectively.

4.2   The First Group Experiment: Simulating the Network Status of Selecting the Preferred Paths

In the first group experiment, we simulate the network status scenario that can obtain the preferred path set. After calculation by Eqs. (1), (2), (3) and (4), the path status is obtained as shown in Table 7.

Table 7. QoS parameters of paths in the first group experiment

Paths	Bandwidth (kb/s)	Delay (ms)	Loss (%)	Jitter (ms)
1st	52	115	17.77	2.29
2nd	70	48	13.04	2.11
3rd	*70*	*6*	*2.68*	*0.61*
4th	*70*	*6*	*2.68*	*0.61*
5th	70	46	9.47	1.51
6th	52	71	8.9	2.55
7th	52	109	13.52	1.45
8th	52	87	10.98	2.85
9th	52	125	15.5	1.75
10th	68	20	4.14	0.79
11th	68	60	10.82	1.69
12th	52	142	19.75	3.09
13th	70	75	15.13	2.91
14th	70	33	5.02	1.41

It can be concluded from Table 7 that the status values of the 3rd and 4th paths meet the condition of Eq. (5), so they are the preferred paths. The paths calculated by other comparison algorithms are the 3rd, the 4th and the 14th paths. WSP and SDP select the 14th path. SWP selects the 4th path. FSEM selects the 3rd path. We respectively inject the simulated flow into three paths and monitors the status of the paths every 5 s. The experimental results are shown in Fig. 3 and Fig. 4. Figure 3 shows the results for throughput, delay, jitter and packet loss rate of the 3rd, the 4th and the 14th paths under the same VoIP traffic load. Figure 4 shows the results for the average status values of three paths under the same load.

We observe that the QoS parameter performance of the 3rd path and the 4th path are better than the 14th path, thereby indicating that SWQoS, SWP and FSEM are more efficient compared with WSP and SDP. However, SWP and FSEM schemes can only select the 3rd path or the 4th path, they can not select all paths that meet the requirements of the network services. Therefore, SWQoS scheme is more better compared with the others because it can not only select the best paths but also select all the paths that meet the network service requirements.

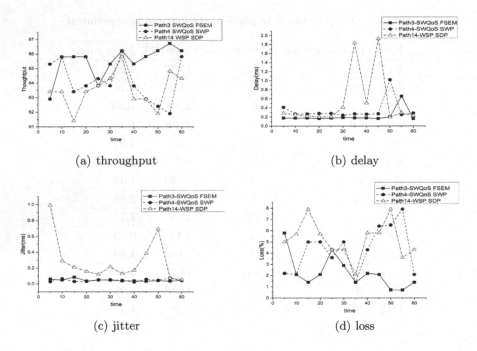

Fig. 3. Comparison of QoS parameters in the first group experiment

4.3   The Second Group Experiment: Simulating the Network Status of Obtaining Satisfied Paths

Similar to the first group experiment above, the purpose of the second group experiment is to select a satisfied path set and compare with the QoS parameter performance of the paths selected by the others schemes. It can be concluded from Table 8 that the status values of the 10th and the 14th paths meet the condition of Eq. (6), thus they are the satisfied paths.

Same as in the first group experiment, the 10th path and the 14th path obtained by SWQoS are better than the 3rd path in QoS parameter performance as shown in Fig. 5 and Fig. 6. Similarly, SWP and FSEM schemes can also obtain the 10th path or the 14th path, both the 10th path and the 14th path cannot be obtained at the same time by two schemes. However, SWQoS can select both the optimal paths of performance and all the paths that meet the requirements of services. Therefore, SWQoS is more effective.

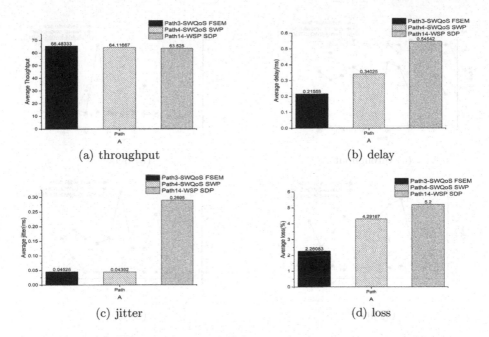

(a) throughput

(b) delay

(c) jitter

(d) loss

Fig. 4. The average values of QoS parameters in the first experiment

Table 8. QoS parameters of paths in the second group experiment

Paths	Bandwidth (kb/s)	Delay (ms)	Loss (%)	Jitter (ms)
1st	52	144	19.24	3.19
2nd	60	77	14.6	3.01
3rd	60	35	6.14	1.51
4th	60	40	7.74	9.52
5th	60	80	15.72	10.42
6th	52	105	13.63	11.46
7th	52	143	19.5	10.36
8th	52	87	7.69	2.85
9th	52	125	13.96	1.75
10th	*60*	*20*	*0.6*	*0.79*
11th	60	60	9.19	1.69
12th	52	129	14.47	2.47
13th	60	62	9.56	2.29
14th	*60*	*20*	*0.6*	*0.79*

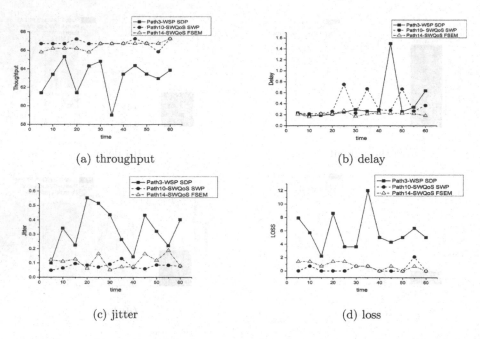

Fig. 5. Comparison of QoS parameters in the second group experiment

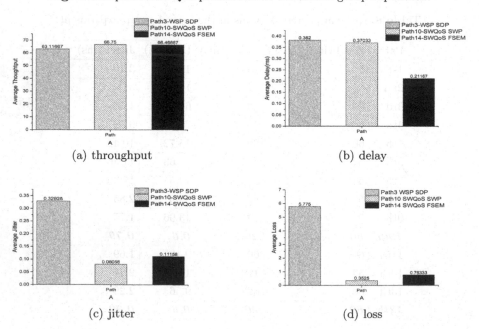

Fig. 6. The average values of QoS parameters in the second experiment

4.4 The Third Group Experiment: Simulating the Network Status of Obtaining Reluctant Paths

In the third group experiment, we first obtain the status of the paths, as shown in Table 9. However, none of the path parameters in the Table 9 match Eqs. (5) or (6). Thus, we implement the step to select the reluctant paths.

Table 9. QoS parameters of paths in the third group experiment

Paths	Bandwidth (kb/s)	Delay (ms)	Loss (%)	Jitter (ms)
1st	32	203	4.32	3.941
2nd	41	106	0.16	0.89
3rd	50	6	0.1	3.689
4th	49	130	18	5.127
5th	41	130	7.2	2.326
6th	32	165	54	5.926
7th	32	143	21.6	2.925
8th	32	178	54	6.764
9th	32	156	21.6	3.763
10th	48	79	2	4.701
11th	41	79	0.8	1.9
12th	30	231	17.28	3.952
13th	30	134	0.64	0.901
14th	30	34	0.4	3.7

1) Settings of Membership Functions. Several membership functions need to be proposed, such as bandwidth, delay, jitter, and packet loss rate. The symbols $bw_{threshold}$, $de_{threshold}$, $jt_{threshold}$ and $lo_{threshold}$ represent the thresholds of the bandwidth, delay, jitter and packet loss rate, respectively. According to the habit of selecting membership functions, three commonly used membership functions are selected, which conform to the characteristics of VoIP protocol. They include the Cauchy function, piecewise function, and normal distribution function. After repeating experiments, the membership functions adopt improved normal distribution functions. The fuzzy membership functions could be defined separately as below:

$$\mathcal{F}_{bw}(x) = \begin{cases} e^{-\left(\dfrac{x - bw_{threshold}}{e}\right)^2} & x < bw_{threshold} \\ 1 & \text{otherwise} \end{cases} \quad (12)$$

$$\mathcal{F}_{de}(x) = \begin{cases} e^{-\left(\dfrac{x - de_{threshold}}{e}\right)^2} & x >= de_{threshold} \\ 1 & \text{otherwise} \end{cases} \quad (13)$$

$$\mathcal{F}_{jt}(x) = \begin{cases} 1 & x < jt_{threshold} \\ e^{-e*(x - jt_{threshold})} & \text{otherwise} \end{cases} \quad (14)$$

$$\mathcal{F}_{lo}(x) = \begin{cases} 1 & x < lo_{threshold} \\ e^{-e*(x-lo_{threshold})} & \text{otherwise} \end{cases} \qquad (15)$$

Adjusting the thresholds of the membership functions are very important to experimental results. The range of threshold values adopts a principle, which is as small as possible. The range of threshold values should be less than the range given in Table 4. For instance, $bw_{threshold}$ value is equal to the preferred bandwidth value in Table 4 plus half of the absolute value of the difference between the preferred bandwidth value and the current maximum bandwidth value in Table 4. In this way, the selected paths can meet the service requirements to the maximum extent. Other thresholds are also determined by the same method. However, the monotonicity of the membership functions needs attention.

2) Result. According to the membership functions, we can compute the fuzzy synthetic evaluation matrix $\mathcal{B}$. Then, the QoS weight values of the service requirements $\mathcal{W}$ are multiplied by the fuzzy synthetic evaluation matrix $\mathcal{B}$ to get the reluctant path according to Eq. (10). $\mathcal{R} = \mathcal{W} \bullet \mathcal{B} = \{0.00012136, 0.41671141, 0.48580011, 0.00000478, 0.00965493, 0.00000054, 0.00194193, 0.00000006, 0.00019796, 0.36397667, 0.39579707, 0.00011811, 0.41666693, 0.41724337\}$. The max value of $\mathcal{R}$ is the 3rd path. Therefore, the reluctant path is the 3rd path in the candidate paths by implementing the SWQoS scheme.

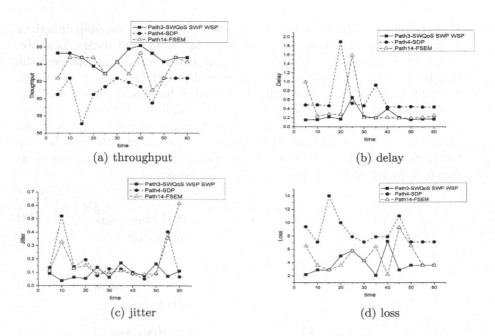

(a) throughput

(b) delay

(c) jitter

(d) loss

Fig. 7. Comparison of QoS parameters in the third experiment

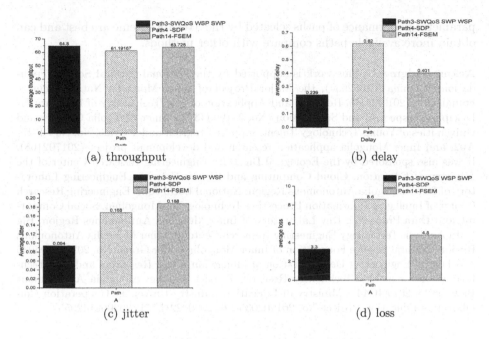

(a) throughput

(b) delay

(c) jitter

(d) loss

Fig. 8. The average values of QoS parameters in the third experiment

Figure 7 and Fig. 8 show the simulation results for network status under VoIP traffic load in the third group experiment. It can be seen that the QoS parameter performance of the 3rd path is better compare with the 4th path and the 14th path most of the time. Although the 3rd path has lower QoS parameter performance than other paths some time, they do not affect the whole QoS guarantee. That's because VoIP protocol is more sensitive to delays and jitters, and the effect of two parameters is more obvious. The results of the average value of QoS parameters shown in Fig. 8 more obviously prove the above situation. Thus the 3rd path can guarantee the QoS requirements of VoIP as much as possible.

Through the above three group experimental results, we can observe that SWQoS can select all the available paths and the performance of the paths are better than other paths compared with other routing methods.

5   Conclusions

In this paper, we propose an stepwise, universal SWQoS scheme to select the available paths from all the candidate paths in SDN. To obtain all the available paths, we fully consider all the performance criteria of the service requirements and current network status. In the absence of preferred path set and satisfied path set, we select the reluctant path set using AHP combined with the fuzzy synthetic evaluation method. Thus, SWQos can meet various Qos requirements of services and obtain the optimal path. Experimental results show that the QoS

518   L. Liu et al.

parameter performance of paths selected by the SWQoS scheme are best and can obtain more available paths compare with other methods.

Acknowledgment. This work is supported by the National Natural Science Foundation of China (61662054), the Major Project of Inner Mongolia Natural Science Foundation (2019ZD15), Research and Application of Key Technology of Big Data for Discipline Inspection and Supervision (No. 2019GG372), Inner Mongolia Colleges and Universities of Young Technology Talent Support Program under Grant No. NJYT-19-A02, and Inner Mongolia application research and development project (201702168). It was also sponsored by the Ecological Big Data Engineering Research Center of the Ministry of Education, Cloud Computing and Service Software Engineering Laboratory of Inner Mongolia Autonomous Region, National Local Joint Engineering Research Center of Intelligent Information Processing Technology for Mongolian, Social Computing and Data Processing Key Laboratory of Inner Mongolia Autonomous Region, Big Data Analysis Technology Engineering Research Center of Inner Mongolia Autonomous Region, Natural Science Foundation of Inner Mongolia under Grand No. 2020MS06030 and Digital Engineering Demonstration of Mongolian Music Resources and Key Technology Research, Key Technology Research Project in Inner Mongolia Autonomous Region No. 2019GG147, Ministry of Education Industry-University Cooperation Collaborative Education Project No. 201902035056, 202002215071, 202002142055.

References

1. Braden, R., Clark, D., Shenker, S.: RFC 1633: integrated services in the internet architecture: an overview (1994)
2. Blake, S., Black, D., Carlson, M., Davies, E., Wang, Z., Weiss, W.: RFC2475: an architecture for differentiated service (1998)
3. Rosen, E., Viswanathan, A., Callon, R.: RFC3031: multiprotocol label switching architecture (2001)
4. Egilmez, H.E., Civanlar, S., Tekalp, A.M.: An optimization framework for QoS-enabled adaptive video streaming over OpenFlow networks. IEEE Trans. Multimedia **15**(3), 710–715 (2012)
5. Karakus, M., Durresi, A.: Quality of service (QoS) in software defined networking (SDN): a survey. J. Netw. Comput. Appl. **80**, 200–218 (2017)
6. Alto, P.: Software-defined networking: the new norm for networks [white paper]. ONF White Paper (2012)
7. Kreutz, D., Ramos, F.M., Verissimo, P.E., Rothenberg, C.E., Azodolmolky, S., Uhlig, S.: Software-defined networking: a comprehensive survey. Proc. IEEE **103**(1), 14–76 (2014)
8. Feamster, N., Rexford, J., Zegura, E.: The road to SDN: an intellectual history of programmable networks. ACM SIGCOMM Comput. Commun. Rev. **44**(2), 87–98 (2014)
9. Nunes, B.A.A., Mendonca, M., Nguyen, X.N., Obraczka, K., Turletti, T.: A survey of software-defined networking: past, present, and future of programmable networks. IEEE Commun. Surv. Tutor. **16**(3), 1617–1634 (2014)
10. Guck, J.W., Bemten, A.V., Reisslein, M., Kellerer, W.: Unicast QoS routing algorithms for SDN: a comprehensive survey and performance evaluation. IEEE Commun. Surv. Tutor. **20**(1), 388–415 (2018)

11. Satty, T.: How to make a decision: the analytic hierarchy process. Eur. J. Oper. Res. **48**(1), 9–26 (1990)
12. Zhang, L., Deering, S., Estrin, D., Shenker, S., Zappala, D.: RSVP: a new resource reservation protocol. IEEE Commun. Mag. **40**(5), 116–127 (2002)
13. Egilmez, H.E., Dane, S.T., Bagci, K.T., Tekalp, A.M.: OpenQoS an OpenFlow controller design for multimedia delivery with end-to-end quality of service over software-defined networks. In: 2012 Asia-Pacific Signal and Information Processing Association Annual Summit and Conference (APSIPA ASC) (2012)
14. Seddiki, M.S., et al.: FlowQoS: QoS for the rest of us. In: Proceedings of the Third Workshop on Hot Topics in Software Defined Networking, pp. 207–208 (2014)
15. Seddiki, M.S., Shahbaz, M., et al.: FlowQoS: per-flow quality of service for broadband access networks. Technical report, Georgia Institute of Technology (2015)
16. Yan, J., Zhang, H., Shuai, Q., Liu, B., Guo, X.: HiQoS: an SDN-based multipath QoS solution. China Commun. **12**(5), 123–133 (2015)
17. Yu, T.F., Wang, K., Hsu, Y.H.: Adaptive routing for video streaming with QoS support over SDN networks. In: International Conference on Information Networking 2015, pp. 318–323, March 2015. https://doi.org/10.1109/ICOIN.2015.7057904
18. Li, J., Chang, X., Ren, Y., Zhang, Z., Wang, G.: An effective path load balancing mechanism based on SDN. In: 2014 IEEE 13th International Conference on Trust, Security and Privacy in Computing and Communications (2014)
19. Gelenbe, E.: Machine learning for network routing. In: 2020 9th Mediterranean Conference on Embedded Computing (MECO), p. 1 (2020). https://doi.org/10.1109/MECO49872.2020.9134073
20. Mao, B., et al.: Routing or computing? The paradigm shift towards intelligent computer network packet transmission based on deep learning. IEEE Trans. Comput. **66**(11), 1946–1960 (2017). https://doi.org/10.1109/TC.2017.2709742
21. Reis, J., Rocha, M., Phan, T.K., Griffin, D., Le, F., Rio, M.: Deep neural networks for network routing. In: 2019 International Joint Conference on Neural Networks (IJCNN), pp. 1–8 (2019). https://doi.org/10.1109/IJCNN.2019.8851733
22. Yu, C., Lan, J., Guo, Z., Hu, Y.: DROM: optimizing the routing in software-defined networks with deep reinforcement learning. IEEE Access **6**, 64533–64539 (2018). https://doi.org/10.1109/ACCESS.2018.2877686
23. ITU-T: ITU-T recommendation G.1010 end-user multimedia QoS categories. ITU-T (2001)
24. Yi, Z.: Research on QoS routing of SDN based on OpenFlow. Master's thesis, University of Electronic Science and Technology of China (2018)
25. Saaty, T.L.: Decision making with the analytic hierarchy process. Int. J. Serv. Sci. **1**(1), 83–98 (2008)
26. Zadeh, L.A.: Fuzzy sets. In: Fuzzy Sets, Fuzzy Logic, and Fuzzy Systems: Selected Papers by Lotfi A Zadeh, pp. 394–432. World Scientific (1996)
27. Octopress: mininet.org. http://mininet.org/. Accessed 12 Feb 2020
28. Fernando, O.A., Xiao, H., Che, X.: Evaluation of underlying switching mechanism for future networks with P4 and SDN (workshop paper). In: Wang, X., Gao, H., Iqbal, M., Min, G. (eds.) CollaborateCom 2019. LNICST, vol. 292, pp. 549–568. Springer, Cham (2019). https://doi.org/10.1007/978-3-030-30146-0_38
29. projectfloodlight.org. http://www.projectfloodlight.org/floodlight/. Accessed 11 Aug 2020
30. iperf.fr. http://iperf.fr/. Accessed 23 Nov 2020

A Novel Approach to Taxi-GPS-Trace-Aware Bus Network Planning

Liangyao Tang[1], Peng Chen[2], Ruilong Yang[1], Yunni Xia[1(✉)], Ning Jiang[3], Yin Li[4], and Hong Xie[1]

[1] School of Computers, Chongqing University, Chongqing 400030, China
[2] School of Computer and Software Engineering, Xihua University, Chengdu 610039, China
[3] Mashang Consumer Finance Co., Ltd. (MSCF), Chongqing 40022, China
[4] Institute of Software Application Technology, Guangzhou and Chinese Academy of Sciences, Guangzhou 511458, China

Abstract. Taxi GPS traces are rich with information regarding the human mobility pattern in metropolitans. In this paper, we aimed at fully exploiting the Taxi GPS traces and addressing the bus network planning problem. Specifically, the proposed framework comprises a method for determining candidate bus stations by utilizing passenger pick-up and drop-off records, a bio-inspired method for yielding bus routes and further for generating the final bus network. To prove the effectiveness of our framework, we conduct simulative studies as well based on a real-world taxi GPS data-set and show that our proposed framework considerably outperforms traditional ones.

Keywords: Taxi GPS traces · Bus routes planning · Bus network planning

1 Introduction

Buses are usually believed to be more energy-efficient, in terms of energy consumed per mile and per person, and less resource requiring than private cars when serving crowded city areas [11]. Nowadays, with the increasing demand for highly-efficient and green public transportation for metropolitan citizens, cost-effective and environment-friendly bus networks shows its great importance in serving versatile transportation demands [15,39].

Liangyao Tang and Peng Chen contribute equally to this work and thus are co-first authors of this paper.

This work is supported by China Scholarship Council Science and Technology Program of Sichuan Province under Grant 2020YFG0326, Talent Program of Xihua University under Grant Z202047, the Chongqing grand research and development project under Grant cstc2019jscx-fxyd0385.

H. Gao and X. Wang (Eds.): CollaborateCom 2021, LNICST 406, pp. 520–533, 2021.
https://doi.org/10.1007/978-3-030-92635-9_30

Traditional method for bus network planning and design are usually based on human surveys [2], which requires high cost and manpower but proved to be inaccurate. Moreover, survey-based planning is incapable of accommodating the fast changes and modifications of the metropolitan networks, especially when metropolitan roads are under continuing construction and optimization. Nowadays, with the wide application of wireless communication technology, taxis are usually equipped with GPS devices [16,20,40,42]. Taxi GPS traces are rich with time and spatial information that can be exploited. Such traces comprise time/positions of pick-up and drop-off events, which can be further used for analyzing the emergence patterns of "hot" areas and predicting real-time transportation needs.

In this paper, we propose a novel approach to planning bus routes and networks by exploiting taxi GPS traces. The propose method is capable of identifying candidate bus stops base on historical taxi passenger pick-up/drop-off distributions, appropriately manage "hot" areas for serving as-many-as-possible passengers and yielding bus networks with maximum passenger flows with the constraint of available candidate stops and serving time. We conduct extensive simulations as well to verify the effectiveness of the proposed approach.

2   Related Work

Existing works in the direction of exploiting automobile traces fall into multiple categories: social dynamics computing, traffic dynamics mining and operational dynamics computing [5,32,35]. The first category relies itself on analysis of collective behavior and movement of a city's population. Related works aim at analyzing the destination places of citizens [19,23], or the distribution patterns of "hot" spots [6,28], or the functions of "hot" spots [24,30]. The second category studies the flow of the population through the city's road network and aims at forecasting traffic flows and travelling durations for drivers [4,21,25,41]. The third category concerns taxi driver's behavior and aims at learning taxi drivers' expert knowledge [7,26,27,29,38].

Bus network planning can be seem as a extended topic of the trace mining ones [1,36,37,44]. It is known as a complex, nonlinear, non-convex, multi-objective NP-hard problem [3,22,34]. Related research objectives comprises planning and optimization of travelling route, travelling durations, travelling cost and throughput of road networks [17,18,43]. Among them, [9] aims to find an optimal bus route for a given origin-destination (OD) pair in a single direction. Similarly, [8] aims to find a bi-directional travelling route for a specified OD pair.

3   Main Steps

3.1   Candidate Bus Stop Identification

The first major stage of our proposed method is to decide candidate bus stops through analyzing the taxi PDRs (taxi passenger pick-up and drop-off spots and times). It comprises two steps: (1) Dividing the bus serving area into small equally-sized grid cells and marking "hot" ones for further processing (A hot grid cell refers to the grid with PDRs greater than zero); (2) Selecting candidate bus stops according to the Algorithm 1.

Hot Grid Cells. We first split the area under study into equally-sized grid cells, each of which covers a 25 m × 25 m square. Thus, the whole area is partitioned into 1500 × 1800 cells. Note that only reachable cells, excluding unreachable ones in terms of, e.g., lakes and mountains, are with PDR and are thus 'hot' ones. Figure 1 is a sample for hot grid cells.

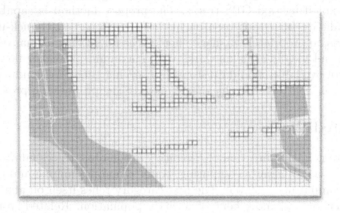

Fig. 1. A sample for hot grid cells.

Selection of Candidate Bus Stops. Traditional methods for deciding candidate stops are performed through clustering PDRs. However, such methods could lead to inappropriate setting of bus stops especially when some clusters are too large for a single stop to cover. To overcome related limitations, we consider an improved strategy, as shown in Algorithm 1 for deciding candidate bus stations:

(1) Taking each "hot" grid cell as the center and a certain distance as the radius to count the PDRs within the range. The above-mentioned distance refers to the service radius of a bus stop, and the service radius must be set properly because it affects the walking distance to the bus stop. According to the "Urban Road Traffic Planning and Design Code" [31], the service radius of the bus stop in this paper is set to 300 m.

(2) Sorting the "hot" grid cells in descending order according to *pdrs* value (It refers to the number of PDRs within 300 m of each "hot" grid.) and select the grid with the largest *pdrs* value as the first candidate stop. Obviously,

---

Algorithm 1: Candidate Stops Selection

Input : The collection of "hot" grid: G
Output: The collection of candidate stops: S
$i = 1$ // Initialization
for *each grid* $\in G$ **do**
    counting the number of PDRs within 300 m of this grid: *pdrs*
end
repeat
    $G = sort(G)$ // sort G according to *pdrs* by descending order
    $S_i = G_1$ // select the grid with the largest *pdrs* as candidate stop
    Deleting the PDRs within 300 m of the G_1
    for *each grid* $\in G$ **do**
        update the *pdrs* of *grid*
    end
    $i = i + 1$
until *Get enough candidate sites*;

---

the larger the *pdrs* value is, the more potential passengers there are around the grid.

(3) Deleting the PDRs within 300 m of the first candidate stop and update the *pdrs* of the remaining "hot" grid cells. This step is mainly to eliminate the impact of the selected candidate sites, because if there are candidate stops around a "hot" grid, the number of potential passengers around the grid will be reduced.

(4) Sorting the remaining "hot" grid cells in descending order according to *pdrs* value, and repeat the above process.

3.2 Bus Network Generation

Taking the bus stations generated above as inputs, the following stage aims at planning bus networks with 2 steps: (1) Estimating the user flow and trip time with any two candidate stops according to trace data. (2) Applying an ant colony algorithm for yielding bus routes and further generating bus networks.

Passenger Flow and Travel Time Evaluation. We employ two matrix, i.e., flow matrix (FM) and travelling time matrix (TM), for capturing travel demand and related time requirement. Every element in the matrix indicates the number of passengers or the required travelling time from one station to another.

In the FM matrix, we count the total passenger flow from the coverage of one stop to that of another. We further calculate the average taxi time between two candidate stops. In addition, in consideration of the speed gap between taxis and buses, we consider the bus travel time between two candidate stops to be the average taxi time multiplied by α ($\alpha = 1.5$). For the paths without taxi trip records, we consider using $dis(i,j)/v$ as an approximation. $dis(i,j)$ is the driving distance between station i and station j, and v is 25 km/h due to the fact that the average speed of taxis is about 25.4 km/h according to taxi GPS traces.

Bus Route Generation. Ant colony algorithm is a general-purpose heuristic algorithm which can be used to solve different combinatorial optimization problems [10,12,13,33]. As illustrated in Algorithm 2, we use ant colony algorithm with elitist strategy for selecting bus routes. This algorithm takes the following intermediate variables as inputs:

(1) Visibility. We define the visibility η_{ij} as passenger flow between station i and station j. $fm(i,j)$ refers to the passenger flow from station i to station j

$$\eta_{ij} = fm(i,j) + fm(j,i) \tag{1}$$

(2) Pheromone update. We update the pheromone τ_{ij} from station i to station j according to the following formula. ρ is pheromone evaporation coefficient, $\Delta\tau_{ij}$ the pheromone increment, m the number of ants, k the k-th ant and Q the pheromone constant.

$$\tau_{ij}(t+1) = (1-\rho) \cdot \tau_{ij}(t) + \Delta\tau_{ij} \tag{2}$$

$$\Delta\tau_{ij} = \sum_{k=1}^{m} \Delta\tau_{ij}^{k} \tag{3}$$

$$\Delta\tau_{ij}^{k} = \begin{cases} Q \cdot Num, & if\ the\ k\text{-}th\ ant\ passes\ edge(i,j) \\ 0, & otherwise \end{cases} \tag{4}$$

(3) Passenger flow. Num represents the passenger flow in both directions of the route. For route $R = [s_1, s_2, ...s_n]$, Num is determined according to the following formula:

$$Num = \sum_{i;j(j>i)}^{n} (fm(s_i, s_j) + fm(s_j, s_i)) \tag{5}$$

(4) Transition probability. Ants randomly select the next stop according to a selection probability affected by visibility and pheromone. We use *allowed* to represent the set of optional next stations and define the transition probability from station i to station j for the k-th ant as:

$$p_{ij}^{k} = \frac{[\tau_{ij}]^{\alpha} \cdot [\eta_{ij}]^{\beta}}{\sum_{k \in allowed}[\tau_{ik}]^{\alpha} \cdot [\eta_{ik}]^{\beta}}, \ if\ j \in allowed \tag{6}$$

(5) Time of current route. T is the total time of the current route in both directions. If T satisfies the time constraint, the corresponding ant continues to search for the next station, otherwise, it stops searching. For current route $R = [s_1, s_2, ...s_n]$, we calculate the T according to the following formula. t_0 is the waiting time of the bus at each stop:

$$T = \sum_{i=1}^{n-1} (tm(s_i, s_{i+1}) + tm(s_{i+1}, s_i)) + (n-2) \cdot t_0 \cdot 2 \tag{7}$$

Algorithm 2: Bus Route Selection

Input : FM: Passenger flow matrix
$\quad\quad\quad$ TM: Travel time matrix
$\quad\quad\quad$ $iter_{max}$: Maximum number of iterations
$\quad\quad\quad$ T_{max}: Time constraints
Output: *Route*

Initialize the parameters of ant colony algorithm.
Initialize the position of the ants.
$count = 0$ // Initialize the number of iterations.
while $count < iter_{max}$ **do**
$\quad$ **for** *each ant* **do**
$\quad\quad$ **repeat**
$\quad\quad\quad$ Choose the next stop j according to Eq. 6.
$\quad\quad\quad$ Calculate the total time T according to Eq. 7.
$\quad\quad$ **until** $T > T_{max}$;
$\quad\quad$ Calculate the passenger flow of the route according to Eq. 5.
$\quad\quad$ Update the best route.
$\quad$ **end**
$\quad$ **for** *each edge*(i, j) **do**
$\quad\quad$ Update pheromone according to Eq. 2–Eq. 4.
$\quad$ **end**
$\quad$ $count = count + 1$
end

Algorithm 3: Bus Network Planning

Input : FM: Passenger flow matrix
$\quad\quad\quad$ TM: Travel time matrix
$\quad\quad\quad$ Num_{min}: Minimum passenger flow
Output: The collection of routes $\{R_1, R_2...R_n\}$
$k = 1$ // Initialization
repeat
$\quad$ Run Bus Route Selection Algorithm, and we get R_k.
$\quad$ //R_k is a route generated by Algorithm 2.
$\quad$ Calculate Num (the passenger flow of the route R_k) according to Eq. 5.
$\quad$ **for** i *in* R_k **do**
$\quad\quad$ **for** j *in* R_k **do**
$\quad\quad\quad$ $fm(i,j) = 0$ //$fm(i,j)$ is the passenger flow from stop i to stop j.
$\quad\quad$ **end**
$\quad$ **end**
$\quad$ $k = k + 1$
until $Num < Num_{min}$;

Bus Network Generation. Algorithm 3 shows the process of bus network generation. It first generates one single bus route using the Bus Route Selection algorithm and then updates passenger flow matrix (FM). For example, we firstly get a route R_k by Algorithm 2. Then, we assume that the passenger flow between the stations of route R_k becomes zero. Finally, we update FM and get a new route based on the new FM.

4   Simulations

In this part, we test our proposed approach with a third-party and real-world taxi GPS data-set. The data-set was generated from taxis in New York City for one week and comprises more than 3.4 million passenger delivering trips formatted as follows:

- Trip_Pickup_Datetime : passenger boarding time
- Trip_Dropoff_Datetime : passenger drop-off time
- Start_Lon : longitude of the place where passengers get on
- Start_Lat : latitude of the place where passengers get on
- End_Lat : longitude of the place where passengers get off
- End_Lat : latitude of the place where passengers get off
- Trip_Distance : total travel distance

	Trip_Pickup_DateTime	Trip_Dropoff_DateTime	Start_Lon	Start_Lat	End_Lon	End_Lat	Trip_Distance
2	2009-12-18 03:09:00	2009-12-18 03:34:00	-73.955745	40.689503	-73.937730	40.737463	14.451909
15	2009-12-19 22:05:00	2009-12-19 22:18:00	-74.026760	40.656373	-74.030567	40.658013	1.271382
17	2009-12-19 23:05:00	2009-12-19 23:38:00	-74.005453	40.740142	-73.967598	40.753637	5.761452
19	2009-12-16 01:31:00	2009-12-16 01:47:00	-74.003327	40.751415	-73.990005	40.690455	8.095000
24	2009-12-16 01:30:00	2009-12-16 01:51:00	-73.981385	40.744072	-73.966352	40.690532	12.665537
...	...	...	...	...	...	...	...

Fig. 2. A sample for data set.

Based on the data-set, we further:

(1) Delete duplicate data and unreasonable data, e.g., the Trip_Distance with values of 0.
(2) Delete invalid data, for example, the place where passengers get on/off the taxi is not in the New York.
(3) Project the longitude and latitude coordinates into plane coordinates according to the UTM (Universal transverse Mercator) projection. It divides the earth's surface into 60 projection zones and New York is located in No.18 projection zone. Since New York is located in the northern hemisphere, the FalseEasting is set to 500 km and FalseNorthing is 0.

We generate 200 stations based on Algorithm 1 and Fig. 3 shows the number of PDRs within the 300 m of these stations. It can be seen from Fig. 3 that the stations ahead contain more PDRs. This indicates that there are more potential passengers at the stations ahead. Among them, the station No.0 contains 43825 PDRs, while the station No.199 only contains 1019 PDRs.

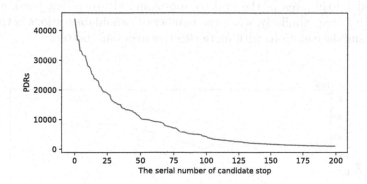

Fig. 3. The number of PDRs within the 300 m of each station.

As shown in the Fig. 4, the PDRs covered by the top 80 stations accounts for 74.29% of the total PDRs, and the PDRs covered by the top 100 stations accounts for 80.12% of the total PDRs. Therefore, we decide to select the top 100 stations as candidate stations.

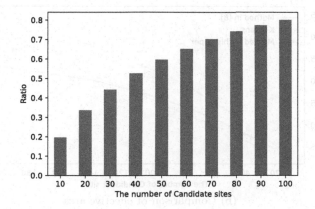

Fig. 4. Proportion of the PDRs covered by top n stops.

In order to compare our method with the baseline methods [8, 14], we propose two strategies to evaluate the rationality of candidate sites:

(1) We compare the number of PDRs covered by the same number of candidate sites. Obviously, with the same number of candidate stations, the more PDRs the candidate stations contain, the more potential passengers there are.

(2) We compare the effective areas covered by the same number of candidate sites. The effective area refers to the regions with taxi passengers pick-up/drop-off records. In this paper, we use the number of "hot" grids multiplied by the area of the grid to approximately represent the size of the effective area. Similarly, when the number of candidate stations is the same, the candidate stations with more effective areas are better.

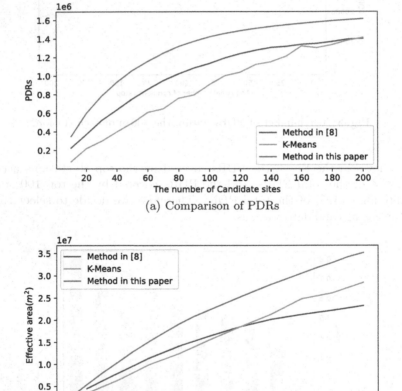

(a) Comparison of PDRs

(b) Comparison of effective area

Fig. 5. Comparison of candidate stops selection methods.

Figure 5 shows the comparison results between our method and baseline methods [8,14], when the number of candidate site is set to 10, 20... 200 respectively. As illustrated in Fig. 5, in the case of the same number of candidate sites, the candidate sites determined by our method can cover more PDRs, which means more potential passengers. Similarly, the candidate sites determined by our method can contain more effective areas, indicating that the stations are more reasonable. Therefore, our method is superior to the baseline methods in many aspects.

In the stage of bus route selection, the methods in [8,9] generate an optimal route for a specified origin-destination (OD) pair. However, our method does not need to specify the origin-destination pair in advance, and can generate a global optimal route.

Parameter-setting of the Bus Route Selection algorithm is shown in Table 1 and the bus network generated in the experiment is shown in Table 2.

Table 1. Experimental parameters

Parameter	Description	Value
stop_num	The number of candidate stops	100
ant_num	The number of ants	500
α	Pheromone weight value	1
β	Visibility weight value	1.5
ρ	Pheromone evaporation rate	0.5
Q	Pheromone constant	1
e	The number of elite ants	10
t	Time constraints	7200 s

Table 2. Bus network

	Path	Bus frequency	Bus type	Passenger flow
1	[15, 10, 2, 6, 0, 4, 11, 8]	20 min	Medium-sized	18652
2	[1, 0, 18, 5, 7, 9, 11, 13]	20 min	Medium-sized	15271
3	[12, 3, 17, 27, 7, 22, 16, 5]	20 min	Medium-sized	12702
4	[23, 20, 4, 13, 24, 8, 28, 25, 21]	20 min	Medium-sized	12308
5	[44, 25, 40, 8, 29, 4, 14, 0, 19]	20 min	Small-sized	9598
6	[30, 6, 33, 1, 34, 26, 5, 14]	20 min	Small-sized	8762
7	[35, 12, 9, 26, 0, 24, 11, 17]	20 min	Small-sized	7806

Next, according to the number of users on the bus at each stop in both directions, we determine the operation frequency and bus capacity for each route in Table 2. For example, for *path*: [15, 10, 2, 6, 0, 4, 11, 8], Fig. 6 illustrates the number of citizens on the bus at each stop, when bus operation frequency is respectively set to 10 min, 20 min, 30 min. Considering the waiting time of passengers, the cost of public transportation system and the capacity of buses, for this path, we set the bus operation frequency to 20 min and choose the medium-sized bus for transportation.

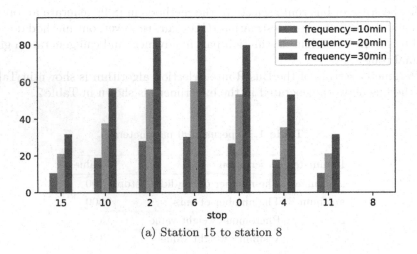

(a) Station 15 to station 8

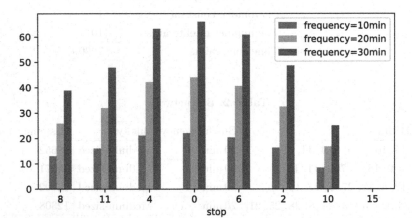

(b) Station 8 to station 15

Fig. 6. The number of passengers on the bus at each stop.

Finally, we compare Bus Route Selection algorithm with the baseline methods [8,9]. As illustrated in Fig. 7, our method achieves higher passenger flow, i.e., the maximum number of passengers being served, than the baseline methods [8,9] at varying time constraints.

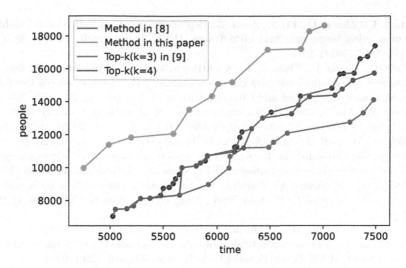

Fig. 7. Comparison of bus route selection methods.

5 Conclusion

In this paper, we propose a novel framework for generating bus routes and networks through exploiting the taxi GPS traces. The proposed framework comprises a method for determining candidate bus stations according to taxi passenger pick-up and drop-off records and a bio-inspired method for yielding bus routes and networks with adaptive strategies for deciding bus frequency and capacity. With a real-world data set of 3.4 million passenger delivery trips, we conduct simulations and show that our proposed framework clearly outperform baseline methods in terms of the number of people served.

References

1. Anagnostopoulos, C., Anagnostopoulos, I., Loumos, V., Kayafas, E.: A license plate-recognition algorithm for intelligent transportation system applications. IEEE Trans. Intell. Transp. Syst. **7**, 377–392 (2006)
2. Aslam, J., Lim, S., Pan, X., Rus, D.: City-scale traffic estimation from a roving sensor network. In: SenSys 2012 (2012)
3. Baaj, M.H., Mahmassani, H.: Trust: a lisp program for the analysis of transit route configurations. Transp. Res. Rec. **1283**(1990), 125–135 (1990)
4. Balan, R., Nguyen, K., Jiang, L.: Real-time trip information service for a large taxi fleet. In: MobiSys 2011 (2011)
5. Castro, P.S., Zhang, D., Chen, C., Li, S., Pan, G.: From taxi GPS traces to social and community dynamics. ACM Comput. Surv. (CSUR) **46**, 1–34 (2013)
6. Chang, H.W., Tai, Y.C., Hsu, J.Y.J.: Context-aware taxi demand hotspots prediction. Int. J. Bus. Intell. Data Min. **5**, 3–18 (2010)
7. Chen, C., et al.: iBOAT: Isolation-based online anomalous trajectory detection. IEEE Trans. Intell. Transp. Syst. **14**, 806–818 (2013)

8. Chen, C., Zhang, D., Li, N., Zhou, Z.: B-planner: planning bidirectional night bus routes using large-scale taxi GPS traces. IEEE Trans. Intell. Transp. Syst. **15**, 1451–1465 (2014)

9. Chen, C., Zhang, D., Zhou, Z., Li, N., Atmaca, T., Li, S.: B-planner: night bus route planning using large-scale taxi gps traces. In: 2013 IEEE International Conference on Pervasive Computing and Communications (PerCom), pp. 225–233 (2013)

10. Deng, W., Xu, J., Zhao, H.: An improved ant colony optimization algorithm based on hybrid strategies for scheduling problem. IEEE Access **7**, 20281–20292 (2019)

11. Dickens, M., Neff, J.: Apta 2011 Public Transportation Fact Book (2011)

12. Dorigo, M., Gambardella, L.: Ant colony system: a cooperative learning approach to the traveling salesman problem. IEEE Trans. Evol. Comput. **1**, 53–66 (1997)

13. Dorigo, M., Maniezzo, V., Colorni, A.: Ant system: optimization by a colony of cooperating agents. IEEE Trans. Syst., Man, Cyber., Part B, Cyber. **26**(1), 29–41 (1996)

14. Hartigan, J., Wong, M.: A k-means clustering algorithm (1979)

15. He, Q., et al.: A game-theoretical approach for user allocation in edge computing environment. IEEE Trans. Parallel Distrib. Syst. **31**, 515–529 (2020)

16. He, Q., et al.: A game-theoretical approach for mitigatingedge ddos attack. IEEE Trans. Dependable Sec. Comput. (2021)

17. Jerby, S., Ceder, A.: Optimal routing design for shuttle bus service. Transp. Res. Rec. **1971**, 14–22 (2006)

18. Jiang, C., Bhat, C., Lam, W.K.: A bibliometric overview of transportation research part b: methodological in the past forty years (1979–2019). Transp. Res. Part B-Method. **138**, 268–291 (2020)

19. Li, B., et al.: Hunting or waiting? discovering passenger-finding strategies from a large-scale real-world taxi dataset. In: 2011 IEEE International Conference on Pervasive Computing and Communications Workshops (PERCOM Workshops), pp. 63–68 (2011)

20. Li, B., He, Q., Chen, F., Jin, H., Xiang, Y., Yang, Y.: Auditing cache data integrity in the edge computing environment. IEEE Trans. Parallel Distrib. Syst. **32**(5), 1210–1223 (2020)

21. Li, X., et al.: Prediction of urban human mobility using large-scale taxi traces and its applications. Front. Comput. Sci. **6**, 111–121 (2011)

22. Liu, C.L., Pai, T., Chang, C.T., Hsieh, C.M.: Path-planning algorithms for public transportation systems. In: ITSC 2001. 2001 IEEE Intelligent Transportation Systems, Proceedings (Cat. No.01TH8585), pp. 1061–1066 (2001)

23. Liu, L., Andris, C., Ratti, C.: Uncovering cabdrivers' behavior patterns from their digital traces. Comput. Environ. Urban Syst. **34**, 541–548 (2010)

24. Liu, Y., Wang, F., Xiao, Y., Gao, S.: Urban land uses and traffic 'source-sink areas': evidence from GPS-enabled taxi data in shanghai. Landscape Urban Plan. **106**, 73–87 (2012)

25. Lu, H., Jin, L., Luo, X., Liao, B., Guo, D., Xiao, L.: RNN for solving perturbed time-varying underdetermined linear system with double bound limits on residual errors and state variables. IEEE Trans. Ind. Inf. **15**(11), 5931–5942 (2019)

26. Luo, X., Liu, H., Gou, G., Xia, Y., Zhu, Q.: A parallel matrix factorization based recommender by alternating stochastic gradient decent. Eng. Appl. Artif. Intell. **25**(7), 1403–1412 (2012)

27. Luo, X., Wang, D., Zhou, M., Yuan, H.: Latent factor-based recommenders relying on extended stochastic gradient descent algorithms. IEEE Trans. Syst., Man, Cyber. Syst. **51**(2), 916–926 (2019)

28. Luo, X., Xia, Y., Zhu, Q., Li, Y.: Boosting the k-nearest-neighborhood based incremental collaborative filtering. Knowl.-Based Syst. **53**, 90–99 (2013)
29. Luo, X., Zhou, M., Li, S., Wu, D., Liu, Z., Shang, M.: Algorithms of unconstrained non-negative latent factor analysis for recommender systems. IEEE Trans. Big Data **7**(1), 227–240 (2019)
30. Pan, G., Qi, G., Wu, Z., Zhang, D., Li, S.: Land-use classification using taxi GPS traces. IEEE Trans. Intell. Transp. Syst. **14**, 113–123 (2013)
31. Shanqing, Z.: Rationality and values of relevant indexes in "code for transport planning on urban road"(gb 50220–95). Transport Standardization (2011)
32. Shen, Y., Zhao, L., Fan, J.: Analysis and visualization for hot spot based route recommendation using short-dated taxi GPS traces. Information **6**, 134–151 (2015)
33. Stützle, T., Hoos, H.: Max-min ant system. Future Gener. Comput. Syst. **16**, 889–914 (2000)
34. Szeto, W.Y., Wu, Y.: A simultaneous bus route design and frequency setting problem for Tin Shui Wai, Hong Kong. Eur. J. Oper. Res. **209**, 141–155 (2011)
35. Tang, J., Liu, F., Wang, Y., Wang, H.: Uncovering urban human mobility from large scale taxi GPS data. Phys. A-stat. Mech. Appl. **438**, 140–153 (2015)
36. Wang, F.: Driving into the future with its. IEEE Intell. Syst. **21**, 94–95 (2006)
37. Wang, F.: Parallel control and management for intelligent transportation systems: Concepts, architectures, and applications. IEEE Trans. Intell. Transp. Syst. **11**, 630–638 (2010)
38. Wu, D., Luo, X., Shang, M., He, Y., Wang, G., Zhou, M.: A deep latent factor model for high-dimensional and sparse matrices in recommender systems. IEEE Trans. Syst., Man, Cyber. Syst. **51**(7), 4285–4296 (2019)
39. Xia, X., Chen, F., He, Q., Grundy, J., Abdelrazek, M., Jin, H.: Cost-effective app data distribution in edge computing. IEEE Trans. Parallel Distrib. Syst. **32**, 31–44 (2021)
40. Xia, X., Chen, F., He, Q., Grundy, J., Abdelrazek, M., Jin, H.: Online collaborative data caching in edge computing. IEEE Trans. Parallel Distrib. Syst. **32**, 281–294 (2021)
41. Yuan, J., Zheng, Y., Xie, X., Sun, G.: T-drive: enhancing driving directions with taxi drivers' intelligence. IEEE Trans. Knowl. Data Eng. **25**, 220–232 (2013)
42. Yuan, L., et al.: Coopedge: a decentralized blockchain-based platform for cooperative edge computing. In: Proceedings of the Web Conference 2021 (2021)
43. Zhao, F., Zeng, X.: Optimization of transit route network, vehicle headways and timetables for large-scale transit networks. Eur. J. Oper. Res. **186**, 841–855 (2008)
44. Zhu, L., Yu, F., Wang, Y., Ning, B., Tang, T.: Big data analytics in intelligent transportation systems: a survey. IEEE Trans. Intell. Transp. Syst. **20**, 383–398 (2019)

Community Influence Maximization Based on Flexible Budget in Social Networks

Mengdi Xiao[1,2], Peng Li[1,2(✉)], Weiyi Huang[1,2], Junlei Xiao[1,2], and Lei Nie[1,2]

[1] College of Computer Science and Technology, Wuhan University of Science and Technology, Wuhan, Hubei, China
lipeng@wust.edu.cn
[2] Hubei Province Key Laboratory of Intelligent Information Processing and Real-Time Industrial System, Wuhan, Hubei, China

Abstract. The influence maximization (IM) problem is a vital issue in social networks. In community structure, community influence maximization (CIM) chooses the seed nodes based on the characteristics of the community structure instead of blindly selecting seed nodes from the entire network. However, it depends on the community size, which results in high influence nodes not being selected due to a lack of budget. In this paper, we propose a budget allocation strategy for the CIM problem. To solve the problem of less influence spread in sparse community structure, we propose a community influence maximization algorithm based on a flexible budget and adopt the reverse influence sampling (RIS) approach to sample the network structure, which reduces the time complexity of the greedy algorithm. Then, we consider the imbalance of influence expansion between communities, and we propose a balanced community influence maximization algorithm, which maintains the relative balance of the influence spread ratio between communities. In addition, we analyze the time complexity of our proposed algorithms and give a theoretical guarantee. Finally, we conduct extensive experiments on three real datasets. Compared with other baseline algorithms, the results show that the proposed algorithms have a good performance in terms of influence maximization and community influence balance.

Keywords: Influence maximization · Reverse influence sampling · Social network

1 Introduction

The problem of influence maximization in social networks has attracted wide attention in recent years, and it has many applications in various aspects, such as rumor control and viral marketing. However, some influence maximization applications (such as public service advertising, safety education) are unbiased,

---

P. Li—This work is partially supported by the NSF of China (No. 61802286).

H. Gao and X. Wang (Eds.): CollaborateCom 2021, LNICST 406, pp. 534–553, 2021.
https://doi.org/10.1007/978-3-030-92635-9_31

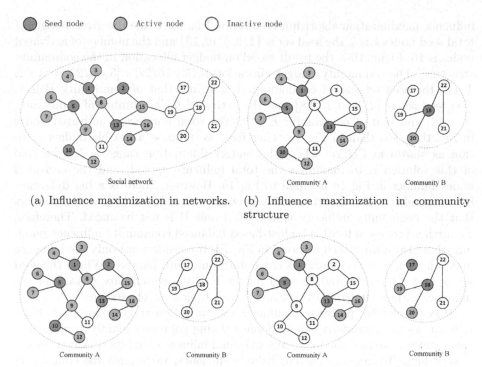

● Seed node   ● Active node   ○ Inactive node

(a) Influence maximization in networks.

Social network

(b) Influence maximization in community structure

Community A   Community B

(c) Flexible budget based community influence maximization

Community A   Community B

(d) Flexible budget based balanced community influence maximization

Community A   Community B

Fig. 1. The distribution of influence under the four conditions.

the balance of influence has become an issue that needs consideration. How to maximize the influence and ensure the balance of influence diffusion has become a critical problem.

Influence maximization aims to select seed nodes S with the maximum influence spread in the social networks. In the traditional methods [8,9], they calculate the marginal influence gain of each node by traversing the entire network, rank the nodes with the influence spread gain in each iteration, and get the maximum influence coverage with seed set S after k iterations. However, the above process may differ somewhat based on its structural characteristics in a community structure. The relationships within communities are tight, and the relationships between communities are spares. So the influence maximization method works within communities, but it does not work well between communities. Therefore, we consider influence maximization in community structure, and we use the budget allocation method between communities.

In community structure, the existing solution of the influence maximization problem is to maximize the influence spread with the predefined budget without considering the network balance. For example, as shown in Fig. 1, there are 22 nodes in the network, which can be divided into two communities A and community B with 16 nodes and 6 nodes, respectively. The result of the traditional influence maximization algorithm is shown in Fig. 1a. While the given budget of total seed nodes k is 5, the seed set is $\{1, 2, 5, 10, 13\}$ and the number of activated nodes is 15. Figure 1b is the result based on budget allocation in the community structure. The community budget allocation is $[5 * 16/22] : [5 * 6/22] = 4 : 1$. Then, the seed set size of community A is 4, and that of community B is 1. The seed set is $\{1, 5, 10, 13, 18\}$ and the activated nodes number of community A is 12, and that of community B is 2. The total number of activated nodes in Fig. 1b is less than in Fig. 1a. Therefore, we propose a flexible budget solution as shown in Fig. 1c, where the budget fluctuation range is 1. The goal of this solution is to maximize the total influence spread, and the activated nodes number in Fig. 1c is equal to Fig. 1a. However, there is a big difference in the number ratio of active nodes between communities A and B. It shows that the community influence spread of A and B is not balanced. Therefore, we further propose a flexible budget-based balanced community influence maximization algorithm, which is shown in Fig. 1d. It considers not only the influence maximization but also considers community influence balance. Where the seed set is $\{1, 5, 13, 17, 18\}$ and the real influence of the community is 14, and the influence ratio is $(9/16):(4/6) = 27:32$, which is close to the balance.

How to solve the problem of influence maximization and community balance is what we are considering now. Some existing solutions mainly traverse the entire network to find the maximum marginal influence, and its time complexity is very high. To improve the algorithm's efficiency, we sample the community structure to obtain approximate influence spread estimates. Although our algorithm reduces the time complexity, it causes new problems. First, the fairness of the budget allocation method needs to be guaranteed. Second, we still need to maximize the whole network influence. Finally, community influence balance and influence maximization need to be weighed.

To solve the above problems, we propose a community influence maximization solution based on a flexible budget. First of all, to ensure the algorithm's fairness, we propose a simple budget allocation strategy based on community size. Second, we propose a flexible budget-based community influence maximization algorithm based on the reverse influence sampling (RIS) method. The RIS algorithm based on community structure estimates the community influence diffusion with ensuring computational efficiency. It selects the seed nodes iteratively with a strategy based on a flexible budget to obtain the maximum influence. Then, we propose a community balance influence algorithm based on RIS to solve the imbalance of community influence caused by a flexible budget. It selects seed nodes iteratively to get a trade-off between influence maximization and influence balance. We evaluate the performance of our proposed solution based on three real-world datasets. Extensive experiments show that the performance of our proposed algorithm is very close to that of traditional influence maximization, and the balance of our proposed algorithm is better than that of the existing algorithms.

In summary, we have made the following contributions:

- We propose an influence maximization algorithm for a flexible budget based on reverse influence sampling to maximum influence within a certain budget fluctuation range.
- Considering the balance of community influence distribution, we improve the above algorithm and propose a balanced influence maximization algorithm with a flexible budget. We analyze the time complexity of our proposed algorithm and give a theoretical guarantee.
- We evaluate our proposed algorithm in three real-world social datasets, and the results show our proposed algorithms have outstanding performance in both influence maximization and influence balancing, respectively, compared with other algorithms.

2 Related Work

Domingos and Richardson et al. [6] are the first to propose the problem of maximizing the influence in social networks. They transform the problem of viral marketing into the problem of maximizing the influence in a formulaic form. Then Kempe et al. [8] study the nature of the influence maximization problem. They propose that the problem is NP-hard, and they find that the greedy algorithm can get a $1 - 1/e$ approximate ratio. The greedy algorithm has very high time complexity due to its dependence on Monte Carlo simulation and the scale of the network. To reduce the time cost of the greedy algorithm, Leskovec et al. [9] propose a lazy-forward method (CELF), which is based on the diminishing marginal influence gain property in each round of node selection process. Talukder A. et al. [17] propose a knapsack-based reverse influence maximization model, where a linear threshold model is used in reverse order to estimate the minimum node cost required for nodes to be activated. Tang Jing et al. [18] propose a new online influence maximization algorithm, where a pause mechanism for the lack of interactivity and flexibility is provided due to the long-running time of the algorithm. The mode of information propagation in an undirected graph is studied by Schoenebeck Grant et al. [14]. Sun, Lichao and Huang et al. [16] study the multi-round influence maximization problem of adaptive and non-adaptive. Becker and R. et al. [3] study the influence distribution and seed node selection in attribute, community, and preference networks, and they weigh the relationship between influence maximization and balance.

Now there are some studies on the influence maximization caused by node attributes or relationships under different scenarios. Chen et al. [5] studies the selection of influence seed nodes when variety of marketing strategies are mixed, Banerjee et al. [1,2] study the complementary and competitive characteristics between projects and propose utility-driven influence maximization model. In the above papers, the marketing scenarios and the relationship between nodes in the network are considered. They all combine the maximization of influence with the possible scenarios in the network but do not consider the structural characteristics of the network. Stoica and Ana-Andreea et al. [15] propose a method to solve the fairness problem in the maximization of social influence. Yang, Yu and Mao *et al.* [19] model the problem of maximizing the sustainable influence, and they propose the coordinate descent framework to consider the project discount within the budget scope. Guo *et al.* studied the influence maximization method in the community structure in [7], and proposed a simplified pipage rounding method based on guaranteeing the approximate ratio, which reduces the complexity of IMCB-Framework further. Lin *et al.* [11] proposes a RIS-based attribute sampling algorithm to tradeoff the maximization of influence and the balance of attributes. Inspired by it, we study the relationship between influence maximization and influence balance on non-attributed networks. In a non-attribute network, community structure can be easily discovered based on the tightness of the relationships between nodes.

Unlike the point of focus in the above papers, this paper focuses on the tradeoff between maximization of influence and balance based on the flexibility of community budget in community structure. In other words, we study the distribution of an unbiased advertisement across communities within the community structure. We propose a community influence maximization algorithm based on upper budget bounds and a balancing algorithm based on upper and lower budget bounds to solve the above problems.

3   System Model and Problem Formulation

In this section, we model the problem and propose a formulaic solution. We summarize the frequently used notations in Table 1.

Table 1. Frequently used notations.

Symbol	Description
$G = (V, E)$	Network G, nodes set V, edges set E
n, m	n is the number of nodes in V, m is the number of edges in E
$N(v_i)$	The neighbor nodes set of node v_i, $v_i in V$
$p_{i,j}$	The probability that node v_i active node $v_j \in N(v_i)$
C, C_i	Community partition in G, C_i is the i-th community in C
k	k is the size of seed nodes
$\sigma(S)$	The number of activated nodes by seed set S
γ	Weighting parameter
$R(v_i)$	The RR set of node v_i

3.1   System Model

We formalize the social network as graph $G = (V, E)$. $V = \{v_1, v_2, ..., v_n\}$ represents the nodes set in the network, where v_i denotes the i-th user node. E represents the edges set in the network, where $e_{i,j} \in E$ denotes the relationship between node v_i and its neighbor node v_j. We define social network nodes size as $n = |V|$, edges size $m = |E|$. In any two nodes v_i, v_j in the network, if there is an edge $e_{i,j}$ between them, it means that node v_i can active its neighbor node v_j with probability $p_{i,j}$. The state of node v_j in graph can be defined as a_i, if node v_i is actived, then $a_i = 1$, otherwise $a_i = 0$. Actived nodes have the potential to activate its neighbor nodes in the next iteration.

In social networks, information propagation follows the influence propagation model. The main propagation process of the influence propagation model is as follows: (1) Given the seed node S and the network graph G, the state of all nodes are unactived $(a_i = 0, \forall i)$ in the network. (2) In the initial time $t = 0$, the nodes v_i of the seed set S are changed to $a_i = 1$. (3) For $\forall t > 0$, all activated nodes in V try to activate their neighbor nodes in a certain probability. Node v_i in time $t - 1$ was not activated and it is activated in t, then in time $t + 1$ node v_i will try to activate the v_j where $v_j \in N(v_i)$ with probability $p_{i,j}$. (4) Iterative spread until there is no active node in the network.

There are two classical influence propagation models in social networks: the independent cascading (IC) and linear threshold (LT) models. Every node has only one chance to activate its neighbor nodes in the IC model, and every node's activation behavior is independent and unrelated. Different from the IC model, the activation behavior of the nodes in the LT model is related. The node v_i activates its neighbor node v_j if and only if $p_{i,j} \geq \delta_j$ in the IC model, where δ_j represents the activation threshold of node v_j. However, in LT model, the node v_j will be activated while $\sum_{v_i \in N(v_j)} p_{i,j} \geq \delta_j$. If the node v_j is not jointly activated by its neighbors at time t, v_j will try to be activated again while any other neighbor node is activated at any time $T > t$.

In this paper, we adopt the trigger model as our influence propagation model, which is based on the independent cascade model and the linear threshold model. Given a directed network graph $G = (V, E)$, any node $v_i \in V$ has a trigger distribution $D(v_i)$, which is the probability distribution based on the subsets of incoming neighbors $N(v_i)$, the trigger set $T(v_i)$ is obtained by randomly sampling $D(v_i)$. Trigger model can be described as: for the trigger set of node v_i, delete any incoming edge from the node not in the trigger set $T(v_i)$ to get the trigger sub-graph G', and activate any node belonging to the seed set S to get the activated nodes set $\sigma(S)$.

Classic influence maximization problem is to find a seed set S from the nodes set V, where $k = |S|$ represents the size of S, the influence of seed set S is the number of nodes activated by S, which is expressed as $\sigma(S)$. The influence of seed set S can be formulated as $\sigma(S) = \sum_{v_i \in V} a_i$, then influence maximization can be formulated as $S^* = \arg \max \sigma(S)$.

3.2 Problem Formulation

In the social networks, there is usually a community structure in Graph G. The node connections within the community are relatively close, while the node connections between communities are relatively sparse. Therefore, we use the classical community partitioning algorithm to divide the entire social network G into t disjoint community subgraphs, which is defined as $C = \{C_1, C_2, ..., C_t\}$.

We split the problem of influence maximization in social networks into two sub-problems, the problem of community influence maximization and the problem of budget allocation. The community budget is defined as $B = \{b_1, b_2, ..., b_t\}$ in the community structure, and the total budget is $k = \sum_{b_i \in B} b_i$. Without loss of generality, we define the cost of each seed as the same and as a unit cost, then the budget $b_i = |S_i|$ represents the seed set size of the i-th community C_i. Based on the budget b_i, we should select the seed set with the maximum influence spread for each community. The influence spread of seed set S_i in community c_i can be expressed as $\sigma_{C_i}(S_i) = \sum_{v_i \in C_i} a_i$. Therefore, the community influence maximization problem can be formulated as:

$$S^* = \arg\max \sum_{\forall C_i \in C} \sigma_{C_i}(S_i) \tag{1}$$

In this paper, we study the problem of community influence maximization based on the flexible budget.

Definition 1 *(Flexible Budget based Community Influence maximization, FBCIM Problem). Given community subgraph $C = \{C_1, C_2, ..., C_t\}$ and budget allocation B. The FBCIM problem is to find a flexible seed budget allocation strategy, where the number of joint active nodes is maximum.*

In Fig. 1, we can see that we solve the FBCIM problem may lead to the unbalanced distribution of influence spread in the community. Therefore, we define the flexible budget based balanced community influence maximization (FBBCIM) problem as follow.

Definition 2 *(Flexible Budget based Balanced Community Influence maximization, FBBCIM Problem). Given a graph G and community distribution subgraph $C = \{C_1, C_2, ..., C_t\}$ and budget allocation B. The FBBCIM problem is to find a flexible seed budget allocation strategy, where the influence maximization and community balance are considered together.*

We use $Q(S)$ represent the set of the ratio of influence and community size in each community, where $Q(S) = \{\sigma_{C_1}(S_1)/n_1, \sigma_{C_2}(S_2)/n_2, ..., \sigma_{C_t}(S_t)/n_t\}$, n_i represents the size of community C_i. We adopt the variance $Var(Q(S))$ of influence proportion as the scale of balance evaluation. At the same time, some high-diffusion communities may allocate few seed nodes, while low-diffusion communities may allocate too many seed nodes. To avoid excessive balanced influence, we have a tradeoff between maximizing influence and balancing influence. According to the above definition, the FBBCIM problem can be expressed as a dual objective optimization problem of maximizing influence and minimizing variance. Therefore, FBBCIM problem can be formulated as:

$$S^* = \arg\max \sum_{\forall C_i \in C} \sigma_{C_i}(S_i) - \gamma n Var(Q(S)) \tag{2}$$

Where γ is the parameter, which describes the importance of variance in the influence function. n is the size of the graph, and we use it to scale up the variance value appropriately.

4   Our Solutions

4.1   General Solution

First, we need to study the rationality of budget allocation. We adopt a simple rounding method to allocate the seed budget based on the ratio of the community size and the size of the seed set. For t communities $C = \{C_1, C_2, ..., C_t\}$, the size of each community is n_i, the total budget is k, and the budget of each community is floating around kn_i/n. We use the community budget rounding method to allocate surplus budget. The steps of the community budget rounding method are as follows:

First, we need to study the rationality of budget allocation. We adopt a simple rounding method to allocate the seed budget based on the ratio of the community size and the size of the seed set. For t communities $C = \{C_1, C_2, ..., C_t\}$, the size of each community is n_i, the total budget is k, and the budget of each community is floating around kn_i/n. We use the community budget rounding method to allocate surplus budget. The steps of the community budget rounding method are as follows:

(1) Let b_i be the integer part of $k * n_i/n$ and let f_i be the fractional part of $k * nc_i/n$. We use $F = \{f_1, f_2, ..., f_t\}$, and $B = \{b_1, b_2, ..., b_t\}$
(2) We sort each element in F from largest to smallest, and the new ordered set denoted by F'.
(3) Let $q = k - \sum_{b_i \in B} b_i$, which represents the number of unallocated seed nodes.
(4) We increment the integer part by 1, that is $b_i + 1$ if f_i is in the top q of F'.

To solve the FBCIM problem, we propose a greedy algorithm with diminishing returns of internal community influence as given in Algorithm 1. We select the seed nodes with the maximum community influence for t communities with a budget allocation of B. In each iteration of each community, we select the node with the maximum influence to be added into the seed set (lines 3–5). To avoid the excessive crossover problem of influence diffusion cover between seed nodes, we sort the marginal influence gain rather than the influence of nodes (line 4), which can be expressed as $\sigma_{C_i}(v|S_i) = \sigma_{C_i}(S_i \cup \{v\}) - \sigma_{C_i}(S_i)$.

Algorithm 1. Algorithm for community influence maximization

Input: social network $G(V, E)$, community $C = C_1, C_2, ..., C_t$, seed budget allocation $B = \{b_1, b_2, ..., b_t\}$
Output: seed set S

1: Initialize: $S \leftarrow \emptyset$
2: **for** $i = 1, ..., t$ **do**
3:     **for** $j = 1, 2, ..., b_i$ **do**
4:         $v = \arg\max \sigma_{c_i}(v|S_i)$
5:         $S_i \leftarrow S_i \cup \{v\}$
6:     Let $S \leftarrow S \cup S_i$
7: return S

We analyze the time complexity of the Algorithm 1. In Algorithm 1, seed nodes are selected for t communities according to their budget settings, and b_i seeding rounds are selected in each community. The marginal influence of nodes in each iteration is calculated. In the greedy algorithm, we most often use the Monte Carlo simulation method to estimate the influence spread of the seed set. In Monte Carlo simulation, the selected seed nodes simulate the information propagation process from the edge relationship according to the corresponding propagation model. The process of a simulation is traversing the edge of the network once, and the time complexity is $O(n_i m_i)$. Assuming we simulate H propagations for each community, the time complexity of Algorithm 1 can be expressed as $O(H \sum_{C_i \in C} b_i n_i m_i) = O(Htknm)$.

4.2   FBCIM Algorithm

It is well known that Monte Carlo greedy algorithm is not scalable and has high computational costs. In this paper, we propose a scalable algorithm based on Reverse Influence Sampling (RIS). It can give an approximate estimate of the influence and has a theoretical guarantee. We use the critical concept Reverse Reachable (RR) set to describe RIS.

Definition 3. *Reverse Reachable (RR) set:* *Given a graph $G = (V, E)$, the RR set $R(v)$ is obtained by sampling the trigger set $T(v)$ with a certain probability.*

According to Definition 3, any node in the RR set $R(v)$ can propagate influence to node v. Any node in the RR set $R(v)$ is activated, then v is activated. Therefore, the reverse reachable (RR) set represents the set of any node when the node is not selected.

Reverse Influence Sampling (RIS) algorithm can be described as follows: First, we randomly sample the trigger set $T(v)$ of any node v to obtain its reverse reachable set $R(v)$, where $R(v)$ represents the nodes set that can activate node v. Then, we sample enough reverse reachable sets to obtain the approximate influence results. When we sample more RR sets, the seed set S will cover more RR sets, the number of activated nodes will be more. Finally, we use a greedy algorithm to calculate the influence spread and get the seeds set with the maximum influence gain.

Lemma 1. *Given disjoint community partitioning $C = \{C_1, C_2, ..., C_t\}$, a RR set $R(v)$, the number of nodes n_i in community C_i, the influence of S can be estimated as: $\mathbb{E}[\sigma(S)] = \sum_{C_i \in C, S_i \in S} n_i \cdot Pr[S_i \cap R(v) \neq \emptyset | v \in C_i]$.*

Proof. For the seed set S_i of community C_i, the proportion that S_i covers the community C_i is equal to the probability that S_i activates random v. The influence of S_i can be estimated as: $\mathbb{E}[\sigma_{C_i}(S_i)] = n_i \cdot Pr[S_i \cap R(v) \neq \emptyset]$, where $Pr[S_i \cap R(v) \neq \emptyset]$ represents the probability that the intersection of the seed set S_i and the RR set $R(v)$ of a random node $v \in C_i$ is not empty. Therefore, the influence of S can be estimated as: $\mathbb{E}[\sigma(S)] = \sum_{C_i \in C, S_i \in S} n_i \cdot Pr[S_i \cap R(v) \neq \emptyset | v \in C_i]$.

For the CIM problem, we propose the Flexible Budget-based Community Influence Maximization (FBCIM) algorithm based on RIS. Our method includes two processes: RR set sampling and seed selection.

RR Set Sampling. First, we generate the RR sets in all community structures. In each community, we sample θ_i nodes and generate θ_i RR sets. Theoretically, the more RR sets are sampled, the higher the approximation of influence results. However, when the number of nodes in the network is relatively large, the number of RR sets is also significant, it will lead to the more calculation cost. So we need to calculate a lower bound on the number of RR sets, that is, to obtain the nearest similar influence result with fewer RR sets.

Unlike traditional influence propagation, node v is activated if and only if the intersection of $R(v)$ and S is not an empty set. In RIS sampling, the activation probability of a node is the probability that the seed set S contains the trigger set of any node. Let $\mathbb{E}$ represent expectation, $\sigma(S_i)$ represent the number of active nodes in the reverse influence sampling model, θ_i and represent the number of RR sets sampled by the community C_i. Based on Lemma 1, the influence of seed set S can be expressed as follows:

$$\sigma(S^*) = \sum_{\forall C_i \in C} \frac{n_i}{\theta_i} \mathbb{E}[\sigma_{C_i \in C}(S_i)] \tag{3}$$

Lemma 2. *The expected influence $\sigma(S)$ is a monotone and submodular function under RIS model.*

Proof. From Lemma 1, we know the influence of S can be expressed as $\sigma(S) = \sum_{C_i \in C, S_i \in S} \sigma_{C_i}(S_i)$. Let $f(S) = \sigma(S)$, $g(S_i) = \sigma_{C_i}(S_i)/n_i = Pr[S_i \cap R(v) \neq \emptyset]$. In the RIS algorithm, we select seed nodes according to the maximum marginal revenue of S covering R, that is, the revenue of nodes selected in each round is diminishing compared with that of the last round. However, with the increase in the number of seed nodes, the overall revenue is increasing. So, $g(S_i)$ is nonnegative, monotonic, and submodular. Because $f(S) = \sum_{S_i \in S} n_i \cdot g(S_i)$, the function $f(S)$ is nonnegative, monotonic, and submodular.

Lemma 3 (Chernoff Bound). *Let $X_i \in [0,1]$ be θ i.i.d random variables with a mean μ. For any $\delta > 0$, $Pr[|\sum X_i - \theta\mu| \geq \delta \cdot \theta\mu] \leq 2\exp(-\frac{\delta^2}{2+\delta} \cdot \theta\mu)$.*

Lemma 3 is the classical Chernoff Bound theoretical analysis. We use it to get the lower bound of θ_i in every community. The low bound means the minimum number of RR sets what the influence estimation and approximation assurance needs.

Lemma 4. *Given community subgraph $C = \{C_1, C_2, ..., C_t\}$, the number of community nodes is n_i, if θ_i satisfies:*

$$\theta_i \geq \frac{n_i(\epsilon + 2)l \log 2n_i}{OPT_i \cdot \epsilon^2} \tag{4}$$

Then, for any set $S_i \in C_i$, the following inequality $|n_i Pr[S_i \cap R \neq \emptyset] - E[\sigma(S_i)]| < \epsilon \cdot OPT_i$ holds with at least $1 - 1/n_i^l$ probability.

Proof. In community C_i, let x_i represents $Pr[S_i \cap R \neq \emptyset]$, which is the probability that the intersection of the seed set with an random RR set is not empty. $E[\sigma(S_i)]$ is the expected influence spread of seed set S_i, μ_i represents the expected activation probability of the node, it can be denoted as $\mu_i = E[\sigma(S_i)]/n_i$. According to Lemma 3 (Chernoff bound), we have:

$$Pr[|n_i x_i - E[\sigma(S_i)]| \geq \epsilon \cdot OPT_i]$$

$$= Pr[|\theta_i x_i - \theta_i \mu_i| \geq \frac{\epsilon \cdot OPT_i}{n_i \mu_i} \cdot \theta_i \mu_i]$$

$$\leq 2 \exp(-\frac{\epsilon^2 \cdot OPT_i^2}{n_i(2n_i \mu_i + \epsilon \cdot OPT_i)} \cdot \theta_i)$$

$$\leq 2 \exp(-\frac{\epsilon^2 \cdot OPT_i}{n_i(2+\epsilon)} \cdot \theta_i) \leq 1/n_i^l$$

Therefore, when $\theta_i \geq \frac{n_i(\epsilon+2)l \log 2n_i}{OPT_i \cdot \epsilon^2}$, the inequality $|n_i x_i - E[\sigma(S_i)]| \geq \epsilon \cdot OPT_i$ holds with probability $1/n_i^l$. Lemma 4 is proved.

Seed Selection. We further optimize Algorithm 1 and propose a FBCIM algorithm based on a flexible budget. Given each community budget and a float parameter, the algorithm allows the actual budget to fluctuate within this parameter. In FBCIM algorithm, the flexible budget of community C_i is defined as $b_i' \leq b_i + \lambda$, where b_i is the budget allocated based on community size, and λ is a flexible parameter, which stipulates the scope of budget fluctuations. As shown in Algorithm 2, under the constraint of total budget k, the node with the maximum influence is iteratively selected to seed set, and the iteration is carried out until the size of seed set within each community is equal to the maximum budget fluctuation upper bound. This seed selection method ensures that the seed nodes selected in each round are optimal. That is, the seed node selection process is a monotonic submodule.

Algorithm 2. FBCIM

Input: social network $G(V, E)$, community subgraph $C = \{C_1, C_2, ..., C_t\}$, seed budget allocation $B = \{b_1, b_2, ..., b_t\}$, θ_i RR sets in random community C_i
Output: seed set S
1: Initialize: $S = \{S_1, S_2, ..., S_t\} \leftarrow \emptyset$
2: **for** $j = 1, 2, ..., k$ **do**
3:      **for** $i = 1, ..., t$ **do**
4:          **if** $|S_i| \leq b_i + \lambda$ **then**
5:              $v^i = \arg\max \sigma_{c_i}(v|S_i)$
6:      Let $\sigma_{v^z}(v^z|S_z)$ be the max in $\{\sigma_{v^1}(v^1|S_1), \sigma_{v^2}(v^2|S_2), ..., \sigma_{v^t}(v^t|S_t)\}$
7:      $S_z \leftarrow S_z \cup \{v^z\}$
8: **return** S

In Algorithm 2, we find the nodes with the most significant influence within each community (Lines 4–5), rank the influence of these nodes, select the nodes with the most influence, and allocate the budget to the community (Line 6). To ensure the algorithm's fairness, we set the size of the community seed set according to the upper bound of the given community budget (Line 4). The above process iterates k rounds and assigns the node with the most significant influence to the community as the seed node (Lines 2–7).

4.3 FBBCIM Algorithm

To alleviate the influence imbalance caused by influence maximization in Algorithm 2, we propose a Flexible Budget-based Balanced Community Influence Maximization (FBBCIM) algorithm as shown in Algorithm 3. Different from Algorithm 2, we restrict the upper and lower bounds of the community budget. The lower bound is set to avoid ignoring the fairness of the community budget for obtaining a more balanced influence. In addition, because the proportion of community influence changes during the seed selection process, we propose redistributing the community budget according to the change of seed size and corresponding influence. The algorithm process is as follows:

Algorithm 3. FBBCIM

Input: social network $G(V, E)$, community subgraph $C = \{C_1, C_2, ..., C_t\}$, seed budget allocation $B = \{b_1, b_2, ..., b_t\}$, θ_i RR sets in random community C_i
Output: seed set S

1: Initialize: $S = \{S_1, S_2, ..., S_t\} \leftarrow \emptyset$
2: **for** $i = 1, ..., t$ **do**
3:      **for** $j = 1, 2, ..., b_i + \lambda$ **do**
4:          $v = \arg\max \sigma_{C_i}(v|S_i)$
5:          $S_i \leftarrow S_i \cup \{v\}$
6:          $f(C_i, S_i) = \sigma_{C_i}(S_i)$
7:          **if** $|S_i| \geq (b_i - \lambda)$ & $|S_i| \leq (b_i + \lambda)$ **then**
8:              append $(j, f(C_i, S_i))$ to $list_i$
9: **if** $\sum_{C_i \in C} |S_i| = k$ **then**
10:      $S^* = \arg\max \sum_{\forall c_i \in C} \sigma_{C_i}(S_i) - \gamma n Var(Q(S))$
11: **return** S^*

First, we find the node with the most significant influence within the upper budget bounds of each community (Lines 3–5) and store the influence change list of upper and lower budget bounds (Lines 7–8). The generated community influence list is selected, and the nodes with the most significant total community influence and the minor total variance are sorted as the seed set (Lines 9–10). Different from Algorithm 2, Algorithm 3 restricts the lower bound of the community budget to avoid excessive transfer of influence.

The time complexity of Algorithm 2 and Algorithm 3 are analyzed as follows. The running process can be divided into two parts in these two algorithms: RR set sampling and seed nodes selection. In Lemma 4, we have got the number of RR sets per community, which is θ_i. Then, we construct the sample subgraph and θ inverse reachable set, the time complexity is $O(\sum \theta_i m_i) = O(\sum_{C_i \in C} b_i m_i \log n_i / \epsilon^2)$. In the node selection stage, we calculate the number of occurrences of each node in the RR set, and each community iterates b_i times, then the time complexity can be expressed as $O(\sum b_i n_i)$. The total time complexity of the algorithm is $O(\sum_{C_i \in C} b_i(m_i + n_i) \log n_i / \epsilon^2) = O(tk(m + n) \log n / \epsilon^2)$, which is much less than the time complexity of Algorithm 1.

5   Performance Evaluation

In this section, we run our proposed algorithms on several real networks and study their distribution in the networks. By comparing our proposed algorithm with some other algorithms, we demonstrate the effectiveness and efficiency of the proposed algorithms.

5.1   Datasets and Parameters Setting

We run our experiment on the following three real-world networks with statistics summarized in Table 2. The Facebook [10] dataset is collected from survey participants using this Facebook app, which is from Stanford Network Analysis Project. LastFM [13] dataset is a music communication platform. The Government [12] dataset is one category of Facebook pages.

Table 2. Dataset statistics.

Dataset	n	m	Direction	Density
Facebook	4039	88234	Undirected	0.001
LastFM	7624	27806	Undirected	0.001
Government	7057	89455	Undirected	0.0036

In the given network datasets above, we use a community discovery algorithm to mine the community structure. In this paper, we use Louvain [4] algorithm to find the community structure of the networks. This algorithm calculates the benefit of modularity from nodes clustering and obtains the community structure by maximizing the network's modularity. We obtained 12, 15, and 19 communities through the community partitioning algorithm in the Facebook, LastFM, and Government datasets, respectively.

In our experiment, we set the propagation probability of the independent cascading model as 0.1. To ensure the accuracy of the influence calculation, we set the Monte Carlo simulation number to 1000, which is used to estimate the influence of the greedy algorithm. In RIS sampling, we set the node sampling probability as 0.1 and set $\varepsilon = 0.1$. In addition, we set the parameter γ in algorithm FBCIM and algorithm FBBCIM to 1. In all experiments, we set the total budget k from 20 to 100, where the unit range is 20. For each experiment, we run five times under the same settings to get the influence estimate of the algorithm.

5.2 Comparison of Algorithms and Metrics

To evaluate our proposed algorithms, we test our proposed algorithms and the baseline algorithms as following:

(1) CELF: CELF is an improved algorithm of the CELF algorithm, which reduces certain time complexity by simplifying the marginal revenue comparison process.
(2) degree: For the heuristic algorithm of degree centrality, the nodes with high degree centrality are selected as seed nodes. In this experiment, we use the DegreeHeuristic algorithm in combination with Algorithm 1 to obtain the comparative data.
(3) FBCIM: Our proposed Flexible Budget-based Community Influence Maximization algorithm considers reasonable budget allocation based on maximizing influence.
(4) FBBCIM: Our proposed Flexible Budget-based Balanced Community Influence Maximization algorithm takes balance into account based on our FBCIM algorithm.

To measure the performance of the algorithm, we use the four criteria:

(1) Expected influence spread: the total number of activated nodes in a graph with seed set S.
(2) Variance of the influence spread ratio between communities: it is denoted by $\gamma n Var(S)$, where n represents the size of the network. This parameter is added to enlarge the variance value. This evaluation metric represents the balance of community influence spread.
(3) The trade-off between maximizing influence spread and community balance: It is denoted by $\sum_{c_i \in C} \sigma_{c_i}(S_i) - \gamma n Var(S)$, where $\sum_{c_i \in C} \sigma_{c_i}(S_i)$ is the influence spread, and $\gamma n Var(Q(S))$ is the community balance.
(4) Running time: The time it takes to run the algorithm. The unit of time is second.

5.3 Evaluation Results

Figure 2 shows the performance of our proposed algorithms on the expected influence spread compared with two other algorithms. Expected influence spread is the most direct way to demonstrate the algorithmic effectiveness in influence maximization. The larger the result of the algorithm, the higher the quality of

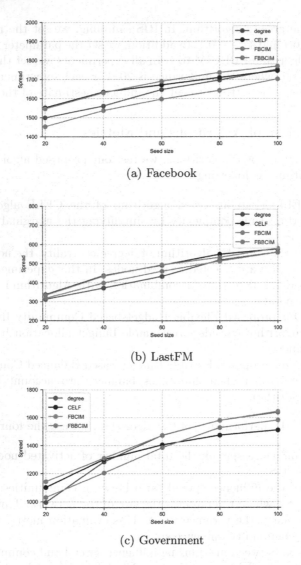

(a) Facebook

(b) LastFM

(c) Government

Fig. 2. Expected influence spread in the three different networks.

the algorithm. In all three subgraphs, FBCIM consistently outperforms all other algorithms. The reason for this is the flexible budget strategy of the FBBCIM algorithm. The performance of the FBBCIM algorithm is similar or slightly less than that of the other algorithms. This is because the FBBCIM will lose some influence to balance the community influence ratio, while the other three algorithms all seek to maximize influence spread. Therefore, it can be concluded from these results that our proposed FBCIM algorithm has outstanding performance in influence maximization, and the FBBCIM algorithm also performs well.

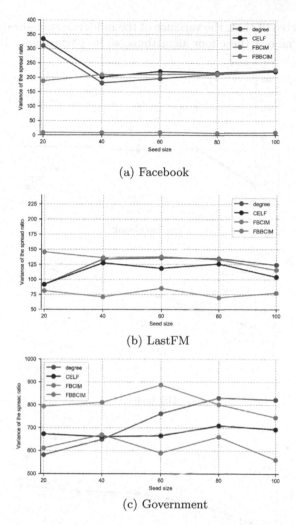

(a) Facebook

(b) LastFM

(c) Government

Fig. 3. Variance of the influence spread ratio between communities in the three different networks.

Figure 3 compares the variance of the influence ratio generated by the four algorithms in the three networks. To show more directly the difference in the variance of influence proportions, we multiply each result by the network size and scaling parameters γ. In this experiment, we set the parameter $\gamma = 10$. In the three subfigures, the variance of community influence ratio in FBCIM, CELF, and Degree algorithms is similar, and higher than that of FBBCIM. The reason is that seed nodes differ in their ability to disseminate information within the community, even if there is a relatively equitable budget allocation strategy. The FBBCIM algorithm that we propose reduces the difference in the proportion of community influence to minimize the difference in the propagation ability of nodes. As depicted of that, the variance of the influence spread ratio in FBBCIM is the smallest among these four algorithms, so its performance is the best.

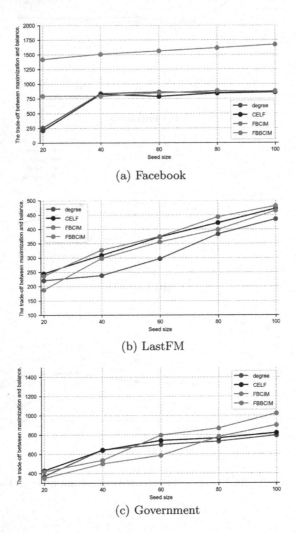

(a) Facebook

(b) LastFM

(c) Government

Fig. 4. The trade-off between community influence maximization and community balance in the three different networks.

Figure 4 draws performance comparison of the trade-off between community influence maximization and community balance. Here, both the algorithm's influence scale and community balance are taken into account. As can be seen from the figure, the performance of FBBCIM proposed by us is better than the other three algorithms. Moreover, the FBCIM algorithm is not the worst among the three networks. Therefore, when community balance is valued, the FBBCIM algorithm is our best choice.

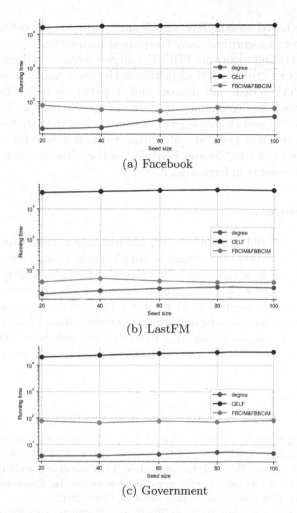

(a) Facebook

(b) LastFM

(c) Government

Fig. 5. Running time in the three different networks.

Figure 5 shows the running times of the four algorithms in three datasets. In the earlier algorithm analysis, we have known that the time complexity of FBCIM and FBBCIM is the equivalent, so their running time is coincident. Therefore, in this figure, we describe both FBCIM and FBBCIM with one line. From the three subfigures, the degree algorithm has the shortest running time in the three subfigures because the degree algorithm is heuristic. Moreover, the degree algorithm has no solid theoretical guarantee. The CELF algorithm has consistently had the most extensive running time, several orders of magnitude more than the FBCIM and Degree algorithms. It is because the CELF algorithm is based on the greedy algorithm. Although there is a theoretical guarantee in CELF, the running time is too great. The FBCIM and FBBCIM algorithms that we propose are based on sampling, significantly reducing the calculation cost. In addition, these two algorithms have theoretical guarantees.

Summary: (1) Our algorithm FBCIM achieves a higher influence spread than degree and CELF. (2) FBBCIM algorithm is the most balanced algorithm compared to FBCIM, CELF and degree, and it performs best when both influence maximization and community balance are considered. (3) The time cost of FBCIM, FBBCIM, and Degree algorithm is much less than that of CELF, while the degree algorithm is a heuristic algorithm without a solid theoretical guarantee. The time cost of FBCIM and FBBCIM is the same and can guarantee the efficiency requirements in large algorithms.

6   Conclusion

In this paper, we study the influence maximization problem based on budget in community structure. We design a simple budget allocation strategy and propose two scalable algorithms: FBCIM and FBBCIM, which solve community influence maximization and balanced community influence maximization, respectively. The RIS method is adopted in CIM to reduce the calculation cost-effectively on guaranteeing the algorithm approximation. FBBCIM is based on FBCIM. We add variance parameters to reduce the imbalance between communities. Finally, we test the proposed algorithms on three real networks and show their effectiveness.

References

1. Banerjee, P., Chen, W., Lakshmanan, L.V.S.: Maximizing social welfare in a competitive diffusion model. Proc. VLDB Endow. **14**(4), 613–625 (2020)
2. Banerjee, P., Chen, W., Lakshmanan, L.V.S.: Maximizing welfare in social networks under a utility driven influence diffusion model. In: Proceedings of the 2019 International Conference on Management of Data (2019)
3. Becker, R., Corò, F., Angelo, G., Gilbert, H.: Balancing spreads of influence in a social network. In: Proceedings of the AAAI Conference on Artificial Intelligence, pp. 3–10 (2020)
4. Blondel, V.D., Guillaume, J.L., Lambiotte, R., Lefebvre, E.: Fast unfolding of communities in large networks. J. Stat. Mech. Theory Exp. **2008**, P10008:1–12 (2008)
5. Chen, W., Wang, C., Wang, Y.: Scalable influence maximization for prevalent viral marketing in large-scale social networks. In: Proceedings of the 16th ACM SIGKDD International Conference on Knowledge Discovery and Data Mining (2010)
6. Domingos, P., Richardson, M.: Mining the network value of customers. In: Proceedings of the Seventh ACM SIGKDD International Conference on Knowledge Discovery and Data Mining (2001)
7. Guo, J., Wu, W.: Influence maximization: seeding based on community structure. ACM Trans. Knowl. Discov. Data **14**(6), 1–22 (2020)
8. Kempe, D., Kleinberg, J., Tardos, E.: Maximizing the spread of influence through a social network. In: Proceedings of the Ninth ACM SIGKDD International Conference on Knowledge Discovery and Data Mining (2003)

9. Leskovec, J., Krause, A., Guestrin, C., Faloutsos, C., VanBriesen, J., Glance, N.: Cost-effective outbreak detection in networks. In: Proceedings of the 13th ACM SIGKDD International Conference on Knowledge Discovery and Data Mining (2007)
10. Leskovec, J., Sosič, R.: SNAP: a general-purpose network analysis and graph-mining library. ACM Trans. Intell. Syst. Technol. (TIST) 8(1), 1–20 (2016)
11. Lin, M., Li, W., Lu, S.: Balanced influence maximization in attributed social network based on sampling. In: Proceedings of the 13th International Conference on Web Search and Data Mining (2020)
12. Rozemberczki, B., Davies, R., Sarkar, R., Sutton, C.: GEMSEC: graph embedding with self clustering. In: Proceedings of the 2019 IEEE/ACM International Conference on Advances in Social Networks Analysis and Mining 2019 (2019)
13. Rozemberczki, B., Sarkar, R.: Characteristic functions on graphs: birds of a feather, from statistical descriptors to parametric models. In: Proceedings of the 29th ACM International Conference on Information and Knowledge Management (CIKM 2020) (2020)
14. Schoenebeck, G., Tao, B.: Influence maximization on undirected graphs: toward closing the (1–1/e) gap. ACM Trans. Econ. Comput. 8(4), 1–36 (2020)
15. Stoica, A.A., Chaintreau, A.: Fairness in social influence maximization. In: Companion Proceedings of the 2019 World Wide Web Conference, pp. 569–574 (2019)
16. Sun, L., Huang, W., Yu, P.S., Chen, W.: Multi-round influence maximization. In: Proceedings of the 24th ACM SIGKDD International Conference on Knowledge Discovery and Data Mining (2018)
17. Talukder, A., Alam, M., Tran, N.H., Niyato, D., Hong, C.S.: Knapsack-based reverse influence maximization for target marketing in social networks. IEEE Access 7, 44182–44198 (2019)
18. Tang, J., Tang, X., Xiao, X., Yuan, J.: Online processing algorithms for influence maximization. In: Proceedings of the 2018 International Conference on Management of Data, pp. 991–1005 (2018)
19. Yang, Y., Mao, X., Pei, J., He, X.: Continuous influence maximization. ACM Trans. Knowl. Discov. Data 14(3), 1–38 (2020)

An Online Truthful Auction for IoT Data Trading with Dynamic Data Owners

Zhenni Feng[1](✉), Junchang Chen[1], and Tong Liu[2]

[1] Donghua University, Shanghai, China
fzn@dhu.edu.cn, 220250@mail.dhu.edu.cn
[2] Shanghai University, Shanghai, China
tong_liu@shu.edu.cn

Abstract. Data is an extremely import asset in modern scientific and commercial society. The life force behind powerful AI or ML algorithms is data, especially lots of data, which makes *data trading* significantly essential to unlocking the power of AI or ML. Data owners who offer personal data and data consumers who request data blocks negotiate with each other to make an agreement on trading prices via a big data trading platform; consequently both sides gain profit from data transactions. A great many existing studies have investigated to trade various kinds of data as well as to protect data privacy, or to construct a decentralized data trading platform due to untrustworthy participants. However, existing studies neglect an important characteristic, *i.e.*, dynamics of both data owners and data requests in IoT data trading. To this end, we first construct an auction-based model to formulate the data trading process and then propose an truthful online data trading algorithm which not only resolves the problem of matching dynamic data owners and randomly generated data requests, but also determines the data trading price of each data block. The proposed algorithm achieves several good properties, such as constant competitive ratio for near-optimal social efficiency, incentive-compatibility, individual rationality of participants, via rigorous theoretical analysis and extensive simulations.

Keywords: Data trading · Dynamic data owners · Near-optimal online auction

1 Introduction

Data is an extremely import asset in modern scientific and commercial society. Predicted by IDC, there will be 55.7 billion connected devices worldwide by 2025, 75% of which will be connected to an IoT platform [10]. Furthermore, data generated by these IoT devices is estimated to be 73.1 ZB then. Most of these data arise from security and video surveillance; industrial IoT applications may also

H. Gao and X. Wang (Eds.): CollaborateCom 2021, LNICST 406, pp. 554–571, 2021.
https://doi.org/10.1007/978-3-030-92635-9_32

take a significant portion of these data. Almost all companies are aggressively turning to artificial intelligence (AI) or machine learning (ML) technology to gain competitive advantages. Nevertheless, the life force behind these AI or ML algorithms is data, especially a vast amount of data. Consequently, *data trading* is significantly essential to unlocking the power of AI or ML.

Data trading is different from the general concept of data sharing. In mobile crowdsensing applications, participants or workers usually share their sensing data and get reward in return [13,14,19]. Application users generally share their data, *e.g.*, web browsing, online shopping orders, to application service providers in order to get accessible to their services. To get things moving, participants, which are generally application users, application service providers and third-party data consumers, go from the general concept of data sharing with others to the specifics of exactly what data they want and what they are willing to give of value in exchange.

Big data has fueled the emergency of data trading platforms which serve as bridges between sellers and buyers. Examples, such as Terbine [2] and GXS TradeWeb [1], have been designed to provide big data trading services. Data owners who offer personal data and data consumers who request data blocks negotiate with each other to make an agreement on trading prices via a big data trading platform; consequently both sides gain profit from data transactions.

Many recent studies [3,6,12,17,22] have paid attention to propose data trading algorithms or to design data trading platforms. Various kinds of datasets, such as raw data samples, range counts, aggregate statistic results, are traded between data owners and data consumers. To negotiate the data trading process between them, data brokers are usually introduced to assist transmission of data trading messages or traded data. To protect data privacy, some recent studies employ encryption algorithms and then to disclose encrypted data to data consumers; other studies introduce privacy-preserving schemes such as differential privacy or its variations to take control of the level of disclosed data privacy. Some other existing studies [4,5,8,9,18,21] consider participants or the data platform are untrustworthy; they propose decentralized data trading platforms based on blockchain technology. Almost all existing studies, however, ignore an important characteristic,*i.e.*, dynamics of both data owners and data requests in IoT data trading, where data owners are end devices and data consumers are application service providers.

In the paper, we model the dynamics of data owners and data requests in the scenario of IoT data trading. For example, when a data request of querying real-time noise level locating at a given block is submitted, a smartphone can serve as a data owner only if the owner of the smartphone passes by the block. Consequently, we assume that data owners are intermittent and only available to trade their data during a specific time period, which is called *active time*, because of limited resources or mobility. Furthermore, data requests are randomly generated by data consumers and then submitted to the data trading platform running on a edge server or a cloud server.

There are several main technical challenges to solve the IoT data trading problem. *First of all*, data owners are allowed to dynamically join in the data trading process and data requests are randomly generated according to the demand of applications. Such uncertain and unpredictable data requests make the data trading process relatively complicated. *Then*, it is very difficult for the data trading platform to make an efficient matching between data owners and data requests, because both the real value of data and active time are private information of data owners. *Finally*, rational and strategic data owners are not willing to offer their valuable data or reveal their private information truthfully except with enough compensation.

To this end, we propose a truthful online auction-based data trading algorithm containing two key components, which resolves two subproblems of how to match data owners with data requests and how to determine trading price of each data block. On the one hand, an auction model is constructed to formulate data trading process and an efficient online matching algorithm based on greedy scheme is further proposed to achieve the near-optimal system efficiency with a constant competitive ration of $1/2$. On the other hand, the trading price of each data block is computed according to a *critical value*, which is the highest bidding price that a data owner would win a bid. Both rigorous theoretical analysis and extensive simulations demonstrate good properties of our proposed online data trading algorithm, *e.g.*, *individual rationality*, *incentive-compatibility*, *near-optimality* on system efficiency. Major technical contributions in this paper are summarized as follows.

- It is the first work, to the best of our knowledge, which takes account of both dynamic behaviors of data owners and randomly generated data requests in the problem formulation of the data trading process.
- We propose a truthful online auction-based data trading algorithm which not only determines the matching rule with incomplete information but also computes proper data trading prices between data owners and data consumers.
- We have demonstrated that the proposed algorithm achieves several good properties via rigorous theoretical analysis and extensive simulations.

The rest of the paper is organized as follows. The system model and problem formulation of data trading is presented in Sect. 2. Then, the proposed online data trading algorithm is discussed in Sect. 3 along with rigorous theoretical analysis. Section 4 provides extensive simulations and numerical results to demonstrate desirable properties of the proposed algorithm. We review relate work in Sect. 5 and finally conclude the paper in Sect. 6.

2   System Model and Problem Formulation

We first introduce participants in the data trading process and describe the data trading model between data owners and data consumers; the mathematical formulation of data trading is then provided.

2.1 System Model

In a data trading market for sharing IoT data, there are mainly two kinds of participants, *end devices* and *edge servers or cloud servers*. End devices who collect data are *data owners*; edge servers or cloud servers who buy data from data owners are *data consumers*.

We divide time into time slots of equal size. Auctions between data owners and data consumers are executed round by round. Data owners are short-sighted so that they expect to get as much profit as possible in the current round. Without loss of generality, we only consider the auction process in a single round.

Data requests are submitted randomly and dynamically. We assume that a practical data requirement can be decomposed into several smaller data requests, each of which can be satisfied by a single data block generated by a single data owner. Let τ_j denote the number of data requests arriving at time slot j. A data request submitted at the time slot j is denoted by $r_{j,k}, k \leq \tau_j$. The set of all data requests is denoted by $R = \{r_{j,k} | j = 1, 2, \cdots, T, k \leq \tau_j\}$, where T is the total number of time slots in each round. For simplicity, we assume that there are a sufficient number of data owners that every data request is matched to a data block at its arriving time slot.

Data owners are dynamic because their data are only accessible during their *active time*. The active time of a data owner i is a time period described by $[s_i, e_i)$, where s_i and e_i are the start time slot and end time slot (not included) of her active time. Each data owner can sell at most one data block or sell her data block at most once during her active time because of limited resources. A data owner i has a valuation v_i for her data block which indicates that she would not trade her data block with a trading price lower than v_i. Similarly, the reported active time is possibly different from her real active time.

2.2 Data Trading Model Based on an Auction Mechanism

The interaction between data owners and data consumers are modeled by an auction mechanism, as shown in Fig. 1. In a data trading market, data owners share their data blocks with others and then they are compensated according to trading prices; data consumers get data blocks and pay to data owners. There exists a third-party trading platform to manage the auction process. The auction process is described as follows:

1. The platform sends data requests to data owners.
2. Each data owner generates a bid B_i which reports the active time and valuation for her data block, and then sends her bid to the platform.
3. The platform matches bids to data requests and determines the trading time slot and trading price (payment) for each selected bid. Then platform return matching results to data owners and data consumers.
4. Each data owner whose bid is selected uploads her data block at a specific time slot.
5. The data consumer pays for the data owner according to the negotiated trading price.

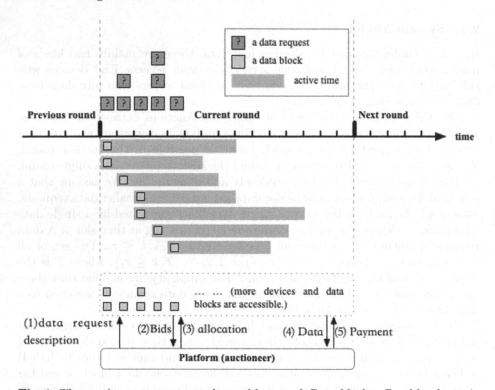

Fig. 1. The auction process repeated round by round. Data blocks offered by dynamic end devices and dynamic data requests arrives at the first five time slots are depicted in the picture.

The platform must determine a *matching rule* and a *trading price rule*. We use $B = \{B_i | i = 1, 2, \cdots, N\}$ to denote the set of all bids submitted by all data owners, where N is the total number of data owners in a round. The matching rule $\mathbf{X}$ is a matrix of indicator variables; each element, $x_{i,j} \in \{1, 0\}, 1 \leq i \leq N, 1 \leq j \leq T$, represents whether bid B_i is selected at time slot j or not. Actually, we should denote the matching rule by $x_{i,j}(B)$ since each $x_{i,j}$ is determined by B. For simplicity, we use $x_{i,j}$ instead of $x_{i,j}(B)$ in the rest of the paper. According to the trading price rule, payment to each bid B_i is denoted by $p_i(B) \in \mathbf{R}$.

In each round, a data owner i submits at most one bid, $B_i = (\tilde{s}_i, \tilde{e}_i, b_i), 1 \leq \tilde{s}_i < \tilde{e}_i \leq T+1, b_i \geq 0$, where $[\tilde{s}_i, \tilde{e}_i]$ is the reported active time and b_i is the bid price. The bid price may be different from real valuation v_i, because data owners are usually *selfish*. Similarly, the reported active time period may be different from her real active time.

Next, we discuss utilities of data owners and explain why data owners are selfish.

Definition 1 (Utility of a data owner). *The utility of a data owner is the difference between the trading price and her valuation if the bid of the data owner is selected; Otherwise her utility is zero. The utility is computed as follows.*

$$u_i^o(B) = \sum_{j=s_i}^{e_i-1} x_{i,j}(p_i(B) - v_i), \tag{1}$$

where $\sum_{j=s_i}^{e_i-1} x_{i,j} \leq 1$ holds because a data owner only trade her data block at most once during her active time.

A data owner is selfish so that she probably selects strategies solely to maximize her utility. The data owner probably misreports start time or end time of her active time as well as to charge a higher price than her valuation. Data owners can not report earlier arrivals or delayed departure, because it is easy to detect their absence and thus they would be punished. Therefore, there are three of strategic behaviors for selfish data owners, i.e., delayed arrival, earlier departure, and misreporting valuation.

Definition 2 (Utility generated by a data request). *The utility of a data requester who publishes a data request is the difference between the amount of profit that the data requester get from data and the trading price. If the k-th data request $r_{j,k}$ at time slot j is matched to the bid B_i of a data owner i, then the utility generated by the data request $r_{j,k}$ is computed as follows:*

$$u_{j,k}^r(B) = x_{i,j}(\varphi_j - p_i(B)), \tag{2}$$

where φ_j is the amount of profit that the data requester get from the traded data block.

Consequently, utilities of all data requests at time slot j is

$$\sum_{k=1}^{\tau_j} u_{j,k}^r(B) = \sum_{i=1}^{N} x_{i,j}(\varphi_j - p_i(B)), \tag{3}$$

where $\sum_{i=1}^{N} x_{i,j} = \tau_j$ holds because a data request can be satisfied by a single data block and there are τ_j data owners selected at time slot j.

Definition 3 (Social efficiency). *The social efficiency is defined as the sum of utilities of all participants. It is computed as follows:*

$$\begin{aligned}
\Delta &= \sum_{i=1}^{N} u_i^o(B) + \sum_{j=1}^{T}\sum_{k=1}^{\tau_j} u_{j,k}^r(B) \\
&= \sum_{i=1}^{N}\sum_{j=s_i}^{e_i-1} x_{i,j}(p_i(B) - v_i) + \sum_{j=1}^{T}\sum_{i=1}^{N} x_{i,j}(\varphi_j - p_i(B)) \\
&\stackrel{(a)}{=} -\sum_{i=1}^{N}\sum_{j=s_i}^{e_i-1} x_{i,j}v_i + \sum_{j=1}^{T}\tau_j\varphi_j + \Big(\sum_{i=1}^{N}\sum_{j=s_i}^{e_i-1} x_{i,j}p_i(B) - \sum_{j=1}^{T}\sum_{i=1}^{N} x_{i,j}p_i(B)\Big) \\
&\stackrel{(b)}{=} C - \sum_{i=1}^{N}\sum_{j=s_i}^{e_i-1} x_{i,j}v_i,
\end{aligned} \tag{4}$$

where (a) uses $\sum_{i=1}^{N} x_{i,j} = \tau_j$ and (b) follows from exchanging two summations and replacing $\sum_{j=1}^{T} \tau_j \varphi_j$ with a constant parameter C whose value is not related to $x_{i,j}$.

2.3 Problem Formulation

In the paper, we aim to design an auction mechanism for the data trading market so that the maximum social efficiency is achieved as well as following properties, i.e., *individual rationality, incentive-compatibility, computation efficiency.*

Definition 4 (Individual rationality). *An auction mechanism satisfies the property of individual rationality if and only if every data owner has a non-negative utility, i.e.,* $u_i^o \geq 0, i = 1, 2 \cdots N$.

Definition 5 (Incentive-compatibility). *An auction mechanism is incentive-compatible if and only if, for each data owner i, she cannot increase her utility by misreporting her private information, i.e.,*

$$u_i^o(B_i \cup B_{-i}) \geq u_i^o(\tilde{B}_i \cup B_{-i}), \tag{5}$$

where $B_i = (s_i, e_i, v_i)$ and $\tilde{B}_i = (\tilde{s}_i, \tilde{e}_i, b_i)$ are not the same, which means any of $s_i \leq \tilde{s}_i$, $e_i \geq \tilde{e}_i$, and $v_i \neq b_i$ holds; B_{-i} denotes the set of all bids except B_i.

In the following, we offer the mathematical formulation of our problem. We need to determine the matching rule of $\mathbf{X} = \{x_{i,j} | i = 1, 2, \cdots, N, j = 1, 2, \cdots, T\}$ by solving the optimization problem defined in (6) as well as the trading price rule $\{p_i | i = 1, 2, \cdots, N\}$ satisfying definitions of (4) and (5).

$$\arg\max_{x_{i,j}} \Delta = \arg\min_{x_{i,j}} \sum_{i=1}^{N} \sum_{j=s_i}^{e_i-1} x_{i,j} v_i,$$

$$s.t. \sum_{j=s_i}^{e_i-1} x_{i,j} \leq 1, \quad \sum_{j \in \{t | t < s_i\} \cup \{t | t \geq e_i\}} x_{i,j} = 0, \forall i,$$

$$\sum_{i=1}^{N} x_{i,j} = \tau_j, \forall j, \tag{6}$$

$$x_{i,j} \in \{0, 1\}, \forall i, \forall j.$$

From the objective of the problem, we can see that maximizing social efficiency is equivalent to minimizing sum of valuation from selected data owners.

Our problem cannot be solved by classic optimization algorithms solving linear programming for several reasons. *Firstly*, valuation v_i for each data owner is not accessible. *Secondly*, both data owners (with their data blocks) and data requests joins the data trading market dynamically, which means $x_{i,j}$ should be determined online without future information. *Thirdly*, the solution of Eq. (6) fails to provide any information about trading prices.

Algorithm 1: Online Matching Algorithm

Input : The set of bids B.

Output: The matching rule $\mathbf{X} = \{x_{i,j} | i = 1, 2, \cdots, N, j = 1, 2, \cdots, T\}$.

1 $A \leftarrow \emptyset, j \leftarrow 1, \mathbf{X} \leftarrow 0;$ // A is the set of all active bids which can
    provide accessible data block at current time slot.

2 **while** $j \leq T$ **do**

3   Remove expired bids (bids with $e_i = j$) from A;

4   Add newly active bids (bids with $s_i = j$) to A;

    /* Greedy select the first τ_j bids with lowest bid price at each
    time slot j.   */

5   **for** $k \leftarrow 0$ **to** τ_j **do**

       // Loop τ_j times.

6   Choose a bid $B_{(i')}$ with the lowest bid price and match it to the k-th
       data request at current time slot, $i.e.$, $x_{(i'),j} \leftarrow 1$;

7         $A \leftarrow A - B_{(i')};$ // Remove $B_{(i')}$ from A.

8   **end for**

9     $j \leftarrow j + 1;$

10 **end while**

11 **return X**

---

3 Online Data Trading Algorithm

In the section, we propose an online data trading algorithm to determine both the matching rule and the trading price rule. Our online data trading algorithm contains two components, which solves the matching subproblem in Scct. 3.1 and trading price subproblem in Sect. 3.2. For simplicity, we assume that every data owner submits a bid which is exactly the same as her private information and then we prove that data owners would honestly report her private information under the given trading price rule in Sect. 3.3.

3.1 Online Matching Algorithm Based on a Greedy Strategy

We propose an online matching algorithm to match data blocks of data owners to data requests using a greedy strategy. The basic idea of this algorithm is to greedily select one with the lowest bid price from current available data blocks to satisfy the newly submitted data request. The selection is executed at the beginning of every time slot. As shown in Algorithm 1, the algorithm maintains a set of active bids which have not been matched to any data request; the set is updated at the beginning of each time slot, $i.e.$, appending or removing bids into the active set according to active time of bids. At each time slot j, the first τ_j bids with lowest bid prices are selected.

3.2 Computing Trading Prices Based on Critical Data Owners

Unfortunately, a VCG-based payment scheme [16] is inapplicable to the online auction mechanism, because the online matching algorithm is not optimal.

Algorithm 2: Trading Price Determination Algorithm (for a single selected bid)

Input : A selected bid $B_i = (s_i, e_i, b_i)$, time slot $\bar{j}$ that B_i is selected (*i.e.*, $x_{i,\bar{j}} = 1$), the set of all bids B.

Output: The trading price p_i of the data block that is associated to B_i.

1   $A \leftarrow \emptyset$, $j \leftarrow 1$, $p_i \leftarrow b_i$;

2   $B \leftarrow B - B_i$; // Remove B_i from the set of all bids.

3   **while** $j < e_i$ **do**

4   Remove expired bids (bids with $e_i = j$) from A;

5   Add newly active bids (bids with $s_i = j$) to A;

6   **for** $k \leftarrow 0$ **to** τ_j **do**

7   Choose a bid $B_{i'} = (s_{i'}, e_{i'}, b_{i'})$ from A with the lowest bid price;

8         $A \leftarrow A - B_{i'}$;

          /* Find the highest bid price from all bids that are selected during $[\bar{j}, e_i - 1]$ */

9   **if** $j \geq \bar{j}$ *and* $b_{i'} > p_i$ **then**

10           $p_i \leftarrow b_{i'}$;

11   **end if**

12   **end for**

13      $j \leftarrow j + 1$;

14 **end while**

15 **return** p_i;

In the paper, we propose a trading price determination scheme based on *critical bid* which guarantees that each data owner reports private information truthfully. The basic idea is to set the trading price of a selected bid B_i as the bid price of the first bid that makes B_i fails. The first bid that makes B_i fails is the critical bid of B_i. Actually, if $B_i = (s_i, e_i, b_i)$ is selected at time slot $\bar{j}$ according to the matching rule, the critical bid $c(B_i)$ of B_i is the bid with the highest bid price which are selected during the time period of $[\bar{j}, e_i - 1]$ other than B_i.

Main steps of computing trading price for a selected bid are shown in Algorithm 2. Firstly, remove $B_i = (s_i, e_i, b_i)$ from the set of all bids B. Secondly, employ the matching rule proposed in Algorithm 1 to find all bids that are selected earlier than $\bar{j}$ and remove all of them from the active set of bids. Finally, find the bid (*i.e.*, the critical bid) with the highest bid price from all bids that are selected during the time period $[\bar{j}, e_i - 1]$ and return the bid price of the critical bid as the trading price. Similarly, we can repeat the procedure in Algorithm 2 for each selected bid. Besides, if a bid is not selected, then the trading price of the data block that is associated to the bid is zero.

3.3 Theoretical Analysis

In the subsection, we prove that the proposed online auction mechanism which contains two components of an online matching algorithm (Algorithm 1) and a trading price determination algorithm (Algorithm 2) satisfies several good properties aforementioned.

To prove the auction mechanism is incentive-compatible, it is equivalent to prove that it satisfies following two conditions: (i) the matching rule in Algorithm 1 is *monotonic*, and (ii) the trading price of the data block associated to each bid is equal to the *critical value*.

Definition 6 (Monotonicity). *The matching rule is monotonic if and only a data owner whose bid $B_i = (s_i, e_i, b_i)$ is selected would also win if she reports a more attractive bid $\tilde{B}_i = (\tilde{s}_i, \tilde{e}_i, \tilde{b}_i)$ with a lower bid price or a longer active time period, i.e., $\tilde{s}_i \leq s_i$, $\tilde{e}_i \geq e_i$, $\tilde{b}_i \leq b_i$.*

Definition 7 (Critical value). *For a data owner whose bid $B_i = (s_i, e_i, b_i)$ is selected, the critical value of the bid is the highest bid price b'_i that the data owner submits a bid $B'_i = (s_i, e_i, b'_i \geq b_i)$ and the new bid B'_i is still selected.*

Theorem 1 (Incentive-compatibility). *The proposed auction mechanism is incentive-compatible, because the matching rule is monotonic and the trading price is set as the critical value.*

Proof. First of all, we show that the matching rule is monotonic. Suppose a bid $B_i = (s_i, e_i, b_i)$ is selected at time slot j according to the matching rule. We replace the bid B_i with another bid $\tilde{B}_i = (\tilde{s}_i, \tilde{e}_i, \tilde{b}_i)$, where $\tilde{s}_i \leq s_i$, $\tilde{e}_i \geq e_i$, $\tilde{b}_i \leq b_i$. Obviously, $\tilde{B}_i$ would be selected at time slot j or earlier. Therefore, the matching rule is proved to be monotonic.

Then we check whether the trading price computed by Algorithm 2 is exactly the critical value. Suppose a data owner whose bid $B_i = (s_i, e_i, b_i)$ is selected at time slot j and the trading price of this bid computed by Algorithm 2 is $\dot{p}_i$. Therefore, there must be another bid B^0 whose bid price is $\dot{p}_i$ and is selected during the time period of $[j, e_i - 1]$. If the data owner submits another bid $\hat{B}_i = (s_i, e_i, \dot{p}_i - \xi)$ instead of B_i, where $\xi > 0$, then $\hat{B}_i$ would be selected during its active time and makes B^0 fails. On the contrary, if the data owner submits another bid $\bar{B}_i = (s_i, e_i, \dot{p}_i + \zeta)$ instead of B_i, where $\zeta > 0$, then $\bar{B}_i$ would not be selected since its bid price is higher than all selected bids during its active time. So we have verified that the trading price computed by Algorithm 2 is the critical value. We therefore conclude that the proposed auction mechanism is incentive-compatible.

Theorem 2 (Individual rationality). *The proposed online auction mechanism is individually rational.*

Proof. For a data owner whose bid fails, her utility is zero. For a data owner whose bid is selected at any time slot, we can compute the trading price of her data block according to Algorithm 2. Suppose a bid $B_i = (s_i, e_i, b_i)$ is selected at time slot j, there must be another bid $B_{i'} = (s_{i'}, d_{i'}, b_{i'})$ chosen at time slot j and p_i is updated to $b_{i'}$ (Line 10 in Algorithm 2). According to the update rule of trading prices, the final trading price would be $p_i \geq b_{i'}$. We can see that $b_{i'} \geq b_i$; otherwise, $B_{i'}$ would be selected at time slot j instead of B_i according to the matching rule in Algorithm 1. Since we have demonstrated that the auction mechanism is incentive-compatible, *i.e.*, every data owner would report their valuation truthfully, we can get that $p_i \geq b_{i'} \geq b_i = v_i$. Therefore, the utility of data owners are always nonnegative.

Theorem 3 (*Competitive ratio*). *The online matching algorithm achieves a competitive ratio of $\frac{1}{2}$, i.e., $\Delta_{online}/\Delta^* \geq \frac{1}{2}$, where Δ_{online} and Δ^* denote the resulting social efficiency of the online matching algorithm and the optimal solution of Eq. (6), respectively.*

Proof. The competitive ratio is computed by introducing a parameter a whose value is 0 initially. For a bid $B_i = (s_i, e_i, v_i)$ that is selected both in the online matching rule and the optimal solution, increase a by $\varphi_j - v_i$ (suppose B_i is selected at time slot j). For a bid B_i that is selected in the optimal solution at time slot j to a data request but not in the online matching rule, suppose this data request is matched to another bid $B_{i'}$ with a bid price of $v_{i'}$ in the online case, i.e., $v_{i'} \leq v_i$, increase a by $\varphi_j - v_{i'}$. Then, we can get $a \geq \Delta^*$.

For each bid that is matched by the online matching rule, its matching value $\varphi_j - v_i$ is added to a at most twice, *i.e.*, $a \leq 2\Delta_{online}$. Therefore, we have $2\Delta_{online} \geq a \geq \Delta^*$.

4   Numerical Illustration

In the section, we perform extensive simulation and report simulation results to show performance of the proposed online auction mechanism for data trading with dynamic data owners.

4.1   Methodology and Simulation Settings

We compare our proposed online auction mechanism with an optimal data trading algorithm with an optimal matching rule based on complete information and a incentive-compatible trading price rule. Suppose that active time of data owners are known and submitted in advance, we employ the Hungarian algorithm to find the optimal matching and then introduce the VCG payment scheme to compute trading prices. That is to say, since we have complete information about all bids and data requests in the auction and make decisions off-line, we can solve the optimization problem in Eq. (6) using the Hungarian algorithm. The VCG scheme can be utilized to determine trading prices, stimulating data owners to report their private information truthfully.

In simulation experiments, we suppose that both data requests and data owners are generated randomly. Specifically, both arrivals of data owners and data requests are generated from Poisson distributions, *i.e.*, the number of data owners newly joining the data trading market at each time slot follows a Poisson distribution with a parameter λ_o; the number of data requests submitted at each time slot follows a Poisson distribution with a parameter λ_r. The active time length, *i.e.*, the number of slots of active time, follows a uniform distribution.

The valuation every data owner or data requester is randomly chosen from a uniform distribution. Settings of all parameters used in simulation experiments are listed in Table 1.

Table 1. Summary of default settings.

Parameter	Default value
Rate of data owners, λ_o	10
Rate of data requests, λ_t	3
Range of active time length	[5, 20]
Range of data owners' valuation	[1, 10]
Number of time slots in a round, T	50

In following pictures, the online auction mechanism and the optimal data trading algorithm are denoted by "Online" and "Opt" in the legends, respectively. Furthermore, these approaches under different valuations of data owners' are evaluated, denoted by "Online, v = [1,5]" and "Opt, v = [1,5]" for example.

These two approaches are evaluated with extensive simulations based on three metrics of *social efficiency*, *competitive ratio*, and *running time*. We conduct several groups of experiments and report comparison results of these two approaches. Each point in these figures is average result over 100 runs.

4.2   Numerical Results

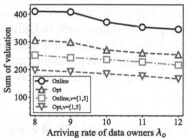 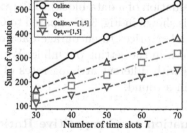

(a) Comparison on sum of valuation with different rates of data owners, λ_o.

(b) Comparison on sum of valuation with different numbers of time slots in a round, T.

Fig. 2. Performance of sum of valuation.

According to Eq. (4), C is a constant which is irrelevant with matching and trading prices of bids, we instead evaluate performance of sum of valuation of all data owners whose have been matching, *i.e.*, $\sum_{i=1}^{N} \sum_{j=s_i}^{e_i-1} x_{i,j} v_i$, in Fig. 2. As shown in Fig. 2a, there is a decrease in sum of valuation for all matched data owners when there are more available data owners with a larger arriving rate λ_o. It is obvious that sum of valuation is on the increase along with the number of time slots in a round, since there are more data requests are satisfied and more data owners are chosen. The performance of Opt outperforms Online in terms of sum of valuation.

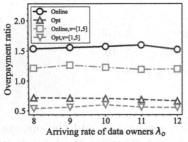 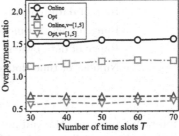

(a) Comparison on overpayment ratio with different rates of data owners, λ_o.

(b) Comparison on overpayment ratio with different numbers of time slots in a round, T.

Fig. 3. Performance of overpayment ratio.

To stimulate data owners to honestly report their valuation, the trading price of a data block is usually no lower than the valuation that data owner claims. We further introduce a metric of *overpayment ratio* to show that data requests should pay extra money to ensure social efficiency. The overpayment ratio is the amount of extra expenditure (*i.e.*, the difference between the trading price and the valuation of a data block) to the valuation. The performance of overpayment ratio is shown in Fig. 3. Compared to Opt, the proposed approach of Online must pay higher prices to encourage data owners' cooperation since less information is known in the setting of Online. We can also get that overpayment ratio keeps stable with different arriving rates of data owners or different numbers of time slots in a round.

Evaluation on Competitive Ratio. We plot empirical CDF of competitive ratio of the proposed online auction with different parameters in Fig. 4. In each parameter setting, simulations are repeated 1,000 times. The result on competitive ratio of each run is regarded as a sample; all these samples are utilized to derive the empirical CDF. When the valuation range of data owners varies, we can see that the competitive ratio is always above the bound of 0.5. To simplify the simulation, we set the valuation of each data request to be the upper bound of the valuation of data owners and then compute social efficiency of Online.

To compare the proposed online auction mechanism with optimal algorithm in detail, we further compare ratio of sum of valuation between different methods

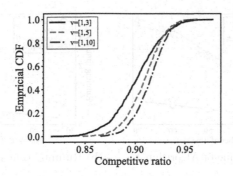

Fig. 4. Empirical CDF of competitive ratio derived from 1,000 simulations.

with different parameter settings; results are shown in Table 2 and Table 3. We can see that the sum of valuation of Online is a little higher than that of Opt. Results of ratio of sum of valuation range from 1.2 to 1.4. Ratio of sum of valuation remains stable, even if number of time slots increases or arriving rate of data owners increases. With smaller range of valuations, *i.e.*, v = [1, 5], the ratio is smaller since data owners participate in a more competitive auction and it is easier to induce their truthfulness.

Table 2. Ratio of sum of valuation when the number of time slots in a round changes

T	30	40	50	60	70
Ratio (Online)	1.35501	1.36143	1.36706	1.37421	1.37441
Ratio (Online, v = [1, 5])	1.27243	1.29097	1.28802	1.28590	1.28542

Table 3. Ratio of sum of valuation when arriving rate of data owners changes.

λ_o	8	9	10	11	12
Ratio (Online)	1.34804	1.37381	1.37681	1.36705	1.36356
Ratio (Online, v = [1, 5])	1.27757	1.26861	1.28496	1.28957	1.29698

Evaluation on Running Time. To show computation efficiency of proposed online auction mechanism, running time of both matching (Algorithm 1) and trading price determination (Algorithm 2) are recorded and plotted in Fig. 5. Plotted lines of Online shows modest growth in running time for both Algorithm 1 and Algorithm 2 as shown Fig. 5; lines of Opt increase dramatically with the number of time slots increases. Consequently, Online is much more computationally efficient than Opt.

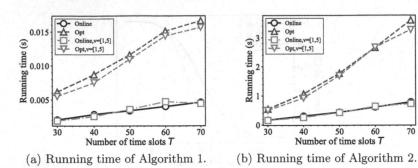

(a) Running time of Algorithm 1.   (b) Running time of Algorithm 2.

Fig. 5. Performance of running time.

5   Related Work

We review related work from the following two aspects and point out that existing solutions cannot be applied to solve our problem.

5.1   Decentralized Data Trading Based on the Blockchain Technology

A great many research papers have paid attention to design a data trading market or platform based on blockchain technology [4,8,9,18,21] because of the absence of a trustworthy and centralized data trading platform, single point of failure and DDoS. Dai et al. claim that both data brokers and buyers are dishonest and none of them is accessible for raw data; data processing and analysis algorithms encoded in smart contracts are deployed in a secure data trading platform supported by the hardware of Intel's Software Guard Extensions (SGX) based secure execution environment [4]. Another trusted data trading platform employing both the blockchain technology and trusted execution environment (TEE) is implemented by Su et al. where the trusted trading platform contains a special kind of nodes each of which supports TEE and serves as a trust exchange for exchanging data or payment between data sellers and data buyers [18]. Ha et al. introduce a decentralized private data trading marketplace called "Digital Me" based on the blockchain technology [8], where data sellers and data buyers trade personal data directly without trustworthy servers. The AI agent included in "Digital Me" serves as a trading assistant to recommend trading prices based on a user's personal data and data transaction history data. He et al. propose a distributed and trusted data trading platform based on blockchain technology to detect misbehavior of participants and a dataset similarity comparison scheme based on MinHash for detecting illegal resale efficiently is then employed [9]. Zheng et al. deploy smart contracts to solve the problem of data matching and reward distribution in a distributed data trading platform and then introduce proxy re-encryption to guarantee secure data transmission, where trading data are encrypted and only valid data requesters are allowed to decrypt trading data

[21]. Nguyen et al. design a distributed ledger based IoT data trading system along with three typical data trading protocols for city-level environmental monitoring using NB-IoT connections and further analyze the cost of data trading in terms of end-to-end transmission latency and energy consumption [15].

5.2 Trading Data with Different Levels of Privacy

Existing studies [3,12,17,22] investigate how to trade private data where the authors propose pricing functions for personal data or other private data to compensate data owners' different levels of privacy loss. Private data with different privacy loss is generally traded at different prices. Furthermore, data in different formats e.g., data samples and range counts, are returned to data consumers. Higher prices should be paid to a higher level of privacy loss caused by traded data undoubtedly. A few desirable properties, e.g., arbitrage-freeness, budget feasibility, performance accuracy based on traded data, are considered when designing pricing functions.

Gao et al. propose a pricing rule based on an auction-based model where both task description and bid prices in bids are possible to disclose sensitive information of data owners; they employ differential privacy schemes at both stages of data collection and trading price determination [7]. Another research paper [11] using the auction-based model introduces geo-indistinguishability to quantify privacy loss of geographical locations and then pays for their sensing cost as well as privacy breach. Zhang et al. point out that disclosure of raw social media data of users probably cause privacy leakage, because anonymous user IDs can be linked to real users and they propose a novel mechanism based on a notion of ϵ-text indistinguishability to guarantee different user privacy as well as to achieve high data utility [20].

All existing studies, however, neglect the dynamic behavior of data owners as well as randomly generated data requests of data consumers in IoT data trading. It is reasonable that end devices only trade their data intermittently due to limited resources or mobility.

6   Conclusion

In the paper we have investigated the data trading problem with dynamic data owners, aiming to share IoT data and to take full advantages of big data. Existing studies have neglected an important observation that a data owner is not always available to trade her data blocks except in her active time period. To this end, we have proposed a truthful and efficient online data trading algorithm which not only resolves the problem of matching dynamic data owners and randomly generated data requests with near-optimal social efficiency, but also determines the trading price of each data block which ensures the incentive-compatibility and individual rationality of participants.

Acknowledgment. This research is partially supported by Shanghai Sailing Program (Grant No. 19YF1402200) and the Fundamental Research Funds for the Central Universities (Grant No. 2232021D-23).

References

1. GXS TradeWeb - A service of GXS. https://gxstradeweb.gxsolc.com/pub-html/EdiServiceInfoFrameset.html. Accessed 11 Apr 2021
2. Terbine: The data exchange for advanced mobility and infrastructure. https://terbine.com/. Accessed 11 Apr 2021
3. Cai, Z., He, Z.: Trading private range counting over big IoT data. In: Proceedings IEEE International Conference Distributed Computing System (ICDCS), pp. 144–153 (2019)
4. Dai, W., Dai, C., Choo, K.K.R., Cui, C., Zou, D., Jin, H.: SDTE: a secure blockchain-based data trading ecosystem. IEEE Trans. Inf. Forensics Secur. **15**, 725–737 (2019)
5. Feng, Z., Chen, J.: Blockchain based mobile crowd sensing for reliable data sharing in IoT systems. In: Proceedings IFIP Networking, pp. 1–3 (2021)
6. Feng, Z., Chen, J., Zhu, Y.: Uncovering value of correlated data: trading data based on iterative combinatorial auction. In: Proceedings IEEE International Conference Mobile Ad-Hoc and Smart System (MASS) (2021, to appear)
7. Gao, G., Xiao, M., Wu, J., Zhang, S., Huang, L., Xiao, G.: DPDT: a differentially private crowd-sensed data trading mechanism. IEEE Internet Things J. **7**(1), 751–762 (2020)
8. Ha, M., Kwon, S., Lee, Y.J., Shim, Y., Kim, J.: Where WTS meets WTB: a blockchain-based marketplace for digital me to trade users' private data. Pervasive Mob. Comput. **59**, 101078 (2019)
9. He, Y., Zhu, H., Wang, C., Xiao, K., Zhou, Y., Xin, Y.: An accountable data trading platform based on blockchain. In: Proceedings IEEE International Conference Computing Communication Workshops (INFOCOM WKSHPS), pp. 1–6 (2019)
10. idc.com: IoT growth demands rethink of long-term storage strategies, says IDC, https://www.idc.com/getdoc.jsp?containerId=prAP46737220. Accessed 11 Apr 2021
11. Jin, W., Xiao, M., Li, M., Guo, L.: If you do not care about it, sell it: trading location privacy in mobile crowd sensing. In: Proceedings IEEE International Conference Computing Communication (INFOCOM), pp. 1045–1053 (2019)
12. Li, C., Li, D.Y., Miklau, G., Suciu, D.: A theory of pricing private data. Commun. ACM **60**(12), 79–86 (2017)
13. Liu, T., Li, D., Cao, C., Gao, H., Li, C., Feng, Z.: Joint location-value privacy protection for spatiotemporal data collection via mobile crowdsensing. In: Proceedings International Conference Collaborative Computing: Networking, Applications and Worksharing (CollaborateCom) (2021, to appear)
14. Liu, T., Wu, W., Zhu, Y., Tong, W.: Accuracy-Guaranteed event detection via collaborative mobile crowdsensing with unreliable users. In: Wang, X., Gao, H., Iqbal, M., Min, G. (eds.) CollaborateCom 2019. LNICST, vol. 292, pp. 729–744. Springer, Cham (2019). https://doi.org/10.1007/978-3-030-30146-0_49
15. Nguyen, L.D., Leyva-Mayorga, I., Lewis, A.N., Popovski, P.: Modeling and analysis of data trading on blockchain-based market in IoT networks. IEEE Internet Things J. **8**(8), 6487–6497 (2021)

16. Nisan, N., Roughgarden, T., Tardos, E.V., Vazirani, V. (eds.): Algorithmic Game Theory. Cambridge University Press, Cambridge (2007)
17. Niu, C., Zheng, Z., Wu, F., Tang, S., Gao, X., Chen, G.: Unlocking the value of privacy: Trading aggregate statistics over private correlated data. In: Proceedings ACM International Conference on Knowledge Discovery and Data Mining (KDD), pp. 2031–2040 (2018)
18. Su, G., Yang, W., Luo, Z., Zhang, Y., Bai, Z., Zhu, Y.: BDTF: a blockchain-based data trading framework with trusted execution environment. CoRR abs/2007.06813 (2020)
19. Yang, C., et al.: Mobile data sharing with multiple user collaboration in mobile crowdsensing (short paper). In: Gao, H., Wang, X., Yin, Y., Iqbal, M. (eds.) CollaborateCom 2018. LNICST, vol. 268, pp. 356–365. Springer, Cham (2019). https://doi.org/10.1007/978-3-030-12981-1_25
20. Zhang, J., Sun, J., Zhang, R., Zhang, Y., Hu, X.: Privacy-preserving social media data outsourcing. In: Proceedings IEEE International Conference Computing Communication (INFOCOM), pp. 1106–1114 (2018)
21. Zheng, S., Pan, L., Hu, D., Li, M., Fan, Y.: A blockchain-based trading platform for big data. In: Proceedings IEEE International Conference Computing Communication Workshops (INFOCOM WKSHPS), pp. 991–996 (2020)
22. Zheng, S., Cao, Y., Yoshikawa, M.: Money cannot buy everything: trading mobile data with controllable privacy loss. In: Proceedings IEEE International Conference Mobile Data Management (MDM), pp. 29–38 (2020)

IoT and Social Networks and Images Handling and Human Recognition

Exploiting Heterogeneous Information for IoT Device Identification Using Graph Convolutional Network

Jisong Yang[1,2], Yafei Sang[1,2(✉)], Yongzheng Zhang[1,2], Peng Chang[1,2], and Chengwei Peng[3]

[1] Institute of Information Engineering, Chinese Academy of Sciences, Beijing, China
{yangjisong,sanyafei,zhangyongzheng,changpeng}@iie.ac.cn
[2] School of Cyber Security, University of Chinese Academy of Sciences, Beijing, China
[3] National Computer Network Emergency Response Technical Team/Coordination Center of China, Beijing, China
pengchengwei@cert.org.cn

Abstract. IoT devices are widely present in production and life. To provide unique resource requirements and Quality of Service for different device types, we are prompted to implement IoT device identification. Existing IoT device identification methods either need to extract features manually or suffer from low effectiveness. In addition, these methods mainly focus on plaintext traffic, and their effectiveness will not work in the encryption era. It remains a challenging task to conduct IoT device identification via TLS encrypted traffic analysis accurately. This work fills the gap by presenting THG-IoT, a novel device identification method using graph convolutional network (GCN). We propose a graph structure named traffic heterogeneous graph (THG), an information-rich representation of encrypted IoT network traffic. The key novelty of THG is two-fold: i) it is a traffic heterogeneous graph containing two kinds of nodes and two kinds of edges. Compared with the sequence model, THG can better model the relationship between the flows and the packets. ii) it implicitly reserves multiple heterogeneous information, including packet length, packet message type, packet context, and flow composition, in the bidirectional packet sequence. Moreover, we utilize THG to convert IoT device identification into a graph node classification problem and design a powerful GCN-based classifier. The experimental results show that THG-IoT achieves excellent performance. The TPR exceeds 95% and the FPR is less than 0.4%, superior to the state-of-the-art methods.

Keywords: IoT device identification · Heterogeneous information · TLS message type · Packet length · Node classification

1 Introduction

With the constant maturity and gradual improvement of Industry 4.0, more and more devices possess unique sensing and communication functions. According to

H. Gao and X. Wang (Eds.): CollaborateCom 2021, LNICST 406, pp. 575–591, 2021.
https://doi.org/10.1007/978-3-030-92635-9_33

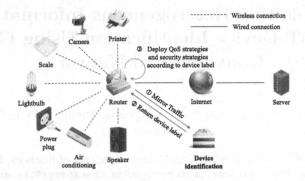

Fig. 1. IoT device identification application scenarios

the development trend of intelligent data systems, it is estimated that 64 billion devices in the world will be connected to the Internet of Things (IoT) by 2025 [14]. Considering the complexity and heterogeneity of the IoT structure itself, it is essential and indispensable to identify IoT devices and formulate different QoS strategies and security strategies according to different types.

The application scenario of IoT device identification is shown in Fig. 1. Various types of IoT devices first connect to the router via a wireless network and then connect to the supplier server via the Internet. The traffic generated by IoT devices will be aggregated in the router. The IoT device identification problem is transformed into a traffic classification problem by deploying a device identification method based on traffic analysis and marking the traffic passing through the router with corresponding device labels. There are many pieces of research on traffic classification, but most of them are based on feature engineering and plaintext payload, which require manual and cumbersome extraction of traffic features. The research by cloud-based security provider Zscaler found that 8.5% were encrypted with TLS, and this percentage will increase rapidly. With the extensive use of the TLS1.2 protocol in IoT device traffic, the payload becomes unresolvable, prompting us to study more robust traffic classification methods.

In this paper, we propose THG-IoT, which is an IoT device identification method using graph convolutional network (GCN). We are motivated by the intuitive observation that each traffic flow, consisting of a series of packets resulting from client-server interactions, can be represented by an information-rich graph structure named *traffic heterogeneous graph* (THG). The benefit of THG lies in that it is capable of reserving as much information as the original flow, such as **packet length, packet message type, packet context**, and **flow composition**, and thus does not require selecting features explicitly by hand. We set flows and packets as nodes to construct the THG and generate flow-packet edges and packet-packet edges. The heterogeneous graph captures the structural information between the flow and the packets it contains and differentiates to characterize the flow.

By using THG, we convert IoT device identification into a graph node classification problem. Moreover, since deep learning has shown its advantages over traditional machine learning techniques, we designed a classifier based on GCN. The classifier automatically extracts features from the input THG and distinguishes different flow nodes by mapping them to different representations in the embedding space. This work fills the gap in the research of IoT device identification based on TLS encrypted traffic.

Our contributions can be briefly summarized as follows:

- We propose the *traffic heterogeneous graph* (THG), a heterogeneous graph containing flow nodes, packet nodes, flow-packet edges, and packet-packet edges. The term heterogeneous means that the graph is composed of multiple nodes with different roles and edges with different meanings. We also designed different weight calculation methods for the two kinds of edges to capture heterogeneous information effectively.
- THG can capture various heterogeneous information between packets and flows, including **packet length**, **packet message type**, **packet context**, and **flow composition**, so as to differentiate the flow.
- We designed THG-IoT, a classification method based on GCN and using THG as input to realize the identification of IoT devices by classifying the flow nodes in THG.
- We evaluate our method based on real IoT device traffic data containing TLS encrypted traffic of 13 real IoT devices. The results show that THG-IoT possesses excellent performance with 95.32% TPR, 0.39% FPR, and 0.9495 FTF. THG-IoT has a 4% improvement at least over the state-of-the-art methods.

2 Preliminaries

In this section, we first introduce the TLS protocol and the basic concepts of graph convolutional networks and then analyze and define the problem we want to solve.

2.1 TLS Basics

The TLS protocol is a popular encryption protocol used to encrypt the transmission data between the client and the server. A typical TLS session is mainly composed of the Handshake layer and the Record layer. The Handshake layer is used to establish a session and negotiate a master key, and the Record layer is used to transfer payload under the secure parameters. Each packet in the TLS protocol communication session has a specific message type. Table 1 presents the message types and their notations of TLS. TLS message types indicate the configuration details of the connection and transmission between the client and the server.

Table 1. Message types of TLS

Value	Message type	Value	Message type	Value	Message type
20	Change Cipher Spec	22:3	Hello Verify Request	22:14	Server Hello Done
22	Handshake	22:4	New Session Ticket	22:15	Certificate Verify
22:0	Hello Request	22:11	Certificate	22:16	Client Key Exchange
22:1	Client Hello	22:12	Server Key Exchange	22:20	Finished
22:2	Server Hello	22:13	Certificate Request	23	Application Data

Table 2. List of notations

Notations	Meaning	Notations	Meaning
$\mathbb{P}$	Flow set	F	Flow nodes set
P	A flow in $\mathbb{P}$	v_i	A node in G
p_i	A packet in flow P	A	The adjacency matrix of G
G	The THG of $\mathbb{P}$	X	The node feature matrix of G
Y_i	Actual label	M_i	The message type of p_i
Z_i	Predicted label	L_i	The packet length of p_i

2.2 Graph Convolutional Networks

Graph Convolutional Network (GCN) [5] is a multi-layer neural network that trains the embedding vector of the node according to the characteristics of the node's neighborhood. For any node $v \in V$ in graph $G = (V, E)$, assuming its feature vector size is m, let $X \in \mathbb{R}^{n \times m}$ be a matrix containing all n nodes with their features, and each row $x_v \in \mathbb{R}^m$ is the feature vector for v. We define an adjacency matrix A of G and the degree matrix D of A, and the diagonal elements of A are set to 1 because of self-loops. GCN captures the information of neighbors in the one-layer convolution relationship. After one-layer GCN, the dimension of the feature vector of the node will change with the dimension of the weight matrix $W_0 \in \mathbb{R}^{m \times k}$. New node feature matrix $X^{(1)} \in \mathbb{R}^{n \times k}$ is computed as:

$$X^{(1)} = \sigma(\widetilde{A} X W_0) \tag{1}$$

where $\widetilde{A} = D^{\frac{1}{2}} A D^{-\frac{1}{2}}$ is the normalized symmetric adjacency matrix and σ is an activation function. The multi-layer GCN iteratively updates the node feature matrix as follow:

$$X^{(i+1)} = \sigma(\widetilde{A} X^i W_i) \tag{2}$$

where i denotes the layer number and $X^0 = X$.

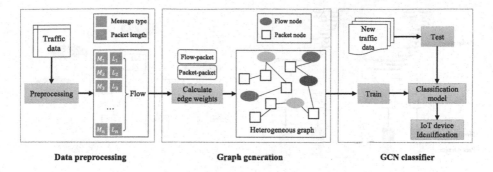

Data preprocessing   **Graph generation**   **GCN classifier**

Fig. 2. The general framework of THG-IoT

2.3 Problem Definition

Given the flow $P = \{p_1, p_2, \cdots, p_n\}, P \in \mathbb{P}$, our goal is to find the objective function $f\{P\} \rightarrow Y_i$, where Y_i represents the classification label of the flow P. In this paper, we construct a heterogeneous graph G based on $\mathbb{P}$, and transform the flow classification problem into a node classification problem. For node v_i, the objective function is $f\{G, A, X, v_i\} \rightarrow Y_i$. The notations used in this paper are shown in Table 2.

3 The THG-IoT Framework

THG-IoT is mainly composed of three modules, as shown in Fig. 2. THG-IoT uses message type and packet length to differentiate the characterization of different classes of flows and further enhance characterization capabilities by establishing heterogeneous graphs between flows and packets. Our method enjoys the advantages of both supervised learning and unsupervised learning. In this section, we will introduce each module in detail.

3.1 Data Preprocessing

We can collect the traffic generated by the communication between IoT devices and the server at the gateway. In network traffic, 5-tuple (source IP, source port, destination IP, destination port, transport protocol) can distinguish different sessions, and the corresponding session is unique. Therefore, we divide the traffic into individual flows based on 5-tuple information, and each flow contains several packets. We take a specific flow as an example to illustrate the particular process of data preprocessing and show it in Fig. 3. Each TLS packet in the flow has a *"tls.record.content_type"* field, and the handshake protocol packet has a *"tls.handshake.type"* field. We extract the values of these fields as the message type of the packet. For example, the message type of the data packet

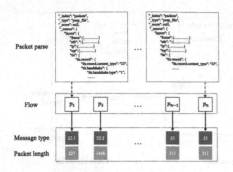 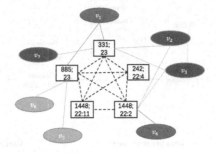

Fig. 3. Data preprocessing   **Fig. 4.** Heterogeneous graph

p_1 is *22:1*, and the message type of the data packet p_n is *23* in Fig. 3. In addition, we extract the packet length from the $['\_source']['layers']['tcp']['tcp.len']$ field. We further connect the message type and packet length of the same packet with a semicolon, such as *"1448;22:12"*, *"311;23"*. Each packet is represented by a string $M_i; L_i$, and each flow is represented by a sequence of strings $\{M_1; L_1, M_2; L_2, \cdots, M_n; L_n\}$.

3.2 Graph Generation

We built a large traffic heterogeneous graph THG from $\mathbb{P}$ as shown in Fig. 4, which contains two kinds of nodes and two kinds of edges.

Nodes. Given a specific flow $P = \{p_1, p_2, \cdots, p_n\}$, we set both the flow P and packets p_i as nodes in THG, which are termed as flow nodes and packet nodes. We scan all the flows in $\mathbb{P}$ to get all the nodes of THG. The total number of nodes of the THG is the total number of flows plus the number of different packets. Each packet node contains the message type and packet length information, and the flow node is the target of classification. In Fig. 4, the ellipse is the flow node, and the rectangle is the data packet node, and different colors represent different classes. When nodes are determined, edges should be added to connect these nodes.

Edges. There are two types of edges in THG: flow-packet edges and packet-packet edges. The flow-packet edge connects the flow node with the packet node belonging to itself. We assume that the flow in Fig. 3 corresponds to the node v_2 in Fig. 4, then the node v_2 will connect the edge between the two packet nodes *"1448;22:12"* and *"311;23"*. We assign flow-packet edge weights to characterize the contribution of a packet to the flow to which it belongs. In this paper, the message type and packet length information $M_i; L_i$ are regarded as a word, the flow is seen as a piece of text. The term frequency-inverse document frequency (TF-IDF) is selected as the flow-packet edge weight calculation method. If the packets owned by two flows are very similar, the two flows are likely to come from the same IoT device.

The packet-packet edge connects any two packet nodes, such as the five packet nodes in Fig. 4, which are connected by a dotted line. The packet-packet edge is also given weights, which are used to characterize the co-occurrence relationship between different data packets and differentiate further and characterize the flow. We use a fixed-size sliding window to collect co-occurrence statistics and employ pointwise mutual information (PMI) to calculate the weight of the edge between two packet nodes. The PMI value of packet nodes v_i, v_j is computed as

$$PMI(v_i, v_j) = log\frac{p(v_i, v_j)}{p(v_i)p(v_j)} \tag{3}$$

where $p(v_i, v_j)$ represents the probability of two packets appearing in the same sliding window, and $p(v_i)$ represents the probability of v_i appearing in the sliding window. The larger the value of PMI, the stronger the correlation between the two packets.

We believe that THG is a powerful representation of traffic and can extract valuable information in the following four aspects.

– **Packet length information.** The packet length is a critical feature in the research of encrypted traffic classification. The packet node in THG contains packet length information, which can be used directly by the GCN classifier.
– **Packet message type information.** The message type is a unique feature of the TLS session packet. Adding it to the packet node in the THG is conducive to the classifier refining and learning the classification features.
– **Packet context information.** The packet-packet edge in THG represents the context information of two packets. The packets appearing in the same flow have a stronger co-occurrence relationship, which enhances the differential representation of the flow.
– **Flow composition information.** The flow-packet edge in THG establishes the contribution relationship between packets and flows. The composition of packets in different flows can vary significantly and act as discriminative features learned by GCN classifiers.

3.3 GCN Classifier

We take the generated heterogeneous graph G as the input of a two-layer GCN network. The adjacency weight matrix $\hat{A}$ is obtained by the weights of the flow-packet edges and the packet-packet edges, and the feature vector matrix X is composed of the One-Hot of all flow nodes. The output of the first layer of GCN passes the ReLU activation function, and the output of the second layer passes through the softmax to get the classification result label.

$$Z = f(\hat{A}, X) = softmax(\tilde{A}ReLU(\tilde{A}XW_0)W_1) \tag{4}$$

where $\tilde{A} = D^{\frac{1}{2}}\hat{A}D^{-\frac{1}{2}}$. Since we only classify flow nodes, the cross-entropy loss function for this problem is defined as:

$$loss = - \sum_{f \in F} \sum_{c=1}^{C} Y_{fc} \ln Z_{fc} \tag{5}$$

where F is the set of flow nodes and C is the number of classes. $Y_{fc} = 1$ when c is equal to the label of f, otherwise $Y_{fc} = 0$. The weight matrix can be trained using gradient descent. Define the output of the second layer of GCN $O_2 = \tilde{A}ReLU(\tilde{A}XW_0)W_1$, and O_2 contains the latest feature vectors of all nodes. We can visualize these vectors and help us intuitively understand how to implement node classification.

For specific application scenarios, more layers of GCN will not significantly improve the final performance [7]. We only use a two-layer GCN network in this work, which ensures the information interaction between the two flow nodes while reducing the complexity of the model.

4   Experimental Evaluation

In this section, we sequentially present the dataset, evaluation metrics, experimental setting, and the abundant experimental results.

4.1   Dataset

We use the dataset [16] to train and test our method THG-IoT. The dataset collects IoT device traffic from a real-world IoT network environment, including switch, echo, printer, etc., which applies the MAC address as the IoT device label. Table 3 shows some statistical results of the dataset, consisting of 373+ thousand traffic from 13 IoT devices. Since the TLS encrypted traffic of some devices in the dataset is relatively small, we randomly selected 1000 flows for each IoT device to ensure a balanced sample number. We divide the dataset into a training set (80%) and a test set (20%); the training set is used to train the parameters of THG-IoT, and the test set is used to test the performance of THG-IoT. 5-fold cross-validation is adopted in our experiment to reduce overfitting and improve the reliability of experiments.

4.2   Evaluation Metrics

For each sample data in the test set, we have the original true label and the predicted label obtained by our method. For each label i, the test set is divided into the following four groups according to the values of the predicted label and the true label.

- TP_i: The predicted label and true label of a payload are both i.
- TN_i: Neither the predicted label nor true label of a payload is i.
- FP_i: The predicted label of a payload is i, and the true label is not i.
- FN_i: The predicted label of a payload is not i, and the true label is i.

Table 3. Summary of dataset in evaluation

ID	IoT device	MAC address	Flows	Packets
0	Amazon echo	44:65:0d:56:cc:d3	55.3K	1.3M
1	Belkin motion sensor	ec:1a:59:83:28:11	41.5K	1.4M
2	Belkin switch	ec:1a:59:79:f4:89	36.0K	1.1M
3	Dropcam	30:8c:fb:2f:e4:b2	152.1K	4.2M
4	HP printer	70:5a:0f:e4:9b:c0	4.6K	248.3K
5	iHome PowerPlug	74:c6:3b:29:d7:1d	2.6K	65.4K
6	Netatmo camera	70:ee:50:18:34:43	16.7K	673.7K
7	Pixstart photo frame	e0:76:d0:33:bb:85	7.2K	68.7K
8	Samsung smart cam	00:16:6c:ab:6b:88	26.1K	1.2M
9	Smart things	d0:52:a8:00:67:5e	8.2K	571.0K
10	TP-Link camera	f4:f2:6d:93:51:f1	11.7K	319.0K
11	Triby speaker	18:b7:9e:02:20:44	5.9K	200.6K
12	Withings sleep sensor	00:24:e4:20:28:c6	6.4K	387.8K

We adopt the True Positive Rate ($TPR_i = \frac{|TP_i|}{|TP_i|+|FN_i|}$) and False Positive Rate ($FPR_i = \frac{|FP_i|}{|TN_i|+|FP_i|}$) to evaluate our method. In addition, we define the following three macro metrics($TPR_{AVE}, FPR_{AVE}, FTF_{AVE}$) to quantitatively evaluate the effectiveness of THG-IoT. The parameter M is the total number of device labels.

$$TPR_{AVE} = \frac{1}{M}\sum_{i=1}^{M} TPR_i \qquad (6)$$

$$FPR_{AVE} = \frac{1}{M}\sum_{i=1}^{M} FPR_i \qquad (7)$$

$$FTF_{AVE} = \frac{1}{M}\sum_{i=1}^{M} \frac{TPR_i}{1+FPR_i} \qquad (8)$$

4.3 Experimental Setting

We set three key hyper-parameters in Table 4 and discuss their optimal values next. Moreover, We take dropout [2] with 0.3 ratio to avoid over-fitting, and the Adam optimizer [4] with learning rate 0.005 is adopted. The core part of THG-IoT is implemented with PyTorch framework. Our experiments were executed on a cluster machine where each node had 32 Octa-core Xeons processors running at 2.10 GHz with 128 GB RAM. We discuss the selection of three key hyper-parameters in the training phase of THG-IoT and analyze relevant experimental results under the optimal parameters.

Table 4. Hyper-parameters of THG-IoT.

Parameter	Meaning	Value
w	Sliding window size	[4, 8, 12]
dim	Embedding vector dimension of nodes	[32, 64, 128]
l	Packet sequence length	[8, 16, 32]

4.4 Parameter Study

Hyper-parameters. We use the grid search method to study three pre-defined hyper-parameters and use the metrics FTF_{AVE} to evaluate the classification performance of different parameter combinations. The results are shown in Fig. 5, and it can be seen that the best performance is obtained when $w = 12, dim = 128, l = 32$, at this time $FTF_{AVE} = 95.07\%$. Moreover, we observe that the FTF_{AVE} of IoT devices has increased as the value of hyper-parameters rises. The reasons for this phenomenon are summarized as follows:

- The increase of the w value allows the packet to pay attention to more other packets near itself, and the information carried by the edge weight formed by packets and packets is more refined, which is beneficial to characterize the flow better. As the value of w increases, it takes to construct a heterogeneous graph also increases quadratic complexity.
- The increase of the dim value makes the representation form of the flow more complicated and makes the representation contain more information of different dimensions, which is beneficial to improve the classification performance of the flow directly. However, a larger dim means that the entire training model becomes more complicated, increasing the training time and reducing training efficiency.
- The increase of the l value allows the flows to be associated with more packets, thereby increasing the difference between different flows. The more packets that the flow pays attention to, the better structure information of the entire flow can be extracted. Similarly, the increase in the value of l makes the construction of the graph and the model's training more complicated.

We finally determine $w = 12, dim = 128, l = 32$ for THG-IoT in view of the aforementioned discussion. Then we compiled the results of device identification, and the confusion matrix of the THG-IoT is presented in Fig. 6. What is striking about the figures is the diagonal that implies correct classification, and the values are almost higher than 90%. Specifically, we observed that among the 13 IoT devices, the recall rate of ID0, ID2, ID8, ID11 reached 98%, which indicates that our method has the excellent ability to identify these IoT devices. Only the classification result of ID6 is not satisfactory, and sometimes it will be classified as ID3, which is presumably related to the fact that they are both "Camera device".

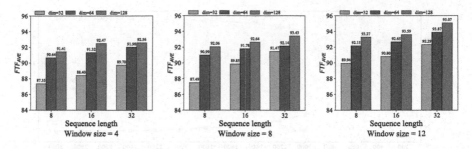

Fig. 5. FTF_{AVE} under different parameter combinations

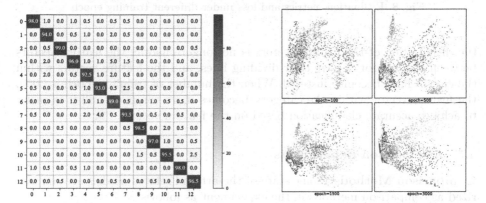

Fig. 6. Confusion matrix   **Fig. 7.** Feature vector visualization

Train Epoch Analysis. In order to find the balance between the classification performance and the number of training epoch, we tested the models trained in different epochs in real-time, and the results obtained are shown in Fig. 8. From the trend of the line graph, it can be found that when the number of training epoch is less than 1000, both TPR_{AVE} and FTF_{AVE} continue to increase, FPR_{AVE} continues to decrease, and the value of loss drops from 2.4230 to 0.4498. When the number of training epochs reaches 1800, TPR_{AVE}, FTF_{AVE} and FPR_{AVE} no longer change significantly, and the corresponding loss value no longer decreases significantly.

In addition, we visualized the feature vector of the flows during training. The high-latitude feature vector is reduced to 2 dimensions by PCA and displayed on the plane, as shown in Fig. 7. Each point in the figure represents a representation vector of a flow after dimensionality reduction, and different colors represent different classes of flows. When the number of training epoch is 100, it can be seen that all the feature vectors are chaotic, and it isn't easy to make a classification. When the number of training epoch is 500, although there are still chaotic local performances, the overall chaotic degree of the feature vector is significantly reduced, and the classification boundary can be easily found for some classes of feature vectors. When the number of training epoch is 1500,

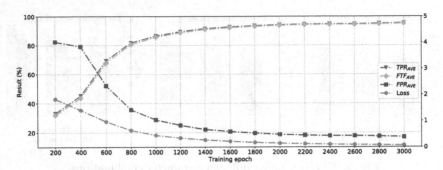

Fig. 8. Evaluation metrics and loss under different training epoch

the distribution of the feature vectors is no longer chaotic, the feature vectors between the categories have clear dividing lines, and the feature vectors within the classes have become clustered. When the number of training epoch is 3000, the feature vector within the category becomes more compact, and it is easier to achieve accurate classification based on the feature vector.

4.5 Comparison Experiments

Comparison Methods. Some state-of-the-art and related models are summarized as comparison methods in the experiment as follows:

- FoSM [6] uses message type sequences of TLS to build first-order Markov model and takes the flow with the maximum probability as the classification result.
- SOB [15] combines the length of the certificate and the length of the first communication data packet into a two-tuple, and determines the final classification result together with the message type sequences. We take 100 as the number of bi-gram clustering.
- MaMPF [8] uses the output probabilities of the message type and the length block Markov models as features to classify encrypted traffic by the random forest. We take 40 as the number of trees in random forest.
- LSTM is an improved recurrent neural network that can solve long-distance dependence problems. The input word vector dimension is specified as 32, and the output vector dimension is specified as 128.
- FS-Net [9] is an end-to-end classification model that only uses packet length information, which adopts a multi-layer encoder-decoder structure to mine the potential sequential characteristics of flows deeply.

Comparison Results. The comparison results are shown in Table 5. We adopt the 5-fold cross-validation experiment for all methods and demonstrate the average and standard deviation of each evaluation metrics. From Table 5, We can obtain the following conclusions.

Table 5. Experimental results with different methods

	FoSM			SOB		
ID	TPR	FPR	FTF	TPR	FPR	FTF
0	13.10 ± 3.02	1.76 ± 0.32	12.88 ± 3.00	54.90 ± 19.19	19.79 ± 3.82	45.44 ± 14.63
1	32.20 ± 2.94	3.18 ± 0.73	31.20 ± 2.79	38.60 ± 8.36	4.35 ± 1.77	37.11 ± 8.42
2	63.50 ± 3.78	5.13 ± 0.51	60.40 ± 3.59	49.60 ± 8.10	3.04 ± 0.90	48.13 ± 7.83
3	36.60 ± 8.24	8.87 ± 2.81	33.45 ± 6.75	49.30 ± 4.71	8.12 ± 1.04	45.61 ± 4.44
4	4.40 ± 2.13	1.19 ± 0.27	4.34 ± 2.09	18.00 ± 6.35	4.48 ± 2.34	17.19 ± 5.92
5	14.60 ± 1.11	1.75 ± 0.40	14.35 ± 1.14	82.80 ± 3.19	1.36 ± 0.34	81.69 ± 3.24
6	52.50 ± 3.30	2.69 ± 0.28	51.12 ± 3.23	57.80 ± 21.63	4.65 ± 1.55	55.54 ± 21.56
7	93.40 ± 1.77	5.26 ± 0.29	88.74 ± 1.88	85.70 ± 1.60	0.61 ± 0.10	85.18 ± 1.59
8	31.50 ± 2.77	8.11 ± 0.54	29.13 ± 2.48	14.80 ± 1.44	1.66 ± 0.82	14.55 ± 1.37
9	79.20 ± 3.33	4.08 ± 0.49	76.09 ± 3.08	24.80 ± 8.44	2.41 ± 1.66	24.12 ± 7.84
10	45.30 ± 3.59	4.45 ± 0.49	43.36 ± 3.29	53.00 ± 9.68	4.05 ± 1.97	50.91 ± 9.09
11	43.00 ± 5.01	11.56 ± 2.66	38.49 ± 3.87	34.70 ± 12.00	2.72 ± 1.07	33.80 ± 11.77
12	30.40 ± 2.50	5.32 ± 0.72	28.85 ± 2.27	20.50 ± 7.71	2.38 ± 1.47	20.05 ± 7.55
AVE	41.52 ± 0.35	4.87 ± 0.03	39.42 ± 0.37	44.96 ± 3.69	4.59 ± 0.31	43.02 ± 3.41

	MaMPF			LSTM		
ID	TPR	FPR	FTF	TPR	FPR	FTF
0	57.30 ± 4.57	5.73 ± 0.33	54.20 ± 4.33	84.40 ± 20.70	**0.11 ± 0.09**	84.32 ± 20.72
1	60.60 ± 2.82	3.05 ± 0.58	58.80 ± 2.61	91.00 ± 5.81	2.02 ± 0.89	89.17 ± 5.09
2	68.20 ± 2.94	3.31 ± 0.39	66.01 ± 2.81	90.40 ± 4.26	1.39 ± 0.91	89.17 ± 4.28
3	35.80 ± 3.28	3.08 ± 0.22	34.72 ± 3.13	74.90 ± 9.12	1.08 ± 0.97	74.16 ± 9.46
4	25.30 ± 2.04	3.84 ± 1.52	24.35 ± 1.72	87.20 ± 9.50	1.75 ± 1.14	85.63 ± 8.66
5	17.30 ± 1.81	0.81 ± 0.18	17.16 ± 1.81	82.50 ± 12.77	0.63 ± 0.46	81.97 ± 12.58
6	64.60 ± 1.28	2.38 ± 0.31	63.10 ± 1.35	**98.00 ± 0.84**	0.36 ± 0.20	**97.65 ± 0.74**
7	93.20 ± 2.11	5.12 ± 0.35	88.66 ± 1.98	93.20 ± 4.81	1.29 ± 0.95	92.00 ± 4.56
8	53.40 ± 5.99	11.68 ± 1.67	47.75 ± 4.72	96.70 ± 2.38	0.64 ± 0.50	96.08 ± 2.15
9	75.10 ± 3.89	2.07 ± 0.26	73.58 ± 3.87	95.80 ± 1.94	0.27 ± 0.17	95.54 ± 1.93
10	62.20 ± 2.01	2.62 ± 0.32	60.61 ± 1.96	97.60 ± 0.86	**0.23 ± 0.07**	97.38 ± 0.90
11	40.50 ± 3.39	4.01 ± 1.81	38.90 ± 2.65	89.80 ± 8.42	1.38 ± 1.33	88.55 ± 7.97
12	44.70 ± 5.14	2.45 ± 0.42	43.61 ± 4.86	78.60 ± 8.26	0.51 ± 0.10	78.21 ± 8.28
AVE	53.71 ± 0.47	3.86 ± 0.04	51.65 ± 0.44	89.24 ± 0.57	0.90 ± 0.04	88.45 ± 0.60

	FS-Net			THG-IoT (Our proposal)		
ID	TPR	FPR	FTF	TPR	FPR	FTF
0	97.80 ± 1.54	0.27 ± 0.17	97.54 ± 1.46	**97.80 ± 0.24**	0.22 ± 0.06	**97.59 ± 0.21**
1	93.40 ± 1.77	**0.38 ± 0.20**	93.04 ± 1.79	**95.30 ± 1.29**	0.41 ± 0.08	**94.91 ± 1.21**
2	97.10 ± 0.86	0.22 ± 0.14	96.89 ± 0.76	**98.80 ± 0.24**	0.09 ± 0.02	**98.71 ± 0.26**
3	93.00 ± 2.07	1.22 ± 0.13	91.88 ± 1.98	**95.40 ± 0.49**	0.74 ± 0.15	**94.70 ± 0.35**
4	81.30 ± 6.27	**0.48 ± 0.25**	80.90 ± 6.10	**94.60 ± 1.96**	0.59 ± 0.05	**94.04 ± 1.96**
5	84.30 ± 6.12	0.49 ± 0.51	83.86 ± 5.68	**92.30 ± 0.40**	0.42 ± 0.05	**91.92 ± 0.39**
6	88.00 ± 1.67	**0.33 ± 0.15**	87.71 ± 1.65	88.00 ± 0.95	0.46 ± 0.03	87.60 ± 0.95
7	91.40 ± 4.45	2.18 ± 1.07	89.42 ± 3.62	**93.60 ± 0.20**	0.61 ± 0.06	**93.01 ± 0.17**
8	95.00 ± 1.58	0.59 ± 0.33	94.44 ± 1.32	**97.30 ± 0.81**	0.38 ± 0.07	**96.94 ± 0.87**
9	95.40 ± 0.58	0.19 ± 0.11	95.22 ± 0.56	**97.70 ± 0.68**	0.13 ± 0.00	**97.58 ± 0.68**
10	91.60 ± 2.97	0.97 ± 0.30	90.72 ± 2.70	94.60 ± 0.73	0.50 ± 0.09	94.13 ± 0.81
11	96.40 ± 1.59	0.53 ± 0.16	95.90 ± 1.61	**97.20 ± 0.68**	0.30 ± 0.04	**96.91 ± 0.71**
12	93.30 ± 3.04	0.66 ± 0.22	92.69 ± 2.92	**96.50 ± 0.00**	0.22 ± 0.02	**96.29 ± 0.02**
AVE	92.15 ± 0.39	0.65 ± 0.03	91.55 ± 0.42	**95.32 ± 0.06**	**0.39 ± 0.01**	**94.95 ± 0.06**

- THG-IoT achieves the best performance and outperforms all other methods. It can be seen from Table 5 that THG-IoT obtains the best TPR_{AVE}, FTF_{AVE} and FPR_{AVE}, and 11 out of 13 IoT devices obtain the best FTF performance. The variance of all categories in the five-fold cross-validation experiment is less than 2%, indicating that THG-IoT has better robustness.
- THG-IoT captures the structural information between packets better than the Markov model. Both FoSM and SOB are typical Markov models. From the results of comparative experiments, we can see that our method is better than these two methods. The Markov model can only capture the information of adjacent packets for the characterization of the sequence, while THG-IoT controls the establishment of the relationship between the packet and other nearby packets according to the parameter w, and captures information from a more extensive range.
- THG-IoT characterizes the heterogeneous information of the flow better than the sequence model. MaMPF, LSTM, and FS-Net take the packet length or message type as the sequence input and model the flow based on the sequence model. In addition, MaMPF uses a random forest classifier with good performance, LSTM uses a cache mechanism to obtain global information better, and FS-Net uses a bi-GRU network to increase the model's functionality. However, the sequence model cannot uniformly capture the structural heterogeneous information between any two packets. THG-IoT compensates for this defect by reconstructing the alternating graph between packets and flows.

4.6   Variant of THG-IoT

Message type sequence and packet length sequence are used as the critical input to characterize the flow. FoSM and SOB only use message type sequence, FS-Net only uses data packet length sequence, and MaMPF and LSTM use the combined sequence of the two. To facilitate comparison and analysis of the performance of different input sequences, we set up three variants for different input sequences and named them THG-IoT-M, THG-IoT-L, and THG-IoT-ML. The results of the experiment are shown in Table 6.

From Table 6, it can be seen that the performance of the combined sequence method is much higher than that of the message type sequence or the packet length sequence alone, which indicates that the more complete the sequence information, the more unique the characterization of different flows. Besides, using the packet length sequence alone is better than using the message type sequence alone, which demonstrates that there are more elements and richer information in the packet length sequence than the message type sequence, thereby increasing the flow of distinction.

Application data is a particular TLS message type that is used to mark the packets transmitted between the client and the server. The *Application data* packets have unique client information, which helps to increase the distinction between different classes of flows. We consider a practical application scenario that the system needs to prevent certain IoT devices from transmitting data as soon as possible, which means that application data packets should be used as

Table 6. Variants of the input sequence

Method variants	TPR_{AVE}	FTF_{AVE}	FPR_{AVE}
THG-IoT-M	64.12	2.81	62.36
THG-IoT-L	86.54	1.10	85.60
THG-IoT-ML	**95.42**	**0.38**	**95.07**

Table 7. Variants of the application data packets

Method variants	TPR_{AVE}	FTF_{AVE}	FPR_{AVE}
THG-IoT-0	87.15	0.92	86.33
THG-IoT-2	88.73	0.93	87.93
THG-IoT-4	91.38	0.71	90.75
THG-IoT-6	93.73	0.51	93.26
THG-IoT-all	**95.42**	**0.38**	**95.07**

little as possible for device identification. In response to this scenario, we set up five variants corresponding to the number of application data packets used as 0, 2, 4, 6, and all.

The experiment results shown in Table 7 are in line with objective application scenario predictions. As the number of application data packets used increases, the performance of IoT device identification gradually becomes more robust. Using all Application data packets improves the performance of TPR_{AVE} by 8% compared to not using Application data packets. Using only 4 Application data packets can make TPR_{AVE} and FTF_{AVE} more than 90%. We adopt specific variants according to different application scenarios.

5   Related Work

Feature Engineering Based. Marchal *et al.* [11] proposed AuDI, a system that identifies IoT devices by analyzing the communication characteristics of device traffic. AuDI analyzes and calculates the periodicity and stability of each device's traffic communication process and uses the kNN clustering algorithm to classify the 33 extracted features. Msadek *et al.* [12] proposed a classification method for IoT devices using statistical characteristics of data packet headers is proposed. Select the relevant flow through the improved sliding window technology. Then, send the header field information and the flow statistics information into the machine learning model to achieve device classification. Thangavelu *et al.* [17] introduced DEFT, a scalable distributed method for fingerprinting IoT devices based on network traffic. DEFT performs device classification at the gateway closer to the IoT device, and the controller and the gateway coordinate to identify new devices in the network. According to the data packet header, the statistical characteristics of common IoT application layer protocols are extracted, and the K-means clustering algorithm and random forest are used for classification.

Deep Learning Based. Aneja *et al.* [1] used Inter-arrival Time (IAT) to construct fingerprints of IoT devices. In the article, the author proposes a new idea for constructing device fingerprints. First, the author draws IAT diagrams of data packets for different IoT devices and processes them into reference images suitable for deep learning libraries. Then the author uses deep learning methods to compare these reference images for further classification, and an accuracy rate of 86.7% was finally achieved based on CNN. Jafari *et al.* [3] used a radio frequency fingerprint as a physical layer authentication method to identify six ZigBee devices from the same manufacturer. This paper uses three deep learning models of deep learning neural network (DNN), convolutional neural network (CNN), and long short-term memory network (LSTM) to model the radio frequency data set collected on the USRP platform. Ortiz *et al.* [13] constructed a probabilistic framework called DeviceMien to provide meaningful feedback in IoT device identification. In the case of unsupervised and unlabeled data, the framework uses a stacked LSTM Autoencoder to automatically learn features and traffic categories from device traffic data packets and probabilistically perform a probabilistic analysis on each device based on the distribution of traffic categories. Lopez-Martin [10] tried to combine recurrent neural networks (RNN) and convolutional neural networks (CNN) to detect the traffic of IoT devices. This method does not require any feature engineering and has better detection results.

6   Conclusions and Future Work

In this paper, we proposed THG-IoT, which can identify IoT devices using the graph convolutional network. We constructed the THG of encrypted flows based on the packet length and TLS message type and turned the IoT device identification problem into the graph node classification problem. THG implicitly reserves multiple heterogeneous information between packets and flows. Then we built a GCN-based classifier that using THG as input for IoT device identification. The abundant experimental results proved the effectiveness of our proposed method, which improved the TPR and FTF by at least 4% over the state-of-the-art.

For future work, we will explore the identification of IoT devices under the new version of TLS traffic. In the TLS1.3 version, all message types after *Server-Hello* are encrypted, and the plaintext is significantly reduced. This poses a great challenge to the method of using the TLS message type sequence.

Acknowledgement. We thank the anonymous reviewers for their insightful comments. This work was supported by the National Key Research and Development Program of China (Grant No. 2019YFB1804504).

References

1. Aneja, S., Aneja, N., Islam, M.S.: IoT device fingerprint using deep learning. In: 2018 IEEE International Conference on Internet of Things and Intelligence System (IOTAIS), pp. 174–179. IEEE (2018)
2. Hinton, G.E., Srivastava, N., Krizhevsky, A., Sutskever, I., Salakhutdinov, R.R.: Improving neural networks by preventing co-adaptation of feature detectors. arXiv preprint arXiv:1207.0580 (2012)
3. Jafari, H., Omotere, O., Adesina, D., Wu, H.H., Qian, L.: IoT devices fingerprinting using deep learning. In: MILCOM 2018-2018 IEEE Military Communications Conference (MILCOM), pp. 1–9. IEEE (2018)
4. Kingma, D.P., Ba, J.: Adam: a method for stochastic optimization. arXiv preprint arXiv:1412.6980 (2014)
5. Kipf, T.N., Welling, M.: Semi-supervised classification with graph convolutional networks. arXiv preprint arXiv:1609.02907 (2016)
6. Korczyński, M., Duda, A.: Markov chain fingerprinting to classify encrypted traffic. In: IEEE INFOCOM 2014-IEEE Conference on Computer Communications, pp. 781–789. IEEE (2014)
7. Li, Q., Han, Z., Wu, X.M.: Deeper insights into graph convolutional networks for semi-supervised learning. In: Proceedings of the AAAI Conference on Artificial Intelligence, vol. 32 (2018)
8. Liu, C., Cao, Z., Xiong, G., Gou, G., Yiu, S.M., He, L.: MaMPF: encrypted traffic classification based on multi-attribute Markov probability fingerprints. In: 2018 IEEE/ACM 26th International Symposium on Quality of Service (IWQoS), pp. 1–10. IEEE (2018)
9. Liu, C., He, L., Xiong, G., Cao, Z., Li, Z.: Fs-Net: a flow sequence network for encrypted traffic classification. In: IEEE INFOCOM 2019-IEEE Conference on Computer Communications, pp. 1171–1179. IEEE (2019)
10. Lopez-Martin, M., Carro, B., Sanchez-Esguevillas, A., Lloret, J.: Network traffic classifier with convolutional and recurrent neural networks for internet of things. IEEE Access 5, 18042–18050 (2017)
11. Marchal, S., Miettinen, M., Nguyen, T.D., Sadeghi, A.R., Asokan, N.: AuDI: toward autonomous IoT device-type identification using periodic communication. IEEE J. Sel. Areas Commun. 37(6), 1402–1412 (2019)
12. Msadek, N., Soua, R., Engel, T.: IoT device fingerprinting: machine learning based encrypted traffic analysis. In: 2019 IEEE Wireless Communications and Networking Conference (WCNC), pp. 1–8. IEEE (2019)
13. Ortiz, J., Crawford, C., Le, F.: DeviceMien: network device behavior modeling for identifying unknown IoT devices. In: Proceedings of the International Conference on Internet of Things Design and Implementation, pp. 106–117 (2019)
14. Riad, K., Huang, T., Ke, L.: A dynamic and hierarchical access control for IoT in multi-authority cloud storage. J. Network Comput. Appl. 160, 102633 (2020)
15. Shen, M., Wei, M., Zhu, L., Wang, M.: Classification of encrypted traffic with second-order Markov chains and application attribute bigrams. IEEE Trans. Inf. Forensics Secur. 12(8), 1830–1843 (2017)
16. Sivanathan, A., et al.: Classifying IoT devices in smart environments using network traffic characteristics. IEEE Trans. Mob. Comput. 18(8), 1745–1759 (2018)
17. Thangavelu, V., Divakaran, D.M., Sairam, R., Bhunia, S.S., Gurusamy, M.: DEFT: a distributed IoT fingerprinting technique. IEEE Internet Things J. 6(1), 940–952 (2018)

Data-Driven Influential Nodes Identification in Dynamic Social Networks

Ye Qian[1] and Li Pan[1,2(✉)]

[1] School of Electronic Information and Electrical Engineering, Shanghai Jiao Tong University, Shanghai, China
panli@sjtu.edu.cn
[2] National Engineering Laboratory for Information Content Analysis Technology, Shanghai, China

Abstract. The identification of influential nodes in social networks has significant commercial and academic value in advertising, information management, and user behavior analysis. Previous work only studies the simple topology of the network without considering the dynamic propagation characteristics of the network, which does not fit the actual scene and hinders wide application. To solve the problem, We develop a data-driven model for the identification of influential nodes in dynamic social networks. Firstly, we introduce an influence evaluation metric BTRank based on user interaction behavior and topic relevance of the information. Combining BTRank, LH-index, and betweenness centrality, we construct a multi-scale comprehensive metric system. Secondly, in order to optimize the metric weights calculated by entropy weight method, we use simulation data to train a regression model and obtain the metric weights by Gradient Descent Algorithm. Thirdly, the weights obtained from training are used in weighted TOPSIS to sort the influence of nodes and identify influential nodes among them. Finally, We compare our model with existing models on four real-world networks. The experimental results have demonstrated significant improvement in both accuracy and effectiveness achieved by our proposed model.

Keywords: Social networks · Influential nodes · Data-driven

1 Introduction

With the development of the mobile internet, the online social network plays a vital role in information dissemination among a vast number of users. The social network can be abstractly regarded as a network topology model composed of nodes and edges, which can reflect the social relations among social individuals. Compared with other nodes, influential nodes in social networks are more likely to affect the state of nearby nodes, which makes the information spread more widely. If we identify influential nodes quickly and accurately, it will be better for the government to achieve the guidance and control of major public feeling's affairs. For business, influential nodes identification applied to precision marketing will effectively improve merchant marketing efficiency and reduce promotional costs. Therefore, the research on influential nodes has excellent theoretical and practical values.

© ICST Institute for Computer Sciences, Social Informatics and Telecommunications Engineering 2021
Published by Springer Nature Switzerland AG 2021. All Rights Reserved
H. Gao and X. Wang (Eds.): CollaborateCom 2021, LNICST 406, pp. 592–607, 2021.
https://doi.org/10.1007/978-3-030-92635-9_34

Previous work proposes a series of models to measure the influence of nodes, which can be summarized as network structure-based methods, topic-based methods, transfer entropy-based methods, etc. [1]. Although these models provide detailed calculation methods for node influence evaluation, all of them use a single metric to measure the influence, which cannot comprehensively and accurately measure the influence of nodes. More recently, researchers regard identifying influential nodes in complex networks as a multi-attribute decision-making problem. Considering the uncertain information fusion, text variables, and other factors, some mathematical tools such as evidence theory and fuzzy set theory are also applied to the identification of influential nodes [2]. According to the similarity between the value of each metric and the ideal value, Liern et al. [3] propose the TOPSIS decision model. While Lu et al. [4] take degree centrality, betweenness centrality, and structural hole as the input of the decision model to measure the node's propagation influence in the complex network. However, these influential node identification models still have three main disadvantages. Firstly, the selection of metrics for the evaluation model is unreasonable. The reason is that they only focus on the metrics related to the network structure without considering the dynamic propagation characteristics of the network, such as degree centrality, betweenness centrality, closeness centrality, and other simple traditional centrality indicators. Secondly, existing models do not maximize the integration of multiple metrics that reflect various aspects of node characteristics. Some decision models assign the same weight to the input metrics, which is unreasonable. Different metrics are obtained by different algorithms, and they play different roles in the network. Other models calculate the internal correlation of indicator value to obtain weights. They only consider the relative importance of the indicators and do not consider the difference between the calculated influence of the nodes and the actual results. Thirdly, the selection of indicators for the decision model lacks theoretical support and ignores the logical derivation process.

To address the above problems, we propose a data-driven influential nodes identification model (DINI). Firstly, we propose a topic-level influence evaluation metric BTRank, which is used to describe the topics of information spread by users and the frequency of interaction between users in the time interval. BTRank is an indicator that takes the dynamic propagation characteristics of social networks into account. In order to comprehensively measure the influence of nodes, we combine BTRank, LH-index, and betweenness centrality to establish a comprehensive metric system, where LH-index reflects node's influence in the local network, and betweenness centrality reflects node's influence in the global network. Then, we propose a data-driven weight optimization algorithm to assign reasonable weights to different metrics, facilitating the integration of multiple metrics. Different from the average distribution method or entropy weight method, the data-driven weight optimization algorithm not only considers the information entropy of each metric but also adds the prior knowledge of ground truth, which optimizes the weight and makes it more reasonable and adaptive. Specifically, we first use the entropy weight method to obtain the initial weights and then set the loss function between the influence results obtained by the proposed model and the actual influence value. Finally, we use Gradient Descent Algorithm to optimize the weight of each metric gradually. We apply the optimized weights to the weighted

TOPSIS model to get the ranking of nodes' influence and select the top 1% nodes as influential nodes. The main contributions of this paper can be summarized as follows:

- We propose a topic-level influence evaluation metric BTRank, which describes the interaction behavior among users and the topic relevance of information dissemination. BTRank is a node influence evaluation metric that takes into account the dynamic propagation characteristics of networks.
- We establish a comprehensive metric system, which considers not only the static and dynamic network characteristics but also the local and global network structure, to make the quantitative analysis of social networks more accurate.
- We propose a data-driven weight optimization algorithm. After using the entropy weight method to calculate the weight of each metric, we add the prior knowledge of ground truth to establish the function of the difference between the estimated value and the actual value. Gradient Descent Algorithm is used to optimize the weights to make them more reasonable and more adaptive.
- We apply the weights obtained by the data-driven weight optimization algorithm to weighted TOPSIS and complete the identification of influential nodes. Experiments conducted on four real-world networks indicate that our proposed DINI model outperforms some baselines in both accuracy and effectiveness.

The rest of the paper is organized as follows. Section 2 reviews related work. We describe our data-driven model for identifying influential nodes in social networks in Sect. 3. Section 4 presents details and results of the experiments. Finally, we conclude our work and describe future research directions in Sect. 5.

2   Related Work

The measurement of nodes' influence in social networks is mainly used to identify influential nodes in social networks. Commonly used methods for studying the influence of nodes are network structure-based method, behavior-based method, transfer entropy-based method, and so on [1].

In the previous study, the centrality of nodes is used to measure the influence of nodes in complex networks [5]. Hu et al. [6] use the concept of degree centrality, which calculates the influence of a node with the node's out-degree. Degree centrality is intuitive and efficient. However, it only reflects the local structural characteristics of nodes and ignores the global characteristics. Considering the global characteristics, Singh et al. [7] propose betweenness centrality and closeness centrality based on a faster algorithm. The difference between the two algorithms is that the former investigates the number of shortest paths through the node, while the latter considers the average length of the shortest paths from the node to other nodes. However, in the actual scene, the scale of social networks is huge, and the computational complexity of these two kinds of centrality is high, which are not very efficient methods to identify influential nodes in social networks. Wang et al. [8] attach importance to those nodes that are more central in the network, even if their degree centrality is not high. Based on such an idea, they propose a K-shell algorithm. In the early stage, social network reflects homogeneity. In other words, the nodes with high out-degree tend to have neighbors with high out-degree. With the gradual evolution of social networks, the number of nodes keeps increasing, and some social networks begin to reflect heterogeneity [9]. Therefore, the simple method based on network topology cannot solve the problem of identifying influential nodes in social networks.

In recent years, some scholars combine information content with network topology. Yu et al. [10] use the sum of three factors, the quality of the user's tweets, the frequency of user's tweets being forwarded, and the similarity of interests between users, as the social influence of the user. Sun et al. [11] combine the amount of information in users' text content, users' emotions, and fans' behaviors to measure user's influence. Some scholars obtain user's influence by studying the behavior of users spreading information. The greedy algorithm proposed by Ren et al. [12] solves the problem of influence transmission in the network and makes the information of users as the initial nodes spread most widely in the network. Zheng et al. [13] consider that users who log on social media frequently and the number of whose neighbor nodes is increasing should have a stronger influence. In actual social media, the user's influence is related to the topics of information [14, 15]. The EIRank algorithm proposed by Bo et al. [16] introduces the topic factor into the measurement of node's influence. EIRank analyzes the information spread by users, summarizes the topics that users are interested in, and finds out the relation between the topics. This relation is combined with the static network structure to obtain the influence of nodes.

The above methods define and measure the influence of nodes in social networks from different perspectives. However, social networks in the real world are more complex than experimental settings, including network topology, user attributes, user behavior, topics of interactive information, and so on. The process of information dissemination among nodes is affected by the above factors. Therefore, in order to make the model more applicable to real-world social networks, it is necessary to combine the network topology and social network propagation elements effectively. Based on this key point, we establish a comprehensive metric system that includes not only static and dynamic network characteristics but also local and global network structure. Moreover, the influential nodes in social networks can be identified more accurately through a more reasonable weight allocation method.

3   Data-Driven Model for Influential Nodes Identification in Social Networks

Considering the dynamic propagation characteristics of real-world social networks, we establish a data-driven influential nodes identification model (DINI), which mainly contains two modules: multi-scale comprehensive metric system and data-driven metric weight optimization, as shown in Fig. 1.

Firstly, we propose an indicator BTRank to describe the influence of nodes at the topic level based on three elements: the topics of blogs posted by users, the structure of social networks, and the interaction behavior among users. Then, we combine BTRank with LH-index and betweenness centrality, which are based on local and global network structure, respectively, to form a comprehensive multi-scale metric system. In the metric weight optimization module, we use the entropy weight method to get the initial weight of each metric and then use the data-driven way to optimize the metric weights. Finally, the metric weights are used in the weighted TOPSIS model to obtain the node influence ranking.

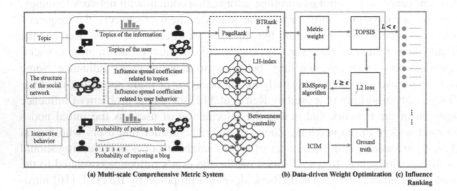

(a) Multi-scale Comprehensive Metric System   (b) Data-driven Weight Optimization (c) Influence Ranking

Fig. 1. The framework of our data-driven influential nodes identification model in dynamic social networks

3.1 Multi-scale Comprehensive Metric System

In order to deal with the inaccurate quantification of nodes' influence in previous models, we establish a multi-scale metric system, as shown in Fig. 1(a). The metric system consists of three metrics. The specific description of each metric is as follows:

(1) Topic-level Influence Measurement Based on User Behavior

In social networks, users are influenced by the content of information from the source users, and users also influence the spread of information from the source users. Then, We can calculate the influence of the users through mining the behavioral characteristics of users. The dissemination of information between users is mainly affected by two factors. The first factor is the user's interest in this information, which varies with users and information. Therefore, it is necessary to model the user's topics and the content of information spread among users. LDA is used to obtain the topic distribution vectors of users and information [17]. According to the distribution vectors, the probability of the user responding to other users will be calculated. The second factor is the interaction level between users. If user v often forwards blogs published by user u, it indicates that the interaction level between user u and user v is high, and user v is likely to continue forwarding blogs published by user u in the future. These two factors are critical in calculating topic-level influence based on user behavior. The influence of user u is denoted by $BR(u)$.

Given a social network $G = (V, E)$, where V represents the set of users and E represents the edges of nodes due to their interaction behavior. The topic-level influence measurement algorithm based on user behavior (BTRank) is divided into three modules:

Based on topic mining technology, the influence propagation coefficient $p(u,v)$ related to topics is calculated. Firstly, the content of blogs published by users is collected, and the unsupervised LDA model is used to obtain the topic distribution vectors of users and blogs. Each user has a topic distribution vector denoted by $\overrightarrow{p_u} = (p_u^1, p_u^2, \ldots, p_u^z)$, where p_u^i denotes the interest level of user u in the ith topic. Information is propagated through the edge, which also has a topic distribution vector denoted by $\overrightarrow{p_{uv}} = (p_{uv}^1, p_{uv}^2, \ldots, p_{uv}^z)$, where p_{uv}^k represents the kth topic's proportion of the information propagated through user u to user v. Finally, the similarity measurement method is used to calculate the influence propagation coefficient $p(u,v)$ related to topics among users.

Based on the temporal information of users publishing or forwarding blogs, the influence propagation coefficient $p(u,v)'$ related to behavior is calculated. We use a statistical method to calculate the probability of users publishing or forwarding blogs in each period. Users publish and forward blogs with obvious periodicity. A day is divided into 24 disjoint time intervals, $p_u(t_i)$ represents the probability of user u publishing blogs in the ith time interval, $\widetilde{p}_v(t_j)$ represents the probability of user v forwarding blogs in the jth time interval. Then, the cumulative probability method is used to obtain influence propagation coefficient $p(u,v)'$ related to behavior.

The influence propagation coefficient $p(u,v|\overrightarrow{p_{uv}})$ is calculated using formula (1). Based on the influence propagation coefficient $p(u,v|\overrightarrow{p_{uv}})$, the PageRank algorithm is used to measure the influence of each user.

$$p(u,v|\overrightarrow{p_{uv}}) = \sum_{i=1}^{z} p_v^i p_{uv}^i + \sum_{i=1}^{23} p_u(t_i) \sum_{j=i+1}^{24} \widetilde{p}_v(t_j) \tag{1}$$

In the PageRank algorithm, the propagation range of a user's information depends on the number of times the information is forwarded by other users. If user v forwards the information of user u, it can be seen as a voting process of user v to user u. User v will contribute its influence to user u with the probability of $p(u,v|\overrightarrow{p_{uv}})$. The influence of u is the accumulation of several users' influence, which is an iterative process. In each iteration, user u's influence is calculated using formula (2).

$$BR(u) = (1-\lambda) + \lambda \sum_{v \in N(u)} p(u,v|\overrightarrow{p_{uv}}) \cdot BR(v) \tag{2}$$

where $N(u)$ represents the set of user u's neighbor users, $p(u,v|\overrightarrow{p_{uv}})$ is the probability of user v responding to user u, $BR(u)$ is the influence of user u. λ is the damping coefficient used to ensure the convergence of the calculation results, which is generally 0.85.

After calculating the influence propagation coefficient $p(u,v|\overrightarrow{p_{uv}})$ by combining the two influence propagation coefficients, we further measure the influence of users using the PageRank algorithm. We show the BTRank algorithm in Algorithm 1.

(2) Influence Measurement Based on Local Network Structure

For each node in the network, the influence of the node can be measured by calculating the influence of its neighbor nodes. The higher the influence of its neighbor nodes are, the higher the influence of the node is [18].

Algorithm 1. Influence Measurement Based on User Behavior and Topics

---

Input: Social network $G(V, E), \overrightarrow{p_u}, \overrightarrow{p_{uv}}, p_u(t_i), \widetilde{p}_v(t_j)$, damping coefficient λ
Output: User k's influence: $BR(k), k = 1, 2, \ldots, n$
1: **for** $(u, v) \in E$ **do**
2:     Compute coefficient $p(u, v)$ related to topics
3:     Compute coefficient $p(u, v)'$ related to behavior
4:     Compute influence propagation coefficient $p(u, v|\overrightarrow{p_{uv}})$ using formula (1)
5: **end for**
6: **repeat**
7:     **for** $k \in V$ **do**
8:         $BR(k) = (1 - \lambda) + \lambda \sum_{v \in N(k)} p(k, v|\overrightarrow{p_{kv}}) \cdot BR(v)$
9:     **end for**
10: **until** Convergence
11: **return** $BR(k)$

$$LH(i) = h(i) + \sum_{v \in N(i)} h(v) \qquad (3)$$

where $h(i)$ is the H-index of node i, that is, node i has at most h neighbors whose degree exceeds h. $N(i)$ is the set of node i's neighbor nodes.

(3) Influence Measurement Based on Global Network Structure

Betweenness centrality can be used to mine the nodes that play a key role in the transmission of information. The betweenness centrality of node i is denoted by $BC(i)$, which is calculated as follows:

$$BC(i) = \sum_{p \neq i \neq q}^{n} \frac{g_{pq}(i)}{g_{pq}} \qquad (4)$$

where g_{pq} represents the number of shortest paths from node p to node q. $g_{pq}(i)$ is the number of paths passing through node i among the g_{pq} shortest paths.

Using the above three indicators, the metric system includes not only static and dynamic network characteristics but also local and global network structure, which makes the quantitative analysis of social networks more accurate.

3.2   Data-Driven Weight Optimization Algorithm

Most of the existing comprehensive evaluation models use uniform distribution or an entropy weight method to assign weight to each metric. Uniform distribution ignores the differences among indicators, which is not in line with the actual situation. The entropy weight method uses the information entropy of indicators to determine the weight [19], ignoring the prior knowledge of ground truth. Therefore, we propose a data-driven metric weight optimization algorithm, as shown in Fig. 1(b). Based on the weights obtained by the entropy weight method, we add the prior knowledge of ground truth to optimize the metric weights. The influence of nodes can be evaluated more accurately with the weighted TOPSIS algorithm. The specific steps are as follows:

The three metrics are combined to generate metric matrix $D(x_{n\times 3})$. After normalization, the normalized metric matrix $D_1 = (a_{ij})$ is obtained.

$$a_{ij} = \frac{x_{ij}}{\sqrt{\sum_{i=1}^{n} x_{ij}^2}}, j = 1, 2, 3 \tag{5}$$

The entropy weight method is used to get the initial metric weights. In general, if the information entropy of a metric is small, which means that the value of the metric changes greatly, provides more information, and plays an important role in a comprehensive evaluation. Then, the weight of the corresponding metric is large. On the contrary, it indicates that the metric value does not change much, provides little information, and plays a small role in the comprehensive evaluation. Then, the weight of the corresponding indicator is small.

Calculate the weight $p_{ij} = \frac{a_{ij}}{\sqrt{\sum_{i=1}^{n} a_{ij}}}$ of the ith user under the jth indicator. The entropy value $e_j = \frac{-p_{ij} \cdot ln p_{ij}}{ln(n)}$ of the jth indicator is obtained with the weight p_{ij}. Then the entropy weight of the jth metric is calculated:

$$w_j = \frac{1 - e_j}{\sum_{j=1}^{3} 1 - e_j} \tag{6}$$

After using the entropy method to obtain the initial weights, we add the prior knowledge of ground truth to optimize the initial weights. The optimization variable is the weights of the three indicators, which are represented by w_1, w_2 and w_3 in turn. The optimization goal is to calculate the difference between ground truth T_{gt} and nodes' influence T_{cal} obtained by the proposed model. We use the $L2$ distance to measure the difference. The optimization problem is defined as follows:

$$\min_{w_1, w_2, w_3} L(w_1, w_2, w_3) = \|T_{gt} - T_{cal}\|_2$$

$$s.t. \begin{cases} 0 < w_1, w_2, w_3 < 1 \\ w_1 + w_2 + w_3 = 1 \end{cases} \tag{7}$$

We choose the commonly used adaptive learning rate RMSprop algorithm that is one of the gradient descent algorithms [20] to solve the optimization problem. The simulation results of independent cascade model based on the influence propagation coefficient $p(u, v | \overrightarrow{p_{uv}})$ are used to form ground truth. Along the gradient direction of the loss function, each metric is updated as follows:

$$w_i^{t+1} = w_i^t - \frac{\eta}{\sqrt{S_t + \phi}} \cdot \frac{\partial L}{\partial w_i^t}, i = 1, 2 \tag{8}$$

$$S_t = \beta S_{t-1} + (1 - \beta) \left[\frac{\partial L}{\partial w_t}\right]^2 \tag{9}$$

where S_t is the sum of the current and past square gradients, the learning rate η is 0.001, the floating-point ϕ is 10^{-7}, and the weight β is 0.9.

In Algorithm 2, we show the method to optimize each indicator's weight. After traversing all the ground truth and updating the weight, the results are the optimized metric weights. The metric weights optimized in this way learn the prior knowledge of ground truth and can be used for comprehensive evaluation more reasonably. The data-driven metric weight optimization algorithm proposed here is a metric weight optimization method that is not only applicable to node influence evaluation models but can also be extended to other comprehensive evaluation models with multiple metrics.

Algorithm 2. Metric Weight Optimization

Input: Influence metric $LH(k)$, $BC(k)$, $BR(k)$, $k = 1, 2, \ldots, n$, threshold ϵ, ground truth T_{gt}
Output: Metric weights w_1, w_2, w_3
1: **for** $k = 1$ to n **do**
2:     Generate metric matrix $D(x_{k \times 3}) = (LH(k), BC(k), BR(k))$
3: **end for**
4: Normalize D to D_1 using formula (5)
5: Generate entropy weight vector $W = (w_1, w_2, w_3)$ using formula (6)
6: Compute T_{cal} by weighted TOPSIS
7: $t = 1$
8: **while** $\left| L^{t+1} - L^t \right| \geq \varepsilon$ **do**
9:     $L(w_1, w_2, w_3) = \|T_{gt} - T_{cal}\|_2$
10:     Update W by RMSprop algorithm
11:     $t = t + 1$
12: **end while**
13: **return** w_1, w_2, w_3

3.3   Influential Nodes Identification Based on Data-Driven Weighted TOPSIS

In order to integrate the above three dimensions of metrics for measuring nodes' influence, we use a weighted TOPSIS algorithm with the metric weights obtained in Sect. 3.2 to rank nodes.

Weighted TOPSIS follows the following steps to measure nodes' influence [21].

Multiply the normalized metric matrix D_1 with the weight vector to get the weighted metric matrix $D_2 = (b_{ij})$:

$$b_{ij} = w_j \cdot a_{ij}, i = 1, 2, \ldots, n; j = 1, 2, 3 \tag{10}$$

When we establish positive ideal solution A^+ and negative ideal solution A^-, the Euclidean distance S_i between node i's value and the two ideal solutions are calculated respectively. Thus, the proximity $K_i = \frac{S_i^-}{S_i^- + S_i^+}$ to the ideal solution is obtained. Node i with a large K_i value means that it is significant, and the information it posts is more widely spread in the network. Conversely, the influence is small.

Complexity analysis: Using three metrics to measure the influence of n nodes in the network takes $O(nn_\theta + nE + n^2 t(\epsilon_1))$, where n_θ denotes the maximum degree of the node, E is the number of edges, $t(\epsilon_1)$ is the number of iterations, and the number of iterations is related to the threshold ϵ_1 for convergence. It takes $O(nlog(n))$ to rank all nodes in descending order. Combining the three metrics using weighted TOPSIS needs $O(nt(\epsilon_2))$. Where ϵ_2 is the threshold for convergence. Therefore, the worst-case time complexity of our algorithm is $O(nE)$.

4 Experiments and Analysis

4.1 Experimental Setup

Datasets: We compare the proposed DINI model with other models on four real-world social networks, including Email network, Facebook network, Enron network, and Gowalla network, where nodes denote users and edges denote interactions between users. Several basic statistics are listed in Table 1. The table contains information about the number of nodes, the number of edges, and the average degree.

Table 1. Statistics of four real-world networks.

Network	Email	Facebook	Enron	Gowalla
$|V|$	1005	3483	36692	137873
$|E|$	16706	65536	183831	661800
Average degree	33.25	37.63	10.02	9.61

Ground Truth: We use the independent cascade model based on the influence propagation coefficient $p\left(u, v | \overrightarrow{p_{uv}}\right)$ mentioned in Sect. 3.1 (ICIM) to simulate the information transmission process in social networks [22]. Each node activates its neighbors with probability $p(u, v | \overrightarrow{p_{uv}})$. The final result is expressed as $F(u)$, representing the total number of nodes that end up in the active state at the end of the propagation process starting from node u.

Evaluation Metric: The influence of nodes is sorted, and the top 1% nodes are selected as influential nodes. We use Spearman's rank correlation coefficient [23] to measure the accuracy of influential nodes identification. Spearman's rank correlation coefficient is defined as follows:

$$\rho = 1 - \frac{6\sum_{i=1}^{N} d_i^2}{N(N^2 - 1)} \tag{11}$$

where ρ is the rank correlation coefficient, d_i is the difference between the rankings of the same node obtained by the model and ground truth, and N is the number of influential nodes. The more accurate the influential nodes are identified, the larger Spearman's rank correlation coefficient ρ is.

Settings: There is a constant parameter ϵ in Algorithm 2, which is the threshold for convergence. We set $\epsilon = 0.002$ for Email, $\epsilon = 0.005$ for Facebook, $\epsilon = 0.135$ for Enron, $\epsilon = 0.215$ for Gowalla. Experiments were conducted on a machine with an Intel Core CPU i7-9700 at 3 GHz and 32 GB memory.

4.2 Performance Comparison

To identify influential nodes in social networks, we sort the influence results of nodes calculated by our proposed DINI model and select the top 1% nodes as influential nodes. Five typical methods are chosen to be baselines. We compare DINI with baselines, including Entropy-based ranking measure (ERM) [24], WVoteRank [25], New Evidential Centrality (NEC) [2], NLPCA [26] and TOPSIS [4]. These methods are comprehensive evaluation models, which make full use of the network structure and the information transmitted between users to measure the influence of users. The experiments are conducted on real-world social networks mentioned in Sect. 4.1 to demonstrate the rationality of weight distribution and the effectiveness of our proposed DINI model.

Influential Nodes Identification Performance. We use Spearman's rank correlation coefficient [23] to measure the accuracy of influential nodes identification by different models. Spearman's rank correlation coefficient ρ is the correlation between the model results and the simulation results, which is higher when the identification of influential nodes is more accurate. Figure 2 shows that ρ changes with the mean value α of the influence propagation coefficient $p\left(u, v | \overrightarrow{p_{uv}}\right)$, where (a) - (d) shows the results of

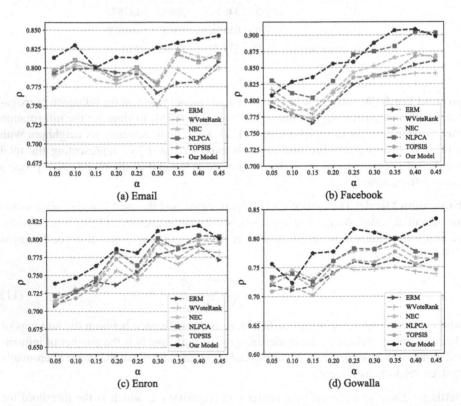

Fig. 2. Correlation between the ranking of influential nodes identified by the influence simulation model and the rankings of the corresponding nodes obtained by our proposed DINI model and other models in (a) Email, (b) Facebook, (c) Enron, (d) Gowalla

various models on four datasets correspondingly. With the increase of α, the overall accuracy of the influential nodes identified by different models trends to increase.

As can be seen from Fig. 2, the model proposed in this paper outperforms other models generally. The most apparent superiority is in (a), when $\alpha = 0.3$, where shows a 5% improvement over the second-place method NEC. Although in rare cases, such as $\alpha = 0.05$ or $\alpha = 0.25$ in (b), The accuracy of our model is 2% lower than that of NLPCA, we still have a far leading efficiency advantage, which is described in Sect. 4.2 in detail. The reason for these cases is that when the value of α is small, the information spreads slowly in the network, and the nodes in the critical position of the network can make the information spread more widely, which makes the model based on the network structure identify influential nodes more effectively. Besides, our proposed model has inherent scalability that is not available in other methods, proving by the consistent advantage across four different data sets. In summary, our model can identify influential nodes more accurately in different data sets and link the propagation characteristics of information with the node's attributes, which can better measure the node's influence.

Comparison on the Activation Capability of Influential Nodes. In general, if a node is considered as an influential node, it should be able to activate more nodes in the independent cascade model based on the influence propagation coefficient $p\left(u, v|\overrightarrow{p_{uv}}\right)$ [22]. Therefore, the top 1% of nodes proposed by each algorithm are selected as the initial

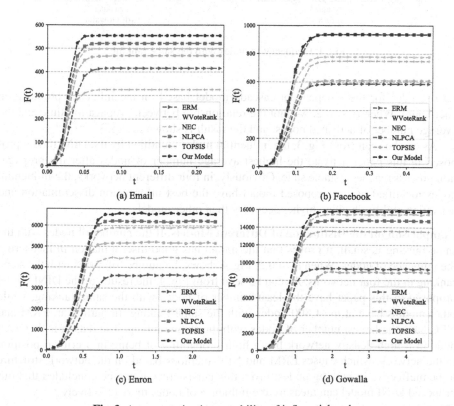

Fig. 3. Average activation capability of influential nodes

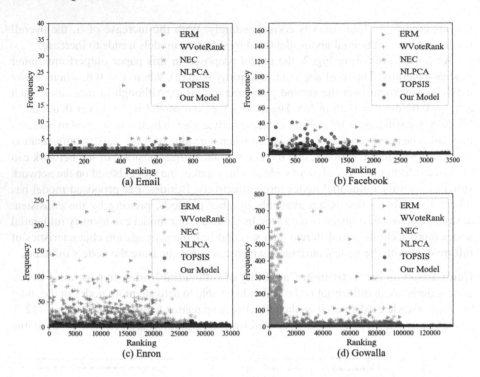

Fig. 4. Frequency of nodes with the same ranking

set of nodes for the independent cascade model based on the influence propagation coefficient. When $\alpha = 0.2$, we plot the relationship between propagation time and the average number of activated nodes, and the result is shown in Fig. 3.

As can be seen from Fig. 3, the influential nodes identified by the DINI model proposed in this paper activate the highest average number of nodes after the propagation process reaches a stable state. Obviously, in four different networks, the influential nodes identified by our proposed model have the best information dissemination and activation ability, showing higher accuracy than other models.

Comparison on the Distinction of Different Models. If the number of nodes with the same ranking is small, it proves that the model has good discrimination in measuring the influence of nodes. When $\alpha = 0.2$, Fig. 4 shows the number of nodes with the same ranking in the four networks. As can be seen from Fig. 4, the results of the DINI model proposed in this paper have the least number of nodes with the same ranking, while other models have at most 800 nodes with the same ranking. In particular, ERM and NEC have the largest number of nodes with the same ranking. The reason is that each node in a high-density network has a high probability of being in a critical position in the network, which causes ERM and NEC methods based on the network structure to be ineffective in ranking nodes. From this perspective, it can be concluded that our proposed DINI model can measure the influence of nodes more effectively.

Comparison on the Running Time of Different Models. The running time of models is a common index to measure the performance of models. When $\alpha = 0.2$, Fig. 5 depicts the running time required to select the influential nodes using six different models on four different networks. It is obvious that the running time of our DINI model is almost the same as that of the WVoteRank, and it is more efficient than NEC and NLPCA. For small networks, such as Email and Facebook, it takes no more than 10 s for our model to complete the identification of influential nodes. For Gowalla network with 137873 nodes and 661800 edges, its running time can also be controlled within 200 s. Though our model is slightly less efficient compared with ERM and WVoteRank, our model shows superiority over them in the accuracy of identifying influential nodes. Therefore, our model has a high time efficiency on the whole.

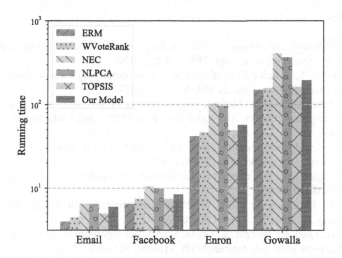

Fig. 5. Running time of six different models

5   Conclusion and Future Work

In this paper, we propose a data-driven model for influential nodes identification in dynamic social networks. To make the measurement of nodes' influence reflect the nodes' real situation in the social network, we mainly focus on three aspects, including the quantitative indicators, the metric evaluation system, and experiment settings. Firstly, we propose a topic-level indicator BTRank to quantify the influence based on user interaction behavior and topics of the information. Combining BTRank, which reflects the dynamic propagation characteristics of the network, with LH-index and betweenness centrality, which reflect the static topology of the network, we establish a multi-scale integrated evaluation metric system. Secondly, We introduce a weight optimization algorithm that computes a set of weights containing prior knowledge and apply the weights to weighted TOPSIS to identify influential nodes. Finally, to simulate the actual propagation process of information in social networks, we use an independent cascade model based on the influence propagation coefficient $p(u, v|\overrightarrow{p_{uv}})$ in our experiments. We do experiments on four real-world social networks and compare the proposed DINI model with five existing advanced models: Entropy-based ranking measure, WVoteRank, New Evidential Centrality, NLPCA, and TOPSIS. Besides, we also compare the effect of weighted TOPSIS and TOPSIS in terms of balancing metrics. The experimental results verify that our model can achieve better performance on both accuracy and effectiveness. In future research, we will make use of spatial and temporal characteristics of information dissemination in social networks to identify influential nodes.

Acknowledgements. This work is supported by National Key Research and Development Plan in China (2018YFC0830500), National Natural Science Foundation of China (62172278).

References

1. Hafiene, N., Karoui, W., Romdhane, L.B.: Influential nodes detection in dynamic social networks: a survey. Expert Syst. Appl. **159**, 113642 (2020)
2. Bian, T., Deng, Y.: A new evidential methodology of identifying influential nodes in complex networks. Chaos Solitons Fractals **103**, 101–110 (2017)
3. Liern, V., Pérez-Gladish, B.: Multiple criteria ranking method based on functional proximity index: un-weighted TOPSIS. Ann. Oper. Res. 1–23 (2020). https://doi.org/10.1007/s10479-020-03718-1
4. Lu, M.: Node importance evaluation based on neighborhood structure hole and improved TOPSIS. Comput. Networks **178**, 107336 (2020)
5. Martin, T., Zhang, X., Newman, M.E.: Localization and centrality in networks. Phys. Rev. E **90**(5), 052808 (2014)
6. Hu, R.J., Li, Q., Zhang, G.Y., Ma, W.C.: Centrality measures in directed fuzzy social networks. Fuzzy Inf. Eng. **7**(1), 115–128 (2015)
7. Singh, R.R., Goel, K., Iyengar, S., Gupta, S.: A faster algorithm to update betweenness centrality after node alteration. Internet Math. **11**(4–5), 403–420 (2015)
8. Wang, Z., Zhao, Y., Xi, J., Du, C.: Fast ranking influential nodes in complex networks using a k-shell iteration factor. Phys. A Stat. Mech. Appl. **461**, 171–181 (2016)
9. Lee, J.K., Choi, J., Kim, C., Kim, Y.: Social media, network heterogeneity, and opinion polarization. J. Commun. **64**(4), 702–722 (2014)
10. Yu, D., Chen, N., Ran, X.: Computational modeling of Weibo user influence based on information interactive network. Online Inf. Rev. (2016)
11. Sun, X., Xie, F.: The three-degree calculation model of microblog users' influence (short paper). In: Gao, H., Wang, X., Yin, Y., Iqbal, M. (eds.) CollaborateCom 2018. LNICST, vol. 268, pp. 151–160. Springer, Cham (2019). https://doi.org/10.1007/978-3-030-12981-1_10
12. Ren, T., et al.: Identifying vital nodes based on reverse greedy method. Sci. Rep. **10**(1), 1–8 (2020)
13. Zheng, Z., Gao, X., Ma, X., Chen, G.: Predicting hot events in the early period through Bayesian model for social networks. IEEE Trans. Knowl. Data Eng. (2020)
14. Riquelme, F., González-Cantergiani, P.: Measuring user influence on twitter: a survey. Inf. Process. Manag. **52**(5), 949–975 (2016)
15. Drakopoulos, G., Kanavos, A., Tsakalidis, A.K.: Evaluating twitter influence ranking with system theory. In: WEBIST (1), pp. 113–120 (2016)
16. Bo, H., McConville, R., Hong, J., Liu, W.: Social network influence ranking via embedding network interactions for user recommendation. In: Companion Proceedings of the Web Conference 2020, pp. 379–384 (2020)

17. Sapul, M.S.C., Aung, T.H., Jiamthapthaksin, R.: Trending topic discovery of twitter tweets using clustering and topic modeling algorithms. In: 2017 14th International Joint Conference on Computer Science and Software Engineering (JCSSE), pp. 1–6. IEEE (2017)
18. Liu, Q., et al.: Leveraging local h-index to identify and rank influential spreaders in networks. Phys. A Stat. Mech. Appl. **512**, 379–391 (2018)
19. Chen, P.: Effects of the entropy weight on TOPSIS. Expert Syst. Appl. **168**, 114186 (2021)
20. Ning, Z., Iradukunda, H.N., Zhang, Q., Zhu, T.: Benchmarking machine learning: how fast can your algorithms go? arXiv preprint arXiv:2101.03219 (2021)
21. Yuan, B., Chang, J.E., Zhang, F.: Influential node identification method of assembly system based on TOPSIS and topology. J. Phys. Conf. Ser. **1605**, 012019 (2020)
22. Li, P., Liu, K., Li, K., Liu, J., Zhou, D.: Estimating user influence ranking in independent cascade model. Phys. A Stat. Mech. Appl. **565**, 125584 (2021)
23. Batyrshin, I.Z., Ramirez-Mejia, I., Batyrshin, I.I., Solovyev, V.: Similarity-Based correlation functions for binary data. In: Martínez-Villaseñor, L., Herrera-Alcántara, O., Ponce, H., Castro-Espinoza, F.A. (eds.) MICAI 2020. LNCS (LNAI), vol. 12469, pp. 224–233. Springer, Cham (2020). https://doi.org/10.1007/978-3-030-60887-3_20
24. Guo, C., Yang, L., Chen, X., Chen, D., Gao, H., Ma, J.: Influential nodes identification in complex networks via information entropy. Entropy **22**(2), 242 (2020)
25. Sun, H., Chen, D., He, J., Ch'ng, E.: A voting approach to uncover multiple influential spreaders on weighted networks. Phys. A Stat. Mech. Appl. **519**, 303–312 (2019)
26. Basu, S., Maulik, U.: Mining important nodes in complex networks using nonlinear PCA. In: 2017 IEEE Calcutta Conference (CALCON), pp. 469–473. IEEE (2017)

Human Motion Recognition Based on Wi-Fi Imaging

Liangliang Lin, Kun Zhao$^{(\boxtimes)}$, Xiaoyu Ma, Wei Xi, Chen Yang, Hui He, and Jizhong Zhao

School of Computer Science and Technology,
Department of Telecommunications, Xi'an Jiatong University,
Xi'an 710004, People's Republic of China

Abstract. The current wireless sensing technology has some problems, such as low resolution caused by narrow signal bandwidth, poor environmental adaptability caused by multipath effect and so on. To solve the above problems, this paper provides a new perception idea, Wi-Fi imaging human actions, and then using image processing method for action recognition. In the Wi-Fi imaging part, according to the different spatial angles of different parts of the body trunk such as head, chest and legs relative to the receiving end, this paper processes the human body reflection signal received by the receiving end, obtains the signal strength corresponding to each azimuth signal in the space, and generates the human body heat map. In the stage of action recognition, firstly, the background interference is removed. According to the characteristics of imaging and action in this paper, a continuous action segmentation method is proposed. The image action features are obtained through intensive sampling. Finally, the SVM is optimized by genetic algorithm to improve the accuracy of action classification under different conditions. Through the analysis of experimental results, the method proposed in this paper can produce high-precision imaging of human body under practical application conditions. The recognition accuracy of different actions in the experiment is more than 90%.

Keywords: Motion recognition · Wi-Fi image · SVM

1 Introduction

Human motion perception and recognition refers to the technology of using devices other than the human naked eye to obtain the current dynamic information of individuals and identify them. With the proposal of virtual reality, augmented reality and other technologies, human motion perception technology is urgently needed in many fields. For example, the motion perception recognition technology is applied to the field of human-computer interaction to get the somatosensory interaction technology, which gets rid of the complex and learning cost equipment and improves the efficiency of life. In recent ten years, with the popularity of the concept of artificial intelligence technology and the rapid development of intelligent devices and the Internet of things, the application scenarios of human motion perception and recognition technology are more extensive, such as directly using simple actions to control the switch and regulation of

© ICST Institute for Computer Sciences, Social Informatics and Telecommunications Engineering 2021
Published by Springer Nature Switzerland AG 2021. All Rights Reserved
H. Gao and X. Wang (Eds.): CollaborateCom 2021, LNICST 406, pp. 608–627, 2021.
https://doi.org/10.1007/978-3-030-92635-9_35

home intelligent appliances in the field of smart home, so as to improve people's quality of family life; In education, conference and other scenarios, use actions directly to control PPT and documents, or intelligently adjust the conference process according to human actions [1, 2].

With the construction of intelligent devices and the Internet of things, the technology of using wireless signals to perceive human actions appears, this technology also belongs to the category of unbound sensing technology. Due to the existence of the communication module in the intelligent device, the wireless signal has the advantages of wide distribution and high signal strength in the indoor space. Because of its original design intention, the penetration of the wireless signal is very superior. When perceiving the human body, it breaks through the restriction that the optical perception must be in the line of sight of the device, and avoids the risk of personal privacy disclosure. However, as an emerging technology, wireless sensing technology also faces some challenges. For example, the main purpose of wireless signal is communication rather than human perception, which needs to be improved in physics or algorithm. Some of the current studies directly use the waveform changes of signals to perceive human actions, but due to the abstraction of waveform changes, it is difficult to distinguish different actions, which reduces the accuracy and has a great impact on the environment. The other part uses the wireless signal to image the human body and then recognize the action, but the imaging effect is poor and the recognition action is single. Therefore, this paper proposes human motion recognition based on Wi-Fi imaging, which aims to improve the imaging algorithm to realize human fine-grained imaging through Wi-Fi perception of human body information, and carry out a variety of motion recognition based on imaging, which has great application scenarios and theoretical significance, and is worthy of in-depth research.

The propagation of electromagnetic signal in space will have a certain degree of attenuation signal, which is related to the length of its propagation path and the reflection and refraction generated by penetrating the object in the propagation process. When the object moves, the generated multipath signal will also change, and the intensity of the received electromagnetic model will also change. Therefore, RSSI is highly sensitive to moving objects in the environment, and different individual movements will have different signal intensity transformation. Therefore, some research work uses RSSI information to roughly image different moving objects, and some work attempts to use RSSI for more accurate human action recognition, but the effect is not good. This is because RSSI is too rough and the granularity is too large to play a role in fine-grained perception [12–15].

Channel state information is the channel attribute in electromagnetic signal transmission, which is used to reflect the linear superposition of the influence of multipath effect caused by reflection, scattering and other reasons on the signal during the propagation of each subcarrier signal. CSI contains the changes of signal propagation to a certain medium (such as air or human body). According to different factors, it can be divided into the influence of attenuation caused by distance, the influence of attenuation of surrounding environment and the influence of object surface scattering. Therefore,

CSI has higher sensitivity to individual movement than RSSI. CSI can therefore play a role in a subtler field of perception. Wihear uses directional antenna to acquire CSI changes caused by slight movement of mouth to perceive different words emitted by individuals; Falldefi uses OFDM as a wireless signal to obtain the changes of CSI and RSSI caused by different human fall actions, and finally identify them [16, 19].

The typical technology of imaging using RF signal is the military radar system and the prohibited article inspection device in subway and train. It uses the characteristics of high reflectivity of metal to electromagnetic signal to search the information of metal objects such as aircraft position and whether to carry knives. The imaging system for human body information and its movement is basically inseparable from professional multi antenna equipment, and the modulation mode is customized to make the transmitted signal ultra-high frequency and ultra-large bandwidth. These systems are essentially different from Wi-Fi based imaging systems because their working scene range is very narrow. The characteristics of high frequency and high penetration make the signal penetrate or be absorbed by the human body rather than reflect the signal. Therefore, it will do varying degrees of harm to human health. These technologies have certain advantages in accuracy, such as CT examination devices in hospitals. In the imaging research using Wi-Fi like signals, researchers have also made some pioneering achievements. RF capture uses the customized waveform of Wi-Fi like signal frequency to achieve a fairly accurate imaging of the human body, which is a big step in the feasibility of Wi-Fi imaging technology. In CARM, the real Wi-Fi signal is used to realize human imaging, but the ideal effect cannot be achieved [20, 30, 32].

The use of wireless RF signals, especially Wi-Fi signals for human motion perception, almost achieves full signal coverage in specific application scenarios, and makes use of existing routing equipment without reinstallation and deployment. Therefore, it is a popular trend in current research. However, in the previous work, some of them directly use Wi-Fi information for motion perception. They need to process the collected continuous waveform data and distinguish the actions through different waveforms, which leads to single recognition action and great difficulty in distinguishing. For example, detecting falling action may be confused with bending, squatting and other actions. The other part uses imaging, the human body is imaged first, and then the action recognition is carried out. However, due to the rough imaging, the detection of actions with large amplitude has a good effect, but the limb movements cannot be detected. This paper improves the previous imaging method, images the human body more finely, obtains more diverse human actions, and proposes a continuous action recognition method to extract different action features and classify actions through SVM to achieve the purpose of action recognition.

Our contribution can be summarized as follows:

1) Wi-Fi imaging using multiple antenna arrays. Visible electromagnetic signals are transmitted to the surface of the object and reflected. The reflected light enters the retina or the optical perceptron such as photosensitive material in the camera, and the inverted image of the object is generated on the retina or photosensitive material. This is the principle of optical sensing imaging. Similarly, the invisible Wi-Fi signal and the

visible light belong to the electromagnetic signal and have the same propagation characteristics. The Wi-Fi signal will also be reflected when it propagates to the human body surface. The reflected signal is received by the receiver, and the image of the object is obtained by processing the signal received by the receiver. In detail, the human trunk and limbs do not belong to the same plane as the signal receiving end, so the reflected signal reflected by the human trunk will form an incident angle with the signal receiving end, and the reflected signals of different parts of the human trunk, such as head, chest and legs, are also different from the incident angle of the signal receiving end, and different individuals are different because of their height and body shape, The intensity and incident angle of the reflected signals of their different body parts are also quite different. All the reflected signals in the environment are obtained by using multiple antennas at the receiving end. Because the filtered reflected signals of the surrounding static objects, the received signals with greater intensity are the reflected signals of the human body. The incident angle of these signals is calculated by the algorithm, the positions of these different body parts can be determined in two-dimensional space, so as to obtain the heat map of the human body. Therefore, this paper uses angle measurement to simplify the complex problem of final imaging using accurate ranging. This paper extends the multi signal classification algorithm applied in one-dimensional space, that is, MUSIC algorithm, to make it applicable in two-dimensional space, and then improves it to improve the human imaging effect.

2) Adding subcarriers to improve the effect of MUSIC algorithm. The angle measurement accuracy of MUSIC algorithm is positively correlated with the number of antennas, but due to the small number of antennas in ordinary home routers in real conditions, the imaging effect will not reach the expectation. There is no doubt that the multipath signals generated by different parts of the human torso will cause measurable phase shift on different physical antennas of the receiver array, which is the physical basis of MUSIC algorithm angle measurement. Due to the characteristics of Wi-Fi transmission, the carrier and subcarrier of the signal will also generate phase shift due to different frequencies, even if the causes of phase shift are different, However, it can be comprehensively calculated mathematically and does not affect the properties of the equation. Therefore, the influence caused by subcarriers can be added to the MUSIC algorithm, because subcarriers play the same role as physical antennas in the algorithm, which essentially increases the number of antennas and improves the measurement accuracy. Here, the space is saved and the algorithm effect is improved without increasing the number of physical antennas. However, the effect of MUSIC algorithm on coherent signals reflected by human body is very poor. Combined with the advantages of different decoherence, this paper proposes a global decoherence method for Wi-Fi human body imaging to further improve the imaging effect, and then verifies each improved effect through simulation.

3) Experimental results and analysis. Firstly, this paper realizes the prototype of the algorithm by using the notebook computer loaded with 9300 network card and multi heel antenna and PicoScenes platform. Then, different volunteers are used for downlink imaging and action recognition in different scenes for the imaging algorithm proposed in this paper, and the influence of other influencing factors such as distance and signal frequency on the model is analyzed to evaluate the effect and robustness of the algorithm.

2   Wi-Fi Imaging Algorithm Based on 3D Virtual Array

The imaging method used in this paper is different from the traditional imaging method, which requires accurate ranging. Its principle is that the Wi-Fi signal is transmitted to different parts of the object and reflected. For the receiving end, the incoming wave directions of the reflected signals of different parts of the object in physical space must be different. If these incoming wave directions can be estimated, the positions of different parts of the object in space can be determined reversely, Combined to get a complete image of the object. As shown in Fig. 1, just as optical perception will not image substances without entities, the surrounding static reflection will be filtered out in this paper, so the image is only a moving human body. The three main different parts of the moving human body (head, trunk and limbs) will reflect the signal, and the receiving end not in the same plane with the human body will receive the reflected signals of these different parts. These reflected signals will have different entry angles with the antenna array plane. Analyze the received signal by some method to obtain the azimuth of the reflected signal from the human body the pitch angle reverses the relative positions of different parts of the human body, and then obtains the heat map of the whole human body.

2.1   Imaging Algorithm Based on Virtual 3D Array

In this paper, the human body needs to be imaged in two-dimensional space. Due to the limitation of antenna, the previous imaging algorithm has poor estimation effect on the similar incident angle in two-dimensional plane, so the performance of the algorithm in human body imaging is still poor, and cannot achieve the goal of human motion imaging.

According to the principle of imaging algorithm, the number of antennas must be greater than the number of sources, and the more antennas, the more accurate the estimation of sources. Therefore, directly increasing the number of physical antennas is the simplest solution, but it is limited to the implementation conditions and application scenarios. If the imaging granularity can be improved without adding physical antennas, it will be of great benefit to the subsequent recognition accuracy and future applications. Therefore, in this section, by analyzing the characteristics of subcarriers, an imaging algorithm based on virtual three-dimensional array is proposed to improve the imaging accuracy. Then, according to the characteristics of 3D virtual array imaging algorithm, a new decoherence algorithm is designed, which improves the effect of imaging algorithm to a greater extent.

It can be seen that the core of the imaging algorithm in this paper is to calculate the guidance vector matrix, which is composed of the phase shift caused by the distance between physical antennas. Wi-Fi technology uses multiple different subcarriers to transmit signals, and the signals are modulated on different subcarriers to achieve the purpose of parallel transmission. Different from the principle that the signal generates phase shift due to time delay on the physical antenna, when the reflected signal of L signal sources is received by the receiving array composed of M antennas, different subcarriers in the same signal reach an array element of the array at the incident angle, and phase shift will also occur due to different frequencies. Take a subcarrier as the origin, and its phase shift is as follows:

$$\Omega = \frac{2\pi(L-1)d(f_1 - f_2)\cos(\varphi)}{c} \tag{1}$$

where f_1 and f_2 represent frequencies of these two different subcarriers, c indicating the speed of light. Because the Wi-Fi signal belongs to a narrow-band signal, the subcarrier frequencies are relatively similar and the denominator order of magnitude is much larger than the numerator. Therefore, even if all 30 subcarriers are calculated into the formula, the maximum phase obtained is less than 0.06 radian, which can hardly participate in the calculation, which has little effect on the improvement of angle measurement algorithm.

However, using the flight time of the incoming wave signal can make the discrimination of subcarrier phase shift more obvious. Similarly, after introducing the flight time, the original phase shift formula is transformed into the following formula

$$\Omega = 2\pi(f_1 - f_2)\tau \tag{2}$$

τ to indicates the flight time. By introducing new parameters, the discrimination of phase difference is greatly enhanced, with a maximum of about 3.5 radians, which meets the measurement standard. As the number of unknowns increases, the difficulty of solving the equations increases. However, under the experimental conditions in this paper, the number of multipath reflected signals generated by the reflection of signals on the human body is generally no more than 7. Because the equation group added with subcarriers is 30 groups, which is much larger than the increased number of unknowns and does not interfere with the solution, the three-dimensional virtual array imaging algorithm is feasible.

We have discussed the phase difference caused by the physical distance of different antennas, but the propagation distance of the same signal to the same antenna will not change. Set the phase of a certain subcarrier as the phase of the original subcarrier, and the phase difference will be caused by the different flight time and frequency of different subcarriers:

$$\Omega(\tau_i) = 2\pi(n-1)f_\delta\tau_1 \tag{3}$$

where f_δ describes the frequency difference of different subcarriers. The phase difference of all N subcarriers is composed of a vector, expressed as:

$$h(\tau_i) = \left[e^{j\Omega_1}, e^{j\Omega_2}, \ldots, e^{j\Omega_N}\right]^T \tag{4}$$

It can also be added into the formula $X(t) = HK(t) + N(t)$, and we get is:

$$
\begin{bmatrix} x_{11}(t) \\ \vdots \\ x_{1N}(t) \\ \vdots \\ x_{M1}(t) \\ \vdots \\ x_{MN}(t) \end{bmatrix} = \begin{bmatrix} e^{j\Phi_{11}}e^{j\Omega_{11}} & \cdots & e^{j\Phi_{1L}}e^{j\Omega_{1L}} \\ \vdots & \ddots & \vdots \\ e^{j\Phi_{L1}}e^{j\Omega_{N1}} & \cdots & e^{j\Phi_{ML}}e^{j\Omega_{NL}} \end{bmatrix} * \begin{bmatrix} s_1(t) \\ \vdots \\ s_L(t) \end{bmatrix} + \begin{bmatrix} n_{11}(t) \\ \vdots \\ n_{1N}(t) \\ \vdots \\ n_{LN}(t) \\ \vdots \\ n_{MN}(t) \end{bmatrix} \tag{5}
$$

where $x_{mn}(t)$ represents the received signal value of the m-th antenna and the n-th subcarrier with a subcarrier of an antenna as the origin received by the antenna array. The physical meaning of Φ_{mi} is the phase difference of the i-th reflected signal on the m-th antenna relative to the origin subcarrier of the origin antenna, and Ω_{ni} represents the phase difference of the i-th reflected signal on the n-th subcarrier relative to the origin subcarrier of the origin antenna. Mathematically, the phase shift caused by subcarriers is equivalent to that caused by physical antennas. In fact, it increases the number of equations and makes it easier to solve. In the physical sense, it increases the number of antennas and improves the accuracy of angle measurement. The next step is still to obtain the guidance vector matrix of the received signal, and then perform eigenvalue decomposition on the matrix, which is consistent with the steps after MUSIC algorithm to obtain the spatial spectral function:

$$
P(\varphi, \theta, \tau) = \frac{h^H(\varphi, \theta, \tau)h(\varphi, \theta, \tau)}{h^H(\varphi, \theta, \tau)UU^H h(\varphi, \theta, \tau)} \tag{6}
$$

A new parameter time of flight τ is introduced, and the physical meaning of the spectral function is that each direction in two-dimensional space combines the signal density of time of flight. As long as the possible azimuth and pitch angles of the source with the strongest signal are found in all flight times, the imaging heat map in two-dimensional space can be obtained.

Although the imaging accuracy is improved according to the measurable phase between subcarriers, the introduction of subcarriers brings new problems. When the source signal is a coherent signal, it is impossible to obtain the guidance vector and then the angle of arrival of the reflected signal according to the principle of the imaging algorithm. There is almost no difference in frequency between subcarriers, so it has coherence and interferes with imaging, so it is necessary to decoherence the signal.

2.2   Description of Improved 3D Decoherence Algorithm

The previous imaging algorithms in this paper mainly estimate the incoming direction of incoherent signals, but through the analysis in Sect. 3.1, the reflected signal of Wi-Fi signal reflected by human body is coherent. According to Eq. (3–9), for coherent signals, complex constant vectors can be extracted, so that the properties of reflected signal matrix will change, Finally, the whole equation cannot use the traditional mathematical method to calculate the guidance vector. Finally, the effect of the imaging algorithm on the coherent source is very poor, so the decoherence algorithm needs to be introduced to decoherence. Among the mainstream decoherence algorithms, the spatial leveling algorithm is simple to implement, but it has estimation defects for a large number of signals. The non-dimensionality reduction processing algorithm is also easy to implement, but the deviation is large [34, 35].

In order to avoid the above defects of smoothing algorithm, a three-dimensional decoherence algorithm based on Toeplitz matrix reconstruction is proposed in this paper. It is still 1 sources. From the derivation, we can get the rank loss of the signal covariance matrix in the case of signal coherence. Here, the subcarrier is also regarded as a virtual array element, and the Toeplitz matrix is constructed by taking the correlation function of the data received by each array element to decoherence the signal. Formula (5) has described that for the received signal values of rows and columns in the virtual antenna array, its correlation function with the received data of the origin array element is:

$$
\begin{aligned}
r_{mn} &= E\left[x_{11}x_{mn}^{\mathrm{H}}\right] \\
&= h_{11}E\left[\mathbf{SS}^{\mathrm{H}}\right]h_{mn}^{\mathrm{H}} + \sigma^2\mathbf{I} \\
&= h_{11}\mathbf{R}_s h_{mn}^{\mathrm{H}} + \sigma^2\mathbf{I}
\end{aligned}
\tag{7}
$$

where $\mathbf{R}_s$ is the source autocovariance matrix, then the correlation function vector satisfies:

$$
[r_{11}, r_{12}, \cdots, r_{MN}] = h_{11}R_s[h_{11}^H, h_{11}^H, h_{12}^H, ..., h_{MN}^H]
\tag{8}
$$

It can be seen that the correlation function vector contains all source information. The Hermitian Toeplitz matrix is constructed from these correlation functions through conjugate operation to restore the rank of the covariance matrix:

$$
\begin{bmatrix}
r_{11}(1) & r_{12}^*(2) & r_{13}^* & \cdots & r_{MN}^* \\
r_{12}(2) & r_{11}(1) & r_{12}^* & \cdots & r_{MN-1}^* \\
r_{13}(3) & r_{12}(2) & r_{11} & \cdots & r_{MN-2}^* \\
\vdots & \vdots & \vdots & \ddots & \vdots \\
r_{MN} & r_{MN-1} & r_{MN-2} & \cdots & r_{11}(1)
\end{bmatrix}
\tag{9}
$$

Due to the nature of Toeplitz matrix, the conjugate operation does not affect the eigenvalue decomposition. Therefore, the eigenvalue decomposition of the new matrix is carried out to obtain the spectral function and estimate the signal angle of arrival. The reconstruction of covariance matrix avoids the damage of smoothing algorithm to array aperture, makes full use of all received information, realizes complete decoherence and improves the imaging effect.

3   Environment Adaptive Human Continuous Motion Recognition

All actions in human daily life are inseparable from the changes of limbs and trunk, and most complex actions are composed of simple actions such as arm waving, leg raising and squatting. They are the basic action unit for human interaction with the outside world. Through different combinations of basic actions, human beings have realized the transformation of the material world. Therefore, this paper sets walking, arm waving, leg raising squat and bend down, and collect the data of individual continuous different movements.

Fig. 1. Schematic diagram: walking, waving, leg lifting, squatting and bending.

3.1   Continuous Action Segmentation

Generally, the human body action will have a process of action start, action reset and then the next action. In this process, the area of the two-dimensional human body heat map will change regularly. Therefore, the area can be used to segment the beginning and end of human action, and a rectangular box can be used to measure the area change in the process of human action, so as to obtain the curve about the area. The maximum point and minimum point in the area transformation are the segmentation points of action. Then the law of area change is more obvious through smoothing.

Fig. 2. Variation of different action areas

In this paper, the ambient noise has been removed in the previous background processing, and a single background is obtained through normalization processing. Therefore, by detecting the value of human body area and background area, select a single threshold a to distinguish the background and human image, and then use a rectangular box to surround the human image as close as possible. The rectangular box is represented by, where t is time, (x,y) represents the coordinates of the upper left corner of the rectangular area box, and W and H are the width and height of the rectangular area. Let represent the rectangular area over time. As shown in Fig. 2 when the human body stands, bends and kicks, the area function will change, and the start and end of each action will cause the maximum and minimum points of the area function. However, due to the instability of human action, the change of area is not stable, and there are many maximum and minimum points of interference, which makes it difficult to find the real action segmentation points. Therefore, smoothing method is used to remove the unstable points.

3.2   Action Feature Extraction

1) Sample dimension alignment

The image sequence of each action can be obtained through continuous action segmentation. Each sequence represents the imaging result of an action, which is called a sample. During action segmentation, various steps are processed to make the number of sequences in each action sample inconsistent. By averaging the number of sequences of different samples, the reference frame number T is determined, which is less than the supplement of the reference, and a certain number is deleted if it is greater than the reference. The dimension of the samples is aligned to facilitate subsequent processing.

2) Get the cigenvector

Through grid division, feature points are intensively sampled at multiple scales of the picture in the action clip. These feature points are tracked over time and the eigenvalues of the autocorrelation matrix of all pixels are obtained. Some feature points lower than the threshold are removed through the threshold.

Calculate the feature point coordinates of the next frame from the feature point coordinates of the previous frame through the formula:

$$P_{t+1} = (x_{t+1}, y_{t+1}) = (x_t, y_t) + (M * \omega_t) * x_t, y_t \tag{10}$$

Use formula (10) to obtain the optical flow median in the neighborhood of the feature point, so as to see the motion vector of the feature point. From the motion vector, the feature point coordinates of the subsequent frame can be calculated. These coordinates form trajectories over time, and the characteristics of actions are extracted from these trajectories. The above track shape can be expressed by $\Delta P_t, \ldots, \Delta P_{t+L-1}$, and the obtained trajectory features are:

$$T = \frac{(\Delta P_t, \ldots, \Delta P_{t+L-1})}{\sum_{j=t}^{t+L-1} \Delta P_j} \tag{11}$$

The features outside the track are further extracted, and the surrounding n existing as track feature points in each frame heat map in a certain track. The region of N constitutes a time-space body for grid division, the space is divided into parts, and the parts are evenly selected for time division. Finally, feature extraction is performed in two regions.

The histogram of gray image gradient is calculated to obtain the hog feature, with a length of 96 (2 * 2 * 3 * 8). The histogram of optical flow direction and amplitude information is calculated to obtain the HOF feature, and the feature length is 108 (2 * 2 * 3 * 9).Calculate the histogram of optical flow image gradient to obtain MBH feature, and the feature length is 192 (2 * 96).

3) Principal component analysis dimensionality reduction

The feature length used for training is 426 dimensions, which is the sum of the above feature dimensions. In this way, there is too much redundant information of feature vector, and the dimension needs to be reduced by principal component analysis.

If there are m samples, there are m features. If the feature dimension is set to N, the feature matrix X of M * n can be formed. The feature normalization of each column of matrix X is that the mean value of each column is 0 and the variance is 1. The covariance matrix C of matrix X is calculated

$$C = \frac{1}{M}X^TX \tag{12}$$

where C is the covariance matrix, and its diagonal elements represent the variance of each feature. The elements of row I, column J and row J, column I are the same, representing the covariance of the I and j features. In order to achieve the purpose of maximum information retention and minimum data repetition, in this matrix, it is the transformation matrix to maximize the diagonal elements and minimize the other elements to achieve orthogonality. Therefore, we need to diagonalize matrix C. Eigenvalue decomposition covariance matrix to obtain the eigenvector and eigenvalue of the matrix. From small to large, the eigenvectors corresponding to each eigenvalue are combined into a matrix, so that the matrix composed of the front row is called linear algebra. It has been proved that $P * C * P^T = \Lambda$, Λ is a diagonal matrix. The covariance matrix C is diagonalized, which is the linear transformation required by PCA. Y = XP, the original data X is changed into k-dimensional data y through the linear change P of k-dimension.

It should be noted that the size of dimension k of the data after dimension reduction is selected, as shown in Figs. 3 and 4. It is necessary to make a trade-off between dimension k and how many data features are retained, and select an appropriate value. In this study, when the reduced dimension is 213, 98% of the information is retained, and K is 213. In this way, the original feature dimension is reduced by half, and then the feature group of each action segment is encoded.

3.3 SVM Classification Based on GA Algorithm Optimization

In this paper, SVM is used to classify and identify the movements of waving, leg raising, bending and squatting. The original purpose of SVM is linear binary classification. Here, There is a multi classification problem and the data distribution is complex, so SVM needs to be modified, so we can maximize the gap between two different class and minimize probability of misclassification. the RBF kernel function be used to get new decision function, and we use SVM in each pair of action. By above method, the problems of non-linear sample space and multi-classification are solved.

Samples are obtained by using 2.4 GHz and 5 GHz Wi-Fi signals respectively. Some of these samples have good effects and some have poor effects, but they all contain features that represent actions, and their corresponding optimal SVM parameters are also different. However, the original SVM uses a deterministic search method to find the optimal parameters with fixed directions and relationships, The search speed is slow and there is a great possibility that the global optimization cannot be obtained. In view of this situation, genetic algorithm is used to optimize SVM, which is mainly to optimize the parameters and improve the adaptability of the model under various conditions.

On the one hand, the stochastic characteristics of genetic algorithm are used to accelerate the speed of SVM in the parameter seeking process; On the other hand, the fitness function is used as the optimization standard to improve the accuracy of action recognition under different conditions. The comparison of fitness function between SVM optimized by genetic algorithm and original SVM is shown in the figure below. Aiming at the weakness of SVM, genetic algorithm is introduced to improve SVM from two aspects: efficient adjustment of parameters and recognition accuracy under different conditions. The training time is about 1/3 of that when it is not optimized, and the accuracy is improved by 8%–10%.

4   Experiment and Result Analysis

In this section, we will introduce the experimental configuration and experimental design, compare different experimental parameters to test the system performance, and finally display and analyze the experimental results.

4.1   Experimental Configuration

In this paper, two 9300 network cards are selected as the transmitter and receiver of the signal. Three antennas are led out from the transmitting end, and six omnidirectional antennas are used at the receiving end to form an antenna array to realize three transmitting and four receiving devices. Six omnidirectional antennas are used to form a uniform circular array to receive reflected signals. The height of the transmitter and receiver is 0.7 m, which is basically located in the middle of the human body and can receive all information about the human body. The experimenter acted in the area where the signal could reach, and collected CSI data using PicoScenes platform. In each case, 10 experimental acquisitions were carried out for different continuous actions, each acquisition time was 5 min, the contracting interval was set to 5 ms, the signal transmission frequency was set to 5800 MHz, the transmission energy was set to 30 dBm, and the subcarrier bandwidth was set to 20 MBps. 60000 data packets can be obtained in each experiment, and each data packet contains CSI data of all subcarriers corresponding to links between antennas.

The experimenter selected two men and two women of different height and body type, with height and weight of [183 cm, 126 kg], number a, [176 cm, 143 kg] number B, [164 cm, 115 kg] number C and [152 cm, 104 kg] number d.

4.2   Human Imaging and Motion Recognition

The action sequence is segmented by using the action segmentation method proposed in Sect. 4, in which the size of the selection matrix is determined based on the threshold A. for the heat map value in this paper is between 0 and 1, and the background is normalized close to 0, take the kick action as an example, select the threshold a as 0.01, 0.05, 0.1 and 0.15 respectively, and the effect is shown in the following picture.

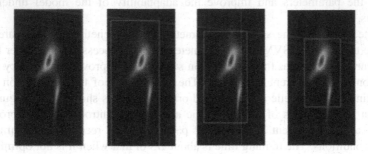

Fig. 3. Threshold selection (a) 0.01, (b) 0.05, (c) 0.1, and (d) 0.15

It can be seen from the above figure that when the threshold is too small, the matrix is large and contains a lot of background. With the increase of the threshold, the bounding box gradually decreases. When the threshold is too large, the bounding box does not contain the weak reflection caused by limbs. The appropriate threshold should be between 0.05 and 0.1, which effectively includes the reflection of limbs, and the change of bounding box with action is obvious. In the subsequent experiments, the threshold is 0.06. In this paper, we need to distinguish five kinds of actions, and the experimenter segments them after continuous different actions. After smoothing, similar to the method of selecting threshold a, select =1 s, =4 s, area difference =100, =300. the SVM will extract a series of features of each action under the stable time sequence.

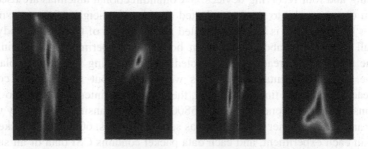

Fig. 4. Imaging results of different actions (a) wave, (b) kick, (c) bend, and (d) squat

4.3 Result Analysis

Figure 5 shows the action recognition results. Generally speaking, the recognition rate of different actions has reached more than 90%, but there is still a gap in the recognition rate between different actions, which is related to the recognition effect of different actions.

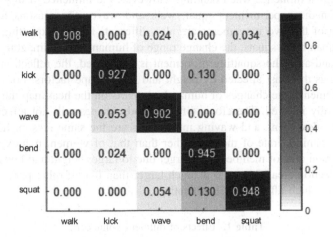

Fig. 5. Different action recognition result

When standing upright, the limbs are closer to the trunk, which is difficult to distinguish. With the extension of the action, the distance between the limbs and the trunk is farther and farther, and it is easier to be distinguished. For squatting and bending movements, the images of the imaging heat map and the real experimental scene are displayed. It can be found that there is a high corresponding relationship. With the experimenter squatting and bending down, the main reflection point is also moving downward, and the pitch angle relative to the receiving antenna array is gradually reduced. For the wave movement, the arm is close to the chest, the arm is thin, and the reflection signal is weak. The wave movement cannot be clearly observed, but when the arm is far away, the change of the azimuth of the reflection point can still be observed. For kicking, the distance between the leg and the chest is far, and the cylindrical structure is easy to generate reflection signals, so the imaging results are better than the waving action, so the recognition rate should be higher.

4.4 Model Test

Individual Factors. Among the experimenters numbered a, B, C and D, the action samples of experimenters numbered B and C are used for training, and the resampled experimental data numbered a, B, C and D are used for testing in the training stage. The recognition effect is shown in Table 1.

It can be found that there is a certain relationship between the recognition rate and the height and body shape of the experimenter. The influence of different height and body shape is more obvious in the bending and leg lifting actions. The larger the reflection area caused by the height and weight of the experimenter, the higher the action recognition rate. When the experimenter carries out the leg lifting action, the leg action amplitude caused by height is quite different. Similarly, Height and body shape also have a great influence when bending. However, the influence of this factor is not obvious when the experimenters squat, walk and wave. The reasons for the same phenomenon of the two movements are also different. For the squatting movement, when the human body squats, the change range of human motion imaging trajectory is the largest, and after the squatting movement is completed, the reflection area of the experimenter is the least affected by height and body shape factors. For walking and waving movements, the changes of human body parts on the heat map during walking can be basically ignored. Therefore, different body shapes will not affect the recognition of this movement, and waving movements are the same reason. In the overall trend, the recognition rate of men is higher than that of women, mainly because the overall reflection area of men becomes larger due to larger height and body shape, and the chest reflection area of women is much larger than that of other parts due to body shape factors, which affects the classification.

Table 1. Effects of different volunteers

Experimenter accuracy/%	A	B	C	D
Walk	90.24	92.33	88.67	89.57
Lift your legs	92.45	91.87	91.34	88.79
Wave	88.21	89.92	89.43	87.94
Bend	94.56	92.74	92.14	90.32
Squat down	94.36	95.48	93.67	92.96

Distance Factor. Since there is walking in the set recognition action, the recognition rate of different actions is also different when the experimenter walks to different distances from the equipment. In order to understand the role of distance factor in model recognition, the experimenter can walk to a distance of 2 m, 4 m, 6 m and 8 m from the equipment to perform other actions. The results are shown in Fig. 6. We observed that the distance factor was roughly negatively correlated with the recognition accuracy. The decrease of recognition rate with distance is mainly because the farther the distance, the higher the degree of signal attenuation and the less useful information in the reflected signal. The experiment here removes the influence of the wall. Therefore, it can be seen that compared with the influence of the wall on the recognition rate, the distance has little effect on the recognition rate. There is no wall at the same distance, even if the distance is far, The reduction of recognition rate is also very gentle, and the signal attenuation with distance is mainly related to the signal propagation frequency. Generally, the maximum space without walls in indoor space will not exceed 8 m * 8 m, and the recognition accuracy of the model for actions at a distance of 8 m is more than 80%, which fully meets the requirements. Therefore, it can be said that the algorithm has high adaptability to distance.

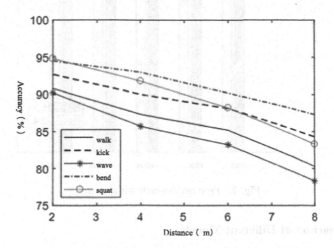

Fig. 6. Influence of different measurement distance

Wi-Fi Signal Frequency Factor. The above experiments lead to the information that the Wi-Fi signal changes with distance and the attenuation degree of wall blocking is related to frequency. In the current Wi-Fi frequency standard, 2.4 GHz is a general frequency, and 5 GHz has been widely popularized. with the development of technology, higher frequency Wi-Fi signal frequencies are also being studied, and high-frequency and low-frequency signals have their own advantages in propagation, therefore, experiments are conducted on the Wi-Fi signals of the above two frequencies. When there is no wall blocking between man and machine, the recognition rates of different Wi-Fi signals are shown in Fig. 7. It can be seen that when the attenuation degree of Wi-Fi signal is small, the recognition rates of the two Wi-Fi signals are very high and similar, and the accuracy of 5g signal model is slightly greater than 2.4G signal, This is mainly because 5g signal has fast transmission rate and contains more information, so the accuracy is slightly higher.

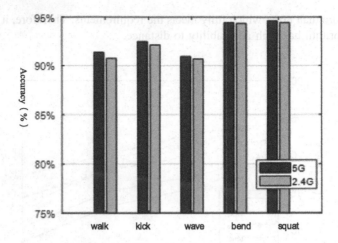

Fig. 7. Propagation path without wall

4.5   Comparison of Different Models

There are other models in the direction of action recognition, such as CARM (an activity recognition scheme), rtfall (a fall detection scheme) and falldefi (several fall action recognition schemes). In order to compare with the model in this paper, the accuracy, recall and F1 score, referred to as PRF, are used to evaluate the performance of the model. The results are shown in the table below (Table 2):

Table 2. Comparison of effects of various models

	Paper model	FallDeFi	CARM	RTFall
F1 score	90.24	86.34	76.41	82.45
Precision	86.65	84.12	73.60	80.01
Recall/TPR	92.33	90.48	80.32	85.20

It can be seen from the table that the effect of the model in this paper is better than other models. On the one hand, the imaging accuracy of the imaging algorithm in this paper is higher, and the information of human limbs is obtained, so the accuracy is higher in some movements about limbs. On the other hand, GA is used to optimize SVM, which has higher resistance to different factors affecting classification, so a certain classification accuracy is improved.

5   Conclusion

For the Wi-Fi imaging part, this paper solves several difficulties of Wi-Fi imaging, transforms the imaging problem into an angle measurement problem, extends the one-dimensional MUSIC algorithm to two-dimensional, and realizes coarse-grained imaging. However, because the number of antennas is too small, it cannot be applied to human height recognition. Directly increasing the number of physical antennas is a direct way to improve the accuracy. However, limited to the implementation conditions and application scenarios, this method is not applicable. Therefore, using the characteristics of Wi-Fi signal transmission using subcarriers, this paper demonstrates the equivalent role of subcarriers and physical antennas in angle measurement theory. Adding subcarriers is actually equivalent to increasing the number of antennas and further improving the imaging accuracy. Combined with the characteristics of various decoherence algorithms, a decoherence algorithm is designed, which first decoherence locally and then decoherence globally. Finally, the effects of these improvements are verified by simulation experiments. Through these improvements, the imaging algorithm can be used in human motion imaging.

In the image processing part, the background interference is removed first, the continuous action is segmented according to the characteristics of imaging and action, the trajectory features and other features of the special action are obtained through intensive sampling, and the dimension of the sample data is reduced by PCA to retain 96% of the information. The classification accuracy of SVM is improved and optimized by genetic algorithm.

Experimental phase, at 9300 The algorithm prototype is implemented on the network card and PicoScenes software platform, and then the imaging and motion recognition are evaluated. In the imaging part, the human imaging effect in CARM is compared to verify the effectiveness of the imaging algorithm and decoherence algorithm in this paper. Then different actions are imaged and displayed. In the classification part, SVM is used to train the model because of the amount of data. For the trained model, several groups of comparative experiments are designed to simulate the impact of factors that may affect the recognition in reality, such as scene, propagation path, human clothing and recognition distance, so as to illustrate the robustness of the system. It is proved that the motion recognition algorithm based on Wi-Fi imaging proposed in this paper is practical and feasible.

Acknowledgment. This work was supported by NSFC 62072367, 61772413, 61802299, 62002284, and Natural Science Foundation of Shaanxi Province 2021JM-025.

References

1. Forbrig, P., Paternó, F., Pejtersen, A.M.: Human-computer interaction. Encycl. Creativity Invention Innovation Entrepreneurship **19**(2), 43–50 (2017)
2. Zesheng, Y.: Based on Kinect Research and Design of Smart Home System Based on Gesture Recognition. Liaoning University of Science and Technology, Liaoning (2017)
3. Yibo, L., Yulin, D.: Indoor abnormal behavior detection of the elderly living alone based on intelligent monitoring. Comput. Appl. Softw. **31**(02), 188–190 (2014)
4. Kwapisz, J.R., Weiss, G.M., Moore, S.: Activity recognition using cell phone accelerometers. ACM SIGKDD Explor. Newsl. **12**(2), 74–82 (2011)
5. Lv, M., et al.: Bi-view semi-supervised learning based semantic human activity recognition using accelerometers. IEEE Trans. Mobile Comput. **17**(9), 1991–2001 (2018)
6. http://www.36kr.com/p/212759.html

7. Jalal, A., Kamal, S., Kim, D.: Shape and motion features approach for activity tracking and recognition from kinect video camera. In: 2015 IEEE 29th International Conference on Advanced Information Networking and Applications Workshops, pp. 445–450. IEEE (2015)
8. Tang, S., Andres, B., Andriluka, M., Schiele, B.: Multi-person tracking by multicut and deep matching. In: Hua, G., Jégou, H. (eds.) ECCV 2016. LNCS, vol. 9914, pp. 100–111. Springer, Cham (2016). https://doi.org/10.1007/978-3-319-48881-3_8
9. Yoshida, T., Taniguchi, Y.: Estimating the number of people using existing WiFi access point in indoor environment. In: Proceedings of the 6th European Conference of Computer Science (ECCS '15), pp. 46–53 (2015)
10. Weng, J., Weng, C., Yuan, J.: Spatio-temporal naive-Bayes nearest-neighbor (ST-NBNN) for skeleton-based action recognition. In: Proceedings of the IEEE Conference on Computer Vision and Pattern Recognition, pp. 4171–4180 (2017)
11. Zou, Y., et al.: Wi-Fi radar: recognizing human behavior with commodity Wi-Fi. IEEE Commun. Mag. 55(10), 105–111 (2017)
12. Wilson, J., Patwari, N.: Radio tomographic imaging with wireless networks. IEEE Trans. Mob. Comput. 9(5), 621–632 (2010)
13. Sigg, S., Blanke, U., Tröster, G.: The telepathic phone: frictionless activity recognition from WiFi-RSSI. In: 2014 IEEE International Conference on Pervasive Computing and Communications (PerCom), pp. 148–155. IEEE (2014)
14. Han, J., et al.: GenePrint: generic and accurate physical-layer identification for UHF RFID tags. IEEE/ACM Trans. Netw. 24(2), 846–858 (2016)
15. Xia, S., et al.: Indoor fingerprint positioning based on Wi-Fi: an overview. ISPRS Int. J. Geo Inf. 6(5), 135 (2017)
16. Sameera, P., et al.: FallDeFi: ubiquitous fall detection using commodity Wi-Fi devices. IEEE Trans. Mob. Comput. 15(11), 2474–9567 (2019)
17. Shangguan, L., et al.: STPP: spatial-temporal phase profiling-based method for relative RFID tag localization. IEEE/ACM Trans. Netw. 25(1), 596–609 (2016)
18. Kotaru, M., et al.: Spotfi: decimeter level localization using WiFi. In: Proceedings of the 2015 ACM Conference on Special Interest Group on Data Communication, pp. 269-282 (2015)
19. Yuw, Z., et al.: ZeroEffort cross-domain gesture recognition with Wi-Fi. In: Proceedings of the 17th Annual International Conference on Mobile Systems, Applications, and Services, 978-1-4503-6661-8. ACM (2020)
20. Stamnes, K., Stamnes, J.: Scattering of Electromagnetic Waves (2001)
21. Liu, Y., et al.: Channel estimation for OFDM. Commun. Surv. Tutorials IEEE 16(4), 1891–1908 (2014)
22. Farhang-Boroujeny, B., Moradi, H.: OFDM inspired waveforms for 5G. IEEE Commun. Surv. Tutorials 18(4), 2474–2492 (2016)
23. Schmidt, R.: Multiple emitter location and signal parameter estimation. IEEE Trans. Antennas Propag. 34(3), 276–280 (1986)
24. Fugui, L., Mingzhen, L.: Structural optimization of depth CNN model based on convolution kernel decomposition and its application in small image recognition. J. Jinggangshan Univ. (Natural Sci. Edition) 39(02), 31–39 (2018)
25. Agarwal, M., et al.: Face recognition using principle component analysis, eigenface and neural network. In: 2010 International Conference on Signal Acquisition and Processing, pp. (310–314). IEEE (2010)
26. Roska, T., Pazienza, G.: Cellular Neural Network (2000)
27. Cortes, C., Vapnik, V.N.: Support vector networks. Mach. Learn. 20(3), 273–297 (1995)
28. Vapnik, V.N., Lerner, A.: Pattern recognition using generalized portrait method. Autom. Remote. Control. 24(6), 774–780 (1963)

29. Kimeldorf, G., Wahba, G.: Some results on Tchebycheffian spline functions. J. Math. Anal. Appl. **33**(1), 82–95 (1971)
30. Huang, D., Nandakumar, R., Gollakota, S.: Feasibility and limits of Wi-Fi imaging. In: Proceedings of the 12th ACM Conference on Embedded Network Sensor Systems, pp. 266–279. ACM (2014)
31. Xiong, J., Jamieson, K.: ArrayTrack: a fine-grained indoor location system. In: Proceedings of the 10th USENIX Conference on Networked Systems Design and Implementation. USENIX Association (2013)
32. Adib, F., et al.: Capturing the human figure through a wall. ACM Trans. Graphics (TOG) **34** (6), 219 (2015)
33. Lei, C., et al.: Principle and architecture analysis of Wi-Fi technology. Telecommun. Sci. **31** (9), 175–181 (2015)
34. Ting, W.: Research on Two-Dimensional DOA Estimation of Coherent Signals. Harbin Engineering University, Harbin (2010)
35. Kang, L., Youbao, X., Zhi, L.: Signal correlation and correction musicTwo dimensional DOA estimation algorithm. Comput. Appl. **32**(02), 592–594 (2012)
36. Ma, Y., Zhou, G., Wang, S.: WiFi sensing with channel state information: a survey. ACM Comput. Surv. (CSUR) **52**(3), 1–36 (2019)
37. Jiang, Z., et al.: PicoScenes: enabling UWB sensing array on COTS Wi-Fi platform. In: Proceedings of the 2019 International Conference on Embedded Wireless Systems and Networks, pp. 264–266 (2019)
38. Mingxing, X.: Human feature extraction and measurement based on image. Zhejiang Univ. Technol (2018)
39. Hang, L.: Statistical Learning Methods. Tsinghua University Press, Beijing (2012)
40. Gao, R.X., Yan, R. Discrete wavelet transform. In: Proceeding of the International Conference on Imaging Science, pp. 193–197 (2016)
41. Kay, S.M.: Modem Spectral Estimation: Theory and Application. Prentice-Hall, Englewood, NJ (1988)

A Pervasive Multi-physiological Signal-Based Emotion Classification with Shapelet Transformation and Decision Fusion

Shichao Zhang[1,2], Xiangwei Zheng[1,2(✉)], Mingzhe Zhang[1,2], Gengyuan Guo[1,2], and Cun Ji[1,2(✉)]

[1] School of Information Science and Engineering, Shandong Normal University, Jinan, China
xwzhengcn@163.com, jicun@sdnu.edu.cn
[2] Shandong Provincial Key Laboratory for Distributed Computer Software Novel Technology, Jinan, China

Abstract. Emotion classification is a hot pot at present. Since physiological signals are objective and difficult to hide, physiological signals are commonly used in emotion classification methods. However, traditional emotional classification based on physiological signals faced the following challenges: low accuracy and low interpretability. For this, this paper proposes a pervasive multi-physiological signal-based emotion classification with shapelet transformation and decision fusion (PMSEC). In PMSEC, the shapelet transformation and feature extraction are carried out for ECG, GSR, and RA, respectively. Following by, six sub-classifiers are constructed for different physiological signals. Lastly, decision-level fusion is implemented to obtain final emotion results. Experimental results show that the proposed PMSEC is not only highly competitive but also has a broad application prospect compared with EEG-based classification method.

Keywords: Emotion classification · Shapelet transformation · Decision fusion

1 Introduction

Emotions play an important role in daily life. Therefore, emotion classification has gradually formed a new research direction [1]. Over the last few decades, a large number of methods have been put forward to recognize human emotions. These methods can be mainly separated into two groups: humans' behaviors based methods and physiological signals based methods. The methods in the first group classify emotion based on humans' behaviors, such as facial expressions, speech, and other signals. However, some people may hide their true emotional state, which leads to errors in the emotional classification results. The other methods classify emotion based on physiological signals for these signals are objective and difficult to hide. The commonly used physiological signals include Electroencephalogram (EEG),

Supplementary Information The online version contains supplementary material available at https://doi.org/10.1007/978-3-030-92635-9_36.

Electrocardiogram (ECG), Galvanic Skin Response (GSR), Respiration Amplitude (RA), Heart Rate signal (HR), etc.

Emotion classification based on physiological signals has also been applied to many fields, such as human-computer interaction, rehabilitation therapy, emotion robot, emotion teaching, and so on [2–4]. However, the traditional physiological signals based emotional classifications rely on feature extraction and classifier training, which results in low accuracy and low interpretability.

For this, this paper proposes a pervasive multi-physiological signal emotion classification with shapelet transformation and decision fusion (PMSEC). In PMSEC, shapelet transformation is introduced to improve interpretability. Shapelet is a discriminative sub-sequence of time series, which can maximally represent a class [5]. The classification principle of a shapelet is illustrated in Fig. 1. To improve the classification accuracy, PMSEC carries out the shapelet transformation and feature extraction process for ECG, GSR, and RA, respectively. And then, six sub-classifiers are constructed and then decision-level fusion is implemented.

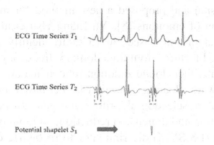

Fig. 1. The classification principle of a shapelet

The main contributions of this paper are as follows:

1) A pervasive multi-physiological signal-based emotion classification with shapelet transformation and decision fusion (PMSEC) method is proposed, which utilizes ECG, GSR, and RA jointly.

2) The concept and algorithm of shapelet transformation are applied to emotion classification based on physiological signals. Shapelet based algorithm has strong interpretability and can explain the differences between different classes.

3) We make experiments to show the effects of our method on the MAHNOB-HCI Multimodal Sentiment Database. The experimental results demonstrate that the proposed PMSEC is not only highly competitive but also has a wider application prospect.

The remainder of this paper is structured as follows. Section 2 reviews some related works on emotion classification based on physiological signals and shapelet-based algorithms. Section 3 describes our method in detail. Experimental results and their analysis are presented in Sect. 4 and our conclusions are given in Sect. 5.

2 Related Works

2.1 Emotion Classification Based on Physiological Signals

In recent years, there have been some meaningful studies on emotion classification. Among these, emotion classification based on physiological signals has attracted great attention in the scientific research field.

Physiological signals are signals that trigger a series of changes in bioelectrical characteristics of the internal organs of the human body due to individual emotion changes. Since physiological signals are objective and difficult to hide, emotion classification based on physiological signals has become a hot topic.

The multimedia affective computing research team led by Professor Picard from Massachusetts Institute of Technology (MIT) is the first research team to put forward a clear definition of "affective computing" [6]. In 2012, according to ECG signals, Agrafioti et al. proposed a dynamically evolving emotion pattern detection method based on empirical mode decomposition, and analyzed the test data of 44 subjects. The results achieved a good effect [7]. Verma G K studied the method of multimodal physiological signal fusion and proposed a new method for multimodal fusion, classification, and prediction of emotions [8]. Yu-Liang Hsu et al. proposed an emotion automatic recognition algorithm based on ECG to identify human emotions. The algorithm firstly extracted features from time domain, frequency domain, and nonlinear analysis for ECG signals, then found emotion-related features in ECG signals, then used a sequential forward floating selection-kernel-based class separability-based (SFFS-KBCSbased) feature selection algorithm and generalized discriminant analysis (GDA) for feature selection and dimension reduction, and finally used the least squares support vector machine (LS-SVM) discriminator to recognize different emotions [9].

2.2 Shapelet-Based Algorithms

Shapelet is a sub-sequence of time series, which can appear in any position of time series. Since shapelet was proposed in 2009 (Ye and Keogh 2009) [10], time series classification algorithms based on shapelet have attracted the attention of many researchers. Shapelet-based classification algorithms are mainly divided into the following three categories: shapelet discovery algorithm [11], shapelet transform algorithm [12], and shapelet learning algorithm [13].

At present, the concept and algorithms of shapalet have been applied in many fields. Ahmed AlDhanhani proposed an automatic event detection framework based on shapelets technology in 2019. This framework introduced shapelet concepts to detect event/congestion patterns and regular traffic conditions. Experiments proved that this framework not only improves interpretability, but also does not affect the classification results [14]. Albert Zorko et al. designed a method to detect sleep and awake based on shapelets of HRV. The algorithm based on shapelet analysis, the time series were divided into two categories by finding the best shapelet [15]. Monica Arul proposed an automatic detection algorithm based on time series shapelet (EQShapelet) for earthquake detection. Compared with other detection methods, the classifier based on

EQShapelet could detect more events and achieved an accuracy rate of 96.3%. In addition, the running speed of the algorithm can be improved in the future [16].

3 Methods

3.1 Overview

The flow chart of PMSEC is shown in Fig. 2. The steps are briefly described as follows: 1) signals extraction, which extracted ECG, GSR, and RA signals from MAHNOB-HCI database, 2) data preprocessing, 3) sub-classification methods, and 4) decision fusion. Next, we will describe the last three steps in detail.

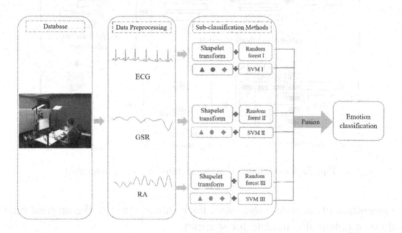

Fig. 2. Flow chart of PMSEC

3.2 Data Preprocessing

Data preprocessing is used to process noise from frequency interference, baseline drift, electrode contact noise, polarization noise, muscle noise, internal amplifier noise and motor artifact noise [6].

PMSEC firstly normalization ECG, GSR, and RA signals. Each signal is normalized as shown in Eq. (1). In Eq. (1), X_n is the data sample of the Nth subject, $\overline{X}_n$ is the mean value of the data under the calm state of the subject, and $\tilde{X}_n$ is the normalized data.

$$\tilde{X}_n = X_n - \overline{X}_n \tag{1}$$

Next, 8-scale wavelet transform based on db6 wavelet basis function was used to denoise ECG [22] and Pan-Tompkins algorithm [23] to detect P, Q, R, S, T wave of ECG. Following by, the second-order Butterworth filter was adopted for GSR, and the low-pass cutoff frequency is set as 0.35 Hz [24]. Considering that the respiratory movement may change when the mood changes, the third-order Butterworth filter was actually used to filter RA signals, and the low-pass cutoff frequency is set as 1 Hz [25].

3.3   Sub-classification Methods

3.3.1   Shapelet Transformation Algorithm

As shown in Fig. 3, there are four main stages of shapelet transformation: 1) generation of candidate shapelets, 2) shapelets distance calculations, 3) quality assessment of shapelets, and 4) discovery of shapelets and data transformation.

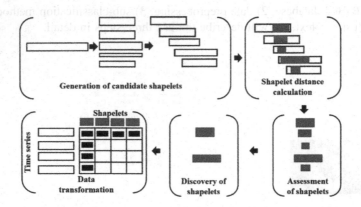

Fig. 3. Flow chart of shapelet transformation algorithm

1) Generation of candidate shapelets. Each subsequence of the original time series *TS* could be a potential candidate for shapelet.

2) Distance calculation between shapelets and time series. Euclidean distance is usually used to measure the similarity between a shapelet and time series. The square of the Euclidean distance between subsequence X and another subsequence Y is defined in Eq. (2), where l is the length of X and Y.

$$d(X, Y) = \sum_{i=1}^{l} (x_i - y_i)^2 \tag{2}$$

3) Quality assessment of shapelets. The standard method for calculating the mass of a shapelet is Information Gain (*IG*). Each segmentation strategy divides dataset T into two sub-datasets, T_I and T_{II}, the *IG* of each segmentation point is calculated as Eq. (3). In Eq. (3), $|T|$ is the instance number of dataset T, and the entropy of each dataset can be calculated as Eq. (4). In Eq. (4), p (1) and p (2) are the proportion of objects in class 1 and class 2, respectively.

$$IG = H(T) - \left(\frac{|T_I|}{|T|} H(T_I) + \frac{|T_{II}|}{|T|} H(T_{II}) \right) \tag{3}$$

$$H(T) = -p(1)\log(p(1)) - p(2)\log(p(2)) \tag{4}$$

4) Discovery of shapelets and data transformation. The candidates with higher information gains are selected as shapelets. Shapelet transformation algorithm calculates the minimum distance between each found shapelets and each time series in the data sets. The algorithm creates a matrix G with n rows and r columns. Each element of the matrix is the minimum Euclidean distance between the time series and each shapelets, and there is a class label at the end of each row. The matrix G can be used as a standard instance data set for machine learning tasks and used in other algorithms.

3.3.2 Feature Extraction
The extracted features of the ECG, GSR, and RA physiological signals are shown in Table 1, Table 2, and Table 3, respectively.

Table 1. ECG features

Time domain	Frequency domain	Nonlinear domain
Mean and standard deviation of R wave, P wave, T wave and HRV wave	The mean value, standard deviation and maximum value of HRV wave power spectral density	Related parameters of Poincaré map, ApEn, SampleEn, the maximum autocorrelation coefficient (ACF_{coef}) and reciprocal of the lag time (ACF_{freq})

Table 2. GSR features

Time domain	Frequency domain	Nonlinear domain
Mean value, standard deviation, maximum value, minimum value, maximum value of GSR, GSR wave first order difference, GSR wave first order difference absolute value, GSR wave second order difference and GSR wave second order difference absolute value	The difference between the median, mean, standard deviation, maximum, minimum and maximum values of the power spectral density of GSR wave	ApEn and SampleEn

Table 3. RA features

Time domain	Frequency domain	Nonlinear domain
The difference of median, mean, standard deviation, maximum, minimum and maximum value of RA wave, the first-order difference of RA wave, the absolute value of first-order difference of RA wave, second order difference of RA wave and second order difference of RA wave	The difference between the median, mean, standard deviation, maximum, minimum and maximum values of the power spectral density of RA wave	ApEn and SampleEn

3.3.3 Sub-classifiers

There are two kinds of sub-classification methods for each physiological signal: shapelet transformation and feature extraction.

For shapelet transformation, the matrix obtained from shapelet transformation is used to train the random forest classifier. In this study, a random forest classifier with 500 trees is used to train the shapelet transform dataset in this study.

For feature extraction, three sub-classifiers based on SVM are established, three groups of emotion features are used to train and test SVM classifiers, and the emotion classification results are obtained.

Through the sub-classification methods, 6 sub-classifiers and 6 emotion classification results are obtained, which will be used for decision-level fusion in the next section.

3.4 Decision-Level Fusion Strategy

After constructing six sub-classifiers of three physiological signals, the weight assignment and optimization of decision-level fusion are implemented by using particle swarm optimization algorithm (PSO) [26]. PSO algorithm is simple and easy to implement and does not have many parameters adjustment and has been widely used in function optimization, neural network training, fuzzy system control and other applications of genetic algorithms.

Let K sub-classifiers established in the fusion model be $E_k, k \in (1, 2, 3, 4, 5, 6)$, and the recognition results of classifiers are output in the form of a posteriori probability as Eq. (5), where $P_k(C_t|x)$ is the result of the normalization of Eq. (6).

$$E_k(x) = (P_k(C_1|x), P_K(C_2|x)) \tag{5}$$

$$P_k(C_t|x) = P(C_t|x, E_k) / \sum_{t=1}^{2} P(C_t|x, E_k), \sum_{t=1}^{2} P_k(C_t|x) = 1 \tag{6}$$

In Eq. (5) and Eq. (6), $P_k(C_t|x, E_k)$ represents the probability that the *kth* classifier recognizes physiological signal x as C_t emotion, and its value range is [0,1], $\{C_t\}_{t=1}^{2}$

representing two types of emotion states, namely, the high or low level of arousal and the positive and negative of value. Therefore, the decision matrix of three physiological signals regarding K sub-classifiers can be constructed as Eq. (7).

$$A_{K\times 2}(x) = \begin{bmatrix} P_1(C_1|x) & P_1(C_2|x) \\ P_2(C_1|x) & P_2(C_2|x) \\ \vdots & \vdots \\ P_K(C_1|x) & P_K(C_2|x) \end{bmatrix} \tag{7}$$

Since the decision performance of a single classifier for different emotions is different, the decision information of each classifier is fused by Eq. (8), where ω_{kt} represents the weight of the kth classifier on C_t emotion, and $\sum_{k=1}^{K} \omega_{kt} = 1$.

$$P_{1.2...,K}(C_t|x) = \sum_{k=1}^{K} \omega_{kt}P_k(C_t|x), t \in (1.2) \tag{8}$$

Aiming at the optimization and assignment of ω_{kt}, the weight solving process is transformed into the maximization of the objective function value in Eq. (9), where C_{xi} and $\widehat{C}_{xi}$ respectively represents the real emotion category value of the ith training sample and the emotion prediction category values calculated by Eq. (10). N represents the number of training samples, and $\omega = (\omega_{11}, \omega_{21}, ..., \omega_{k1}, ..., \omega_{k2})$ represents the $D(D = K \times 2)$ weight to be optimized.

$$J(\omega) = \frac{\sum_{i=1}^{N} 1_{C_{xi}=\widehat{C}_{xi}}}{N} = \frac{\sum_{i=1}^{N} 1_{C_{xi}=\arg\max_{C_t}\{\sum_{k=1}^{K} \omega_{kt}P_k(C_t|)x\}_{t=1}^2}}{N} \tag{9}$$

$$\widehat{C}_{xi} = \arg\max_{C_t}\left\{\sum_{k=1}^{K} \omega_{kt}P_k(C_t|x)\right\}_{t=1}^2 \tag{10}$$

The inputs of PSO algorithm is the training set (Decision matrix $\{A_{K\times 2}(x_i)\}_{i=1}^{N}$ and real emotion category $\{C_{xi}\}_{i=1}^{N}$), the inertial weight of the particle α (A fixed weight of 0.5 is used), learning factor β_1, β_2 (Usually the value of $\beta_1 = \beta_2 = 2$ is taken), number of particle S, maximum number of iterations Q (The default value is 1000), recognition rate threshold value η (Default value is 85%), the output of the algorithm is the optimal weight coefficient ω. The PSO algorithm steps are as follows.

Step 1: Initialize a particle swarm and randomly generate S particles $\{\omega^j\}_{j=1}^{S}$ in the D-dimensional space, and generate the speed $\{v^j\}_{j=1}^{S}$ of the particle as shown in Eq. (11). Denoted p_j is the historical optimal solution searched so far for the jth particle, and the global optimal position is calculated at the same time. Denoted g as the historical optimal solution searched so far for the whole particle swarm as shown in Eq. (12).

$$\begin{cases} \omega^j = (\omega_{11}^j, \omega_{21}^j, \cdots, \omega_{k1}^j, \cdots, \omega_{k2}^j) \\ v^j = (v_{11}^j, v_{21}^j, \cdots, v_{k1}^j, \cdots, v_{k2}^j) \end{cases} \qquad (11)$$

$$\begin{cases} p^j = (p_{11}^j, p_{21}^j, \cdots, p_{k1}^j, \cdots, p_{k2}^j) \\ g = (g_{11}, g_{21}, \cdots, g_{k1}, \cdots, g_{k2}) \end{cases} \qquad (12)$$

Step 2: Two accelerated weighted particles, r_1 and r_2, are randomly generated in the interval [0,1]. Each particle updates its speed and position according to Eq. (13) according to its own optimal solution and historical optimal solution.

$$\begin{cases} v_{kt}^{j+1} = (\alpha v_{kt}^j + \beta_1 r_1 (p_{kt}^j - \omega_{kt}^j) + \beta_2 r_2 (g_{kt}^j - \omega_{kt}^j)) \\ \omega_{kt}^{j+1} = \omega_{kt}^j + v_{kt}^{j+1} \end{cases} \qquad (13)$$

Step 3: Each particle is normalized according to Eq. (14).

$$\omega_{kt}^j = \omega_{kt}^j / \sum_{k=1}^{K} \omega_{kt}^j, \ 1 \le k \le K, \ 1 \le t \le 2 \qquad (14)$$

Step 4: Find the global optimal solution. Take Eq. (9) as the objective function, update the optimal position $p^j(1 \le j \le S)$ of each particle and the optimal position g of the whole particle swarm, and update the current optimal recognition rate η'.

Step 5: Check the termination condition, end the algorithm when the number of iterations q reaches the preset maximum number of iterations Q or the current optimal recognition rate η' reaches the recognition rate threshold $\eta(\eta' < \eta)$, or when the optimal solution is no longer changing, otherwise go to Step 2.

4 Experimental Results and Analysis

4.1 Database

MAHNOB-HCI Multimodal Sentiment Database is a popular public standard multimodal emotion database [33]. It is widely used in non-commercial academic research, and many articles use this data to test and validate their algorithms. The emotional data modes of this dataset are 32 channels of EEG, 3 channels of ECG, 1 channel of RA, 1 channel of GSR, Skin Temperature, Face and Body Video, Eye Gaze signal, and Audio. In this study, 1 channel of ECG signal (the 35th channel), 1 channel of RA signal and 1 channel of GSR signal are selected from this dataset. The model we used is a two-dimensional emotion model of high/low arousal and positive/negative valence.

4.2 Results of Emotion Classification of a Single Physiological Signal

In this section, we respectively calculated the precision, recall, accuracy and F1-score of the six sub-classifiers and decision level fusion in the experimental process, as shown in Table 4, and then analyzed the experimental results.

(1) Results analysis of the shapelet transformation

We conducted shapelet transformation experiments on the three physiological signals and classified them. In the classification results of shapelet transformation, ECG achieved the best results in both arousal and valence, which were 0.728 and 0.752 respectively. It shows that the shapelet in the ECG contains richer emotional information than GSR and RA.

(2) Results analysis of feature extraction

Traditional emotion classification is mostly based on extracting different features and using the extracted features to train the classifier for classification. The experimental results are shown in Table 4. In the experiment of feature extraction, ECG has achieved the best results in terms of both arousal and valence. This is partly because ECG contains richer emotional information and extracts more nonlinear features, which are more conducive to improving the accuracy of classification.

(3) Results comparison of shapelet transformation and feature extraction

In terms of arousal, the accuracy of shapelet transformation of ECG is lower than that of feature extraction, while the classification accuracy of shapelet transformation of GSR and RA are higher than that of feature extraction. In terms of valence, the classification accuracy of shapelet transformation of ECG and GSR are higher than that of feature extraction, and the accuracy of shapelet transformation of RA is lower than that of feature extraction.

The results showed that the shapelet transformation in the classification of ECG, GSR and RA with high/low arousal and positive/negative valence was at the same level as the traditional method of extracting features and training the SVM classifier, and guaranteed a high classification accuracy. Therefore, shapelet transformation can be used as an effective tool for classifying high/low arousal and positive/negative valence.

Table 4. The experimental results of emotion classification on multi-physiological signals with 6 sub-classifiers and decision level fusion

Signal	Method	Precision	Recall	F1-score	Accuracy
ECG	Shapelet transform	0.640(A)	0.800(A)	0.711(A)	0.728(A)
	+ Random forest	0.651(V)	0.871(V)	0.745(V)	0.752(V)
	Feature extraction	0.575(A)	0.767(A)	0.657(A)	0.777(A)
	+ SVM	0.655(V)	0.853(V)	0.735(V)	0.733(V)
GSR	Shapelet transform	0.697(A)	0.825(A)	0.736(A)	0.689(A)
	+ Random forest	0.680(V)	0.862(V)	0.760(V)	0.712(V)
	Feature extraction	0.734(A)	0.850(A)	0.788(A)	0.675(A)
	+ SVM	0.639(V)	0.793(V)	0.708(V)	0.699(V)
RA	Shapelet transform	0.687(A)	0.842(A)	0.757(A)	0.714(A)
	+ Random forest	0.647(V)	0.776(V)	0.706(V)	0.675(V)
	Feature extraction	0.671(A)	0.817(A)	0.737(A)	0.701(A)
	+ SVM	0.660(V)	0.802(V)	0.724(V)	0.684(V)
PSO		0.608(A)	0.892(A)	0.723(A)	0.854(A)
		0.593(A)	0.879(A)	0.708(A)	0.835(V)

4.3 Results Comparisons

This section compares the results of other methods using the MAHNOB-HCI database in terms of the high/low levels of arousal and the positive/negative levels of valence. Table 5 provides a comparison of the proposed algorithm with those in other literature.

According to Table 5, compared with the methods without EEG, the proposed method achieved the best results. Compared with [9, 37, 38] which adopts a single signal, our method is in the leading position in both arousal and valence dimensions. Compared with the two literatures that use EEG and facial expression fusion, our algorithm is higher than [39] in terms of arousal and valence, and is basically at the same level as [40]. However, compared with EEG signals, the ECG, GSR and RA signals we use are simpler and easier to collect, so our method have a wider application prospect. Compared with the two methods [40, 41] using deep learning, the accuracy of our method is indeed lower than that of deep learning, but deep learning often requires a large amount of data and a large amount of computation.

Table 5. The comparison of the proposed algorithm with other methods in other literature

Reference	Signals	Method	Accuracy	
			Arousal	Valence
M. Ben and Z. Lachiri [36]	ECG, RESP, SC, GSR	Statistical parameters & SVM	64.23%	68.75%
Ferdinando et al. [37]	ECG	Statistical parameters & KNN	59.70%	55.80%
Hsu et al. [9]	ECG	Time Domain, Frequency Domain, Nonlinear Analysis & SFFS-KBCS + GDA + LS-SVM	49.20%	44.10%
Baghizadeh et al. [38]	ECG (use of 5-different time-series generated from ST Intervals Poincaré map)	Time Domain, Frequency Domain, Time Frequency Domain, Nonlinear Domain Analysis & SVM-Linear, SVM Polynomial	82.17%	78.07%
Huang et al. [39]	EEG, Face	Features Extraction& CNN + SVM & Decision-level Fusion	75.63% 74.17%	75.00% 75.21%
Siddharth et al. [40]	EEG, Face	LSTM	86.31%	83.70%
Zhang et al. [41]	EEG, GSR, RESP, TEMP, PLET	HFCNN, RF	88.28%	89.00%
Our method	ECG, GSR, RA	Shapelet transformation + Random Forest & Features Extraction + SVM & Decision-level Fusion	85.44%	83.50%

5 Conclusion and Future Work

In this paper, a decision level fusion model of multiple physiological signals is presented. Firstly, ECG, GSR and RA signals in MAHNOB-HCI database are preprocessed, and then two independent sub-classification methods, shapelet transformation and feature extraction, are carried out on the processed signals, and a total of six sub-classifiers are created. Finally, the particle swarm optimization algorithm (PSO) is used to find the optimal weight for the six sub-classifiers, and the decision level fusion is implemented. The results of this study indicate that ECG contains more emotional information than GSR and RA. In addition, shapelet algorithm as a time series classification method, has demonstrated that it can be used in the classification of physiological signals with high/low levels of arousal and positive/negative levels of valence. Among the methods for emotion classification without EEG signal, the accuracy of our method is almost the best. Compared with the method using EEG signal for emotion classification, the signals (ECG, GSR, and RA) we used are simpler and easier to collect, so it has a wider application prospect in daily life.

In future, we will consider mining deeper features of a single physiological signal and consider shapelet as a feature to conduct feature level fusion. In addition, we will adopt deep learning to achieve good results.

Acknowledgments. We are grateful for the support of the Natural Science Foundation of Shandong Province (No. ZR2020LZH008, ZR2020QF112, ZR2019MF071).

References

1. Cavallo, F., Semeraro, F., Fiorini, L., Magyar, G., Sinčák, P., Dario, P.: Emotion modelling for social robotics applications: a review. J. Bionic Eng. **15**(2), 185–203 (2018). https://doi.org/10.1007/s42235-018-0015-y
2. Tojo, T., Ono, O., Noh, N.B., Yusof, R.: Interactive Tutor Robot for collaborative e-learning system. Electr. Eng. Jap. **203**(3), 22–29 (2018). https://doi.org/10.1541/ieejeiss.137.1373
3. Basiri, M., Schill, F., Lima, P.U., Floreano, D.: Localization of emergency acoustic sources by micro aerial vehicles. J. Field Robot. **35**(2), 187–201 (2018). https://doi.org/10.1002/rob.21733
4. Choe, Y.: Meaning Versus Information, Prediction Versus Memory, and Question Versus Answer (2021). arXiv e-prints. https://arxiv.org/abs/2107.13393
5. Bagnall, A., Lines, J., Bostrom, A., Large, J., Keogh, E.: The great time series classification bake off: a review and experimental evaluation of recent algorithmic advances. Data Min. Knowl. Disc. **31**(3), 606–660 (2016). https://doi.org/10.1007/s10618-016-0483-9
6. Picard, R.W., Vyzas, E., Healey, J.: Toward machine emotional intelligence: analysis of affective physiological state. IEEE Trans. Pattern Anal. Mach. Intell. **23**(10), 1175–1191 (2001). https://doi.org/10.1109/34.954607
7. Agrafioti, F., Hatzinakos, D., Anderson, A.K.: ECG pattern analysis for emotion detection. IEEE Trans. Affect. Comput. **3**(1), 102–115 (2012). https://doi.org/10.1109/T-AFFC.2011.28
8. Verma, G.K., Tiwary, U.S.: Multimodal fusion framework: a multiresolution approach for emotion classification and recognition from physiological signals. Neuroimage **102**, 162–172 (2014). https://doi.org/10.1016/j.neuroimage.2013.11.007
9. Hsu, Y., Wang, J., Chiang, W., Hung, C.: Automatic ECG-based emotion recognition in music listening. IEEE Trans. Affect. Comput. **11**(1), 85–99 (2020). https://doi.org/10.1109/TAFFC.2017.2781732
10. Ye, L., Keogh, E.: Time series shapelets: a novel technique that allows accurate, interpretable and fast classification. Data Min. Knowl. Disc. **22**, 149–182 (2011). https://doi.org/10.1007/s10618-010-0179-5

11. Rakthanmanon, T., Keogh, E.: Fast shapelets: a scalable algorithm for discovering time series shapelets. In: Proceedings of the 2013 SIAM International Conference on Data Mining, pp. 668–676. Society for Industrial and Applied Mathematics, Philadelphia (2013). https://doi.org/10.1137/1.9781611972832.74

12. Hills, J., Lines, J., Baranauskas, E., Mapp, J., Bagnall, A.: Classification of time series by shapelet transformation. Data Min. Knowl. Disc. **28**(4), 851–881 (2013). https://doi.org/10.1007/s10618-013-0322-1

13. Grabocka, J., Schilling, N., Wistuba, M., et al.: Learning time-series shapelets. In: Proceedings of the 20th ACM SIGKDD International Conference on Knowledge Discovery and Data Mining, pp. 392–401 (2014). https://doi.org/10.1145/2623330.2623613

14. Aldhanhani, A., Damiani, E., Mizouni, R., et al.: Framework for traffic event detection using Shapelet Transform. Eng. Appl. Artif. Intell. **82**, 226–235 (2019). https://doi.org/10.1016/j.engappai.2019.04.002

15. Albert, Z., Matthias, F., et al.: Heart rhythm analyzed via shapelets distinguishes sleep from awake. Front. Physiol. **10**, 1554−1554 (2019)https://doi.org/10.3389/fphys.2019.01554

16. Arul, M., Kareem, A.: Shapelets for Earthquake Detection (2019). arXiv e-prints. https://arxiv.org/abs/1911.09086

17. Zhao, C., Wang, T., Liu, S., et al.: A fast time series shapelet discovery algorithm combining selective extraction and subclass clustering. J. Softw. **31**(03), 763–777 (2020). https://doi.org/10.13328/j.cnki.jos.005912

18. Renard, X., et al.: Random-shapelet: an algorithm for fast shapelet discovery. In: 2015 IEEE International Conference on Data Science and Advanced Analytics (DSAA), pp. 1–10 (2015). https://doi.org/10.1109/DSAA.2015.7344782

19. Lin, J., et al.: Experiencing SAX: a novel symbolic representation of time series. Data Min. Knowl. Disc. **15**(2), 107–144 (2007). https://doi.org/10.1007/s10618-007-0064-z

20. Ji, C., et al.: A fast shapelet discovery algorithm based on important data points. Int. J. Web Serv. Res. **14**(2), 67–80 (2017). https://doi.org/10.4018/IJWSR.2017040104

21. Ji, C., et al.: A fast shapelet selection algorithm for time series classification. Comput. Netw. **148**, 231–240 (2019). https://doi.org/10.1016/j.conet.2018.11.031

22. Goshvarpour, A., Abbasi, A., Goshvarpour, A.: An accurate emotion recognition system using ECG and GSR signals and matching pursuit method. Biomed. J. **40**, 355–368 (2017). https://doi.org/10.1016/j.bj.2017.11.001

23. Pan, J., Tompkins, W.J.: A real-time QRS detection algorithm. IEEE Trans. Biomed. Eng. **32**(3), 230–236 (1985). https://doi.org/10.1109/TBME.1985.325532

24. Shukla, J., et al.: Feature extraction and selection for emotion recognition from electrodermal activity. IEEE Trans. Affect. Comput. **12**(4), 857–869 (2019). https://doi.org/10.1109/TAFFC.2019.2901673

25. Wei, W., et al.: Emotion recognition based on weighted fusion strategy of multichannel physiological signals. Comput. Intell. Neurosci. **1–9**, 2018 (2018). https://doi.org/10.1155/2018/5296523

26. Kennedy, J., Eberhart, R.: Particle swarm optimization. In: Proceedings of ICNN'95 - International Conference on Neural Networks, vol. 4, pp. 1942–1948 (1995). https://doi.org/10.1109/ICNN.1995.488968

27. Udhaya Kumar, S., Hannah Inbarani, H.: PSO-based feature selection and neighborhood rough set-based classification for BCI multiclass motor imagery task. Neural Comput. Appl. **28**(11), 3239–3258 (2016). https://doi.org/10.1007/s00521-016-2236-5

28. Khamis, H., et al.: QRS detection algorithm for telehealth electrocardiogram recordings. IEEE Trans. Biomed. Eng. **63**(7), 1377–1388 (2016). https://doi.org/10.1109/TBME.2016.2549060

29. Kim, K.H., Bang, S.W., Kim, S.R.: Emotion recognition system using short-term monitoring of physiological signals. Med. Biol. Eng. Comput. **42**, 419–427 (2004). https://doi.org/10. 1007/BF02344719

30. Arul, M., Kareem, A.: Data Anomaly Detection for Structural Health Monitoring of Bridges using Shapelet Transform (2020). arXive-prints. https://doi.org/10.13140/RG.2.2.34565. 78567

31. Yan, W., Li, G.: Research on time series classification based on shapelet. Comput. Sci. **046** (001), 29–35 (2019). 10.11896%EF%BC%8Fj.issn.1002-137X.2019.01.005

32. Yin, Z., et al.: Recognition of emotions using multimodal physiological signals and an ensemble deep learning mode. Comput. Methods Programs Biomed. **140**, 93–110 (2017). https://doi.org/10.1016/j.cmpb.2016.12.005

33. Soleymani, M.: A multimodal database for affect recognition and implicit tagging. IEEE Trans. Affect. Comput. **3**(1), 42–55 (2012). https://doi.org/10.1109/T-AFFC.2011.25

34. Ebrahimi, Z., et al.: A review on deep learning methods for ECG arrhythmia classification. Expert Syst. Appl. X **7**, 100033 (2020). https://doi.org/10.1016/j.eswax.2020.100033

35. Li, C., Xu, C., Feng, Z.: Analysis of physiological for emotion recognition with IRS model. Neurocomputing **178**, 103–111 (2016). https://doi.org/10.1016/j.neucom.2015.07.112

36. Mimoun, B.H.W., Zied, L.: Emotion classification in arousal valence model using MAHNOB-HCI database. Int. J. Adv. Comput. Sci. Appl. **8**(3) (2017). https://doi.org/10. 14569/IJACSA.2017.080344

37. Ferdinando, H., Seppänen, T., Alasaarela, E.: Comparing features from ECG pattern and HRV analysis for emotion recognition system. In: 2016 IEEE Conference on Computational Intelligence in Bioinformatics and Computational Biology (CIBCB), pp. 1–6 (2016). https:// doi.org/10.1109/CIBCB.2016.7758108

38. Maryam, B., Keivan, M., Fardad, F., Nader, J.D.: A new emotion detection algorithm using extracted features of the different time-series generated from ST intervals Poincaré map. Biomed. Signal Process. Control **59**, 101902 (2020). https://doi.org/10.1016/j.bspc. 2020.101902

39. Huang, Y., Yang, J., Liu, S., Pan, J.: Combining facial expressions and electroencephalography to enhance emotion recognition. Future Internet **11**, 105 (2019). https://doi.org/10. 3390/fi11050105

40. Siddharth, S., Jung, T.-P., Sejnowski, T.J.: Utilizing Deep Learning Towards Multi-modal Bio-sensing and Vision-based Affective Computing (2019). arXiv e-prints. https://doi.org/ 10.1109/TAFFC.2019.2916015

41. Zhang, Y., Cheng, C., Zhang, Y.: Multimodal emotion recognition using a hierarchical fusion convolutional neural network. IEEE Access **9**, 7943–7951 (2021). https://doi.org/10. 1109/ACCESS.2021.3049516

A Novel and Efficient Distance Detection Based on Monocular Images for Grasp and Handover

Dianwen Liu[1], Pengfei Yi[1(✉)], Dongsheng Zhou[1,2], Qiang Zhang[1,2],
Xiaopeng Wei[2], Rui Liu[1], and Jing Dong[1]

[1] Dalian University, Dalian, People's Republic of China
[2] Dalian University of Technology, Dalian, People's Republic of China

Abstract. Robot grasping and human-robot handover (HRH) tasks can significantly facilitate people's production and life. In these tasks, robots need to obtain the real-time 3D position of the object, and the distance from the object to the camera plane is the critical information to get the object position. Currently, depth camera-based distance detection methods always need additional equipment, which results in more complexity and cost. In contrast, RGB camera-based methods often assume that the object's size is known or the object is at a fixed height. To make distance detection more adaptive and with low cost, a novel and efficient distance detection method based on monocular RGB images is proposed in this paper. With a simple marker, the method can estimate the object's distance in real-time from the pixel information obtained by a general, lightweight target detector. Experiments on the Baxter robot platform show the effectiveness of the proposed method, where the success rate of the grasping test reaches 87.5%, and the success rate of the HRH test goes 84.7%.

Keywords: Monocular RGB image · Distance detection · Grasping · Human-robot handover

1 Introduction

Grasping and human-robot handover (HRH) have a broad prospect of application. Grasping is one of the classic tasks in robotics, widely used in various industries of people's production and life. For example, industrial robots can accomplish the pick-and-place task, which is laborious for human laborers, and domestic robots can assist disabled or older people in their daily grasping tasks [17]. In addition, HRH [11] has become a research hotspot of cooperative robotics in recent years [2]. For example, industrial robot assistants can fetch or pass tools to human workers to increase efficiency in factories. Service robots can fetch or pass objects that older or disabled adults are needed to help them live independently.

© ICST Institute for Computer Sciences, Social Informatics and Telecommunications Engineering 2021
Published by Springer Nature Switzerland AG 2021. All Rights Reserved
H. Gao and X. Wang (Eds.): CollaborateCom 2021, LNICST 406, pp. 642–658, 2021.
https://doi.org/10.1007/978-3-030-92635-9_37

To achieve grasping or HRH, robots must locate the 3D position of the object quickly and accurately. It's worth noting that in HRH tasks, unlike grasping tasks where the object is placed on a table of constant height, the object will be held in the human hand, and its height will change, which will make it more difficult for robots to locate the object.

The distance from the object to the camera plane, namely depth information of the object, is the key information for robots to locate the object. Currently, most robots are equipped with RGB cameras. Still, the object is difficult to be located from RGB cameras directly because 2D images obtained by RGB cameras lack depth information in space. Existing analytic-based object distance detection algorithms [3,7,8] are fast but restricted, which are difficult to adapt to complex and changeable robot applications. For example: (1) The object must be placed on a fixed horizontal plane to calculate the distance, which makes the algorithm unsuitable for HRH tasks; (2) In some algorithms, the size of the object must be known, which limits its applicability to a variety of objects. As a result, researchers need to provide additional distance sensors to robots in practical tasks. With the development of depth cameras, the current mainstream method is to use the depth information obtained by depth cameras to calculate the distance between the object and robot [14,19,20]. This approach is practical, but there are some problems: (1) Due to the high price of depth cameras, large-scale deployment in a factory environment will significantly increase the cost; (2) The hardware performance and technical level of depth cameras will seriously restrict its practical application effect; (3) When additional depth cameras are loaded, their coordinate system needs to be embedded very precisely into the robot system, which adds the complexity of device deployment.

For human beings, they can easily judge the distance of the object and complete grasp or handover tasks, which is the effect of complex human visual mechanism: (1) For objects of the same size, the size of the object perceived by human eyes is inversely proportional to the distance. (2) For multiple objects of different sizes, people can quickly judge the distance between different objects and themselves, which benefits from the rich prior knowledge acquired by birth, enabling people to know the size difference between objects in advance judge the distance. (3) Rich prior knowledge can help people quickly estimate the size of the unknown object by comparing it with the known object to judge the distance of the unknown object.

Inspired by human vision, this paper proposes a novel two-stage unknown object distance detection algorithm based on monocular RGB images. In the first stage, a known size marker is used as prior knowledge to calculate the distance of the unknown object on the desk. In the second stage, the object's distance on the desk is used to calculate the object's distance at any position. The algorithm uses a general, lightweight object detector to obtain the pixel information of the object in real-time, establishes a geometric-imaging model, and calculates the distance of the object on the desk and in the air successively. It meets precision and real-time performance; simultaneously, it is cheap and easy to be deployed.

2   Related Works

There are three ways to estimate the distance from the object to the camera plane: RGB-D-based methods, analytics-based methods, and model-based methods. Firstly, recent works of grasping and HRH tasks are introduced in Sect. 2.1, where the distance is obtained by RGB-D camera. Afterward, the traditional analytics-based distance detects methods are discussed in Sect. 2.2. Lastly, the model-based methods are introduced in Sect. 2.3, which use deep neural networks to detect the distance of the object in RGB images directly.

2.1   RGB-D-Based Methods

Robot grasping is a classic task. In recent years, most studies have completed the grasping tasks by connecting the external RGB-D camera to obtain the 6D target pose [18] or infer the 6D grasping pose of the end-effector [9,10,16]. In these works, the object distance is calculated from point cloud information by the k-means clustering algorithm. Zhang et al. [21] estimated grasping points through RGB images and realized grasping on the Baxter robot platform, however in actual grasping tasks, instead of using the RGB camera on the robot's head, they use an external depth camera.

HRH has become an important research topic. Rosenberger et al. [14] used object detection, human segmentation, hand segmentation, grasping point detection to realize the handover of a variety of objects while ensuring the safety of human partners. Yang et al. [20] divided the gestures of people grasping objects into seven categories, and each category corresponds to a grasping trajectory. In their subsequent research [19], they proposed an HRH method for transferring objects of arbitrary shape, size, and stiffness. In these implementations, objects are also located by the RGB-D camera.

2.2   Analytic-Based Methods

The analytical-based algorithms use the principle of optical imagery and geometry to directly solve the distance between the object and the camera plane from RGB images. The distance can be calculated from a simple optical imaging model when the object's size is known. For example, Pathi et al. [12] implemented a distance detection algorithm for people, in which the pixel distance from human ear to clavicle is obtained through a human pose estimation algorithm, and then using optical imaging model to calculate the distance between human and cameras. The process assumes that the actual distance from the ear to the clavicle is the same for different adults and takes it as a known size.

The challenge is to detect the distance of unknown objects using RGB images. Ricardo et al. [13] proposed a handover method based on binocular vision. This method configures two RGB cameras to obtains the distance of the object through multi-view geometry [1]. Krishnan et al. [8] proposed a complex logarithmic mapping (CLM) method, but it required objects to be placed on and perpendicular to the optical axis, so it is not widely used. Chiang et al. [3] proposed a geometric triangulation method for measuring the object distance, in which some hard-to-measure information in the workspace needs to be taken as known quantities. Jmzad [7] implemented a soccer robot that uses a distance detection algorithm with variable pitch angles to locate the ball. Still, the algorithm also needed to measure some hard-to-measure information in advance, and the ball must be on the ground.

2.3   Model-Based Methods

With the development of deep learning, some researchers use neural networks to solve monocular distance detection problems. Zhu et al. [22] proposed a distance detection model for specific objects, which is used for vehicle distance detection in the process of autonomous driving. This method is only suitable for specific objects (size range is fixed) and unsuitable for indoor multi-object robot tasks. Haseeb et al. [6] also proposed a distance detection model for road scenes and set the average size of different categories of targets. This method requires manual measurement to mark large datasets. Griffin et al. [5] proposed a target depth estimation model, which requires the relative movement of the target and camera and maps the pixel transformation of the target to distance. Still, this method is only suitable for the circumstances of a comparatively fixed object pose when object location is unchanged. Posture changes, the pixel will also undergo significant changes, resulting in a drastic change of detection distance.

To sum up, the model-based distance detection method is more flexible. It has high accuracy but slow reasoning speed and poor generalization, making it difficult to use for complex and changeable robot tasks. Moreover, it needs to take up extra space, train the network, and download many data sets, which is inefficient. Using RGB-D cameras to obtain the object distance can meet the requirements of robot tasks in speed and accuracy. Still, the cost is high, and the RGB-d camera's hardware performance and technical level seriously restrict the practical application effect. Analytic-based methods have some advantages of high efficiency, low cost, and easy deployment. In theory, the effect of these methods is less affected by the camera hardware performance and more depends on the rationality of the algorithm. Unfortunately, existing analytics-based algorithms have significant limitations and cannot be applied to robotic tasks. Therefore, our goal is to design a novel analysis-based algorithm to detect the distance between the unknown object and the camera plane using the monocular image. It can be successfully applied to grasping or HRH tasks while retaining the advantages of traditional algorithms overcoming some limitations.

3   Method

The implementation process of our method is as follows (see Fig. 1): Firstly, the robot obtains the image of the workspace through a head-fixed camera (internal parameters are known). Then a general detector is used to obtain the 2D

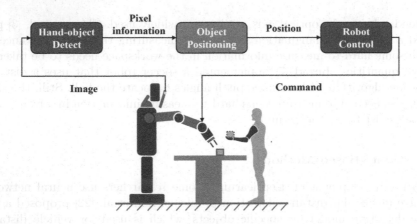

Fig. 1. General diagram of our method

pixel information of the hand and the object. Next, the 2D pixel information is transformed into the 3D position of the object through the object positioning module. Finally, the robot control module receives the 3D position of the object and sends commands to control the manipulator to move towards the target, to implement grasping or HRH.

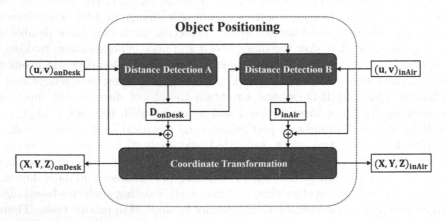

Fig. 2. Overview of object positioning module, where $(u, v)_{onDesk}$, D_{onDesk}, $(X, Y, Z)_{onDesk}$ respectively represent the pixel coordinates of the object on the desk, the distance to the camera, and world coordinates; $(u, v)_{inAir}$, D_{inAir}, $(X, Y, Z)_{inAir}$ respectively represent the pixel coordinates of the object in the air, the distance to the camera, and world coordinates.

In this section, we describe the object positioning module (see Fig. 2) in detail, which is composed of three sub-modules: (1) Distance Detection A, which is used to detect the distance of the object on the desk; (2) Distance Detection B, which is used to detect the distance of the object in the air. (3) Coordinate

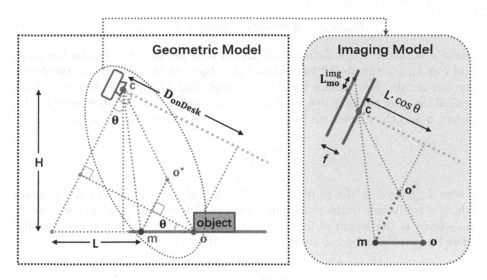

Fig. 3. Geometric-imaging model of workspace, in which the object is on the desk

conversion, which converts the object pixel coordinates to world coordinates using the acquired distance between the object and the camera plane. We focus on Distance Detection A (Sect. 3.1) and Distance Detection B (Sect. 3.2). In the grasping task, only Distance Detection A and Coordinate Transformation are needed to locate the object. In HRH, Distance Detection A is first used to get the object's distance on the desk, and then Distance Detection B and Coordinate Transformation are used to locate the object's location in the air.

3.1 Distance Detection A

This sub-module calculates the distance of the object to the camera plane when the object is on the desk (D_{onDesk}). We abstract the side view of the workspace as a geometer-imaging model (see Fig. 3): the height of camera c to the tabletop is H, the horizontal distance of the marker m to the camera plane is L, the camera inclination angle is θ, the focal length is f, and object is abstracted as a particle o. Before calculation, workspace is simply marked: the pixel information of marker $Pix_{mark}(\Delta u_{mark}, v_{mark})$ is obtained and taken as a known quantity, where v_{mark} represents the v-axis pixel coordinates at the bottom of marker and Δu_{mark} represents the pixel width at the bottom of the marker. The essence of this process is, at the position where the vertical distance to the camera is H, and the horizontal distance to the camera plane is L, the pixel information projected by a line segment of length W in the camera is fixed. This pixel information can be regarded as a size prior information in the current workspace and used as a reference of the unknown object. Therefore, only one mark is required when the workbench height is fixed; it needs to be re-marked when its height is changed. The distance of the object on the desk can be expressed as:

$$D_{onDesk} = distanceA(Pix_{mark}, Pix_{object}) \tag{1}$$

where $distanceA(\cdot)$ is a distance calculation function of the object on the desk, and Pix_{object} is the pixel information of the object on the desk, which the detector can obtain. Next comes the concrete reasoning of Eq. 1. To be specific, an optical imaging model is first established according to the distance between the object o and the marker m: Next comes the concrete reasoning of Eq. 1.

$$\frac{L_{mo^*}}{L \cdot \cos\theta} = \frac{L_{mo}^{img}}{f} \tag{2}$$

where $L_{mo}^{img}(cm)$ is the projection length of line segment mo in the camera, which is equivalent to the projection length of virtual image mo^* in the camera; $L_{mo^*}(cm)$ is the length of line segment mo^*. In addition, an optical imaging model is established according to the width W of the marker:

$$\frac{W}{L \cdot \cos\theta} = \frac{W^{img}}{f} \tag{3}$$

where $W(cm)$ is the actual width of the marker and $W^{img}(cm)$ is the projection width of the marker in the camera. In the geometric model, triangle similarity theorem is used to obtain:

$$\frac{L_{mo'}}{H \cdot \sec\theta} = \frac{D_{onDesk} \cdot \sec\theta - L}{D_{onDesk} \cdot \sec\theta} \tag{4}$$

From Eqs. 2, 3 and 4, it can be deduced that:

$$D_{onDesk} = \frac{L}{\sec\theta - \frac{W}{H} \cdot \frac{L_{mo}^{img}}{W^{img}}} \tag{5}$$

In addition, we can easily find that:

$$\frac{L_{mo}^{img}}{W^{img}} = \frac{\Delta v_{mo} \cdot dv}{\Delta u_{mark} \cdot du} \tag{6}$$

where Δv_{mo} is the pixel length of line segment mo in image, Δu_{mark} is the pixel width of marker. du, $dv(cm)$ is the length of each pixel in direction of u and v axis respectively. In general, $du = dv(cm)$, so Eq. 5 can be changed to:

$$D_{onDesk} = \frac{L}{\sec\theta - \frac{W}{H} \cdot \frac{\Delta v_{mo}}{\Delta u_{mark}}} \tag{7}$$

wherein, θ can be set by itself or returned by robot topic, Δv_{mo} and Δu_{mark} can be directly obtained by pixel information Pix_{mark} and Pix_{object}. So D_{onDesk} can be solved. Equation 7 is equivalent to Eq. 1.

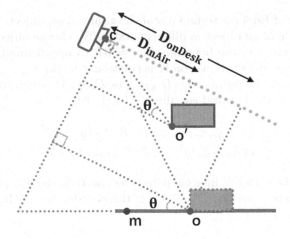

Fig. 4. Geometric-imaging model of workspace

3.2 Distance Detection B

This submodule uses the previous information to calculate the distance from the object in the air to the camera plane (D_{inAir}):

$$D_{inAir} = distanceB(Pix'_{object}, Pix_{object}, D_{onDesk}) \tag{8}$$

where $distanceB(\cdot)$ is a function of the distance of the object in the air, Pix'_{object} is the pixel information when the object is in the air. Next comes the concrete reasoning of Eq. 8. Specifically, we use the previously obtained information of the object on the desk to calculate the distance D_{inAir} (see Fig. 4). We set up optical imaging models for the two states of the object on the workbench and in the air:

$$\frac{W_{object}}{D_{onDesk}} = \frac{W_{object}^{img}}{f} \tag{9}$$

$$\frac{W_{object}}{D_{inAir}} = \frac{W_{object'}^{img}}{f} \tag{10}$$

where $W_{object}(cm)$ represents the actual width of object, $W_{object}^{img}(cm)$ is projection width in the camera of the object on the workbench, and $W_{object}^{img} = \Delta u_{object} \cdot du$, where Δu_{object} is pixel width of object on the workbench. $W_{object'}^{img}(cm)$ is the projection width in camera of the object in the air: $W_{object'}^{img} = \Delta u'_{object} \cdot du$, in which u'_{object} is the pixel width of object in the air. According to Eq. 9 and Eq. 10:

$$D_{inAir} = D_{onDesk} \cdot \frac{\Delta u'_{object}}{\Delta u_{object}} \tag{11}$$

The essence of Eq. 11 is to find a scale-invariant of an object, to connect the pixel information of an object in different states. We choose object width as the scale-invariant and use the bounding box width to approximately replace the pixel width of the object. Equation 11 is equivalent to Eq. 8.

After obtaining the distance D (D_{onDesk} or D_{inAir}), we can convert 2D pixel coordinates (u, v) to 3D world coordinates (X, Y, Z):

$$\begin{cases} (x_c, y_c, z_c)^T & = D \cdot K^{-1} \cdot (u, v, 1)^T \\ (X, Y, Z, 1)^T & = T^{-1} \cdot (x_c, y_c, z_c, 1)^T \end{cases} \tag{12}$$

where $K_{3\times3}$ is the camera's internal parameters matrix, and $T_{4\times4}$ is the camera's external parameters matrix, (x_c, y_c, z_c) is the coordinates of the object in the camera coordinate system.

4   Experiments and Results

To evaluate the effectiveness of our approach, a set of experiments are performed on a Baxter robot platform. First, introduce our experimental equipment in Sect. 4.1 and preliminary work experiments in Sect. 4.2. Afterward, the process and results of the robot grasping experiment are reported in Sect. 4.3. Then the process and results of HRH experiments are reported in Sect 4.4. After that, the time cost of our method is fully described in Sect. 4.5. Lastly, the weaknesses of the proposed distance detection algorithm are qualitatively discussed in Sect. 4.6.

4.1   Experimental Equipment

A Baxter robot is used in the experiment, with an RGB camera (1280 × 800) on its head, a laptop computer with an NVIDIA 1050 Ti graphics for image processing, and robot control. A workbench is placed within the effective working range of the manipulator in front of the robot. A worker is present in the scene. Ten objects with different materials, shapes, and sizes (see Fig. 5) are used as targets, including five large-size objects and five small-size objects to verify the applicability of our method to objects with different attributes.

4.2   Preliminary Work

For the Baxter workspace (see Fig. 6), actually, we measured the distance L' from the marker to the Y-O-Z plane of the world coordinates instead of measuring L. Because L' is easier to be measured and L can be calculated by L':

$$L = L' - (X_{camera} - H \cdot \tan\theta) \tag{13}$$

where X_{camera} is the distance from the camera to the world coordinate Y-O-Z plane.

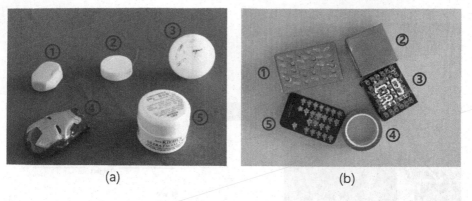

(a)   (b)

Fig. 5. Objects used in the task, the unit of size is cm. (a) Large objects (L-objects), in order: plastic box ($15.3 \times 10.8 \times 4.3$), square carton ($10.0 \times 10.0 \times 6.0$), long carton ($11.6 \times 10.0 \times 4.6$), adhesive tape ($10.4 \times 5.0^2$), Long tin box ($15.4 \times 9.4 \times 3.0$). (b) Small objects (S-objects), in order: rubber ($3.1 \times 2.1 \times 1.2$), small cap ($0.9 \times 2.3^2$), small ball ($3.8^3$), U-disk ($4.3 \times 2.9 \times 1.6$), cosmetic box ($3.5 \times 3.0^2$).

We get the camera's parameter matrix ($K_{3\times 3}$ and $T_{4\times 4}$) and correct the camera's distortion. Then the workspace is measured and marked to obtain $H = 58(cm)$, $W = 6.3(cm)$, $L' = 52(cm)$, $Pix_{mark}(\Delta u_{mark}, \Delta v_{mark}) = (52, 790)$.

For the object detector, we chose YOLOv5 [4] to meet the real-time requirements and trained it with a 100K Frame-level dataset [15], in which only objects in contact with hands are labeled. After training, the detector takes a single RGB image as input, outputs the bounding box of the hand and objects that contact the hand. Objects that do not contact the hand cannot be detected. In addition, we filter the detection results to exclude the cases of the hand touching the tabletop, touching itself, and misdirecting the robotic arm as the hand, which can lead to task failure or cause danger (see Fig. 7).

4.3   Grasping Tests

The diagram of grasping tasks is shown in Fig. 8. Clamping jaws of different specifications (9.5 cm–13.5 cm, 2 cm–6 cm) are used to perform grasping tests on five large and five small objects (L-object and S-object) successfully. Objects are placed on the workbench at any location. The worker randomly touches the object and drives the robot to grab it; When the manipulator is moving, the worker constantly adjusts the location of other objects (only change the location, not change the object posture). 200 experiments are conducted on large objects and small objects respectively (total of 400), and each of which resulted in one of the following four conditions:

– Success: Baxter successfully grabs the object and puts it back where it was.
– Drop: Baxter grabs the object, but it drops out and is not put back in place.

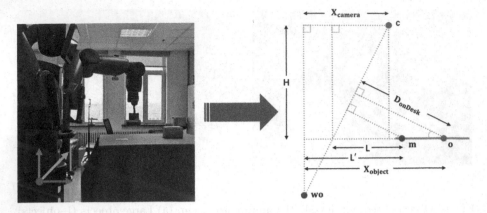

Fig. 6. Real workspace is abstracted as a geometric model, where Colored labels are world coordinates

Table 1. Success rates of the grasping tests about grasp L-objects (Large objects) and S-objects (Small objects).

	Success	Drop	Inaccuracy	Unreachable
L-objects	91.5%	0%	6.5%	2%
S-objects	83.5%	5%	9%	2.5%
Total	87.5%	2.5%	7.75%	2.25%

– Inaccuracy: Baxter cannot accurately move its arm to the object location and fails to grasp.
– Unreachable: Baxter gets the object's location but can't move the arm to it.

Table 1 shows the result of grasping: The overall grasping success rate is 87.5%, among which the grasping success rate of large objects reaches 91.5%, while that of small objects is only 83.5%, this is because small objects are much smaller in size than large objects and are difficult to grab. Large objects do not fall, while small objects fall 5% of the time, this is because when there is a slight deviation, large objects can be grabbed firmly, but small objects cannot be grabbed firmly. Failed to fetch in 7.75% of cases: (1) Most of the time is after the start moving, the change in the object location; (2) Part of the situation because of the two objects get too close, makes the detector produce error detection, which results in location errors; (3) For small objects, some deviations make the robot arm cannot reach the object grasping range, resulting in fetching failure. In 2.25% of cases, the manipulator cannot reach the target position because only the left manipulator is used, and the angle of the end-effector is fixed perpendicular to the working platform. Still, such grasping failure has nothing to do with the accuracy of our algorithm.

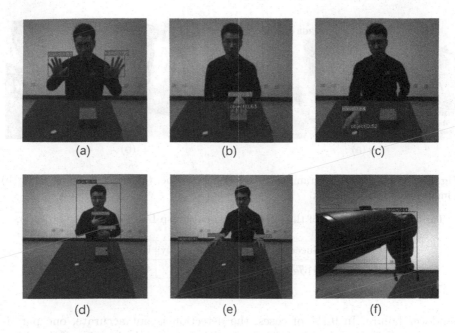

Fig. 7. Detection effect. (a) The hand does not make contact with the object, only detects the hand. (b) When hands touch objects, both hands, and objects are detected. (c) Even small objects can be detected. (d) Hands touching the body makes people the target of detection, which will lead to dangerous interactions (e) Touch the workbench, and the workbench will be detected, which will cause the task to fail. (f) The robot arms are wrongly detected as the human hand.

4.4   Human-Robot Handover Tests

The diagram of Human-robot handover tests is shown in Fig. 8. Only large objects are used (small objects cannot be used for handover). Three objects are placed on the workbench, and 10 participants in the experiment (including six males and four females). Workers randomly grab objects on the workbench for handover, take the objects to different positions in the workspace and keep the object posture relatively fixed. Each worker carried out 30 handover tests a total of 300 times, and the results of each test can be divided into the following three conditions:

- Success: Baxter successfully catches the object and returns it to the human.
- No hand: The detector cannot detect a hand, end the handover in advance.
- Inaccuracy: Mechanical cannot move to the object location accurately and get it.
- Unreachable: Baxter is unable to move its arm to the positions of the object.

HRH results are shown in Table 2: The success rate of experiments is 84.7%; 1.3% of the cases does not detect the hand, which is because the hand is blocked by the object and cannot obtain the interactive information, which led to the

(a)   (b)

Fig. 8. The schematic diagram of the experiment scene. (a) robot grasping task. (b) human-robot handover task.

Table 2. Success rates of the HRH tests about grasp L-objects (Large objects).

	Success	No hand	Inaccuracy	Unreachable
L-objects	84.7%	1.3%	9.7%	4.3%

handover failure. In 9.7% of cases, the detection is not accurate, one part is because the failure of the handover due to the change of the location of the object when the manipulator moved to the object, another part is due to the inaccurate positioning, which causes the failure of the handover due to the inability of the manipulator to move to the handover accurately. In 4.3% of cases, the robotic arm cannot reach the location. Here are two possible situations: (a) The object is located correctly but beyond the effective workspace and cannot be reached; (b) The object is in the effective workspace but is wrongly positioned outside the effective workspace and cannot be reached. There is no excellent criterion to distinguish between the two conditions.

4.5   Time Cost

To prove that our method is efficient, has a low delay, and can meet the real-time requirements of the robot, we conducted an additional 40 experiments (20 grasping experiments and 20 HRH experiments). We counted the following five time indicators respectively:

- IAP-ATM: Image acquisition average time, the average time required to pull an image from a video stream and preprocess it.
- MD-ATM: Model detection average time, the average time from image input to YOLOv5 model to output detection results.
- DC-ATM: Distance calculation average time, the average time required to calculate the distance of the object and locate its position in space.
- RM-ATM: Robot movement average time, the average time it takes for the robot to move from start to finish the task.
- T-ATM: Total average time, the total average time it takes from the moment the image is pulled to the robot completes the task.

Table 3. Average time cost of IAP-ATM (image acquisition average time), MD-ATM (model detection average time), DC-ATM (distance calculation average time), RM-ATM (robot movement average time), and T-ATM (Total average time). Time measured in s.

	IAP-ATM	MD-ATM	DC-ATM	RM-ATM	T-ATM
Grasping	4.416 s	8.353×10^{-2} s	7.328×10^{-5} s	28.83 s	33.33 s
HRH	3.595 s	8.306×10^{-2} s	7.419×10^{-5} s	35.02 s	38.70 s
Total	4.006 s	8.330×10^{-2} s	7.374×10^{-5} s	31.93 s	36.02 s

Table 3 shows the time cost of our approach. We can see obviously that the main factors affecting time cost are robot moving time, the network delay to some extent, also affect the time cost. The model of reasoning and distance calculation takes very little time, hardly can be ignored, this shows that our proposed distance detection algorithm is low latency, can satisfy the real-time requirements of robot tasks. The factors that restrict the real-time completion of tasks are the robot's own moving speed and network delay. In grasping and HRH tasks: (1) The distance detection time is almost the same, which indicates that our distance detection method is very stable. (2) The difference in image pulling time is nearly 1s, caused by network fluctuation. (3) The difference in the moving time of the robots is about 7 s, which is because the trajectories and distances of the two robots are different.

4.6 Qualitative Results and Future Work

Besides the reported evaluation, additional internal tests are conducted to identify weaknesses that should be addressed in future work. During these tests, (1) Change the position of the object to traverse the entire workspace. (2) Change the posture of the object. These internal tests do not have a strict evaluation procedure but are intended to verify the adaptability of the proposed algorithm to object position and pose.

The further tests reveal some limitations of our approach. (see Fig. 9): (1) Because the height information of the object is ignored, when the object is very high, the success rate of grasping and transferring will decrease. (2) When calculating the distance of objects in the air, we choose the width at the lowest point of the object as the size-invariant, so the object must be facing the robot and keep its relative posture unchanged. When the object rotates, the accuracy of distance detection will be affected; (3) We simply use the width of the object bounding box from the detector to replace the object width approximately. When the object deviates from the central axis of the robot, there will be an error.

These limitations are essentially caused by the loss of spatial information in 2D RGB images. The object detection model can only obtain 2D position information of the object, cannot obtain the size and pose information of the object. In future work, we will obtain more abundant information of the object through a real-time monocular-based object 6D pose estimation model of the

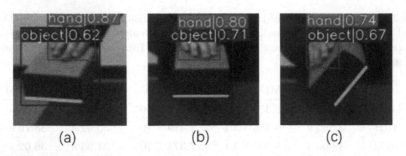

(a)   (b)   (c)

Fig. 9. The green line is the object's width, and the blue box is the bounding box. (a) When the object is facing the robot and on the central axis, the width of the bounding box is approximately equal to the object's width. (b) After changing the object's pose, the width of the bounding box is not equal to the object's width. (c) When the object is on the optical axis, the width of the bounding box is not equal to the object's width. (Color figure online)

object based on RGB image to further improve the current distance detection algorithms and the practicability of the algorithm in practical tasks.

5   Conclusion

This paper proposes a novel two-stage unknown object distance detection algorithm based on monocular RGB images, and a marker of known size is used as prior knowledge. Compared with existing RGB-D-based and analytics-based methods, the proposed algorithm has more adaptive, higher efficiency with low cost meets the requirements of real-time, accuracy, and effectiveness in robot tasks. It is successfully used for robot grasping and HRH tasks.

But the algorithm also has some limitations in actual use. For example, the object posture needs to be relatively fixed, and the algorithm is unsuitable for irregular objects. The root cause of these limitations is that the RGB camera loses the 3D information. A simple object detector can only obtain the pixel position information, which cannot determine object size and pose. There are some other restrictions: two-stage although distance detection algorithm is effective to solve the monocular RGB image the distance of the object under the air, the flexibility to drop, and the space parameter measurement method is more simple than before, but still affect the practicability of the proposed algorithm. These two points, to a certain extent, restrict the practical application of the algorithm. We will focus on real-time monocular-based object 6D pose estimation technology and grab point estimation technology. We will use them to obtain more rich information about objects from RGB images to address the limitations of the need for fixed object orientation and the inability to grasp irregular objects. We will also further improve the proposed distance detection algorithm to improve the practicability of practical tasks.

Acknowledgement. This work was supported in part by the Key Program of NSFC (Grant No. U1908214), Special Project of Central Government Guiding Local Science and Technology Development (Grant No. 2021JH6/10500140), Program for the Liaoning Distinguished Professor, Program for Innovative Research Team in University of Liaoning Province, Dalian and Dalian University, the Scientific Research fund of Liaoning Provincial Education Department (No. L2019606), Dalian University Scientific Research Platform Project (No. 202101YB03), and in part by the Science and Technology Innovation Fund of Dalian (Grant No. 2020JJ25CY001).

References

1. Andrew, A.M.: Multiple view geometry in computer vision. Kybernetes **30**(9/10), 1333–1341 (2001)
2. Bauer, A., Wollherr, D., Buss, M.: Human-robot collaboration: a survey. Int. J. Hum. Robot. **5**(1), 47–66 (2008)
3. Chiang, Y.-M., Hsu, N.-Z., Lin, K.-L.: Driver assistance system based on monocular vision. In: Nguyen, N.T., Borzemski, L., Grzech, A., Ali, M. (eds.) IEA/AIE 2008. LNCS (LNAI), vol. 5027, pp. 1–10. Springer, Heidelberg (2008). https://doi.org/10.1007/978-3-540-69052-8_1
4. Glenn, J., Alex, S., Jirka, B.: Ultralytics/YOLOv5: v5.0 - YOLOv5-P6 1280 models, AWS, Supervise.ly and YouTube integrations (2021). https://doi.org/10.5281/zenodo.4679653
5. Chiang, Y.-M., Hsu, N.-Z., Lin, K.-L.: Driver assistance system based on monocular vision. In: Nguyen, N.T., Borzemski, L., Grzech, A., Ali, M. (eds.) IEA/AIE 2008. LNCS (LNAI), vol. 5027, pp. 1–10. Springer, Heidelberg (2008). https://doi.org/10.1007/978-3-540-69052-8_1
6. Haseeb, M., Guan, J., Ristic-Durrant, D.: Disnet: a novel method for distance estimation from monocular camera. In: 10th Planning. Perception and Navigation for Intelligent Vehicles (PPNIV18), IROS (2018)
7. Jamzad, M., Foroughnassiraei, A., Chiniforooshan, E.: Middle sized soccer robots: Arvand. In: Robot Soccer World Cup (RoboCup) 1999. vol. 1856, pp. 61–73. Springer (1999). https://doi.org/10.1007/3-540-45327-X_4
8. Krishnan, J.V.G., Manoharan, N., Rani, B.S.: Estimation of distance to texture surface using complex log mapping. J. Comput. Appl. **3**(3), 16 (2010)
9. Mousavian, A., Eppner, C., Fox, D.: 6-DoF GraspNet: variational grasp generation for object manipulation. In: International Conference on Computer Vision (ICCV) 2019, pp. 2901–2910. IEEE/CVF (2019). https://doi.org/10.1109/ICCV.2019.00299
10. Murali, A., Mousavian, A., Eppner, C.: 6-DOF grasping for target-driven object manipulation in clutter. In: International Conference on Robotics and Automation (ICRA) 2020, pp. 6232–6238. IEEE (2020). https://doi.org/10.1109/ICRA40945.2020.9197318
11. Ortenzi, V., Cosgun, A., Pardi, T.: Object handovers: a review for robotics. IEEE Trans. Robot. **37**, 1855–1873 (2021)
12. Pathi, S.K., Kiselev, A., Kristoffersson, A.: A novel method for estimating distances from a robot to humans using egocentric RGB camera. Sensors **19**(14), 3142 (2019)
13. Ricardo, S.M., Konstantinos, C., Apostolos, M.: Benchmark for human-to-robot handovers of unseen containers with unknown filling. IEEE Robot. Autom. Lett. **5**(2), 1642–1649 (2020)

14. Rosenberger, P., Cosgun, A., Newbury, R.: Object-independent human-to-robot handovers using real time robotic vision. IEEE Robot. Autom. Lett. **6**(1), 17–23 (2021)

15. Shan, D., Geng, J., Shu, M.: Understanding human hands in contact at internet scale. In: 2020 IEEE/CVF Conference on Computer Vision and Pattern Recognition (CVPR) (2020). https://doi.org/10.1109/CVPR42600.2020.00989

16. Vohra, M., Prakash, R., Behera, L.: Real-time grasp pose estimation for novel objects in densely cluttered environment. In: International Conference on Robot and Human Interactive Communication (RO-MAN) 2019, pp. 1–6. IEEE (2019). https://doi.org/10.1109/RO-MAN46459.2019.8956438

17. Vyas, D.R., Markana, A., Padhiyar, N.: Robotic grasp synthesis using deep learning approaches: a survey. In: Sahni, M., Merigó, J.M., Jha, B.K., Verma, R. (eds.) Mathematical Modeling, Computational Intelligence Techniques and Renewable Energy. AISC, vol. 1287, pp. 117–130. Springer, Singapore (2021). https://doi.org/10.1007/978-981-15-9953-8_11

18. Wang, C., Xu, D., Zhu, Y.: Densefusion: 6d object pose estimation by iterative dense fusion. In: Conference on Computer Vision and Pattern Recognition (CVPR) 2019, pp. 3338–3347. IEEE/CVF (2019). https://doi.org/10.1109/CVPR.2019.00346

19. Yang, W., Paxton, C., Arsalan, M.: Reactive human-to-robot handovers of arbitrary objects (2020)

20. Yang, W., Paxton, C., Cakmak, M.: Human grasp classification for reactive human-to-robot handovers. In: International Conference on Intelligent Robots and Systems (IROS) 2020, pp. 11123–11130. IEEE/RSJ (2020). https://doi.org/10.1109/IROS45743.2020.9341004

21. Zhang, H., Lan, X., Bai, S.: Roi-based robotic grasp detection for object overlapping scenes. In: International Conference on Intelligent Robots and Systems (IROS) 2019. pp. 4768–4775. IEEE/RSJ (2019). https://doi.org/10.1109/IROS40897.2019.8967869

22. Zhu, J., Fang, Y.: Learning object-specific distance from a monocular image. In: International Conference on Computer Vision (ICCV) 2019, pp. 3838–3847. IEEE/CVF (2019). https://doi.org/10.1109/ICCV.2019.00394

Images Handling and Human Recognition and Edge Computing

A Novel Gaze-Point-Driven HRI Framework for Single-Person

Wei Li[1], Pengfei Yi[1], Dongsheng Zhou[1,2(✉)], Qiang Zhang[1,2], Xiaopeng Wei[2], Rui Liu[1], and Jing Dong[1]

[1] Dalian University, Dalian, People's Republic of China
zhouds@dlu.edu.cn
[2] Dalian University of Technology, Dalian, People's Republic of China

Abstract. Human-robot interaction (HRI) is a required method of information interaction in the age of intelligence. The new human-robot collaboration work mode is based on this information interaction method. Most of the existing HRI strategies have some limitations: Firstly, limb-based HRI relies heavily on the user's physical movements, making interaction impossible when physical activity is limited. Secondly, voice-based HRI is vulnerable to noise in the interaction environment. Lastly, while gaze-based HRI reduces the reliance on physical movements and the impact of noise in the interaction environment, external wearables result in a less convenient and natural interaction process and increase costs. This paper proposed a novel gaze-point-driven interaction framework using only RGB cameras to provide a more convenient and less restricted way of interaction. At first, gaze points are estimated from images captured by cameras. Then, targets can be determined by matching these points and positions of objects. At last, objects gazed at by an interactor can be grabbed by the robot. Experiments under conditions of different lighting, distances, and different users on the Baxter robot show the robustness of this framework.

Keywords: Human-robot interaction · Gaze point · Grab

1 Introduction

HRI is ubiquitous, and especially the new human-robot collaborative work model is inseparable from the interaction between workers and robots. There is a wide range of interaction methods such as touch (touch display control), voice (voice input control), gesture-based, gaze-based, etc. Touch [1], voice [15], and gesture-based interaction [13,22] are currently the most common methods of HRI. However, these three interaction methods have some limitations, such as being susceptible to the interaction environment (e.g., interaction distances, noise in the

© ICST Institute for Computer Sciences, Social Informatics and Telecommunications Engineering 2021
Published by Springer Nature Switzerland AG 2021. All Rights Reserved
H. Gao and X. Wang (Eds.): CollaborateCom 2021, LNICST 406, pp. 661–677, 2021.
https://doi.org/10.1007/978-3-030-92635-9_38

environment), relying heavily on the user's body movements, and contact interaction increases the risk of spreading some diseases. In the actual human-robot collaboration process, there may be situations where workers and robots cannot interact through language due to the noise in the work environment, and busy workers cannot operate the robot through their limbs. To improve the efficiency of human-robot collaboration and make human-robot interaction more natural, smooth, and safer, so nowadays, gaze-point-based human-robot interaction is increasingly used.

Most current gaze-point-based HRI methods require the help of external wearables (e.g., eye-tracking devices). Although external wearables, such as eye-tracking devices, are inexpensive, they have achieved good experimental results. However, there are still many disadvantages such as calibration problems, inconvenience in carrying, and influence by other glasses (e.g., myopic glasses). Collaborators wear eye-tracking for collaboration, which not only increases costs but also inconveniences collaboration.

To overcome the drawback of external wearables in HRI, improve the convenience of HRI based on gaze points, and make HRI more intelligent and humanized. Inspired by human gaze behavior, we build a framework for gaze-point-driven interaction without the assistance of any external wearables. In this framework, the robot uses only RGB cameras to infer the gaze target from the user's gaze points and then actively grasp it. Our main work in this paper is as follows:

- We build an HRI framework in which the robot infers the gaze target by gaze points and actively grasps it without the aid of any external wearables, significantly improving the ease and reducing the cost of interaction.
- We design a filter to filter invalid gaze points, thus enabling the robot to reason more accurately about the user's gaze target.
- We propose an efficient sorting-based entity matching method that does not rely on object features and does not require the processing of object features.

2   Related Work

In the gaze-point-based HRI approach, gaze point estimation is the first step in the interaction. Then uses the estimated gaze points to drive the robot through the various tasks in HRI. This section provides an overview of the methods commonly used for gaze point estimation and their application in HRI.

2.1   Gaze Point Estimation

In the current research, the methods of acquitting gaze points can divide into those that rely on external wearables and those that do not. Among the methods with external wearables, the main one is the use of eye-tracking. While among the methods without external wearables, deep learning-based methods are the dominant ones.

Methods with External Wearables: [7] proposed a mobile VR eye-tracking method, which can acquire eye images through a head-mounted camera without the assistance of additional equipment. [12] developed a head-mounted device to track line of sight and estimate user's gaze point. The head-mounted device can use in combination with a large endoscope camera, infrared light, and mobile phone. [10] built a system that can track high-speed eye movements, and the system can achieve a working frequency far 500 Hz. [16] proposed an automatic calibration method for 3D gaze estimation. With the development of deep learning, some scholars combine deep learning with eye trackers. [11] developed a pupil tracking technology based on deep learning for wearable device eye trackers.

Methods Without External Wearables: [17] designed a deep neural network architecture specifically for gaze estimation based on monocular input. [5] proposed a method to estimate people's overall attention in images. [6] took the relative position of key-points of the face as input for gaze estimation and used a confidence gated unit in the input layer of the network to deal with problems such as occlusion and incomplete key-points of the face. [3] proposed a face-based asymmetric regression evaluation network (FARE-Net). [4] proposed a model that simulates the dynamic interaction between scene and head features and infers the attention targets that change over time. [20] used a deep network structure to enable the robot to acquire human gaze or determine whether the user is making eye contact with the robot; still, it does not generate grasping and other interactive actions.

The above review shows that researchers have favored both gaze point estimation methods with external wearables and without external wearables, and the research on gaze point estimation has been increasing. Similarly, in HRI, researchers are keen to incorporate gaze points into HRI tasks to improve the naturalness of interaction. The following sections provide an overview of the usages of gaze points in HRI.

2.2 Application of Gaze Points in HRI

Modern-day approaches incorporating gaze points in human-robot interactions vary widely, and the completed tasks are also different. Based on the "GT3D" binocular eye tracker, [21] used Gaussian process regression to deal with the uncertainty of gaze point estimation to control the robotic arm better to complete the grasping task. [8] robot arm control by line-of-sight tracking so that the robotic arm can perform writing and painting artistic creations in 3D space. [23] developed a robot assistance system with intuitive and free gaze interaction, which estimates the user's gaze point in 3D space and controls the robotic arm for grasping and placing operations. [18] proposed a framework for grounding and learning examples through demonstration and eye-tracking (GLIDE). [9] built a platform based on a gaze tracking control robot, and users were able to navigate the mobile robot using only gazes as the system input. [24] proposed to use an automatic target location method combined with human gaze to extract relevant object positions, and this method can simplify the entire interaction process and improve the accuracy of interaction.

As seen from the above review, the methods to estimate gaze points can divide into those that rely on external wearables and those that do not. The method with external wearables is highly accurate but has the disadvantages of being inconvenient and costly to use. The way without external wearables uses deep learning techniques to enhance gaze point estimation accuracy greatly and is easy and inexpensive to use. In recent years, gaze points have also been used in a variety of HRI tasks. Still, most HRI tasks use external wearables (e.g., eye tracking devices) for estimation, which reduces the ease and naturalness of interaction and increases the cost of interaction. We combine the gaze point estimation method that does not rely on an external wearable and build an HRI framework that does not require the help of any external wearables and only requires RGB cameras.

3   Methods

This paper built a framework for human-robot interaction using only RGB cameras without the aid of other external wearables. This interaction framework requires inference of what object is gazed at by persons from global information, acquires the position of the object from local information in preparation for the mobile grabbing of the robot arm, entity matching through correspondence between global and regional information. In this framework, we, therefore, use the global view to obtain global information and use the local perspective to obtain local information.

3.1   Overview

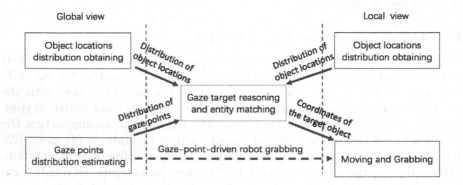

Fig. 1. Gaze-point-driven HRI framework.

As shown in Fig. 1, the global view mainly consists of object locations distribution obtaining and gaze points distribution estimating. The object locations distribution obtaining is used to obtain the position distribution of objects. The gaze points distribution estimating is used to estimate the number of valid gaze points falling around each object. The gaze target reasoning and entity matching use gaze points distribution information from the global view to infer the gaze target and find the target object in the local view through a matching algorithm. The object locations distribution obtaining and the moving and grabbing are used in the local perspective. The object locations distribution obtaining has the same function as the object locations distribution obtaining module in the global view. The moving and grabbing convert the pixel coordinates of the objects in the local view into coordinates in the robot's coordinate system. It generates commands to drive the robot arm to move and grasp. The implementation details of the framework are described below.

3.2   Object Locations Distribution Obtaining

We use a vision-based object detection model to detect objects and determine the position distribution of each object, then filter and sort the detected object bounding boxes. Figure 2 is a flow chart of this module, showing the process of obtaining the distribution of object positions. The detailed implementations of duplicate border filters and reordering of border positions are described below.

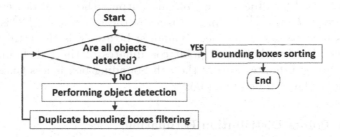

Fig. 2. Flowchart of the object locations distribution obtaining

Duplicate Bounding Boxes Filtering. In object detection, non-maximum suppression is used to determine the unique bounding box of an object. However, there is still the problem of duplicate bounding boxes for the same object during the experiments. Inspired by the non-maximum suppression algorithm, we solve this problem by suppressing the minimum distance between the center point of the object bounding boxes, as shown in filter f_d:

$$f_d(b_0) = \begin{cases} 0, \exists\ f_p(b_0, b_i) < \mathbb{N}, b_i \in s \\ 1, otherwise \end{cases} \tag{1}$$

where $s = \{b_1, b_2, b_3, \ldots, b_l | b_i = <(x_{min}, y_{min}), (x_{max}, y_{max}), w, h>\}$, $s \in \mathbb{R}^{1 \times l}$ is a storage unit that is used to save the object bounding boxes, l is the number of saved object bounding boxes, $< (x_{min}, y_{min}), (x_{max}, y_{max}), w, h >$ denotes the top left coordinate, bottom right coordinate, width and height of the object bounding box, respectively. f_p is used to find the Euclidean distance between two points, b_0 is bounding box of the object to be saved currently, b_i is bounding box of the object saved in the s memory cell, and threshold N is the minimum distance. If f_d is 0, it means that there is a bounding box in s that overlaps with b_0 then b_0 is discarded. If it is 1, b_0 is saved to s.

Bounding Boxes Sorting. It means all objects have been detected when $f_n = 1$, but the position of the object bounding boxes saved in s may not correspond to the position of objects on the workbench. To ensure that the position order of the bounding boxes in the s corresponds to the object space position order, we use $sort(.)$ for sorting. The $sort(.)$ is a function that sorts objects according to the center point coordinates of their bounding boxes. Use f_n to verify that all objects have been detected, expressed as follows:

$$f_n(I) = \begin{cases} 1, if \ \sum_{b_i \in D(I)} f_d(b_i) = N \\ 0, otherwise \end{cases} \tag{2}$$

$$sort(s) = \{c(s_i) < c(s_j), i < j\} \tag{3}$$

where $D(.)$ is used for object detection and outputs the set of detected objects bounding boxes, $I \in \mathbb{R}^{H \times W \times 3}$ image taken by the robot, $s \in \mathbb{R}^{1 \times N}$ is a storage unit that is used to save the object bounding boxes, N is the total number of objects, $c(s_i)$ denotes the center point coordinates of the i-th bounding box, When $i < j$, the coordinate value of the i-th bounding box is less than the coordinate value of the j-th bounding box.

3.3 Gaze Points Distribution Estimating

The distribution of valid gaze points around the object is obtained by first estimating the gaze points in the global view based on head information and scene information, processing the invalid gaze points, and recording the number of valid gaze points around each object. While estimating the gaze points, the robot also adjusts its interaction state based on the header information. Figure 3 is a flow chart of the gaze points distribution estimating, showing the process of obtaining a valid gaze points distribution, and the corresponding pseudocode is shown in Algorithm 1. The implementations of the robot's interaction mode adjustment and gaze point filtering are described in detail below.

Interaction Model Adjusting. In this framework, the robot may enter an invalid interaction mode for two reasons: 1) when a user with no interaction intent appears within the robot vision and wakes the robot into interaction mode, this is an invalid interaction mode. 2) when a user leaves during the interaction and does not have enough information about the gaze point to complete the subsequent interaction, the robot is in an invalid interaction mode.

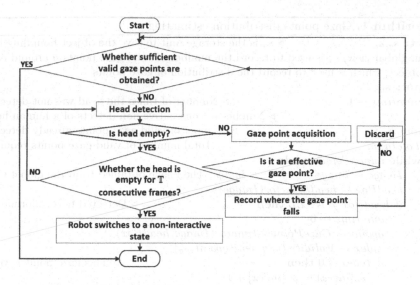

Fig. 3. Flowchart of the gaze points distribution estimating

The robot adjusts its mode according to the header information to avoid invalid interaction modes. If a person's head is not detected in $\mathbb{T}$ consecutive frames during the gaze point acquisition phase, $f_t = 0$, the robot will move from the interaction model to a non-interaction mode. f_t is an interaction status detector, which represents as follows:

$$f_\Phi = \begin{cases} 1, \, if \; \Phi(I) = 0 \\ 0, \, otherwise \end{cases} \tag{4}$$

$$f_t = \begin{cases} 0, \, if \; \sum_t^{t+\mathbb{T}} f_\Phi = \mathbb{T} \\ 1, \, otherwise \end{cases} \tag{5}$$

where t is the first time a video frame without head information is detected, $\Phi(.)$ is a head detector and returns the number of heads detected.

Gaze Points Filtering. To acquire the distribution of valid gaze points around each object, we start with determining whether the estimated gaze points are valid and filtering invalid gaze points, and finally, recording the distribution of valid gaze points around each object. This framework builds on the work of [4] to perform gaze point estimation based on the head pose.

Our filter focuses on two types of invalid gaze points in gaze point estimation: remote gaze points and saccade gaze points.

Remote Gaze Points: We call points far from the object above the workbench invalid remote gaze points, and we use maximum distance suppression to filter such points. We find the minimum distance of the gaze point from the nearest

---

Algorithm 1. Gaze points distribution estimating

---

input: s_g, s_n ▷ s_g is the storage unit to save the object bounding box in global view; s_n is used to record the number of valid gaze points g around each object, which is used to record the distribution of gaze points

output: s_n

 1: $numtims \leftarrow 0$ ▷ Number of times the head was not detected

 2: T ▷ Number of consecutive non-detects of a human head

 3: $num \leftarrow 0$ ▷ Number of valid gaze points already detected

 4: $Totalnum$ ▷ Total number of valid gaze points required

 5: **while** $num < Totalnum$ **do**

 6:      $Image \leftarrow image$ ▷ $image$: The image obtained from the global view

 7:      $headBox \leftarrow headDetector(Image)$

 8:      **if** $headBox \neq NULL$ **then**   ▷ Detected head information

 9:          $numtims \leftarrow 0$

10:          $gpoint \leftarrow GazePointEstimator(Image, headBox)$

11:          $index \leftarrow ValidityJudgment(gpoint, s_g)$

12:          **if** $index \neq 0$ **then**   ▷ This gaze point is valid

13:             $s_n[index] \leftarrow s_n[index] + 1$

14:             $num \leftarrow num + 1$

15:          **else**   ▷ This gaze point is invalid and discarded

16:             continue

17:          **end if**

18:      **else**   ▷ Head information is not detected

19:          $numtims \leftarrow numtims + 1$

20:          **if** $numtims \neq T$ **then**

21:             continue

22:          **else**   ▷ Head information is not detected in consecutive T-frames

23:             Robot switches to a non-interactive state

24:             **return** $null$

25:          **end if**

26:      **end if**

27: **end while**

28: **return** s_n

---

object and then apply maximum suppression of the minimum distance to filter the remote invalid gaze points. As indicated by f_{dis} :

$$f_g(g, s_g) = \begin{cases} 0, & if\ g \in b_i\ and\ b_i \in s_g \\ \min_{b_i \in s_g}(f_{ps}(g, b_i)), & otherwise \end{cases} \qquad (6)$$

$$f_{dis}(g, s_g, s_n) = \begin{cases} 0, if\ f_g(g, s_g) > \mathbb{D} \\ 1, otherwise \end{cases} \qquad (7)$$

where $s_g = \{b_1, b_2, b_3, \ldots, b_N | b_i =< (x_{min}, y_{min}), (x_{max}, y_{max}), w, h >\}$ is the storage unit used to save the object bounding box in global view, g is the gaze point (x, y), $s_n = \{n_1, n_2, n_3, \ldots, n_N | n_i \in \{0, 1, 2, 3, \ldots\}\}$ and s_g have the same size, and s_n is used to record the number of valid gaze points g around each object, indicating the number of times each object was gazed, N is the total number of objects, b_i is the bounding box of the object saved in the s_g memory cell, n_i is the number of times the object at the corresponding position was gazed at, $< (x_{min}, y_{min}), (x_{max}, y_{max}), w, h >$ denotes the top left coordinate, bottom right coordinate, width and height of the object bounding box, respectively. f_{ps} is a function that calculates the shortest distance from point g to the rectangular region b_i formed by the bounding box, f_g is used to find the bounding box of the object in s_g that is closest to the gaze point g and the minimum distance, the threshold $\mathbb{D}$ is the maximum distance. If the distance returned by f_g is greater than $\mathbb{D}$, the gaze point g is invalid and the value of f_{dis} is 0, otherwise, f_{dis} is 1 and 1 is added to the position of s_n in relation to the nearest object, indicating that the corresponding object is gazed once, as illustrated in Fig. 4.

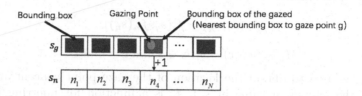

Fig. 4. The correspondence between s_g and s_n. where $n_1, n_2, n_3, ..., n_N$ denote the number of times the bounding box at the corresponding position has been gazed at. If the bounding box in s_g is gazed, 1 will be added to the position corresponding to it in s_n. Like n_4 will be added by 1.

Saccade Gaze Points: The gaze points detected during the user's movement are unstable gaze points, which we call saccade gaze points, and the small number of saccade gaze points cannot be used as a solid basis for reasoning about the gaze target. We solve the saccade gaze points problem by adopting a constraint on the minimum number of times that valid gaze points are detected, as denoted by f_s below:

$$f_s(g, s_g, s_n) = \begin{cases} 1, & if \ \sum_t^e f_{dis}(g, s_g, s_n) = \mathbb{Q} \\ -1, & otherwise \end{cases} \tag{8}$$

where the threshold $\mathbb{Q}$ is the total number of valid gaze points required, t is the starting frame, e is considered to be the end frame when $\mathbb{Q}$ valid gaze points are detected. Suppose the value of f_s is 1. In that case, it means that a sufficient number of valid gaze points have been obtained, and the gaze target can be inferred based on the distribution of the obtained valid gaze points; otherwise, the detection of valid gaze points will continue.

3.4   Gaze Target Reasoning and Entity Matching

The gaze target reasoning and entity matching module infers gaze targets based on the distribution of valid gaze points in the global view and object positions. It uses a sorting-based entity matching method to match gaze targets in the local perspective. The implementations of gaze target reasoning and sorting-based entity matching will be described in detail below.

Gaze Target Reasoning. In this framework, we use an approach similar to human gaze reasoning, considering the location where the gaze point is most concentrated as the location where the gaze target is located. In Eq. 7, we can know the distribution of the valid gaze point g according to s_n. We consider the location of the maximum value in s_n as the location where the gaze points are most concentrated and consider this location as the location of the gaze target. With f_{ob}, we get the gaze target object in s_g based on this position. The expression is as follows:

$$f_{ob}(s_g, s_n) = f_v(s_g, f_m(s_n)) \tag{9}$$

$$f_v(s, index) = b_i, b_i \in s \text{ and } i = index \tag{10}$$

where f_m is used to obtain the location of the gaze point concentration (the *index* of the maximum value in s_n), f_v is a function for inferring the gaze target object from the storage cell based on the location of the gaze point concentration (Retrieve the object at *index* position in the storage cell), $s = \{b_1, b_2, b_3, \ldots, b_N | b_i = < (x_{min}, y_{min}), (x_{max}, y_{max}), w, h >\}, s \in \mathbb{R}^{1 \times N}$ is a storage unit that is used to save the object bounding boxes, N is the number of saved object bounding boxes, b_i is the bounding box of the object saved in the s memory cell, $< (x_{min}, y_{min}), (x_{max}, y_{max}), w, h >$ denotes the top left coordinate, bottom right coordinate, width and height of the object bounding box, respectively.

Entity Matching. The entity matching method is a matching method that does not rely on object features but only on the order of object storage locations, which requires that the positions between objects in different views are relatively invariant and the position of the bounding boxes in the storage unit corresponds to each other. The expressions are as follows:

$$\Theta_g = \{\theta_1^g, \theta_2^g, \theta_3^g, \ldots, \theta_N^g | x(\theta_i^g) < x(\theta_j^g), i < j\} \tag{11}$$

$$\Theta_l = \{\theta_1^l, \theta_2^l, \theta_3^l, \ldots, \theta_N^l | x(\theta_i^l) < x(\theta_j^l), i < j\} \tag{12}$$

$$Match : \Theta_g \approx \Theta_l \tag{13}$$

where both θ_i^g and θ_i^l are composed of $< (x_{min}, y_{min}), (x_{max}, y_{max}), w, h >$, Θ_g and Θ_l are ordered sets that store the bounding boxes of objects in the global and local views. $x(\theta_i^g)$ and $x(\theta_i^l)$ denote the x-value of the coordinates of the centre point of the i-th bounding box in the global view and local view, respectively. When $i < j$, the x-value of the coordinate of the centre point of the i-th bounding box is less than the x-value of the coordinate of the centre point of the j-th bounding box. N is the number of entities, $< (x_{min}, y_{min}), (x_{max}, y_{max}), w, h >$

denotes the top left coordinate, bottom right coordinate, width and height of the object bounding box, respectively. In the matching process, we consider that the same location of Θ_g and Θ_l stores the bounding box of the same object under different views. Figure 5 provides a more intuitive representation of the sorting-based entity matching method.

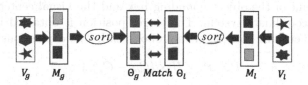

$$V_g \qquad M_g \qquad\qquad \Theta_g \text{ Match } \Theta_l \qquad\qquad M_l \qquad V_l$$

Fig. 5. Sort-based entity matching: V_g, V_l are the object positions in the global and local views, and M_g and M_l store the bounding boxes of the objects in the global and local views, respectively, which are sorted by *sort* to obtain the ordered sets Θ_g and Θ_l. The bounding boxes of the same object are stored in the same order in both Θ_g and Θ_l, so that during the matching process, we assume that Θ_g and Θ_l store the same object at the same location. *sort* is a sorting function, like Eq. 3.

3.5 Moving and Grabbing

The moving and grabbing module focuses on moving the arm to the target object and grabbing it. Figure 6 shows the flow of the operation. Converting the pixel coordinates of the target object to world coordinates in the Baxter robot's coordinate system is the key to the moving procedure, and we will describe the coordinate conversion in more detail below.

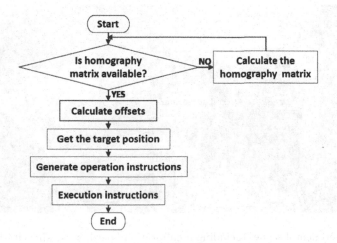

Fig. 6. Flowchart of the moving and grabbing

Coordinate Conversion. The coordinate conversion is primarily used to convert pixel coordinates to robot coordinates. The pixel position information of the bounding box of the target object obtained using Eq. 9 is in the pixel coordinate system. It cannot be used for the control of the robot directly. We use the calibration board to calculate the homography matrix from the camera image plane to the working plane. The homography matrix is used to find the offset between the center point of the object bounding box and the visual center point in the robot's base coordinate system. The target position is obtained based on this offset and the current position of the robot's end. The expressions show below:

$$P = P_r + H \times (C_{ob} - C_v) + \gamma \tag{14}$$

where $H \in \mathbb{R}^{3 \times 3}$ is the homography matrix. $C_{ob} \in \mathbb{R}^{3 \times 1}$, $C_v \in \mathbb{R}^{3 \times 1}$ are the homogeneous coordinates of the object bounding box centre point and the visual centre point, respectively. P_r is the current position at the end of the robot arm, γ is the deviation compensation, P is the target position at the end of the robot arm.

4   Results

Experiments were conducted using the Baxter robot under different lighting conditions, different interaction distances, and different persons to verify the robustness of the interaction framework. In Subsect. 4.1, we will describe our experimental equipment and experimental setup. In Subsect. 4.2, we will analyze the experimental results. Presentation videos are listed as following: daytime [Youku] or [Youtube], nighttime [Youku] or [Youtube].

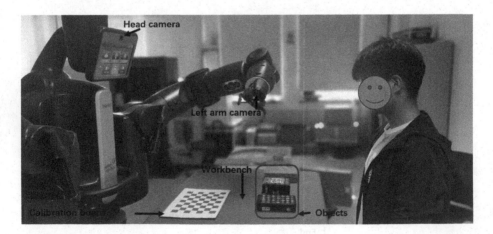

Fig. 7. Experimental scene. Including a calibration board, a workbench, the object to be grasped, and the Baxter robot head and left arm camera

4.1   Experimental Equipment and Setup

Throughout the experiment, we used a computer with an RTX2080 GPU, a workbench, a calibration board, the object to be grasped, a Baxter robot with parallel grippers, and the Baxter robot head camera and left arm camera as hardware devices, as shown in Fig. 7. The Baxter head camera is a global view camera, and the Baxter left arm camera is a local view camera. We implement the program using a mixture of python 3.7 and the C++ programming language.

In this experiment, we need the user's head information, and we trained Yolov4 [2] as a head detector based on the head dataset provided in [19]. During the training process, we set max_batches to 5000 depending on the amount of data and the number of categories. We performed the training on a computer with an RTX2060 and achieved mAP of 95.5%. In this interaction framework, we also use Yolov4 trained on the Microsoft coco [14] dataset as an object detector. The separation distance between objects on the workbench is controlled to be around 10 cm.

As shown in Fig. 8, we set the interaction distance (the distance of the experimenter from the workbench) into three distance bands: near (within 60 cm), medium (60 cm–120 cm), and far (120 cm–200 cm), and calculate the mission success rate for different distances.

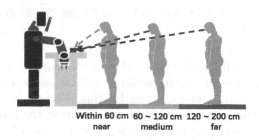

Within 60 cm   60 ~ 120 cm   120 ~ 200 cm
near   medium   far

Fig. 8. Interaction distance. Including three distance bands: near, medium, and far

4.2   Experimental Results

To verify the robustness of this interaction framework under different lighting conditions, we conduct experiments under natural daylight exposure and evening lighting conditions. The data in Table 1 shows that the average success rates are 90.0% and 91.1% for near range, 87.8% and 82.2% for medium range, 76.7% and 74.4% for far range under natural daylight and evening lighting conditions, respectively. Although there is a difference in the average success rate of this interaction framework under different lighting conditions, the difference is not significant. The average success rate for each person in daytime natural lighting conditions is also similar to their average success rate in evening lighting conditions. This all proves that this interaction framework is robust under different lighting conditions.

Table 1. Task success rate and average success rate of different experimenters under different illumination and different interaction distance conditions. U_*^c (this experimenter wears glasses), $*\_f$ (this experimenter is a female), $*\_N$ (near), $*\_M$ (medium), $*\_F$ (far), Nl_* (natural daylight exposure conditions), El_* (evening lighting conditions). The data format for each row is several successes/total number of executions (s/t). AVGSR1 (the average success rate of each experimenter under natural lighting conditions during the day and night lighting conditions), AVGSR2 (the average success rate of different interaction distances under different lighting conditions). 83.7% (the overall average success rate).

Experimenters	Nl_N	Nl_M	Nl_F	El_N	El_M	El_F	AVGSR1
U_1	10/10	9/10	10/10	9/10	9/10	9/10	93.3%
U_2^c	8/10	10/10	8/10	9/10	8/10	8/10	85.0%
U_{3_f}	10/10	9/10	8/10	10/10	10/10	7/10	90.0%
U_4^c	9/10	10/10	9/10	10/10	8/10	8/10	90.0%
U_{5_f}	9/10	7/10	8/10	9/10	8/10	6/10	78.3%
U_6^c	9/10	8/10	5/10	9/10	6/10	7/10	73.3%
U_7^c	8/10	9/10	9/10	8/10	7/10	8/10	81.7%
$U_{8\_f}^c$	9/10	9/10	5/10	10/10	9/10	8/10	83.3%
U_{9_f}	9/10	8/10	7/10	8/10	9/10	6/10	78.3%
AVGSR2	90.0%	87.8%	76.7%	91.1%	82.2%	74.4%	**83.7%**

The AVGSR2 data in Table 1 shows that, under natural daylight conditions, the average success rates at a near, medium, and far distance are 90.0%, 87.8%, and 76.7%, respectively; under evening lighting conditions, are 91.9%, 82.2% and 74.4%, respectively. It is clear from the data that the average success rate decreases as the distance gets further. The individual data for each person also shows that the success rate decreases as the distance gets further. Although the average success rate is gradually decreasing, the average success rate exceeds 90.0%, 82.2%, 74.4% for near, medium, and far distances, respectively, which shows that this interaction framework is robust and effective at different interaction distances.

To verify the robustness of this interactive framework across different body types, heights, clothing, genders, and in the presence of other glasses. Nine experimenters (five male and four female, different body types, heights, clothing, and some wearing glasses) are invited to perform the same interactive task under natural daylight exposure and evening lighting conditions. To exclude the effects of light and interaction distance, we analyze the data in Table 1 under the same light and interaction distance conditions. There is little difference in the number of successes between the near and medium-distance person. The number of wins in the far distance interactions is not significantly different, except for a person's relatively low number of successes. In Table 1, the average success rate is 84.66% for males and 82.48% for females, with a slight difference between the two percentages. The average success rate is 82.66% for those wearing glasses and 84.98% for those not wearing glasses, with a bit of difference between the two percentages. The data suggest that this interaction framework is robust across different body types, heights, clothing, and gender conditions, and that person can still interact efficiently when wearing other glasses.

The above experimental results show that this interaction framework is highly robust and can successfully perform interaction tasks at different interaction distances, in different lighting environments, and with users of various sizes. Wearing other glasses does not affect the efficient use of this interaction framework. However, this interaction framework can only be used in a single-person situation, which is a limitation of this framework and a problem we will address in the future.

5   Conclusion

In this work, the human-robot interaction framework using only RGB cameras without the assistance of any external wearables is built. In this framework, the robot uses only RGB cameras to infer the target of the person's gaze and grasps it. The interaction framework does not require external wearables, makes interaction more accessible and reduces interaction costs. We conducted several sets of experiments based on the Baxter robot and demonstrated the high robustness of our interaction framework.

As this framework uses only RGB cameras during the interaction, a calibration board is required to assist in the conversion of coordinates. The entity matching algorithm converts the target object from the global view to the local perspective to convert coordinates more accurately, resulting in a longer interaction time. In future work, we will focus on solving the conversion problem and implementing and extending the framework to accommodate multi-player human-robot interaction.

Acknowledgment. This work was supported in part by the Key Program of NSFC (Grant No. U1908214), Special Project of Central Government Guiding Local Science and Technology Development (Grant No. 2021JH6/10500140), Program for the Liaoning Distinguished Professor, Program for Innovative Research Team in University of Liaoning Province, Dalian and Dalian University, the Scientific Research fund of Liaoning Provincial Education Department (No. L2019606), Dalian University Scientific Research Platform project, and in part by the Science and Technology Innovation Fund of Dalian (Grant No. 2020JJ25CY001).

References

1. Acien, A., Morales, A., Vera-Rodriguez, R., Fierrez, J.: Smartphone sensors for modeling human-computer interaction: general outlook and research datasets for user authentication. In: 2020 IEEE 44th Annual Computers, Software, and Applications Conference (COMPSAC), pp. 1273–1278. IEEE Computer Society, Los Alamitos (2020)

2. Bochkovskiy, A., Wang, C.Y., Liao, H.Y.M.: Yolov4: optimal speed and accuracy of object detection. arXiv preprint arXiv:2004.10934 (2020)
3. Cheng, Y., Zhang, X., Lu, F., Sato, Y.: Gaze estimation by exploring two-eye asymmetry. IEEE Trans. Image Process. **29**, 5259–5272 (2020)
4. Chong, E., Wang, Y., Ruiz, N., Rehg, J.M.: Detecting attended visual targets in video. In: 2020 IEEE/CVF Conference on Computer Vision and Pattern Recognition (CVPR), pp. 5395–5405. IEEE Computer Society, Los Alamitos (2020)
5. Chong, E., Ruiz, N., Wang, Y., Zhang, Y., Rozga, A., Rehg, J.M.: Connecting Gaze, scene, and attention: generalized attention estimation via joint modeling of gaze and scene saliency. In: Ferrari, V., Hebert, M., Sminchisescu, C., Weiss, Y. (eds.) ECCV 2018. LNCS, vol. 11209, pp. 397–412. Springer, Cham (2018). https://doi.org/10.1007/978-3-030-01228-1_24
6. Dias, P.A., Malafronte, D., Medeiros, H., Odone, F.: Gaze estimation for assisted living environments. In: 2020 IEEE Winter Conference on Applications of Computer Vision (WACV), pp. 279–288. IEEE, Snowmass Village, Colorado (2020)
7. Drakopoulos, P., Koulieris, G.A., Mania, K.: Front camera eye tracking for mobile VR. In: 2020 IEEE Conference on Virtual Reality and 3D User Interfaces Abstracts and Workshops (VRW), pp. 642–643. Atlanta (2020)
8. Dziemian, S., Abbott, W.W., Faisal, A.A.: Gaze-based teleprosthetic enables intuitive continuous control of complex robot arm use: writing & drawing. In: 2016 6th IEEE International Conference on Biomedical Robotics and Biomechatronics (BioRob), pp. 1277–1282. IEEE, University Town, Singapore (2016)
9. Gêgo, D., Carreto, C., Figueiredo, L.: Teleoperation of a mobile robot based on eye-gaze tracking. In: 2017 12th Iberian Conference on Information Systems and Technologies (CISTI), pp. 1–6. IEEE, Lisbon, Portugal (2017)
10. Hosp, B., Eivazi, S., Maurer, M., Fuhl, W., Geisler, D., Kasneci, E.: Remoteeye: an open-source high-speed remote eye tracker:implementation insights of a pupil- and glint-detection algorithm for high-speed remote eye tracking. Behav. Res. Methods **52**(3), 1387–1401 (2020)
11. Kuo, T.L., Fan, C.P.: Design and implementation of deep learning based pupil tracking technology for application of visible-light wearable eye tracker. In: 2020 IEEE International Conference on Consumer Electronics (ICCE), pp. 1–2. IEEE, Las Vegas (2020)
12. Lee, K.F., Chen, Y.L., Yu, C.W., Chin, K.Y., Wu, C.H.: Gaze tracking and point estimation using low-cost head-mounted devices. Sensors **20**(7) (2020)
13. Li, X.: Human-robot interaction based on gesture and movement recognition. Sig. Process. Image Commun. **81**, 115686 (2020)
14. Lin, T.-Y., et al.: Microsoft COCO: common objects in context. In: Fleet, D., Pajdla, T., Schiele, B., Tuytelaars, T. (eds.) ECCV 2014. LNCS, vol. 8693, pp. 740–755. Springer, Cham (2014). https://doi.org/10.1007/978-3-319-10602-1_48
15. Liu, J., Chang, W., Li, J., Wang, J.: Design and implementation of human-computer interaction intelligent system based on speech control. Comput.-Aid. Des. Appl. **17**, 22–34 (2020)
16. Liu, M., Fu Li, Y., Liu, H.: 3D gaze estimation for head-mounted devices based on visual saliency. In: 2020 IEEE/RSJ International Conference on Intelligent Robots and Systems (IROS), pp. 10611–10616. IEEE, Las Vegas (2020)
17. Park, S., Spurr, A., Hilliges, O.: Deep pictorial gaze estimation. In: Ferrari, V., Hebert, M., Sminchisescu, C., Weiss, Y. (eds.) ECCV 2018. LNCS, vol. 11217, pp. 741–757. Springer, Cham (2018). https://doi.org/10.1007/978-3-030-01261-8_44

18. Penkov, S., Bordallo, A., Ramamoorthy, S.: Physical symbol grounding and instance learning through demonstration and eye tracking. In: 2017 IEEE International Conference on Robotics and Automation (ICRA), pp. 5921–5928. IEEE, Marina Bay Sands (2017)
19. Radhakrishnan, P.: Head-detection-using-yolo. https://github.com/pranoyr/head-detection-using-yolo. Accessed 4 Oct 2020
20. Saran, A., Majumdar, S., Short, E.S., Thomaz, A., Niekum, S.: Human gaze following for human-robot interaction. In: 2018 IEEE/RSJ International Conference on Intelligent Robots and Systems (IROS), pp. 8615–8621. IEEE, Madrid (2018)
21. Tostado, P.M., Abbott, W.W., Faisal, A.A.: 3D gaze cursor: continuous calibration and end-point grasp control of robotic actuators. In: 2016 IEEE International Conference on Robotics and Automation (ICRA), pp. 3295–3300. IEEE, Stockholm (2016)
22. Tsai, T.H., Huang, C.C., Zhang, K.L.: Design of hand gesture recognition system for human-computer interaction. Multimedia Tools Appl. **79**(9), 5989–6007 (2020)
23. Wang, M.Y., Kogkas, A.A., Darzi, A., Mylonas, G.P.: Free-view, 3D gaze-guided, assistive robotic system for activities of daily living. In: 2018 IEEE/RSJ International Conference on Intelligent Robots and Systems (IROS), pp. 2355–2361. IEEE, Madrid (2018)
24. Weber, D., Santini, T., Zell, A., Kasneci, E.: Distilling location proposals of unknown objects through gaze information for human-robot interaction. In: 2020 IEEE/RSJ International Conference on Intelligent Robots and Systems (IROS), pp. 11086–11093. IEEE, Las Vegas (2020)

Semi-automatic Segmentation of Tissue Regions in Digital Histopathological Image

Xin He[1]📷, Kairun Chen[2]📷, and Mengning Yang[1(✉)]📷

[1] School of Big Data and Software Engineering, Chongqing University, Chongqing, China
mnyang@cqu.edu.cn
[2] Chongqing Meehoo Technology Co., Ltd., Chongqing, China

Abstract. Segmentation of tissue regions in the digital histopathological images refers to the identification and segmentation of tissues such as epithelium, glandular cavity, fibers, etc. Precise segmentation of tissues is key to pre-determining the regions with the greatest diagnostic value, which can support clinical diagnosis, particularly with regard to etiology and severity. In view of the uneven quality of histopathological images and the difficulty of manual segmentation. In this paper, an approach based on weakly supervised learning and deep learning has been proposed to build a semi-automatic segmentation model of tissue regions. The model uses superpixel classification to pre-segment the tissues, the tissue region boundary is preserved, and the automatic segmentation of the tissues is finally achieved based on the deep convolutional neural network. The effectiveness of the model is evaluated on 600 cervical histopathology images provided by the hospital. The results show that the proposed method achieves 82.52% mean IoU of epithelial segmentation and 81.67% mean IoU of glandular lumen segmentation in cervical histopathological images. The model is superior to traditional manual feature representation methods and classical deep convolution neural network methods in segmentation accuracy and efficiency.

Keywords: Digital histopathological image · Tissues segmentation · Superpixels · Deep convolutional neural network

1 Introduction

Digital histopathological images are obtained by making stained sections of patients' suspected lesion tissues and then photographing them in high resolution using whole slide imaging (WSI) technology. They are analyzed and judged by the pathologists, and this histopathological examination process has become the

© ICST Institute for Computer Sciences, Social Informatics and Telecommunications Engineering 2021
Published by Springer Nature Switzerland AG 2021. All Rights Reserved
H. Gao and X. Wang (Eds.): CollaborateCom 2021, LNICST 406, pp. 678–696, 2021.
https://doi.org/10.1007/978-3-030-92635-9_39

"gold standard" of cancer detection and diagnosis. Histopathological images can be used by doctors to know about the status of tumor cells, which is of great significance for the diagnosis, classification, and prognosis of tumors. These images include many tissues such as stroma, epithelium, and glandular lumen, etc. Different tissues are closely related to the type and severity of the disease. For example, the diagnosis of cervical squamous cell carcinoma is often based on the proliferation and arrangement of cells in cervical epithelial tissue [1]. Therefore, it is important for pathologists to focus on the tissue region with the greatest diagnostic value in histopathological images before judging the disease. However, the traditional manual segmentation method is limited by time-consuming, unstable, strong subjectivity, and high error rate due to visual fatigue. Developing an intelligent automated algorithm to efficiently segment different tissues is crucial for the development of pathology-assisted diagnosis [2, 3].

The field of histopathological image segmentation faces the following challenges: (1) Compared with natural images, histopathological images have unclear semantic regions, unobvious boundaries and high similarity among tissues, and the analysis of histopathological images is more challenging. (2) Histopathological images are usually super-resolution images, which have greater computational complexity. (3) Lack of large-scale annotated datasets [4], the lesion dataset changes greatly and needs to be labeled by clinical experts. Therefore, classical segmentation algorithms for natural images, such as threshold segmentation, region growth and edge detection, are not fully applicable to histopathological images. In order to segment histopathological image tissue regions, this paper starts with the accurate dividing of the tissue boundaries, superpixels [5] have been shown to be able to efficiently and completely segment local regions. Convolutional neural networks with excellent performance in image processing are used to classify superpixel in an inexact supervised learning way. To simplify operations and further improve segmentation accuracy, the obtained tissue images are used to make deep learning datasets, an automatic segmentation model is trained end-to-end based on a deep convolutional neural network, so as to achieve faster and more accurate segmentation of tissue regions. The innovation of this article lies in:

1. Improve the accuracy and interpretability of the model through the key human-computer interaction.
2. Superpixel-level labeling replaces the pixel-level labeling, which improves the labeling efficiency and preserves tissue boundaries.
3. The semi-automatic segmentation model has strong universality and can be applied to segmentation tasks of various tumor histopathological images.

The organizational structure of the paper is as follows: The second section reviews the related work of tissue region segmentation in histopathological images. The third section introduces preliminaries, including image preprocessing and semi-automatic segmentation methods. In the fourth section, the experimental results are discussed and analyzed, and the effectiveness of the method is verified. The last section gives conclusion and future research contents.

2   Related Work

In the past few decades, with the mature development of image scanning technology and the improvement of computing power, as well as the emergence of automatic analysis algorithms, significant progress has been made in histopathology image segmentation. To deeply explore the tissue segmentation methods of histopathological images, this section sorts out and analyzes the effective techniques which are used to segment different interest objects in histopathological images. Research on these techniques can be broadly divided into two main categories: methods based on hand-crafted features and methods based on deep learning.

2.1   Methods Based on Hand-Crafted Features

Support vector machine (SVM) model based on local binary patterns (LBP) was used to automatically distinguish epithelial and stromal in digitized tumor tissue microarrays (TMAs) of colorectal cancer [6], the image is narrowed and divided into square blocks, and the blocks are then independently classified using an SVM model. Also a bayesian model was used to automatically segment the stromal tissue in the immunohistochemical (IHC) image based on color and texture features [7]. The patch was classified by the deep learning method [8], and patch level statistics and morphological characteristics were input into the random forest (RF) regression model to classify the whole slide image. All of the above approaches are based on hand-crafted feature representation, the segmentation precision is unsatisfactory due to high computation and limited feature extraction. In recent years, artificial intelligence technologies such as deep learning have made breakthroughs in various fields. More and more researchers have turned their attention to the application of deep learning in histopathological images and achieved outstanding results.

2.2   Methods Based on Deep Learning

Convolutional Neural Networks (CNNs) perform well in histopathological image processing [9]. The supervised classification of a CNN was combined with unsupervised image segmentation to distinguish the epithelial and stromal tissues of H&E images [10], the combination of deep learning and boundary localization improved boundary segmentation accuracy, but the performance of this method was limited for images with fuzzy boundaries. De [11] proposed a method for segmenting renal tissue using CNN, experiments were conducted using three different network architectures, with about 90% accuracy. Nirschl [12] provided a deep learning framework for the segmentation of muscle cells and stroma in H&E stained heart biopsy samples. The framework uses AlexNet architecture to train pixel-level classifiers for segmentation. Compared with a random forest classifier with 333 intensity and texture features, the framework is superior in AUC and F-score.

Fully Convolutional Networks (FCN) has made significant progress in image semantic segmentation with the advantages of unlimited input size and end-to-end training of the model [13]. Chen [14] segmented the colon glands based on the Deep Contour-Aware Networks (DCAN). DCAN adopts an auxiliary supervision mechanism to overcome the problem of gradient disappearance when training. This method ranked first in the 2015 MICCAI glandular segmentation challenge and 2015 MICCAI nuclear segmentation challenge. Lahiani [15] used an end-to-end color deconvolution deep learning method to segment tissues in multi-staining immunohistochemical images, digital histopathological images with multiple staining effects can be automatically segmented into the tumor, healthy tissue, necrotic region and background based on a FCN, however, the scheme is difficult to judge when the image source is unknown or the imaging quality is poor.

U-Net as the baseline of medical image processing methods [16], many scholars have processed and analyzed medical images based on U-Net or its improved version. A group of sparsely annotated histopathological images was used to train U-Net and FCNs of different depths [17], and the pixel-based AUC score was 0.97. However, this method has a large network scale, many parameters and a long training time. Based on the U-Net architecture, a positive predictive value of 0.89 ± 0.16 and a sensitivity of 0.92 ± 0.1 were obtained in the epidermal or non-epidermal pixel classification task [18], but the entire epidermis region cannot be divided.

Other methods based on deep learning are also used in tissue segmentation of histopathological images. In [19], HistoSegNet is proposed for the semantic segmentation of tissues in the histopathological image. It is superior to the more complex weakly supervised semantic segmentation method and can be extended to other datasets without retraining. A general neural network method is designed to segment disease-related regions in medical images [20], it requires only two types of tags at the sample level. Using the labeled samples to train a meta-network, which deduces a segmented neural network to segment the disease-related regions in the image, and identify tumor regions or tumor-free regions reliably. However, the diversity of training data needs to be improved.

It can be known from literature research that dataset, segmentation accuracy, and model portability are still the difficulties in tissue segmentation. Therefore, we constructed a semi-automatic segmentation model suitable for various tumor image tissues based on pathological knowledge and deep learning. Semi-automatic refers to obtaining high-quality segmentation samples through the interactive pre-segmentation model, which can be used to train the tissue automatic segmentation model end to end.

3   Preliminaries

This section introduces the image preprocessing, and describes in detail the semi-automatic segmentation methods, including pre-segmentation and automatic segmentation.

3.1   Methodology Overview

The semi-automatic segmentation model of tissues is divided into two stages: pre-segmentation based on superpixel classification and automatic segmentation based on deep CNNs. The pre-segmentation process yields a large number of high-quality tissue segmentation results, providing a dataset quickly for automatic segmentation model in the next stage. Automatic segmentation utilizes the deep learning method to learn the segmented image dataset end-to-end and ultimately achieves the automatic segmentation of tissues. The semi-automatic segmentation model greatly reduces the burden of manual segmentation and improves the accuracy of the automatic segmentation of tissue regions. The segmentation process is shown in Fig. 1.

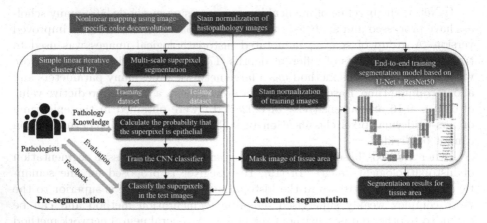

Fig. 1. Segmentation process of tissue region in histopathological image.

3.2   Histopathological Images Preprocessing: Staining Normalization

The imaging of histopathological images is influenced by factors such as staining degree and scanning equipment. Deep learning algorithms are extremely sensitive to the color structure of images, so it is necessary to normalize image to the color distribution of the template image to reduce the variance. Based on the representation derived from color deconvolution, the nonlinear mapping from the source image to the template image is found, and the staining normalization is realized [21]. It is a spectral normalization method that converts all images into a spectral distribution of the template image (see Fig. 2).

3.3   Pre-segmentation of Tissue Regions

Multi-scale Superpixel Segmentation. Superpixel [5], first proposed by Ren and Malik, is a clustering-based segmentation algorithm, which clusters a series

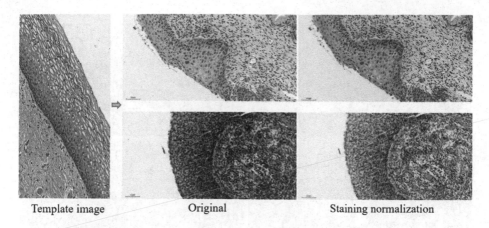

Template image   Original   Staining normalization

Fig. 2. Stain normalization results of histopathological images.

of pixels that are adjacent to each other and have similar color, brightness, texture into small regions. The simple linear iterative clustering (SLIC) method proposed by Achanta [22] is simple, fast in running, and capable of generating uniformly distributed and compact superpixels. SLIC method is used to segment superpixels with different scales in this paper, the multi-scale facilitates the model to reconcile the labeling cost and segmentation effect. The multi-scale superpixel segmentation is shown in Fig. 3.

Rectangularization and Labeling of Superpixel. Superpixels are rectangularized to be able to feed them into the CNN for classification. The external rectangle of the superpixel is obtained by topologically analyzing the superpixel mask. According to Table 1, the preset size of the superpixel blocks is cropped out in the external rectangle, where SP_{number} indicates the number of superpixels segmented in an image (1430×712), SP_{size} indicates the size of the superpixels after rectangularization.

Table 1. Predefined size for cropping superpixel rectangle.

SP_{number}	SP_{size}
250	64
180	86
130	100
50	128
40	156

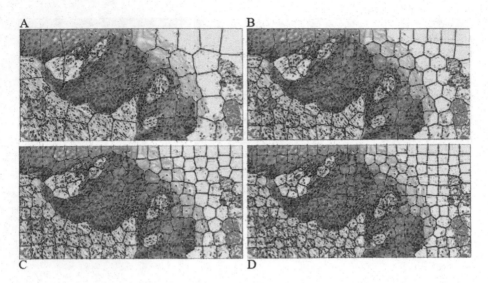

Fig. 3. Superpixel segmentation. The images are divided into 50(A), 130(B), 180(C) and 250(D) superpixels.

Based on Eqs. 1 and 2, the coordinates (x_-, y_-) of the top-left point of the rectangularized superpixel block are calculated. x, y are the top-left point coordinates of the external rectangle, and W, H are the width and height of the external rectangle. Starting from the top-left point, pixels are taken to the right and down, and reverses when the boundary is encountered. The pseudocode is shown in Table 2. The results are shown in Fig. 4.

$$x_- = x - \lceil 1/2(SP_{size} - W)\rceil \tag{1}$$

$$y_- = y - \lceil 1/2(SP_{size} - H)\rceil \tag{2}$$

Superpixel labeling is done by people trained in basic pathology, the probability that each superpixel rectangular belongs to a specific tissue is determined, which is used for inexact supervised learning. As shown in Eq. 3, where K is the number of superpixels.

$$p_i \approx area(target\ tissue)/area(superpixel)$$
$$p_i \in [0,1], i \in K \tag{3}$$

Pre-segmentation Architecture. CNN is constructed to train the superpixel classification model in an inexact supervised learning way, the network structure is shown in Fig. 5. The classification of superpixels is the initial segmentation of tissues. The process is shown from 4 to 12.

$$a^1 = Relu(X * W^1 + b^1) \qquad n\_kernel = 20, size\_kernel = 5 * 5 \tag{4}$$

$$a^2 = pool(a^1) \qquad size\_pool\_kernel = 2 * 2, stride = 2 * 2 \tag{5}$$

Table 2. Pseudocode for the method of superpixel rectangularization.

Input: A histopathological image img, the top-left point coordinates (x,y), width W and height H of the superpixel external rectangle.

Output: Rectangularized superpixel roi.

$x_- \leftarrow x - \lceil SP_{size} - W)/2 \rceil$

if $x_- < 0$:

    $x_- \leftarrow 0$

elif $x_- + SP_{size} > img.shape[0]$:

    $x_- \leftarrow img.shape[0] - SP_{size}$

$y_- \leftarrow y - \lceil (SP_{size} - H)/2 \rceil$

if $y_- < 0$:

    $y_- \leftarrow 0$

elif $y_- + SP_{size} > img.shape[1]$:

    $y_- \leftarrow img.shape[1] - SP_{size}$

$roi \leftarrow img[x_- : x_- + SP_{size}, y_- : y_- + SP_{size}]$

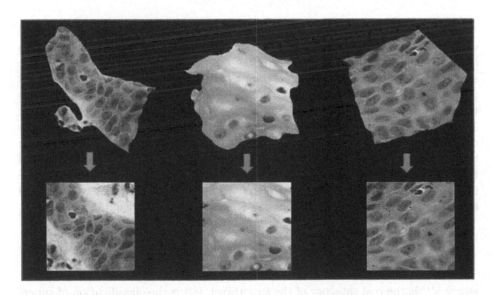

Fig. 4. The results of superpixel rectangularization.

$$a^3 = Relu(a^2 * W^3 + b^3) \qquad n\_kernel = 40, size\_kernel = 4 * 4 \qquad (6)$$

$$a^4 = pool(a^3) \qquad size\_pool\_kernel = 2 * 2, stride = 2 * 2 \qquad (7)$$

$$a^5 = flatten(a^4) \tag{8}$$

$$a^6 = softmax(W^6 * a^5 + b^6) \tag{9}$$

$$a^7 = dropout(a^6) \tag{10}$$

$$a^8 = W^8 * a^7 + b^7 \tag{11}$$

$$\hat{y} = argmax(a^8) \tag{12}$$

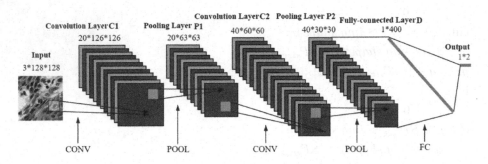

Fig. 5. Convolutional neural network structure.

where the input X is a three-dimensional superpixel image, and a^1,a^3 are the outputs of the convolutional layers, respectively. *Relu* is the activation function of the convolutional layer, n_kernel is the number of convolutional kernels, the size of convolutional kernel is $size\_kernel$. *flatten* function makes the a^4 matrix into a one-dimensional vector a^5 for input to the fully connected layer. a^2, a^4 is the output of max-pooling layer, $size\_pool\_kernel$ is the size of pooling kernel, *stride* is the step size. a^6 is the fully connected layer output after activation, the activation function is the *softmax* function, a^7 is the result of discard units from the network with a certain probability to prevent overfitting. a^8 is the output of fully connected layer with no activation, finally, it returns the subscript $\hat{y}$ of the maximum value in a^8, which is the output. The cross-entropy loss is minimized to train superpixel classification model, which is defined as Eq. 13.

$$L = -\sum_{i=1}^{N} y^{(i)} \log \hat{y}^{(i)} + \left(1 - y^{(i)}\right) \log \left(1 - \hat{y}^{(i)}\right) \tag{13}$$

where $y^{(i)}$ is the real category of the superpixel, $\hat{y}^{(i)}$ is the classification of superpixel by model, L represents the difference between the predicted output and the real category.

3.4 Automatic Segmentation of Tissue Regions

Obtaining Ground Truth for Automatic Segmentation. The mask obtained by binarization of the pre-segmented result is taken as ground truth, the tissue part of the RGB is 255. The input and label images are data enhanced. Each image is cut to half size of the original image. The staining normalized image is the input and the mask of the corresponding tissue is the label.

Automatic Segmentation Architecture. Both low-level and high-level features of images are important for tissue segmentation. Skip-connection and U-structure of the U-Net enable learning of both high-level and low-level features. An improved U-Net architecture for replacing the VGG Net [16] with the first four layers of the ResNet50 is used to avoid gradient disappearance for deep network training, which is called U-Net+ResNet50 in this paper, U-Net+ResNet50 structure is shown in Fig. 6.

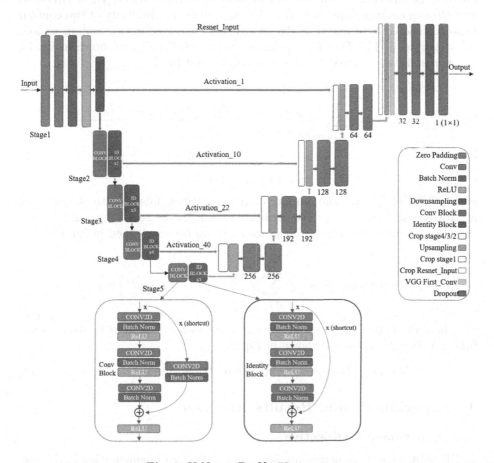

Fig. 6. U-Net + ResNet50 structure.

The ResNet50 downsampling process includes Conv Block and Identity Block, the role of Conv Block is to change the dimensionality of the feature vector. Identity Block with consistent input and output dimensions for identity mapping. The bottom left Conv Block is used to handle mismatches between input images and output dimensions, where the convolutional layers on the shortcut connection is used to adjust the dimensionality of the input, ensuring that inputs and outputs on the main path can be summed. The shortcut connection in the bottom right Identity Block spans three hidden layers. In addition, the U-Net+ResNet50 adopts the "same" mode for zero-padding filling at the edge of the image. Batch normalization is performed between each convolutional operation and the activation function to prevent overfitting to some extent. Maxpooling is used for the downsampling of pathological features and bilinear interpolation for upsampling.

Automatic segmentation model is trained based on the U-Net + ResNet50, the input is staining normalized images, the output is tissue masks. The loss function Bce_Dice_Loss for end-to-end training consists of two parts: $Dice\ loss$ and $Binary\ crossentropy\ loss$. $Dice\ loss$ describes the similarity of two contour regions, denoted by A and B as the set of pixels contained in two contour regions of category y, $\hat{y}$. The $Dice\ loss$ is derived from $DSC(A, B)$, as shown in Eq. 14, where p and r are defined as shown in Eqs. 15 and 16.

$$Dice\ loss = 1 - DSC(A, B)$$

$$= 1 - \frac{\sum_{n=1}^{N} p_n r_n + \varepsilon}{\sum_{n=1}^{N} p_n + r_n + \varepsilon} - \frac{\sum_{n=1}^{N}(1 - p_n)(1 - r_n) + \varepsilon}{\sum_{n=1}^{N} 2 - p_n - r_n + \varepsilon} \tag{14}$$

$$p = TP/(TP + FP) \tag{15}$$

$$r = TP/(TP + FN) \tag{16}$$

where TP, FP, FN are the number of true positives, false positives, and false negatives, respectively, p_n is the accuracy rate, r_n is the recall rate, and ε is the smoothing parameter. The $Binary\ crossentropy\ loss$ is defined in Eq. 17.

$$Binary\ crossentropy\ loss = - \sum_{i=1}^{N} y^{(i)} \log \hat{y}^{(i)} + \left(1 - y^{(i)}\right) \log \left(1 - \hat{y}^{(i)}\right) \tag{17}$$

The sum of the Dice loss and the $Binary\ crossentropy\ loss$ is taken as loss function, Bce_Dice_Loss is shown in Eq. 18.

$$Bce\_Dice\_Loss = Binary\ crossentropy\ loss + Dice\ loss \tag{18}$$

4   Experiments and Results Analysis

4.1   Experimental Objective

1. Different hyperparameters are set to optimize pre-segmentation and automatic segmentation models respectively.

2. Four widely used methods are compared with our proposed method to measure the segmentation performance of our method.

4.2   Dataset

Pre-segmentation Dataset. Cervical histopathological image dataset provided and authorized by the hospital. All personal information is withheld to protect patient privacy. Pre-process 600 cropped images, they are divided into superpixels. Data enhancement of labeled superpixels by rotating, flipping, adding noise. Finally, 22,032 superpixels are obtained as a superpixel classification dataset (SCD).

Automatic Segmentation Dataset. The automatic segmentation dataset (ASD) is constructed based on the pre-segmentation results. The ASD consists of 1662 staining normalized images and corresponding tissue masks (ground truth).

4.3   Experimental Setup

The experimental settings of pre-segmentation and automatic segmentation are the same, and the positive and negative samples of the dataset are balanced, of which 90% is taken as a training set, the remaining 10% is taken as a validation set, real-time samples from the hospital as a test set. Each evaluation metric is cross-validated by a 5-fold cross-validation and the final results averaged. All experiments were performed on an Amax NVIDIA Titan V server with a 12G GPU.

The methods widely used in image segmentation are selected for comparison with our method, SVM-RBF [23] and Random Forest (RF) [24], as the most commonly used and better performing segmentation methods based on manual feature, are used to verify the limitations of classical traditional methods in pathological image processing. FCN [13] and U-Net [16] are widely used deep learning segmentation methods, and U-Net is also the baseline of the proposed method.

4.4   Experimental Results and Analysis

Metrics of Pre-segmentation Model. Superpixel classification is evaluated by Mean Cross-Entropy (MCE) Loss and Accuracy. *MCE Loss* and *Accuracy* are defined as shown in Eqs. 19 and 20. *MCE loss* characterizes the difference between the predicted output and the true label. TP, TN, FP, FN in Eq. 20 are elements of the confusion matrix. *Accuracy* can partly indicate whether the classifier is effective.

$$MCE\ Loss = -(1/N) \sum_{i=1}^{N} y^{(i)} \log\ \hat{y}^{(i)} + \left(1 - y^{(i)}\right) \log\ \left(1 - \hat{y}^{(i)}\right) \tag{19}$$

$$Accuracy = (TP + TN) / (TP + FP + TN + FN) \tag{20}$$

Pathology-trained personnel performs a rapid evaluation of the superpixel classification results, and makes continuous improvements in the input superpixel size and network hyperparameters until an optimal classification model is obtained. Finally, the superpixel classification results are reorganized into tissue segmentation results according to the division rules and categories.

Pre-segmentation includes superpixel segmentation and rectangularization. In addition, a histopathological image is divided into superpixels for independent classification, ignoring the correlation between the superpixels in close positions. In view of this, the contents of the image are directly learned based on deep learning to avoid superpixel segmentation. It simplifies the production of the dataset and improves the accuracy of tissue segmentation by learning the overall features.

Pre-segmentation Results of Tissue Region. The classification of the superpixels based on the CNN is shown in Table 3, where *Epochs* is the number of training rounds and Batch Size (BS) is the batch size of each input data during training. It is clear that at *Epochs* of 24 and *Batch Size* of 20, *Mean Loss* and *Accuracy* are iterated to optimal, achieving 85% classification accuracy.

Table 3. Classification results of superpixels with size 128 * 128.

	Epochs = 30	Epochs = 24	Epochs = 23		Epochs = 19
	BS = 23	BS = 20	BS = 10	BS = 25	BS = 30
Mean loss	0.2404	**0.2350**	0.3200	0.2620	0.2683
Accuracy	0.8472	**0.8489**	0.8358	0.8466	0.8471

Experiments also verify the effect of different sizes of superpixels. If a superpixel is too small, and contains too little contextual information, which will lead to poor classification accuracy. While it is too large to properly segment tissue boundaries. As shown in Table 4, the classification model achieved more accurate results when the superpixel size is 128 × 128, that is, each pathological image is divided into 50 superpixels.

Table 4. Classification results of superpixels with different sizes.

	Size = 64 × 64	Size = 86 × 86	Size = 100 × 100	Size = 128 × 128	Size = 156 × 156
Mean loss	0.2855	0.2820	0.2649	**0.2350**	0.2417
Accuracy	0.8213	0.8275	0.8311	**0.8489**	0.8401

The superpixels are recombined base on categories to obtain the segmentation result of the tissue, which is visually shown in Fig. 7. Pre-segmentation provides a better segmentation of images with clear borders. However, for images with more complex staining distribution and more disordered cell arrangement, the pre-segmentation model has a greater error.

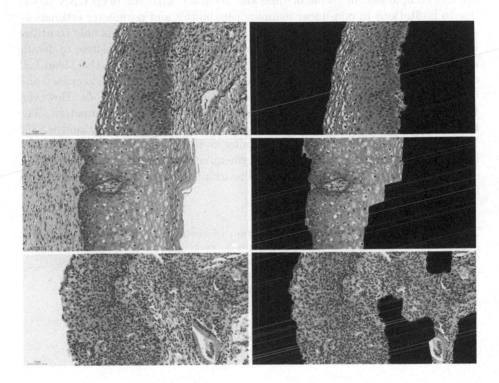

Fig. 7. Pre-segmentation results of tissue regions in histopathological images.

Metrics of Automatic Segmentation Model. *Accuracy* and the *Mean_IoU* as evaluation metrics of models. They are defined as shown in Eqs. 21 and 22, respectively. The Intersection-over-Union (IoU) refers to the ratio of intersection and union between the target region generated by the model and the originally marked region. In Eq. 22, P and G represent the predicted and ground truth, and N is the number of samples.

$$Accuracy = (TP + TN)/((TP + FP) + (TN + FN)) \tag{21}$$

$$Mean\_IoU = (1/N) * (area(P) \cap area(G))/((area(P) \cup area(G)) \\ = (1/N) * TP/(FN + TP + FP) \tag{22}$$

Automatic Segmentation Results of Tissue Region. Table 5 shows the *Accuracy* and *Mean_IoU* of the five models for segmenting the tissues. For such massive and dense data as histopathological images, SVM-RBF and RF are not comparable in terms of time and accuracy with the deep CNN model due to limitations in non-linear mapping capabilities and parameter estimation, achieving only about 86% but taking up to 15 h or more. FCN is able to utilize information from multiple layers simultaneously, but it is not sensitive to details and lacks spatial consistency, achieving only 89% accuracy and the Mean_IoU of 0.6341. The baseline network U-Net achieves 94% segmentation accuracy and 0.7866 Mean_IoU, which is more accurate than the above methods. However, U-Net usually needs random initialization and has many parameters. The improved U-net +ResNet50 in this paper achieves 95% accuracy and 0.8252 Mean_IoU, compared to other deep learning models, it achieved better segmentation results, and each image can be segmented in less than one second, which satisfies the need for fast segmentation. The intuitive segmentation of each model is shown in Fig. 8.

Table 5. Results of automatic segmentation of tissue regions in histopathological images.

Methods	Accuracy	Mean_IoU	Training time (min)
SVM-RBF	0.8603	0.5269	839
RF	0.8578	0.5537	924
FCN	0.8944	0.6341	56
U-Net	0.9416	0.7866	**53**
U-Net+ResNet50	**0.9542**	**0.8252**	70

The results of pre-segmentation and automatic segmentation of epithelial tissue are visually compared, as shown in Fig. 9. The automatic segmentation model based on deep learning learns the unique cell arrangement patterns and specific pathological features of the epithelial tissue and thus performs well even on cervical pathology images with disordered cell alignment and uneven staining.

The semi-automatic segmentation model consists of a pre-segmentation model and an automatic segmentation model. To further validate the generality of the model, it is also used to segment glandular cavity tissues in cervical pathology images. 620 histopathological images are taken as datasets, each of which contains one or more glandular cavity tissues. The visual display of the segmentation results is shown in Fig. 10, which achieves medically acceptable results. This further demonstrates that the semi-automatic segmentation model has a strong generalization ability and provides a versatile solution to segment lesion regions in various tumor images.

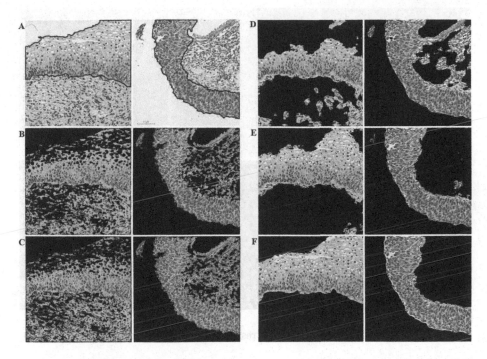

Fig. 8. Automatic segmentation results of tissue regions. Orginal (A), SVM-RBF (B), RF (C), FCN (D), U-Net (E), U-Net+ResNet50 (F). In the original image, the red outline is the epithelial tissue, the white region is the background, and the rest is fibrous tissue. (Color figure online)

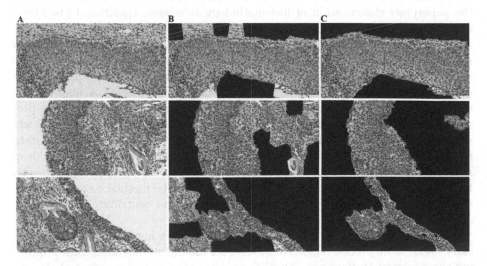

Fig. 9. Comparison of pre-segmentation results and automatic segmentation results. Original (A) and pre-segmentation results (B), automatic segmentation results (C).

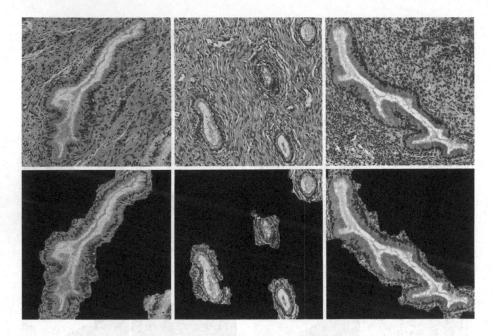

Fig. 10. Segmentation results of glandular cavity.

5   Conclusion and Future Work

In this study, a semi-automatic segmentation model for tissue regions is constructed to accurately segment tissues in small datasets. The model starts with the superpixel classification of histopathological images, traditional pixel-level labeling is replaced by superpixel-level labeling. Superpixels preserve the boundary of tissues, and can be labeled quickly. This inexact supervised learning approach greatly reduces physician burden and improves labeling efficiency. The guidance of pathological knowledge makes the results of tissue segmentation more accurate and improves the interpretability of the model. A high-quality training dataset for the deep learning model is constructed based on tissue pre-segmentation results, and it is learned end-to-end to segment tissues more quickly and accurately. The model provides a versatile solution for rapid and accurate segmentation of various tissue regions, and the techniques for constructing deep model datasets greatly reduce the reliance of medical research on public datasets. Individual physicians or small teams can also follow the method to annotate the data in their field, so that they can start research and contribute to the development of the medical field.

Accurate segmentation of tissues like epithelium in histopathological images can provide more precise regions of interest for intelligent diagnosis, thus improving the accuracy of diagnosis. Another member of our team learns pathological features in the segmented epithelial tissue, such as color, texture and cell density characteristics, and then uses deep learning techniques for lesion grading, demonstrating that focusing on the tissue regions yields more accurate results than diagnosing on the whole image.

Moreover, there are still some issues to be resolved in this paper. First, the rewards and penalties of reinforcement learning can be used for superpixel classification in order to obtain an optimal model. Second, immunohistochemical images such as ki67 and p16 can be combined to provide richer pathological features, thus improving the accuracy of tissue segmentation and disease diagnosis. Finally, our research cannot be limited to the segmentation of tissues, microscopic cell segmentation and morphological analysis, and macroscopic studies of tissue structure and location can further aid computerized diagnosis.

Funding Information. This study is supported by Fundamental Research Funds for the Central Universities under Grant 2020CDCGRJ013 and the Science and Technology innovation ability enhancement project of Third Military Medical University under Grant 2019XQY14.

References

1. Doorbar, J., Griffin, H.: Refining our understanding of cervical neoplasia and its cellular origins. Papillomavirus Res. **7**, 176–179 (2019)
2. Zhang, X., Liu, W., Dundar, M., et al.: Towards large-scale histopathological image analysis: hashing-based image retrieval. IEEE Trans. Med. Imag. **34**, 496–506 (2015)
3. Xu, J., Luo, X., Wang, G., et al.: A deep convolutional neural network for segmenting and classifying epithelial and stromal regions in histopathological images. Neurocomputing **191**, 214–223 (2016)
4. Ren, X., Malik, J.: Learning a classification model for segmentation. In: Tomaszewski, J.E., Gurcan, M.N. (eds.) ICCV 2003, LNCS, vol. 1, pp. 10–17. IEEE (2003). https://doi.org/10.1109/ICCV.2003.1238308
5. Lin, B., Deng, S., et al.: FocAnnot: patch-wise active learning for intensive cell image segmentation. In: Gao, H., Wang, X., Iqbal, M., Yin, Y., Yin, J., Gu, N. (eds.) Collaborative computing: networking, applications and worksharing. CollaborateCom 2020, LNCS, Social Informatics and Telecommunications Engineering, vol 350. Springer, Cham (2020). https://doi.org/10.1007/978-3-030-67540-0_21
6. Linder, N., Konsti, J., Turkki, R., et al.: Identification of tumor epithelium and stroma in tissue microarrays using texture analysis. Diagn. Pathol. **7**(1), 22 (2012)
7. Hiary, H., Alomari, R.S., Saadah, M., Chaudhary, V.: Automated segmentation of stromal tissue in histology images using a voting Bayesian model. Signal, Image Video Process. **7**(6), 1229–1237 (2012). https://doi.org/10.1007/s11760-012-0393-2
8. Vu, Q.D., Graham, S., Kurc, T., et al.: Methods for segmentation and classification of digital microscopy tissue images. Front. Bioeng. Biotechnol. **7**, 53 (2019)
9. Krizhevsky, A., Sutskever, I., Hinton, G.E.: Imagenet classification with deep convolutional neural networks. Commun. ACM **60**, 84–90 (2017)
10. Al-Milaji, Z., Ersoy, I., Hafiane, A., et al.: Integrating segmentation with deep learning for enhanced classification of epithelial and stromal tissues in H&E images. Patt. Recogn. Lett. **119**(MAR.), 214–221 (2017)

11. De Bel, T., Hermsen, M., Smeets, B., et al.: Automatic segmentation of histopathological slides of renal tissue using deep learning. Medical Imaging 2018. In: Proceedings of the SPIE, LNCS, vol. 10581, pp. 1058112 (2018). https://doi.org/10.1117/12.2293717

12. Nirschl, J.J., Janowczyk, A., Peyster, E.G., et al.: Deep learning tissue segmentation in cardiac histopathology images. In: Deep Learning for Medical Image Analysis, pp. 179–195 (2017)

13. Long, J., Shelhamer, E., Darrell, T.: Fully convolutional networks for semantic segmentation. IEEE Trans. Patt. Anal. Mach. Intell. **39**(4), 640–651 (2015)

14. Chen, H., Qi, X., Yu, L., et al.: DCAN: deep contour-aware networks for object instance segmentation from histology images. Med. Image Anal. **36**, 135–146 (2017)

15. Lahiani, A., Gildenblat, J., Klaman, I., et al.: Generalising multistain immunohistochemistry tissue segmentation using end-to-end colour deconvolution deep neural networks. IET Image Process. **13**(7), 1066–1073 (2019)

16. Ronneberger, O., Fischer, P., Brox, T.: U-Net: convolutional networks for biomedical image segmentation. In: Navab, N., Hornegger, J., Wells, W.M., Frangi, A.F. (eds.) MICCAI 2015. LNCS, vol. 9351, pp. 234–241. Springer, Cham (2015). https://doi.org/10.1007/978-3-319-24574-4_28

17. Bulten, W., Hulsbergen-van, d.K.C., van d.L.J., et al.: Automated segmentation of epithelial tissue in prostatectomy slides using deep learning. Medical Imaging 2018. In: Proceedings of the SPIE, LNCS, vol. 10581 (2018). https://doi.org/10.1117/12.2292872

18. Oskal, K.R.J., Risdal, M., Janssen, E.A.M., Undersrud, E.S., Gulsrud, T.O.: A U-net based approach to epidermal tissue segmentation in whole slide histopathological images. SN Appl. Sci. **1**(7), 1–12 (2019). https://doi.org/10.1007/s42452-019-0694-y

19. Chan, L., Hosseini, M.S., Rowsell, C., et al.: Histosegnet: semantic segmentation of histological tissue type in whole slide images. In: Proceedings of the IEEE International Conference on Computer Vision 2019, LNCS, pp. 10661–10670. IEEE (2019). https://doi.org/10.1109/ICCV.2019.01076

20. Schuhmacher, D., Gerwert, K., Mosig, A.: A generic neural network approach to infer segmenting classifiers for disease-associated regions in medical image data. medRxiv (2020). https://doi.org/10.1101/2020.02.27.20028845

21. Khan, A.M., Rajpoot, N., Treanor, D., et al.: A nonlinear mapping approach to stain normalization in digital histopathology images using image-specific color deconvolution. IEEE Trans. Biomed. Eng. **61**(6), 1729–1738 (2014)

22. Achanta, R., Shaji, A., Smith, K., et al.: SLIC superpixels compared to state-of-the-art superpixel methods. IEEE Trans. Patt. Anal. Mach. Intell. **34**(11), 2274–2282 (2012)

23. Cruz-Roa, A., Díaz, G., Romero, E., et al.: Automatic annotation of histopathological images using a latent topic model based on non-negative matrix factorization. J. Pathol. Inform. **2**(2), S4 (2011)

24. Urbán, S., Tanács, A.: Atlas-based global and local RF segmentation of head and neck organs on multimodal MRI images. In: Proceedings of the 10th International Symposium on Image and Signal Processing and Analysis 2017, pp. 99–103. IEEE. https://doi.org/10.1109/ISPA.2017.8073577

T-UNet: A Novel TC-Based Point Cloud Super-Resolution Model for Mechanical LiDAR

Lu Ren[2,3,4], Deyi Li[2,3,4], Zhenchao Ouyang[1,3(✉)], Jianwei Niu[2,3], and Wen He[4,5]

[1] Zhongfa Aviation University, Hangzhou 310000, Zhejiang, China
[2] State Key Laboratory of Software Development Environment, BeiHang University, Beijing 100191, China
[3] Hangzhou Innovation Institute, BeiHang University, Hangzhou 310000, Zhejiang, China
[4] Nanhu Laboratory, Jiaxin 314000, Zhejiang, China
ouyangkid@buaa.edu.cn
[5] Chinese Academy of Military Science, Beijing, China

Abstract. Mechanical LiDAR is one of the most crucial perception sensors for autonomous vehicles. However, the vertical angular resolution of low-cost multi-beam LiDAR is small, limiting the perception and movement range of mobile agents. This paper presents a novel temporal convolutional (TC)-based U-Net model for point cloud super-resolution, which can optimize the point cloud of low-cost LiDAR based on fusing spatiotemporal features of the point cloud. We project the 3D point cloud on a 2D image plane and extend a U-Net convolutional neural network model with a temporal convolutional (TC) module for processing consecutive frames. Each time the model generates one dense/up-sampled image from low-end LiDAR consecutive frames and projects it back into the 3D space as the final result. Considering the intrinsic noise of LiDAR, the structural similarity index measure (SSIM) is introduced as the loss function. Experiments are carried out on both datasets generated by the CARLA simulator and a small-scale dataset collected from actual road conditions with a local vehicle platform. Results show that the proposed model achieves a high peak signal to noise ratio (PSNR). It means the T-UNet model can effectively upsample the sparse point cloud of low-cost LiDAR to a dense point cloud which is almost indistinguishable from the high-end LiDAR point cloud. The source code can be accessed at https://github.com/donkeyofking/lidar-sr.git

Keywords: LiDAR · Point cloud upsampling · Super-resolution · Temporal convolution · U-Net

© ICST Institute for Computer Sciences, Social Informatics and Telecommunications Engineering 2021
Published by Springer Nature Switzerland AG 2021. All Rights Reserved
H. Gao and X. Wang (Eds.): CollaborateCom 2021, LNICST 406, pp. 697–712, 2021.
https://doi.org/10.1007/978-3-030-92635-9_40

1   Introduction

LiDAR (Light Detection and Ranging) is one of the most fundamental and crucial sensors for intelligent robots, self-driving vehicles, and unmanned aerial vehicles (UAV), for it can obtain accurate distance information of the surrounding environments [10,11]. Moreover, LiDAR adopts an active sensing mode that is not affected by ambient light so that it can work at night and dark underground scenes (e.g., caves, mines, and tunnels). Based on the relatively accurate environmental information obtained by LiDAR, sundry robot perception tasks (e.g., target detection and tracking [14,22], segmentation [3,6], simultaneous localization and mapping [13,21,23,31], and navigation [4]) can be performed.

Currently, the off-the-shelf LiDAR can be divided into mechanical LiDAR, and solid-state LiDAR [1]. During the past few decades, we widely use the former in different applications and scenes, and it is designed with 360°of the horizontal field of view (FOV). Furthermore, its vertical field of view is determined by the number of included laser beams and the angle (uniform or non-uniform) between adjacent laser beams (also called angular resolution). The design allows mechanical LiDAR to obtain all the surrounding information within a single scan quickly. However, the predefined angular resolutions (in both horizontal and vertical directions) limit the resolution of perception, especially the horizontal angular resolution. Besides, as the distance increases, the spacing between each laser beam gradually increases. It also results in sparse features in the final 3D point cloud. The solid-state LiDAR fixes this problem with non-repetitive scanning technology by concentrating all the laser beams in a limited field of view, which can scan densely during a certain time interval [2]. Nevertheless, this design sacrifices the field of view range. Furthermore, the complex movement of the mechanical structure to achieve non-repetitive scanning also reduces the final ranging accuracy of the sensor, which is why the FOV of solid-state LiDAR is usually relatively limited. To achieve 360-°horizontal sensing, multiple LiDAR, sophisticated installation structures and synchronization algorithms are usually required [8,18,19].

To avoid the defects involving mechanical LiDAR, the researchers consider the recent progress in point cloud [9,30] and image-based [25] super-resolution technology with deep learning-based modeling. A novel TC [15,16] (Temporal Convolutional) based U-Net model for the point cloud of mechanical LiDAR is proposed. With our model, the sparse point cloud captured by low-cost LiDAR (i.e., Ouster-16/Robosense-32) can be enhanced in real-time and lightweight-edge computing equipment. Moreover, the point cloud upsampled can achieve similar performance compared with the point cloud of high-cost LiDAR (i.e., Ouster-64/Robosense-128). Our T-UNet model can easily extend sparse point cloud from low-cost LiDAR sensor to dense point cloud with dense laser beams, improving the performance of subsequent point cloud-based perception modules. The proposed model can also be treated as a pre-processing module that can be easily inserted into the current workflow of single LiDAR perceptual system or multi-sensor fusion system.

Our research converts the 3D point cloud densification problem into a 2D plane image super-resolution one. The TC module performs like a memory mechanism and tries to capture the inter-frame information to densify and interpolate sparse point clouds. We first project the raw 3D point cloud from the LiDAR with fewer laser beams on a unique 2D plane and enhance the 2D range image with a super-resolution model–T-UNet. Considering the projected point cloud 2D image is much sparser than the camera images and lacks texture information, the enhancement mainly concentrates on the spatial information. As the LiDAR scanning is continuous in the time domain, we extend the typical convolution network into a temporal convolutional one with shared weights. This operation helps fuse information of the time domain while ensuring the lightweight of the model. The structural similarity index measure (SSIM) is introduced as the loss function and ensures that the model can maintain the spatial consistency of the enhanced point cloud. Two scales of enhancement are considered, i.e., upsampling the 16 laser beams into 64 laser beams and upsampling 32 laser beams into 128 laser beams. The final results show that our T-UNet model can achieve a high peak signal to noise ratio (PSNR) on both synthetic data and actual sensor data.

The contributions of this paper are summarized as follows:

- A real-world dataset for 32-to-128 laser beams LiDAR point cloud super-resolution task is released.
- By combining temporal convolutional with U-Net, the proposed T-UNet model can deal with continuous frames and capture the spatiotemporal patterns.
- The dilation convolution is used to replace the pooling layer, and it helps to improve the receptive field without losing information.

The rest of this paper is organized as follows. Sect. 2 briefly reviews the early works on deep learning-based super-resolution tasks and temporal convolutional networks. The detailed design of the T-UNet model is presented in Sect. 3 and the model is evaluated on both the synthetic data captured in the CARLA simulator [7] and the real-world data collected of a local vehicle platform in Sect. 4. Section 5 summarizes the current work and several possible improvements for future research.

2 Related Works

This section introduces the geometric heuristic-based, and deep learning-based progresses on both point cloud and image upsampling or super-resolution tasks. Early research on point cloud upsampling or super-resolution mainly concentrates on point clouds calculated by structured light and a stereo camera. This kind of sensor can only offer approximate 3D distances covering a very close range (i.e., $\leq 10\,\text{m}$) and capture a dense point cloud with both distance and color information. Weinmann et al. [27] refined the resolution limitations of the individual projector-based structured light system with multiple cameras and projectors and used the iterated bundle adjustment registration of the point cloud from different sensors. This framework aims at reconstructing the whole shape of an object with limited size in an indoor environment. By mining the local triangles relationship built from low-resolution point cloud data, Dinesh et al. [5] proposed a novel bipartite graph approximation-based method with the piecewise-smooth to refine the up-sampled point cloud. Their work can preserve the piecewise smoothness of an object's surface after increasing the point cloud density. Without using additional ancillary data, such as RGB (Red, Green, Blue) color, multiple aligned depth maps, or a database of high-resolution depth exemplars, Michael et al. [12] introduced a depth super-resolution method with the reasoning in terms of patches of 3D points, such as repetition of geometric primitives or object symmetry. With the motion information of the 6-DoF (Degree of Freedom) rigid body, this method can achieve super-resolution of a depth map.

Due to the nonlinear fitting ability and data-driven automatic convergence of the deep learning-based model, recent works try to solve the 3D point cloud super-resolution task with neural networks. For overcoming the limitations of deep learning-based 3D objects super-resolution, the 3D appearance SR (3DASR) dataset [12] is published. The researchers then extended the 2D learning-based SR methods into 3D multi-view tasks by utilizing both the coordinates in 3D space and the texture of color at different layers of the two sub-neural networks. Their work projects the 3D object into a 2D texture map and a normal map, generates a super-resolution texture map, projects it back to 3D space. An adversarial residual graph network [28] is proposed to learn the local similarity and the analogy between low-resolution input and high-resolution output. The residual graph convolution, skip connection design, and a novel loss that combines Chamfer distance and graph adversarial loss extend the normal GCN (Graph Convolution Network). PU-Net [30] tries to learn multi-level features with a multi-branch convolution unit with a joint loss function (i.e., reconstruction loss and repulsion loss). PU-GAN [17] further extends the PU-Net with a generative adversarial network (GAN) and self-attention mechanism. The additional discriminator module helps the generator converge faster. The designers also test PU-GAN on outdoor dataset KITTI [8], but with no ground truth. However, the model parameters and calculations also increase for the additional module. Previous works are mainly designed to generate a high-resolution point cloud of a single object for 3D reconstruction. The super-resolution task for outdoor scenes is rarely considered because LiDAR point clouds are long distances and have too much noise in the outdoor environments.

Shan et al. [24] first projected the 3D point cloud into a 2D image and used an image super-resolution U-Net model to generate a high-resolution image, and back-projected the image into 3D space. The Monte-Carlo dropout is combined to remove noisy points learned by the model. They evaluated their model on a dataset generated from a self-driving simulator. The fundamental problem for designing the outdoor point cloud super-resolution model for self-driving applications with data-driven-based deep learning is the shortage of relevant datasets. Moreover, the contextual information provided by the spatiotemporal constraint of LiDAR equipped on a vehicle has not been fully exploited. Therefore, we extend this work with temporal convolutional network [15,16,29] and a novel loss function that can guide the model to learn the spatial structure of the environment. We do not simply deploy the model with common used RNN (Recurrent Neural Network) or LSTM (Long Short Term Memory) because they are neither flexible nor suitable for the high-dimensional LiDAR point cloud.

3   Model Architecture

One of the key factors restricting the popularization of autonomous driving and mobile intelligent robots is the cost of multi-beam LiDAR. The point cloud from a low-cost LiDAR is usually sparse and has low resolution, making it difficult for human annotators to recognize objects. Considering the methods mentioned earlier are not designed for real-time situations for on-road 3D LiDAR point cloud densification on self-driving vehicles, we introduce the T-UNet model designed for upsampling the point cloud from a low-cost LiDAR. We aim to generate stable and accurate dense point clouds of the road scenes from consecutive sparse point clouds.

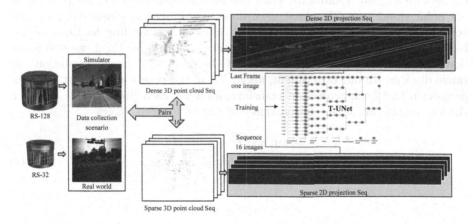

Fig. 1. The proposed T-UNet model for point cloud densification for onboard LiDAR of self-driving vehicle.

Due to the highly correlated consecutive frames from a LiDAR, we take a point cloud sequence instead of one frame pair and learn the inter-frame correlation (temporal and spatial relationships) with a novel TCN framework. To further reduce the computational load of the deep learning model, we transform the 3D space interpolation problem into the super-resolution of the 2D image problem. In this way, we first project the 3D sparse point cloud into a panoramic image according to the coordinating mapping. And then, we encode the feature maps with transposed convolutional layers with dilated units. Each time we connect the features (only the last frame) with deeper layers (upsampling part) before downsampling the scales of the feature map with a high-way (concatenate) module, helping us combine the TCN with the traditional U-Net architecture. Simultaneously, we combine the feature maps from two adjacent frames to integrate data flows while downsampling. The TCN-based U-Net model tries to recover the blocks with short memories from continuous frames and generates a 2D feature image that can be back-projected to the 3D Cartesian coordinate. This model guarantees the extraction of Spatial-temporal information and hierarchically boosts the processing speed by accelerating continuous point cloud frames in a pipeline. The whole process is illustrated in Fig. 1. We evaluate the present model on CARLA simulator [7] and our platform with a RoboSense Ruby (with 128 laser beams: RS-128).

3.1   Point Cloud Projection and Back-Projection

For processing point cloud of the mechanical LiDAR, uniform or non-uniform laser beams are prefined. The laser angle determines the sensor's vertical field of view (FOV), and its horizontal field of view is 360°. The point cloud density is related to the laser beams (vertical angular resolution) and the rotating speed (horizontal angular resolution). With this information, we can easily project the 3D point cloud on a 2D plane, according to the 3D coordinate (x, y, z), and vertical angle (ω), offset (δ) of azimuth angle (α) according to Eq. 1. When back-projecting the 2D image into 3D spaces according to Eq. 2, we will abandon some points due to the limitation of the predefined image resolution and range distance. In our experiment, about 7% of the points will be lost after the projection and back-projection transformation for each point cloud. However, the lost points take a tiny proportion and can be ignored.

$$
\begin{aligned}
r &= \sqrt{x^2 + y^2 + z^2} \\
\alpha + \delta &= arctan(y/x) \\
\omega &= arcsin(z/r)
\end{aligned}
\tag{1}
$$

$$
\begin{aligned}
x &= rcos(\omega)sin(\alpha + \delta) \\
y &= rcos(\omega)cos(\alpha + \delta) \\
z &= rsin(\omega)
\end{aligned}
\tag{2}
$$

According to azimuth and vertical angles, projecting the point cloud is much flexible when dealing with the LiDAR with non-uniform laser beams used in our platform. Figure 2 illustrates the experimental vehicle platform with an RS-32/RS-128 and other equipment. We also illustrate a comparison between Ouster-64 with uniform laser beams and RS-128 with non-uniform laser beams. RS-128's unique design enables the point cloud to be concentrated in the middle of the vertical FOV, thereby helping point clouds obtain more features of on-road targets (such as cars, pedestrians) instead of the ground surface.

Fig. 2. The vehicle platform with RS-128 (left) and a comparison of bird-view images between RS-128 (non-uniform) and Ouster-64 (uniform).

Figure 3 illustrates the projecting results of the 2D images from RS-32 and RS-128. It can be seen that the widths of the two range images are the same due to the two sensors share the same horizontal FOV and angular resolution. However, the difference is four times the height between the two frames (Fig. 3(a)

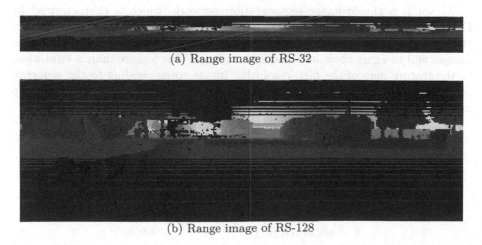

(a) Range image of RS-32

(b) Range image of RS-128

Fig. 3. Comparison between range images of RS-32 and RS-128.

and 3(b)). Therefore, our T-UNet model aims to recover a dense range image of RS-128 from sparse range image of RS-32.

We also collected LiDAR data from the CARLA self-driving simulator for model training. However, we only defined uniform LiDAR of Ouster-16 and Ouster-64 with the same vertical FOV but different angle resolutions. Table 1 lists the detailed sensor parameters related to this study.

Table 1. LiDAR parameters used in this study.

Names	Laser beams	FOV (vertical)	Uniform
CARLA16	16	$(-15.8°, 15.8°)$	Yes
CARLA64	64	$(-15.8°, 15.8°)$	Yes
Ouster-16	16	$(-15.8°, 15.8°)$	Yes
Ouster-64	64	$(-15.8°, 15.8°)$	Yes
CARLA32	32	$(-25°, 15°)$	No
CARLA128	128	$(-25°, 15°)$	No
RS-32	32	$(-25°, 15°)$	No
RS-128	128	$(-25°, 15°)$	No

3.2   T-UNet Model

To further exploit the temporal association between consecutive point cloud frames, we combine the temporal convolutional network with U-Net architecture Fig. 4.

U-Net is a classic image segmentation network, which is characterized by U-Shape encoder-decoder structure and skip-connection. We use TCN to modify the encoder module by extracting features from a sequence of low-resolution images and merging them into one feature map. This feature map is equivalent to the feature map of the high-resolution image corresponding to the sequence of low-resolution images. Then the decoder module up samples the feature map and generates the high-resolution image as output.

$$I_t^{low} = \mathcal{P}(PC_t^{low}) \tag{3}$$

$$I_t^{high} = \mathcal{P}(PC_t^{high}) \tag{4}$$

Then we define the object of our model as follows:

$$\hat{\theta} = \arg\min_{\theta} \mathcal{L}_{SSIM}(\hat{I}_t^{high}, I_t^{high}) + \lambda\Phi(\theta) \tag{5}$$

$$\hat{I}_t^{high} = \mathcal{F}(I_t^{low}, I_{t-1}^{low}, I_{t-2}^{low} \cdots I_{t-(l-1)}^{low}, \theta) \tag{6}$$

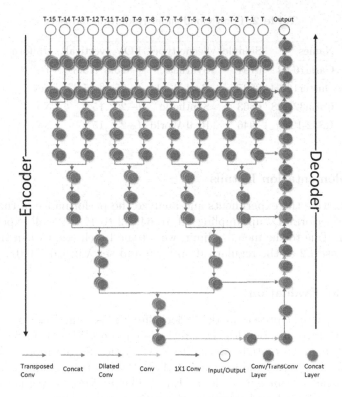

Fig. 4. The architecture of T-UNet model.

where l is the length of low-resolution point cloud sequence, $\mathcal{F}$ defines the function of our model and θ is the parameters of the model. $\mathcal{L}$ represents the SSIM loss function between the generated high-resolution image $\hat{I}_t^{high}$ and the ground truth image I_t^{high}, $\Phi(\theta)$ is the regularization term and λ is the trade-off factor.

4   Experimental Study

4.1   Datasets

Considering the number of existing point cloud super-resolution datasets for the open scene rare, we prepare two datasets from the CARLA simulator and one dataset from the real world. In our experiments, the model up samples I^{low} and generates I^{high}. Therefore, we separately generate I^{low} and I^{high} data with different pre-defined LiDAR in the simulator and on-board LiDAR. Two different upsampling scales are considered, i.e., 16–64 and 32–128; the detailed information is listed in Table 2. The Ouster16-64 is a public released from [24], the CARLA16-64 and CARLA32-128 are synthetic data generated from CARLA, and the RS32-128 is the dataset collected based on our local platform, respectively. For convenience, all datasets are stored in a uniform format.

Table 2. Point cloud super-resolution datasets

Names	Samples	Secne type	FOV (vertical)	Uniform
Carla16-64	6791	Synthetic	$(-15.8°, 15.8°)$	Yes
Ouster16-64	8825	Real-World	$(-15.8°, 15.8°)$	Yes
Carla32-128	6863	Synthetic	$(-25°, 15°)$	No
RS32-128	946	Real-World	$(-25°, 15°)$	No

4.2 Implementation Details

Now we describe the experiments and analyze the performance of the proposed model. We perform 4× upsampling (16 to 64, 32 to 128) for all experiments in this section. Due to the memory limit, we set the batch size to 1 in the training phase. We use $L2$ as the regularization term and set λ in Eq. 5 0.01.

4.3 Model Evaluation

We first compare the present model under different loss functions on the synthetic dataset of Carla16-64. Besides the SSIM, we choose MSE (Mean Square Error) and MAE (Mean Absolute Error) as the loss function for our model. Each model is trained for 20 epochs, and we calculated the PSNR of the ground truth and the generated results from our model as Eq. 7. The PSNR is extended from MSE (Mean Square Error) and describes the ratio between the maximum possible power of a signal and the power of corrupting noise that affects the fidelity of its representation. The higher PSNR, the better consistency between the generated data and ground truth.

$$PSNR = 10 \cdot \log_{10}(\frac{MAX_I^2}{MSE})$$

$$SSIM = [l(I_1, I_2)]^\alpha [c(I_1, I_2)]^\beta [s(I_1, I_2)]^\gamma \tag{7}$$

$$MSE = \frac{1}{mn} \sum_{i=0}^{m-1} \sum_{j=0}^{n-1} [I_1(i,j) - I_2(i,j)]^2$$

where MAX_I is the maximum pixel value of the image, $I_1(i,j)$, $I_2(i,j)$ are two images at the same size. Considering the project images are stored using 8 bits, $MAX_I = 255$. The m and n are the width and height of the projected image. The details about SSIM can be found in [26].

Figure 5 illustrates the PSNRs of the different models during training epochs. It can be seen that the model quickly converged after about six epochs, and the value of PSNR periodically jitter because we use sequence data to train the T-UNet model. However, all the losses of models were finally reduced to an acceptable value under 0.05. Among the three models, the one with the SSIM loss function achieves the highest PSNR from the beginning and can guarantee

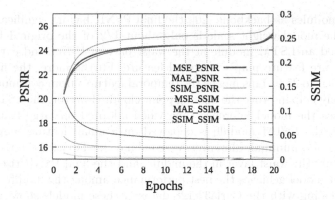

Fig. 5. The PSNR of the models with different loss functions during the training phase.

continued advantages along the whole sequence. Therefore, we use the SSIM as the loss function for the rest evaluation.

We further compare the U-Net model [24] with three different extended versions on both Carla16-64 and Carla32-128. The comparison result also helps us measure how the models deal with data under different scales. Usually the number of laser beams determines the size of a point cloud. Ideally, a LiDAR with 32 laser beams is about twice the point number than 16 laser beams. However, when a laser beam disappears without returning, there is an inevitable fluctuation in the number of points.

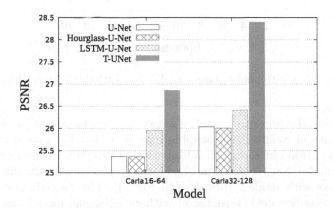

Fig. 6. The PSNR of the four selected models on Carla16-64 and Carla32-128.

Figure 6 illustrates the basic U-Net model and three improved versions. The basic U-Net model only contains an encoder and decoder module with several highway connections between deep (decoder) and shallow (encoder) layers. It is easy to see that when replacing the unstructured U-Net with three stacked

Hourglass modules as backbone [20], the final PSNR has no significant change. However, the model weight scale is only about 1/2 of the original U-Net. We further added an LSTM module at the waist part of U-Net and extracted the most miniature feature map's temporal feature. Furthermore, the final PSNR rises from 25.4 to 25.9. This means the temporal features from continuous frames can effectively help improve the missing point cloud. However, the LSTM module heavily raises the model inference time when adopted on large feature maps. In our case, the LSTM module is deployed on a small feature map, and the promotion is also minimal. Moreover, we combine the TC with the U-Net model instead of only the waist layer in the middle, and the final PSNR rises to 26.83. The T-UNet model achieves the best performance among the modified versions.

When dealing with the Carla32-128 dataset, these models show a very similar distribution as on Carla16-64 but with a higher PSNR value because when dealing with the dense point cloud, the observed information is sufficient for recovering the environment with our model.

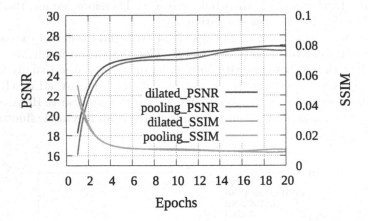

Fig. 7. The T-UNet with pooling layers vs. T-UNet with dilated convolutional layers.

Considering the previous hourglass network can achieve similar performance with fewer model weights, one possible reason is that the max-pooling layers in U-Net may drop out too much information. Therefore, all pooling layers are removed from T-UNet (*pooling_* in Fig. 7), and replaced with the dilated convolutional layers with stride 2 ((*dilated_* in Fig. 7)). The dilated convolution can enlarge the receptive field (kernel size) without enlarging model weights. It can be seen that the T-UNet model with dilated convolution layer achieves more stable PSNR during training epochs than the T-UNet model with max-pooling. Moreover, both models achieve very similar losses on SSIM.

Finally, we test our model on the four datasets mentioned above (two synthetic data and two real-world sensor data) in Fig. 8. It shows that when dealing with a more dense point cloud, the final up-sampled result shows higher PSNR, which means the more information the model gets, the better the final point

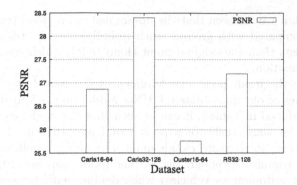

Fig. 8. The performances of the dilated T-UNet models when dealing with different datasets.

cloud can be restored. The PSNR can achieve nearly 28.5 when using a point cloud of 32 laser beams to generate a point cloud of LiDAR with 128 laser beams. Simultaneously, when dealing with the actual data with more vital randomness, both results reduce the overall PSNR (Ouster16-64 and RS32-128).

Table 3. Time consumptions of the final model on different point scales (on Nvidia GTX2080Ti)

Mode	Processing Time			
	Carla16-64	Ouster16-64	Carla32-128	RS32-128
Single frame	48.6 ms	50.1 ms	152.3 ms	161.2 ms
1000 frames	51.59 s	53.17 s	160.07 s	164.02 s

(a) Ouster16-64: Input (left), GT (middle) and model output(right)

(b) RS32-128: Input (left), GT (middle) and model output(right)

Fig. 9. The super-resolution results. The above pictures are the Ouster16-64 experiment. The below are the RS32-128 experiment. Left: Low-resolution point cloud; Middle: Ground truth; Right: Our model generated.

From Fig. 9, it can be seen that the upsampled point cloud (right) has some noise points compared with ground truth (middle), but it still generates lots of effective points than the original point cloud (left), which can provide more abundant information.

Considering the input point cloud influences the model, we estimated the calculation times of our final dilated T-UNet model on the four different scales of datasets. As listed in Table 3, it can be seen that the model can achieve near 53.0 ms/frame for small scales of a point cloud and 164.0 ms/frame for large scales of the point cloud (point clouds number is about 2 million for a LiDAR with 128 laser beams). Although the current model can not 10 Hz (a typical requirement for autonomous vehicles) when dealing with large-scale data, we think it still has room for improvement.

5   Conclusion

Dense point cloud from high-cost LiDAR supports stable perception for autonomous vehicles and robots. However, the high cost also severely restricts related applications. To achieve better perception performance based on sparse low-end LiDAR, we design a novel TC-based U-Net model for point cloud super-resolution. We first project the 3D point cloud into a 2D image plane, process the continuous frames with dilated convolutional layers, encode the temporal feature with a TC module, and generate super-resolution point feature maps. By re-projecting it back to the 3D space, the dense point cloud is achieved. With the SSIM loss function, the model can capture a more stable spatial consistency of the point cloud. Furthermore, dilated convolutional also helps enlarge the reception field of each kernel without reducing details. The final model can achieve higher PSNR through the improvements as mentioned earlier. In the future, we plan to compress the model with knowledge distillation for better deployment on edge computing devices and real-time processing. The generated result is also considered to be quantitatively evaluated through point cloud-based detection or tracking models.

Acknowledgment. This work has been supported by China Postdoctoral Science Foundation (2020M 681798), Qianjiang Excellent Post-Doctoral Program (2020Y4A001) and 2020 Zhejiang Postdoctoral Research Project (ZJ2020011). JITRI Suzhou Automotive Research Institute Project (CEC20190404). The authors would like to thank Plusgo for their cooperation during data collection.

References

1. Aijazi, A., Malaterre, L., Trassoudaine, L., Checchin, P.: Systematic evaluation and characterization of 3d solid state lidar sensors for autonomous ground vehicles. Int. Arch. Photogrammetry, Remote Sens. Spatial Inf. Sci. **43**, 199–203 (2020)
2. Atanacio-Jiménez, G., et al.: Lidar velodyne hdl-64e calibration using pattern planes. Int. J. Adv. Robot. Syst. **8**(5), 59 (2011)

3. Behley, J., et al.: Semantickitti: a dataset for semantic scene understanding of lidar sequences. In: Proceedings of the IEEE International Conference on Computer Vision, pp. 9297–9307 (2019)
4. Christian, J.A., Cryan, S.: A survey of lidar technology and its use in spacecraft relative navigation. In: AIAA Guidance, Navigation, and Control (GNC) Conference, p. 4641 (2013)
5. Dinesh, C., Cheung, G., Bajić, I.V.: 3d point cloud super-resolution via graph total variation on surface normals. In: 2019 IEEE International Conference on Image Processing (ICIP), pp. 4390–4394. IEEE (2019)
6. Dong, X., Niu, J., Cui, J., Fu, Z., Ouyang, Z.: Fast segmentation-based object tracking model for autonomous vehicles. In: Qiu, M. (ed.) ICA3PP 2020. LNCS, vol. 12453, pp. 259–273. Springer, Cham (2020). https://doi.org/10.1007/978-3-030-60239-0_18
7. Dosovitskiy, A., Ros, G., Codevilla, F., Lopez, A., Koltun, V.: Carla: an open urban driving simulator. arXiv preprint arXiv:1711.03938 (2017)
8. Geiger, A., Lenz, P., Urtasun, R.: Are we ready for autonomous driving? the kitti vision benchmark suite. In: 2012 IEEE Conference on Computer Vision and Pattern Recognition, pp. 3354–3361. IEEE (2012)
9. Gevrekci, M., Pakin, K.: Depth map super resolution. In: 2011 18th IEEE International Conference on Image Processing, pp. 3449–3452. IEEE (2011)
10. Guo, Y., Wang, H., Hu, Q., Liu, H., Liu, L., Bennamoun, M.: Deep learning for 3d point clouds: a survey. IEEE Trans. Pattern Anal. Mach. Intell. **43**(12), 4338–4364 (2020)
11. Hanson, W., Jones, R., Jones, R.: The roman military presence at dalswinton, dumfriesshire: a reassessment of the evidence from aerial, geophysical and lidar survey. Britannia **50**, 285–320 (2019)
12. Hornacek, M., Rhemann, C., Gelautz, M., Rother, C.: Depth super resolution by rigid body self-similarity in 3d. In: Proceedings of the IEEE Conference on Computer Vision and Pattern Recognition, pp. 1123–1130 (2013)
13. Huang, B., Zhao, J., Liu, J.: A survey of simultaneous localization and mapping. arXiv preprint arXiv:1909.05214 (2019)
14. Huang, R., et al.: An lstm approach to temporal 3d object detection in lidar point clouds. arXiv preprint arXiv:2007.12392 (2020)
15. Lea, C., Flynn, M.D., Vidal, R., Reiter, A., Hager, G.D.: Temporal convolutional networks for action segmentation and detection. In: proceedings of the IEEE Conference on Computer Vision and Pattern Recognition, pp. 156–165 (2017)
16. Lea, C., Vidal, R., Reiter, A., Hager, G.D.: Temporal convolutional networks: a unified approach to action segmentation. In: Hua, G., Jégou, H. (eds.) ECCV 2016. LNCS, vol. 9915, pp. 47–54. Springer, Cham (2016). https://doi.org/10.1007/978-3-319-49409-8_7
17. Li, R., Li, X., Fu, C.W., Cohen-Or, D., Heng, P.A.: Pu-gan: a point cloud upsampling adversarial network. In: Proceedings of the IEEE International Conference on Computer Vision, pp. 7203–7212 (2019)
18. Lin, J., Liu, X., Zhang, F.: A decentralized framework for simultaneous calibration, localization and mapping with multiple lidars. arXiv preprint arXiv:2007.01483 (2020)
19. Milella, A., Reina, G., Nielsen, M.: A multi-sensor robotic platform for ground mapping and estimation beyond the visible spectrum. Precision Agric. **20**(2), 423–444 (2018). https://doi.org/10.1007/s11119-018-9605-2

20. Newell, A., Yang, K., Deng, J.: Stacked hourglass networks for human pose estimation. In: Leibe, B., Matas, J., Sebe, N., Welling, M. (eds.) ECCV 2016. LNCS, vol. 9912, pp. 483–499. Springer, Cham (2016). https://doi.org/10.1007/978-3-319-46484-8_29

21. Ouyang, Z., Liu, Y., Zhang, C., Niu, J.: A cgans-based scene reconstruction model using lidar point cloud. In: 2017 IEEE International Symposium on Parallel and Distributed Processing with Applications and 2017 IEEE International Conference on Ubiquitous Computing and Communications (ISPA/IUCC), pp. 1107–1114. IEEE (2017)

22. Ouyang, Z., Wang, C., Liu, Yu., Niu, J.: Multiview CNN model for sensor fusion based vehicle detection. In: Hong, R., Cheng, W.-H., Yamasaki, T., Wang, M., Ngo, C.-W. (eds.) PCM 2018. LNCS, vol. 11166, pp. 459–470. Springer, Cham (2018). https://doi.org/10.1007/978-3-030-00764-5_42

23. Shan, T., Englot, B.: Lego-loam: Lightweight and ground-optimized lidar odometry and mapping on variable terrain. In: 2018 IEEE/RSJ International Conference on Intelligent Robots and Systems (IROS), pp. 4758–4765. IEEE (2018)

24. Shan, T., Wang, J., Chen, F., Szenher, P., Englot, B.: Simulation-based lidar super-resolution for ground vehicles. arXiv preprint arXiv:2004.05242 (2020)

25. Wang, Z., Chen, J., Hoi, S.C.: Deep learning for image super-resolution: a survey. IEEE Trans. Pattern Anal. Mach. Intell. 43(10), 3365–3387 (2020)

26. Wang, Z., Bovik, A., Sheikh, H., Simoncelli, E.: Image quality assessment: from error visibility to structural similarity. IEEE Trans. Image Process. 13(4), 600–612 (2004). https://doi.org/10.1109/TIP.2003.819861

27. Weinmann, M., Schwartz, C., Ruiters, R., Klein, R.: A multi-camera, multi-projector super-resolution framework for structured light. In: 2011 International Conference on 3D Imaging, Modeling, Processing, Visualization and Transmission, pp. 397–404. IEEE (2011)

28. Wu, H., Zhang, J., Huang, K.: Point cloud super resolution with adversarial residual graph networks. arXiv preprint arXiv:1908.02111 (2019)

29. Yan, J., Mu, L., Wang, L., Ranjan, R., Zomaya, A.Y.: Temporal convolutional networks for the advance prediction of ENSO. Sci. Rep. 10(1), 1–15 (2020)

30. Yu, L., Li, X., Fu, C.W., Cohen-Or, D., Heng, P.A.: Pu-net: Point cloud upsampling network. In: Proceedings of the IEEE Conference on Computer Vision and Pattern Recognition, pp. 2790–2799 (2018)

31. Zhang, J., Singh, S.: Loam: lidar odometry and mapping in real-time. In: Robotics: Science and Systems. vol. 2 (2014)

Computation Offloading for Multi-user Sequential Tasks in Heterogeneous Mobile Edge Computing

Huanhuan Xu[1], Jingya Zhou[1,2(✉)], and Fei Gu[1]

[1] School of Computer Science and Technology, Soochow University,
Suzhou 215006, China
20195227020@stu.suda.edu.cn, {jy_zhou,gufei}@suda.edu.cn
[2] State Key Laboratory of Mathematical Engineering and Advanced Computing,
Wuxi 214125, China

Abstract. Currently, many computation offloading studies in Mobile Edge Computing (MEC) mainly focus on the multi-user and multi-task offloading, where the tasks are independent and inseparable. However, nowadays many mobile applications, such as Augmented Reality (AR) glasses, face recognition, *etc.*, can be classified into sequential tasks. Dependencies among tasks and resource competition among multiple users in the heterogeneous edge environment make the offloading problem very challenging. In this paper, we propose a new multi-user sequential task (MUST) framework to address the above challenge. Specifically, we present a comprehensive analysis of the time cost of the task offloading process in the MUST framework and define the multi-user sequential task offloading problem. Moreover, we prove the problem is NP-hard and propose a reMUST algorithm based on regular expression to obtain the approximate optimal solution. Numerous experiments have shown that the proposed method is superior to existing alternatives in terms of cost and system scalability.

Keywords: Mobile edge computing · Computation offloading · Sequential tasks · Regular expression

This work was supported by National Natural Science Foundation of China (Grant Nos. 61972272, U1905211 and 62072321), the Open Project Program of the State Key Laboratory of Mathematical Engineering and Advanced Computing (Grant No. 2019A04), Jiangsu Planned Projects for Postdoctoral Research Funds (Grant No. 1701173B), Jiangsu Overseas Visiting Scholar Program for University Prominent Young & Middle-aged Teachers and Presidents.

H. Gao and X. Wang (Eds.): CollaborateCom 2021, LNICST 406, pp. 713–727, 2021.
https://doi.org/10.1007/978-3-030-92635-9_41

1   Introduction

With the iterative evolution of mobile device (MD) technology, users' demand for complex mobile applications is soaring. These applications, such as mobile AR, live video, collaborative editing, *etc.*, require massive computational resources of MDs to provide low-latency services [1,2]. However, since the performance of MDs is constrained by many issues, such as physical size, battery capacity, and system architecture, their computational capacities may not be able to meet the latency and computing requirements of these applications. By deploying cloud computing services to the edge of the network, MEC effectively provides low-latency services for the users [3]. MEC uses a computation offloading technology to offload computation-intensive tasks to nearby MEC servers, which expands the computing and storage capacity of MDs. Meanwhile, MEC overcomes the disadvantages of the traditional cloud computing system, such as network congestion and long transmission delay caused by the large number of task processes and the long physical distance between the data center and the network edge.

As a key technology of MEC, computation offloading distributes the computational tasks of MDs to the edge environment, which alleviates the shortages of the MDs in resource storage, computational performance, and energy efficiency [4]. Recently, many researchers work on computation offloading in MEC. These studies can reduce energy consumption and makespan effectively, or make a balance between them, by selecting task offloading decisions flexibly. However, in real mobile application scenarios, there are still some problems to be solved:

(1) **Heterogeneity**: MEC servers are heterogeneous and deployed on a large scale. There may be multiple service providers at the edge network. Even for the same service provider, different periods of deployment and equipment updates result in heterogeneity. This situation leads to great uncertainty about the cost of executing the task, which will seriously affect the efficiency of task execution.
(2) **Multi-User**: MEC is multi-user oriented, which is different from the single-user scenario that has been widely studied. Multi-user scenario has different user requirements, heterogeneous tasks, and resource competition. Especially, the competition between MDs leads to low wireless rates in the network and long queuing time on edge servers.
(3) **Dependency**: At present, many mobile applications are not independent tasks, but consist of sequential dependent tasks. For instance, Fig. 1 shows that Google Project Glass with AR technology, including the following function modules: video capture, video parsing, target recognition, content mapping, video synthesis, and real-time display. Among them, video parsing and content mapping require massive computational resources and offloading them to edge servers can effectively reduce the processing delay of the overall task.

To solve these problems, we propose a Multi-User Sequential Tasks (MUST) framework for offloading users' sequential tasks in MEC. Our contributions are summarized as follows:

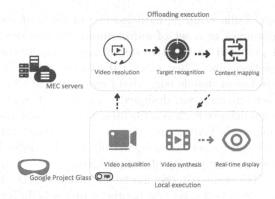

Fig. 1. The AR application execution flow

(1) We propose a framework named MUST to solve the multi-user sequential tasks offloading problem in heterogeneous MEC. When it comes to offloading the sequential tasks, we adopt the MUST framework to reduce the system cost according to the current network environment.

(2) Based on the MUST framework, we give a formal definition of the problem and further prove that it is NP-hard. By relaxing the integer limit of offloading decision and using the penalty function to consider other constraints, we propose a heuristic method based on Regular Expression (RE) to get the approximate optimal solution to the original problem.

(3) We make extensive numerical experiments to evaluate the effectiveness of the proposed scheme. The results demonstrate that the proposed scheme not only reduces the average delay but also improves the scalability of the system, compared with the conventional offloading strategies.

2   Related Work

A great deal of work has been done on the computation offloading problem and various offloading strategies have been proposed. Among them, the dependent task offloading problem in MEC is very complicated. Only a few studies pay attention to the sequential dependent constraint.

Kao *et al.* [5] designed a polynomial-time approximation algorithm that minimizes the maximum completion time when offloading dependent tasks under resource constraints. Zhao *et al.* [6] jointly considered dependent task offloading and service caching placement to minimize application completion time. Liu *et al.* [7] considered the problem of dependent task placement and scheduling with on-demand function configuration on the server, and proposed an approximate algorithm to minimize the application completion time. However, the above works did not take the real multi-user edge environment into account, where multiple users compete for marginal limited resources fiercely.

Fan *et al.* [8] proposed the offloading problem of multi-user dependent tasks to minimize the total cost of all applications within the time limit for completion of each application. Jošilo *et al.* [9] modeled the problem of allocating wireless and computing resources to a set of autonomous wireless devices as a Stackelberg game and proposed an effective decentralized equilibrium algorithm to reduce task completion time. Chen *et al.* [10] studied the multi-user computation offloading problem of mobile edge cloud computing under a multi-channel wireless interference environment, designed a distributed computing offloading algorithm that could achieve nash equilibrium, and obtained good computation offloading performance. Whereas, the above works ignored the heterogeneous characters of the edge. Edge servers are generally heterogeneous in reality, which will further enhance the complexity of the problem.

Habak *et al.* [11] considered the model of using a MD cluster to accept offloaded tasks and proposed a general heuristic task scheduling scheme, whose goal is to maximize the effective execution of offloading tasks by the cluster. Lin *et al.* [12] considered the scheduling problem of an application composed of dependent tasks and proposed a heuristic algorithm. However, the authors assumed that only the native processors of a single MD are considered. Similarly, Sundar *et al.* [13] examined the problem of dependent task offloading in heterogeneous networks to minimize cost under application deadline. Sundar *et al.* [14] studied the scheduling of an application composed of dependent tasks in a general heterogeneous edge computing system and proposed a heuristic algorithm to obtain an effective solution.

In this work, edge servers have a limited workload, and there is latency cost associated with task execution, data communication, and task queuing on them, which leads to a unique formulation of the problem that has not been studied in the existing literature.

3   MUST Model and Problem Formulation

In this section we first introduce the system model, which includes the network model, task model, and cost model. Then we formally define the offloading problem of multi-user sequential tasks (MUST) with delay and load limits, and prove that there is no constant solution to this problem.

3.1   System Model

A. Network Model

As shown in Fig. 2, the typical MEC scenario includes two parts, *i.e.*, the user and the edge. The user side has N MDs, represented by a set $\mathbf{N} = \{1, 2, \ldots, N\}$. In the edge, we define $\mathbf{K} = \{1, 2, \ldots, K\}$ as the set of K access points with different computational capabilities, which we call computational access points (CAPs) for simplicity. CAPs are interconnected at a high speed via wired links [3], while MD transmits data with CAP via the single-carrier wireless channel of orthogonal frequency division multiple access (OFDMA). The channel gain is affected by shadow fading and path loss caused by the route. It is assumed that

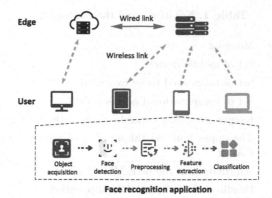

Fig. 2. An example of MEC scenario with multiple users

each CAP can cover all MDs and has an upper limit c on the number of accepted tasks, which is similar to the heterogeneous edge environment with dense MDs.

Each MD has an application to be processed, and the set $\mathbf{D} = \{d_1, d_2, \ldots, d_N\}$ represents the applications on all MDs. Each application $d \in \mathbf{D}$ consists of a set $\mathbf{M} = \{m_1^d, m_2^d, \ldots, m_M^d\}$ of sequential tasks. The m-th task on the n-th MD is denoted by $\mathcal{T}_{m,n}(s_{m,n}, \eta_{m,n})$, where $s_{m,n}$ (bits) represents the size of input data in the current task, which is also the size of output data in the previous task, and $\eta_{m,n}$ represents the number of CPU cycles spent by each bit of the task data.

Each task can be offloaded to a nearby CAP. According to the offloading decision of the previous task, the data transmission process of the current task may be different. The CAP may not process the task immediately after receiving the task data but puts it in a queue.

B. Task Dependency Model

The application of the n-th MD has a deadline T_n^{max}, and can be divided into several tasks with sequential dependent as mentioned before [15]. We use an example of face recognition to illustrate the execution flow of the application. As shown in Fig. 2, the MD needs to collect relevant face data, preprocess the collected data at the beginning, and send the processed data to the CAP, where the application service has been deployed and the model has been preloaded. The CAP can perform match detection, and return the match result to the MD. In this paper, we mainly focus on the offloading of such sequential tasks. Notations and their meanings are listed in Table 1.

We assume that the first task of the application is started from local MD and the final result must be returned to MD. Each task can be executed on one of the CAPs. We define the offloading decision variable of the task m on MD n as $x_{m,n}$, which means whether the system offloads this task to CAP k.

Further, we define $\mathbf{X} = \{\mathrm{x}_{m,n} \mid m \in \mathbf{M}, n \in \mathbf{N}\}$ as the offloading strategy of all tasks, $\mathbf{N}_k$ as the set of MDs connected to CAP k and $\mathbf{M}_k$ as the set of tasks offloaded to CAP k. For a given $\mathbf{X}$, we calculate the wireless rate as:

Table 1. Notations and their meanings

Notation	Meaning
N	Set of mobile devices
M	Set of tasks need to be executed
K	Set of computational access points
$x_{m,n}$	Integer offloading variable of the n-th MD's task m
c	The upper limit on the number of accepted tasks on CAP k
$\mathcal{T}_{m,n}$	Information of the n-th MD's task m
T_n^{max}	Deadline of the n-th MD's application
T_n	Completing time of the n-th MD's application
$T_{m,n}$	Completing time of the n-th MD's task m
W	Wireless channel bandwidth of CAPs
β	Data rate of wire link between CAPs
p_n	Device power of MD n
$h_{n,k}$	Channel gain of transmitting tasks from MD n to CAP k
$r_{n,k}$	Wireless transmission rate of the n-th MD's tasks between MD n and CAP k
$t_{m,n}^{exec}$	Executing time of the n-th MD's task m
$t_{m,n}^{tran}$	Data transmission time of the n-th MD's task m
$t_{m,n}^{queu}$	Queuing time of the n-th MD's task to be executed on CAP k

$$r_{n,k}(\mathbf{X}) = W \log_2 \left(1 + \frac{p_n h_{n,k}}{\varpi + \sum_{i \in \mathbf{N}_k \backslash \{n\}} p_i h_{i,k}} \right), \quad (1)$$

where p_n is the power of the n-th MD, $h_{n,k}$ is the channel gain from MD n to CAP k based on the shadow attenuation and path loss, ϖ is the thermal noise of link between MD n and CAP k and W is the wireless channel bandwidth [16].

C. Cost Model

In our scenario, the goal of computation offloading is to minimize the average cost of the applications by offloading different sequential tasks from MDs to CAPs. Next we introduce our cost model where the process of each task consists of three parts: execution, communication, and queuing.

Task Execution: To illustrate the delay caused by execution, we define the execution time of the task m on CAP k is $t_{m,n}^{exec} = \frac{s_{m,n} \eta_{m,n}}{f_k}$.

Task Communication: Considering different communication processes, the transmission time of the n-th MD's task m can be divided into three cases:

(1) Task m is the first task and needs to be offloaded from MD n to CAP k for execution. Considering the wireless transmission process of the task, the time it consumed is $t_{m,.n}^{tran} = \frac{s_{m,n}}{r_{n,k}}$.

(2) Task m is an intermediate task and its offloading decision is different from its previous task. The task data requires to be transmitted from one CAP to another for execution. Since the fixed wired bandwidth between CAPs is β, the transmission delay is $t_{m,n}^{tran} = \frac{s_{m,n}}{\beta}$.

(3) Task m is an intermediate task and its offloading decision is identical to its previous task. It means that adjacent tasks are offloaded to the same CAP. Since there is no data transmission process, the transmission delay is $t_{m,n}^{tran} = 0$.

Since the amount of data returned by the application's final task is generally small, we ignore the transmission delay of these data like the article [10].

Task Queuing: The task potentially cannot be executed immediately when it arrives at CAP, it needs to wait until the current CAP finishes its tasks. Each CAP handles the tasks on it as an M/G/1 queue by utilizing queuing theory. It is assumed that v_k is the arrival rate of the task on the CAP k. Let θ denote the variable of computation time for any task. $E_k[\theta]$ and $E_k[\theta^2]$ represent its first and second moments. According to the Pollaczek-Khinchin formula [17], the expected queuing time for a task m of the n-th MD on CAP k is $t_{m,n}^{queu} = \frac{v_k E_k[\theta^2]}{2(1-v_k E_k[\theta])}$.

Since the application completion consists of the execution, communication and queuing of all tasks, the total delay of the n-th MD's application is:

$$T_n(\mathbf{X}_n) = \sum_{m=1}^{M} T_{m,n}(x_{m,n}) = \sum_{m=1}^{M} \left(t_{m,n}^{exec} + t_{m,n}^{tran} + t_{m,n}^{queu} \right). \tag{2}$$

3.2   Problem Formulation

Given a set of resource-constrained CAPs and a set of MDs (each MD's application consisting of several dependent tasks and has a deadline), we define the multi-user sequential dependent task offloading (MUSTO) problem. Since the goal is to minimize the average application delay in the system while satisfying the delay limit of each application and the load capacity of CAPs, the problem of MUST can be formulated as problem **P1** below:

$$\underset{\{\mathbf{X}\}}{\text{minimize}} \quad \frac{1}{N} \sum_{m=1}^{M} \sum_{n=1}^{N} T_{m,n}(x_{m,n}), \tag{3}$$

$$\text{s. t.} \quad C_1: \quad |\mathbf{D}_k| \leq c, \forall k \in \mathbf{K},$$

$$C_2: \quad x_{m,n} \in \{1, \dots, K\}, \forall m \in \mathbf{M}, \forall n \in \mathbf{N},$$

$$C_3: \quad \sum_{m=1}^{M} T_{m,n} \leq T_n^{max}, \forall n \in \mathbf{N}.$$

C_1 indicates that there is a limit to the number of tasks that each CAP can accept, C_2 represents that the offloading decision is an integer variable, and C_3 indicates that the overall task execution time of MD's tasks cannot exceed its deadline. Then we rephrase the offloading decision variable to the one-hot type $\mathbf{x}_{m,n} = [0, \ldots, 1, \ldots]^\top$, where $(\cdot)^\top$ represents the transpose of a matrix or a vector. Hence, the offloading decision variable can be represented as follows:

$$x_{m,n,k} = \begin{cases} 1 & \text{if MD } n' \text{ s task } m \text{ is offloaded to CAP} k, \\ 0 & \text{otherwise.} \end{cases} \tag{4}$$

Since each task can only be assigned to one CAP for execution, it must satisfy $\sum_{k=1}^{K} x_{m,n,k} = 1$. We define the potential cost of offloading the n-th MD's task m to CAP k as $C_{m,n,k}$. Therefore, the problem **P1** can be rewritten as **P2**:

$$\underset{\mathbf{X}}{\text{minimize}} \sum_{m=1}^{M} \sum_{n=1}^{N} \sum_{k=1}^{K} C_{m,n,k} x_{m,n,k},$$

$$\text{s. t.} \quad C_3 : \quad \sum_{m=1}^{M} T_{m,n} \leq T_n^{\max}, \forall n \in \mathbf{N},$$

$$C_4 : \quad \sum_{m=1}^{M} \sum_{n=1}^{N} x_{m,n,k} \leq c, \forall m \in \mathbf{M}, \forall n \in \mathbf{N}, \forall k \in \mathbf{K}, \tag{5}$$

$$C_5 : \quad x_{m,n,k} \in \{0,1\}, \forall m \in \mathbf{M}, \forall n \in \mathbf{N}, \forall k \in \mathbf{K},$$

$$C_6 : \quad \sum_{k=1}^{K} x_{m,n,k} = 1, \forall m \in \mathbf{M}, \forall n \in \mathbf{N}.$$

C_3 and C_4 indicate that the total task execution time of the MD cannot exceed its delay limit and there is a limit to tasks each CAP can undertake respectively.

Theorem 1. *The problem of MUSTO is NP-hard.*

Proof. GAP problem is NP-hard, which is described as follows [18]. The system assigns mutually independent works to the limited-resource agents. A job can only be served by one agent, an agent can serve multiple jobs, and the total amount of resources required by the agent to complete the work cannot exceed its limit. The mathematical model of GAP is expressed in the following form:

$$\text{minimize} \quad F(x) = \sum_{i=1}^{m} \sum_{j=1}^{n} c_{i,j} x_{i,j},$$

$$\text{s. t.} \quad \sum_{j=1}^{n} r_{i,j} x_{i,j} \leqslant b_i, \forall i \in \mathbf{I}, \tag{6}$$

$$\sum_{i=1}^{m} x_{i,j} = 1, \forall j \in \mathbf{J},$$

$$x_{i,j} \in \{0,1\}, \forall i \in \mathbf{I}, \forall j \in \mathbf{J},$$

where $\mathbf{I} = \{1, 2, \ldots, m\}$ is the agent set, $\mathbf{J} = \{1, 2, \ldots, n\}$ is the working set, b_i represents the number of resources owned by the i-th agent, $r_{i,j}$ represents the number of resources consumed by the i-th agent to serve the work j and $c_{i,j}$ represents the cost of the i-th agent to serve the work j.

The objective function is to minimize the system cost. Equation (6) ensures that the total amount of resources consumed by the work j which is assigned to the i-th agent does not exceed the resource limit, each work j can only be assigned to one agent, and $x_{i,j}$ is the binary decision variable respectively.

From the above analysis, we find that when every application does not exceed its deadline, the MUSTO problem is a set of N standard GAP problems. Therefore, the MUSTO problem is NP-hard and cannot obtain a specific solution.

4   Regular Expression Based Algorithm for MUST

In this section, we present a regular expression-based algorithm for MUST, called reMUST. The workflow of reMUST is shown in Fig. 3. Specifically, reMUST offloads sequential tasks by conducting four steps: (1) *Relaxing the MUSTO Problem* to construct an ordinary least square problem (OLS) with constraints. (2) *Using penalty function* method to remove constraints for this relaxed problem. (3) *Using regular expression* for the standard OLS problem. (4) *Offloading tasks* by following the approximate optimal strategy.

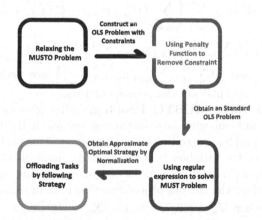

Fig. 3. The workflow of the reMUST algorithm: 1) relaxing the MUSTO problem to construct a limited OLS problem, 2) using penalty function to remove constraints, 3) using regular expression for the rewritten MUSTO problem to get the approximate optimal solution, and 4) offloading tasks according to the strategy.

Relaxing the MUSTO Problem. To reformulate problem in Eq. (5) in the OLS form, the offloading decisions must take binary values. Nevertheless, the offloading decisions may not be integer vectors under matrix operations.

Therefore, we relax the offloading decisions. By combining one-hot vector with relaxing offloading decisions, we simplify **P2** as follows:

$$\underset{\mathbf{X}}{\text{minimize}} \quad \frac{1}{N} \sum_{m=1}^{M} \sum_{n=1}^{N} \mathbf{C}_{m,n}^{\top} \mathbf{x}_{m,n}, \tag{7}$$

$$\text{s. t. } C_3, C_4.$$

Using Penalty Function. After relaxing the variables, we use the penalty functions to replace the constraints of the problem. We make a two-layer optimization with the ordinary least square form. We further rewrite the objective function and add the formulation limitations of application delay and CAP load to **P2**. By using the definition of sums of squared deviations, the objective function of average system cost can be replaced as follows:

$$\underset{\mathbf{X}}{\text{minimize}} \quad \frac{1}{N} \sum_{n=1}^{N} \sum_{m=1}^{M} \left(\mathbf{C}_{m,n}^{\top} \mathbf{x}_{m,n} - \overline{\mathbf{C}^{\top} \mathbf{x}} \right)^2, \tag{8}$$

$$\text{s.t.} \quad C_3, C_4,$$

where $\overline{\mathbf{C}^{\top} \mathbf{x}}$ represents the average task delay. Since the least square method cannot be applied to the optimization problem with limited conditions, we add C_3 and C_4 to the objective function by using the penalty functions and get:

$$\underset{\mathbf{X}}{\text{minimize}} \quad \mathrm{E} = \frac{1}{N} \sum_{n=1}^{N} \left(\sum_{m=1}^{M} \left(\mathbf{C}_{m,n}^{\top} \mathbf{x}_{m,n} - \overline{\mathbf{C}^{\top} \mathbf{x}} \right)^2 + \rho_1 \right) + \rho_2, \tag{9}$$

where $\rho_1 = \mu \left(\max \left(0, \sum_{m=1}^{M} \mathbf{C}_{m,n,k}^{\top} \mathbf{x}_{m,n,k} - T_n^{\max} \right) \right)$ and

$\rho_2 = \sum_{k=1}^{K} \phi \left(\max \left(0, \sum_{m=1}^{M} \sum_{n=1}^{N} \mathbf{x}_{m,n,k} - c \right) \right)$ are penalty functions of constraints C_3, C_4, and μ, ϕ are the penalty factors.

Using RE to Solve MUSTO Problem. After getting the form of the relaxed MUSTO problem with no constraints, we use a RE-based method to solve the MUSTO problem according to its properties. This problem is a function of mapping some vector variables to a number. By mapping offloading strategy to cost, the first-order partial derivatives of cost concerning offloading strategy can be obtained. Since the current object function is to minimize $E(\mathbf{X}_n)$, it gets optimal solution when $\nabla_{\mathbf{x}_n} E(\mathbf{X}_n) = 0$, i.e., $\mathbf{X}_n = \left(\mathbf{C}_n^{\top} \mathbf{C}_n \right)^{-1} \mathbf{C}_n^{\top} \overline{\rho_1}$ according to Theorem 2. After using normalization we can finally obtain our approximate optimal offloading strategy $\mathbf{X}_n^*$ for tasks on the n-th MD.

Theorem 2. *For a least square problem, i.e., minimize $\sum_{i=1}^{I} \left(\mathbf{A}^{\top} x_i - b_i \right)^2$, its solution satisfies $\mathbf{X} = \left(\mathbf{A}^{\top} \mathbf{A} \right)^{-1} \mathbf{A}^{\top} \mathbf{B}$.*

Proof. For a matrix $\mathbf{A} \in \mathbb{R}^{m \times n}$, a constant vector $\mathbf{B} \in \mathbb{R}^{m \times 1}$ and the vector variable $\mathbf{X} \in \mathbb{R}^{n \times 1}$ in the ordinary least square problem, we have:

$$\nabla_{\mathbf{x}} f(\mathbf{X}) = \nabla_{\mathbf{x}} (\mathbf{A}\mathbf{X} - \mathbf{B})^{\top} (\mathbf{A}\mathbf{X} - \mathbf{B})$$

$$\Leftrightarrow \nabla_{\mathbf{x}} f(\mathbf{X}) = \nabla_{\mathbf{x}} \left(\mathbf{X}^{\top} \mathbf{A}^{\top} \mathbf{A} \mathbf{X} - \mathbf{X}^{\top} \mathbf{A}^{\top} \mathbf{B} - \mathbf{B}^{\top} \mathbf{A} \mathbf{X} + \mathbf{B}^{\top} \mathbf{B} \right).$$

Algorithm 1. Offloading Algorithm for MUST

Input:. MDs information, CAPs information, network parameters
Output:. optimal offloading strategy $\mathbf{X}^*$, average cost E^*
Initial:. Random offloading strategy $\mathbf{X}$

1: **while** there is application in $\mathbf{N}'$ **do**
2:     Calculate current wireless rate r
3:     **for** task in $\mathbf{M}$ **do**
4:         Calculate execution $t_{m,n}^{exec}$ and queuing cost $t_{m,n}^{queu}$
5:         **if** normalized offloading strategy $\mathbf{x}_{m,n} ==$ prior **then**
6:             Communication cost $= 0$
7:         **else**
8:             Calculate communication cost $t_{m,n}^{tran}$
9:         Compute cost vector $\mathbf{C}_n$
10:     Construct optimization function in (9)
11:     Compute partial optimal strategy $\mathbf{X}_n^*$
12: **return** $\mathbf{X}^*$, E^*

Since $\mathbf{B}^\top\mathbf{B}$ is a real number, the optimization can be further converted to:

$$\nabla_{\mathbf{x}}f(\mathbf{X}) = \nabla_{\mathbf{x}}\left(\mathbf{X}^\top\mathbf{A}^\top\mathbf{A}\mathbf{X} - \mathbf{X}^\top\mathbf{A}^\top\mathbf{B} - \mathbf{B}^\top\mathbf{A}\mathbf{X} + \mathbf{B}^\top\mathbf{B}\right)$$

$$\Leftrightarrow \nabla_{\mathbf{x}}f(\mathbf{X}) = \nabla_{\mathbf{x}}tr\left(\mathbf{X}^\top\mathbf{A}^\top\mathbf{A}\mathbf{X}\right) - 2\nabla_{\mathbf{x}}\left(tr\left(\mathbf{B}^\top\mathbf{A}\mathbf{X}\right)\right)$$

$$\Leftrightarrow \nabla_{\mathbf{x}}f(\mathbf{X}) = \nabla_{\mathbf{x}}tr\left(\mathbf{X}^\top\mathbf{A}^\top\mathbf{A}\mathbf{X}\right) - 2\mathbf{A}^\top\mathbf{B},$$

where $tr\left(\cdot\right)$ represents the trace of the matrix. According to the matrix property, we can get $\nabla_{\mathbf{x}}tr\left(\mathbf{X}^\top\mathbf{A}^\top\mathbf{A}\mathbf{X}\right)\mathbf{A}\mathbf{A}^\top\mathbf{X} + \mathbf{A}^\top\mathbf{A}\mathbf{X} = 2\mathbf{A}^\top\mathbf{A}\mathbf{X}$. Then, we have

$$\nabla_{\mathbf{x}}f(\mathbf{X}) = \nabla_{\mathbf{x}}\left(\mathbf{X}^\top\mathbf{A}^\top\mathbf{A}\mathbf{X} - \mathbf{X}^\top\mathbf{A}^\top\mathbf{B} - \mathbf{B}^\top\mathbf{A}\mathbf{X} + \mathbf{B}^\top\mathbf{B}\right)$$

$$\Leftrightarrow \nabla_{\mathbf{x}}f(\mathbf{X}) = 2\mathbf{A}^\top\mathbf{A}\mathbf{X} - 2\mathbf{A}^\top\mathbf{B}.$$

Therefore, we get the optimal solution when $2\mathbf{A}^\top\mathbf{A}\mathbf{X} - 2\mathbf{A}^\top\mathbf{B} = 0$, *i.e.* , $\mathbf{X} = \left(\mathbf{A}^\top\mathbf{A}\right)^{-1}\mathbf{A}^\top\mathbf{B}$.

Offloading Tasks. We offload tasks according to the offloading strategy until all tasks are offloaded. The reMUST algorithm is formally described in Algorithm 1. It starts from a strategy profile in which all tasks are offloaded randomly. We first denote $\mathbf{N}'$ by the set of MDs that have never changed their strategy, noting that at the beginning $\mathbf{N}=\mathbf{N}'$. The reMUST algorithm for each task consists of two phases that are executed sequentially. In the first phase, it calculates execution, communication, and queuing cost by offloading tasks to potential CAPs. In the second strategy judgment phase, we calculate the solution of **P2**.

5   Performance Evaluation

In this section, we perform extensive simulation experiments to evaluae the performance of our algorithm. The simulation is carried out on a MacBook Pro with a 2 GHz Intel Core i5 quad-core processor.

5.1   Setup

The edge environment is a 250×250 m^2 area with 50 MDs and 9 CAPs embedded edge servers. Each MD has an application that consists of 5 sequential tasks. The size of the initial computing task follows the uniform distribution of $[600, 1200]$ KB. The number of CPU cycles required for completing the task follows the random distribution of $[500, 1000]$ Megacycles, and the computing capacity of CAP is selected from $\{15, 20, 30\}$ GHz [19]. According to the channel gain model in [19], the channel loss is set as $140. 7 + 36. 7\log_{10}(d)$, where d represents the distance between the MD and CAP. The positions of MD and CAP follow a random and uniform distribution, and CAPs interconnect with each other via wired links. For the communication model, the noise power is set as $\varpi = -100$ dBm, $p_n = 23$ dBm, $\beta = 1$ Gbps [16,20]. The wireless channel is a single carrier channel, and the channel bandwidth is $W = 36$ MHz. We compare our reMUST algorithm with the following benchmark schemes:

- *Random Offloading Scheme (ROS)*. It randomly assigns tasks to arbitrary CAPs for execution.
- *Greedy Offloading with Load Balance (GOLB)*. It always assigns tasks to the CAP with the lowest workload.

5.2   Results

In the first group of experiments, we evaluate the average application delay of all task offloading algorithms. Figure 4 (a) shows that all algorithms have a sustained increasing trend along with the growth of the number of MDs. Among them, reMUST always achieves the lowest average latency. We notice that reMUST's queuing time is longer than GOLB when the number of MDs comes to 100. It is because GOLB tends to offload tasks to CAPs with low workloads, which leads to high transmission delay, while reMUST takes both data transmission and task queuing into account. Although reMUST has a long queuing process, its transmission delay is significantly lower than other algorithms. Thus, reMUST achieves the lowest overall delay compared with other algorithms. Figure 4 (b) illustrates that the average delays of all algorithms decrease as CAPs increase. We find that the average delay falls slowly when the number of CAPs exceeds 12. Then the queuing time decreases significantly. In this case, when the number of CAP reaches 12, the system resources are sufficient for the current task scale and the resource competition weakens. Compared to other algorithms, the advantage of reMUST is mainly reflected in the reduction of data transmission delay.

In the second group of experiments, we evaluate the impact on the completion rate of applications, where completion rate refers to the proportion of completed applications within their deadlines. Figure 5 illustrates that all algorithms perform worse while the number of MDs is larger. Among them, only reMUST has a nearly 100% completion rate when the number of MDs is 60, while the completion rate of *ROS* and *GOLB* is about 50%. When there are

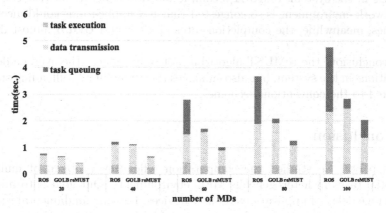

(a) Impact of the number of MDs on delay

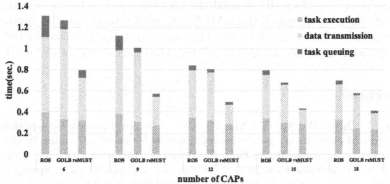

(b) Impact of the number of CAPs on delay

Fig. 4. Number of MDs & CAPs vs. average delay

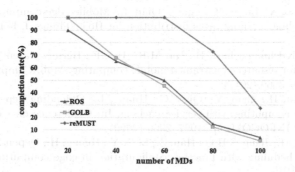

Fig. 5. Number of MDs vs. completion rate

100 MDs in the system, reMUST's completion rate is nearly 30%, it is because the network environment is so congested that few applications could meet their deadlines, meanwhile, the completion rates of *ROS* and *GOLB* almost drop to zero.

In conclusion, the reMUST algorithm not only reduces the average delay of applications in the system, but also enhances the scalability of offloading strategy compared to the conventional schemes.

6   Conclusion

In this paper, we studied the computation offloading problem of multi-user sequential tasks in heterogeneous MEC environments. Our goal is to minimize the average delay of applications on user devices. For this fundamental issue, we proposed the MUST framework to complete multi-user sequential tasks offloading. Specifically, we formally defined the problem and further proved that it is NP-hard. Then we proposed the reMUST algorithm to obtain the approximate optimal solution based on solid theoretical analysis. Finally, we conducted extensive simulation experiments to demonstrate the effectiveness of the reMUST.

References

1. Lei, Y., Zheng, W., Ma, Y., Xia, Y., Xia, Q.: A novel probabilistic-performance-aware and evolutionary game-theoretic approach to task offloading in the hybrid cloud-edge environment. In: Gao, H., Wang, X., Iqbal, M., Yin, Y., Yin, J., Gu, N. (eds.) CollaborateCom 2020. LNICST, vol. 349, pp. 255–270. Springer, Cham (2021). https://doi.org/10.1007/978-3-030-67537-0_16
2. Liu, F., Lv, B., Huang, J., Ali, S.: Towards mobility-aware dynamic service migration in mobile edge computing. In: Gao, H., Wang, X., Iqbal, M., Yin, Y., Yin, J., Gu, N. (eds.) CollaborateCom 2020. LNICST, vol. 349, pp. 115–131. Springer, Cham (2021). https://doi.org/10.1007/978-3-030-67537-0_8
3. Chen, N., Yang, Y., Zhang, T., Zhou, M.T., Luo, X., Zao, J.K.: Fog as a service technology. IEEE Commun. Mag. **56**(11), 95–101 (2018)
4. Yang, B., Cao, X., Li, X., Zhang, Q., Qian, L.: Mobile-edge-computing-based hierarchical machine learning tasks distribution for IIoT. Things J. IoT-J **7**(3), 2169–2180 (2020)
5. Kao, Y.H., Krishnamachari, B., Ra, M.R., Bai, F.: Hermes: latency optimal task assignment for resource-constrained mobile computing. Mob. Comput. INFOCOM **16**(11), 3056–3069 (2017)
6. Zhao, G., Xu, H., Zhao, Y., Qiao, C., Huang, L.: Offloading dependent tasks in mobile edge computing with service caching. In: Conference on Computer Communications INFOCOM, pp. 1997–2006 (2020)
7. Liu, L., Tan, H., Jiang, S.H.C., Han, Z., Li, X.Y., Huang, H.: Dependent task placement and scheduling with function configuration in edge computing. In: IWQoS, pp. 1–10 (2019)
8. Fan, Y., Zhai, L., Wang, H.: Cost-efficient dependent task offloading for multiusers. IEEE Access **7**, 115843–115856 (2019)

9. Jošilo, S., Dán, G.: Wireless and computing resource allocation for selfish computation offloading in edge computing. In: Conference on Computer Communications INFOCOM, pp. 2467–2475 (2019)
10. Chen, X., Jiao, L., Li, W., Fu, X.: Efficient multi-user computation offloading for mobile-edge cloud computing. Netw. TON **24**(5), 2795–2808 (2016)
11. Habak, K., Ammar, M., Harras, K.A., Zegura, E.: Femto clouds: leveraging mobile devices to provide cloud service at the edge. In: International Conference on Cloud Computing CLOUD, pp. 9–16 (2015)
12. Lin, X., Wang, Y., Xie, Q., Pedram, M.: Task Scheduling with Dynamic Voltage and Frequency Scaling for Energy Minimization in the Mobile Cloud Computing Environment. IEEE Trans. Services Comput. **8**(2), 175–186 (2015)
13. Sundar, S., Liang, B.: Communication Augmented Latest Possible Scheduling for cloud computing with delay constraint and task dependency. In: Conference on Computer Communications Workshops (INFOCOM WKSHPS) INFOCOM WKSHPS, pp. 1009–1014 (2016)
14. Sundar, S., Liang, B.: Offloading dependent tasks with communication delay and deadline constraint. In: INFOCOM, pp. 37–45 (2018)
15. Yang, L., Cao, J., Cheng, H., Ji, Y.: Multi-user computation partitioning for latency sensitive mobile cloud applications. TOC **64**(8), 2253–2266 (2015)
16. Chen, X.: Decentralized computation offloading game for mobile cloud computing. IEEE Trans. Parallel Distrib. Syst. **26**(4), 974–983 (2015)
17. Chan, W.C.: Pollaczek-Khinchin formula for the M/G/1 queue in discrete time with vacations. IEE Proc. Comput. Digital Tech. **144**(4), 222–226 (2002)
18. Author, R., Review, H., Source, H., Logic, S., Url, S.: Review of Computers and Intractability. A Guide to the Theory of NP-Completeness (1983)
19. Tran, T.X., Pompili, D.: Joint task offloading and resource allocation for multi-server mobile-edge computing networks. Technol. TVT **68**(1), 856–868 (2019)
20. Guo, H., Liu, J.: Collaborative computation offloading for multiaccess edge computing over fiber-wireless networks. Technol. TVT **67**(5), 4514–4526 (2018)

Model-Based Evaluation and Optimization of Dependability for Edge Computing Systems

Jingyu Liang, Bowen Ma, Sikandar Ali, and Jiwei Huang[✉]

Beijing Key Laboratory of Petroleum Data Mining, China University of Petroleum - Beijing, Beijing 102249, China
2020310701@student.cup.edu.cn, {sikandar,huangjw}@cup.edu.cn

Abstract. Edge computing moves part of the computing tasks to the edge of the network to improve service capabilities while reducing latency. It has been successfully applied in Internet of Things (IoT) and mobile computing systems. With the increasing popularity of edge computing, the ability of an edge computing system continuously providing services to users without interruptions and failures, which is also known as the dependability, has become an important issue. However, the evaluation and optimization of dependability attributes of an edge computing system still remains an largely unexplored problem. In this paper, we study this issue from a model-based viewpoint. We propose an atomic dependability model of a server and provide quantitative analyses of dependability attributes with Markov chain techniques. In order to facilitate the analyses of multiple attributes in large-scale environments, we adopt a state aggregation method for model simplification, and present its corresponding theoretical proof. Considering the edge-cloud collaboration, we put forward the dependability model of an edge computing system, and provide an evaluation approach using the state aggregation technique. Furthermore, taking task offloading as an example, we formulate the dependability optimization as a continuous-time Markov decision problem (CTMDP), and propose an efficient approach of solving the problem with reinforcement learning. Finally, we use a real-world dataset to conduct simulation experiments, and the experimental results validate the efficacy of our approach.

Keywords: Dependability · Edge computing · Continue-time Markov decision process · State aggregation

1 Introduction

Cloud computing, as a business model that provides services on demand through the network, has been widely applied in IT industry [1,2]. However, with the emergence of mobile smart devices and wireless communication technologies such as Internet of Things (IoT) and 5G, the requirements for low latency have raised

© ICST Institute for Computer Sciences, Social Informatics and Telecommunications Engineering 2021
Published by Springer Nature Switzerland AG 2021. All Rights Reserved
H. Gao and X. Wang (Eds.): CollaborateCom 2021, LNICST 406, pp. 728–747, 2021.
https://doi.org/10.1007/978-3-030-92635-9_42

dramatically, which brings significant challenges to cloud computing. Especially for date-intensive services, a extremely large amount of data needs to be uploaded from users to the cloud, which costs plenty of time for data transmission through the Internet. To attack this challenge, edge computing as a novel computing paradigm has been proposed. Servers are deployed at the edges of the networks very close to the terminal devices, and they are able to provide services directly to users for reducing latency by avoiding all service requests being uploaded to the cloud [3,4]. Also, with edge computing, the bandwidth pressure within the backbone network as well as energy consumption within the cloud can be reduced.

Meanwhile, the computational capability of an edge server is quite limited comparing with a cloud data center, and thus some computation-intensive tasks still need to be handled by the cloud. Instead of replacing the cloud, edge computing needs to cooperate with cloud computing to better fulfill user requests [5,6]. For example, in factories, edge servers can process data generated by equipment more quickly and detect abnormal situations in time. The edge nodes periodically upload the processed data to the cloud for storage and management. At the same time, the cloud is also responsible for monitoring data transmission. With the close cooperation between the edge and the cloud, the factories can be operated and managed in a very high efficiency. Within such a complex multi-layered architecture, how to design an advanced task offloading scheme has become an urgently required issue to be addressed.

As the IoT and mobile computing have been widely applied in various scenarios especially for some critical applications such as medical health-care, electric power management, military applications, etc. Besides high performance, dependability is also one of the most important requirements, which is the degree of stability to which a system can provide fault-free services without interruption [7]. In order to guarantee the dependability of a computing system, its evaluation and optimization is a critical research problem. Although there have been several works dedicating to study the dependability issue in traditional HPC systems and cloud computing by analyzing system log data or constructing mathematical models [8], the dependability research of systems with close collaboration between edge computing and cloud computing has still not been explored to a large extent.

In this paper, we make an attempt at filling this gap by studying the dependability evaluation and optimization for edge computing systems from a mathematical modeling aspect. By analyzing large-scale system failure data and surveying existing work, we propose a state transition model of edge computing systems. With detailed mathematical analyses, the dependability attributes can be quantitatively evaluated. A state aggregation approach is designed, with which a comprehensive system model of an edge-cloud collaboration system can be established and theoretically analyzed with very high efficiency. Furthermore, we formulate the dependability optimization for task offloading in edge computing as a continuous-time Markov decision process (CTMDP), and introduce reinforcement learning techniques to solve the problem efficiently. Finally, we conduct simulation experiments based on failure data collected from real-life computing systems, and validate the effectiveness of our approach.

The remainder of this paper is organized as follows. In Sect. 2, we survey the related work most pertinent to this paper. In Sect. 3, the dependability state transition model of the edge and the cloud is established, and dependability attributes are analyzed and quantified. In Sect. 4, a state aggregation technique is put forward, with which the dependability model of an edge-cloud collaboration system is proposed. In Sect. 5, the dependability optimization problem is formulated by a continuous-time Markov decision process. To solve such problem with high efficiency in large-scale systems, the time difference method and reinforcement learning are introduced and algorithms are presented. In Sect. 6, we conduct simulation experiments based on real-life dataset, and validate our approach. Finally, Sect. 7 concludes this paper.

2   Related Work

Research on dependability has always been of profound significance. Pan et al. provided a general definition of cloud computing system dependability, and discussed some methods for establishing cloud computing reliability models [9]. Guan *et al.* proposed a cloud dependability analysis (CDA) framework and designed a failure metric directed a cyclic graph (DAG) to analyze the correlation between various performance indicators and failure events in virtualized and non-virtualized systems [10]. In [11], authors research and set goals to connect different kinds of servers with private clouds and designate the servers as hosts to establish a multi-cloud dependability system. Qiu *et al.* considered parallel redundant big data tasks, and proposed a joint modeling method to consider the correlation between reliability-performance and reliability-energy consumption during failures, and proposed a genetic algorithm to find the optimal solution [12]. In [13], firstly according to the priority of the service requested by the user, the use of redundant resources is analyzed from the perspective of dependability index and service cost. Then the reliability of the proposal received from the cloud provider was evaluated based on the negotiation strategy suggested for the requester. Bai *et al.* proposed a method for constructing a complex network model that regards cloud services as nodes and finds the relationships between services to construct a complex network model. Through relationships key services can be found in a large number of services, thereby improving cloud services reliability [14]. Zhang *et al.* proposed an optimization model based on queuing theory. Using this optimization model, energy can be minimized under the constraints of average response time, response time reliability, stability of the queuing system, and resource maximization. Optimal resource allocation for cloud computing strategies [15]. Luo *et al.* proposed a resource scheduling algorithm that can optimize reliability, system performance, and energy consumption based on the Semi-Markov model modeling method. Furthermore, Laplace-Stieltjes transform and Bayesian method are used to analyze the correlation between reliability and performance and reliability and Energy [16]. In [17], the author proposes a mechanism that can continuously update the reliability of cloud resources and provide users with reliable resource scheduling in the cloud computing environment. In [18], the author provides a newest solution to automate the negotiation process in the cloud environment.

With the rapid growth of user service requests, many people have begun to research edge computing. For instance Song et al. [19] proposed regular assignment of tasks in the edge computing in order to increase the number of tasks that can be processed in the edge computing network and at the same time to ensure efficient task management. Some main technologies that can support the edge paradigm are introduced in [20]. Further, authors discussed some existing problems and proposed relevant solutions along with the advantages and precautions of each solutions in different scenarios [20]. An edge computing architecture to overcome the network bottleneck caused by the massive increase in the internet of Things (IoT) devices and data with the development of cloud computing are proposed in [21]. Their architecture is based on the based on the $\lambda - CoAP$ architecture, covering edge computing and cloud infrastructure. Regarding the placement of edge servers in an edge computing environment, Chen et al. [22], proposed an effective method of placing edge servers to reduce access latency, optimize load balance, and prevent the big data generated at the edge from making the cloud computing center unable to process effectively. In [23], a risk-based mobile edge computing system resource configuration optimization method is proposed. The method is capable to tackle the risk of edge server overload while performing resource allocation, and further improve system efficiency and service quality. Xiao et al. proposed a predictive mapping optimization heuristic method to place edge servers to meet the resource demand forecasting problem, divide tasks into subtasks, map the positions of servers and subtasks, and complete the information interaction between server and data, and proposed a cross-regional resource optimization model to reduce costs [24].

Although numerous published articles address [13,19–22], various Edge/Cloud computing subjects such as design and architectures [19], optimal resource allocation strategies [13], resource configuration optimization [21], resource demand forecasting [22], and task offloading [23]. However, the dependability of edge computing and cloud collaboration is not considered in none of the above work, which is one of the important factors to ensure the normal service provided by the system.

3 Dependability Modeling and Analysis of Edge/Cloud Server

This section builds a state model based on the service computing system, and introduces dependability and its attributes. In order to analyze and quantify these attributes, consider discrete-time Markov Chain (DTMC) for embedding.

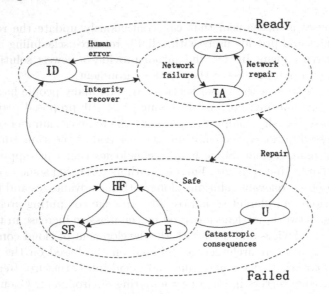

Fig. 1. Service computing system dependability mode

3.1 System State Transition Model

During the operation of the system, there might be a series of problems, such as failure, recovery and so on. Since this series of events will affect the state of the whole service system and have a decisive impact on the processing capacity of the whole system. It is important to discuss the state transition of the system according to the failure in the service computing system for the evaluation of dependability [25]. In this paper, in addition to the analysis of the types of failures, the correlation between failures within the system is also considered, that is, some failures of the system will affect the system, and lead to new failures before the recovery of the failure, which is called the correlation of failure.

According to a large-scale study of failures in high-performance computing systems [8], the fault is roughly divided into Human error; Environment; Network failure; Software failure; Hardware failure. In our previous work, we studied the state transition transition of the service computing system [26]. On this basis, we considered the correlation between faults, and considered the correlation between different faults to help us further analyze the dependability, the system dependability model is shown in Fig. 1. The status is hierarchical, divided into two parts as *Ready* or *Failed*. In the ready state to further divide the system, whether the system has a state can be divided into the Integrity to ensure Integrity-Maintained and Integrity-Destroyed (ID) two kinds of state. Integrity under the guarantee of Accessible (A) on behalf of all the components in the system are normal state, but in the process of system processes the request, may be due to failure of the network system can't receive the user request, unable to complete the user request, this state is defined as Inaccessible (IA). Integrity-Destroyed state means that when the system is processing a service request, the system may be modified by inappropriate modules or data due to human error, but this change has not caused system failure or service failure.

In this paper, referring to the failed state of large-scale service systems, hardware failures and software failures account for a large proportion. Therefore, hardware failures and software failures are the biggest causes of failures, but environmental failures are also very important. We regard these types of failures as repairable failures and classify them as Safe. Unsafe (U) means that the failure may lead to catastrophic consequences. In different situations, users or service providers are required to define the level of catastrophic consequences.

The CTMC has a transition probability matrix P_{ij}, and uses the transfer rate of the CTMC to calculate its transfer probability, as

$$P_{ij} = \frac{q_{i,j}}{\sum_{k \neq i} q_{i,k}} \cdot (1 - e^{-\sum_{k \neq i} q_{i,k}\xi}), i \neq j \tag{1}$$

$$P_{ij} = e^{-\sum_{k \neq i} q_{i,k}\xi}, i = j \tag{2}$$

By solving a system of linear equations, steady-state probability $v = [v_{ID}, v_A, v_{IA}, v_{HF}, v_{SF}, v_E, v_U]$ embedded a discrete-time Markov chain (DTMC) in each state.

$$\begin{cases} v \cdot P = \pi \\ \sum_i v_i = 1 \end{cases} \tag{3}$$

Assuming that the average waiting time of each state in the CTMC model is t_i and the steady-state probability of DTMC is π_i, the steady-state probability of any state $i \in \{ID, IA, A, HF, SF, E, U\}$ in the CTMC model can be obtained as V:

$$\pi_i = \frac{v_i t_i}{\sum_j v_j t_j}, \forall i. \tag{4}$$

As it has been assumed that each epoch has time period of equal length $t_i = \xi$ for any i, it is obvious that

$$\pi_i = \frac{v_i t_i}{\sum_i v_j t_j} = \frac{v_i}{\sum_i v_j} = v_i. \tag{5}$$

Therefore, the embedded DTMC steady-state probability and the original CTMC steady-state probability are equal. The expected DTMC reward in steady-state is expressed ,which is proved equivalent to the CTMC reward function shown in

$$\bar{R} = \sum_{i \subseteq s} = v_i \cdot r_i = \pi_i \cdot r_i \tag{6}$$

3.2   Analysis of Dependability Attributes

This paper introduces a widely recognized dependability, which is an aggregate concept that contains five attributes. Based on the definition of dependability attribute and the proposed state model of service computation, this paper makes a mathematical analysis of some attributes of dependability attribute, so as to evaluate the service system more directly [27].

From the perspective of system design and optimization, generally speaking, steady-state dependability analysis is more meaningful than transient analysis (that is, to analyze the dependability attributes of the system at a certain time t). In view of aforementioned definition of dependability attributes, based on the steady-state probability of states, quantitative analysis of different attributes can be carried out.

Availability represents the ability of the system to provide a working service to the user when the user needs it, and is expressed as:

$$A = Pr(Accessible(t)) = \pi_A \tag{7}$$

Integrity reflects the ability of the service computing system to resist inappropriate systems, use the following expression to analyze the attribute.

$$I = \pi_A + \pi_{IA} \tag{8}$$

Reliability is the probability that services can work continuously at a given time. In steady-state, reliability is the steady-state probability of the ready state, and is expressed as:

$$R = \pi_A + \pi_{IA} + \pi_{ID} \tag{9}$$

Safety provides services without catastrophic consequences, it can be expressed as follows:

$$S = \pi_{ID} + \pi_A + \pi_{IA} + \pi_{SF} + \pi_{HF} + \pi_E \tag{10}$$

Maintainability describes the ability to repair and modify the system. This article considers that all faults can be repaired, so the expression of Safety and Maintainability in this paper are the same.

4   Dependability Modeling and Analysis of Edge Computing System

To simplify the complex system model, in this section, a state aggregation method is proposed, and further prove that the models before and after aggregation are equivalent. Additionally, on the basis of this an edge-cloud collaboration model is proposed.

4.1   State Aggregation Technique

In the service computing system, due to the large number of servers and complex distribution, it is difficult to directly obtain a complete model, the solution is to aggregate complex models. The purpose of state aggregation is to reduce the state space and aggregate it into a two-state Markov reward model. The reward values of Ready (R) state and Failed (F) state are set to 1 and 0 respectively. The aggregation model is shown in Fig. 2, here the red transition (RT) is define as a state transition from the ready state to the failed state, while the transition from the failed state to the ready state as green transition (GT) [28].

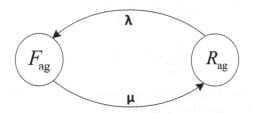

Fig. 2. The aggregated model

The transition rate from up to down state is set as λ and is denoted as:

$$\lambda = \sum_{t_{i,j} \subseteq RT} Pr\{s_i \mid R\} \cdot q_{i,j} = \frac{\sum\limits_{t_{i,j} \subseteq RT} \pi_i q_{i,j}}{\pi_{RT}} \tag{11}$$

$$\pi_{RT} = \sum_{s_k \subseteq R} \pi_k \tag{12}$$

s_i is the state i, π_k is the CTMC steady-state probability in state k, $t_{i,j}$ shows the transition from state i to state j. μ is similar:

$$\mu = \sum_{t_{i,j} \subseteq GT} Pr\{s_i \mid F\} \cdot q_{i,j} = \frac{\sum\limits_{t_{i,j} \subseteq GT} \pi_i q_{i,j}}{\pi_F} \tag{13}$$

$$\pi_F = \sum_{s_k \subseteq F} \pi_k \tag{14}$$

The aggregation model expressed in $\pi = [\pi_{R_{ag}}, \pi_{F_{ag}}]$ steady-state probability [29], and the steady-state probabilities are:

$$\pi_{R_{ag}} = \frac{\mu}{\lambda + \mu} \tag{15}$$

$$\pi_{F_{ag}} = \frac{\lambda}{\lambda + \mu} \tag{16}$$

The steady-state probability of the aggregate model is equivalent to that of the original model.

Proof. Step 1: Consider the Kolmogorov differential equation in steady-state:

$$\pi \times Q = 0 \tag{17}$$

π is the steady-state probability vector, Q is the transfer rate matrix, we have:

$$\pi_i \cdot q_i = \sum_{j \neq i} \pi_j \cdot q_{j,i} \tag{18}$$

$$\pi_i (\sum_{k \neq i} q_{i,k}) = \sum_{j \neq i} \pi_j \cdot q_{j,i} \tag{19}$$

Step 2: Without loss of generality, for analysis and proof, the up-states are numbered from 1 to h and the set of down-states from $h+1$ to n;

$$\sum_{i=1}^{h} \pi_i (\sum_{k \neq i} q_{i,k}) = \sum_{h+1}^{n} \sum_{j \neq i} \pi_j \cdot q_{j,i} \tag{20}$$

$$\sum_{i=1}^{h} \pi_i \cdot (\sum_{k=1, k \neq i}^{h} q_{i,k} + \sum_{k=h+1}^{n} q_{i,k})$$
$$= \sum_{i=1}^{h} (\sum_{j=1, j \neq i}^{h} \pi_j q_{j,i} + \sum_{j=h+1}^{n} \pi_j q_{j,i}) \tag{21}$$

then:

$$\sum_{i=1}^{h} \pi_i \cdot \sum_{k=1, k \neq i}^{h} q_{i,k} + \sum_{i=1}^{h} \pi_i \cdot \sum_{k=h+1}^{n} q_{i,k}$$
$$= \sum_{i=1}^{h} \sum_{j=1, j \neq i}^{h} \pi_j \cdot q_{i,j} + \sum_{i=1}^{h} \sum_{j=h+1}^{n} \pi_j \cdot q_{j,i} \tag{22}$$

To make the first item on the left and the first item on the right equal, the order of summation can be changed, thus

$$\sum_{i=1}^{h} \sum_{j=h+1}^{n} \pi_j \cdot q_{j,i} = \sum_{i=1}^{h} \pi_i \cdot \sum_{k=h+1}^{n} q_{i,k} \tag{23}$$

then:

$$\sum_{t_{i,j} \subseteq RT} \pi_i \cdot q_{i,j} = \sum_{t_{i,j} \subseteq GT} \pi_i \cdot q_{i,j} \tag{24}$$

Therefore, we have

$$\frac{\lambda}{\mu} = \frac{\frac{1}{\pi_R}(\sum_{t_{i,j} \subseteq RT} \pi_i \cdot q_{i,j})}{\frac{1}{\pi_F}(\sum_{t_{i,j} \subseteq GT} \pi_i \cdot q_{i,j})} = \frac{1/\pi_R}{1/\pi_F} = \frac{\pi_F}{\pi_R} \tag{25}$$

$$\lambda = \frac{\pi_F}{\pi_R} \cdot \mu \qquad (26)$$

From (11)–(13), and $\pi_R + \pi_F = 1$, it is shown

$$\pi_{R_{ag}} = \frac{\mu}{\frac{\pi_F}{\pi_R} \cdot \mu + \mu} = \frac{1}{\frac{\pi_F}{\pi_R} + 1} = \frac{\pi_R}{\pi_R + \pi_F} = \pi_R \qquad (27)$$

The proof for showing $\pi_{F_{ag}} = \pi_F$ is similar.

4.2 Dependability Model of Edge Computing Systems

Edge computing applications use the computational, sensor, and networking resources of nearby mobile and stationary computing devices. Therefore, in contrast to cloud computing, edge computing processes data locally near to its origin [30], there by reducing the network transmission load and communication latency. Edge computing and cloud computing cooperate, and the proposed model is shown in Fig. 3. This article assumes that as long as either the edge or

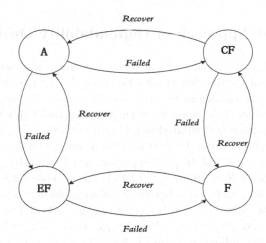

Fig. 3. Dependability model of edge-cloud system

the cloud is in a normal state, the system is reliable at this time. If both the edge and the cloud fail, the system fails. According to the hypothesis, we can draw four possible states of the service system: The state A means that both the edge server and the cloud server can process user requests normally; State EF means that the cloud server can process user requests normally, and the edge server is faulty, and the system is reliable; State CF means that the cloud server is faulty, but the edge server is operating normally and the system is reliable; The state F represents that both the edge server and the cloud server are faulty, and the system fails at this time.

In order to evaluate the dependability attributes, a state diagram based on a continuous-time Markov chain (CTMC) is established. The CTMC transition rate matrix is represented by Q, as shown below:

$$Q = \begin{bmatrix} q_{A,A} & q_{CF} & q_{EF} & 0 \\ q_A & q_{CF,CF} & 0 & q_F \\ q_A & 0 & q_{EF,EF} & q_F \\ 0 & q_{CF} & q_{EF} & q_{F,F} \end{bmatrix},$$

where,

$$q_{i,i} = -\sum_{j \neq i} q_{i,j}, i \subseteq A, CF, EF, F \tag{28}$$

The steady-state probability in each state, saying $\pi = [\pi_A, \pi_{CF}, \pi_{EF}, \pi_F]$, can be obtained using the following expressions:

$$\begin{cases} \pi \cdot Q = 0 \\ \sum_i \pi_i = 1 \end{cases} \tag{29}$$

5   Model and Approach of Dependability Optimization

In the design and management of the service computing system, reliability is one of the important factors that cannot be ignored. In a real service computing system, the system is continuous in service, rather than discrete. When tasks are offloaded to the server, it is guaranteed that the system can effectively process user requests. This issue is simulated as a Continuous-time Markov decision process (CTMDP) problem through the state transition model [31]. A five tuple can be used to describe the CTMDP problem typically: $\{S, S_0, A(i), q^\mu(j \mid i), r^\mu(i)\}$, where the state space S is a finite set of fully observable states of the system. S_0 is the initial state, and $A(i)$ a family of measurable subsets of actions applicable in $i \subseteq S$. The policy μ is the set of all actions selected in every state, i.e., $\mu = [a_1, a_2...a_{|S|}]$ and $\mu(i) = a_i \subseteq A(i)$, $q^\mu(j|i)$ is an infinitesimal generator of transition probability following a specific formula, the formula is as follows:

$$P^\mu(t) = e^{tq^\mu(j|i)} \tag{30}$$

In order to maximize the expected average reward function $\bar{J}^\mu$ under the policy μ, it is necessary to find the optimal policy $\mu = \mu^*$ and define it as the sum of future rewards within an infinite horizon.

$$\bar{J}^\mu(i) = \lim_{T \to \infty} \frac{1}{T} E\{\int_0^T r^\mu(i)(X_t)dt|X_0 = i\} \tag{31}$$

The $\bar{J}^\mu$ converges to a value denoted as η^μ for any initial state i is

$$\eta^\mu = \bar{J}^\mu(i) = \pi^\mu r \tag{32}$$

Algorithm 1 :The Policy Iteration Algorithm

1: Let the initial policy to μ_0, and $k = 0$.
2: Using (32) and (35), get the performance potential function g^{μ_k}.
3: Improve the policy by:

$$\mu_{k+1} = argmax\{r^{\mu} + Q^{\mu}g^{\mu_k}\}$$

4: if $\mu_{k+1} = \mu_k$ then
5:     Set $\mu^* = \mu_{k+1}$.
6:     Stop.
7: else
8:     Set $k = k + 1$.
9:     Go to Step 2.
10: end if.

---

where π^{μ} is the steady state probability, and η^{μ} is shown as

$$\eta_k = \frac{1}{T_k} \int_0^{T_k} r(X_t)dt \tag{33}$$

The relationship between steady-state probability and the transfer rate matrix is shown in (29). Markov processes can be divided into embedded Markov chains and the sojourn time of each state, the transition probabilities of the embedded Markov chain P^{μ} satisfies the equation

$$Q_{\mu} = \Lambda^{\mu}(P^{\mu} - I) \tag{34}$$

where $\Lambda^{\mu} = diag[\lambda_1^{\mu}, \lambda_2^{\mu}, ...]$, $\lambda_i^{\mu} > 0$ represents the transfer rate of state i. The length of time represented by the state i can satisfy the exponential probability distribution. The CTMDP model will support Poisson equation:

$$Q^{\mu}g^{\mu} = -r^{\mu} + \eta^{\mu}e \tag{35}$$

where $g^{\mu}(i)$ is expressed as the performance potential function of state i. The performance potential function $g^{\mu}(i)$ is the difference between the instantaneous return and the long-term average return, the expectation in the infinite time domain. For different state paths, the long-term average return of the system is only related to the initial state. If a state is selected as the initial state and its performance potential is defined as $g(i) = 0$, then the long-term of other states and states i is the difference between average returns in performance potential. In addition, for the CTMDP model, the performance potential function can be expressed by the following formula:

$$g^{\mu}(i) = \lim_{T \to \infty} \frac{1}{T} E\{ \int_0^T \{r^{\mu}(X_t) - \eta^{\mu}\}dt | X_0 = i\} \tag{36}$$

In this paper, we only consider the case of policy iteration algorithms. The policy algorithm used to solve the CTMDP problem will be introduced in this section. Firstly, the reward difference formula between the two strategies, namely μ_1 and μ_2, we have

$$\eta^{\mu_1} - \eta^{\mu_2}$$
$$= \pi^{\mu_1} r^{\mu_1} - \pi^{\mu_2} e \eta^{\mu_2} \tag{37}$$
$$= \pi^{\mu_1} [(r^{\mu_1} + Q^{\mu_1} g^{\mu_2}) - (r^{\mu_2} + Q^{\mu_2} g^{\mu_2})]$$

Definition 1. *If there is at least one vector with the same dimension, which can satisfy $u(i) \preceq v(i)$, and the rest are equal, it can be expressed as $u \preceq v$. Similarly, $u \succeq v$ can be defined in the same way as above.*

Theorem 1. *μ_1 and μ_2 are two policies of the CTMDP problem, if $r^{\mu_1} + Q^{\mu_1} g^{\mu_2} \succeq r^{\mu_2} + Q^{\mu_2} g^{\mu_2}$, then $\eta^{\mu_1} \succeq \eta^{\mu_2}$; Moreover, if $r^{\mu_1} + Q^{\mu_1} g^{\mu_2} \preceq r^{\mu_2} + Q^{\mu_2} g^{\mu_2}$, then $\eta^{\mu_1} \preceq \eta^{\mu_2}$.*

Proof. The theorem can be easily obtained based on Definition 1, (37) and the fact that the steady-state probability of an ergodic Markov process is greater than 0.

Theorem 2. *Consider a CTMDP problem. For any $\mu \subseteq D$, the strategy μ^* can be called the optimal strategy.*

$$r^{\mu^*} + Q^{\mu^*} g^{\mu^*} \succeq r^{\mu} + Q^{\mu} g^{\mu} \tag{38}$$

Proof. We first need to prove the necessary conditions. Assume that μ^* is the optimal strategy. We need to prove the validity of (38). Suppose there is a policy μ^ξ that makes (38) invalid, that is, there is at least one state i that can satisfy the following

$$r^{\mu^*}(i) + \sum_{j=1}^{S} q^{\mu^*}(j|i) g^{\mu^*}(j) < r^{\mu^\xi}(i) + \sum_{j=1}^{S} q^{\mu^\xi}(j|i) g^{\mu^*}(j)$$

Construct a new policy $\hat{\mu}$ in this way $\hat{\mu}(j) = \mu^*(j)$ while $j \neq i$; $\hat{\mu}(j) = \mu^*(j)$ while $j = i$. Thus, we can get $r^{\mu^*} + Q^{\mu^*} g^{\mu^*} \succeq r^{\mu} + Q^{\mu} g^{\mu}$. By applying Theorem 2, we see that $\eta^{\hat{\mu}} > \eta^{\mu^*}$ which contradicts to the previous assumption.

Next, we will give sufficient conditions to carry out a proof. Assume that (38) holds. Based on the application of Theorem 2, we can find that for any D, $\eta^{\mu^*} > \eta^{\mu}$, therefore, μ^* is the optimal strategy.

The previous description proves that as the theoretical basis, the strategy iteration algorithm is as Algorithm 1. In practical problems, due to the uncertainty or large scale of the state space, it is almost impossible to solve the Poisson equation. The structure of the evaluation strategy is evaluated through multiple sample paths and return values using the temporal differences (TD) method, which is a very effective method [32]. This section will introduce the TD algorithm used to evaluate the performance potential function.

Let $\{T_0, T_1, ...\}$ be a transition time sequence in CTMDP, when $t \subseteq [T_k, T_{k+1}]$, $x_t = X_k, k = \{0, 1, 2, 3...\}$. The sojourn time of state X_k is expressed

---

Algorithm 2 :g^μ evaluation algorithm

1: Let $k = 0$, $\eta^\mu = 0$, $g^\mu(i) = 0$, $I_i = 0$, where $I(i)$ represents the occurrence time of state i, $i \subseteq S$
2: Select a small constant $\xi > 0$
3: $k \leftarrow k + 1$
4: $I_{X_k} \leftarrow I_{X_k+1}$.
5: $\eta_\mu \leftarrow \eta_\mu + \frac{T_{k+1}-T_k}{T_{k+1}}[r^\mu(X_k) - \eta^\mu]$
6: $g^\mu \leftarrow g^\mu + s_k\{[r^\mu(X_t) - \eta^\mu]\tau_k\} + g^\mu(X_{k+1} - g^\mu(X_k)\})$
7: if $|g_{k+1} - g_k|_{max} < \xi$, then
8:     output: g^μ
9:     STOP.
10: else
11:     Go to Step 2.
12: end if

---

as $\tau(X_k) = T_{k+1} - T_k$, use the (37) to get the approximate solution of g^μ, where s_k indicates the number of iterations:

$$g^\mu \approx g^\mu + s_k\{[r^\mu(X_t) - \eta^\mu]\tau_k\} + g^\mu(X_{k+1} - g^\mu(X_k)\}) \tag{39}$$

Using (32), we get

$$\eta^\mu_{k+1} = \frac{1}{T_{k+1}} \int_0^{T_{k+1}} r^\mu(X_t)dt$$

$$= \frac{1}{T_{k+1}}[\int_0^{T_k} r^\mu(X_t)dt + r^\mu(X_t)(T_{k+1} - T_k)]$$

$$= \frac{T_k}{T_{k+1}}\eta^\mu_k + \frac{T_{k+1} - T_k}{T_{k+1}}r^\mu(X_k) \tag{40}$$

$$= \eta^\mu_k + \frac{T_{k+1} - T_k}{T_{k+1}}[r^\mu(X_k) - \eta^\mu_k]$$

Then, the average average reward value η_μ can be approximately expressed as

$$\eta_\mu \approx \eta_\mu + \frac{T_{k+1} - T_k}{T_{k+1}}[r^\mu(X_k) - \eta^\mu] \tag{41}$$

Use the (39) and (41) to get the Algorithm 2 for evaluating the performance potential function.

6   Empirical Evaluation

6.1   Data Set and Experimental Settings

In this article, a large high-performance service system is used, which is provided by $LANL$ [8]. This data set contains the failure and repair times of more than 20

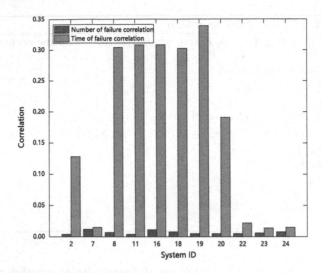

Fig. 4. The correlation analysis of failures in the system

different systems collected from 1996 to 2005. When the system fault is detected, the operator will establish a fault record, and the system administrator will notify the operator to fill in the end time of the fault after the successful repair. Since the number of nodes and processors in some systems differs by hundreds or thousands of times, they share the same characteristics with the ability of different edge computing and cloud computing to provide services in reality. Therefore, this data set is used to confirm our method.

When analyzing computer system failures, it is necessary to consider the correlation between the failures, and use the large-scale service computing system the $LANL$ data set for analysis. Through the analysis, it can be concluded that some systems have failure correlations, as shown in Fig. 4. Although there is not much difference in the proportion of the number of fault correlation to the total number of faults, there is a big difference in the time of fault correlation. Therefore, the consideration of fault correlation cannot be ignored. There are six types of failures in the data set, which are software, human error, network, facilities, hardware and undetermined, facility failures are caused by the environments. By analyzing the correlation between failures, the dependability of the system can be further accurately evaluated. Through the analysis of typical failures in the data set, as shown in the Fig. 5, the human failures destroy the integrity of the system, and the network failures cause the system to change from accessible state to inaccessible state. Under certain circumstances, this part of the failures can be accepted to avoid system failure. In addition, the analysis considers software, facilities, hardware, and make the system enter to the failed, and they are considered to be repairable in this paper. As users and suppliers define the level of catastrophe, in the case, we believe the undetermined failures are regarded as catastrophic consequences.

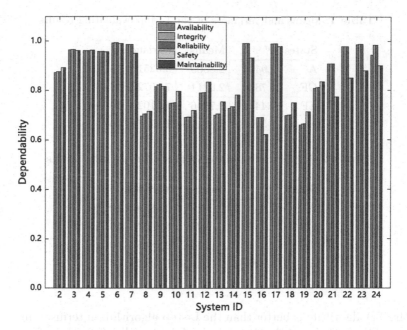

Fig. 5. Dependability attributes in the system

6.2 Experimental Results

The theorem is used to prove the correctness of the Algorithm 1, since in the real system the state space is usually large, therefore, it is almost impossible to solve the Poisson equation, Algorithm 2 is used to evaluate the performance potential function. This article uses the proposed edge and cloud collaboration model to confirm Algorithm 2. The return value for the fault state is set to 0, and the return value for the normal state is set to 1, the stay time of the system satisfies the exponential distribution, we have repeated the simulation 10 times, and each state transition is 10,000 times as the path, and the average is taken. The dynamic process of the performance potential function is shown in Fig. 6.

We first regularize the obtained approximate g^μ value and then calculate the mean and variance of the deviation from the theoretical value, the outcome are presented in Table 1. In order to further evaluate the performance of Algorithm 2, the L-step algorithm of CTMDP performance potentials used for comparison with algorithm. Since the L-step algorithm cannot obtain a graph of the change in performance potential with the number of transitions, the two are compared. The result obtained by the algorithm and the mean and variance of the theoretical deviation are used to evaluate the performance of our algorithm. The results of the L-step algorithm are shown in Table 2. The experimental results show

Table 1. Means and variances of the L-step evaluating results

States	g^μ	Means	Variances
A	2.9407	2.8509	9.3251×10^{-3}
CF	2.7870	2.6211	7.1072×10^{-3}
EF	2.4443	1.7946	1.0102×10^{-2}
F	−3.8787	−3.6148	1.2947×10^{-2}

Table 2. Means and variances of the TD evaluating results

States	g^μ	Means	Variances
A	2.9407	0.0718	6.6579×10^{-3}
CF	2.7870	0.0675	6.4457×10^{-3}
EF	2.4443	0.0676	9.1805×10^{-3}
F	−3.8787	−0.0959	1.2130×10^{-2}

that the TD algorithm is better than the L-step algorithm in terms of mean and variance. The deviation of the theoretical value is smaller than that of the L-step method.

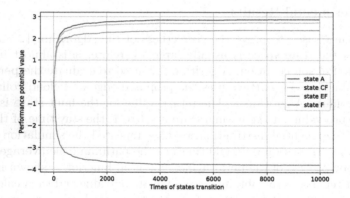

Fig. 6. Simulation state process

7   Conclusion

With the increasing demand on low latency and high privacy, computing models that combine edge computing and traditional cloud computing have been widely used, and therefore, dependability has become one of the key challenges. In this paper, we establish a dependability model, which takes into account the failure correlation, embeds the discrete-time Markov chain, and quantitatively analyzes the attributes of dependability. For a cloud-edge computing system, we propose a state aggregation approach, and prove the equivalence of the models before and after aggregation. Moreover, we formulate the dependability optimization problem as a continuous-time Markov decision process, and present effective algorithms to solve the problem. Finally, we evaluate our approach by simulation experiments based on real-world dataset. In future, we will continue to analyze the large-scale open-source data, propose new models capturing the dynamics of edge computing systems, and design new methods with higher efficiency.

Acknowledgment. This work is supported by Beijing Nova Program (No. Z201100006820082), National Natural Science Foundation of China (No. 61972414), National Key Research and Development Plan (No. 2016YFC0303700), Beijing Natural Science Foundation (No. 4202066), and the Fundamental Research Funds for Central Universities (Nos. 2462018YJRC040 and 2462020YJRC001).

References

1. Ozcan, M.O., Odaci, F., Ari, I.: Remote debugging for containerized applications in edge computing environments. In: 2019 IEEE International Conference on Edge Computing (EDGE), pp. 30–32 (2019). https://doi.org/10.1109/EDGE.2019.00021
2. Amanatullah, Y., Lim, C., Ipung, H.P., Juliandri, A.: Toward cloud computing reference architecture: cloud service management perspective. In: International Conference on ICT for Smart Society, pp. 1–4 (2013). https://doi.org/10.1109/ICTSS.2013.6588059
3. Wei, X., et al.: MVR: an architecture for computation offloading in mobile edge computing. In: 2017 IEEE International Conference on Edge Computing (EDGE), pp. 232–235 (2017). https://doi.org/10.1109/IEEE.EDGE.2017.42
4. Xu, J., Palanisamy, B., Ludwig, H., Wang, Q.: Zenith: utility-aware resource allocation for edge computing. In: 2017 IEEE International Conference on Edge Computing (EDGE), pp. 47–54 (2017). https://doi.org/10.1109/IEEE.EDGE.2017.15
5. Loghin, D., Ramapantulu, L., Teo, Y.M.: Towards analyzing the performance of hybrid edge-cloud processing. In: 2019 IEEE International Conference on Edge Computing (EDGE), pp. 87–94 (2019). https://doi.org/10.1109/EDGE.2019.00029
6. Jain, R., Tata, S.: Cloud to edge: distributed deployment of process-aware IoT applications. In: 2017 IEEE International Conference on Edge Computing (EDGE), pp. 182–189 (2017). https://doi.org/10.1109/IEEE.EDGE.2017.32
7. Esteves-Verissimo, P., Völp, M., Decouchant, J., Rahli, V., Rocha, F.: Meeting the challenges of critical and extreme dependability and security. In: 2017 IEEE 22nd Pacific Rim International Symposium on Dependable Computing (PRDC), pp. 92–97 (2017). https://doi.org/10.1109/PRDC.2017.21
8. Schroeder, B., Gibson, G.A.: A large-scale study of failures in high-performance computing systems. IEEE Trans. Dependable Secure Comput. **7**(4), 337–350 (2010). https://doi.org/10.1109/TDSC.2009.4
9. Pan, Y., Hu, N.: Research on dependability of cloud computing systems. In: 2014 10th International Conference on Reliability, Maintainability and Safety (ICRMS), pp. 435–439 (2014). https://doi.org/10.1109/ICRMS.2014.7107234

10. Guan, Q., Chiu, C., Fu, S.: CDA: a cloud dependability analysis framework for characterizing system dependability in cloud computing infrastructures. In: 2012 IEEE 18th Pacific Rim International Symposium on Dependable Computing, pp. 11–20 (2012). https://doi.org/10.1109/PRDC.2012.10

11. Walunj, S.G., Nagrare, T.H.: Dependability issues on cloud environment and analyzing server responsibilities. In: 2018 2nd International Conference on Inventive Systems and Control (ICISC), pp. 926–928 (2018). https://doi.org/10.1109/ICISC.2018.8398936

12. Qiu, X., Luo, L., Dai, Y.: Reliability-performance-energy joint modeling and optimization for a big data task. In: 2016 IEEE International Conference on Software Quality, Reliability and Security Companion (QRS-C), pp. 334–338 (2016). https://doi.org/10.1109/QRS-C.2016.51

13. Mondal, S.K., Sabyasachi, A.S., Muppala, J.K.: On dependability, cost and security trade-off in cloud data centers. In: 2017 IEEE 22nd Pacific Rim International Symposium on Dependable Computing (PRDC), pp. 11–19 (2017). https://doi.org/10.1109/PRDC.2017.12

14. Bai, Y., Zhang, H., Fu, Y.: Reliability modeling and analysis of cloud service based on complex network. In: 2016 Prognostics and System Health Management Conference (PHM-Chengdu), pp. 1–5 (2016). https://doi.org/10.1109/PHM.2016.7819907

15. Zhang, N., Li, R.: Resource optimization with reliability consideration in cloud computing. In: 2016 Annual Reliability and Maintainability Symposium (RAMS), pp. 1–6 (2016). https://doi.org/10.1109/RAMS.2016.7447982

16. Luo, L., Li, H., Qiu, X., Tang, Y.: A resource optimization algorithm of cloud data center based on correlated model of reliability, performance and energy. In: 2016 IEEE International Conference on Software Quality, Reliability and Security Companion (QRS-C), pp. 416–417 (2016). https://doi.org/10.1109/QRS-C.2016.69

17. Chowdhury, A., Tripathi, P.: Enhancing cloud computing reliability using efficient scheduling by providing reliability as a service. In: 2014 International Conference on Parallel, Distributed and Grid Computing, pp. 99–104 (2014). https://doi.org/10.1109/PDGC.2014.7030723

18. Dastjerdi, A.V., Buyya, R.: An autonomous reliability-aware negotiation strategy for cloud computing environments. In: 2012 12th IEEE/ACM International Symposium on Cluster, Cloud and Grid Computing (CCGrid 2012), pp. 284–291 (2012). https://doi.org/10.1109/CCGrid.2012.101

19. Song, Y., Yau, S.S., Yu, R., Zhang, X., Xue, G.: An approach to QoS-based task distribution in edge computing networks for IoT applications. In: 2017 IEEE International Conference on Edge Computing (EDGE), pp. 32–39 (2017). https://doi.org/10.1109/IEEE.EDGE.2017.50

20. Caprolu, M., Di Pietro, R., Lombardi, F., Raponi, S.: Edge computing perspectives: architectures, technologies, and open security issues. In: 2019 IEEE International Conference on Edge Computing (EDGE), pp. 116–123 (2019). https://doi.org/10.1109/EDGE.2019.00035

21. Martín Fernández, C., Díaz Rodríguez, M., Rubio Muñoz, B.: An edge computing architecture in the internet of things. In: 2018 IEEE 21st International Symposium on Real-Time Distributed Computing (ISORC), pp. 99–102 (2018). https://doi.org/10.1109/ISORC.2018.00021

22. Chen, X., Liu, W., Chen, J., Zhou, J.: An edge server placement algorithm in edge computing environment. In: 2020 12th International Conference on Advanced Infocomm Technology (ICAIT), pp. 85–89 (2020). https://doi.org/10.1109/ICAIT51223.2020.9315526

23. Badri, H., Bahreini, T., Grosu, D., Yang, K.: Risk-based optimization of resource provisioning in mobile edge computing. In: 2018 IEEE/ACM Symposium on Edge Computing (SEC), pp. 328–330 (2018). https://doi.org/10.1109/SEC.2018.00033
24. Xiao, K., Gao, Z., Wang, Q., Yang, Y.: A heuristic algorithm based on resource requirements forecasting for server placement in edge computing. In: 2018 IEEE/ACM Symposium on Edge Computing (SEC), pp. 354–355 (2018). https://doi.org/10.1109/SEC.2018.00043
25. Ribeiro, R., Favarim, F., Barbosa, M.A.C., Koerich, A.L., Enembreck, F.: Combining learning algorithms: an approach to Markov decision processes. In: Cordeiro, J., Maciaszek, L.A., Filipe, J. (eds.) ICEIS 2012. LNBIP, vol. 141, pp. 172–188. Springer, Heidelberg (2013). https://doi.org/10.1007/978-3-642-40654-6_11
26. Huang, J., Lin, C., Kong, X., Wei, B., Shen, X.: Modeling and analysis of dependability attributes for services computing systems. IEEE Trans. Serv. Comput. 7(4), 599–613 (2014). https://doi.org/10.1109/TSC.2013.8
27. Avizienis, A., Laprie, J., Randell, B., Landwehr, C.: Basic concepts and taxonomy of dependable and secure computing. IEEE Trans. Dependable Secure Comput. 1(1), 11–33 (2004). https://doi.org/10.1109/TDSC.2004.2
28. Lanus, M., Yin, L., Trivedi, K.S.: Hierarchical composition and aggregation of state-based availability and performability models. IEEE Trans. Reliab. 52, 44–52 (2003)
29. Stewart, W.J.: Introduction to the Numerical Solution of Markov Chains. Princeton University Press, Princeton (1994)
30. Zheng, S., Tilevich, E.: A programming model for reliable and efficient edge-based execution under resource variability. In: 2019 IEEE International Conference on Edge Computing (EDGE) (2019)
31. Huang, J., Lin, C., Wan, J.: Modeling, analysis and optimization of dependability-aware energy efficiency in services computing systems. In: 2013 IEEE International Conference on Services Computing, pp. 683–690. IEEE (2013)
32. Jia, S., Shen, L., Xue, H.: Continuous-time Markov decision process with average reward: using reinforcement learning method. In: 2015 34th Chinese Control Conference (CCC), pp. 3097–3100 (2015). https://doi.org/10.1109/ChiCC.2015.7260117

Author Index

Ali, Sikandar   I-728, II-65

Bai, Xu   II-416
Bernard, Ngounou   I-20

Cai, Guanyu   II-276
Cai, Haini   I-318
Cao, Bin   II-316
Cao, Buqing   I-213
Cao, Chenhong   II-87
Cao, Zhihui   II-3
Chang, Mengmeng   I-299
Chang, Peng   I-575
Chao, Meiling   II-435
Chen, Dajiang   I-156
Chen, Dong   II-104
Chen, Gaojian   I-38
Chen, Jun   II-385
Chen, Junchang   I-554
Chen, Junjie   I-213
Chen, Kairun   I-678
Chen, Maojian   II-190
Chen, Peng   I-520
Chen, Rong   II-276
Chen, Xin   II-50, II-144
Chen, Yige   I-117, I-348
Chen, Yuqing   I-54
Chiu, Brian   II-104, II-124

Dai, Hong-Ning   I-368
Dai, Qiong   II-416
Deng, Zhidong   I-479
Ding, Bo   II-158
Ding, Weilong   II-385
Ding, Xu   I-99, II-335
Ding, Zhiming   I-299
Dong, Jing   I-642, I-661
Dou, Wanchun   II-33
Du, Bowen   II-104, II-124
Du, Jiajie   II-401
Du, Yongkang   II-385

Fan, Guijun   II-18
Fan, Guisheng   I-244, I-420

Fan, Hongfei   II-104, II-124
Fan, Jing   II-316
Fan, Yuqi   II-335
Fang, Mohan   I-20
Feng, Haodong   I-54
Feng, Meiqi   I-368
Feng, Qilong   II-352
Feng, Zhenni   I-554, II-87
Fu, Xiang   II-158

Gao, Honghao   II-87
Gao, Min   I-138, I-229, II-435
Ge, Weimin   I-479
Gu, Fei   I-713
Gu, Tianyi   I-259
Guo, Gengyuan   I-628
Guo, Xiao-Yong   I-498

Han, Jianmin   I-20
Han, Yanbo   I-38
He, Fei   II-175
He, Hui   I-608
He, Wen   I-697
He, Xin   I-678
He, Xionghui   II-206
He, Yulin   II-435
Hou, Yuexian   I-277
Hu, Rong   I-213
Huang, Hongyu   I-389, II-18
Huang, Jiahui   II-316
Huang, Jiwei   I-728, II-65
Huang, Kaiwen   I-259
Huang, Teng   I-244
Huang, Weiyi   I-534
Huang, Zijie   I-420
Huang, Ziyang   II-190
Huang, Zunfu   I-277

Ji, Cun   I-628
Jia, Weixing   I-171
Jia, Nannan   I-299
Jiang, Jinfeng   II-104, II-124
Jiang, Lei   II-416
Jiang, Ning   I-520

Jiang, Qinxue   I-277
Jiang, Xutong   II-33
Jiao, Boyang   I-38
Jiao, Libo   II-50, II-144
Jiao, Litao   I-368
Jin, Canghong   II-175
Jin, Hui   I-479

Ke, Yuxian   I-460
Kuang, Li   II-276, II-352

Li, Chengfan   II-87
Li, Dan   II-87
Li, Deyi   I-697
Li, Juanru   I-439
Li, Lutong   I-299
Li, Ning   I-191
Li, Peng   I-534
Li, Ping   I-259
Li, Qianwen   I-38
Li, Wei   I-661
Li, Wenxiong   II-241
Li, Xiaohong   I-479
Li, Yantao   I-389, II-18
Li, Yin   I-520
Li, Zhen   II-158
Liang, Jingyu   I-728
Liang, Yan   II-50
Liang, Ying   II-385
Lin, Liangliang   I-608
Liu, Bowen   II-33
Liu, Changzheng   II-261
Liu, Dianwen   I-642
Liu, Hongtao   I-277
Liu, Jian   I-368
Liu, Jianxiao   I-67
Liu, Jianxun   I-213
Liu, Junrong   I-439
Liu, Lin   I-498
Liu, Rui   I-642, I-661
Liu, Sihang   I-67
Liu, Tong   I-554, II-87
Liu, Yu   II-451
Liu, Zunhao   I-299
Lu, Jianfeng   I-20
Luo, Jie   II-158
Luo, Xiong   II-190
Lv, Qing   I-3
Lv, Xiang   II-3

Ma, Bowen   I-728
Ma, Shengcheng   II-50
Ma, Xiaoyu   I-608
Ma, Yifan   II-124
Mateen, Muhammad   I-318
Mu, Nankun   II-18

Nie, Lei   I-534
Niu, Jianwei   I-697

Ouyang, Zhenchao   I-697, II-451

Pan, Li   I-592, II-401
Panaousis, Emmanouil   I-368
Peng, Chengwei   I-575
Peng, Hao   I-20
Peng, Huailiang   II-416
Peng, Mi   I-213
Peng, Qiaojuan   II-190

Qi, Xiuxiu   II-241
Qian, Ye   I-592
Qin, Zhen   I-156

Ren, Lu   I-697, II-451
Ren, Yuheng   II-241

Sang, Yafei   I-575
Shang, Siyuan   II-335
Shen, Hailun   II-190
Sheng, Yu   II-276
Shi, Dianxi   II-370
Shu, Xiao   II-158
Su, Yaqianwen   II-206, II-370
Sun, Haifeng   II-3
Sun, Song   I-318

Tan, Jiefu   I-191, II-206
Tan, Yufu   II-65
Tang, Liangyao   I-520
Tian, Changbo   I-405

Wang, Bo   I-277
Wang, Ding   I-368
Wang, Guiling   I-171
Wang, Haotian   II-65
Wang, Huan   I-191
Wang, Jia   I-138
Wang, Jiahao   II-241
Wang, Jiaxing   II-316
Wang, Jing   I-38

Wang, Miao   I-3
Wang, Shiqi   I-138
Wang, Yang   I-99
Wang, Yifeng   II-297
Wang, Yipeng   I-348
Wang, Zongwei   I-138, II-435
Wang, Zuohua   I-83
Wei, Xiaopeng   I-642, I-661
Wen, Junhao   I-138, I-318
Wu, Bilian   II-144
Wu, Fan   I-138
Wu, Minghui   II-175
Wu, Wendi   I-191
Wu, Yingbo   II-225
Wu, Yunlong   I-191

Xi, Wei   I-608
Xia, Yunni   I-520
Xiao, Junlei   I-534
Xiao, Mengdi   I-534
Xie, Hong   I-520
Xie, Min   II-158
Xie, Yujian   I-460
Xing, Hai-Feng   I-498
Xing, Mengda   II-385
Xiong, Naixue   I-277
Xu, Chi   I-439
Xu, Guangquan   I-368
Xu, Huanhuan   I-713
Xu, Jiachi   I-191, II-206
Xu, Ruihong   I-67
Xu, Wenhua   II-104
Xue, Chao   II-206

Yang, Bowen   I-299
Yang, Chen   I-608
Yang, Huan   I-318
Yang, Jisong   I-575
Yang, Kang   I-420
Yang, Mengning   I-678
Yang, Peng   I-117, I-348
Yang, Ruilong   I-520
Yang, Senqiao   II-297
Yang, Shaowu   II-370
Yang, Xingguang   I-420
Yang, Xuankai   I-171
Yang, Yao   I-389
Yang, Yin   II-297
Yang, Zemeng   I-67
Yang, Zichao   II-124

Yao, Juan   II-225
Yao, Xin   I-20
Ye, Peng   I-333
Yi, Pengfei   I-642, I-661
Yin, Tao   I-405
Ying, Weizhi   I-83
Yu, Huiqun   I-244, I-420
Yu, Juan   I-20
Yu, Lei   I-54
Yu, Qing   I-83
Yu, Yang   I-229, II-225
Yuan, Lei   I-299
Yuan, Limengzi   I-460, II-261
Yuan, Yunjing   I-38

Zang, Tianning   I-117, I-348
Zeng, Jun   I-229, II-225
Zhang, Benzhuang   II-261
Zhang, Changjie   II-451
Zhang, Chaokui   I-333
Zhang, Fengquan   I-171
Zhang, Hanwen   I-156
Zhang, Junwei   II-435
Zhang, Mingzhe   I-628
Zhang, Ning   II-3
Zhang, Qiang   I-642, I-661
Zhang, Qingwang   I-67
Zhang, Shichao   I-628
Zhang, Song   II-33
Zhang, Tianpu   II-385
Zhang, Wang   II-416
Zhang, Xiang   I-333
Zhang, Yao   I-479
Zhang, Yaowen   II-370
Zhang, Yiteng   II-104, II-124
Zhang, Yongjun   II-206, II-370
Zhang, Yongzheng   I-117, I-348, I-405, I-575
Zhang, Yue   II-352
Zhao, Chenran   II-370
Zhao, Chong   I-99
Zhao, Jizhong   I-608
Zhao, Kun   I-608
Zhao, Qingzhan   I-460
Zhao, Quanwu   II-435
Zhao, Xingbiao   II-261
Zhao, Xuan   II-33
Zhao, Yizhu   I-229
Zheng, Hang   I-99
Zheng, Limin   I-3

Zheng, Xi   I-368
Zheng, Xiangwei   I-628
Zheng, Yuchen   I-460, II-261
Zhong, Zhenyang   I-333
Zhou, Dongsheng   I-642, I-661

Zhou, Jian-Tao   I-498
Zhou, Jingya   I-713
Zhou, Wei   I-229
Zhou, Yuan   I-117, I-348
Zhou, Ziyi   I-420

Printed in the United States
by Baker & Taylor Publisher Services